Stability and Control of Large-Scale Dynamical Systems

PRINCETON SERIES IN APPLIED MATHEMATICS

Edited by

Ingrid Daubechies, *Princeton University*
Weinan E, *Princeton University*
Jan Karel Lenstra, *Eindhoven University*
Endre Süli, *University of Oxford*

The Princeton Series in Applied Mathematics publishes high quality advanced texts and monographs in all areas of applied mathematics. Books include those of a theoretical and general nature as well as those dealing with the mathematics of specific applications areas and real-world situations.

Stability and Control of Large-Scale Dynamical Systems

A Vector Dissipative Systems Approach

Wassim M. Haddad

Sergey G. Nersesov

PRINCETON UNIVERSITY PRESS

PRINCETON AND OXFORD

Library of Congress Cataloging-in-Publication Data

Haddad, Wassim M., 1961–
    Stability and control of large-scale dynamical systems: a vector dissipative systems approach. / Wassim M. Haddad, Sergey G. Nersesov.
        p. cm. — (Princeton series in applied mathematics)
    Includes bibliographical references and index.
    ISBN: 978-0-691-15346-9 (alk. paper)

    1. Lyapunov stability. 2. Energy dissipation. 3. Dynamics. 4. Large scale systems.
I. Nersesov, Sergey G., 1976– II. Title. III. Series.

QA871.H15 2011
003′.71—dc23   2011019426

British Library Cataloging-in-Publication Data is available

This book has been composed in Times Roman in LaTeX

The publisher would like to acknowledge the authors of this volume for providing the camera-ready copy from which this book was printed.

Printed on acid-free paper. ∞

press.princeton.edu

Printed in the United States of America

10 9 8 7 6 5 4 3 2 1

Εἷς ἐμοὶ μύριοι, ἐὰν ἄριστος ἦι.

To me one is worth ten thousand if he is truly outstanding.

—Herakleitos of Ephesus, Ionia, Greece

Χρόνον τε γενέσθαι εἰκόνα τοῦ ἀιδίου. Κἀκεῖνον μὲν ἀεὶ μένειν, τὴν δὲ τοῦ οὐρανοῦ φορὰν χρόνον εἶναι· καὶ γὰρ νύκτα καὶ ἡμέραν καὶ μῆνα καὶ τὰ τοιαῦτα πάντα χρόνου μέρη εἶναι. Διόπερ ἄνευ τῆς τοῦ κόσμου φύσεως οὐκ εἶναι χρόνον· ἅμα γὰρ ὑπάρχειν αὐτῷ καὶ χρόνον εἶναι.

Time was created as an image of the eternal. While time is everlasting, time is the outcome of change (motion) in the universe. And as night and day and month and the like are all part of time, without the physical universe time ceases to exist. Thus, the creation of the universe has spawned the arrow of time.

—Plato of Athens, Attiki, Greece

Ἄτοπον εἶναι εν μεγάλω πεδίω ἕνα στάχυν γενηθῆναι και ἕνα κόσμον εν τω απείρω. Ὅτι δ᾽ άπειρος κατά το πλῆθος, δῆλον εκ του άπειρα τα αίτια είναι· ει γάρ ο μέν κόσμος πεπερασμένος, τα δ᾽ αιτια πάντα άπειρα, εξ ών όδε ο κόσμος γέγονεν, ανάγκη απείρους είναι. Οπου γάρ τα πάντα αίτια, εκεί και τα αποτελέσματα· αίτια δ᾽ ήτοι αι άτομοι ή τα στοιχεία.

To consider the earth as the only inhabited world in the infinite universe is as absurd as to assert that in an entire field sown with millet, only one grain will grow. That the universe is infinite with an infinite number of worlds follows from the infinite number of causalities that govern it. If the universe were finite and the causes that caused it infinite, then the universe would be comprised of an infinite number of worlds. For where all causes concur by the blending and altering of atoms or elements in the physical universe, there their effects must also appear.

—Metrodoros of Chios, Chios, Greece

Εἶναι τε ὥσπερ γενέσεις κόσμου, οὕτω καί αυξήσεις καί φθίσεις καί φθοράς, κατά τινά ανάγκην.

From its genesis, the cosmos has spawned multitudinous worlds that evolve in accordance to a supreme law that is responsible for their expansion, enfeeblement, and eventual demise.

—Leukippos of Miletus, Ionia, Greece

Contents

Preface xiii

Chapter 1. Introduction 1

1.1 Large-Scale Interconnected Dynamical Systems 1
1.2 A Brief Outline of the Monograph 5

Chapter 2. Stability Theory via Vector Lyapunov Functions 9

2.1 Introduction 9
2.2 Notation and Definitions 9
2.3 Quasi-Monotone and Essentially Nonnegative Vector Fields 10
2.4 Generalized Differential Inequalities 14
2.5 Stability Theory via Vector Lyapunov Functions 18
2.6 Discrete-Time Stability Theory via Vector
Lyapunov Functions 34

**Chapter 3. Large-Scale Continuous-Time Interconnected
Dynamical Systems** 45

3.1 Introduction 45
3.2 Vector Dissipativity Theory for Large-Scale Nonlinear Dynamical Systems 46
3.3 Extended Kalman-Yakubovich-Popov Conditions for Large-Scale Nonlinear Dynamical Systems 61
3.4 Specialization to Large-Scale Linear Dynamical Systems 68
3.5 Stability of Feedback Interconnections of Large-Scale Nonlinear Dynamical Systems 71

**Chapter 4. Thermodynamic Modeling of Large-Scale
Interconnected Systems** 75

4.1 Introduction 75
4.2 Conservation of Energy and the First Law
of Thermodynamics 75
4.3 Nonconservation of Entropy and the Second Law
of Thermodynamics 79
4.4 Semistability and Large-Scale Systems 82
4.5 Energy Equipartition 86

4.6   Entropy Increase and the Second Law of Thermodynamics   88
4.7   Thermodynamic Models with Linear Energy Exchange   90

Chapter 5. Control of Large-Scale Dynamical Systems via Vector Lyapunov Functions   **93**

5.1   Introduction   93
5.2   Control Vector Lyapunov Functions   94
5.3   Stability Margins, Inverse Optimality, and Vector Dissipativity   99
5.4   Decentralized Control for Large-Scale Nonlinear Dynamical Systems   102

Chapter 6. Finite-Time Stabilization of Large-Scale Systems via Control Vector Lyapunov Functions   **107**

6.1   Introduction   107
6.2   Finite-Time Stability via Vector Lyapunov Functions   108
6.3   Finite-Time Stabilization of Large-Scale Dynamical Systems 114
6.4   Finite-Time Stabilization for Large-Scale Homogeneous Systems   119
6.5   Decentralized Control for Finite-Time Stabilization of Large-Scale Systems   121

Chapter 7. Coordination Control for Multiagent Interconnected Systems   **127**

7.1   Introduction   127
7.2   Stability and Stabilization of Time-Varying Sets   129
7.3   Control Design for Multivehicle Coordinated Motion   135
7.4   Stability and Stabilization of Time-Invariant Sets   141
7.5   Control Design for Static Formations   144
7.6   Obstacle Avoidance in Multivehicle Coordination   145

Chapter 8. Large-Scale Discrete-Time Interconnected Dynamical Systems   **153**

8.1   Introduction   153
8.2   Vector Dissipativity Theory for Discrete-Time Large-Scale Nonlinear Dynamical Systems   154
8.3   Extended Kalman-Yakubovich-Popov Conditions for Discrete-Time Large-Scale Nonlinear Dynamical Systems   168
8.4   Specialization to Discrete-Time Large-Scale Linear Dynamical Systems   173
8.5   Stability of Feedback Interconnections of Discrete-Time Large-Scale Nonlinear Dynamical Systems   177

Chapter 9. Thermodynamic Modeling for Discrete-Time Large-Scale Dynamical Systems

9.1	Introduction	181
9.2	Conservation of Energy and the First Law of Thermodynamics	182
9.3	Nonconservation of Entropy and the Second Law of Thermodynamics	187
9.4	Nonconservation of Ectropy	189
9.5	Semistability of Discrete-Time Thermodynamic Models	191
9.6	Entropy Increase and the Second Law of Thermodynamics	198
9.7	Thermodynamic Models with Linear Energy Exchange	200

Chapter 10. Large-Scale Impulsive Dynamical Systems

10.1	Introduction	211
10.2	Stability of Impulsive Systems via Vector Lyapunov Functions	213
10.3	Vector Dissipativity Theory for Large-Scale Impulsive Dynamical Systems	224
10.4	Extended Kalman-Yakubovich-Popov Conditions for Large-Scale Impulsive Dynamical Systems	249
10.5	Specialization to Large-Scale Linear Impulsive Dynamical Systems	259
10.6	Stability of Feedback Interconnections of Large-Scale Impulsive Dynamical Systems	264

Chapter 11. Control Vector Lyapunov Functions for Large-Scale Impulsive Systems

11.1	Introduction	271
11.2	Control Vector Lyapunov Functions for Impulsive Systems	272
11.3	Stability Margins and Inverse Optimality	279
11.4	Decentralized Control for Large-Scale Impulsive Dynamical Systems	284

Chapter 12. Finite-Time Stabilization of Large-Scale Impulsive Dynamical Systems

12.1	Introduction	289
12.2	Finite-Time Stability of Impulsive Dynamical Systems	289
12.3	Finite-Time Stabilization of Impulsive Dynamical Systems	297
12.4	Finite-Time Stabilizing Control for Large-Scale Impulsive Dynamical Systems	300

Chapter 13. Hybrid Decentralized Maximum Entropy Control for Large-Scale Systems

		305

13.1   Introduction   305

13.2   Hybrid Decentralized Control and Large-Scale Impulsive Dynamical Systems   306

13.3   Hybrid Decentralized Control for Large-Scale Dynamical Systems   313

13.4   Interconnected Euler-Lagrange Dynamical Systems   319

13.5   Hybrid Decentralized Control Design   323

13.6   Quasi-Thermodynamic Stabilization and Maximum Entropy Control   327

13.7   Hybrid Decentralized Control for Combustion Systems   335

13.8   Experimental Verification of Hybrid Decentralized Controller   341

Chapter 14. Conclusion   **351**

Bibliography   **353**

Index   **367**

Preface

Modern complex large-scale dynamical systems arise in virtually every aspect of science and engineering and are associated with a wide variety of physical, technological, environmental, and social phenomena. Such systems include large-scale aerospace systems, power systems, communications systems, network systems, transportation systems, large-scale manufacturing systems, integrative biological systems, economic systems, ecological systems, and process control systems. These systems are strongly interconnected and consist of interacting subsystems exchanging matter, energy, or information with the environment. In addition, the subsystem interactions often exhibit remarkably complex system behaviors. Complexity here refers to the quality of a system wherein interacting subsystems form multiechelon hierarchical evolving structures exhibiting emergent system properties.

The sheer size, or dimensionality, of large-scale dynamical systems necessitates decentralized analysis and control system synthesis methods for their analysis and control design. Specifically, in analyzing complex large-scale interconnected dynamical systems it is often desirable to treat the overall system as a collection of interacting subsystems. The behavior and properties of the aggregate large-scale system can then be deduced from the behaviors of the individual subsystems and their interconnections. Often the need for such an analysis framework arises from computational complexity and computer throughput constraints. In addition, for controller design the physical size and complexity of large-scale systems impose severe constraints on the communication links among system sensors, processors, and actuators, which can render centralized control architectures impractical. This problem leads to consideration of decentralized controller architectures involving multiple sensor-processor-actuator subcontrollers without real-time intercommunication. The design and implementation of decentralized controllers is a nontrivial task involving control-system architecture determination and actuator-sensor assignments for a particular subsystem, as well as processor software design for each subcontroller of a given architecture.

In this monograph, we develop a unified stability analysis and control design framework for nonlinear large-scale interconnected dynamical systems based on vector Lyapunov function methods and vector dissipativity theory. The use of vector Lyapunov functions in dynamical system theory offers a very flexible framework for stability analysis since each component of the vector Lyapunov function can satisfy less rigid requirements as compared to a single scalar Lyapunov function. Moreover, in the analysis of large-scale interconnected nonlinear dynamical systems, several Lyapunov functions arise naturally from the stability properties of each individual subsystem. In addition, since large-scale dynamical systems have numerous input, state, and output properties related to conservation, dissipation, and transport of energy, matter, or information, extending classical dissipativity theory to capture conservation and dissipation notions on the subsystem level provides a natural energy flow model for large-scale dynamical systems. Aggregating the dissipativity properties of each of the subsystems by appropriate storage functions and supply rates allows us to study the dissipativity properties of the composite large-scale system using the newly developed notions of vector storage functions and vector supply rates. The monograph is written from a system-theoretic point of view and can be viewed as a contribution to dynamical system and control system theory.

After a brief introduction to large-scale interconnected dynamical systems in Chapter 1, fundamental stability theory for nonlinear dynamical systems using vector Lyapunov functions is developed in Chapter 2. In Chapter 3, we extend classical dissipativity theory to vector dissipativity for addressing large-scale systems using vector storage functions and vector supply rates. Chapter 4 develops connections between thermodynamics and large-scale dynamical systems. A detailed treatment of control design for large-scale systems using control vector Lyapunov functions is given in Chapter 5, whereas extensions of these results for addressing finite-time stability and stabilization are given in Chapter 6. Next, in Chapter 7 we develop a stability and control design framework for coordination control of multiagent interconnected systems. Chapters 8 and 9 present discrete-time extensions of vector dissipativity theory and system thermodynamic connections of large-scale systems, respectively. A detailed treatment of stability analysis and vector dissipativity for large-scale impulsive dynamical systems is given in Chapter 10. Chapters 11 and 12 provide extensions of finite-time stabilization and stabilization of large-scale impulsive dynamical systems. In Chapter 13, a novel class of fixed-order, energy- and entropy-based hybrid decentralized controllers is developed for large-scale dynamical systems. Finally, in Chapter 14 we present conclusions.

The first author would like to thank Dennis S. Bernstein and David C. Hyland for their valuable discussions on large-scale vibrational systems over the years. The first author would also like to thank Paul Katinas for several insightful and enlightening discussions on the statements quoted in ancient Greek on page vii. In some parts of the monograph we have relied on work we have done jointly with Jevon M. Avis, VijaySekhar Chellaboina, Qing Hui, and Rungun Nathan; it is a pleasure to acknowledge their contributions.

The results reported in this monograph were obtained at the School of Aerospace Engineering, Georgia Institute of Technology, Atlanta, and the Department of Mechanical Engineering of Villanova University, Villanova,

Pennsylvania, between January 2004 and February 2011. The research support provided by the Air Force Office of Scientific Research and the Office of Naval Research over the years has been instrumental in allowing us to explore basic research topics that have led to some of the material in this monograph. We are indebted to them for their support.

Atlanta, Georgia, June 2011, *Wassim M. Haddad*
Villanova, Pennsylvania, June 2011, *Sergey G. Nersesov*

Chapter One

Introduction

1.1 Large-Scale Interconnected Dynamical Systems

Modern complex dynamical systems[1] are highly interconnected and mutually interdependent, both physically and through a multitude of information and communication network constraints. The sheer size (i.e., dimensionality) and complexity of these large-scale dynamical systems often necessitates a hierarchical decentralized architecture for analyzing and controlling these systems. Specifically, in the analysis and control-system design of complex large-scale dynamical systems it is often desirable to treat the overall system as a collection of interconnected subsystems. The behavior of the aggregate or composite (i.e., large-scale) system can then be predicted from the behaviors of the individual subsystems and their interconnections. The need for decentralized analysis and control design of large-scale systems is a direct consequence of the physical size and complexity of the dynamical model. In particular, computational complexity may be too large for model analysis while severe constraints on communication links between system sensors, actuators, and processors may render centralized control architectures impractical. Moreover, even when communication constraints do not exist, decentralized processing may be more economical.

In an attempt to approximate high-dimensional dynamics of large-scale structural (oscillatory) systems with a low-dimensional diffusive (nonoscillatory) dynamical model, structural dynamicists have developed thermodynamic energy flow models using stochastic energy flow techniques. In particular, statistical energy analysis (SEA) predicated on averaging system states over the statistics of the uncertain system parameters have been extensively developed for mechanical and acoustic vibration problems [109,119,129,163,173]. Thermodynamic models are derived from large-scale dynamical systems of discrete subsystems involving stored energy flow among subsystems based on the assumption of weak subsystem coupling or identical subsystems. However, the ability of SEA to predict the dynamic behavior of a complex large-scale dynamical system in terms of pairwise subsystem interactions is severely limited by the coupling strength of the remaining subsystems on the subsystem pair. Hence, it is not surprising

---

[1]Here we have in mind large flexible space structures, aerospace systems, electric power systems, network systems, communications systems, transportation systems, economic systems, and ecological systems, to cite but a few examples.

that SEA energy flow predictions for large-scale systems with strong coupling can be erroneous.

Alternatively, a deterministic thermodynamically motivated energy flow modeling for large-scale structural systems is addressed in [113–115]. This approach exploits energy flow models in terms of thermodynamic energy (i.e., ability to dissipate heat) as opposed to stored energy and is not limited to weak subsystem coupling. Finally, a stochastic energy flow *compartmental model* (i.e., a model characterized by conservation laws) predicated on averaging system states over the statistics of stochastic system exogenous disturbances is developed in [21]. The basic result demonstrates how compartmental models arise from second-moment analysis of state space systems under the assumption of weak coupling. Even though these results can be potentially applicable to large-scale dynamical systems with weak coupling, such connections are not explored in [21].

An alternative approach to analyzing large-scale dynamical systems was introduced by the pioneering work of Šiljak [159] and involves the notion of *connective stability*. In particular, the large-scale dynamical system is decomposed into a collection of subsystems with local dynamics and uncertain interactions. Then, each subsystem is considered independently so that the stability of each subsystem is combined with the interconnection constraints to obtain a *vector Lyapunov function* for the composite large-scale dynamical system, guaranteeing connective stability for the overall system.

Vector Lyapunov functions were first introduced by Bellman [14] and Matrosov[2] [133] and further developed by Lakshmikantham *et al.* [118], with [65, 127, 131, 132, 136, 159, 160] exploiting their utility for analyzing large-scale systems. Extensions of vector Lyapunov function theory that include matrix-valued Lyapunov functions for stability analysis of large-scale dynamical systems appear in the monographs by Martynyuk [131,132]. The use of vector Lyapunov functions in large-scale system analysis offers a very flexible framework for stability analysis since each component of the vector Lyapunov function can satisfy less rigid requirements as compared to a single scalar Lyapunov function. Weakening the hypothesis on the Lyapunov function enlarges the class of Lyapunov functions that can be used for analyzing the stability of large-scale dynamical systems. In particular, each component of a vector Lyapunov function need not be positive definite with a negative or even negative-semidefinite derivative. The time derivative

---

[2]Even though the theory of vector Lyapunov functions was discovered independently by Bellman and Matrosov, their formulation was quite different in the way that the components of the Lyapunov functions were defined. In particular, in Bellman's formulation the components of the vector Lyapunov functions correspond to disjoint subspaces of the state space, whereas Matrosov allows for the components to be defined in the entire state space. The latter formulation allows for the components of the vector Lyapunov functions to capture the whole state space and, hence, account for interconnected dynamical systems with overlapping subsystems.

of the vector Lyapunov function need only satisfy an element-by-element vector inequality involving a vector field of a certain comparison system. Moreover, in large-scale systems several Lyapunov functions arise naturally from the stability properties of each subsystem. An alternative approach to vector Lyapunov functions for analyzing large-scale dynamical systems is an input-output approach, wherein stability criteria are derived by assuming that each subsystem is either finite gain, passive, or conic [5, 122, 123, 168].

In more recent research, Šiljak [161] developed new and original concepts for modeling and control of large-scale complex systems by addressing system dimensionality, uncertainty, and information structure constraints. In particular, the formulation in [161] develops control law synthesis architectures using decentralized information structure constraints while addressing multiple controllers for reliable stabilization, decentralized optimization, and hierarchical and overlapping decompositions. In addition, decomposition schemes for large-scale systems involving system inputs and outputs as well as dynamic graphs defined on a linear space as one-parameter groups of invariant transformations of the graph space are developed in [178].

Graph theoretic concepts have also been used in stability analysis and decentralized stabilization of large-scale interconnected systems [34, 45]. In particular, graph theory [51, 63] is a powerful tool in investigating structural properties and capturing connectivity properties of large-scale systems. Specifically, a directed graph can be constructed to capture subsystem interconnections wherein the subsystems are represented as nodes and energy, matter, or information flow is represented by edges or arcs. A related approach to graph theory for modeling large-scale systems is bond-graph modeling [35, 107], wherein connections between a pair of subsystems are captured by a bond and energy, matter, or information is exchanged between subsystems along connections. More recently, a major contribution to the analysis and design of interconnected systems is given in [172]. This work builds on the work of bond graphs by developing a modeling behavioral methodology wherein a system is viewed as an interconnection of interacting subsystems modeled by tearing, zooming, and linking.

In light of the fact that energy flow modeling arises naturally in large-scale dynamical systems and vector Lyapunov functions provide a powerful stability analysis framework for these systems, it seems natural that dissipativity theory [170, 171] on the subsystem level, can play a key role in unifying these analysis methods. Specifically, dissipativity theory provides a fundamental framework for the analysis and design of control systems using an input, state, and output description based on system energy[3] related considerations [70, 170]. The dissipation hypothesis on dynamical systems results in a fundamental constraint on their dynamic behavior wherein a dissipative dynamical system can deliver to its surroundings only a fraction of its energy

---

[3]Here the notion of energy refers to abstract energy for which a physical system energy interpretation is not necessary.

and can store only a fraction of the work done to it. Such conservation laws are prevalent in large-scale dynamical systems such as aerospace systems, power systems, network systems, structural systems, and thermodynamic systems.

Since these systems have numerous input, state, and output properties related to conservation, dissipation, and transport of energy, extending dissipativity theory to capture conservation and dissipation notions on the subsystem level would provide a natural energy flow model for large-scale dynamical systems. Aggregating the dissipativity properties of each of the subsystems by appropriate storage functions and supply rates would allow us to study the dissipativity properties of the composite large-scale system using *vector storage functions* and *vector supply rates*. Furthermore, since vector Lyapunov functions can be viewed as generalizations of composite energy functions for all of the subsystems, a generalized notion of dissipativity, namely, *vector dissipativity*, with appropriate vector storage functions and vector supply rates, can be used to construct vector Lyapunov functions for nonlinear feedback large-scale systems by appropriately combining vector storage functions for the forward and feedback large-scale systems. Finally, as in classical dynamical system theory [70], vector dissipativity theory can play a fundamental role in addressing robustness, disturbance rejection, stability of feedback interconnections, and optimality for large-scale dynamical systems.

The design and implementation of control law architectures for large-scale interconnected dynamical systems is a nontrivial control engineering task involving considerations of weight, size, power, cost, location, type, specifications, and reliability, among other design considerations. All these issues are directly related to the properties of the large-scale system to be controlled and the system performance specifications. For conceptual and practical reasons, the control processor architectures in systems composed of interconnected subsystems are typically distributed or decentralized in nature. Distributed control refers to a control architecture wherein the control is distributed via multiple computational units that are interconnected through information and communication networks, whereas decentralized control refers to a control architecture wherein local decisions are based only on local information. In a decentralized control scheme, the large-scale interconnected dynamical system is controlled by multiple processors operating independently, with each processor receiving a subset of the available subsystem measurements and updating a subset of the subsystem actuators. Although decentralized controllers are more complicated to design than distributed controllers, their implementation offers several advantages. For example, physical system limitations may render it uneconomical or impossible to feed back certain measurement signals to particular actuators.

Since implementation constraints, cost, and reliability considerations often require decentralized controller architectures for controlling large-scale systems, decentralized control has received considerable attention in the literature [17, 22, 48, 96–99, 104, 125, 126, 145, 150, 154, 158–160, 162]. A straightforward decentralized control design technique is that of *sequential optimization* [17, 48, 104], wherein a sequential centralized subcontroller design procedure is applied to an augmented closed-loop plant composed of the actual plant and the remaining subcontrollers. Clearly, a key difficulty with decentralized control predicated on sequential optimization is that of dimensionality. An alternative approach to sequential optimization for decentralized control is based on *subsystem decomposition* with centralized design procedures applied to the individual subsystems of the large-scale system [96–99, 125, 126, 145, 150, 154, 158–160]. Decomposition techniques exploit subsystem interconnection data and in many cases, such as in the presence of very high system dimensionality, are absolutely essential for designing decentralized controllers.

1.2 A Brief Outline of the Monograph

The main objective of this monograph is to develop a general stability analysis and control design framework for nonlinear large-scale interconnected dynamical systems, with an emphasis on vector Lyapunov function methods and vector dissipativity theory. The main contents of the monograph are as follows. In Chapter 2, we establish notation and definitions and develop stability theory for large-scale dynamical systems. Specifically, stability theorems via vector Lyapunov functions are developed for continuous-time and discrete-time nonlinear dynamical systems. In addition, we extend the theory of vector Lyapunov functions by constructing a generalized comparison system whose vector field can be a function of the comparison system states as well as the nonlinear dynamical system states. Furthermore, we present a generalized convergence result which, in the case of a scalar comparison system, specializes to the classical Krasovskii-LaSalle invariant set theorem.

In Chapter 3, we extend the notion of dissipative dynamical systems to develop an energy flow modeling framework for large-scale dynamical systems based on vector dissipativity notions. Specifically, using vector storage functions and vector supply rates, dissipativity properties of a composite large-scale system are shown to be determined from the dissipativity properties of the subsystems and their interconnections. Furthermore, extended Kalman-Yakubovich-Popov conditions, in terms of the subsystem dynamics and interconnection constraints, characterizing vector dissipativeness via vector system storage functions, are derived. In addition, these results are used to develop feedback interconnection stability results for large-scale nonlinear dynamical systems using vector Lyapunov functions. Specialization of these results to passive and nonexpansive large-scale dynamical systems is also provided.

In Chapter 4, we develop connections between thermodynamics and large-scale dynamical systems. Specifically, using compartmental dynamical system theory, we develop energy flow models possessing energy conservation and energy equipartition principles for large-scale dynamical systems. Next, we give a deterministic definition of *entropy* for a large-scale dynamical system that is consistent with the classical definition of entropy and show that it satisfies a Clausius-type inequality leading to the law of nonconservation of entropy. Furthermore, we introduce a new and dual notion to entropy, namely, *ectropy*, as a measure of the tendency of a dynamical system to do useful work and grow more organized, and show that conservation of energy in an isolated thermodynamic large-scale system necessarily leads to nonconservation of ectropy and entropy. In addition, using the system ectropy as a Lyapunov function candidate, we show that our large-scale thermodynamic energy flow model has convergent trajectories to Lyapunov stable equilibria determined by the system initial subsystem energies.

In Chapter 5, we introduce the notion of a *control vector Lyapunov function* as a generalization of *control Lyapunov functions* [6], and show that asymptotic stabilizability of a nonlinear dynamical system is equivalent to the existence of a control vector Lyapunov function. Moreover, using control vector Lyapunov functions, we construct a universal decentralized feedback control law for a decentralized nonlinear dynamical system that possesses guaranteed gain and sector margins in each decentralized input channel. Furthermore, we establish connections between the notion of vector dissipativity developed in Chapter 3 and optimality of the proposed decentralized feedback control law. The proposed control framework is then used to construct decentralized controllers for large-scale nonlinear systems with robustness guarantees against full modeling uncertainty. In Chapter 6, we extend the results of Chapter 5 to develop a general framework for finite-time stability analysis based on vector Lyapunov functions. Specifically, we construct a vector comparison system whose solution is finite-time stable and relate this finite-time stability property to the stability properties of a nonlinear dynamical system using a vector comparison principle. Furthermore, we design a universal decentralized finite-time stabilizer for large-scale dynamical systems that is robust against full modeling uncertainty.

Next, using the results of Chapter 5, in Chapter 7 we develop a stability and control design framework for time-varying and time-invariant sets of nonlinear dynamical systems. We then apply this framework to the problem of coordination control for multiagent interconnected systems. Specifically, by characterizing a moving formation of vehicles as a time-varying set in the state space, a distributed control design framework for multivehicle coordinated motion is developed by designing stabilizing controllers for time-varying sets of nonlinear dynamical systems. In Chapters 8 and 9, we present discrete-time extensions of vector dissipativity theory and system thermodynamic connections of large-scale systems developed in Chapters 3 and 4, respectively.

In Chapter 10, we provide generalizations of the stability results developed in Chapter 2 to address stability of impulsive dynamical systems via vector Lyapunov functions. Specifically, we provide a generalized comparison principle involving hybrid comparison dynamics that are dependent on the comparison system states as well as the nonlinear impulsive dynamical system states. Furthermore, we develop stability results for impulsive dynamical systems that involve vector Lyapunov functions and hybrid comparison inequalities. In addition, we develop vector dissipativity notions for large-scale nonlinear impulsive dynamical systems. In particular, we introduce a generalized definition of dissipativity for large-scale nonlinear impulsive dynamical systems in terms of a hybrid vector inequality, a vector hybrid supply rate, and a vector storage function. Dissipativity properties of the large-scale impulsive system are shown to be determined from the dissipativity properties of the individual impulsive subsystems making up the large-scale system and the nature of the system interconnections. Using the concepts of dissipativity and vector dissipativity, we also develop feedback interconnection stability results for impulsive nonlinear dynamical systems. General stability criteria are given for Lyapunov, asymptotic, and exponential stability of feedback impulsive dynamical systems. In the case of quadratic hybrid supply rates corresponding to net system power and weighted input-output energy, these results generalize the positivity and small gain theorems to the case of nonlinear large-scale impulsive dynamical systems.

Using the concepts developed in Chapter 10, in Chapter 11 we extend the notion of control vector Lyapunov functions to impulsive dynamical systems. Specifically, using control vector Lyapunov functions, we construct a universal hybrid decentralized feedback stabilizer for a decentralized affine in the control nonlinear impulsive dynamical system that possesses guaranteed gain and sector margins in each decentralized input channel. These results are then used to develop hybrid decentralized controllers for large-scale impulsive dynamical systems with robustness guarantees against full modeling and input uncertainty. Finite-time stability analysis and control design extensions for large-scale impulsive dynamical systems are addressed in Chapter 12.

In Chapter 13, a novel class of fixed-order, energy-based hybrid decentralized controllers is proposed as a means for achieving enhanced energy dissipation in large-scale vector lossless and vector dissipative dynamical systems. These dynamic decentralized controllers combine a logical switching architecture with continuous dynamics to guarantee that the system plant energy is strictly decreasing across switchings. The general framework leads to hybrid closed-loop systems described by impulsive differential equations [82]. In addition, we construct hybrid dynamic controllers that guarantee that each subsystem-subcontroller pair of the hybrid closed-loop system is consistent with basic thermodynamic principles. Special cases of energy-based hybrid controllers involving state-dependent switching are described, and several illustrative examples are given as well as an experimental test bed is designed to demonstrate the efficacy of the proposed approach. Finally, we draw conclusions in Chapter 14.

Stability Theory via Vector Lyapunov Functions

2.1 Introduction

In this chapter, we introduce the notion of vector Lyapunov functions for stability analysis of nonlinear dynamical systems. The use of vector Lyapunov functions in dynamical system theory offers a flexible framework for stability analysis because each component of the vector Lyapunov function can satisfy less rigid requirements as compared to a single scalar Lyapunov function. Specifically, since for many nonlinear dynamical systems constructing a system Lyapunov function can be a difficult task, weakening the hypothesis on the Lyapunov function enlarges the class of Lyapunov functions that can be used for analyzing system stability. Moreover, in the analysis of large-scale interconnected nonlinear dynamical systems, several Lyapunov functions arise naturally from the stability properties of each individual subsystem.

2.2 Notation and Definitions

In this section, we introduce notation and several definitions needed for developing the main results of this monograph. In a definition or when a word is defined in the text, the concept defined is italicized. Italics in the running text is also used for emphasis. The definition of a word, phrase, or symbol is to be understood as an "if and only if" statement. Lower-case letters such as x denote vectors, upper-case letters such as A denote matrices, upper-case script letters such as $\mathcal{S}$ denote sets, and lower-case Greek letters such as α denote scalars; however, there are a few exceptions to this convention. The notation $\mathcal{S}_1 \subset \mathcal{S}_2$ means that $\mathcal{S}_1$ is a proper subset of $\mathcal{S}_2$, whereas $\mathcal{S}_1 \subseteq \mathcal{S}_2$ means that either $\mathcal{S}_1$ is a proper subset of $\mathcal{S}_2$ or $\mathcal{S}_1$ is equal to $\mathcal{S}_2$. Throughout the monograph we use two basic types of mathematical statements, namely, *existential* and *universal* statements. An existential statement has the form: there exists $x \in \mathcal{X}$ such that a certain condition C is satisfied; whereas a universal statement has the form: condition C holds for all $x \in \mathcal{X}$. For universal statements we often omit the words "for all" and write: condition C holds, $x \in \mathcal{X}$.

The notation used in this monograph is fairly standard. Specifically, $\mathbb{R}$ (respectively, $\mathbb{C}$) denotes the set of real (respectively, complex) numbers,

$\overline{\mathbb{Z}}_+$ denotes the set of nonnegative integers, $\mathbb{Z}_+$ denotes the set of positive integers, $\mathbb{R}^n$ (respectively, $\mathbb{C}^n$) denotes the set of $n \times 1$ real (respectively, complex) column vectors, $\mathbb{R}^{n \times m}$ (respectively, $\mathbb{C}^{n \times m}$) denotes the set of real (respectively, complex) $n \times m$ matrices, $\mathbb{S}^n$ denotes the set of $n \times n$ symmetric matrices, $\mathbb{N}^n$ (respectively, $\mathbb{P}^n$) denotes the set of $n \times n$ nonnegative-definite (respectively, positive-definite) matrices, $(\cdot)^{\mathrm{T}}$ denotes transpose, $(\cdot)^+$ denotes the Moore-Penrose generalized inverse, $(\cdot)^{\#}$ denotes the group generalized inverse, $(\cdot)^{\mathrm{D}}$ denotes the Drazin inverse, $\otimes$ denotes Kronecker product, $\oplus$ denotes Kronecker sum, I_n or I denotes the $n \times n$ identity matrix, and $\mathbf{e}$ denotes the ones vector of order n, that is, $\mathbf{e} = [1, \ldots, 1]^{\mathrm{T}}$. For $x \in \mathbb{R}^q$ we write $x \geq\geq 0$ (respectively, $x >> 0$) to indicate that every component of x is nonnegative (respectively, positive). In this case, we say that x is *nonnegative* or *positive*, respectively. Likewise, $A \in \mathbb{R}^{p \times q}$ is *nonnegative* or *positive* if every entry of A is nonnegative or positive, respectively, which is written as $A \geq\geq 0$ or $A >> 0$, respectively. In addition, $\overline{\mathbb{R}}_+^q$ and $\mathbb{R}_+^q$ denote the nonnegative and positive orthants of $\mathbb{R}^q$, that is, if $x \in \mathbb{R}^q$, then $x \in \overline{\mathbb{R}}_+^q$ and $x \in \mathbb{R}_+^q$ are equivalent, respectively, to $x \geq\geq 0$ and $x >> 0$. Furthermore, $\mathcal{L}_2$ denotes the space of square-integrable Lebesgue measurable functions on $[0, \infty)$ and $\mathcal{L}_\infty$ denotes the space of bounded Lebesgue measurable functions on $[0, \infty)$. Finally, we denote the boundary, the interior, and the closure of the set $\mathcal{S}$ by $\partial\mathcal{S}$, $\overset{\circ}{\mathcal{S}}$, and $\overline{\mathcal{S}}$, respectively.

We write $\| \cdot \|$ for the Euclidean vector norm, $\mathcal{R}(A)$ and $\mathcal{N}(A)$ for the range space and the null space of a matrix A, respectively, $\mathrm{spec}(A)$ for the spectrum of the square matrix A including multiplicity, $\alpha(A)$ for the spectral abscissa of A (that is, $\alpha(A) = \max\{\mathrm{Re}\,\lambda : \lambda \in \mathrm{spec}(A)\}$), $\rho(A)$ for the spectral radius of A (that is, $\rho(A) = \max\{|\lambda| : \lambda \in \mathrm{spec}(A)\}$), and $\mathrm{ind}(A)$ for the index of A (that is, the size of the largest Jordan block of A associated with $\lambda = 0$, where $\lambda \in \mathrm{spec}(A)$). For a matrix $A \in \mathbb{R}^{p \times q}$, $\mathrm{row}_i(A)$ and $\mathrm{col}_j(A)$ denote the ith row and jth column of A, respectively. Furthermore, we write $V'(x)$ for the Fréchet derivative of V at x, $\mathcal{B}_\varepsilon(x)$, $x \in \mathbb{R}^n$, $\varepsilon > 0$, for the *open ball centered* at x with *radius* ε, $M \geq 0$ (respectively, $M > 0$) to denote the fact that the Hermitian matrix M is nonnegative (respectively, positive) definite, inf to denote infimum (that is, the greatest lower bound), sup to denote supremum (that is, the least upper bound), and $x(t) \to \mathcal{M}$ as $t \to \infty$ to denote that $x(t)$ approaches the set $\mathcal{M}$ (that is, for each $\varepsilon > 0$ there exists $T > 0$ such that $\mathrm{dist}(x(t), \mathcal{M}) < \varepsilon$ for all $t > T$, where $\mathrm{dist}(p, \mathcal{M}) \triangleq \inf_{x \in \mathcal{M}} \|p - x\|$). Finally, the notions of openness, convergence, continuity, and compactness that we use throughout the monograph refer to the topology generated on $\mathbb{R}^q$ by the norm $\| \cdot \|$.

2.3 Quasi-Monotone and Essentially Nonnegative Vector Fields

To develop the fundamental results of vector Lyapunov stability theory for nonlinear dynamical systems, we begin by considering the general nonlinear autonomous dynamical system

$$\dot{x}(t) = f(x(t)), \qquad x(0) = x_0, \qquad t \in \mathcal{I}_{x_0}, \tag{2.1}$$

where $x(t) \in \mathcal{D} \subseteq \mathbb{R}^n$, $t \in \mathcal{I}_{x_0}$, is the system state vector, $\mathcal{D}$ is an open set, $f : \mathcal{D} \to \mathbb{R}^n$ is continuous on $\mathcal{D}$, and $\mathcal{I}_{x_0} = [0, \tau_{x_0})$, $0 \leq \tau_{x_0} \leq \infty$, is the *maximal interval of existence* for the solution $x(\cdot)$ of (2.1). A continuously differentiable function $x : \mathcal{I}_{x_0} \to \mathcal{D}$ is said to be a *solution* to (2.1) on the interval $\mathcal{I}_{x_0} \subseteq \mathbb{R}$ with *initial condition* $x(0) = x_0$ if and only if $x(t)$ satisfies (2.1) for all $t \in \mathcal{I}_{x_0}$. We assume that for every initial condition $x(0) \in \mathcal{D}$ and every $\tau_{x_0} > 0$, the dynamical system (2.1) possesses a unique solution $x : [0, \tau_{x_0}) \to \mathcal{D}$ on the interval $[0, \tau_{x_0})$. We denote the solution to (2.1) with initial condition $x(0) = x_0$ by $s(\cdot, x_0)$, so that the *flow* of the dynamical system (2.1) given by the map $s : [0, \tau_{x_0}) \times \mathcal{D} \to \mathcal{D}$ is continuous in x and continuously differentiable in t and satisfies the *consistency* property $s(0, x_0) = x_0$ and the *semigroup* property $s(\tau, s(t, x_0)) = s(t + \tau, x_0)$, for all $x_0 \in \mathcal{D}$ and $t, \tau \in [0, \tau_{x_0})$ such that $t + \tau \in [0, \tau_{x_0})$. Unless otherwise stated, we assume $f(\cdot)$ is Lipschitz continuous on $\mathcal{D}$. Furthermore, $x_{\mathrm{e}} \in \mathcal{D}$ is an *equilibrium point* of (2.1) if and only if $f(x_{\mathrm{e}}) = 0$. In addition, a subset $\mathcal{D}_{\mathrm{c}} \subseteq \mathcal{D}$ is an *invariant set* relative to (2.1) if $\mathcal{D}_{\mathrm{c}}$ contains the orbits of all its points. Finally, recall that if all solutions to (2.1) are bounded, then it follows from the Peano-Cauchy theorem [70, p. 76] that $\mathcal{I}_{x_0} = \mathbb{R}$.

The following definition introduces the notion of Z-, M-, essentially nonnegative, compartmental, and nonnegative matrices.

Definition 2.1. Let $W \in \mathbb{R}^{q \times q}$. W is a Z-*matrix* if $W_{(i,j)} \leq 0$, $i, j = 1, \ldots, q$, $i \neq j$. W is an M-*matrix* (respectively, a *nonsingular* M-*matrix*) if W is a Z-matrix and all the principal minors of W are nonnegative (respectively, positive). W is *essentially nonnegative* if $-W$ is a Z-matrix, that is, $W_{(i,j)} \geq 0$, $i, j = 1, \ldots, q$, $i \neq j$. W is *compartmental* if W is essentially nonnegative and $\sum_{i=1}^{q} W_{(i,j)} \leq 0$, $j = 1, \ldots, q$. Finally, W is *nonnegative*[1] (respectively, *positive*) if $W_{(i,j)} \geq 0$ (respectively, $W_{(i,j)} > 0$), $i, j = 1, \ldots, q$.

A fundamental concept in the stability analysis of large-scale dynamical systems is the comparison principle, which invokes *quasi-monotone increasing* functions. The following definition adopted from [159] introduces such a class of functions.

Definition 2.2. A function $w = [w_1, \ldots, w_q]^{\mathrm{T}} : \mathbb{R}^q \times \mathcal{V} \to \mathbb{R}^q$, where $\mathcal{V} \subseteq \mathbb{R}^s$, is of *class* $\mathcal{W}$ if for every fixed $y \in \mathcal{V} \subseteq \mathbb{R}^s$, $w_i(z', y) \leq w_i(z'', y)$, $i = 1, \ldots, q$, for all $z', z'' \in \mathbb{R}^q$ such that $z'_j \leq z''_j$, $z'_i = z''_i$, $j = 1, \ldots, q$, $i \neq j$, where z_i denotes the ith component of z.

---

[1] In this monograph, it is important to distinguish between a square nonnegative (respectively, positive) matrix and a nonnegative-definite (respectively, positive-definite) matrix.

If $w(\cdot, y) \in \mathcal{W}$, then we say that w satisfies the *Kamke condition* [106, 169]. Note that if $w(z, y) = W(y)z$, where $W : \mathcal{V} \to \mathbb{R}^{q \times q}$, then the function $w(\cdot, y)$ is of class $\mathcal{W}$ if and only if $W(y)$ is essentially nonnegative for all $y \in \mathcal{V}$, that is, all the off-diagonal entries of the matrix function $W(\cdot)$ are nonnegative. Furthermore, note that it follows from Definition 2.2 that every scalar ($q = 1$) function $w(z, y)$ is of class $\mathcal{W}$.

The following definition introduces the notion of essentially nonnegative functions [19, 69].

Definition 2.3. Let $w = [w_1, \dots, w_q]^{\mathrm{T}} : \mathcal{V} \subseteq \overline{\mathbb{R}}_+^q \to \mathbb{R}^q$. Then w is *essentially nonnegative* if $w_i(r) \geq 0$ for all $i = 1, \dots, q$ and $r \in \overline{\mathbb{R}}_+^q$ such that $r_i = 0$, where r_i denotes the ith component of r.

Note that if $w : \mathbb{R}^q \to \mathbb{R}^q$ is such that $w(\cdot) \in \mathcal{W}$ and $w(0) \geq\geq 0$, then w is essentially nonnegative; the converse, however, is not generally true. However, if $w(r) = Wr$, where $W \in \mathbb{R}^{q \times q}$ is essentially nonnegative, then $w(\cdot)$ is essentially nonnegative and $w(\cdot) \in \mathcal{W}$.

Proposition 2.1 ([72]). Suppose $\overline{\mathbb{R}}_+^q \subset \mathcal{V}$. Then $\overline{\mathbb{R}}_+^q$ is an invariant set with respect to

$$\dot{r}(t) = w(r(t)), \quad r(t_0) = r_0, \quad t \geq t_0, \tag{2.2}$$

if and only if $w : \mathcal{V} \to \mathbb{R}^q$ is essentially nonnegative.

Proof. Define $\mathrm{dist}(r, \overline{\mathbb{R}}_+^q) \triangleq \inf_{y \in \overline{\mathbb{R}}_+^q} \|r - y\|$, $r \in \mathbb{R}^q$. Now, suppose $w : \mathcal{D} \to \mathbb{R}^q$ is essentially nonnegative and let $r \in \overline{\mathbb{R}}_+^q$. For every $i \in \{1, \dots, q\}$, if $r_i = 0$, then $r_i + hw_i(r) = hw_i(r) \geq 0$ for all $h \geq 0$, whereas, if $r_i > 0$, then $r_i + hw_i(r) > 0$ for all $|h|$ sufficiently small. Thus, $r + hw(r) \in \overline{\mathbb{R}}_+^q$ for all sufficiently small $h > 0$, and hence, $\lim_{h \to 0^+} \mathrm{dist}(r + hw(r), \overline{\mathbb{R}}_+^q)/h = 0$. It now follows from Lemma 2.1 of [72], with $r(0) = r_0$, that $r(t) \in \overline{\mathbb{R}}_+^q$ for all $t \in [0, \tau_{r_0})$.

Conversely, suppose that $\overline{\mathbb{R}}_+^q$ is invariant with respect to (2.2), let $r(0) \in \overline{\mathbb{R}}_+^q$, and suppose, *ad absurdum*, r is such that there exists $i \in \{1, \dots, q\}$ such that $r_i(0) = 0$ and $w_i(r(0)) < 0$. Then, since w is continuous, there exists sufficiently small $h > 0$ such that $w_i(r(t)) < 0$ for all $t \in [0, h)$, where $r(t)$ is the solution to (2.2). Hence, $r_i(t)$ is strictly decreasing on $[0, h)$, and thus, $r(t) \notin \overline{\mathbb{R}}_+^q$ for all $t \in (0, h)$, which leads to a contradiction. $\square$

The following corollary to Proposition 2.1 is immediate.

Corollary 2.1. Let $W \in \mathbb{R}^{q \times q}$. Then W is essentially nonnegative if and only if $e^{W(t - t_0)}$ is nonnegative for all $t \geq t_0$.

Proof. The proof is a direct consequence of Proposition 2.1 with

$w(r) = Wr$. For completeness of exposition, we provide a proof here based on matrix mathematics. To prove necessity, note that, since W is essentially nonnegative, it follows that $W_\alpha \triangleq W + \alpha I$ is nonnegative, where $\alpha \triangleq -\min\{W_{(1,1)}, \ldots, W_{(q,q)}\}$. Hence, $e^{W_\alpha(t-t_0)} = e^{(W+\alpha I)(t-t_0)} \geq\geq 0$, $t \geq t_0$, and hence, $e^{W(t-t_0)} = e^{-\alpha(t-t_0)}e^{W_\alpha(t-t_0)} \geq\geq 0$, $t \geq t_0$.

Conversely, suppose $e^{W(t-t_0)} \geq\geq 0$, $t \geq t_0$, and assume, *ad absurdum*, that there exist i, j such that $i \neq j$ and $W_{(i,j)} < 0$. Now, since $e^{W(t-t_0)} = \sum_{k=0}^{\infty}(k!)^{-1}W^k(t-t_0)^k$, it follows that

$$[e^{W(t-t_0)}]_{(i,j)} = I_{(i,j)} + (t-t_0)W_{(i,j)} + \mathcal{O}(t-t_0)^2, \tag{2.3}$$

where $\mathcal{O}(t-t_0)^2/(t-t_0) \to 0$ as $t \to t_0$. Thus, as $t \to t_0$ and $i \neq j$, it follows that $[e^{W(t-t_0)}]_{(i,j)} < 0$ for some t sufficiently close to t_0, which leads to a contradiction. Hence, W is essentially nonnegative. $\square$

The following definition and lemma are needed for developing several of the results in later sections.

Definition 2.4. The equilibrium solution $r(t) \equiv r_\mathrm{e}$ of (2.2) is *Lyapunov stable (with respect to $\overline{\mathbb{R}}_+^q$)* if, for every $\varepsilon > 0$, there exists $\delta = \delta(\varepsilon) > 0$ such that if $r_0 \in \mathcal{B}_\delta(r_\mathrm{e}) \cap \overline{\mathbb{R}}_+^q$, then $r(t) \in \mathcal{B}_\varepsilon(r_\mathrm{e}) \cap \overline{\mathbb{R}}_+^q$, $t \geq t_0$. The equilibrium solution $r(t) \equiv r_\mathrm{e}$ of (2.2) is *semistable (with respect to $\overline{\mathbb{R}}_+^q$)* if it is Lyapunov stable (with respect to $\overline{\mathbb{R}}_+^q$) and there exists $\delta > 0$ such that if $r_0 \in \mathcal{B}_\delta(r_\mathrm{e}) \cap \overline{\mathbb{R}}_+^q$, then $\lim_{t\to\infty} r(t)$ exists and converges to a Lyapunov stable equilibrium point. The equilibrium solution $r(t) \equiv r_\mathrm{e}$ of (2.2) is *asymptotically stable (with respect to $\overline{\mathbb{R}}_+^q$)* if it is Lyapunov stable (with respect to $\overline{\mathbb{R}}_+^q$) and there exists $\delta > 0$ such that if $r_0 \in \mathcal{B}_\delta(r_\mathrm{e}) \cap \overline{\mathbb{R}}_+^q$, then $\lim_{t\to\infty} r(t) = r_\mathrm{e}$. Finally, the equilibrium solution $r(t) \equiv r_\mathrm{e}$ of (2.2) is *globally asymptotically stable (with respect to $\overline{\mathbb{R}}_+^q$)* if the previous statement holds for all $r_0 \in \overline{\mathbb{R}}_+^q$.

Definition 2.4 introduces several types of stability notions of dynamical systems with respect to *relatively open* subsets of the nonnegative orthant of the state space containing the system equilibrium point [72]. In the case where the system trajectories are not restricted to the nonnegative orthant, the stability definitions introduced in Definition 2.4 reduce to the usual stability definitions [70]. In this monograph we do *not* distinguish between stability notions with respect to $\mathbb{R}^q$ versus $\overline{\mathbb{R}}_+^q$ as it is clear from the context which stability definition is meant. For the statement of the next result, recall that a matrix $W \in \mathbb{R}^{q \times q}$ is *semistable* if and only if $\lim_{t\to\infty} e^{Wt}$ exists [21, 69], whereas W is *asymptotically stable* if and only if $\lim_{t\to\infty} e^{Wt} = 0$.

Lemma 2.1. Suppose $W \in \mathbb{R}^{q \times q}$ is essentially nonnegative. If W is semistable (respectively, asymptotically stable), then there exist a scalar

$\alpha \geq 0$ (respectively, $\alpha > 0$) and a nonnegative vector $p \in \overline{\mathbb{R}}_+^q$, $p \neq 0$, (respectively, positive vector $p \in \mathbb{R}_+^q$) such that

$$W^{\mathrm{T}}p + \alpha p = 0. \tag{2.4}$$

Proof. Since W is semistable if and only if $\lambda = 0$ or $\mathrm{Re}\,\lambda < 0$, where $\lambda \in \mathrm{spec}(W)$, and $\mathrm{ind}(W) \leq 1$, it follows from Theorem 4.6 of [15] that $-W^{\mathrm{T}}$ is an M-matrix. Now, recalling that (see [93], p. 119) $-W^{\mathrm{T}}$ is an M-matrix if and only if there exist a scalar $\beta > 0$ and an $n \times n$ nonnegative matrix $B \geq\geq 0$ such that $\beta \geq \rho(B)$ and $-W^{\mathrm{T}} = \beta I_q - B$, it follows that W^{T} can be written as $W^{\mathrm{T}} = B - \beta I_q$, where $\beta > 0$. Now, since $B \geq\geq 0$, it follows from Theorem 8.3.1 of [92] that $\rho(B) \in \mathrm{spec}(B)$ and there exists $p \geq\geq 0$, $p \neq 0$, such that $Bp = \rho(B)p$. Hence, $W^{\mathrm{T}}p = Bp - \beta p = (\rho(B) - \beta)p = -\alpha p$, where $\alpha \triangleq \beta - \rho(B) \geq 0$, which proves that there exist $p \geq\geq 0$, $p \neq 0$, and $\alpha \geq 0$ such that (2.4) holds. In the case where W is asymptotically stable, the result is a direct consequence of the Perron-Frobenius theorem. $\qquad\square$

Finally, we introduce the notion of class $\mathcal{W}_{\mathrm{d}}$ functions involving *non-decreasing* functions.

Definition 2.5. A function $w = [w_1, \ldots, w_q]^{\mathrm{T}} : \mathbb{R}^q \times \mathcal{V} \to \mathbb{R}^q$, where $\mathcal{V} \subseteq \mathbb{R}^s$, is of *class* $\mathcal{W}_{\mathrm{d}}$ if for every fixed $y \in \mathcal{V} \subseteq \mathbb{R}^s$, $w(z', y) \leq\leq w(z'', y)$ for all $z', z'' \in \mathbb{R}^q$ such that $z' \leq\leq z''$.

Note that if $w(z, y) = W(y)z$, where $W : \mathcal{V} \to \mathbb{R}^{q \times q}$, then the function $w(\cdot, y)$ is of class $\mathcal{W}_{\mathrm{d}}$ if and only if $W(y)$ is nonnegative for all $y \in \mathcal{V}$, that is, all entries of the matrix function $W(\cdot)$ are nonnegative. Furthermore, note that if $w(\cdot, y) \in \mathcal{W}_{\mathrm{d}}$, then $w(\cdot, y) \in \mathcal{W}$.

2.4 Generalized Differential Inequalities

In this section, we develop a generalized comparison principle involving differential inequalities, wherein the underlying *comparison system* is partially dependent on the state of a dynamical system. Specifically, we consider the nonlinear comparison system given by

$$\dot{z}(t) = w(z(t), y(t)), \quad z(t_0) = z_0, \quad t \in \mathcal{I}_{z_0}, \tag{2.5}$$

where $z(t) \in \mathcal{Q} \subseteq \mathbb{R}^q$, $t \in \mathcal{I}_{z_0}$, is the comparison system state vector, $y : \mathcal{T} \to \mathcal{V} \subseteq \mathbb{R}^s$ is a *given* continuous function, $\mathcal{I}_{z_0} \subseteq \mathcal{T} \subseteq \overline{\mathbb{R}}_+$ is the maximal interval of existence of a solution $z(t)$ of (2.5), $\mathcal{Q}$ is an open set, $0 \in \mathcal{Q}$, and $w : \mathcal{Q} \times \mathcal{V} \to \mathbb{R}^q$. We assume that $w(\cdot, y(t))$ is continuous in t and satisfies the Lipschitz condition

$$\|w(z', y(t)) - w(z'', y(t))\| \leq L\|z' - z''\|, \quad t \in \mathcal{T}, \tag{2.6}$$

for all $z', z'' \in \mathcal{B}_\delta(z_0)$, where $\delta > 0$ and $L > 0$ is a Lipschitz constant. Hence, it follows from Theorem 2.2 of [110] that there exists $\tau > 0$ such that (2.5) has a unique solution over the time interval $[t_0, t_0 + \tau]$.

Theorem 2.1. Consider the nonlinear comparison system (2.5). Assume that the function $w : \mathcal{Q} \times \mathcal{V} \to \mathbb{R}^q$ is continuous and $w(\cdot, y)$ is of class $\mathcal{W}$. If there exists a continuously differentiable vector function $V = [v_1, \ldots, v_q]^{\mathrm{T}} : \mathcal{I}_{z_0} \to \mathcal{Q}$ such that

$$\dot{V}(t) << w(V(t), y(t)), \quad t \in \mathcal{I}_{z_0}, \tag{2.7}$$

then

$$V(t_0) << z_0, \quad z_0 \in \mathcal{Q}, \tag{2.8}$$

implies

$$V(t) << z(t), \quad t \in \mathcal{I}_{z_0}, \tag{2.9}$$

where $z(t)$, $t \in \mathcal{I}_{z_0}$, is the solution to (2.5).

Proof. Since $V(t)$, $t \in \mathcal{I}_{z_0}$, is continuous it follows that for sufficiently small $\tau > 0$,

$$V(t) << z(t), \quad t \in [t_0, t_0 + \tau]. \tag{2.10}$$

Now, suppose, *ad absurdum*, that inequality (2.9) does not hold on the entire interval $\mathcal{I}_{z_0}$. Then there exists $\hat{t} \in \mathcal{I}_{z_0}$ such that $V(t) << z(t)$, $t \in [t_0, \hat{t})$, and for at least one $i \in \{1, \ldots, q\}$,

$$v_i(\hat{t}) = z_i(\hat{t}) \tag{2.11}$$

and

$$v_j(\hat{t}) \le z_j(\hat{t}), \quad j \ne i, \quad j = 1, \ldots, q. \tag{2.12}$$

Since $w(\cdot, y) \in \mathcal{W}$, it follows from (2.7), (2.11), and (2.12) that

$$\dot{v}_i(\hat{t}) < w_i(V(\hat{t}), y(\hat{t})) \le w_i(z(\hat{t}), y(\hat{t})) = \dot{z}_i(\hat{t}), \tag{2.13}$$

which, along with (2.11), implies that for sufficiently small $\hat{\tau} > 0$,

$$v_i(t) > z_i(t), \quad t \in [\hat{t} - \hat{\tau}, \hat{t}). \tag{2.14}$$

This result contradicts the fact that $V(t) << z(t)$, $t \in [t_0, \hat{t})$, and hence establishes (2.9). $\qquad\square$

Next, we present a stronger version of Theorem 2.1 where the strict inequalities are replaced by soft inequalities.

Theorem 2.2. Consider the nonlinear comparison system (2.5). Assume that the function $w : \mathcal{Q} \times \mathcal{V} \to \mathbb{R}^q$ is continuous and $w(\cdot, y)$ is of class $\mathcal{W}$. Let $z(t)$, $t \in \mathcal{I}_{z_0}$, be the solution to (2.5) and $[t_0, t_0 + \tau] \subseteq \mathcal{I}_{z_0}$. If there exists a continuously differentiable vector function $V : [t_0, t_0 + \tau] \to \mathcal{Q}$ such that

$$\dot{V}(t) \leq\leq w(V(t), y(t)), \quad t \in [t_0, t_0 + \tau], \tag{2.15}$$

then

$$V(t_0) \leq\leq z_0, \quad z_0 \in \mathcal{Q}, \tag{2.16}$$

implies

$$V(t) \leq\leq z(t), \quad t \in [t_0, t_0 + \tau]. \tag{2.17}$$

Proof. Consider the family of comparison systems given by

$$\dot{z}(t) = w(z(t), y(t)) + \tfrac{\varepsilon}{n}\mathbf{e}, \quad z(t_0) = z_0 + \tfrac{\varepsilon}{n}\mathbf{e}, \tag{2.18}$$

where $\varepsilon > 0$, $n \in \overline{\mathbb{Z}}_+$, and $t \in \mathcal{I}_{z_0 + \frac{\varepsilon}{n}\mathbf{e}}$, and let the solution to (2.18) be denoted by $s_{(n)}(t, z_0 + \tfrac{\varepsilon}{n}\mathbf{e})$, $t \in \mathcal{I}_{z_0 + \frac{\varepsilon}{n}\mathbf{e}}$. Now, it follows from Theorem 3 of [44, p. 17] that there exists a compact interval $[t_0, t_0 + \tau] \subseteq \mathcal{I}_{z_0}$ such that $s_{(n)}(t, z_0 + \tfrac{\varepsilon}{n}\mathbf{e})$, $t \in [t_0, t_0 + \tau]$, is defined for all sufficiently large n. Moreover, it follows from Theorem 2.1 that

$$V(t) << s_{(n)}(t, z_0 + \tfrac{\varepsilon}{n}\mathbf{e}) << s_{(m)}(t, z_0 + \tfrac{\varepsilon}{m}\mathbf{e}), \quad n > m, \quad t \in [t_0, t_0 + \tau], \tag{2.19}$$

for all sufficiently large $m \in \overline{\mathbb{Z}}_+$.

Since the functions $s_{(n)}(t, z_0 + \tfrac{\varepsilon}{n}\mathbf{e})$, $t \in [t_0, t_0 + \tau]$, $n \in \overline{\mathbb{Z}}_+$, are continuous in t, decreasing in n, and bounded from below, it follows that the sequence of functions $s_{(n)}(\cdot, z_0 + \tfrac{\varepsilon}{n}\mathbf{e})$ converges uniformly on the compact interval $[t_0, t_0 + \tau]$ as $n \to \infty$; that is, there exists a continuous function $\hat{z} : [t_0, t_0 + \tau] \to \mathcal{Q}$ such that

$$s_{(n)}(t, z_0 + \tfrac{\varepsilon}{n}\mathbf{e}) \to \hat{z}(t), \quad n \to \infty, \tag{2.20}$$

uniformly on $[t_0, t_0 + \tau]$. Hence, it follows from (2.19) and (2.20) that

$$V(t) \leq\leq \hat{z}(t), \quad t \in [t_0, t_0 + \tau]. \tag{2.21}$$

Next, note that it follows from (2.18) that

$$s_{(n)}(t, z_0 + \tfrac{\varepsilon}{n}\mathbf{e}) = z_0 + \tfrac{\varepsilon}{n}\mathbf{e} + \int_{t_0}^{t} w(s_{(n)}(\sigma, z_0 + \tfrac{\varepsilon}{n}\mathbf{e}), y(\sigma)) \mathrm{d}\sigma,$$
$$t \in [t_0, t_0 + \tau], \tag{2.22}$$

which implies that $\hat{z}(t_0) = z_0$ and, since $y(\cdot)$ and $w(\cdot,\cdot)$ are continuous, $w(s_{(n)}(t, z_0 + \frac{\varepsilon}{n}\mathbf{e}), y(t)) \to w(\hat{z}(t), y(t))$ as $n \to \infty$ uniformly on $[t_0, t_0 + \tau]$. Hence, taking the limit as $n \to \infty$ on both sides of (2.22) yields

$$\hat{z}(t) = z_0 + \int_{t_0}^{t} w(\hat{z}(\sigma), y(\sigma))\mathrm{d}\sigma, \quad t \in [t_0, t_0 + \tau], \qquad (2.23)$$

which implies that $\hat{z}(t)$ is the solution to (2.5) on the interval $[t_0, t_0 + \tau]$. Hence, by uniqueness of solutions of (2.5) we obtain that $\hat{z}(t) = z(t)$, $[t_0, t_0 + \tau]$. This, along with (2.21), proves the result. $\qquad\square$

Next, consider the nonlinear dynamical system given by

$$\dot{x}(t) = f(x(t)), \quad x(t_0) = x_0, \quad t \in \mathcal{I}_{x_0}, \qquad (2.24)$$

where $x(t) \in \mathcal{D} \subseteq \mathbb{R}^n$, $t \in \mathcal{I}_{x_0}$, is the system state vector, $\mathcal{I}_{x_0}$ is the maximal interval of existence of a solution $x(t)$ of (2.24), $\mathcal{D}$ is an open set, $0 \in \mathcal{D}$, and $f(\cdot)$ is Lipschitz continuous on $\mathcal{D}$. The following result is a direct consequence of Theorem 2.2.

Corollary 2.2. Consider the nonlinear dynamical system (2.24). Assume that there exists a continuously differentiable vector function $V : \mathcal{D} \to \mathcal{Q} \subseteq \mathbb{R}^q$ such that

$$V'(x)f(x) \leq\leq w(V(x), x), \quad x \in \mathcal{D}, \qquad (2.25)$$

where $w : \mathcal{Q} \times \mathcal{D} \to \mathbb{R}^q$ is a continuous function, $w(\cdot, x) \in \mathcal{W}$, and

$$\dot{z}(t) = w(z(t), x(t)), \quad z(t_0) = z_0, \quad t \in \mathcal{I}_{z_0, x_0}, \qquad (2.26)$$

has a unique solution $z(t)$, $t \in \mathcal{I}_{z_0, x_0}$, where $x(t)$, $t \in \mathcal{I}_{x_0}$, is a solution to (2.24). If $[t_0, t_0 + \tau] \subseteq \mathcal{I}_{x_0} \cap \mathcal{I}_{z_0, x_0}$, then

$$V(x_0) \leq\leq z_0, \quad z_0 \in \mathcal{Q}, \qquad (2.27)$$

implies

$$V(x(t)) \leq\leq z(t), \quad t \in [t_0, t_0 + \tau]. \qquad (2.28)$$

Proof. For every given $x_0 \in \mathcal{D}$, the solution $x(t)$, $t \in \mathcal{I}_{x_0}$, to (2.24) is a well-defined function of time. Hence, define $\eta(t) \triangleq V(x(t))$, $t \in \mathcal{I}_{x_0}$, and note that (2.25) implies

$$\dot{\eta}(t) \leq\leq w(\eta(t), x(t)), \quad t \in \mathcal{I}_{x_0}. \qquad (2.29)$$

Moreover, since $[t_0, t_0 + \tau] \subseteq \mathcal{I}_{x_0} \cap \mathcal{I}_{z_0, x_0}$ is a compact interval, it follows from Theorem 2.2, with $y(t) \equiv x(t)$ and $V(x_0) = \eta(t_0) \leq\leq z_0$, that

$$V(x(t)) = \eta(t) \leq\leq z(t), \quad t \in [t_0, t_0 + \tau], \qquad (2.30)$$

which establishes the result. $\square$

If in (2.24) $f : \mathbb{R}^n \to \mathbb{R}^n$ is globally Lipschitz continuous, then (2.24) has a unique solution $x(t)$ for all $t \geq t_0$. A more restrictive sufficient condition for global existence and uniqueness of solutions to (2.24) is continuous differentiability of $f : \mathbb{R}^n \to \mathbb{R}^n$ and uniform boundedness of $f'(x)$ on $\mathbb{R}^n$. Note that if the solutions to (2.24) and (2.26) are globally defined for all $x_0 \in \mathcal{D}$ and $z_0 \in \mathcal{Q}$, then the result of Corollary 2.2 holds for any arbitrarily large but compact interval $[t_0, t_0 + \tau] \subset \overline{\mathbb{R}}_+$. For the remainder of this chapter we assume that the solutions to the systems (2.24) and (2.26) are defined for all $t \geq t_0$. Continuous differentiability of $f(\cdot)$ and $w(\cdot, \cdot)$ provides a sufficient condition for the existence and uniqueness of solutions to (2.24) and (2.26) for all $t \geq t_0$.

2.5 Stability Theory via Vector Lyapunov Functions

In this section, we develop a generalized vector Lyapunov function framework for the stability analysis of nonlinear dynamical systems using the generalized comparison principle developed in Section 2.4. Specifically, consider the cascade nonlinear dynamical system given by

$$\dot{z}(t) = w(z(t), x(t)), \quad z(t_0) = z_0, \quad t \geq t_0, \tag{2.31}$$

$$\dot{x}(t) = f(x(t)), \quad x(t_0) = x_0, \tag{2.32}$$

where $z_0 \in \mathcal{Q} \subseteq \mathbb{R}^q$, $x_0 \in \mathcal{D} \subseteq \mathbb{R}^n$, $[z^{\mathrm{T}}(t), x^{\mathrm{T}}(t)]^{\mathrm{T}}$, $t \geq t_0$, is the solution to (2.31) and (2.32), $w : \mathcal{Q} \times \mathcal{D} \to \mathbb{R}^q$ is continuous, $w(\cdot, x) \in \mathcal{W}$, $w(0,0) = 0$, $f : \mathcal{D} \to \mathbb{R}^n$ is Lipschitz continuous on $\mathcal{D}$, and $f(0) = 0$.

The following definition involving the notion of partial stability is needed for the next result.

Definition 2.6 ([70]). i) The nonlinear dynamical system (2.31) and (2.32) is *Lyapunov stable with respect to z* if, for every $\varepsilon > 0$ and $x_0 \in \mathcal{D}$, there exists $\delta = \delta(\varepsilon, x_0) > 0$ such that $\|z_0\| < \delta$ implies that $\|z(t)\| < \varepsilon$ for all $t \geq t_0$.

ii) The nonlinear dynamical system (2.31) and (2.32) is *Lyapunov stable with respect to z uniformly in x_0* if, for every $\varepsilon > 0$, there exists $\delta = \delta(\varepsilon) > 0$ such that $\|z_0\| < \delta$ implies that $\|z(t)\| < \varepsilon$ for all $t \geq t_0$ and for all $x_0 \in \mathcal{D}$.

iii) The nonlinear dynamical system (2.31) and (2.32) is *asymptotically stable with respect to z* if it is Lyapunov stable with respect to z and, for every $x_0 \in \mathcal{D}$, there exists $\delta = \delta(x_0) > 0$ such that $\|z_0\| < \delta$ implies that $\lim_{t\to\infty} z(t) = 0$.

iv) The nonlinear dynamical system (2.31) and (2.32) is *asymptotically stable with respect to z uniformly in x_0* if it is Lyapunov stable with respect to z uniformly in x_0 and there exists $\delta > 0$ such that $\|z_0\| < \delta$ implies that

$\lim_{t\to\infty} z(t) = 0$ uniformly in z_0 and x_0 for all $x_0 \in \mathcal{D}$.

v) The nonlinear dynamical system (2.31) and (2.32) is *globally asymptotically stable with respect to z* if it is Lyapunov stable with respect to z and $\lim_{t\to\infty} z(t) = 0$ for all $z_0 \in \mathbb{R}^q$ and $x_0 \in \mathbb{R}^n$.

vi) The nonlinear dynamical system (2.31) and (2.32) is *globally asymptotically stable with respect to z uniformly in x_0* if it is Lyapunov stable with respect to z uniformly in x_0 and $\lim_{t\to\infty} z(t) = 0$ uniformly in z_0 and x_0 for all $z_0 \in \mathbb{R}^q$ and $x_0 \in \mathbb{R}^n$.

vii) The nonlinear dynamical system (2.31) and (2.32) is *exponentially stable with respect to z uniformly in x_0* if there exist positive scalars α, β, and δ such that $\|z_0\| < \delta$ implies that $\|z(t)\| \leq \alpha\|z_0\|e^{-\beta(t-t_0)}$, $t \geq t_0$, for all $x_0 \in \mathcal{D}$.

$viii$) The nonlinear dynamical system (2.31) and (2.32) is *globally exponentially stable with respect to z uniformly in x_0* if there exist positive scalars α and β such that $\|z(t)\| \leq \alpha\|z_0\|e^{-\beta(t-t_0)}$, $t \geq t_0$, for all $z_0 \in \mathbb{R}^q$ and $x_0 \in \mathbb{R}^n$.

Theorem 2.3. Consider the nonlinear dynamical system (2.24). Assume that there exist a continuously differentiable vector function $V : \mathcal{D} \to \mathcal{Q} \cap \overline{\mathbb{R}}_+^q$ and a positive vector $p \in \mathbb{R}_+^q$ such that $V(0) = 0$, the scalar function $v : \mathcal{D} \to \overline{\mathbb{R}}_+$ defined by $v(x) \triangleq p^{\mathrm{T}} V(x)$, $x \in \mathcal{D}$, is such that $v(x) > 0$, $x \neq 0$, and

$$V'(x)f(x) \leq\leq w(V(x), x), \quad x \in \mathcal{D}, \tag{2.33}$$

where $w : \mathcal{Q} \times \mathcal{D} \to \mathbb{R}^q$ is continuous, $w(\cdot, x) \in \mathcal{W}$, and $w(0,0) = 0$. Then the following statements hold:

i) If the nonlinear dynamical system (2.31) and (2.32) is Lyapunov stable with respect to z uniformly in x_0, then the zero solution $x(t) \equiv 0$ to (2.24) is Lyapunov stable.

ii) If the nonlinear dynamical system (2.31) and (2.32) is asymptotically stable with respect to z uniformly in x_0, then the zero solution $x(t) \equiv 0$ to (2.24) is asymptotically stable.

iii) If $\mathcal{D} = \mathbb{R}^n$, $\mathcal{Q} = \mathbb{R}^q$, $v : \mathbb{R}^n \to \overline{\mathbb{R}}_+$ is radially unbounded, and the nonlinear dynamical system (2.31) and (2.32) is globally asymptotically stable with respect to z uniformly in x_0, then the zero solution $x(t) \equiv 0$ to (2.24) is globally asymptotically stable.

iv) If there exist constants $\nu \geq 1$, $\alpha > 0$, and $\beta > 0$ such that $v : \mathcal{D} \to \overline{\mathbb{R}}_+$ satisfies

$$\alpha\|x\|^\nu \leq v(x) \leq \beta\|x\|^\nu, \quad x \in \mathcal{D}, \tag{2.34}$$

and the nonlinear dynamical system (2.31) and (2.32) is exponentially stable with respect to z uniformly in x_0, then the zero solution $x(t) \equiv 0$ to (2.24) is exponentially stable.

v) If $\mathcal{D} = \mathbb{R}^n$, $\mathcal{Q} = \mathbb{R}^q$, there exist constants $\nu \geq 1$, $\alpha > 0$, and $\beta > 0$ such that $v : \mathbb{R}^n \to \overline{\mathbb{R}}_+$ satisfies (2.34), and the nonlinear dynamical system (2.31) and (2.32) is globally exponentially stable with respect to z uniformly in x_0, then the zero solution $x(t) \equiv 0$ to (2.24) is globally exponentially stable.

Proof. Assume there exist a continuously differentiable vector function $V : \mathcal{D} \to \mathcal{Q} \cap \overline{\mathbb{R}}_+^q$ and a positive vector $p \in \mathbb{R}_+^q$ such that $v(x) = p^{\mathrm{T}}V(x)$, $x \in \mathcal{D}$, is positive definite, that is, $v(0) = 0$ and $v(x) > 0$, $x \neq 0$. Note that since $v(x) = p^{\mathrm{T}}V(x) \leq \max_{i=1,\ldots,q}\{p_i\}\mathbf{e}^{\mathrm{T}}V(x)$, $x \in \mathcal{D}$, the function $\mathbf{e}^{\mathrm{T}}V(x)$, $x \in \mathcal{D}$, is also positive definite. Thus, there exist $r > 0$ and class $\mathcal{K}$ functions [85] α, $\beta : [0, r] \to \overline{\mathbb{R}}_+$ such that $\mathcal{B}_r(0) \subset \mathcal{D}$ and

$$\alpha(\|x\|) \leq \mathbf{e}^{\mathrm{T}}V(x) \leq \beta(\|x\|), \quad x \in \mathcal{B}_r(0). \tag{2.35}$$

i) Let $\varepsilon > 0$ and choose $0 < \hat{\varepsilon} < \min\{\varepsilon, r\}$. It follows from Lyapunov stability of the nonlinear dynamical system (2.31) and (2.32) with respect to z uniformly in x_0 that there exists $\mu = \mu(\hat{\varepsilon}) = \mu(\varepsilon) > 0$ such that if $\|z_0\|_1 < \mu$, where $\|\cdot\|_1$ denotes the absolute sum norm, then $\|z(t)\|_1 < \alpha(\hat{\varepsilon})$, $t \geq t_0$, for every $x_0 \in \mathcal{D}$. Now, choose $z_0 = V(x_0) \geq\geq 0$, $x_0 \in \mathcal{D}$. Since $V(x)$, $x \in \mathcal{D}$, is continuous, the function $\mathbf{e}^{\mathrm{T}}V(x)$, $x \in \mathcal{D}$, is also continuous. Hence, for $\mu = \mu(\hat{\varepsilon}) > 0$ there exists $\delta = \delta(\mu(\hat{\varepsilon})) = \delta(\varepsilon) > 0$ such that $\delta < \hat{\varepsilon}$, and if $\|x_0\| < \delta$, then $\mathbf{e}^{\mathrm{T}}V(x_0) = \mathbf{e}^{\mathrm{T}}z_0 = \|z_0\|_1 < \mu$, which implies that $\|z(t)\|_1 < \alpha(\hat{\varepsilon})$, $t \geq t_0$.

Now, with $z_0 = V(x_0) \geq\geq 0$, $x_0 \in \mathcal{D}$, and the assumption that $w(\cdot, x) \in \mathcal{W}$, $x \in \mathcal{D}$, it follows from (2.33) and Corollary 2.2 that $0 \leq\leq V(x(t)) \leq\leq z(t)$ on every compact interval $[t_0, t_0 + \tau]$, and hence, $\mathbf{e}^{\mathrm{T}}z(t) = \|z(t)\|_1$, $t \in [t_0, t_0 + \tau]$. Let $\tau > t_0$ be such that $x(t) \in \mathcal{B}_r(0)$, $t \in [t_0, t_0 + \tau]$, for all $x_0 \in \mathcal{B}_\delta(0)$. Thus, using (2.35), if $\|x_0\| < \delta$, then

$$\alpha(\|x(t)\|) \leq \mathbf{e}^{\mathrm{T}}V(x(t)) \leq \mathbf{e}^{\mathrm{T}}z(t) < \alpha(\hat{\varepsilon}), \quad t \in [t_0, t_0 + \tau], \tag{2.36}$$

which implies $\|x(t)\| < \hat{\varepsilon} < \varepsilon$, $t \in [t_0, t_0 + \tau]$.

Next, suppose, *ad absurdum*, that for some $x_0 \in \mathcal{B}_\delta(0)$ there exists $\hat{t} > t_0 + \tau$ such that $\|x(\hat{t})\| = \hat{\varepsilon}$. Then, for $z_0 = V(x_0)$ and the compact interval $[t_0, \hat{t}]$ it follows from (2.33) and Corollary 2.2 that $V(x(\hat{t})) \leq\leq z(\hat{t})$, which implies that $\alpha(\hat{\varepsilon}) = \alpha(\|x(\hat{t})\|) \leq \mathbf{e}^{\mathrm{T}}V(x(\hat{t})) \leq \mathbf{e}^{\mathrm{T}}z(\hat{t}) < \alpha(\hat{\varepsilon})$. This is a contradiction, and hence, for a given $\varepsilon > 0$ there exists $\delta = \delta(\varepsilon) > 0$ such that for all $x_0 \in \mathcal{B}_\delta(0)$, $\|x(t)\| < \varepsilon$, $t \geq t_0$, which implies Lyapunov stability of the zero solution $x(t) \equiv 0$ to (2.24).

ii) It follows from i) and the asymptotic stability of the nonlinear dynamical system (2.31), (2.32) with respect to z uniformly in x_0 that the zero solution to (2.24) is Lyapunov stable and there exists $\mu > 0$ such that if $\|z_0\|_1 < \mu$, then $\lim_{t \to \infty} z(t) = 0$ for every $x_0 \in \mathcal{D}$. As in i), choose $z_0 = V(x_0) \geq\geq 0$, $x_0 \in \mathcal{D}$. It follows from Lyapunov stability of the zero solution to (2.24) and the continuity of $V : \mathcal{D} \to \mathcal{Q} \cap \overline{\mathbb{R}}_+^q$ that there exists $\delta = \delta(\mu) > 0$ such that if $\|x_0\| < \delta$, then $\|x(t)\| < r$, $t \geq t_0$, and $\mathbf{e}^{\mathrm{T}} V(x_0) = \mathbf{e}^{\mathrm{T}} z_0 = \|z_0\|_1 < \mu$. Thus, by asymptotic stability of (2.31) and (2.32) with respect to z uniformly in x_0, for every arbitrary $\varepsilon > 0$ there exists $T = T(\varepsilon) > t_0$ such that $\|z(t)\|_1 < \alpha(\varepsilon)$, $t \geq T$. Thus, it follows from (2.33) and Corollary 2.2 that $0 \leq\leq V(x(t)) \leq\leq z(t)$ on every compact interval $[t_0, T + \tau]$, and hence, $\mathbf{e}^{\mathrm{T}} z(t) = \|z(t)\|_1$, $t \in [t_0, T + \tau]$, and, by (2.35),

$$\alpha(\|x(t)\|) \leq \mathbf{e}^{\mathrm{T}} V(x(t)) \leq \mathbf{e}^{\mathrm{T}} z(t) < \alpha(\varepsilon), \quad t \in [T, T + \tau]. \qquad (2.37)$$

Now, suppose, *ad absurdum*, that for some $x_0 \in \mathcal{B}_\delta(0)$, $\lim_{t \to \infty} x(t) \neq 0$, that is, there exists a sequence $\{t_k\}_{k=1}^\infty$, with $t_k \to \infty$ as $k \to \infty$, such that $\|x(t_k)\| \geq \hat{\varepsilon}$, $k \in \mathbb{Z}_+$, for some $0 < \hat{\varepsilon} < r$. Choose $\varepsilon = \hat{\varepsilon}$ and the interval $[T, T + \tau]$ such that at least one $t_k \in [T, T + \tau]$. Then it follows from (2.37) that $\alpha(\varepsilon) \leq \alpha(\|x(t_k)\|) < \alpha(\varepsilon)$, which is a contradiction. Hence, there exists $\delta > 0$ such that for all $x_0 \in \mathcal{B}_\delta(0)$, $\lim_{t \to \infty} x(t) = 0$ which, along with Lyapunov stability, implies asymptotic stability of the zero solution $x(t) \equiv 0$ to (2.24).

iii) Suppose $\mathcal{D} = \mathbb{R}^n$, $\mathcal{Q} = \mathbb{R}^q$, $v : \mathbb{R}^n \to \overline{\mathbb{R}}_+$ is a radially unbounded function, and the nonlinear dynamical system (2.31) and (2.32) is globally asymptotically stable with respect to z uniformly in x_0. In this case, for $V : \mathbb{R}^n \to \overline{\mathbb{R}}_+^q$ the inequality (2.35) holds for all $x \in \mathbb{R}^n$, where the functions $\alpha, \beta : \overline{\mathbb{R}}_+ \to \overline{\mathbb{R}}_+$ are of class $\mathcal{K}_\infty$ [85]. Furthermore, Lyapunov stability of the zero solution $x(t) \equiv 0$ to (2.24) follows from i). Next, for every $x_0 \in \mathbb{R}^n$ and $z_0 = V(x_0) \in \overline{\mathbb{R}}_+^q$, identical arguments as in ii) can be used to show that $\lim_{t \to \infty} x(t) = 0$, which proves global asymptotic stability of the zero solution $x(t) \equiv 0$ to (2.24).

iv) Suppose (2.34) holds. Since $p \in \mathbb{R}_+^q$, then

$$\hat{\alpha}\|x\|^\nu \leq \mathbf{e}^{\mathrm{T}} V(x) \leq \hat{\beta}\|x\|^\nu, \quad x \in \mathcal{D}, \qquad (2.38)$$

where $\hat{\alpha} \triangleq \alpha / \max_{i=1,\ldots,q}\{p_i\}$ and $\hat{\beta} \triangleq \beta / \min_{i=1,\ldots,q}\{p_i\}$. It follows from the exponential stability of the nonlinear dynamical system (2.31) and (2.32) with respect to z uniformly in x_0 that there exist positive constants γ, μ, and η such that if $\|z_0\|_1 < \mu$, then

$$\|z(t)\|_1 \leq \gamma\|z_0\|_1 e^{-\eta(t-t_0)}, \quad t \geq t_0, \qquad (2.39)$$

for all $x_0 \in \mathcal{D}$. Choose $z_0 = V(x_0) \geq\geq 0$, $x_0 \in \mathcal{D}$. By continuity of $V : \mathcal{D} \to \mathcal{Q} \cap \overline{\mathbb{R}}_+^q$, there exists $\delta = \delta(\mu) > 0$ such that for all $x_0 \in \mathcal{B}_\delta(0)$, $\mathbf{e}^{\mathrm{T}} V(x_0) = \mathbf{e}^{\mathrm{T}} z_0 = \|z_0\|_1 < \mu$. Furthermore, it follows from (2.33), (2.38), (2.39), and Corollary 2.2 that, for all $x_0 \in \mathcal{B}_\delta(0)$, the inequality

$$\hat{\alpha}\|x(t)\|^\nu \leq \mathbf{e}^{\mathrm{T}} V(x(t)) \leq \mathbf{e}^{\mathrm{T}} z(t)$$

$$\leq \gamma \|z_0\|_1 e^{-\eta(t-t_0)}$$
$$\leq \gamma \hat{\beta} \|x_0\|^{\nu} e^{-\eta(t-t_0)} \tag{2.40}$$

holds on every compact interval $[t_0, t_0 + \tau]$. This in turn implies that, for every $x_0 \in \mathcal{B}_\delta(0)$,

$$\|x(t)\| \leq \left(\frac{\gamma \hat{\beta}}{\hat{\alpha}} \right)^{\frac{1}{\nu}} \|x_0\| e^{-\frac{\eta}{\nu}(t-t_0)}, \quad t \in [t_0, t_0 + \tau]. \tag{2.41}$$

Now, suppose, *ad absurdum*, that for some $x_0 \in \mathcal{B}_\delta(0)$ there exists $\hat{t} > t_0 + \tau$ such that

$$\|x(\hat{t})\| > \left(\frac{\gamma \hat{\beta}}{\hat{\alpha}} \right)^{\frac{1}{\nu}} \|x_0\| e^{-\frac{\eta}{\nu}(\hat{t}-t_0)}. \tag{2.42}$$

Then for the compact interval $[t_0, \hat{t}]$, it follows from (2.41) that

$$\|x(\hat{t})\| \leq \left(\frac{\gamma \hat{\beta}}{\hat{\alpha}} \right)^{\frac{1}{\nu}} \|x_0\| e^{-\frac{\eta}{\nu}(\hat{t}-t_0)}, \tag{2.43}$$

which is a contradiction. Thus, inequality (2.41) holds for all $t \geq t_0$ establishing exponential stability of the zero solution $x(t) \equiv 0$ to (2.24).

v) The proof is identical to the proof of *iv*). $\qquad\qquad\square$

If $V : \mathcal{D} \to \mathcal{Q} \cap \overline{\mathbb{R}}_+^q$ satisfies the conditions of Theorem 2.3 we say that $V(x)$, $x \in \mathcal{D}$, is a *vector Lyapunov function* [159]. Note that for stability analysis each component of a vector Lyapunov function need not be positive definite with a negative definite or negative-semidefinite time derivative along the trajectories of (2.31) and (2.32). This provides more flexibility in searching for a vector Lyapunov function as compared to a scalar Lyapunov function for addressing the stability of nonlinear dynamical systems.

It is important to note here that comparison systems with vector fields dependent on the states of both the system dynamics and the comparison system have been addressed in the literature [61, 65, 66], with [65] providing stability analysis using partial stability notions. However, a key difference between our formulation and the results given in [65] is in the definitions of partial stability used to analyze the stability of the generalized comparison system.

Specifically, the partial stability definitions used in [65] (see Definitions 2 and 3 on pages 161 and 162) require that the entire initial system state of the generalized comparison system lie in a neighborhood of the origin, whereas in our definition of partial stability the initial system state corresponding to (2.32) can be arbitrary. This weaker assumption leads to stronger results. Furthermore, in Theorem 2.3 each component of the vector

Lyapunov function is dependent on the entire state x of the dynamical system, while in Theorem 29 of [65] (see p. 210) the vector Lyapunov function is component-decoupled, that is, $V(x) = [v_1(x_1), \ldots, v_q(x_q)]^{\mathrm{T}}$, $x_i \in \mathbb{R}^{n_i}$, $i = 1, \ldots, q$. In addition, in Theorem 2.3 we only require that the scalar function $v(x) \triangleq p^{\mathrm{T}} V(x)$, $x \in \mathcal{D}$, be positive definite, whereas in Theorem 29 of [65] each component of a vector Lyapunov function $V(x)$, $x \in \mathcal{D}$, is assumed to be a positive-definite function of its argument.

Sufficient conditions for partial stability of the nonlinear dynamical system (2.31) and (2.32) are given in [70]. Specifically, Theorem 1 of [41] establishes partial stability of (2.31) and (2.32) in terms of a scalar Lyapunov function that is dependent on both the states z and x. Alternatively, Corollary 1 of [41] provides partial stability of (2.31) and (2.32) in terms of a scalar Lyapunov function that is only dependent on the comparison system state z, which can simplify the stability analysis. In this case, the expanded dimension of the system (2.31) and (2.32) does not introduce additional complexity for the partial stability analysis of the generalized comparison system. As in standard vector Lyapunov theory, this ensures a reduced dimension for the analysis of the comparison system while addressing a more general class of nonlinear systems. This point is further discussed in Section 5.4.

The following corollary to Theorem 2.3 is immediate and corresponds to the standard vector Lyapunov theorem addressed in the literature [159].

Corollary 2.3. Consider the nonlinear dynamical system (2.24). Assume that there exist a continuously differentiable vector function $V : \mathcal{D} \to \mathcal{Q} \cap \overline{\mathbb{R}}_+^q$ and a positive vector $p \in \mathbb{R}_+^q$ such that $V(0) = 0$, the scalar function $v : \mathcal{D} \to \overline{\mathbb{R}}_+$ defined by $v(x) \triangleq p^{\mathrm{T}} V(x)$, $x \in \mathcal{D}$, is such that $v(x) > 0$, $x \neq 0$, and

$$V'(x) f(x) \leq\leq w(V(x)), \quad x \in \mathcal{D், \tag{2.44}$$

where $w : \mathcal{Q} \to \mathbb{R}^q$ is continuous, $w(\cdot) \in \mathcal{W}$, and $w(0) = 0$. Then the stability properties of the zero solution $z(t) \equiv 0$ to

$$\dot{z}(t) = w(z(t)), \quad z(t_0) = z_0, \quad t \geq t_0, \tag{2.45}$$

where $z_0 \in \mathcal{Q}$, imply the corresponding stability properties of the zero solution $x(t) \equiv 0$ to (2.24). In particular, if the zero solution $z(t) \equiv 0$ to (2.45) is Lyapunov (respectively, asymptotically) stable, then the zero solution $x(t) \equiv 0$ to (2.24) is Lyapunov (respectively, asymptotically) stable. If, in addition, $\mathcal{D} = \mathbb{R}^n$, $\mathcal{Q} = \mathbb{R}^q$, and $v : \mathbb{R}^n \to \overline{\mathbb{R}}_+$ is a positive-definite, radially unbounded function, then global asymptotic stability of the zero solution $z(t) \equiv 0$ to (2.45) implies global asymptotic stability of the zero solution $x(t) \equiv 0$ to (2.24). Moreover, if there exist constants $\nu \geq 1$, $\alpha > 0$, and $\beta > 0$ such that $v : \mathcal{D} \to \overline{\mathbb{R}}_+$ satisfies (2.34), then exponential stability of the zero solution $z(t) \equiv 0$ to (2.45) implies exponential stability of the zero solution $x(t) \equiv 0$ to (2.24). Finally, if $\mathcal{D} = \mathbb{R}^n$, $\mathcal{Q} = \mathbb{R}^q$, there exist constants $\nu \geq 1$, $\alpha > 0$, and $\beta > 0$ such that $v : \mathbb{R}^n \to \overline{\mathbb{R}}_+$ satisfies (2.34), then global exponential stability of the zero solution $z(t) \equiv 0$ to (2.45) implies global exponential stability of the zero solution $x(t) \equiv 0$ to (2.24).

Proof. The proof is a direct consequence of Theorem 2.3 with $w(z, x) \equiv w(z)$. $\qquad\square$

Example 2.1. Consider the nonlinear dynamical system given by

$$\dot{x}_1(t) = -x_1(t) - x_1^2(t)x_2^3(t), \quad x_1(0) = x_{10}, \quad t \geq 0, \qquad (2.46)$$
$$\dot{x}_2(t) = -x_2^3(t) + x_1^2(t)x_2^2(t), \quad x_2(0) = x_{20}. \qquad (2.47)$$

Note that Lyapunov's indirect method fails to yield any information on the stability of the zero solution $x(t) \triangleq [x_1(t), x_2(t)]^{\mathrm{T}} \equiv 0$ of (2.46) and (2.47). To examine the stability of (2.46) and (2.47) consider the vector Lyapunov function candidate $V(x) = [v_1(x), v_2(x)]^{\mathrm{T}}$, $x \in \mathbb{R}^2$, with $v_1(x) = \frac{1}{2}x_1^2$ and $v_2(x) = \frac{1}{4}x_2^4$. Clearly, $V(0) = 0$ and $\mathbf{e}^{\mathrm{T}}V(x)$, $x \in \mathbb{R}^2$, is a positive-definite function.

Next, consider the domain $\mathcal{D} \triangleq \{x \in \mathbb{R}^2 : |x_1| \leq 1, |x_2| \leq c_2\}$, where $c_2 > 0$, and note that

$$\dot{v}_1(x(t)) = x_1(t)(-x_1(t) - x_1^2(t)x_2^3(t)) \leq -x_1^2(t) + |x_1^3(t)x_2^3(t)|$$
$$\leq (-2 + 2c_2^3)v_1(x(t)), \qquad (2.48)$$
$$\dot{v}_2(x(t)) = x_2^3(t)(-x_2^3(t) + x_1^2(t)x_2^2(t)) \leq -x_2^6(t) + |x_1^2(t)x_2^5(t)|$$
$$\leq 2c_2^5 v_1(x(t)) - 8v_2^{\frac{3}{2}}(x(t)), \qquad (2.49)$$

for all $x(t) \in \mathcal{D}$, $t \geq 0$. Thus, the comparison system (2.45) is given by

$$\dot{z}_1(t) = (-2 + 2c_2^3)z_1(t), \quad z_1(0) = z_{10}, \quad t \geq 0, \qquad (2.50)$$
$$\dot{z}_2(t) = 2c_2^5 z_1(t) - 8z_2^{\frac{3}{2}}(t), \quad z_2(0) = z_{20}, \qquad (2.51)$$

where $(z_{10}, z_{20}) \in \overline{\mathbb{R}}_+ \times \overline{\mathbb{R}}_+$. Note that $w(z) \triangleq [(-2+2c_2^3)z_1, 2c_2^5 z_1 - 8z_2^{\frac{3}{2}}]^{\mathrm{T}} \in \mathcal{W}$, where $z \triangleq [z_1, z_2]^{\mathrm{T}}$.

Next, to show stability of the zero solution $z(t) \equiv 0$ to the comparison system, consider the linear Lyapunov function candidate $v(z) = z_1 + z_2$, $z \in \overline{\mathbb{R}}_+^2$. Clearly, $v(0) = 0$ and $v(z) > 0$, $z \in \overline{\mathbb{R}}_+^2 \setminus \{0\}$. Moreover,

$$\dot{v}(z(t)) = 2(-1 + c_2^3 + c_2^5)z_1(t) - 8z_2^{\frac{3}{2}}(t), \quad t \geq 0. \qquad (2.52)$$

In order to ensure asymptotic stability of the zero solution $z(t) \equiv 0$, it suffices to take $c_2 = 0.83$. In this case, it follows from Corollary 2.3 that the zero solution $x(t) \equiv 0$ to (2.46) and (2.47) is asymptotically stable. $\qquad\triangle$

Next, we present a convergence result via vector Lyapunov functions that allows us to establish asymptotic stability of the nonlinear dynamical system (2.24) using weaker conditions than those assumed in Theorem 2.3.

Theorem 2.4. Consider the nonlinear dynamical system (2.24), assume that there exist a continuously differentiable vector function $V = [v_1, \ldots, v_q]^{\mathrm{T}} : \mathcal{D} \to \mathcal{Q} \cap \overline{\mathbb{R}}_+^q$ and a positive vector $p \in \mathbb{R}_+^q$ such that $V(0) = 0$, the scalar function $v : \mathcal{D} \to \overline{\mathbb{R}}_+$ defined by $v(x) \triangleq p^{\mathrm{T}} V(x)$, $x \in \mathcal{D}$, is such that $v(x) > 0$, $x \neq 0$, and

$$V'(x)f(x) \leq\leq w(V(x), x), \quad x \in \mathcal{D}, \tag{2.53}$$

where $w : \mathcal{Q} \times \mathcal{D} \to \mathbb{R}^q$ is continuous, $w(\cdot, x) \in \mathcal{W}_{\mathrm{d}}$, and $w(0, 0) = 0$, such that the nonlinear dynamical system (2.31) and (2.32) is Lyapunov stable with respect to z uniformly in x_0. Let $\mathcal{R}_i \triangleq \{x \in \mathcal{D} : v_i'(x)f(x) - w_i(V(x), x) = 0\}$, $i = 1, \ldots, q$. Then there exists $\mathcal{D}_{\mathrm{c}} \subset \mathcal{D}$ such that $x(t) \to \mathcal{R} \triangleq \cap_{i=1}^q \mathcal{R}_i$ as $t \to \infty$ for all $x(t_0) = x_0 \in \mathcal{D}_{\mathrm{c}}$. Moreover, if $\mathcal{R}$ contains no trajectory other than the trivial trajectory, then the zero solution $x(t) \equiv 0$ to (2.24) is asymptotically stable.

Proof. Since the nonlinear dynamical system (2.31) and (2.32) is Lyapunov stable with respect to z uniformly in x_0, it follows that there exists $\hat{\delta} > 0$ such that if $\|z_0\|_1 < \hat{\delta}$, then the partial system trajectories $z(t)$, $t \geq t_0$, of (2.31) and (2.32) are bounded for all $x_0 \in \mathcal{D}$. Furthermore, since $V(x)$, $x \in \mathcal{D}$, is continuous, it follows that there exists $\delta_1 = \delta_1(\hat{\delta}) > 0$ such that $\mathbf{e}^{\mathrm{T}} V(x_0) < \hat{\delta}$ for all $x_0 \in \mathcal{B}_{\delta_1}(0)$. In addition, it follows from Theorem 2.3 that the zero solution $x(t) \equiv 0$ to (2.24) is Lyapunov stable, and hence, for a given $\varepsilon > 0$ such that $\mathcal{B}_\varepsilon(0) \subset \overset{\circ}{\mathcal{D}}$ there exists $\delta_2 = \delta_2(\varepsilon) > 0$ such that if $x_0 \in \mathcal{B}_{\delta_2}(0)$, then $x(t) \in \mathcal{B}_\varepsilon(0)$, $t \geq t_0$, where $x(t)$, $t \geq t_0$, is the solution to (2.24). Choose $\delta = \min\{\delta_1, \delta_2\}$ and define $\mathcal{D}_{\mathrm{c}} \triangleq \mathcal{B}_\delta(0) \subset \mathcal{D}$. Then for all $z_0 = V(x_0)$ and $x_0 \in \mathcal{D}_{\mathrm{c}}$, it follows that $x(t) \in \mathcal{B}_\varepsilon(0)$, $t \geq t_0$, and $z(t)$, $t \geq t_0$, is bounded.

Next, consider the function

$$W_i(x, t) \triangleq v_i(x) - \int_{t_0}^t w_i(V(x(s)), x(s))\mathrm{d}s, \quad t \geq t_0, \quad x \in \mathcal{D}, \quad i = 1, \ldots, q. \tag{2.54}$$

It follows from (2.53) that

$$\dot{W}_i(x(t), t) = v_i'(x(t))f(x(t)) - w_i(V(x(t)), x(t)) \leq 0, \quad t \geq t_0, \quad x_0 \in \mathcal{D}, \tag{2.55}$$

which implies that $W_i(x(t), t)$, $i \in \{1, \ldots, q\}$, is a nonincreasing function of time, and hence, $\lim_{t \to \infty} W_i(x(t), t)$, $i \in \{1, \ldots, q\}$, exists. Moreover, $W_i(x(t_0), t_0) = v_i(x(t_0)) < +\infty$ for all $x(t_0) = x_0 \in \mathcal{D}$ since $v_i(x)$, $x \in$

$\mathcal{D}$, $i \in \{1, \ldots, q\}$, is continuous. Now, suppose, *ad absurdum*, that for some initial condition $x(t_0) = x_0 \in \mathcal{D}_{\mathrm{c}}$,

$$\lim_{t \to \infty} W_i(x(t), t) = -\infty, \quad i \in \{1, \ldots, q\}. \tag{2.56}$$

Since the function $v_i(x)$, $x \in \mathcal{D}$, $i \in \{1, \ldots, q\}$, is continuous on the compact set $\overline{\mathcal{B}_\varepsilon(0)}$, it follows that $v_i(x(t))$, $t \geq t_0$, is bounded, and hence,

$$\lim_{t \to \infty} \int_{t_0}^{t} w_i(V(x(s)), x(s)) \mathrm{d}s = +\infty, \quad i \in \{1, \ldots, q\}. \tag{2.57}$$

Now, it follows from (2.53) and Corollary 2.2 that $V(x(t)) \leq\leq z(t)$, $t \geq t_0$, for $z(t_0) = V(x(t_0))$. Note that since $x_0 \in \mathcal{D}_{\mathrm{c}}$ it follows that $z(t)$, $t \geq t_0$, is bounded.

Furthermore, since $w(\cdot, x) \in \mathcal{W}_{\mathrm{d}}$ it follows that

$$\begin{aligned}
v_i(x(t)) &\leq v_i(x(t_0)) + \int_{t_0}^{t} w_i(V(x(s)), x(s)) \mathrm{d}s \\
&\leq z_i(t_0) + \int_{t_0}^{t} w_i(z(s), x(s)) \mathrm{d}s \\
&= z_i(t),
\end{aligned} \tag{2.58}$$

for all $t \geq t_0$. Since $z(t)$, $t \geq t_0$, is bounded and $v_i(x)$, $x \in \mathcal{D}$, $i \in \{1, \ldots, q\}$, is continuous, it follows that there exists $M > 0$ such that

$$\left| \int_{t_0}^{t} w_i(V(x(s)), x(s)) \mathrm{d}s \right| < M < +\infty, \quad t \geq t_0, \quad i \in \{1, \ldots, q\}. \tag{2.59}$$

This is a contradiction, and hence, $\lim_{t \to \infty} W_i(x(t), t)$, $i \in \{1, \ldots, q\}$, exists and is finite for every $x_0 \in \mathcal{D}_{\mathrm{c}}$. Thus, for every $x_0 \in \mathcal{D}_{\mathrm{c}}$, it follows that

$$\begin{aligned}
\int_{t_0}^{t} \dot{W}_i(x(s), s) \mathrm{d}s &= \int_{t_0}^{t} [v_i'(x(s)) f(x(s)) - w_i(V(x(s)), x(s))] \mathrm{d}s \\
&= W_i(x(t), t) - W_i(x_0, t_0), \quad t \geq t_0,
\end{aligned} \tag{2.60}$$

and hence,

$$\lim_{t \to \infty} \int_{t_0}^{t} [v_i'(x(s)) f(x(s)) - w_i(V(x(s)), x(s))] \mathrm{d}s, \quad i \in \{1, \ldots, q\}, \tag{2.61}$$

exists and is finite.

Next, since $f(\cdot)$ is Lipschitz continuous on $\mathcal{D}$ and $x(t) \in \mathcal{B}_\varepsilon(0)$ for all $x_0 \in \mathcal{D}_{\mathrm{c}}$ and $t \geq t_0$, it follows that

$$\|x(t_2) - x(t_1)\| = \left\| \int_{t_1}^{t_2} f(x(s)) \mathrm{d}s \right\| \leq L \int_{t_1}^{t_2} \|x(s)\| \mathrm{d}s \leq L\varepsilon(t_2 - t_1),$$
$$t_2 \geq t_1 \geq t_0, \tag{2.62}$$

where L is the Lipschitz constant on $\mathcal{D}_{\mathrm{c}}$. Thus, it follows from (2.62) that for every $\gamma > 0$ there exists $\mu = \mu(\gamma) = \frac{\gamma}{L\varepsilon}$ such that

$$\|x(t_2) - x(t_1)\| < \gamma, \quad |t_2 - t_1| < \mu, \tag{2.63}$$

which shows that $x(t)$, $t \geq t_0$, is uniformly continuous. Next, since $x(t)$ is uniformly continuous and $v_i'(x)f(x) - w_i(V(x), x)$, $x \in \mathcal{D}$, $i \in \{1, \ldots, q\}$, is continuous, it follows that $v_i'(x(t))f(x(t)) - w_i(V(x(t)), x(t))$, $i \in \{1, \ldots, q\}$, is uniformly continuous at every $t \geq t_0$. Hence, it follows from Barbalat's lemma [70, p. 221] that $v_i'(x(t))f(x(t)) - w_i(V(x(t)), x(t)) \to 0$ as $t \to \infty$ for all $x_0 \in \mathcal{D}_{\mathrm{c}}$ and $i \in \{1, \ldots, q\}$. Repeating the above analysis for all $i = 1, \ldots, q$, it follows that $x(t) \to \mathcal{R} = \cap_{i=1}^{q} \mathcal{R}_i$ for all $x_0 \in \mathcal{D}_{\mathrm{c}}$. Finally, if $\mathcal{R}$ contains no trajectory other than the trivial trajectory, then $\mathcal{R} = \{0\}$, and hence, $x(t) \to 0$ as $t \to \infty$ for all $x_0 \in \mathcal{D}_{\mathrm{c}}$, which proves asymptotic stability of the zero solution $x(t) \equiv 0$ to (2.24). $\qquad\square$

Note that $\mathcal{R} = \cap_{i=1}^{q} \mathcal{R}_i \neq \varnothing$ since $0 \in \mathcal{R}$. Furthermore, recall that for every bounded solution $x(t)$, $t \geq t_0$, to (2.24) with initial condition $x(t_0) = x_0$, the positive limit set $\omega(x_0)$ of (2.24) is a nonempty, compact, invariant, and connected set with $x(t) \to \omega(x_0)$ as $t \to \infty$ [70]. If $q = 1$ and $w(V(x), x) \equiv 0$, then it can be shown that the Lyapunov derivative $\dot{V}(x)$ vanishes on the positive limit set $\omega(x_0)$, $x_0 \in \mathcal{D}_{\mathrm{c}}$, so that $\omega(x_0) \in \mathcal{R}$. Moreover, since $\omega(x_0)$ is a positively invariant set with respect to (2.24), it follows that for all $x_0 \in \mathcal{D}_{\mathrm{c}}$, the trajectory of (2.24) converges to the largest invariant set $\mathcal{M}$ contained in $\mathcal{R}$. In this case, Theorem 2.4 specializes to the classical Krasovskii-LaSalle invariant set theorem [70].

If for some $k \in \{1, \ldots, q\}$, $w_k(V(x), x) \equiv 0$ and $v_k'(x)f(x) < 0$, $x \in \mathcal{D}$, $x \neq 0$, then $\mathcal{R} = \mathcal{R}_k = \{0\}$. In this case, it follows from Theorem 2.4 that the zero solution $x(t) \equiv 0$ to (2.24) is asymptotically stable. Note that even though for $k \in \{1, \ldots, q\}$ the time derivative $\dot{v}_k(x)$, $x \in \mathcal{D}$, is negative definite, the function $v_k(x)$, $x \in \mathcal{D}$, can be nonnegative definite, in contrast to classical Lyapunov stability theory, to ensure asymptotic stability of (2.24).

Next, we use the vector Lyapunov stability results of Theorem 2.3 to develop partial stability analysis results for nonlinear dynamical systems [41]. Specifically, consider the nonlinear dynamical system (2.24) with partitioned dynamics[2] given by

$$\dot{x}_{\mathrm{I}}(t) = f_{\mathrm{I}}(x_{\mathrm{I}}(t), x_{\mathrm{II}}(t)), \quad x_{\mathrm{I}}(t_0) = x_{\mathrm{I}0}, \quad t \geq t_0, \tag{2.64}$$

$$\dot{x}_{\mathrm{II}}(t) = f_{\mathrm{II}}(x_{\mathrm{I}}(t), x_{\mathrm{II}}(t)), \quad x_{\mathrm{II}}(t_0) = x_{\mathrm{II}0}, \tag{2.65}$$

where $x_{\mathrm{I}}(t) \in \mathcal{D}_{\mathrm{I}}$, $t \geq t_0$, $\mathcal{D}_{\mathrm{I}} \subseteq \mathbb{R}^{n_{\mathrm{I}}}$ is an open set such that $0 \in \mathcal{D}_{\mathrm{I}}$, $x_{\mathrm{II}}(t) \in \mathbb{R}^{n_{\mathrm{II}}}$, $t \geq t_0$, $f_{\mathrm{I}} : \mathcal{D}_{\mathrm{I}} \times \mathbb{R}^{n_{\mathrm{II}}} \to \mathbb{R}^{n_{\mathrm{I}}}$ is such that for all $x_{\mathrm{II}} \in \mathbb{R}^{n_{\mathrm{II}}}$,

---

[2]Here we use the Roman subscripts I and II as opposed to Arabic subscripts 1 and 2 for denoting the partial states of x not to confuse the partial states with the component states of the vector Lyapunov function.

$f_\mathrm{I}(0, x_\mathrm{II}) = 0$ and $f_\mathrm{I}(\cdot, x_\mathrm{II})$ is locally Lipschitz in x_I, $f_\mathrm{II} : \mathcal{D}_\mathrm{I} \times \mathbb{R}^{n_\mathrm{II}} \to \mathbb{R}^{n_\mathrm{II}}$ is such that for every $x_\mathrm{I} \in \mathcal{D}_\mathrm{I}$, $f_\mathrm{II}(x_\mathrm{I}, \cdot)$ is locally Lipschitz in x_II, $x(t) \triangleq [x_\mathrm{I}^\mathrm{T}(t), x_\mathrm{II}^\mathrm{T}(t)]^\mathrm{T} \in \mathcal{D} = \mathcal{D}_\mathrm{I} \times \mathbb{R}^{n_\mathrm{II}} \subseteq \mathbb{R}^n$, $t \geq t_0$, $x_0 \triangleq [x_\mathrm{I0}^\mathrm{T}, x_\mathrm{II0}^\mathrm{T}]^\mathrm{T}$, and $n_\mathrm{I} + n_\mathrm{II} = n$. For the nonlinear dynamical system (2.64) and (2.65) the definitions of partial stability given in Definition 2.6 hold with (2.31) and (2.32) replaced by (2.64) and (2.65). Note that for the dynamical system (2.24), $f(x_\mathrm{I}, x_\mathrm{II}) = [f_\mathrm{I}^\mathrm{T}(x_\mathrm{I}, x_\mathrm{II}), f_\mathrm{II}^\mathrm{T}(x_\mathrm{I}, x_\mathrm{II})]^\mathrm{T}$, $(x_\mathrm{I}, x_\mathrm{II}) \in \mathcal{D}_\mathrm{I} \times \mathbb{R}^{n_\mathrm{II}}$.

Corollary 2.4. Consider the nonlinear dynamical system (2.64) and (2.65). Assume that there exist a continuously differentiable vector function $V : \mathcal{D}_\mathrm{I} \times \mathbb{R}^{n_\mathrm{II}} \to \mathcal{Q} \cap \overline{\mathbb{R}}_+^q$, a positive vector $p \in \mathbb{R}_+^q$, and class $\mathcal{K}$ functions $\alpha(\cdot)$ and $\beta(\cdot)$ such that the scalar function $v : \mathcal{D}_\mathrm{I} \times \mathbb{R}^{n_\mathrm{II}} \to \overline{\mathbb{R}}_+$ defined by $v(x_\mathrm{I}, x_\mathrm{II}) \triangleq p^\mathrm{T} V(x_\mathrm{I}, x_\mathrm{II})$ satisfies

$$\alpha(\|x_\mathrm{I}\|) \leq v(x_\mathrm{I}, x_\mathrm{II}) \leq \beta(\|x_\mathrm{I}\|), \quad (x_\mathrm{I}, x_\mathrm{II}) \in \mathcal{D}_\mathrm{I} \times \mathbb{R}^{n_\mathrm{II}}, \tag{2.66}$$

and

$$V'(x_\mathrm{I}, x_\mathrm{II}) f(x_\mathrm{I}, x_\mathrm{II}) \leq\leq w(V(x_\mathrm{I}, x_\mathrm{II}), x_\mathrm{I}, x_\mathrm{II}), \quad (x_\mathrm{I}, x_\mathrm{II}) \in \mathcal{D}_\mathrm{I} \times \mathbb{R}^{n_\mathrm{II}}, \tag{2.67}$$

where $w : \mathcal{Q} \times \mathcal{D}_\mathrm{I} \times \mathbb{R}^{n_\mathrm{II}} \to \mathbb{R}^q$ is continuous, $w(\cdot, x_\mathrm{I}, x_\mathrm{II}) \in \mathcal{W}$, and $w(0, x_\mathrm{I}, x_\mathrm{II}) = 0$, $(x_\mathrm{I}, x_\mathrm{II}) \in \mathcal{D}_\mathrm{I} \times \mathbb{R}^{n_\mathrm{II}}$. Then the following statements hold:

i) If the nonlinear dynamical system (2.31), (2.64), and (2.65) is Lyapunov (respectively, asymptotically) stable with respect to z uniformly in $(x_\mathrm{I0}, x_\mathrm{II0})$, then the nonlinear dynamical system (2.64) and (2.65) is Lyapunov (respectively, asymptotically) stable with respect to x_I uniformly in x_II0.

ii) If $\mathcal{D}_\mathrm{I} = \mathbb{R}^{n_\mathrm{I}}$, $\mathcal{Q} = \mathbb{R}^q$, the functions $\alpha(\cdot)$ and $\beta(\cdot)$ are class $\mathcal{K}_\infty$, and the nonlinear dynamical system (2.31), (2.64), and (2.65) is globally asymptotically stable with respect to z uniformly in $(x_\mathrm{I0}, x_\mathrm{II0})$, then the nonlinear dynamical system (2.64) and (2.65) is globally asymptotically stable with respect to x_I uniformly in x_II0.

iii) If there exist constants $\nu \geq 1$, $\alpha > 0$, and $\beta > 0$ such that $v : \mathcal{D}_\mathrm{I} \times \mathbb{R}^{n_\mathrm{II}} \to \overline{\mathbb{R}}_+$ satisfies

$$\alpha\|x_\mathrm{I}\|^\nu \leq v(x_\mathrm{I}, x_\mathrm{II}) \leq \beta\|x_\mathrm{I}\|^\nu, \quad (x_\mathrm{I}, x_\mathrm{II}) \in \mathcal{D}_\mathrm{I} \times \mathbb{R}^{n_\mathrm{II}}, \tag{2.68}$$

and the nonlinear dynamical system (2.31), (2.64), and (2.65) is exponentially stable with respect to z uniformly in $(x_\mathrm{I0}, x_\mathrm{II0})$, then the nonlinear dynamical system (2.64) and (2.65) is exponentially stable with respect to x_I uniformly in x_II0.

iv) If $\mathcal{D}_\mathrm{I} = \mathbb{R}^{n_\mathrm{I}}$, $\mathcal{Q} = \mathbb{R}^q$, there exist constants $\nu \geq 1$, $\alpha > 0$, and $\beta > 0$ such that $v : \mathbb{R}^{n_\mathrm{I}} \times \mathbb{R}^{n_\mathrm{II}} \to \overline{\mathbb{R}}_+$ satisfies (2.68), and the nonlinear dynamical system (2.31), (2.64), and (2.65) is globally exponentially stable with respect to z uniformly in $(x_{\mathrm{I}0}, x_{\mathrm{II}0})$, then the nonlinear dynamical system (2.64) and (2.65) is globally exponentially stable with respect to x_I uniformly in $x_{\mathrm{II}0}$.

Proof. To show Lyapunov stability of the nonlinear dynamical system (2.64) and (2.65) with respect to x_I uniformly in $x_{\mathrm{II}0}$, recall that $w(\cdot, x_\mathrm{I}, x_\mathrm{II}) \in \mathcal{W}$ and $w(0, x_\mathrm{I}, x_\mathrm{II}) = 0$, $(x_\mathrm{I}, x_\mathrm{II}) \in \mathcal{D}_\mathrm{I} \times \mathbb{R}^{n_\mathrm{II}}$, ensures that the partial solution $z(t)$, $t \geq t_0$, to the nonlinear dynamical system (2.31), (2.64), and (2.65) remains in $\overline{\mathbb{R}}_+^q$ for all $z_0 \in \overline{\mathbb{R}}_+^q$ and $(x_{\mathrm{I}0}, x_{\mathrm{II}0}) \in \mathcal{D}_\mathrm{I} \times \mathbb{R}^{n_\mathrm{II}}$. Since $p \in \mathbb{R}_+^q$ is a positive vector it follows from (2.66) that

$$\frac{\alpha(\|x_\mathrm{I}\|)}{\max_{i=1,\dots,q}\{p_i\}} \leq \mathbf{e}^\mathrm{T} V(x_\mathrm{I}, x_\mathrm{II}) \leq \frac{\beta(\|x_\mathrm{I}\|)}{\min_{i=1,\dots,q}\{p_i\}}, \quad (x_\mathrm{I}, x_\mathrm{II}) \in \mathcal{D}_\mathrm{I} \times \mathbb{R}^{n_\mathrm{II}}.$$

$$(2.69)$$

Next, let $\varepsilon > 0$ and note that it follows from Lyapunov stability of the nonlinear dynamical system (2.31), (2.64), and (2.65) with respect to z uniformly in $(x_{\mathrm{I}0}, x_{\mathrm{II}0})$ that there exists $\mu = \mu(\varepsilon) > 0$ such that if $\|z_0\|_1 < \mu$ and $z_0 \in \overline{\mathbb{R}}_+^q$, then $\|z(t)\|_1 < \alpha(\varepsilon)/\max_{i=1,\dots,q}\{p_i\}$ and $z(t) \in \overline{\mathbb{R}}_+^q$, $t \geq t_0$, for every $(x_{\mathrm{I}0}, x_{\mathrm{II}0}) \in \mathcal{D}_\mathrm{I} \times \mathbb{R}^{n_\mathrm{II}}$. Now, choose $z_0 = V(x_{\mathrm{I}0}, x_{\mathrm{II}0}) \geq\geq 0$, $(x_{\mathrm{I}0}, x_{\mathrm{II}0}) \in \mathcal{D}_\mathrm{I} \times \mathbb{R}^{n_\mathrm{II}}$. Since $V(\cdot, \cdot)$ is continuous, the function $\mathbf{e}^\mathrm{T} V(\cdot, \cdot)$ is also continuous. Moreover, it follows from the continuity of $\beta(\cdot)$ that for $\mu = \mu(\varepsilon)$ there exists $\delta = \delta(\mu(\varepsilon)) = \delta(\varepsilon) > 0$ such that $\delta < \varepsilon$ and if $\|x_{\mathrm{I}0}\| < \delta$, then $\beta(\|x_{\mathrm{I}0}\|)/\min_{i=1,\dots,q}\{p_i\} < \mu$ which, by (2.69), implies that $\mathbf{e}^\mathrm{T} V(x_{\mathrm{I}0}, x_{\mathrm{II}0}) = \mathbf{e}^\mathrm{T} z_0 = \|z_0\|_1 < \mu$ for all $x_{\mathrm{II}0} \in \mathbb{R}^{n_\mathrm{II}}$, and hence, $\mathbf{e}^\mathrm{T} z(t) = \|z(t)\|_1 < \alpha(\varepsilon)/\max_{i=1,\dots,q}\{p_i\}$, $t \geq t_0$. In addition, it follows from (2.67) and Corollary 2.2 that $V(x_\mathrm{I}(t), x_\mathrm{II}(t)) \leq\leq z(t)$ on every compact interval $[t_0, t_0 + \tau]$. Thus, it follows from (2.69) that for all $\|x_{\mathrm{I}0}\| < \delta$, $x_{\mathrm{II}0} \in \mathbb{R}^{n_\mathrm{II}}$, and $t \in [t_0, t_0 + \tau]$,

$$\frac{\alpha(\|x_\mathrm{I}(t)\|)}{\max_{i=1,\dots,q}\{p_i\}} \leq \mathbf{e}^\mathrm{T} V(x_\mathrm{I}(t), x_\mathrm{II}(t)) \leq \mathbf{e}^\mathrm{T} z(t) < \frac{\alpha(\varepsilon)}{\max_{i=1,\dots,q}\{p_i\}}, \quad (2.70)$$

which implies that $\|x_\mathrm{I}(t)\| < \varepsilon$, $t \in [t_0, t_0 + \tau]$.

Next, suppose, *ad absurdum*, that for some $x_{\mathrm{I}0} \in \mathcal{D}_\mathrm{I}$ with $\|x_{\mathrm{I}0}\| < \delta$ and for some $x_{\mathrm{II}0} \in \mathbb{R}^{n_\mathrm{II}}$ there exists $\hat{t} > t_0$ such that $\|x_\mathrm{I}(\hat{t})\| = \varepsilon$. Then, for $z_0 = V(x_{\mathrm{I}0}, x_{\mathrm{II}0})$ and the compact interval $[t_0, \hat{t}]$ it follows from Corollary 2.2 that $V(x_\mathrm{I}(\hat{t}), x_\mathrm{II}(\hat{t})) \leq\leq z(\hat{t})$, which implies that

$$\frac{\alpha(\varepsilon)}{\max_{i=1,\dots,q}\{p_i\}} = \frac{\alpha(\|x_\mathrm{I}(\hat{t})\|)}{\max_{i=1,\dots,q}\{p_i\}}$$

$$\leq \mathbf{e}^\mathrm{T} V(x_\mathrm{I}(\hat{t}), x_\mathrm{II}(\hat{t}))$$

$$\leq \mathbf{e}^{\mathrm{T}} z(\hat{t})$$

$$< \frac{\alpha(\varepsilon)}{\max_{i=1,\dots,q}\{p_i\}}. \tag{2.71}$$

This is a contradiction, and hence, for a given $\varepsilon > 0$ there exists $\delta = \delta(\varepsilon) > 0$ such that for all $x_{\mathrm{I0}} \in \mathcal{D}_{\mathrm{I}}$ with $\|x_{\mathrm{I0}}\| < \delta$ and for all $x_{\mathrm{II0}} \in \mathbb{R}^{n_{\mathrm{II}}}$, $\|x_{\mathrm{I}}(t)\| < \varepsilon$, $t \geq t_0$, which implies Lyapunov stability of the nonlinear dynamical system (2.64) and (2.65) with respect to x_{I} uniformly in x_{II0}.

The remainder of the proof involves similar arguments as given above and as in the proof of parts $ii) - v)$ of Theorem 2.3 and, hence, is omitted.

$\square$

Next, we provide a time-varying extension of Corollary 2.3. In particular, we consider the nonlinear time-varying dynamical system

$$\dot{x}(t) = f(t, x(t)), \quad x(t_0) = x_0, \quad t \geq t_0, \tag{2.72}$$

where $x(t) \in \mathcal{D} \subseteq \mathbb{R}^n$, $0 \in \mathcal{D}$, $f : [t_0, \infty) \times \mathcal{D} \to \mathbb{R}^n$ is such that $f(\cdot, \cdot)$ is jointly continuous in t and x, for every $t \in [t_0, \infty)$, $f(t, 0) = 0$, and $f(t, \cdot)$ is locally Lipschitz in x uniformly in t for all t in compact subsets of $[0, \infty)$.

Theorem 2.5. Consider the nonlinear time-varying dynamical system (2.72). Assume that there exist a continuously differentiable vector function $V : [0, \infty) \times \mathcal{D} \to \mathcal{Q} \cap \overline{\mathbb{R}}_+^q$, a positive vector $p \in \mathbb{R}_+^q$, and class $\mathcal{K}$ functions $\alpha, \beta : [0, r] \to \overline{\mathbb{R}}_+$ such that $V(t, 0) = 0$, $t \in [0, \infty)$, the scalar function $v : [0, \infty) \times \mathcal{D} \to \overline{\mathbb{R}}_+$ defined by $v(t, x) \triangleq p^{\mathrm{T}} V(t, x)$, $(t, x) \in [0, \infty) \times \mathcal{D}$, is such that

$$\alpha(\|x\|) \leq v(t, x) \leq \beta(\|x\|), \quad (t, x) \in [0, \infty) \times \mathcal{B}_r(0), \quad \mathcal{B}_r(0) \subseteq \mathcal{D}, \tag{2.73}$$

and

$$\frac{\partial V}{\partial t}(t, x) + \frac{\partial V}{\partial x}(t, x) f(t, x) \leq\leq w(t, V(t, x)), \quad (t, x) \in [0, \infty) \times \mathcal{D}, \tag{2.74}$$

where $w : [0, \infty) \times \mathcal{Q} \to \mathbb{R}^q$ is continuous, $w(t, \cdot) \in \mathcal{W}$, and $w(t, 0) = 0$, $t \in [0, \infty)$. Then the stability properties of the zero solution $z(t) \equiv 0$ to

$$\dot{z}(t) = w(t, z(t)), \quad z(t_0) = z_0, \quad t \geq t_0, \tag{2.75}$$

where $z_0 \in \mathcal{Q} \cap \overline{\mathbb{R}}_+^q$, imply the corresponding stability properties of the zero solution $x(t) \equiv 0$ to (2.72). In particular, if the zero solution $z(t) \equiv 0$ to (2.75) is uniformly Lyapunov (respectively, uniformly asymptotically) stable, then the zero solution $x(t) \equiv 0$ to (2.72) is uniformly Lyapunov (respectively, uniformly asymptotically) stable. If, in addition, $\mathcal{D} = \mathbb{R}^n$, $\mathcal{Q} = \mathbb{R}^q$, and $\alpha(\cdot), \beta(\cdot)$ are class $\mathcal{K}_\infty$ functions, then global uniform asymptotic stability of the zero solution $z(t) \equiv 0$ to (2.75) implies global uniform asymptotic stability of the zero solution $x(t) \equiv 0$ to (2.72). Moreover, if there exist constants $\nu \geq 1$, $\alpha > 0$, and $\beta > 0$ such that $v : [0, \infty) \times \mathcal{D} \to \overline{\mathbb{R}}_+$ satisfies

$$\alpha \|x\|^{\nu} \leq v(t, x) \leq \beta \|x\|^{\nu}, \quad (t, x) \in [0, \infty) \times \mathcal{D}, \tag{2.76}$$

then exponential stability of the zero solution $z(t) \equiv 0$ to (2.75) implies exponential stability of the zero solution $x(t) \equiv 0$ to (2.72). Finally, if $\mathcal{D} = \mathbb{R}^n$, $\mathcal{Q} = \mathbb{R}^q$, there exist constants $\nu \geq 1$, $\alpha > 0$, and $\beta > 0$ such that $v : [0, \infty) \times \mathbb{R}^n \to \overline{\mathbb{R}}_+$ satisfies (2.76), then global exponential stability of the zero solution $z(t) \equiv 0$ to (2.75) implies global exponential stability of the zero solution $x(t) \equiv 0$ to (2.72).

Proof. The proof is a direct consequence of Corollary 2.4 with $x_{\mathrm{I}}(t) \equiv x(t)$, $x_{\mathrm{II}}(t) \equiv t$, $n_{\mathrm{II}} = 1$, and $f_{\mathrm{II}}(x_{\mathrm{I}}(t), x_{\mathrm{II}}(t)) \equiv 1$. $\qquad\square$

Example 2.2. Consider the nonlinear time-varying dynamical system given by

$$\dot{x}_1(t) = -2(5 + \sin t)x_1(t) + (5 + \sin t)\left[\frac{[x_1(t) - x_2(t)]^3 + [x_1(t) + x_2(t)]^3}{x_1^2(t) + x_2^2(t)}\right]$$
$$-(1 + e^{-t})[x_1(t) + x_2(t)]^5, \quad x_1(t_0) = x_{10}, \quad t \geq t_0, \tag{2.77}$$
$$\dot{x}_2(t) = -2(5 + \sin t)x_2(t) + (5 + \sin t)\left[\frac{[x_1(t) - x_2(t)]^3 - [x_1(t) + x_2(t)]^3}{x_1^2(t) + x_2^2(t)}\right]$$
$$-(1 + e^{-t})[x_1(t) + x_2(t)]^5, \quad x_2(t_0) = x_{20}. \tag{2.78}$$

To analyze (2.77) and (2.78), consider the vector Lyapunov function $V(x) = [V_1(x), V_2(x)]^{\mathrm{T}}$, $x \in \mathbb{R}^2$, $x \triangleq [x_1, x_2]^{\mathrm{T}}$, where

$$V_1(x) = \frac{1}{2}(x_1 - x_2)^2, \quad x \in \mathbb{R}^2,$$
$$V_2(x) = \frac{1}{2}(x_1 + x_2)^2, \quad x \in \mathbb{R}^2.$$

Note that each component of $V(x)$ is not positive definite, whereas $\mathbf{e}^{\mathrm{T}}V(x) = V_1(x) + V_2(x) = x_1^2 + x_2^2$, $x \in \mathbb{R}^2$, is positive definite and radially unbounded. Using the fact that $\frac{a^2 - b^2}{a^2 + b^2} \leq 1$, for all $a, b \in \mathbb{R}$, it follows that

$$\dot{V}_1(t, x) \leq -2(5 + \sin t)(x_1 - x_2)^2 + 2(5 + \sin t)(x_1 + x_2)^2$$
$$= -4(5 + \sin t)V_1(x) + 4(5 + \sin t)V_2(x) \tag{2.79}$$

and

$$\dot{V}_2(t, x) \leq -2(5 + \sin t)(x_1 + x_2)^2 + 2(5 + \sin t)(x_1 - x_2)^2$$
$$-2(1 + e^{-t})(x_1 + x_2)^6$$
$$= 4(5 + \sin t)V_1(x) - 4(5 + \sin t)V_2(x) - 16(1 + e^{-t})V_2^3(x). \tag{2.80}$$

Next, consider the comparison system given by

$$\dot{z}(t) = w(t, z(t)), \quad z(t_0) = z_0, \quad t \geq t_0, \tag{2.81}$$

where

$$w(t, z) = \begin{bmatrix} -4(5 + \sin t)z_1 + 4(5 + \sin t)z_2 \\ 4(5 + \sin t)z_1 - 4(5 + \sin t)z_2 - 16(1 + e^{-t})z_2^3 \end{bmatrix},$$

and note that $w(t, \cdot) \in \mathcal{W}$. To show global uniform asymptotic stability of the comparison system (2.81), consider the scalar Lyapunov function

$$v(z) = \frac{1}{2}z_1^2 + \frac{1}{2}z_2^2, \quad z \in \overline{\mathbb{R}}_+^2,$$

and note that

$$\frac{1}{3}\|z\|^2 < v(z) < \|z\|^2, \quad z \in \overline{\mathbb{R}}_+^2.$$

Furthermore, note that

$$\dot{v}(z) \leq -16(z_1 - z_2)^2 - 16z_2^4 < 0, \quad z \in \overline{\mathbb{R}}_+^2, \quad z \neq 0. \tag{2.82}$$

Since the right-hand side of (2.82) is negative-definite, it follows from Theorem 2.5 that the zero solution $x(t) \equiv 0$ to (2.77) and (2.78) is globally uniformly asymptotically stable.

Alternatively, consider the quadratic scalar Lyapunov function for the system (2.77) and (2.78) given by

$$v(x) = \mathbf{e}^{\mathrm{T}}V(x) = V_1(x) + V_2(x) = x_1^2 + x_2^2, \quad x \in \mathbb{R}^2, \tag{2.83}$$

and note that it follows from (2.79) and (2.80) that

$$\dot{v}(t, x) \leq -2(1 + e^{-t})(x_1 + x_2)^6 \leq 0, \quad x \in \mathbb{R}^2, \tag{2.84}$$

which only proves uniform Lyapunov stability of the zero solution $x(t) \equiv 0$ to (2.77) and (2.78). In addition, note that the Krasovskii-LaSalle invariance principle *cannot* be used in this case since (2.77) and (2.78) is time-varying.

Finally, consider the more general quadratic scalar Lyapunov function given by

$$v(x) = p_1 x_1^2 + p_2 x_2^2, \quad x \in \mathbb{R}^2, \tag{2.85}$$

where $p_1, p_2 > 0$, and note that

$$\begin{aligned}
\dot{v}(t, x) = &-4(5 + \sin t)(p_1 x_1^2 + p_2 x_2^2) \\
&+2(5 + \sin t)(p_1 x_1 + p_2 x_2)\frac{(x_1 - x_2)^3}{x_1^2 + x_2^2} \\
&+2(5 + \sin t)(p_1 x_1 - p_2 x_2)\frac{(x_1 + x_2)^3}{x_1^2 + x_2^2} \\
&-2(1 + e^{-t})(p_1 x_1 + p_2 x_2)(x_1 + x_2)^5, \quad x \in \mathbb{R}^2.
\end{aligned} \tag{2.86}$$

It is clear that if $p_1 \neq p_2$, then $\dot{v}(t, x)$ is sign indefinite, whereas if we set $p_1 = p_2 = p > 0$, then $\dot{v}(t, x)$ satisfies inequality (2.84), modulo p, which again only proves uniform Lyapunov stability of the zero solution $x(t) \equiv 0$ to (2.77) and (2.78). $\triangle$

Next, we give a generalization of the converse Lyapunov theorem that establishes the existence of a vector Lyapunov function for an asymptotically stable nonlinear dynamical system. This result is used in Chapter 5 to establish the equivalence between asymptotic stabilizability and the existence of a control vector Lyapunov function.

Theorem 2.6. Consider the nonlinear dynamical system (2.24). Let $\delta > 0$ and $\mathcal{D}_0 = \mathcal{B}_\delta(0) \subset \mathcal{D}$, and assume that $f : \mathcal{D} \to \mathbb{R}^n$ is continuously differentiable and the zero solution $x(t) \equiv 0$ to (2.24) is asymptotically stable. Then there exist a continuously differentiable componentwise positive definite vector function $V = [v_1, \ldots, v_q]^{\mathrm{T}} : \mathcal{D}_0 \to \overline{\mathbb{R}}_+^q$ and a continuous function $w = [w_1, \ldots, w_q]^{\mathrm{T}} : \overline{\mathbb{R}}_+^q \to \mathbb{R}^q$ such that $V(0) = 0$, $w(\cdot) \in \mathcal{W}$, $w(0) = 0$, $V'(x)f(x) \leq\leq w(V(x))$, $x \in \mathcal{D}_0$, and the zero solution $z(t) \equiv 0$ to

$$\dot{z}(t) = w(z(t)), \quad z(t_0) = z_0, \quad t \geq t_0, \tag{2.87}$$

where $z_0 \in \overline{\mathbb{R}}_+^q$, is asymptotically stable.

Proof. Since the zero solution $x(t) \equiv 0$ to (2.24) is asymptotically stable it follows from Theorem 3.14 of [110] that there exist a continuously differentiable positive definite function $\tilde{v} : \mathcal{D}_0 \to \overline{\mathbb{R}}_+$ and class $\mathcal{K}$ functions [85] $\alpha(\cdot)$, $\beta(\cdot)$, and $\gamma(\cdot)$ such that

$$\alpha(\|x\|) \leq \tilde{v}(x) \leq \beta(\|x\|), \quad x \in \mathcal{D}_0, \tag{2.88}$$
$$\tilde{v}'(x)f(x) \leq -\gamma(\|x\|), \quad x \in \mathcal{D}_0. \tag{2.89}$$

Furthermore, it follows from (2.88) and (2.89) that

$$\tilde{v}'(x)f(x) \leq -\gamma \circ \beta^{-1}(\tilde{v}(x)), \quad x \in \mathcal{D}_0, \tag{2.90}$$

where "$\circ$" denotes the composition operator and $\beta^{-1} : [0, \beta(\delta)] \to \overline{\mathbb{R}}_+$ is the inverse function of $\beta(\cdot)$, and hence, $\beta^{-1}(\cdot)$ and $\gamma \circ \beta^{-1}(\cdot)$ are class $\mathcal{K}$ functions.

Next, define $V = [v_1, \ldots, v_q]^{\mathrm{T}} : \mathcal{D}_0 \to \overline{\mathbb{R}}_+^q$ such that $v_i(x) \triangleq \tilde{v}(x)$, $x \in \mathcal{D}_0$, $i = 1, \ldots, q$. Then it follows that $V(0) = 0$ and $V'(x)f(x) \leq\leq w(V(x))$, $x \in \mathcal{D}_0$, where $w = [w_1, \ldots, w_q]^{\mathrm{T}} : \overline{\mathbb{R}}_+^q \to \mathbb{R}^q$ is such that $w_i(V(x)) = -\gamma \circ \beta^{-1}(v_i(x))$, $x \in \mathcal{D}_0$. Note that $w(\cdot) \in \mathcal{W}$ and $w(0) = 0$. To show that the zero solution $z(t) \equiv 0$ to (2.87) is asymptotically stable, consider the Lyapunov function candidate $\hat{v}(z) \triangleq \mathbf{e}^{\mathrm{T}} z$, $z \in \overline{\mathbb{R}}_+^q$. Note that $\hat{v}(0) = 0$, $\hat{v}(z) > 0$, $z \in \overline{\mathbb{R}}_+^q$, $z \neq 0$, and $\dot{\hat{v}}(z) = -\sum_{i=1}^{q} \gamma \circ \beta^{-1}(z_i) < 0$, $z \in \overline{\mathbb{R}}_+^q$, $z \neq 0$. Thus, the zero solution $z(t) \equiv 0$ to (2.87) is asymptotically stable, which completes the proof. $\square$

Finally, to elucidate how to use the vector Lyapunov functions framework to address the problem of control design for nonlinear dynamical systems consider the controlled nonlinear dynamical system given by

$$\dot{x}(t) = F(x(t), u(t)), \quad x(t_0) = x_0, \quad t \geq t_0, \tag{2.91}$$

where $x_0 \in \mathcal{D}$, $\mathcal{D} \subseteq \mathbb{R}^n$ is an open set, $0 \in \mathcal{D}$, $u(t) \in U \subseteq \mathbb{R}^m$, $t \geq t_0$, is the control input, U is the set of all admissible control inputs, $F : \mathcal{D} \times U \to \mathbb{R}^n$ is Lipschitz continuous for all $(x, u) \in \mathcal{D} \times U$, and $F(0, 0) = 0$. Moreover, assume that for every $x_0 \in \mathcal{D}$ and $u \in U$ the solution $x(t)$ to (2.91) is unique and defined for all $t \geq t_0$.

Next, assume there exist a continuously differentiable vector function $V : \mathcal{D} \to \mathcal{Q} \cap \overline{\mathbb{R}}_+^q$ and a positive vector $p \in \mathbb{R}_+^q$ such that $V(0) = 0$, $v(x) \triangleq p^{\mathrm{T}} V(x)$, $x \in \mathcal{D}$, is positive definite, and

$$V'(x)F(x, u) \leq\leq w(V(x), x, u), \quad x \in \mathcal{D}, \quad u \in U, \tag{2.92}$$

where $w : \mathcal{Q} \times \mathcal{D} \times U \to \mathbb{R}^q$ is continuous. Furthermore, define the feedback control law $\phi : \mathcal{Q} \times \mathcal{D} \to U$ given by $u = \phi(V(x), x)$, $x \in \mathcal{D}$, so that $\phi(0, 0) = 0$ and the closed-loop system (2.91) is given by

$$\dot{x}(t) = F(x(t), \phi(V(x(t)), x(t))), \quad x(t_0) = x_0, \quad t \geq t_0. \tag{2.93}$$

Now, if $\phi(\cdot, \cdot)$ is such that the system

$$\dot{z}(t) = \tilde{w}(z(t), x(t)), \quad z(t_0) = z_0, \quad t \geq t_0, \tag{2.94}$$

$$\dot{x}(t) = \tilde{f}(x(t)), \quad x(t_0) = x_0, \tag{2.95}$$

where $\tilde{w}(z, x) \triangleq w(z, x, \phi(z, x))$, $z \in \mathcal{Q}$, $x \in \mathcal{D}$, $\tilde{f}(x) \triangleq F(x, \phi(V(x), x))$, $x \in \mathcal{D}$, $\tilde{w}(\cdot, x) \in \mathcal{W}$, $\tilde{w}(0, x) = 0$, $x \in \mathcal{D}$, $\tilde{f}(0) = 0$, $z_0 \in \mathcal{Q} \cap \overline{\mathbb{R}}_+^q$, and $x_0 \in \mathcal{D}$, is asymptotically stable with respect to z uniformly in x_0, then the zero solution $x(t) \equiv 0$ to the closed-loop system (2.93) is asymptotically stable. To ensure partial asymptotic stability of the system (2.94) and (2.95) the results of Theorem 1 and Corollary 1 of [41] can be used.

2.6 Discrete-Time Stability Theory via Vector Lyapunov Functions

In this section, we introduce several definitions and some key results needed for analyzing discrete-time, large-scale nonlinear dynamical systems.

Definition 2.7. Let $w = [w_1, \cdots, w_q]^{\mathrm{T}} : \mathcal{V} \subseteq \overline{\mathbb{R}}_+^q \to \mathbb{R}^q$. Then w is *nonnegative* if $w(r) \geq\geq 0$ for all $r \in \overline{\mathbb{R}}_+^q$.

Recall from Definition 2.5 that a function $w = [w_1, \ldots, w_q]^{\mathrm{T}} : \mathbb{R}^q \to \mathbb{R}^q$ is of class $\mathcal{W}_{\mathrm{d}}$ or is nondecreasing, if $w(z') \leq\leq w(z'')$ for all $z', z'' \in \mathbb{R}^q$ such that $z' \leq\leq z''$. Note that if $w : \mathbb{R}^q \to \mathbb{R}^q$ is such that $w(\cdot) \in \mathcal{W}_{\mathrm{d}}$ and

$w(0) \geq\geq 0$, then w is nonnegative. Furthermore, note that, if $w(z) = Wz$, then $w(\cdot)$ is nonnegative if and only if $W \in \mathbb{R}^{q \times q}$ is nonnegative.

Proposition 2.2 ([72]). Suppose $\overline{\mathbb{R}}_+^q \subset \mathcal{V}$. Then $\overline{\mathbb{R}}_+^q$ is an invariant set with respect to

$$r(k+1) = w(r(k)), \quad r(0) = r_0, \quad k \in \overline{\mathbb{Z}}_+, \tag{2.96}$$

if and only if $w : \mathcal{V} \to \mathbb{R}^q$ is nonnegative.

Proof. Suppose $w : \mathcal{D} \to \mathbb{R}^q$ is nonnegative and let $r(0) \in \overline{\mathbb{R}}_+^q$. Then, for every $i \in \{1, \ldots, q\}$ it follows that $r_i(k+1) = w_i(r(k)) \geq 0$. Thus, $r(k) \in \overline{\mathbb{R}}_+^q$, $k \in \overline{\mathbb{Z}}_+$. Conversely, suppose $r(k) \in \overline{\mathbb{R}}_+^q$, $k \in \overline{\mathbb{Z}}_+$, for all $r(0) \in \overline{\mathbb{R}}_+^q$ and assume, *ad absurdum*, that there exists $i \in \{1, \ldots, n\}$ and $r_0 \in \overline{\mathbb{R}}_+^q$ such that $w_i(r_0) < 0$. In this case, with $r(0) = r_0$, $r_i(1) = w_i(r(0)) = w_i(r_0) < 0$, which is a contradiction. $\qquad\square$

Definition 2.8. Let $W \in \mathbb{R}^{q \times q}$. W is (*discrete-time*) *compartmental* if W is nonnegative and $\sum_{i=1}^q W_{(i,j)} \leq 1$, $j = 1, \ldots, q$.

The following definition and lemma are needed for developing several of the results of this section.

Definition 2.9. The equilibrium solution $r(k) \equiv r_{\mathrm{e}}$ of (2.96) is *Lyapunov stable* if, for every $\varepsilon > 0$, there exists $\delta = \delta(\varepsilon) > 0$ such that if $r_0 \in \mathcal{B}_\delta(r_{\mathrm{e}}) \cap \overline{\mathbb{R}}_+^q$, then $r(k) \in \mathcal{B}_\varepsilon(r_{\mathrm{e}}) \cap \overline{\mathbb{R}}_+^q$, $k \in \overline{\mathbb{Z}}_+$. The equilibrium solution $r(k) \equiv r_{\mathrm{e}}$ of (2.96) is *semistable* if it is Lyapunov stable and there exists $\delta > 0$ such that if $r_0 \in \mathcal{B}_\delta(r_{\mathrm{e}}) \cap \overline{\mathbb{R}}_+^q$, then $\lim_{k \to \infty} r(k)$ exists and converges to a Lyapunov stable equilibrium point. The equilibrium solution $r(k) \equiv r_{\mathrm{e}}$ of (2.96) is *asymptotically stable* if it is Lyapunov stable and there exists $\delta > 0$ such that if $r_0 \in \mathcal{B}_\delta(r_{\mathrm{e}}) \cap \overline{\mathbb{R}}_+^q$, then $\lim_{k \to \infty} r(k) = r_{\mathrm{e}}$. Finally, the equilibrium solution $r(k) \equiv r_{\mathrm{e}}$ of (2.96) is *globally asymptotically stable* if the previous statement holds for all $r_0 \in \overline{\mathbb{R}}_+^q$.

Recall that a matrix $W \in \mathbb{R}^{q \times q}$ is (*discrete-time*) *semistable* if and only if $\lim_{k \to \infty} W^k$ exists [72] while W is (*discrete-time*) *asymptotically stable* if and only if $\lim_{k \to \infty} W^k = 0$.

Lemma 2.2. Suppose $W \in \mathbb{R}^{q \times q}$ is nonsingular and nonnegative. If W is semistable (respectively, asymptotically stable), then there exist a scalar $\alpha \geq 1$ (respectively, $\alpha > 1$) and a nonnegative vector $p \in \overline{\mathbb{R}}_+^q$, $p \neq 0$, (respectively, positive vector $p \in \mathbb{R}_+^q$) such that

$$W^{-\mathrm{T}}p = \alpha p. \tag{2.97}$$

Proof. Since W is semistable, it follows from Theorem 3.3 of [71] that $|\lambda| < 1$ or $\lambda = 1$ and $\lambda = 1$ is semisimple, where $\lambda \in \mathrm{spec}(W)$. Since $W^{\mathrm{T}} \geq\geq$

0, it follows from the Perron-Frobenius theorem that $\rho(W) \in \text{spec}(W)$, and hence, there exists $p \geq\geq 0$, $p \neq 0$, such that $W^{\mathrm{T}}p = \rho(W)p$. In addition, since W is nonsingular, $\rho(W) > 0$. Hence, $W^{\mathrm{T}}p = \alpha^{-1}p$, where $\alpha \triangleq 1/\rho(W)$, which proves that there exist $p \geq\geq 0$, $p \neq 0$, and $\alpha \geq 1$ such that (2.97) holds. In the case where W is asymptotically stable, the result is a direct consequence of the Perron-Frobenius theorem. $\qquad\square$

Next, we consider the method of vector Lyapunov functions for stability analysis of discrete-time nonlinear dynamical systems. To develop the theory of vector Lyapunov functions, we first introduce some results on vector difference inequalities and the *vector comparison principle*. Consider the discrete-time nonlinear dynamical system given by

$$z(k+1) = w(z(k), y(k)), \quad z(k_0) = z_0, \quad k \in \mathcal{I}_{z_0}, \tag{2.98}$$

where $z(k) \in \mathcal{Q} \subseteq \mathbb{R}^q$, $k \in \mathcal{I}_{z_0}$, is the system state vector, $y : \mathcal{T} \to \mathcal{V} \subseteq \mathbb{R}^s$ is a given sequence, $\mathcal{I}_{z_0} \subseteq \mathcal{T} \subseteq \mathbb{Z}$ is the maximal interval of existence of a solution $z(k)$ to (2.98), $\mathcal{Q}$ is an open set, $0 \in \mathcal{Q}$, and $w : \mathcal{Q} \times \mathcal{V} \to \mathbb{R}^q$ is a continuous function on $\mathcal{Q} \times \mathcal{V}$.

Theorem 2.7. Consider the discrete-time nonlinear dynamical system (2.98). Assume that the function $w : \mathcal{Q} \times \mathcal{V} \to \mathbb{R}^q$ is continuous and $w(\cdot, y)$ is of class $\mathcal{W}_{\mathrm{d}}$. If there exists a continuous vector function $V : \mathcal{I}_{z_0} \to \mathcal{Q}$ such that

$$V(k+1) \leq\leq w(V(k), y(k)), \quad k \in \mathcal{I}_{z_0}, \tag{2.99}$$

then

$$V(k_0) \leq\leq z_0, \quad z_0 \in \mathcal{Q}, \tag{2.100}$$

implies

$$V(k) \leq\leq z(k), \quad k \in \mathcal{I}_{z_0}, \tag{2.101}$$

where $z(k)$, $k \in \mathcal{I}_{z_0}$, is the solution to (2.98).

Proof. Suppose, *ad absurdum*, that inequality (2.101) does not hold on the entire interval $\mathcal{I}_{z_0}$. Then there exists $\hat{k} \in \mathcal{I}_{z_0}$ such that $V(k) \leq\leq z(k), k_0 \leq k < \hat{k}$, and for at least one $i \in \{1, \ldots, q\}$,

$$v_i(\hat{k}) > z_i(\hat{k}) \tag{2.102}$$

and

$$v_j(\hat{k}) \leq z_j(\hat{k}), \quad j \neq i, \quad j = 1, \ldots, q. \tag{2.103}$$

Since $w(\cdot, y) \in \mathcal{W}_d$ it follows from (2.98), (2.99), and (2.102) that

$$\begin{aligned}
w_i(z(\hat{k} - 1), y(\hat{k} - 1)) &= z_i(\hat{k}) \\
&< v_i(\hat{k}) \\
&\leq w_i(V(\hat{k} - 1), y(\hat{k} - 1)) \\
&\leq w_i(z(\hat{k} - 1), y(\hat{k} - 1)),
\end{aligned} \qquad (2.104)$$

which is a contradiction. $\qquad\square$

Theorem 2.8. Consider the discrete-time nonlinear dynamical system (2.98). Assume that the function $w : \mathcal{Q} \times \mathcal{V} \to \mathbb{R}^q$ is continuous and $w(\cdot, y)$ is of class $\mathcal{W}_d$. Let $z(k)$, $k \in \mathcal{I}_{z_0}$, be the solution to (2.98) and $\{k_0, \dots, k_0 + \tau\} \subseteq \mathcal{I}_{z_0}$, where $\tau \in \overline{\mathbb{Z}}_+$. If there exists a continuous vector function $V : \{k_0, \dots, k_0 + \tau\} \to \mathcal{Q}$ such that

$$V(k+1) \leq\leq w(V(k), y(k)), \quad k \in \{k_0, \dots, k_0 + \tau\}, \qquad (2.105)$$

then

$$V(k_0) \leq\leq z_0, \quad z_0 \in \mathcal{Q}, \qquad (2.106)$$

implies

$$V(k) \leq\leq z(k), \quad k \in \{k_0, \dots, k_0 + \tau\}. \qquad (2.107)$$

Proof. The proof is similar to that of Theorem 2.7 and, hence, is omitted. $\qquad\square$

Next, consider the discrete-time nonlinear dynamical system given by

$$x(k+1) = f(x(k)), \quad x(k_0) = x_0, \quad k \in \mathcal{I}_{x_0}, \qquad (2.108)$$

where $x(k) \in \mathcal{D} \subseteq \mathbb{R}^n$, $k \in \mathcal{I}_{x_0}$, is the system state vector, $\mathcal{I}_{x_0} \subset \mathbb{Z}$ is the maximal interval of existence of a solution $x(k)$ to (2.108), $\mathcal{D}$ is an open set, $0 \in \mathcal{D}$, and $f(\cdot)$ is continuous on $\mathcal{D}$. The following result is a direct corollary of Theorem 2.8.

Corollary 2.5. Consider the discrete-time nonlinear dynamical system (2.108). Assume there exists a continuous vector function $V : \mathcal{D} \to \mathcal{Q} \subseteq \mathbb{R}^q$ such that

$$V(x(k+1)) \leq\leq w(V(x(k)), x(k)), \quad x_0 \in \mathcal{D}, \quad k \in \mathcal{I}_{x_0}, \qquad (2.109)$$

where $w : \mathcal{Q} \times \mathcal{D} \to \mathbb{R}^q$ is a continuous function, $w(\cdot, x) \in \mathcal{W}_d$, $x(k)$, $k \in \mathcal{I}_{x_0}$, is the solution to (2.108), and the equation

$$z(k+1) = w(z(k), x(k)), \quad z(k_0) = z_0, \quad k \in \mathcal{I}_{z_0, x_0}, \qquad (2.110)$$

has a unique solution $z(k)$, $k \in \mathcal{I}_{z_0, x_0}$. If $\{k_0, \ldots, k_0 + \tau\} \subseteq \mathcal{I}_{x_0} \cap \mathcal{I}_{z_0, x_0}$, then

$$V(x_0) \leq\leq z_0, \quad z_0 \in \mathcal{Q}, \tag{2.111}$$

implies

$$V(x(k)) \leq\leq z(k), \quad k \in \{k_0, \ldots, k_0 + \tau\}. \tag{2.112}$$

Proof. For every given $x_0 \in \mathcal{D}$, the solution $x(k)$, $k \in \mathcal{I}_{x_0}$, to (2.108) is well defined. With $\eta(k) \triangleq V(x(k))$, $k \in \mathcal{I}_{x_0}$, it follows from (2.109) that

$$\eta(k+1) \leq\leq w(\eta(k), x(k)), \quad k \in \mathcal{I}_{x_0}. \tag{2.113}$$

Moreover, if $\{k_0, \ldots, k_0 + \tau\} \subseteq \mathcal{I}_{x_0} \cap \mathcal{I}_{z_0, x_0}$, then it follows from Theorem 2.8 with $y(k) \equiv x(k)$, that $V(x_0) = \eta(k_0) \leq\leq z_0$ implies

$$V(x(k)) = \eta(k) \leq\leq z(k), \quad k \in \{k_0, \ldots, k_0 + \tau\}, \tag{2.114}$$

which establishes the result. $\qquad\square$

Note that if the solutions to (2.108) and (2.110) are globally defined for all $x_0 \in \mathcal{D}$ and $z_0 \in \mathcal{Q}$, then the result of Corollary 2.5 holds for every $k \geq k_0$. For the remainder of this section we assume that the solutions to the systems (2.108) and (2.110) are defined for all $k \geq k_0$. Furthermore, consider the cascade discrete-time nonlinear dynamical system given by

$$z(k+1) = w(z(k), x(k)), \quad z(k_0) = z_0, \quad k \geq k_0, \tag{2.115}$$
$$x(k+1) = f(x(k)), \quad x(k_0) = x_0, \tag{2.116}$$

where $z_0 \in \mathcal{Q} \cap \overline{\mathbb{R}}_+^q$, $0 \in \mathcal{Q}$, $x_0 \in \mathcal{D}$, $[z^{\mathrm{T}}(k), x^{\mathrm{T}}(k)]^{\mathrm{T}}$, $k \geq k_0$, is the solution to (2.115), (2.116), $w : \mathcal{Q} \times \mathcal{D} \to \mathbb{R}^q$ is continuous, $w(\cdot, x) \in \mathcal{W}_{\mathrm{d}}$, $w(0, x) = 0$, $x \in \mathcal{D}$, $f : \mathcal{D} \to \mathbb{R}^n$, and $f(0) = 0$. Note that since $w(\cdot, x) \in \mathcal{W}_{\mathrm{d}}$ and $w(0, x) = 0$, $x \in \mathcal{D}$, then for every $x \in \mathcal{D}$ and $z \in \mathcal{Q} \cap \overline{\mathbb{R}}_+^q$ it follows that $w(z, x) \geq\geq 0$, which implies that for every $x_0 \in \mathcal{D}$ and $z_0 \in \mathcal{Q} \cap \overline{\mathbb{R}}_+^q$ the solution $z(k)$, $k \geq k_0$, remains in $\overline{\mathbb{R}}_+^q$.

The following definition involving the notion of partial stability for discrete-time systems is needed for the next result.

Definition 2.10. i) The nonlinear dynamical system (2.115) and (2.116) is *Lyapunov stable with respect to z* if, for every $\varepsilon > 0$ and $x_0 \in \mathcal{D}$, there exists $\delta = \delta(\varepsilon, x_0) > 0$ such that $\|z_0\| < \delta$ implies that $\|z(k)\| < \varepsilon$ for all $k \in \overline{\mathbb{Z}}_+$.

ii) The nonlinear dynamical system (2.115) and (2.116) is *Lyapunov stable with respect to z uniformly in x_0* if, for every $\varepsilon > 0$, there exists $\delta = \delta(\varepsilon) > 0$ such that $\|z_0\| < \delta$ implies that $\|z(k)\| < \varepsilon$ for all $k \in \overline{\mathbb{Z}}_+$ and for all $x_0 \in \mathcal{D}$.

iii) The nonlinear dynamical system (2.115) and (2.116) is *asymptotically stable with respect to z* if it is Lyapunov stable with respect to z and, for every $x_0 \in \mathcal{D}$, there exists $\delta = \delta(x_0) > 0$ such that $\|z_0\| < \delta$ implies that $\lim_{k \to \infty} \|z(k)\| = 0$.

iv) The nonlinear dynamical system (2.115) and (2.116) is *asymptotically stable with respect to z uniformly in x_0* if it is Lyapunov stable with respect to z uniformly in x_0 and there exists $\delta > 0$ such that $\|z_0\| < \delta$ implies that $\lim_{k \to \infty} \|z(k)\| = 0$ uniformly in z_0 and x_0 for all $x_0 \in \mathcal{D}$.

v) The nonlinear dynamical system (2.115) and (2.116) is *globally asymptotically stable with respect to z* if it is Lyapunov stable with respect to z and $\lim_{k \to \infty} \|z(k)\| = 0$ for all $z_0 \in \mathbb{R}^q$ and $x_0 \in \mathbb{R}^n$.

vi) The nonlinear dynamical system (2.115) and (2.116) is *globally asymptotically stable with respect to z uniformly in x_0* if it is Lyapunov stable with respect to z uniformly in x_0 and $\lim_{k \to \infty} \|z(k)\| = 0$ uniformly in z_0 and x_0 for all $z_0 \in \mathbb{R}^q$ and $x_0 \in \mathbb{R}^n$.

vii) The nonlinear dynamical system (2.115) and (2.116) is *geometrically stable with respect to z uniformly in x_0* if there exist scalars $\alpha, \beta > 1$, and $\delta > 0$ such that $\|z_0\| < \delta$ implies that $\|z(k)\| \leq \alpha \|z_0\| \beta^{-k}$, $k \in \overline{\mathbb{Z}}_+$, for all $x_0 \in \mathcal{D}$.

viii) The nonlinear dynamical system (2.115) and (2.116) is *globally geometrically stable with respect to z uniformly in x_0* if there exist scalars $\alpha, \beta > 1$ such that $\|z(k)\| \leq \alpha \|z_0\| \beta^{-k}$, $k \in \overline{\mathbb{Z}}_+$, for all $z_0 \in \mathbb{R}^q$ and $x_0 \in \mathbb{R}^n$.

Theorem 2.9. Consider the discrete-time nonlinear dynamical system (2.108). Assume that there exist a continuous function $V : \mathcal{D} \to \mathcal{Q} \cap \overline{\mathbb{R}}_+^q$ and a positive vector $p \in \mathbb{R}_+^q$ such that $V(0) = 0$, the scalar function $v : \mathcal{D} \to \overline{\mathbb{R}}_+$ defined by $v(x) \triangleq p^{\mathrm{T}} V(x)$, $x \in \mathcal{D}$, is such that $v(0) = 0$, $v(x) > 0$, $x \neq 0$, and

$$V(x(k+1)) \leq\leq w(V(x(k)), x(k)), \quad x_0 \in \mathcal{D}, \quad k \geq k_0, \qquad (2.117)$$

where $w : \mathcal{Q} \times \mathcal{D} \to \mathbb{R}^q$ is continuous, $w(\cdot, x) \in \mathcal{W}_{\mathrm{d}}$, and $w(0, x) = 0$, $x \in \mathcal{D}$. Then the following statements hold:

i) If the system (2.115) and (2.116) is Lyapunov stable with respect to z uniformly in x_0, then the zero solution $x(k) \equiv 0$ to (2.108) is Lyapunov stable.

ii) If the system (2.115) and (2.116) is asymptotically stable with respect to z uniformly in x_0, then the zero solution $x(k) \equiv 0$ to (2.108) is asymptotically stable.

iii) If $\mathcal{D} = \mathbb{R}^n$, $\mathcal{Q} = \mathbb{R}^q$, $v : \mathbb{R}^n \to \overline{\mathbb{R}}_+$ is positive definite and radially unbounded, and the system (2.115) and (2.116) is globally asymptotically stable with respect to z uniformly in x_0, then the zero solution $x(k) \equiv 0$ to (2.108) is globally asymptotically stable.

iv) If there exist constants $\nu \geq 1$, $\alpha > 0$, and $\beta > 0$ such that $v : \mathcal{D} \to \overline{\mathbb{R}}_+$ satisfies

$$\alpha \|x\|^{\nu} \leq v(x) \leq \beta \|x\|^{\nu}, \quad x \in \mathcal{D}, \tag{2.118}$$

and the system (2.115) and (2.116) is geometrically stable with respect to z uniformly in x_0, then the zero solution $x(k) \equiv 0$ to (2.108) is geometrically stable.

v) If $\mathcal{D} = \mathbb{R}^n$, $\mathcal{Q} = \mathbb{R}^q$, there exist constants $\nu \geq 1$, $\alpha > 0$, and $\beta > 0$ such that $v : \mathcal{D} \to \overline{\mathbb{R}}_+$ satisfies (2.118), and the system (2.115) and (2.116) is globally geometrically stable with respect to z uniformly in x_0, then the zero solution $x(k) \equiv 0$ to (2.108) is globally geometrically stable.

Proof. Assume there exist a continuous function $V : \mathcal{D} \to \mathcal{Q} \cap \overline{\mathbb{R}}_+^q$ and a positive vector $p \in \mathbb{R}_+^q$ such that $v(x) = p^{\mathrm{T}} V(x)$, $x \in \mathcal{D}$, is positive definite, that is, $v(0) = 0$, $v(x) > 0$, $x \neq 0$. Note that $v(x) = p^{\mathrm{T}} V(x) \leq \max_{i=1,\ldots,q}\{p_i\} \mathbf{e}^{\mathrm{T}} V(x)$, $x \in \mathcal{D}$, and hence, the function $\mathbf{e}^{\mathrm{T}} V(x)$, $x \in \mathcal{D}$, is also positive definite. Thus, there exist $r > 0$ and class $\mathcal{K}$ functions α, $\beta : [0, r] \to \overline{\mathbb{R}}_+$ such that $\mathcal{B}_r(0) \subset \mathcal{D}$ and

$$\alpha(\|x\|) \leq \mathbf{e}^{\mathrm{T}} V(x) \leq \beta(\|x\|), \quad x \in \mathcal{B}_r(0). \tag{2.119}$$

i) Let $\varepsilon > 0$ and choose $0 < \hat{\varepsilon} < \min\{\varepsilon, r\}$. It follows from uniform Lyapunov stability of (2.115) and (2.116) with respect to z that there exists $\mu = \mu(\hat{\varepsilon}) = \mu(\varepsilon) > 0$ such that if $\|z_0\|_1 < \mu$ and $z_0 \in \overline{\mathbb{R}}_+^q$, then $\|z(k)\|_1 < \alpha(\hat{\varepsilon})$ and $z(k) \in \overline{\mathbb{R}}_+^q$, $k \geq k_0$, for every $x_0 \in \mathcal{D}$. Now, choose $z_0 = V(x_0) \geq\geq 0$, $x_0 \in \mathcal{D}$. Since $V(x)$, $x \in \mathcal{D}$, is continuous, then the function $\mathbf{e}^{\mathrm{T}} V(x)$, $x \in \mathcal{D}$, is also continuous. Hence, for $\mu = \mu(\hat{\varepsilon}) > 0$ there exists $\delta = \delta(\mu(\hat{\varepsilon})) = \delta(\varepsilon) > 0$ such that $\delta < \hat{\varepsilon}$ and if $\|x_0\| < \delta$, then $\mathbf{e}^{\mathrm{T}} V(x_0) = \mathbf{e}^{\mathrm{T}} z_0 < \mu$, which implies that $\mathbf{e}^{\mathrm{T}} z(k) = \|z(k)\|_1 < \alpha(\hat{\varepsilon})$, $k \geq k_0$. In addition, it follows from (2.117) and Corollary 2.5 that $V(x(k)) \leq\leq z(k)$, $k \geq k_0$. Thus, using (2.119), if $\|x_0\| < \delta$, then

$$\alpha(\|x(k)\|) \leq \mathbf{e}^{\mathrm{T}} V(x(k)) \leq \mathbf{e}^{\mathrm{T}} z(k) < \alpha(\hat{\varepsilon}), \quad k \geq k_0, \tag{2.120}$$

which implies $\|x(k)\| < \hat{\varepsilon} < \varepsilon$, $k \geq k_0$. Hence, for a given $\varepsilon > 0$ there exists $\delta = \delta(\varepsilon) > 0$ such that for all $x_0 \in \mathcal{B}_{\delta}(0)$, $\|x(k)\| < \varepsilon$, $k \geq k_0$, which implies Lyapunov stability of the zero solution $x(k) \equiv 0$ to (2.108).

ii) It follows from *i)* and the uniform asymptotic stability of (2.115) and (2.116) with respect to z that the zero solution to (2.108) is Lyapunov stable and there exists $\mu > 0$ such that if $\|z_0\|_1 < \mu$ and $z_0 \in \overline{\mathbb{R}}_+^q$, then $\lim_{k \to \infty} z(k) = 0$ for every $x_0 \in \mathcal{D}$. As in *i)*, choose $z_0 = V(x_0) \geq\geq 0$, $x_0 \in \mathcal{D}$. It follows from Lyapunov stability of the zero solution to (2.108) and the continuity of $V : \mathcal{D} \to \mathcal{Q} \cap \overline{\mathbb{R}}_+^q$ that there exists $\delta = \delta(\mu) > 0$ such that if $\|x_0\| < \delta$, then $\|x(k)\| < r$, $k \geq k_0$, and $\mathbf{e}^{\mathrm{T}}V(x_0) = \mathbf{e}^{\mathrm{T}}z_0 = \|z_0\|_1 < \mu$. Thus, by uniform asymptotic stability of (2.115) and (2.116) with respect to z, for any arbitrary $\varepsilon > 0$ there exists $K = K(\varepsilon) > k_0$ such that $\mathbf{e}^{\mathrm{T}}z(k) = \|z(k)\|_1 < \alpha(\varepsilon)$, $k \geq K$. Thus, it follows from (2.117) and Corollary 2.5 that $V(x(k)) \leq\leq z(k)$, $k \geq K$, and hence, by (2.119),

$$\alpha(\|x(k)\|) \leq \mathbf{e}^{\mathrm{T}}V(x(k)) \leq \mathbf{e}^{\mathrm{T}}z(k) < \alpha(\varepsilon), \quad k \geq K. \qquad (2.121)$$

Hence, there exists $\delta > 0$ such that for all $x_0 \in \mathcal{B}_\delta(0)$, $\lim_{k\to\infty} x(k) = 0$, which, along with Lyapunov stability, implies asymptotic stability of the zero solution $x(k) \equiv 0$ to (2.108).

iii) Suppose $\mathcal{D} = \mathbb{R}^n$, $\mathcal{Q} = \mathbb{R}^q$, $v : \mathbb{R}^n \to \overline{\mathbb{R}}_+$ is a positive-definite, radially unbounded function, and the system (2.115) and (2.116) is globally asymptotically stable with respect to z uniformly in x_0. In this case for $V : \mathbb{R}^n \to \overline{\mathbb{R}}_+^q$ the inequality (2.119) holds for all $x \in \mathbb{R}^n$ where the functions α, $\beta : \overline{\mathbb{R}}_+ \to \overline{\mathbb{R}}_+$ are of class $\mathcal{K}_\infty$. Furthermore, Lyapunov stability of the zero solution $x(k) \equiv 0$ to (2.108) follows from *i)*. Next, for every $x_0 \in \mathbb{R}^n$ and $z_0 = V(x_0) \in \overline{\mathbb{R}}_+^q$ the identical arguments as in *ii)* can be used to show that $\lim_{k\to\infty} x(k) = 0$, which proves global asymptotic stability of the zero solution $x(k) \equiv 0$ to (2.108).

iv) Suppose (2.118) holds. Since $p \in \mathbb{R}_+^q$, then

$$\hat{\alpha}\|x\|^\nu \leq \mathbf{e}^{\mathrm{T}}V(x) \leq \hat{\beta}\|x\|^\nu, \quad x \in \mathcal{D}, \qquad (2.122)$$

where $\hat{\alpha} \triangleq \alpha/\max_{i=1,\dots,q}\{p_i\}$ and $\hat{\beta} \triangleq \beta/\min_{i=1,\dots,q}\{p_i\}$. It follows from the geometric stability of (2.115) and (2.116) with respect to z uniformly in x_0 that there exist positive constants γ, μ, and $\eta > 1$ such that if $\|z_0\|_1 < \mu$ and $z_0 \in \overline{\mathbb{R}}_+^q$, then $z(k) \in \overline{\mathbb{R}}_+^q$, $k \geq k_0$, and

$$\|z(k)\|_1 \leq \gamma\|z_0\|_1\eta^{-(k-k_0)}, \quad k \geq k_0, \qquad (2.123)$$

for all $x_0 \in \mathcal{D}$. Choose $z_0 = V(x_0) \geq\geq 0$, $x_0 \in \mathcal{D}$. By continuity of $V : \mathcal{D} \to \mathcal{Q} \cap \overline{\mathbb{R}}_+^q$, there exists $\delta = \delta(\mu) > 0$ such that for all $x_0 \in \mathcal{B}_\delta(0)$, $\mathbf{e}^{\mathrm{T}}V(x_0) = \mathbf{e}^{\mathrm{T}}z_0 = \|z_0\|_1 < \mu$. Furthermore, it follows from (2.122), (2.123), and Corollary 2.5 that for all $x_0 \in \mathcal{B}_\delta(0)$,

$$\begin{aligned}
\hat{\alpha}\|x(k)\|^\nu &\leq \mathbf{e}^{\mathrm{T}}V(x(k)) \\
&\leq \mathbf{e}^{\mathrm{T}}z(k) \\
&\leq \gamma\|z_0\|_1\eta^{-(k-k_0)} \\
&\leq \gamma\hat{\beta}\|x_0\|^\nu\eta^{-(k-k_0)}, \quad k \geq k_0. \qquad (2.124)
\end{aligned}$$

This in turn implies that for every $x_0 \in \mathcal{B}_\delta(0)$,

$$\|x(k)\| \leq \left(\frac{\gamma\hat{\beta}}{\hat{\alpha}}\right)^{\frac{1}{\nu}}\|x_0\|\eta^{-\frac{k-k_0}{\nu}}, \quad k \geq k_0, \qquad (2.125)$$

which establishes geometric stability of the zero solution $x(k) \equiv 0$ to (2.108).

v) The proof is identical to the proof of *iv*). $\square$

If $V : \mathcal{D} \to \mathcal{Q} \cap \overline{\mathbb{R}}_+^q$ satisfies the conditions of Theorem 2.9 we say that $V(x)$, $x \in \mathcal{D}$ is a (*discrete-time*) *vector Lyapunov function*. Note that for stability analysis each component of a vector Lyapunov function need not be positive definite, nor does it need to have a negative-definite time difference along the trajectories of (2.115) and (2.116). This provides more flexibility in searching for a vector Lyapunov function as compared to a scalar Lyapunov function for addressing the stability of discrete-time nonlinear dynamical systems. The following corollary to Theorem 2.9 is immediate.

Corollary 2.6. Consider the discrete-time nonlinear dynamical system (2.108). Assume that there exist a continuous function $V : \mathcal{D} \to \mathcal{Q} \cap \overline{\mathbb{R}}_+^q$ and a positive vector $p \in \mathbb{R}_+^q$ such that $V(0) = 0$, the scalar function $v : \mathcal{D} \to \overline{\mathbb{R}}_+$ defined by $v(x) = p^{\mathrm{T}}V(x)$, $x \in \mathcal{D}$, is such that $v(0) = 0$, $v(x) > 0$, $x \neq 0$, and

$$V(f(x)) \leq\leq w(V(x)), \quad x \in \mathcal{D}, \tag{2.126}$$

where $w : \mathcal{Q} \to \mathbb{R}^q$ is continuous, $w(\cdot) \in \mathcal{W}_{\mathrm{d}}$, and $w(0) = 0$. Then the stability properties of the zero solution $z(k) \equiv 0$ to

$$z(k+1) = w(z(k)), \quad z(k_0) = z_0, \quad k \geq k_0, \tag{2.127}$$

where $z_0 \in \mathcal{Q} \cap \overline{\mathbb{R}}_+^q$, imply the corresponding stability properties of the zero solution $x(k) \equiv 0$ to (2.108). In particular, if the zero solution $z(k) \equiv 0$ to (2.127) is Lyapunov (respectively, asymptotically) stable, then the zero solution $x(k) \equiv 0$ to (2.108) is Lyapunov (respectively, asymptotically) stable. If, in addition, $\mathcal{D} = \mathbb{R}^n$, $\mathcal{Q} = \mathbb{R}^q$, and $v : \mathbb{R}^n \to \overline{\mathbb{R}}_+$ is a positive-definite, radially unbounded function, then global asymptotic stability of the zero solution $z(k) \equiv 0$ to (2.127) implies global asymptotic stability of the zero solution $x(k) \equiv 0$ to (2.108). Moreover, if there exist constants $\nu \geq 1$, $\alpha > 0$, and $\beta > 0$ such that $v : \mathcal{D} \to \overline{\mathbb{R}}_+$ satisfies (2.118), then geometric stability of the zero solution $z(k) \equiv 0$ to (2.127) implies geometric stability of the zero solution $x(k) \equiv 0$ to (2.108). Finally, if $\mathcal{D} = \mathbb{R}^n$, $\mathcal{Q} = \mathbb{R}^q$, there exist constants $\nu \geq 1$, $\alpha > 0$, and $\beta > 0$ such that $v : \mathcal{D} \to \overline{\mathbb{R}}_+$ satisfies (2.118), then global geometric stability of the zero solution $z(k) \equiv 0$ to (2.127) implies global geometric stability of the zero solution $x(k) \equiv 0$ to (2.108).

Proof. The proof is a direct consequence of Theorem 2.9 with $w(z, x) \equiv w(z)$. $\square$

Finally, we provide a time-varying extension of Corollary 2.6. In particular, we consider the discrete-time nonlinear time-varying dynamical system

$$x(k+1) = f(k, x(k)), \quad x(k_0) = x_0, \quad k \geq k_0, \tag{2.128}$$

where $x(k) \in \mathcal{D} \subseteq \mathbb{R}^n$, $0 \in \mathcal{D}$, $f : \{k_0, \ldots, k_1\} \times \mathcal{D} \to \mathbb{R}^n$ is such that $f(\cdot, \cdot)$ is continuous in k and x, and for every $k \in \{k_0, \ldots, k_1\}$, $f(k, 0) = 0$.

Theorem 2.10. Consider the discrete-time nonlinear time-varying dynamical system (2.128). Assume that there exist a continuous vector function $V : \overline{\mathbb{Z}}_+ \times \mathcal{D} \to \mathcal{Q} \cap \overline{\mathbb{R}}_+^q$, a positive vector $p \in \mathbb{R}_+^q$, and class $\mathcal{K}$ functions $\alpha, \beta : [0, r] \to \overline{\mathbb{R}}_+$ such that $V(k, 0) = 0$, $k \in \overline{\mathbb{Z}}_+$, the scalar function $v : \overline{\mathbb{Z}}_+ \times \mathcal{D} \to \overline{\mathbb{R}}_+$ defined by $v(k, x) \triangleq p^{\mathrm{T}} V(k, x)$, $(k, x) \in \overline{\mathbb{Z}}_+ \times \mathcal{D}$, is such that

$$\alpha(\|x\|) \leq v(k, x) \leq \beta(\|x\|), \quad (k, x) \in \overline{\mathbb{Z}}_+ \times \mathcal{B}_r(0), \quad \mathcal{B}_r(0) \subseteq \mathcal{D}, \quad (2.129)$$

and

$$V(k + 1, x(k + 1)) \leq\leq w(k, V(k, x(k))), \quad x_0 \in \mathcal{D}, \quad k \geq k_0, \quad (2.130)$$

where $w : \overline{\mathbb{Z}}_+ \times \mathcal{Q} \to \mathbb{R}^q$ is continuous, $w(k, \cdot) \in \mathcal{W}_{\mathrm{d}}$, and $w(k, 0) = 0$, $k \in \overline{\mathbb{Z}}_+$. Then the stability properties of the zero solution $z(k) \equiv 0$ to

$$z(k + 1) = w(k, z(k)), \quad z(k_0) = z_0, \quad k \geq k_0, \quad (2.131)$$

where $z_0 \in \mathcal{Q} \cap \overline{\mathbb{R}}_+^q$, imply the corresponding stability properties of the zero solution $x(k) \equiv 0$ to (2.128). In particular, if the zero solution $z(k) \equiv 0$ to (2.131) is uniformly Lyapunov (respectively, uniformly asymptotically) stable, then the zero solution $x(k) \equiv 0$ to (2.128) is uniformly Lyapunov (respectively, uniformly asymptotically) stable. If, in addition, $\mathcal{D} = \mathbb{R}^n$, $\mathcal{Q} = \mathbb{R}^q$, and $\alpha(\cdot), \beta(\cdot)$ are class $\mathcal{K}_\infty$ functions, then global uniform asymptotic stability of the zero solution $z(k) \equiv 0$ to (2.131) implies global uniform asymptotic stability of the zero solution $x(k) \equiv 0$ to (2.128). Moreover, if there exist constants $\nu \geq 1$, $\alpha > 0$, and $\beta > 0$ such that $v : \overline{\mathbb{Z}}_+ \times \mathcal{D} \to \overline{\mathbb{R}}_+$ satisfies

$$\alpha\|x\|^\nu \leq v(k, x) \leq \beta\|x\|^\nu, \quad (k, x) \in \overline{\mathbb{Z}}_+ \times \mathcal{D}, \quad (2.132)$$

then geometric stability of the zero solution $z(k) \equiv 0$ to (2.131) implies geometric stability of the zero solution $x(k) \equiv 0$ to (2.128). Finally, if $\mathcal{D} = \mathbb{R}^n$, $\mathcal{Q} = \mathbb{R}^q$, there exist constants $\nu \geq 1$, $\alpha > 0$, and $\beta > 0$ such that $v : \overline{\mathbb{Z}}_+ \times \mathbb{R}^n \to \overline{\mathbb{R}}_+$ satisfies (2.132), then global geometric stability of the zero solution $z(k) \equiv 0$ to (2.131) implies global geometric stability of the zero solution $x(k) \equiv 0$ to (2.128).

Proof. The proof is similar to the proof of Theorem 2.9 and is left as an exercise for the reader. $\qquad\square$

Large-Scale Continuous-Time Interconnected Dynamical Systems

3.1 Introduction

In this chapter, we develop vector dissipativity notions for large-scale nonlinear dynamical systems, a notion not previously considered in the literature. In particular, we introduce a generalized definition of dissipativity for large-scale nonlinear dynamical systems in terms of a *vector dissipation inequality* involving a vector supply rate, a vector storage function, and an essentially nonnegative, semistable dissipation matrix. The vector dissipation inequality reflects the fact that some of the supplied generalized energy of the large-scale system is stored, and some is dissipated. The dissipated generalized energy is nonnegative and is given by the difference of what is supplied and what is stored. Generalized notions of vector available storage and vector required supply are also defined and shown to be element-by-element ordered, nonnegative, and finite. On the subsystem level, the proposed approach provides an energy flow balance in terms of the stored subsystem energy, the supplied subsystem energy, the subsystem energy gained from all other subsystems independent of the subsystem coupling strengths, and the subsystem energy dissipated. Furthermore, for large-scale dynamical systems decomposed into interconnected subsystems, dissipativity of the composite system is shown to be determined from the dissipativity properties of the individual subsystems and the nature of the system interconnections.

In addition, we develop extended Kalman-Yakubovich-Popov conditions, in terms of the local subsystem dynamics and the interconnection constraints, for characterizing vector dissipativeness via vector storage functions for large-scale dynamical systems. Finally, using the concepts of vector dissipativity and vector storage functions as candidate vector Lyapunov functions, we develop feedback interconnection stability results for large-scale nonlinear dynamical systems. Specifically, by appropriately combining vector storage functions and vector supply rates, general stability criteria are given for Lyapunov and asymptotic stability of feedback interconnections of large-scale dynamical systems. In the case of vector quadratic supply rates involving net subsystem powers and input-output subsystem energies, these results provide a generalization to the positivity and small gain theorems for large-scale systems predicated on vector Lyapunov functions.

3.2 Vector Dissipativity Theory for Large-Scale Nonlinear Dynamical Systems

In this section, we extend the notion of dissipative dynamical systems to develop the generalized notion of vector dissipativity for large-scale nonlinear dynamical systems. First, however, we recall the standard notions of dissipativity [70, 170] and exponential dissipativity [42, 70] for nonlinear dynamical systems $\mathcal{G}$ of the form

$$\dot{x}(t) = f(x(t)) + G(x(t))u(t), \quad x(t_0) = x_0, \quad t \geq t_0, \tag{3.1}$$

$$y(t) = h(x(t)) + J(x(t))u(t), \tag{3.2}$$

where $x \in \mathcal{D} \subseteq \mathbb{R}^n$, $u \in U \subseteq \mathbb{R}^m$, $y \in Y \subseteq \mathbb{R}^l$, $f : \mathcal{D} \to \mathbb{R}^n$ and satisfies $f(0) = 0$, $G : \mathcal{D} \to \mathbb{R}^{n \times m}$, $h : \mathcal{D} \to \mathbb{R}^l$ and satisfies $h(0) = 0$, and $J : \mathcal{D} \to \mathbb{R}^{l \times m}$.

For the dynamical system $\mathcal{G}$ given by (3.1) and (3.2) defined on the state space $\mathbb{R}^n$, let $\mathcal{U}$ and $\mathcal{Y}$ define input and output spaces, respectively, consisting of continuous bounded U-valued and Y-valued functions on the semi-infinite interval $[0, \infty)$. The set $U \subseteq \mathbb{R}^m$ contains the set of input values, that is, for every $u(\cdot) \in \mathcal{U}$ and $t \in [0, \infty)$, $u(t) \in U$. The set $Y \subseteq \mathbb{R}^l$ contains the set of output values, that is, for every $y(\cdot) \in \mathcal{Y}$ and $t \in [0, \infty)$, $y(t) \in Y$. The spaces $\mathcal{U}$ and $\mathcal{Y}$ are assumed to be closed under the shift operator, that is, if $u(\cdot) \in \mathcal{U}$ (respectively, $y(\cdot) \in \mathcal{Y}$), then the function defined by $u_T \triangleq u(t + T)$ (respectively, $y_T \triangleq y(t + T)$) is contained in $\mathcal{U}$ (respectively, $\mathcal{Y}$) for all $T \geq 0$. A similar notation also holds for discrete-time systems discussed in later chapters.

For the nonlinear dynamical system $\mathcal{G}$ we assume that the required properties for the existence and uniqueness of solutions are satisfied; that is, $u(\cdot)$ satisfies sufficient regularity conditions such that (3.1) has a unique solution forward in time. For the nonlinear dynamical system $\mathcal{G}$ given by (3.1) and (3.2) a function $s : U \times Y \to \mathbb{R}$ such that $s(0, 0) = 0$ is called a *supply rate* [70, 170] if it is locally integrable for all input-output pairs satisfying (3.1) and (3.2); that is, for all input-output pairs $u(\cdot) \in \mathcal{U}$ and $y(\cdot) \in \mathcal{Y}$ satisfying (3.1) and (3.2), $s(\cdot, \cdot)$ satisfies $\int_{t_1}^{t_2} |s(u(\sigma), y(\sigma))| d\sigma < \infty$, $t_2 \geq t_1 \geq t_0$.

Definition 3.1 ([70])**.** The nonlinear dynamical system $\mathcal{G}$ given by (3.1) and (3.2) is *exponentially dissipative* (respectively, *dissipative*) with respect to the supply rate $s(u, y)$ if there exist a continuous nonnegative definite function $v_s : \mathbb{R}^n \to \overline{\mathbb{R}}_+$, called a *storage function*, and a scalar $\varepsilon > 0$ (respectively, $\varepsilon = 0$) such that $v_s(0) = 0$ and the *dissipation inequality*

$$e^{\varepsilon t_2} v_s(x(t_2)) \leq e^{\varepsilon t_1} v_s(x(t_1)) + \int_{t_1}^{t_2} e^{\varepsilon t} s(u(t), y(t)) dt, \quad t_2 \geq t_1, \tag{3.3}$$

is satisfied for all $t_1, t_2 \geq t_0$, where $x(t), t \geq t_1$, is the solution of (3.1) with $u(\cdot) \in \mathcal{U}$. The nonlinear dynamical system $\mathcal{G}$ given by (3.1) and (3.2) is *lossless with respect to the supply rate* $s(u, y)$ if the dissipation inequality is satisfied as an equality with $\varepsilon = 0$ for all $t_2 \geq t_1 \geq t_0$ and all $u(\cdot) \in \mathcal{U}$.

If $v_{\mathrm{s}}(\cdot)$ is continuously differentiable, then an equivalent statement for exponential dissipativity (respectively, dissipativity) of the dynamical system (3.1) and (3.2) is

$$\dot{v}_{\mathrm{s}}(x(t)) + \varepsilon v_{\mathrm{s}}(x(t)) \leq s(u(t), y(t)), \quad t \geq t_0, \quad u(\cdot) \in \mathcal{U}, \quad y(\cdot) \in \mathcal{Y}, \quad (3.4)$$

where $\varepsilon > 0$ (respectively, $\varepsilon = 0$) and $\dot{v}_{\mathrm{s}}(x(t))$ denotes the total derivative of $v_{\mathrm{s}}(x)$ along the state trajectories $x(t), t \geq t_0$, of (3.1).

Next, to develop vector dissipativity theory, consider large-scale nonlinear dynamical systems $\mathcal{G}$ of the form

$$\dot{x}(t) = F(x(t), u(t)), \quad x(t_0) = x_0, \quad t \geq t_0, \quad (3.5)$$

$$y(t) = H(x(t), u(t)), \quad (3.6)$$

where $x \in \mathcal{D} \subseteq \mathbb{R}^n$, $u \in U \subseteq \mathbb{R}^m$, $y \in Y \subseteq \mathbb{R}^l$, $F : \mathcal{D} \times U \to \mathbb{R}^n$, $H : \mathcal{D} \times U \to Y$, $\mathcal{D}$ is an open set with $0 \in \mathcal{D}$, and $F(0,0) = 0$. Here, we assume that $\mathcal{G}$ represents a large-scale dynamical system composed of q interconnected controlled subsystems $\mathcal{G}_i$ so that, for all $i = 1, \ldots, q$,

$$F_i(x, u_i) = f_i(x_i) + \mathcal{I}_i(x) + G_i(x_i)u_i, \quad (3.7)$$

$$H_i(x_i, u_i) = h_i(x_i) + J_i(x_i)u_i, \quad (3.8)$$

where $x_i \in \mathcal{D}_i \subseteq \mathbb{R}^{n_i}$, $u_i \in U_i \subseteq \mathbb{R}^{m_i}$, $y_i \triangleq H_i(x_i, u_i) \in Y_i \subseteq \mathbb{R}^{l_i}$, $(u_i(\cdot), y_i(\cdot)) \in \mathcal{U}_i \times \mathcal{Y}_i$ is the input-output pair for the ith subsystem, where $\mathcal{U}_i$ and $\mathcal{Y}_i$ denote the ith subsystem input and output spaces, $f_i : \mathbb{R}^{n_i} \to \mathbb{R}^{n_i}$ and $\mathcal{I}_i : \mathcal{D} \to \mathbb{R}^{n_i}$ are Lipschitz continuous and satisfy $f_i(0) = 0$ and $\mathcal{I}_i(0) = 0$, $G_i : \mathbb{R}^{n_i} \to \mathbb{R}^{n_i \times m_i}$ is continuous, $h_i : \mathbb{R}^{n_i} \to \mathbb{R}^{l_i}$ and satisfies $h_i(0) = 0$, $J_i : \mathbb{R}^{n_i} \to \mathbb{R}^{l_i \times m_i}$, $\sum_{i=1}^{q} n_i = n$, $\sum_{i=1}^{q} m_i = m$, and $\sum_{i=1}^{q} l_i = l$.

Furthermore, for the system $\mathcal{G}$ we assume that the required properties for the existence and uniqueness of solutions are satisfied; that is, for every $i \in \{1, \ldots, q\}$, $u_i(\cdot)$ satisfies sufficient regularity conditions such that the system (3.5) has a unique solution forward in time. We define the composite input and composite output for the large-scale system $\mathcal{G}$ as $u \triangleq [u_1^{\mathrm{T}}, \ldots, u_q^{\mathrm{T}}]^{\mathrm{T}}$ and $y \triangleq [y_1^{\mathrm{T}}, \ldots, y_q^{\mathrm{T}}]^{\mathrm{T}}$, respectively. Note that in this case the set $U = U_1 \times \cdots \times U_q$ contains the set of input values and $Y = Y_1 \times \cdots \times Y_q$ contains the set of output values whereas $\mathcal{U} = \mathcal{U}_1 \times \cdots \times \mathcal{U}_q$ and $\mathcal{Y} = \mathcal{Y}_1 \times \cdots \times \mathcal{Y}_q$ define the input and output spaces for (3.5) and (3.6).

Definition 3.2. For the large-scale nonlinear dynamical system $\mathcal{G}$ given by (3.5) and (3.6) a vector function $S = [s_1, \ldots, s_q]^{\mathrm{T}} : U \times Y \to \mathbb{R}^q$ such that $S(u, y) \triangleq [s_1(u_1, y_1), \ldots, s_q(u_q, y_q)]^{\mathrm{T}}$ and $S(0, 0) = 0$ is called a *vector* *supply rate* if it is componentwise locally integrable for all input-output pairs satisfying (3.5) and (3.6); that is, for every $i \in \{1, \ldots, q\}$ and for all input-output pairs $(u_i, y_i) \in \mathcal{U}_i \times \mathcal{Y}_i$ satisfying (3.5) and (3.6), $s_i(\cdot, \cdot)$ satisfies $\int_{t_1}^{t_2} |s_i(u_i(s), y_i(s))| \mathrm{d}s < \infty$, $t_2 \geq t_1 \geq t_0$.

Definition 3.3. The large-scale nonlinear dynamical system $\mathcal{G}$ given by (3.5) and (3.6) is *vector dissipative* (respectively, *exponentially vector dissipative*) *with respect to the vector supply rate* $S(u, y)$ if there exist a continuous, nonnegative definite vector function $V_{\mathrm{s}} = [v_{\mathrm{s}1}, \ldots, v_{\mathrm{s}q}]^{\mathrm{T}} : \mathcal{D} \to \overline{\mathbb{R}}_+^q$, called a *vector storage function*, and an essentially nonnegative *dissipation matrix* $W \in \mathbb{R}^{q \times q}$ such that $V_{\mathrm{s}}(0) = 0$, W is semistable (respectively, asymptotically stable), and the *vector dissipation inequality*

$$V_{\mathrm{s}}(x(T)) \leq\leq e^{W(T-t_0)} V_{\mathrm{s}}(x(t_0)) + \int_{t_0}^{T} e^{W(T-t)} S(u(t), y(t)) \mathrm{d}t, \quad T \geq t_0,$$

$$(3.9)$$

is satisfied, where $x(t)$, $t \geq t_0$, is the solution to (3.5) with $u(\cdot) \in \mathcal{U}$. The large-scale nonlinear dynamical system $\mathcal{G}$ given by (3.5) and (3.6) is *vector lossless with respect to the vector supply rate* $S(u, y)$ if the vector dissipation inequality is satisfied as an equality with W semistable.

Note that if the subsystems $\mathcal{G}_i$ of $\mathcal{G}$ are *disconnected*, that is, $\mathcal{I}_i(x) \equiv 0$ for all $i = 1, \ldots, q$, and $-W \in \mathbb{R}^{q \times q}$ is diagonal and nonnegative definite, then it follows from Definition 3.3 that each of disconnected subsystems $\mathcal{G}_i$ is dissipative (respectively, exponentially dissipative) in the sense of Definition 3.1. A similar remark holds in the case where $q = 1$.

Next, define the *vector available storage* of the large-scale nonlinear dynamical system $\mathcal{G}$ by

$$V_{\mathrm{a}}(x_0) \triangleq \sup_{T \geq t_0, \, u(\cdot)} \left[-\int_{t_0}^{T} e^{-W(t-t_0)} S(u(t), y(t)) \mathrm{d}t \right], \qquad (3.10)$$

where $x(t)$, $t \geq t_0$, is the solution to (3.5) with $x(t_0) = x_0$ and admissible inputs $u \in \mathcal{U}$. The supremum in (3.10) is taken over all admissible inputs $u(\cdot)$, all time $t \geq t_0$, and all system trajectories with initial value $x(t_0) = x_0$ and terminal value left free. In addition, the supremum in (3.10) is taken componentwise, which implies that for each component of $V_{\mathrm{a}}(\cdot)$ the supremum is calculated separately. Note that $V_{\mathrm{a}}(x_0) \geq\geq 0$, $x_0 \in \mathcal{D}$, since $V_{\mathrm{a}}(x_0)$ is the supremum over a set of vectors containing the zero vector ($T = t_0$).

To state the main results of this section the following definition is required.

Definition 3.4. The large-scale nonlinear dynamical system $\mathcal{G}$ given by (3.5) and (3.6) is *completely reachable* if for all $x_0 \in \mathcal{D} \subseteq \mathbb{R}^n$ there exist a finite time $t_{\mathrm{i}} < t_0$ and a square integrable input $u(\cdot)$ defined on $[t_{\mathrm{i}}, t_0]$ such that the state $x(t)$, $t \geq t_{\mathrm{i}}$, can be driven from $x(t_{\mathrm{i}}) = 0$ to $x(t_0) = x_0$. A large-scale nonlinear dynamical system $\mathcal{G}$ is *zero-state observable* if $u(t) \equiv 0$ and $y(t) \equiv 0$ imply $x(t) \equiv 0$.

Theorem 3.1. Consider the large-scale nonlinear dynamical system $\mathcal{G}$ given by (3.5) and (3.6), and assume that $\mathcal{G}$ is completely reachable. Let $W \in \mathbb{R}^{q \times q}$ be essentially nonnegative and semistable (respectively, asymptotically stable). Then

$$\int_{t_0}^{T} e^{-W(t-t_0)} S(u(t), y(t)) \mathrm{d}t \geq\geq 0, \quad T \geq t_0, \quad u \in \mathcal{U}, \qquad (3.11)$$

for $x(t_0) = 0$ if and only if $V_{\mathrm{a}}(0) = 0$ and $V_{\mathrm{a}}(x)$ is finite for all $x \in \mathcal{D}$. Moreover, if (3.11) holds, then $V_{\mathrm{a}}(x)$, $x \in \mathcal{D}$, is a vector storage function for $\mathcal{G}$, and hence, $\mathcal{G}$ is vector dissipative (respectively, exponentially vector dissipative) with respect to the vector supply rate $S(u, y)$.

Proof. Suppose $V_{\mathrm{a}}(0) = 0$ and $V_{\mathrm{a}}(x)$, $x \in \mathcal{D}$, is finite. Then

$$0 = V_{\mathrm{a}}(0) = \sup_{T \geq t_0, u(\cdot)} \left[-\int_{t_0}^{T} e^{-W(t-t_0)} S(u(t), y(t)) \mathrm{d}t \right], \qquad (3.12)$$

which implies (3.11).

Next, suppose (3.11) holds. Then, for $x(t_0) = 0$,

$$\sup_{T \geq t_0, u(\cdot)} \left[-\int_{t_0}^{T} e^{-W(t-t_0)} S(u(t), y(t)) \mathrm{d}t \right] \leq\leq 0, \qquad (3.13)$$

which implies that $V_{\mathrm{a}}(0) \leq\leq 0$. However, since $V_{\mathrm{a}}(x_0) \geq\geq 0$, $x_0 \in \mathcal{D}$, it follows that $V_{\mathrm{a}}(0) = 0$. Moreover, since $\mathcal{G}$ is completely reachable it follows that for every $x_0 \in \mathcal{D}$ there exist $\hat{t} > t_0$ and an admissible input $u(\cdot)$ defined on $[t_0, \hat{t}]$ such that $x(\hat{t}) = x_0$. Now, since (3.11) holds for $x(t_0) = 0$ it follows that for all admissible $u(\cdot) \in \mathcal{U}$,

$$\int_{t_0}^{T} e^{-W(t-t_0)} S(u(t), y(t)) \mathrm{d}t \geq\geq 0, \quad T \geq \hat{t}, \qquad (3.14)$$

or, equivalently, multiplying (3.14) by the nonnegative matrix $e^{W(\hat{t}-t_0)}$, $\hat{t} \geq t_0$, yields

$$-\int_{\hat{t}}^{T} e^{-W(t-\hat{t})} S(u(t), y(t)) \mathrm{d}t \leq\leq \int_{t_0}^{\hat{t}} e^{-W(t-\hat{t})} S(u(t), y(t)) \mathrm{d}t$$

$$\leq\leq Q(x_0)$$

$$<< \infty, \quad T \geq \hat{t}, \quad u \in \mathcal{U}, \qquad (3.15)$$

where $Q : \mathcal{D} \to \mathbb{R}^q$. Hence,

$$V_{\mathrm{a}}(x_0) = \sup_{T \geq \hat{t},\, u(\cdot)} \left[-\int_{\hat{t}}^{T} e^{-W(t-\hat{t})} S(u(t), y(t)) \mathrm{d}t \right] \leq\leq Q(x_0) << \infty,$$

$$x_0 \in \mathcal{D}, \qquad (3.16)$$

which implies that $V_{\mathrm{a}}(x_0)$, $x_0 \in \mathcal{D}$, is finite.

Finally, since (3.11) implies that $V_{\mathrm{a}}(0) = 0$ and $V_{\mathrm{a}}(x)$, $x \in \mathcal{D}$, is finite it follows from the definition of the vector available storage that

$$-V_{\mathrm{a}}(x_0) \leq\leq \int_{t_0}^{T} e^{-W(t-t_0)} S(u(t), y(t)) \mathrm{d}t$$

$$= \int_{t_0}^{t_{\mathrm{f}}} e^{-W(t-t_0)} S(u(t), y(t)) \mathrm{d}t$$

$$+ \int_{t_{\mathrm{f}}}^{T} e^{-W(t-t_0)} S(u(t), y(t)) \mathrm{d}t, \quad T \geq t_0. \qquad (3.17)$$

Now, multiplying (3.17) by the nonnegative matrix $e^{W(t_{\mathrm{f}}-t_0)}$, $t_{\mathrm{f}} \geq t_0$, it follows that

$$e^{W(t_{\mathrm{f}}-t_0)} V_{\mathrm{a}}(x_0) + \int_{t_0}^{t_{\mathrm{f}}} e^{W(t_{\mathrm{f}}-t)} S(u(t), y(t)) \mathrm{d}t$$

$$\geq\geq \sup_{T \geq t_{\mathrm{f}},\, u(\cdot)} \left[-\int_{t_{\mathrm{f}}}^{T} e^{-W(t-t_{\mathrm{f}})} S(u(t), y(t)) \mathrm{d}t \right]$$

$$= V_{\mathrm{a}}(x(t_{\mathrm{f}})), \qquad (3.18)$$

which implies that $V_{\mathrm{a}}(x)$, $x \in \mathcal{D}$, is a vector storage function, and hence, $\mathcal{G}$ is vector dissipative (respectively, exponentially vector dissipative) with respect to the vector supply rate $S(u, y)$. $\qquad \square$

It follows from Lemma 2.1 that if $W \in \mathbb{R}^{q \times q}$ is essentially nonnegative and semistable (respectively, asymptotically stable), then there exist a scalar $\alpha \geq 0$ (respectively, $\alpha > 0$) and a nonnegative vector $p \in \overline{\mathbb{R}}_+^q, p \neq 0$, (respectively, $p \in \mathbb{R}_+^q$) such that (2.4) holds. In this case,

$$p^{\mathrm{T}} e^{Wt} = p^{\mathrm{T}} [I_q + Wt + \tfrac{1}{2} W^2 t^2 + \cdots]$$

$$= p^{\mathrm{T}} [I_q - \alpha t I_q + \tfrac{1}{2} \alpha^2 t^2 I_q + \cdots]$$

$$= e^{-\alpha t} p^{\mathrm{T}}, \ t \in \mathbb{R}. \qquad (3.19)$$

Using (3.19), we define the (scalar) *available storage* for the large-scale nonlinear dynamical system $\mathcal{G}$ by

$$v_{\mathrm{a}}(x_0) \triangleq \sup_{T \geq t_0,\, u(\cdot)} \left[-\int_{t_0}^{T} p^{\mathrm{T}} e^{-W(t-t_0)} S(u(t), y(t)) \mathrm{d}t \right]$$

$$= \sup_{T \geq t_0,\, u(\cdot)} \left[-\int_{t_0}^{T} e^{\alpha(t-t_0)} s(u(t), y(t)) \mathrm{d}t \right], \qquad (3.20)$$

where $s : U \times Y \to \mathbb{R}$ defined as $s(u,y) \triangleq p^{\mathrm{T}} S(u,y)$ is the (scalar) supply rate for the large-scale nonlinear dynamical system $\mathcal{G}$. Clearly, $v_{\mathrm{a}}(x) \geq 0$ for all $x \in \mathcal{D}$. As in standard dissipativity theory, the available storage $v_{\mathrm{a}}(x)$, $x \in \mathcal{D}$, denotes the maximum amount of (scaled) energy that can be extracted from the large-scale nonlinear dynamical system $\mathcal{G}$ at any finite time T.

The following theorem relates vector storage functions and vector supply rates to scalar storage functions and scalar supply rates of large-scale dynamical systems.

Theorem 3.2. Consider the large-scale nonlinear dynamical system $\mathcal{G}$ given by (3.5) and (3.6). Suppose $\mathcal{G}$ is vector dissipative (respectively, exponentially vector dissipative) with respect to the vector supply rate $S : U \times Y \to \mathbb{R}^q$ and with vector storage function $V_{\mathrm{s}} : \mathcal{D} \to \overline{\mathbb{R}}_+^q$. Then there exists $p \in \overline{\mathbb{R}}_+^q$, $p \neq 0$, (respectively, $p \in \mathbb{R}_+^q$) such that $\mathcal{G}$ is dissipative (respectively, exponentially dissipative) with respect to the scalar supply rate $s(u,y) = p^{\mathrm{T}} S(u,y)$ and with storage function $v_{\mathrm{s}}(x) \triangleq p^{\mathrm{T}} V_{\mathrm{s}}(x)$, $x \in \mathcal{D}$. Moreover, in this case $v_{\mathrm{a}}(x)$, $x \in \mathcal{D}$, is a storage function for $\mathcal{G}$ and

$$0 \leq v_{\mathrm{a}}(x) \leq v_{\mathrm{s}}(x), \quad x \in \mathcal{D}. \qquad (3.21)$$

Proof. Suppose $\mathcal{G}$ is vector dissipative (respectively, exponentially vector dissipative) with respect to the vector supply rate $S(u,y)$. Then there exist an essentially nonnegative, semistable (respectively, asymptotically stable) dissipation matrix W and a vector storage function $V_{\mathrm{s}} : \mathcal{D} \to \overline{\mathbb{R}}_+^q$ such that the dissipation inequality (3.9) holds. Furthermore, it follows from Lemma 2.1 that there exist $\alpha \geq 0$ (respectively, $\alpha > 0$) and a nonzero vector $p \in \overline{\mathbb{R}}_+^q$ (respectively, $p \in \mathbb{R}_+^q$) satisfying (2.4). Hence, premultiplying (3.9) by p^{T} and using (3.19) it follows that

$$e^{\alpha T} v_{\mathrm{s}}(x(T)) \leq e^{\alpha t_0} v_{\mathrm{s}}(x(t_0)) + \int_{t_0}^{T} e^{\alpha t} s(u(t), y(t)) \mathrm{d}t, \quad T \geq t_0, \quad u \in \mathcal{U},$$

$$(3.22)$$

where $v_{\mathrm{s}}(x) = p^{\mathrm{T}} V_{\mathrm{s}}(x)$, $x \in \mathcal{D}$, which implies dissipativity (respectively, exponential dissipativity) of $\mathcal{G}$ with respect to the supply rate $s(u,y)$ and with storage function $v_{\mathrm{s}}(x)$, $x \in \mathcal{D}$.

Moreover, since $v_{\mathrm{s}}(0) = 0$, it follows from (3.22) that for $x(t_0) = 0$,

$$\int_{t_0}^{T} e^{\alpha(t-t_0)} s(u(t), y(t)) \mathrm{d}t \geq 0, \quad T \geq t_0, \quad u \in \mathcal{U}, \qquad (3.23)$$

which, using (3.20), implies that $v_a(0) = 0$. Now, it can be easily shown that $v_a(x)$, $x \in \mathcal{D}$, satisfies (3.22), and hence, the available storage defined by (3.20) is a storage function for $\mathcal{G}$.

Finally, it follows from (3.22) that

$$v_s(x(t_0)) \geq e^{\alpha(T-t_0)} v_s(x(T)) - \int_{t_0}^{T} e^{\alpha(t-t_0)} s(u(t), y(t)) \mathrm{d}t$$

$$\geq - \int_{t_0}^{T} e^{\alpha(t-t_0)} s(u(t), y(t)) \mathrm{d}t, \quad T \geq t_0, \quad u \in \mathcal{U}, \qquad (3.24)$$

which implies

$$v_s(x(t_0)) \geq \sup_{T \geq t_0, u(\cdot)} \left[- \int_{t_0}^{T} e^{\alpha(t-t_0)} s(u(t), y(t)) \mathrm{d}t \right] = v_a(x(t_0)), \qquad (3.25)$$

and hence, (3.21) holds. $\qquad \square$

It follows from Theorem 3.1 that if (3.11) holds for $x(t_0) = 0$, then the vector available storage $V_a(x)$, $x \in \mathcal{D}$, is a vector storage function for $\mathcal{G}$. In this case, it follows from Theorem 3.2 that there exists $p \in \overline{\mathbb{R}}_+^q$, $p \neq 0$, such that $v_s(x) \triangleq p^{\mathrm{T}} V_a(x)$ is a storage function for $\mathcal{G}$ that satisfies (3.22), and hence, by (3.21), $v_a(x) \leq p^{\mathrm{T}} V_a(x)$, $x \in \mathcal{D}$. It is important to note that it follows from Theorem 3.2 that if $\mathcal{G}$ is vector dissipative, then $\mathcal{G}$ can either be (scalar) dissipative or (scalar) exponentially dissipative.

The following theorem provides sufficient conditions guaranteeing that all scalar storage functions defined in terms of vector storage functions, that is, $v_s(x) = p^{\mathrm{T}} V_s(x)$, of a given vector dissipative large-scale nonlinear dynamical system are positive definite.

Theorem 3.3. Consider the large-scale nonlinear dynamical system $\mathcal{G}$ given by (3.5) and (3.6) and assume that $\mathcal{G}$ is zero-state observable. Furthermore, assume that $\mathcal{G}$ is vector dissipative (respectively, exponentially vector dissipative) with respect to the vector supply rate $S(u, y)$ and there exist $\alpha \geq 0$ and $p \in \mathbb{R}_+^q$ such that (2.4) holds. In addition, assume that there exist functions $\kappa_i : Y_i \to U_i$ such that $\kappa_i(0) = 0$ and $S_i(\kappa_i(y_i), y_i) < 0$, $y_i \neq 0$, for all $i = 1, \ldots, q$. Then for all vector storage functions $V_s : \mathcal{D} \to \overline{\mathbb{R}}_+^q$ the storage function $v_s(x) \triangleq p^{\mathrm{T}} V_s(x)$, $x \in \mathcal{D}$, is positive definite, that is, $v_s(0) = 0$ and $v_s(x) > 0$, $x \in \mathcal{D}$, $x \neq 0$.

Proof. It follows from Theorem 3.2 that $v_a(x)$, $x \in \mathcal{D}$, is a storage function for $\mathcal{G}$ that satisfies (3.22). Next, suppose, *ad absurdum*, that there exists $x \in \mathcal{D}$ such that $v_a(x) = 0$, $x \neq 0$. Then it follows from the definition of $v_a(x)$, $x \in \mathcal{D}$, that for $x(t_0) = x$,

$$\int_{t_0}^{T} e^{\alpha(t-t_0)} s(u(t), y(t)) \mathrm{d}t \geq 0, \quad T \geq t_0, \quad u \in \mathcal{U}. \qquad (3.26)$$

However, for $u_i = k_i(y_i)$ we have $s_i(\kappa_i(y_i), y_i) < 0$, $y_i \neq 0$, for all $i = 1, \ldots, q$ and since $p >> 0$ it follows that $y_i(t) = 0$, $t \geq t_0$, $i = 1, \ldots, q$, which further implies that $u_i(t) = 0$, $t \geq t_0$, $i = 1, \ldots, q$. Since $\mathcal{G}$ is zero-state observable it follows that $x = 0$, and hence, $v_a(x) = 0$ if and only if $x = 0$. The result now follows from (3.21).

Finally, for the exponentially vector dissipative case it follows from Lemma 2.1 that $p >> 0$, with the rest of the proof identical to that given above. $\qquad\square$

Next, we introduce the concept of *vector required supply* of a large-scale nonlinear dynamical system. Specifically, define the vector required supply of the large-scale dynamical system $\mathcal{G}$ by

$$V_r(x_0) \triangleq \inf_{T \geq -t_0,\, u(\cdot)} \int_{-T}^{t_0} e^{-W(t-t_0)} S(u(t), y(t)) dt, \qquad (3.27)$$

where $x(t)$, $t \geq -T$, is the solution to (3.5) with $x(-T) = 0$ and $x(t_0) = x_0$. The infimum in (3.27) is taken over all system trajectories starting from $x(-T) = 0$ at time $t = -T$ and ending at $x(t_0) = x_0$ at time $t = t_0$, and all times $t \geq t_0$ or, equivalently, over all admissible inputs $u(\cdot)$ which drive the dynamical system $\mathcal{G}$ from the origin to x_0 over the time interval $[-T, t_0]$. If the system is not reachable from the origin, then we define $V_r(x_0) = \infty$. Note that since, with $x(t_0) = 0$, the infimum in (3.27) is the zero vector, it follows that $V_r(0) = 0$. Moreover, since $\mathcal{G}$ is completely reachable it follows that $V_r(x) << \infty$, $x \in \mathcal{D}$. Using the notion of the vector required supply we present necessary and sufficient conditions for dissipativity of a large-scale dynamical system with respect to a vector supply rate.

Theorem 3.4. Consider the large-scale nonlinear dynamical system $\mathcal{G}$ given by (3.5) and (3.6), and assume that $\mathcal{G}$ is completely reachable. Then $\mathcal{G}$ is vector dissipative (respectively, exponentially vector dissipative) with respect to the vector supply rate $S(u, y)$ if and only if

$$0 \leq\leq V_r(x) << \infty, \quad x \in \mathcal{D}. \qquad (3.28)$$

Moreover, if (3.28) holds, then $V_r(x)$, $x \in \mathcal{D}$, is a vector storage function for $\mathcal{G}$. Finally, if the vector available storage $V_a(x)$, $x \in \mathcal{D}$, is a vector storage function for $\mathcal{G}$, then

$$0 \leq\leq V_a(x) \leq\leq V_r(x) << \infty, \quad x \in \mathcal{D}. \qquad (3.29)$$

Proof. Suppose (3.28) holds and let $x(t)$, $t \in \mathbb{R}$, satisfy (3.5) with admissible inputs $u(\cdot) \in \mathcal{U}$ and $x(t_0) = x_0$. Then it follows from the definition of $V_r(\cdot)$ that for $-T \leq t_f \leq t_0$ and $u(\cdot) \in \mathcal{U}$,

$$V_r(x_0) \leq\leq \int_{-T}^{t_0} e^{-W(t-t_0)} S(u(t), y(t)) dt$$

$$= \int_{-T}^{t_{\mathrm{f}}} e^{-W(t-t_0)} S(u(t), y(t)) \mathrm{d}t$$

$$+ \int_{t_{\mathrm{f}}}^{t_0} e^{-W(t-t_0)} S(u(t), y(t)) \mathrm{d}t, \tag{3.30}$$

and hence,

$$V_{\mathrm{r}}(x_0) \leq\leq e^{W(t_0 - t_{\mathrm{f}})} \inf_{T \geq -t_{\mathrm{f}},\, u(\cdot)} \left[\int_{-T}^{t_{\mathrm{f}}} e^{-W(t-t_{\mathrm{f}})} S(u(t), y(t)) \mathrm{d}t \right]$$

$$+ \int_{t_{\mathrm{f}}}^{t_0} e^{-W(t-t_0)} S(u(t), y(t)) \mathrm{d}t$$

$$= e^{W(t_0 - t_{\mathrm{f}})} V_{\mathrm{r}}(x(t_{\mathrm{f}})) + \int_{t_{\mathrm{f}}}^{t_0} e^{W(t_0 - t)} S(u(t), y(t)) \mathrm{d}t, \tag{3.31}$$

which shows that $V_{\mathrm{r}}(x)$, $x \in \mathcal{D}$, is a vector storage function for $\mathcal{G}$, and hence, $\mathcal{G}$ is vector dissipative with respect to the vector supply rate $S(u, y)$.

Conversely, suppose that $\mathcal{G}$ is vector dissipative with respect to the vector supply rate $S(u, y)$. Then there exists a nonnegative vector storage function $V_{\mathrm{s}}(x)$, $x \in \mathcal{D}$, such that $V_{\mathrm{s}}(0) = 0$. Since $\mathcal{G}$ is completely reachable it follows that for $x(t_0) = x_0$ there exist $T > -t_0$ and $u(t)$, $t \in [-T, t_0]$, such that $x(-T) = 0$. Hence, it follows from the vector dissipation inequality (3.9) that

$$0 \leq\leq V_{\mathrm{s}}(x(t_0)) \leq\leq e^{W(t_0 + T)} V_{\mathrm{s}}(x(-T)) + \int_{-T}^{t_0} e^{W(t_0 - t)} S(u(t), y(t)) \mathrm{d}t, \tag{3.32}$$

which implies that for all $T \geq -t_0$ and $u(\cdot) \in \mathcal{U}$,

$$0 \leq\leq \int_{-T}^{t_0} e^{W(t_0 - t)} S(u(t), y(t)) \mathrm{d}t \tag{3.33}$$

or, equivalently,

$$0 \leq\leq \inf_{T \geq -t_0,\, u(\cdot)} \int_{-T}^{t_0} e^{W(t_0 - t)} S(u(t), y(t)) \mathrm{d}t = V_{\mathrm{r}}(x_0). \tag{3.34}$$

Since, by complete reachability, $V_{\mathrm{r}}(x) << \infty$, $x \in \mathcal{D}$, it follows that (3.28) holds.

Finally, suppose that $V_{\mathrm{a}}(x)$, $x \in \mathcal{D}$, is a vector storage function. Then for $x(-T) = 0$, $x(t_0) = x_0$, and $u(\cdot) \in \mathcal{U}$ it follows that

$$V_{\mathrm{a}}(x(t_0)) \leq\leq e^{W(t_0 + T)} V_{\mathrm{a}}(x(-T)) + \int_{-T}^{t_0} e^{W(t_0 - t)} S(u(t), y(t)) \mathrm{d}t, \tag{3.35}$$

which implies that, for all $x(t_0) = x_0 \in \mathcal{D}$,

$$0 \leq\leq V_{\mathrm{a}}(x(t_0)) \leq\leq \inf_{T \geq -t_0,\, u(\cdot)} \int_{-T}^{t_0} e^{W(t_0-t)} S(u(t), y(t))dt = V_{\mathrm{r}}(x(t_0)).$$

$$(3.36)$$

Since $x(t_0) = x_0 \in \mathcal{D}$ is arbitrary and, by definition, $V_{\mathrm{r}}(x) << \infty$, $x \in \mathcal{D}$, (3.36) implies (3.29). $\qquad\square$

The next result is a direct consequence of Theorems 3.1 and 3.4.

Proposition 3.1. Consider the large-scale nonlinear dynamical system $\mathcal{G}$ given by (3.5) and (3.6). Let $M = \mathrm{diag}\,[\mu_1, \ldots, \mu_q]$ be such that $0 \leq \mu_i \leq 1$, $i = 1, \ldots, q$. If $V_{\mathrm{a}}(x)$, $x \in \mathcal{D}$, and $V_{\mathrm{r}}(x)$, $x \in \mathcal{D}$, are vector storage functions for $\mathcal{G}$, then

$$V_{\mathrm{s}}(x) = M V_{\mathrm{a}}(x) + (I_q - M)V_{\mathrm{r}}(x), \quad x \in \mathcal{D}, \qquad (3.37)$$

is a vector storage function for $\mathcal{G}$.

Proof. Note that $M \geq\geq 0$ and $I_q - M \geq\geq 0$ if and only if $M = \mathrm{diag}\,[\mu_1, \ldots, \mu_q]$ and $\mu_i \in [0,1]$, $i = 1, \ldots, q$. Now, the result is a direct consequence of the vector dissipation inequality (3.9) by noting that if $V_{\mathrm{a}}(x)$ and $V_{\mathrm{r}}(x)$ satisfy (3.9), then $V_{\mathrm{s}}(x)$ satisfies (3.9). $\qquad\square$

Next, recall that if $\mathcal{G}$ is vector dissipative (respectively, exponentially vector dissipative), then there exist $p \in \overline{\mathbb{R}}_+^q$, $p \neq 0$, and $\alpha \geq 0$ (respectively, $p \in \mathbb{R}_+^q$ and $\alpha > 0$) such that (2.4) and (3.19) hold. Now, define the (scalar) *required supply* for the large-scale nonlinear dynamical system $\mathcal{G}$ by

$$v_{\mathrm{r}}(x_0) \triangleq \inf_{T \geq -t_0,\, u(\cdot)} \int_{-T}^{t_0} p^{\mathrm{T}} e^{-W(t-t_0)} S(u(t), y(t))dt$$

$$= \inf_{T \geq -t_0,\, u(\cdot)} \int_{-T}^{t_0} e^{\alpha(t-t_0)} s(u(t), y(t))dt, \quad x_0 \in \mathcal{D}, \qquad (3.38)$$

where $s(u, y) = p^{\mathrm{T}} S(u, y)$ and $x(t)$, $t \geq -T$, is the solution to (3.5) with $x(-T) = 0$ and $x(t_0) = x_0$. It follows from (3.38) that the required supply of a large-scale nonlinear dynamical system is the minimum amount of generalized energy that can be delivered to the large-scale system to transfer it from a state of minimum storage $x(-T) = 0$ to a given state $x(t_0) = x_0$. Using the same arguments as in the case of the vector required supply, it follows that $v_{\mathrm{r}}(0) = 0$ and $v_{\mathrm{r}}(x) < \infty$, $x \in \mathcal{D}$.

Next, using the notion of required supply, we show that all storage functions of the form $v_{\mathrm{s}}(x) = p^{\mathrm{T}} V_{\mathrm{s}}(x)$, where $p \in \overline{\mathbb{R}}_+^q$, $p \neq 0$, are bounded from above by the required supply and bounded from below by the available storage. Hence, a dissipative large-scale nonlinear dynamical system can deliver to its surroundings only a fraction of all of its stored subsystem energies and can store only a fraction of the work done to all of its subsystems.

Corollary 3.1. Consider the large-scale nonlinear dynamical system $\mathcal{G}$ given by (3.5) and (3.6). Assume that $\mathcal{G}$ is vector dissipative with respect to a vector supply rate $S(u,y)$ and with vector storage function $V_{\mathrm{s}} : \mathcal{D} \to \overline{\mathbb{R}}_+^q$. Then $v_{\mathrm{r}}(x)$, $x \in \mathcal{D}$, is a storage function for $\mathcal{G}$. Moreover, if $v_{\mathrm{s}}(x) \triangleq p^{\mathrm{T}} V_{\mathrm{s}}(x)$, $x \in \mathcal{D}$, where $p \in \overline{\mathbb{R}}_+^q$, $p \neq 0$, then

$$0 \leq v_{\mathrm{a}}(x) \leq v_{\mathrm{s}}(x) \leq v_{\mathrm{r}}(x) < \infty, \quad x \in \mathcal{D}. \tag{3.39}$$

Proof. It follows from Theorem 3.2 that if $\mathcal{G}$ is vector dissipative with respect to the vector supply rate $S(u,y)$ and with a vector storage function $V_{\mathrm{s}} : \mathcal{D} \to \overline{\mathbb{R}}_+^q$, then there exists $p \in \overline{\mathbb{R}}_+^q$, $p \neq 0$, such that $\mathcal{G}$ is dissipative with respect to the supply rate $s(u,y) = p^{\mathrm{T}} S(u,y)$ and with storage function $v_{\mathrm{s}}(x) = p^{\mathrm{T}} V_{\mathrm{s}}(x)$, $x \in \mathcal{D}$. Hence, it follows from (3.22), with $x(-T) = 0$ and $x(t_0) = x_0$, that

$$\int_{-T}^{t_0} e^{\alpha(t-t_0)} s(u(t), y(t)) \mathrm{d}t \geq 0, \quad T \geq -t_0, \quad u \in \mathcal{U}, \tag{3.40}$$

which implies that $v_{\mathrm{r}}(x_0) \geq 0$, $x_0 \in \mathcal{D}$. Furthermore, it is easy to see from the definition of a required supply that $v_{\mathrm{r}}(x)$, $x \in \mathcal{D}$, satisfies the dissipation inequality (3.22). Hence, $v_{\mathrm{r}}(x)$, $x \in \mathcal{D}$, is a storage function for $\mathcal{G}$.

Next, it follows from the dissipation inequality (3.22), with $x(-T) = 0$, $x(t_0) = x_0$, and $u(\cdot) \in \mathcal{U}$, that

$$e^{\alpha t_0} v_{\mathrm{s}}(x(t_0)) \leq e^{-\alpha T} v_{\mathrm{s}}(x(-T)) + \int_{-T}^{t_0} e^{\alpha t} s(u(t), y(t)) \mathrm{d}t$$

$$= \int_{-T}^{t_0} e^{\alpha t} s(u(t), y(t)) \mathrm{d}t, \tag{3.41}$$

which implies that

$$v_{\mathrm{s}}(x(t_0)) \leq \inf_{T \geq -t_0, \, u(\cdot)} \int_{-T}^{t_0} e^{\alpha(t-t_0)} s(u(t), y(t)) \mathrm{d}t = v_{\mathrm{r}}(x(t_0)). \tag{3.42}$$

Finally, it follows from Theorem 3.2 that $v_{\mathrm{a}}(x)$, $x \in \mathcal{D}$, is a storage function for $\mathcal{G}$, and hence, using (3.21) and (3.42), (3.39) holds. $\square$

It follows from Theorem 3.4 that if $\mathcal{G}$ is vector dissipative with respect to the vector supply rate $S(u,y)$, then $V_{\mathrm{r}}(x)$, $x \in \mathcal{D}$, is a vector storage function for $\mathcal{G}$ and, by Theorem 3.2, there exists $p \in \overline{\mathbb{R}}_+^q$, $p \neq 0$, such that $v_{\mathrm{s}}(x) \triangleq p^{\mathrm{T}} V_{\mathrm{r}}(x)$, $x \in \mathcal{D}$, is a storage function for $\mathcal{G}$ satisfying (3.22). Hence, it follows from Corollary 3.1 that $p^{\mathrm{T}} V_{\mathrm{r}}(x) \leq v_{\mathrm{r}}(x)$, $x \in \mathcal{D}$.

The next result relates vector (respectively, scalar) available storage and vector (respectively, scalar) required supply for vector lossless large-scale dynamical systems.

Theorem 3.5. Consider the large-scale nonlinear dynamical system $\mathcal{G}$ given by (3.5) and (3.6). Assume that $\mathcal{G}$ is completely reachable to and from the origin. If $\mathcal{G}$ is vector lossless with respect to the vector supply rate $S(u, y)$ and $V_{\mathrm{a}}(x)$, $x \in \mathcal{D}$, is a vector storage function, then $V_{\mathrm{a}}(x) = V_{\mathrm{r}}(x)$, $x \in \mathcal{D}$. Moreover, if $V_{\mathrm{s}}(x)$, $x \in \mathcal{D}$, is a vector storage function for $\mathcal{G}$, then all (scalar) storage functions of the form $v_{\mathrm{s}}(x) = p^{\mathrm{T}} V_{\mathrm{s}}(x)$, $x \in \mathcal{D}$, where $p \in \overline{\mathbb{R}}_{+}^{q}$, $p \neq 0$, are given by

$$
\begin{aligned}
v_{\mathrm{s}}(x_0) = v_{\mathrm{a}}(x_0) = v_{\mathrm{r}}(x_0) &= -\int_{t_0}^{T_+} e^{\alpha(t-t_0)} s(u(t), y(t)) \mathrm{d}t \\
&= \int_{-T_-}^{t_0} e^{\alpha(t-t_0)} s(u(t), y(t)) \mathrm{d}t, \qquad (3.43)
\end{aligned}
$$

where $x(t)$, $t \geq t_0$, is the solution to (3.5) with $u(\cdot) \in \mathcal{U}$, $x(t_0) = x_0 \in \mathcal{D}$, and $s(u, y) = p^{\mathrm{T}} S(u, y)$ for every T_-, T_+ such that $x(-T_-) = 0$ and $x(T_+) = 0$.

Proof. Suppose $\mathcal{G}$ is vector lossless with respect to the vector supply rate $S(u, y)$. Since $\mathcal{G}$ is completely reachable to and from the origin, it follows that, for every $x_0 = x(t_0) \in \mathcal{D}$, there exist $T_+ > t_0$, $-T_- < t_0$, and $u(t) \in U$, $t \in [-T_-, T_+]$, such that $x(-T_-) = 0$, $x(T_+) = 0$, and $x(t_0) = x_0$. Now, it follows from the dissipation inequality (3.9), which is satisfied as an equality, that

$$
0 = \int_{-T_-}^{T_+} e^{W(T_+ - t)} S(u(t), y(t)) \mathrm{d}t, \qquad (3.44)
$$

or, equivalently,

$$
\begin{aligned}
0 &= \int_{-T_-}^{T_+} e^{-W(t-t_0)} S(u(t), y(t)) \mathrm{d}t \\
&= \int_{-T_-}^{t_0} e^{-W(t-t_0)} S(u(t), y(t)) \mathrm{d}t + \int_{t_0}^{T_+} e^{-W(t-t_0)} S(u(t), y(t)) \mathrm{d}t \\
&\geq\!\geq \inf_{u(\cdot), \, T \geq -t_0} \int_{-T}^{t_0} e^{-W(t-t_0)} S(u(t), y(t)) \mathrm{d}t \\
&\quad + \inf_{u(\cdot), \, T \geq t_0} \int_{t_0}^{T} e^{-W(t-t_0)} S(u(t), y(t)) \mathrm{d}t \\
&= V_{\mathrm{r}}(x_0) - V_{\mathrm{a}}(x_0), \qquad (3.45)
\end{aligned}
$$

which implies that $V_{\mathrm{r}}(x_0) \leq\!\leq V_{\mathrm{a}}(x_0)$, $x_0 \in \mathcal{D}$. However, it follows from Theorem 3.4 that if $\mathcal{G}$ is vector dissipative and $V_{\mathrm{a}}(x)$, $x \in \mathcal{D}$, is a vector storage function, then $V_{\mathrm{a}}(x) \leq\leq V_{\mathrm{r}}(x)$, $x \in \mathcal{D}$, which along with (3.45) implies that $V_{\mathrm{a}}(x) = V_{\mathrm{r}}(x)$, $x \in \mathcal{D}$. Furthermore, since $\mathcal{G}$ is vector lossless there exist a nonzero vector $p \in \overline{\mathbb{R}}_+^q$ and a scalar $\alpha \geq 0$ satisfying (2.4).

Next, it follows from (3.44) that

$$
\begin{aligned}
0 &= \int_{-T_-}^{T_+} p^{\mathrm{T}} e^{-W(t-t_0)} S(u(t), y(t)) \mathrm{d}t \\
&= \int_{-T_-}^{T_+} e^{\alpha(t-t_0)} s(u(t), y(t)) \mathrm{d}t \\
&= \int_{-T_-}^{t_0} e^{\alpha(t-t_0)} s(u(t), y(t)) \mathrm{d}t + \int_{t_0}^{T_+} e^{\alpha(t-t_0)} s(u(t), y(t)) \mathrm{d}t \\
&\geq \inf_{u(\cdot),\, T \geq -t_0} \int_{-T}^{t_0} e^{\alpha(t-t_0)} s(u(t), y(t)) \mathrm{d}t \\
&\quad + \inf_{u(\cdot),\, T \geq t_0} \int_{t_0}^{T} e^{\alpha(t-t_0)} s(u(t), y(t)) \mathrm{d}t \\
&= v_{\mathrm{r}}(x_0) - v_{\mathrm{a}}(x_0), \quad x_0 \in \mathcal{D},
\end{aligned}
\tag{3.46}
$$

which along with (3.39) implies that for every (scalar) storage function of the form $v_{\mathrm{s}}(x) = p^{\mathrm{T}} V_{\mathrm{s}}(x)$, $x \in \mathcal{D}$, the equality $v_{\mathrm{a}}(x) = v_{\mathrm{s}}(x) = v_{\mathrm{r}}(x)$, $x \in \mathcal{D}$, holds. Moreover, since $\mathcal{G}$ is vector lossless the inequalities (3.22) and (3.41) are satisfied as equalities and

$$
v_{\mathrm{s}}(x_0) = -\int_{t_0}^{T_+} e^{\alpha(t-t_0)} s(u(t), y(t)) \mathrm{d}t = \int_{-T_-}^{t_0} e^{\alpha(t-t_0)} s(u(t), y(t)) \mathrm{d}t,
$$

$$
\tag{3.47}
$$

where $x(t)$, $t \geq t_0$, is the solution to (3.5) with $u(\cdot) \in \mathcal{U}$, $x(-T_-) = 0$, $x(T_+) = 0$, and $x(t_0) = x_0 \in \mathcal{D}$. $\qquad\square$

The next proposition presents a characterization for vector dissipativity of large-scale nonlinear dynamical systems in the case where $V_{\mathrm{s}}(\cdot)$ is continuously differentiable.

Proposition 3.2. Consider the large-scale nonlinear dynamical system $\mathcal{G}$ given by (3.5) and (3.6), and assume $V_{\mathrm{s}} = [v_{\mathrm{s}1}, \ldots, v_{\mathrm{s}q}]^{\mathrm{T}} : \mathcal{D} \to \overline{\mathbb{R}}_+^q$ is a continuously differentiable vector storage function for $\mathcal{G}$. Then $\mathcal{G}$ is vector dissipative with respect to the vector supply rate $S(u, y)$ if and only if

$$
\dot{V}_{\mathrm{s}}(x(t)) \leq\leq W V_{\mathrm{s}}(x(t)) + S(u(t), y(t)), \quad t \geq t_0, \quad u(t) \in U, \tag{3.48}
$$

where $\dot{V}_{\mathrm{s}}(x(t))$, $t \geq t_0$, denotes the total time derivative of each component of $V_{\mathrm{s}}(\cdot)$ along the state trajectories $x(t)$, $t \geq t_0$, of $\mathcal{G}$.

Proof. Suppose $\mathcal{G}$ is vector dissipative with respect to the vector supply rate $S(u, y)$ and with a continuously differentiable vector storage function $V_{\mathrm{s}} : \mathcal{D} \to \overline{\mathbb{R}}_+^q$. Then, with $T = t_2$ and $t_0 = t_1$, it follows from (3.9) that there exists a nonnegative vector function $l(t_1, t_2, x_0, u(\cdot)) \geq\geq 0$, $t_2 > t_1 \geq t_0$, $x_0 \in \mathcal{D}$, $u(\cdot) \in \mathcal{U}$, such that

$$V_{\mathrm{s}}(x(t_2)) = e^{W(t_2 - t_1)} V_{\mathrm{s}}(x(t_1)) + \int_{t_1}^{t_2} e^{W(t_2 - t)} S(u(t), y(t)) \mathrm{d}t$$
$$- l(t_1, t_2, x_0, u(\cdot)), \tag{3.49}$$

or, equivalently,

$$e^{-Wt_2} V_{\mathrm{s}}(x(t_2)) - e^{-Wt_1} V_{\mathrm{s}}(x(t_1)) = \int_{t_1}^{t_2} e^{-Wt} S(u(t), y(t)) \mathrm{d}t$$
$$- e^{-Wt_2} l(t_1, t_2, x_0, u(\cdot)). \tag{3.50}$$

Now, dividing (3.50) by $t_2 - t_1$ and letting $t_2 \to t_1$, (3.50) is equivalent to

$$\frac{\mathrm{d}}{\mathrm{d}t} \left[e^{-Wt} V_{\mathrm{s}}(x(t)) \right]\Big|_{t=t_1} = e^{-Wt_1} S(u(t_1), y(t_1))$$
$$- e^{-Wt_1} \lim_{t_2 \to t_1} \frac{l(t_1, t_2, x_0, u(\cdot))}{t_2 - t_1}, \tag{3.51}$$

where the limit in (3.51) exists since $V_{\mathrm{s}}(\cdot)$ is assumed to be continuously differentiable. Next, premultiplying (3.51) by e^{Wt_1}, $t_1 \geq 0$, yields

$$\dot{V}_{\mathrm{s}}(x(t_1)) - W V_{\mathrm{s}}(x(t_1)) = S(u(t_1), y(t_1)) - \lim_{t_2 \to t_1} \frac{l(t_1, t_2, x_0, u(\cdot))}{t_2 - t_1}, \tag{3.52}$$

which, since $\lim_{t_2 \to t_1} \frac{l(t_1, t_2, x_0, u(\cdot))}{t_2 - t_1} \geq\geq 0$ and t_1 is arbitrary, gives (3.48). The converse is immediate and, hence, is omitted. $\square$

Recall that if a disconnected subsystem $\mathcal{G}_i$ (i.e., $\mathcal{I}_i(x) \equiv 0$, $i \in \{1, \ldots, q\}$) of a large-scale dynamical system $\mathcal{G}$ is exponentially dissipative (respectively, dissipative) with respect to a supply rate $s_i(u_i, y_i)$, then there exist a storage function $v_{\mathrm{s}i} : \mathbb{R}^{n_i} \to \overline{\mathbb{R}}_+$ and a constant $\varepsilon_i > 0$ (respectively, $\varepsilon_i = 0$) such that the dissipation inequality (3.3) holds. In the case where $v_{\mathrm{s}i} : \mathbb{R}^{n_i} \to \overline{\mathbb{R}}_+$ is continuously differentiable, (3.3) yields

$$v_{\mathrm{s}i}'(x_i)(f_i(x_i) + G_i(x_i)u_i) \leq -\varepsilon_i v_{\mathrm{s}i}(x_i) + s_i(u_i, y_i), \quad x_i \in \mathbb{R}^{n_i}, \quad u_i \in U_i. \tag{3.53}$$

The next result relates exponential dissipativity with respect to a scalar supply rate of each disconnected subsystem $\mathcal{G}_i$ of $\mathcal{G}$ with vector dissipativity (or, possibly, exponential vector dissipativity) of $\mathcal{G}$ with respect to a vector supply rate.

Proposition 3.3. Consider the large-scale nonlinear dynamical system $\mathcal{G}$ given by (3.5) and (3.6). Assume that each disconnected subsystem $\mathcal{G}_i$ of $\mathcal{G}$ is exponentially dissipative with respect to the supply rate $s_i(u_i, y_i)$ and with a continuously differentiable storage function $v_{si} : \mathbb{R}^{n_i} \to \overline{\mathbb{R}}_+$, $i = 1, \ldots, q$. Furthermore, assume that the interconnection functions $\mathcal{I}_i : \mathcal{D} \to \mathbb{R}^{n_i}$, $i = 1, \ldots, q$, of $\mathcal{G}$ are such that

$$v'_{si}(x_i)\mathcal{I}_i(x) \leq \sum_{j=1}^{q} \xi_{ij}(x)v_{sj}(x_j), \quad x \in \mathcal{D}, \quad i = 1, \ldots, q, \qquad (3.54)$$

where $\xi_{ij} : \mathcal{D} \to \mathbb{R}$, $i, j = 1, \ldots, q$, are given bounded functions. If $W \in \mathbb{R}^{q \times q}$ is semistable (respectively, asymptotically stable), with

$$W_{(i,j)} = \left\{ \begin{array}{ll} -\varepsilon_i + \alpha_{ii}, & i = j, \\ \alpha_{ij}, & i \neq j, \end{array} \right. \qquad (3.55)$$

where $\varepsilon_i > 0$ and $\alpha_{ij} \triangleq \max\{0, \sup_{x \in \mathcal{D}} \xi_{ij}(x)\}$, $i, j = 1, \ldots, q$, then $\mathcal{G}$ is vector dissipative (respectively, exponentially vector dissipative) with respect to the vector supply rate $S(u, y) \triangleq [s_1(u_1, y_1), \ldots, s_q(u_q, y_q)]^{\mathrm{T}}$ and with vector storage function $V_s(x) \triangleq [v_{s1}(x_1), \ldots, v_{sq}(x_q)]^{\mathrm{T}}$, $x \in \mathcal{D}$.

Proof. Since each disconnected subsystem $\mathcal{G}_i$ of $\mathcal{G}$ is exponentially dissipative with respect to the supply rate $s_i(u_i, y_i)$, $i = 1, \ldots, q$, it follows from (3.53) and (3.54) that, for all $u_i(\cdot) \in \mathcal{U}_i$ and $i = 1, \ldots, q$,

$$\dot{v}_{si}(x_i(t)) = v'_{si}(x_i(t))[f_i(x_i(t)) + \mathcal{I}_i(x(t)) + G_i(x_i(t))u_i(t)]$$

$$\leq -\varepsilon_i v_{si}(x_i(t)) + s_i(u_i(t), y_i(t)) + \sum_{j=1}^{q} \xi_{ij}(x(t))v_{sj}(x_j(t))$$

$$\leq -\varepsilon_i v_{si}(x_i(t)) + s_i(u_i(t), y_i(t)) + \sum_{j=1}^{q} \alpha_{ij}v_{sj}(x_j(t)), \quad t \geq t_0.$$

$$(3.56)$$

Now, the result follows from Proposition 3.2 by noting that for all subsystems $\mathcal{G}_i$ of $\mathcal{G}$,

$$\dot{V}_s(x(t)) \leq\leq W V_s(x(t)) + S(u(t), y(t)), \quad t \geq t_0, \quad u(t) \in U, \qquad (3.57)$$

where W is essentially nonnegative and, by assumption, semistable (respectively, asymptotically stable) and $V_s(x) \triangleq [v_{s1}(x_1), \ldots, v_{sq}(x_q)]^{\mathrm{T}}$, $x \in \mathcal{D}$, is a vector storage function for $\mathcal{G}$. $\square$

As a special case of vector dissipativity theory we can analyze the stability of large-scale nonlinear dynamical systems. Specifically, assume that the large-scale dynamical system $\mathcal{G}$ is vector dissipative (respectively, exponentially vector dissipative) with respect to the vector supply rate $S(u, y)$

and with a continuously differentiable vector storage function $V_{\mathrm{s}} : \mathcal{D} \to \overline{\mathbb{R}}_+^q$. Moreover, assume that the conditions of Theorem 3.3 are satisfied. Then it follows from Proposition 3.2, with $u(t) \equiv 0$ and $y(t) \equiv 0$, that

$$\dot{V}_{\mathrm{s}}(x(t)) \leq\leq W V_{\mathrm{s}}(x(t)), \quad t \geq t_0, \tag{3.58}$$

where $x(t)$, $t \geq t_0$, is a solution to (3.5) with $x(t_0) = x_0$ and $u(t) \equiv 0$. Now, it follows from Corollary 2.3, with $w(r) = Wr$, that the zero solution $x(t) \equiv 0$ to (3.5), with $u(t) \equiv 0$, is Lyapunov (respectively, asymptotically) stable.

More generally, the problem of control system design for large-scale nonlinear dynamical systems can be addressed within the framework of vector dissipativity theory. In particular, suppose that there exists a continuously differentiable vector function $V_{\mathrm{s}} : \mathcal{D} \to \overline{\mathbb{R}}_+^q$ such that $V_{\mathrm{s}}(0) = 0$ and

$$\dot{V}_{\mathrm{s}}(x(t)) \leq\leq \mathcal{F}(V_{\mathrm{s}}(x(t)), u(t)), \quad t \geq t_0, \quad u(t) \in U, \tag{3.59}$$

where $\mathcal{F} : \overline{\mathbb{R}}_+^q \times \mathbb{R}^m \to \mathbb{R}^q$ and $\mathcal{F}(0,0) = 0$. Then the control system design problem for a large-scale dynamical system reduces to constructing an *energy* feedback control law $\phi : \overline{\mathbb{R}}_+^q \to U$ of the form

$$u = \phi(V_{\mathrm{s}}(x)) \triangleq [\phi_1^{\mathrm{T}}(V_{\mathrm{s}}(x)), \ldots, \phi_q^{\mathrm{T}}(V_{\mathrm{s}}(x))]^{\mathrm{T}}, \quad x \in \mathcal{D}, \tag{3.60}$$

where $\phi_i : \overline{\mathbb{R}}_+^q \to U_i$, $\phi_i(0) = 0$, $i = 1, \ldots, q$, such that the zero solution $r(t) \equiv 0$ to the comparison system

$$\dot{r}(t) = w(r(t)), \quad r(t_0) = V_{\mathrm{s}}(x(t_0)), \quad t \geq t_0, \tag{3.61}$$

is rendered asymptotically stable, where $w(r) \triangleq \mathcal{F}(r, \phi(r))$ is of class $\mathcal{W}$. In this case, if there exists $p \in \mathbb{R}_+^q$ such that $v_{\mathrm{s}}(x) \triangleq p^{\mathrm{T}} V_{\mathrm{s}}(x)$, $x \in \mathcal{D}$, is positive definite, then it follows from Corollary 2.3 that the zero solution $x(t) \equiv 0$ to (3.5), with u given by (3.60), is asymptotically stable.

As can be seen from the above discussion, using an energy feedback control architecture and exploiting the comparison system within the control design for large-scale nonlinear dynamical systems can significantly reduce the dimensionality of a control synthesis problem in terms of a number of states that need to be stabilized. It should be noted, however, that for stability analysis of large-scale dynamical systems the comparison system need not be linear as implied by (3.58). A nonlinear comparison system would still guarantee stability of a large-scale dynamical system provided that the conditions of Corollary 2.3 are satisfied.

3.3 Extended Kalman-Yakubovich-Popov Conditions for Large-Scale Nonlinear Dynamical Systems

In this section, we show that vector dissipativeness (respectively, exponential vector dissipativeness) of a large-scale nonlinear dynamical system $\mathcal{G}$

of the form (3.5) and (3.6) can be characterized in terms of the local subsystem functions $f_i(\cdot)$, $G_i(\cdot)$, $h_i(\cdot)$, and $J_i(\cdot)$, along with the interconnection structures $\mathcal{I}_i(\cdot)$ for $i = 1, \ldots, q$. For the results in this section we consider the special case of dissipative systems with quadratic vector supply rates and set $\mathcal{D} = \mathbb{R}^n$, $U_i = \mathbb{R}^{m_i}$, and $Y_i = \mathbb{R}^{l_i}$. Specifically, let $R_i \in \mathbb{S}^{m_i}$, $S_i \in \mathbb{R}^{l_i \times m_i}$, and $Q_i \in \mathbb{S}^{l_i}$ be given and assume $S(u,y)$ is such that $s_i(u_i, y_i) = y_i^{\mathrm{T}} Q_i y_i + 2 y_i^{\mathrm{T}} S_i u_i + u_i^{\mathrm{T}} R_i u_i$, $i = 1, \ldots, q$. Furthermore, for the remainder of this chapter we assume that there exists a continuously differentiable vector storage function $V_{\mathrm{s}}(x)$, $x \in \mathbb{R}^n$, for the large-scale nonlinear dynamical system $\mathcal{G}$.

For the statement of the next result recall that $x = [x_1^{\mathrm{T}}, \ldots, x_q^{\mathrm{T}}]^{\mathrm{T}}$, $u = [u_1^{\mathrm{T}}, \ldots, u_q^{\mathrm{T}}]^{\mathrm{T}}$, $y = [y_1^{\mathrm{T}}, \ldots, y_q^{\mathrm{T}}]^{\mathrm{T}}$, $x_i \in \mathbb{R}^{n_i}$, $u_i \in \mathbb{R}^{m_i}$, $y_i \in \mathbb{R}^{l_i}$, $i = 1, \ldots, q$, $\sum_{i=1}^{q} n_i = n$, $\sum_{i=1}^{q} m_i = m$, and $\sum_{i=1}^{q} l_i = l$. Furthermore, for (3.5) and (3.6) define $\mathcal{F} : \mathbb{R}^n \to \mathbb{R}^n$, $G : \mathbb{R}^n \to \mathbb{R}^{n \times m}$, $h : \mathbb{R}^n \to \mathbb{R}^l$, and $J : \mathbb{R}^n \to \mathbb{R}^{l \times m}$ by $\mathcal{F}(x) \triangleq [\mathcal{F}_1^{\mathrm{T}}(x), \ldots, \mathcal{F}_q^{\mathrm{T}}(x)]^{\mathrm{T}}$, where $\mathcal{F}_i(x) \triangleq f_i(x_i) + \mathcal{I}_i(x)$, $i = 1, \ldots, q$, $G(x) \triangleq \mathrm{block\text{–}diag}[G_1(x_1), \ldots, G_q(x_q)]$, $h(x) \triangleq [h_1^{\mathrm{T}}(x_1), \ldots, h_q^{\mathrm{T}}(x_q)]^{\mathrm{T}}$, and $J(x) \triangleq \mathrm{block\text{–}diag}[J_1(x_1), \ldots, J_q(x_q)]$. Moreover, for all $i = 1, \ldots, q$, define $\hat{R}_i \in \mathbb{S}^m$, $\hat{S}_i \in \mathbb{R}^{l \times m}$, and $\hat{Q}_i \in \mathbb{S}^l$ such that each of these block matrices consists of zero blocks except, respectively, for the matrix blocks $R_i \in \mathbb{S}^{m_i}$, $S_i \in \mathbb{R}^{l_i \times m_i}$, and $Q_i \in \mathbb{S}^{l_i}$ on (i, i) position. Finally, we introduce a more general definition of vector dissipativity involving an underlying nonlinear comparison system.

Definition 3.5. The large-scale nonlinear dynamical system $\mathcal{G}$ given by (3.5) and (3.6) is *vector dissipative* (respectively, *exponentially vector dissipative*) *with respect to the vector supply rate* $S(u, y)$ if there exists a continuous, nonnegative definite vector function $V_{\mathrm{s}} = [v_{\mathrm{s}1}, \ldots, v_{\mathrm{s}q}]^{\mathrm{T}} : \mathcal{D} \to \overline{\mathbb{R}}_+^q$, called a *vector storage function*, and a class $\mathcal{W}$ function $w : \overline{\mathbb{R}}_+^q \to \mathbb{R}^q$ such that $V_{\mathrm{s}}(0) = 0$, $w(0) = 0$, the zero solution $r(t) \equiv 0$ to the comparison system

$$\dot{r}(t) = w(r(t)), \quad r(t_0) = r_0, \quad t \geq t_0, \tag{3.62}$$

is Lyapunov (respectively, asymptotically) stable, and the *vector dissipation inequality*

$$V_{\mathrm{s}}(x(T)) \leq\leq V_{\mathrm{s}}(x(t_0)) + \int_{t_0}^{T} w(V_{\mathrm{s}}(x(t)))\mathrm{d}t + \int_{t_0}^{T} S(u(t), y(t))\mathrm{d}t, \tag{3.63}$$

is satisfied for all $T \geq t_0$, where $x(t)$, $t \geq t_0$, is the solution to (3.5) with $u(\cdot) \in \mathcal{U}$. The large-scale nonlinear dynamical system $\mathcal{G}$ given by (3.5) and (3.6) is *vector lossless with respect to the vector supply rate* $S(u, y)$ if the vector dissipation inequality is satisfied as an equality with the zero solution $r(t) \equiv 0$ to (3.62) being Lyapunov stable.

If $V_{\mathrm{s}}(\cdot)$ is continuously differentiable, then (3.63) can be equivalently written as

$$\dot{V}_{\mathrm{s}}(x(t)) \leq\leq w(V_{\mathrm{s}}(x(t))) + S(u(t), y(t)), \quad t \geq t_0, \quad u(t) \in U. \tag{3.64}$$

If in Definition 3.5 the function $w : \overline{\mathbb{R}}_+^q \to \mathbb{R}^q$ is such that $w(r) = Wr$, where $W \in \mathbb{R}^{q \times q}$, then W is essentially nonnegative and Definition 3.5 collapses to Definition 3.3.

Theorem 3.6. Consider the large-scale nonlinear dynamical system $\mathcal{G}$ given by (3.5) and (3.6). Let $R_i \in \mathbb{S}^{m_i}$, $S_i \in \mathbb{R}^{l_i \times m_i}$, and $Q_i \in \mathbb{S}^{l_i}$, $i = 1, \ldots, q$. Then $\mathcal{G}$ is vector dissipative (respectively, exponentially vector dissipative) with respect to the vector quadratic supply rate $S(u, y)$, where $s_i(u_i, y_i) = u_i^{\mathrm{T}} R_i u_i + 2 y_i^{\mathrm{T}} S_i u_i + y_i^{\mathrm{T}} Q_i y_i$, $i = 1, \ldots, q$, if and only if there exist functions $V_{\mathrm{s}} = [v_{\mathrm{s}1}, \ldots, v_{\mathrm{s}q}]^{\mathrm{T}} : \mathbb{R}^n \to \overline{\mathbb{R}}_+^q$, $w = [w_1, \ldots, w_q]^{\mathrm{T}} : \overline{\mathbb{R}}_+^q \to \mathbb{R}^q$, $\ell_i : \mathbb{R}^n \to \mathbb{R}^{s_i}$, and $\mathcal{Z}_i : \mathbb{R}^n \to \mathbb{R}^{s_i \times m}$, such that $v_{\mathrm{s}i}(\cdot)$ is continuously differentiable, $v_{\mathrm{s}i}(0) = 0$, $i = 1, \ldots, q$, $w \in \mathcal{W}$, $w(0) = 0$, the zero solution $r(t) \equiv 0$ to (3.62) is Lyapunov (respectively, asymptotically) stable, and, for all $x \in \mathbb{R}^n$ and $i = 1, \ldots, q$,

$$0 = v_{\mathrm{s}i}'(x)\mathcal{F}(x) - h^{\mathrm{T}}(x)\hat{Q}_i h(x) - w_i(V_{\mathrm{s}}(x)) + \ell_i^{\mathrm{T}}(x)\ell_i(x), \tag{3.65}$$

$$0 = \tfrac{1}{2} v_{\mathrm{s}i}'(x)G(x) - h^{\mathrm{T}}(x)(\hat{S}_i + \hat{Q}_i J(x)) + \ell_i^{\mathrm{T}}(x)\mathcal{Z}_i(x), \tag{3.66}$$

$$0 = \hat{R}_i + J^{\mathrm{T}}(x)\hat{S}_i + \hat{S}_i^{\mathrm{T}} J(x) + J^{\mathrm{T}}(x)\hat{Q}_i J(x) - \mathcal{Z}_i^{\mathrm{T}}(x)\mathcal{Z}_i(x). \tag{3.67}$$

Proof. First, suppose that there exist functions $v_{\mathrm{s}i} : \mathbb{R}^n \to \overline{\mathbb{R}}_+$, $\ell_i : \mathbb{R}^n \to \mathbb{R}^{s_i}$, $\mathcal{Z}_i : \mathbb{R}^n \to \mathbb{R}^{s_i \times m}$, $w : \overline{\mathbb{R}}_+^q \to \mathbb{R}^q$, such that $v_{\mathrm{s}i}(\cdot)$ is continuously differentiable and nonnegative-definite, $v_{\mathrm{s}i}(0) = 0$, $i = 1, \ldots, q$, $w(0) = 0$, $w \in \mathcal{W}$, the zero solution $r(t) \equiv 0$ to (3.62) is Lyapunov (respectively, asymptotically) stable, and (3.65)–(3.67) are satisfied. Then, for every admissible input $u(\cdot) \in \mathcal{U}$, $t_1, t_2 \in \mathbb{R}$, $t_2 \geq t_1 \geq t_0$, and $i = 1, \ldots, q$, it follows from (3.65)–(3.67) that

$$\int_{t_1}^{t_2} s_i(u_i(t), y_i(t)) \mathrm{d}t = \int_{t_1}^{t_2} [u^{\mathrm{T}}(t)\hat{R}_i u(t) + 2 y^{\mathrm{T}}(t)\hat{S}_i u(t)$$

$$+ y^{\mathrm{T}}(t)\hat{Q}_i y(t)]\mathrm{d}t$$

$$= \int_{t_1}^{t_2} [h^{\mathrm{T}}(x(t))\hat{Q}_i h(x(t))$$

$$+ 2 h^{\mathrm{T}}(x(t))(\hat{S}_i + \hat{Q}_i J(x(t)))u(t)$$

$$+ u^{\mathrm{T}}(t)(J^{\mathrm{T}}(x(t))\hat{Q}_i J(x(t))$$

$$+ J^{\mathrm{T}}(x(t))\hat{S}_i + \hat{S}_i^{\mathrm{T}} J(x(t)) + \hat{R}_i)u(t)]\mathrm{d}t$$

$$= \int_{t_1}^{t_2} [v_{\mathrm{s}i}'(x(t))(\mathcal{F}(x(t)) + G(x(t))u(t))$$

$$+ \ell_i^{\mathrm{T}}(x(t))\ell_i(x(t)) + 2\ell_i^{\mathrm{T}}(x(t))\mathcal{Z}_i(x(t))u(t)$$

$$+u^{\mathrm{T}}(t)\mathcal{Z}_i^{\mathrm{T}}(x(t))\mathcal{Z}_i(x(t))u(t) - w_i(V_{\mathrm{s}}(x(t)))]\mathrm{d}t$$

$$= \int_{t_1}^{t_2} [\dot{v}_{\mathrm{s}i}(x(t)) + [\ell_i(x(t)) + \mathcal{Z}_i(x(t))u(t)]^{\mathrm{T}}$$

$$\cdot [\ell_i(x(t)) + \mathcal{Z}_i(x(t))u(t)] - w_i(V_{\mathrm{s}}(x(t)))]\mathrm{d}t$$

$$\geq v_{\mathrm{s}i}(x(t_2)) - v_{\mathrm{s}i}(x(t_1)) - \int_{t_1}^{t_2} w_i(V_{\mathrm{s}}(x(t)))\mathrm{d}t,$$

$$(3.68)$$

where $x(t)$, $t \geq t_0$, satisfies (3.5). Now, the result follows from (3.68) with vector storage function $V_{\mathrm{s}}(x) = [v_{\mathrm{s}1}(x), \ldots, v_{\mathrm{s}q}(x)]^{\mathrm{T}}$, $x \in \mathbb{R}^n$.

Conversely, suppose that $\mathcal{G}$ is vector dissipative (respectively, exponentially vector dissipative) with respect to the vector supply rate $S(u,y)$, where $s_i(u_i, y_i) = u_i^{\mathrm{T}} R_i u_i + 2y_i^{\mathrm{T}} S_i u_i + y_i^{\mathrm{T}} Q_i y_i$, $i = 1, \ldots, q$. Then there exist a vector storage function $V_{\mathrm{s}} = [v_{\mathrm{s}1}, \ldots, v_{\mathrm{s}q}]^{\mathrm{T}} : \mathbb{R}^n \to \overline{\mathbb{R}}_+^q$ and a class $\mathcal{W}$ function $w = [w_1, \ldots, w_q]^{\mathrm{T}} : \overline{\mathbb{R}}_+^q \to \mathbb{R}^q$ such that $V_{\mathrm{s}}(0) = 0$, $w(0) = 0$, the zero solution $r(t) \equiv 0$ to (3.62) is Lyapunov (respectively, asymptotically) stable, and, for all $i = 1, \ldots, q$, $t_2 \geq t_1$, and $u \in \mathcal{U}$,

$$v_{\mathrm{s}i}(x(t_2)) \leq v_{\mathrm{s}i}(x(t_1)) + \int_{t_1}^{t_2} s_i(u_i(t), y_i(t))\mathrm{d}t + \int_{t_1}^{t_2} w_i(V_{\mathrm{s}}(x(t)))\mathrm{d}t. \quad (3.69)$$

Since, by assumption, $v_{\mathrm{s}i}(\cdot)$ is continuously differentiable, (3.69) is equivalent to

$$\dot{v}_{\mathrm{s}i}(x(t)) \leq s_i(u_i(t), y_i(t)) + w_i(V_{\mathrm{s}}(x(t))), \quad t \geq t_0, \quad (3.70)$$

where $x(t)$, $t \geq t_0$, satisfies (3.5). Now, with $t = t_0$ it follows from (3.70) that

$$v_{\mathrm{s}i}'(x_0)(\mathcal{F}(x_0) + G(x_0)u(t_0)) \leq s_i(u_i(t_0), y_i(t_0)) + w_i(V_{\mathrm{s}}(x_0)), \quad (3.71)$$

for all $u(t_0) \in \mathbb{R}^m$, $y(t_0) \in \mathbb{R}^l$, $x_0 \in \mathbb{R}^n$, and $i = 1, \ldots, q$.

Next, for all $i = 1, \ldots, q$, let $d_i : \mathbb{R}^n \times \mathbb{R}^m \to \mathbb{R}$ be such that

$$d_i(x,u) \triangleq -v_{\mathrm{s}i}'(x)(\mathcal{F}(x) + G(x)u) + s_i(u_i, h_i(x_i) + J_i(x_i)u_i) + w_i(V_{\mathrm{s}}(x)).$$

$$(3.72)$$

Now, since $x_0 \in \mathcal{D}$ is arbitrary, it follows from (3.71) that $d_i(x,u) \geq 0$, $x \in \mathbb{R}^n$, $u \in \mathbb{R}^m$, $i = 1, \ldots, q$. Furthermore, note that $d_i(x,u)$ given by (3.72) is quadratic in u, and hence, there exist functions $\ell_i : \mathbb{R}^n \to \mathbb{R}^{s_i}$ and $\mathcal{Z}_i : \mathbb{R}^n \to \mathbb{R}^{s_i \times m}$ such that, for all $i = 1, \ldots, q$, $x \in \mathbb{R}^n$, and $u \in \mathbb{R}^m$,

$$\begin{aligned} d_i(x,u) &= [\ell_i(x) + \mathcal{Z}_i(x)u]^{\mathrm{T}}[\ell_i(x) + \mathcal{Z}_i(x)u] \\ &= -v_{\mathrm{s}i}'(x)(\mathcal{F}(x) + G(x)u) + s_i(u_i, h_i(x_i) + J_i(x_i)u_i) \\ &\quad + w_i(V_{\mathrm{s}}(x)) \end{aligned}$$

$$= -v'_{si}(x)(\mathcal{F}(x) + G(x)u) + u^{\mathrm{T}}\hat{R}_i u + 2(h(x) + J(x)u)^{\mathrm{T}}\hat{S}_i u$$
$$+ (h(x) + J(x)u)^{\mathrm{T}}\hat{Q}_i(h(x) + J(x)u) + w_i(V_s(x)). \tag{3.73}$$

Now, equating coefficients of equal powers yields (3.65)–(3.67). $\qquad\square$

Using (3.65)–(3.67) it follows that for $T \geq t_0 \geq 0$ and $i = 1, \ldots, q$,

$$\int_{t_0}^{T} s_i(u_i(t), y_i(t))\mathrm{d}t + \int_{t_0}^{T} w_i(V_s(x(t)))\mathrm{d}t = v_{si}(x(T)) - v_{si}(x(t_0))$$

$$+ \int_{t_0}^{T} [\ell_i(x(t)) + \mathcal{Z}_i(x(t))u(t)]^{\mathrm{T}}[\ell_i(x(t)) + \mathcal{Z}_i(x(t))u(t)]\mathrm{d}t, \tag{3.74}$$

where $V_s(x) = [v_{s1}(x), \ldots, v_{sq}(x)]^{\mathrm{T}}$, $x \in \mathbb{R}^n$, which can be interpreted as a *generalized energy* balance equation for the ith subsystem of $\mathcal{G}$ where $v_{si}(x(T)) - v_{si}(x(t_0))$ is the stored or accumulated generalized energy of the ith subsystem, the two path-dependent terms on the left are, respectively, the external supplied energy to the ith subsystem and the energy gained by the ith subsystem from the net energy flow between all subsystems due to subsystem coupling, and the second path-dependent term on the right corresponds to the dissipated energy from the ith subsystem.

Equivalently, (3.74) can be rewritten for all $i = 1, \ldots, q$ as

$$\dot{v}_{si}(x(t)) = s_i(u_i(t), y_i(t)) + w_i(V_s(x(t)))$$
$$- [\ell_i(x(t)) + \mathcal{Z}_i(x(t))u(t)]^{\mathrm{T}}[\ell_i(x(t)) + \mathcal{Z}_i(x(t))u(t)], \quad t \geq t_0,$$
$$\tag{3.75}$$

which yields a set of q generalized energy conservation equations for the large-scale dynamical system $\mathcal{G}$. Specifically, (3.75) shows that the rate of change in generalized energy, or generalized power, of the ith subsystem of $\mathcal{G}$ is equal to the generalized system power input to the ith subsystem plus the instantaneous rate of energy supplied to the ith subsystem from the net energy flow between all subsystems minus the internal generalized system power dissipated from the ith subsystem.

Note that if $\mathcal{G}$ with $u(t) \equiv 0$ is vector dissipative (respectively, exponentially vector dissipative) with respect to the vector quadratic supply rate where $Q_i \leq 0$, $i = 1, \ldots, q$, then it follows from the vector dissipation inequality that

$$\dot{V}_s(x(t)) \leq\leq w(V_s(x(t))) + S(0, y(t)) \leq\leq w(V_s(x(t))), \quad t \geq t_0, \tag{3.76}$$

where $S(0, y) = [s_1(0, y_1), \ldots, s_q(0, y_q)]^{\mathrm{T}}$, $s_i(0, y_i(t)) = y_i^{\mathrm{T}}(t)Q_i y_i(t) \leq 0$, $t \geq t_0$, $i = 1, \ldots, q$, and $x(t)$, $t \geq t_0$, is the solution to (3.5) with $u(t) \equiv 0$. If, in addition, there exists $p \in \mathbb{R}_+^q$ such that $p^{\mathrm{T}}V_s(x)$, $x \in \mathbb{R}^n$, is positive definite, then it follows from Corollary 2.3 that the undisturbed ($u(t) \equiv 0$) large-scale nonlinear dynamical system (3.5) is Lyapunov (respectively, asymptotically) stable.

Next, we extend the notions of passivity and nonexpansivity to vector passivity and vector nonexpansivity.

Definition 3.6. The large-scale nonlinear dynamical system $\mathcal{G}$ given by (3.5) and (3.6) with $m_i = l_i$, $i = 1, \ldots, q$, is *vector passive* (respectively, *vector exponentially passive*) if it is vector dissipative (respectively, exponentially vector dissipative) with respect to the vector supply rate $S(u, y)$, where $s_i(u_i, y_i) = 2y_i^{\mathrm{T}} u_i$, $i = 1, \ldots, q$.

Definition 3.7. The large-scale nonlinear dynamical system $\mathcal{G}$ given by (3.5) and (3.6) is *vector nonexpansive* (respectively, *vector exponentially nonexpansive*) if it is vector dissipative (respectively, exponentially vector dissipative) with respect to the vector supply rate $S(u, y)$, where $s_i(u_i, y_i) = \gamma_i^2 u_i^{\mathrm{T}} u_i - y_i^{\mathrm{T}} y_i$, $i = 1, \ldots, q$, and $\gamma_i > 0$, $i = 1, \ldots, q$, are given.

Note that a mixed vector passive-nonexpansive formulation of $\mathcal{G}$ can also be considered. Specifically, one can consider large-scale nonlinear dynamical systems $\mathcal{G}$ which are vector dissipative with respect to vector supply rates $S(u, y)$, where $s_i(u_i, y_i) = 2y_i^{\mathrm{T}} u_i$, $i \in \mathcal{N}_{\mathrm{p}}$, $s_j(u_j, y_j) = \gamma_j^2 u_j^{\mathrm{T}} u_j - y_j^{\mathrm{T}} y_j$, $\gamma_j > 0$, $j \in \mathcal{N}_{\mathrm{ne}}$, $\mathcal{N}_{\mathrm{p}} \cap \mathcal{N}_{\mathrm{ne}} = \emptyset$, and $\mathcal{N}_{\mathrm{p}} \cup \mathcal{N}_{\mathrm{ne}} = \{1, \ldots, q\}$. Furthermore, supply rates for vector input strict passivity, vector output strict passivity, and vector input-output strict passivity generalizing the passivity notions given in [89] can also be considered.

The next result presents constructive sufficient conditions guaranteeing vector dissipativity of $\mathcal{G}$ with respect to a vector quadratic supply rate for the case where the vector storage function $V_{\mathrm{s}}(x)$, $x \in \mathbb{R}^n$, is component decoupled, that is, $V_{\mathrm{s}}(x) = [v_{\mathrm{s}1}(x_1), \ldots, v_{\mathrm{s}q}(x_q)]^{\mathrm{T}}$, $x \in \mathbb{R}^n$.

Theorem 3.7. Consider the large-scale nonlinear dynamical system $\mathcal{G}$ given by (3.5) and (3.6). Assume that there exist functions $V_{\mathrm{s}} = [v_{\mathrm{s}1}, \ldots, v_{\mathrm{s}q}]^{\mathrm{T}} : \mathbb{R}^n \to \overline{\mathbb{R}}_+^q$, $w = [w_1, \ldots, w_q]^{\mathrm{T}} : \overline{\mathbb{R}}_+^q \to \mathbb{R}^q$, $\ell_i : \mathbb{R}^n \to \mathbb{R}^{s_i}$, $\mathcal{Z}_i : \mathbb{R}^n \to \mathbb{R}^{s_i \times m_i}$ such that $v_{\mathrm{s}i}(\cdot)$ is continuously differentiable, $v_{\mathrm{s}i}(0) = 0$, $i = 1, \ldots, q$, $w \in \mathcal{W}$, $w(0) = 0$, the zero solution $r(t) \equiv 0$ to (3.62) is Lyapunov (respectively, asymptotically) stable, and, for all $x \in \mathbb{R}^n$ and $i = 1, \ldots, q$,

$$0 \geq v_{\mathrm{s}i}'(x_i)[f_i(x_i) + \mathcal{I}_i(x)] - h_i^{\mathrm{T}}(x_i) Q_i h_i(x_i) - w_i(V_{\mathrm{s}}(x)) + \ell_i^{\mathrm{T}}(x_i) \ell_i(x_i), \tag{3.77}$$

$$0 = \tfrac{1}{2} v_{\mathrm{s}i}'(x_i) G_i(x_i) - h_i^{\mathrm{T}}(x_i)(S_i + Q_i J_i(x_i)) + \ell_i^{\mathrm{T}}(x_i) \mathcal{Z}_i(x_i), \tag{3.78}$$

$$0 \leq R_i + J_i^{\mathrm{T}}(x_i) S_i + S_i^{\mathrm{T}} J_i(x_i) + J_i^{\mathrm{T}}(x_i) Q_i J_i(x_i) - \mathcal{Z}_i^{\mathrm{T}}(x_i) \mathcal{Z}_i(x_i). \tag{3.79}$$

Then $\mathcal{G}$ is vector dissipative (respectively, exponentially vector dissipative) with respect to the vector supply rate $S(u, y)$, where $s_i(u_i, y_i) = u_i^{\mathrm{T}} R_i u_i + 2y_i^{\mathrm{T}} S_i u_i + y_i^{\mathrm{T}} Q_i y_i$, $i = 1, \ldots, q$.

Proof. For every admissible input $u = [u_1^{\mathrm{T}}, \ldots, u_q^{\mathrm{T}}]^{\mathrm{T}}$ such that $u_i \in \mathbb{R}^{m_i}$, $t_1, t_2 \in \mathbb{R}$, $t_2 \geq t_1 \geq t_0$, and $i = 1, \ldots, q$, it follows from (3.77)–(3.79) that

$$\int_{t_1}^{t_2} s_i(u_i(t), y_i(t))\mathrm{d}t = \int_{t_1}^{t_2} [u_i^{\mathrm{T}}(t)R_i u_i(t) + 2y_i^{\mathrm{T}}(t)S_i u_i(t)$$
$$+ y_i^{\mathrm{T}}(t)Q_i y_i(t)]\mathrm{d}t$$
$$= \int_{t_1}^{t_2} [h_i^{\mathrm{T}}(x_i(t))Q_i h_i(x_i(t))$$
$$+ 2h_i^{\mathrm{T}}(x_i(t))(S_i + Q_i J_i(x_i(t)))u_i(t)$$
$$+ u_i^{\mathrm{T}}(t)(J_i^{\mathrm{T}}(x_i(t))Q_i J_i(x_i(t)) + J_i^{\mathrm{T}}(x_i(t))S_i$$
$$+ S_i^{\mathrm{T}} J_i(x_i(t)) + R_i)u_i(t)]\mathrm{d}t$$
$$\geq \int_{t_1}^{t_2} [v'_{\mathrm{s}i}(x_i(t))[f_i(x_i(t)) + \mathcal{I}_i(x(t))$$
$$+ G_i(x_i(t))u_i(t)] + \ell_i^{\mathrm{T}}(x_i(t))\ell_i(x_i(t))$$
$$+ 2\ell_i^{\mathrm{T}}(x_i(t))\mathcal{Z}_i(x_i(t))u_i(t)$$
$$+ u_i^{\mathrm{T}}(t)\mathcal{Z}_i^{\mathrm{T}}(x_i(t))\mathcal{Z}_i(x_i(t))u_i(t) - w_i(V_{\mathrm{s}}(x(t)))]\mathrm{d}t$$
$$= \int_{t_1}^{t_2} [\dot{v}_{\mathrm{s}i}(x_i(t)) + [\ell_i(x_i(t)) + \mathcal{Z}_i(x_i(t))u_i(t)]^{\mathrm{T}}$$
$$\cdot [\ell_i(x_i(t)) + \mathcal{Z}_i(x_i(t))u_i(t)] - w_i(V_{\mathrm{s}}(x(t)))]\mathrm{d}t$$
$$\geq v_{\mathrm{s}i}(x_i(t_2)) - v_{\mathrm{s}i}(x_i(t_1)) - \int_{t_1}^{t_2} w_i(V_{\mathrm{s}}(x(t)))\mathrm{d}t,$$

$$(3.80)$$

where $x(t)$, $t \geq t_0$, satisfies (3.5). Now, the result follows from (3.80) with vector storage function $V_{\mathrm{s}}(x) = [v_{\mathrm{s}1}(x_1), \dots, v_{\mathrm{s}q}(x_q)]^{\mathrm{T}}$, $x \in \mathbb{R}^n$. $\qquad\square$

Finally, we provide necessary and sufficient conditions for the case where the large-scale nonlinear dynamical system $\mathcal{G}$ is vector lossless with respect to a vector quadratic supply rate.

Theorem 3.8. Consider the large-scale nonlinear dynamical system $\mathcal{G}$ given by (3.5) and (3.6). Let $R_i \in \mathbb{S}^{m_i}$, $S_i \in \mathbb{R}^{l_i \times m_i}$, and $Q_i \in \mathbb{S}^{l_i}$, $i = 1, \dots, q$. Then $\mathcal{G}$ is vector lossless with respect to the vector quadratic supply rate $S(u, y)$, where $s_i(u_i, y_i) = u_i^{\mathrm{T}} R_i u_i + 2y_i^{\mathrm{T}} S_i u_i + y_i^{\mathrm{T}} Q_i y_i$, $i = 1, \dots, q$, if and only if there exist functions $V_{\mathrm{s}} = [v_{\mathrm{s}1}, \dots, v_{\mathrm{s}q}]^{\mathrm{T}} : \mathbb{R}^n \to \overline{\mathbb{R}}_+^q$ and $w = [w_1, \dots, w_q]^{\mathrm{T}} : \overline{\mathbb{R}}_+^q \to \mathbb{R}^q$ such that $v_{\mathrm{s}i}(\cdot)$ is continuously differentiable, $v_{\mathrm{s}i}(0) = 0$, $i = 1, \dots, q$, $w \in \mathcal{W}$, $w(0) = 0$, the zero solution $r(t) \equiv 0$ to (3.62) is Lyapunov stable, and, for all $x \in \mathbb{R}^n$ and $i = 1, \dots, q$,

$$0 = v'_{\mathrm{s}i}(x)\mathcal{F}(x) - h^{\mathrm{T}}(x)\hat{Q}_i h(x) - w_i(V_{\mathrm{s}}(x)), \qquad (3.81)$$
$$0 = \tfrac{1}{2}v'_{\mathrm{s}i}(x)G(x) - h^{\mathrm{T}}(x)(\hat{S}_i + \hat{Q}_i J(x)), \qquad (3.82)$$
$$0 = \hat{R}_i + J^{\mathrm{T}}(x)\hat{S}_i + \hat{S}_i^{\mathrm{T}} J(x) + J^{\mathrm{T}}(x)\hat{Q}_i J(x). \qquad (3.83)$$

Proof. The proof is analogous to the proof of Theorem 3.6. $\qquad\square$

3.4 Specialization to Large-Scale Linear Dynamical Systems

In this section, we specialize the results of Section 3.3 to the case of large-scale linear dynamical systems. Specifically, we assume that $w \in \mathcal{W}$ is linear so that $w(r) = Wr$, where $W \in \mathbb{R}^{q \times q}$ is essentially nonnegative, and consider the large-scale linear dynamical system $\mathcal{G}$ given by

$$\dot{x}(t) = Ax(t) + Bu(t), \quad x(t_0) = x_0, \quad t \geq t_0, \tag{3.84}$$
$$y(t) = Cx(t) + Du(t), \tag{3.85}$$

where $A \in \mathbb{R}^{n \times n}$ and A is partitioned as $A \triangleq [A_{ij}], i, j = 1, \ldots, q, A_{ij} \in \mathbb{R}^{n_i \times n_j}, \sum_{i=1}^{q} n_i = n, B = \text{block--diag}[B_1, \ldots, B_q], C = \text{block--diag}[C_1, \ldots, C_q], D = \text{block--diag}[D_1, \ldots, D_q], B_i \in \mathbb{R}^{n_i \times m_i}, C_i \in \mathbb{R}^{l_i \times n_i}, D_i \in \mathbb{R}^{l_i \times m_i}$, and $i = 1, \ldots, q$.

Theorem 3.9. Consider the large-scale linear dynamical system $\mathcal{G}$ given by (3.84) and (3.85). Let $R_i \in \mathbb{S}^{m_i}$, $S_i \in \mathbb{R}^{l_i \times m_i}$, and $Q_i \in \mathbb{S}^{l_i}$, $i = 1, \ldots, q$. Then $\mathcal{G}$ is vector dissipative (respectively, exponentially vector dissipative) with respect to the vector supply rate $S(u, y)$, where $s_i(u_i, y_i) = u_i^{\mathrm{T}} R_i u_i + 2 y_i^{\mathrm{T}} S_i u_i + y_i^{\mathrm{T}} Q_i y_i$, $i = 1, \ldots, q$, if and only if there exist $W \in \mathbb{R}^{q \times q}$, $P_i \in \mathbb{N}^n$, $L_i \in \mathbb{R}^{s_i \times n}$, and $Z_i \in \mathbb{R}^{s_i \times m}$, $i = 1, \ldots, q$, such that W is essentially nonnegative and semistable (respectively, asymptotically stable), and, for all $i = 1, \ldots, q$,

$$0 = A^{\mathrm{T}} P_i + P_i A - C^{\mathrm{T}} \hat{Q}_i C - \sum_{j=1}^{q} W_{(i,j)} P_j + L_i^{\mathrm{T}} L_i, \tag{3.86}$$

$$0 = P_i B - C^{\mathrm{T}} (\hat{S}_i + \hat{Q}_i D) + L_i^{\mathrm{T}} Z_i, \tag{3.87}$$

$$0 = \hat{R}_i + D^{\mathrm{T}} \hat{S}_i + \hat{S}_i^{\mathrm{T}} D + D^{\mathrm{T}} \hat{Q}_i D - Z_i^{\mathrm{T}} Z_i. \tag{3.88}$$

Proof. Sufficiency follows from Theorem 3.6 with $\mathcal{F}(x) = Ax$, $G(x) = B$, $h(x) = Cx$, $J(x) = D$, $w(r) = Wr$, $\ell_i(x) = L_i x$, $\mathcal{Z}_i(x) = Z_i$, and $v_{\mathrm{s}i}(x) = x^{\mathrm{T}} P_i x$, $i = 1, \ldots, q$.

To show necessity, suppose $\mathcal{G}$ is vector dissipative with respect to the vector supply rate $S(u, y)$, where $s_i(u_i, y_i) = u_i^{\mathrm{T}} R_i u_i + 2 y_i^{\mathrm{T}} S_i u_i + y_i^{\mathrm{T}} Q_i y_i$, $i = 1, \ldots, q$. Then it follows from Theorem 3.6, with $w(r) = Wr$, that there exist $V_{\mathrm{s}} : \mathbb{R}^n \to \overline{\mathbb{R}}_+^q$, $\ell_i : \mathbb{R}^n \to \mathbb{R}^{s_i}$, and $\mathcal{Z}_i : \mathbb{R}^n \to \mathbb{R}^{s_i \times m}$, such that W is essentially nonnegative and semistable (respectively, asymptotically stable), $V_{\mathrm{s}}(x) \triangleq [v_{\mathrm{s}1}(x), \ldots, v_{\mathrm{s}q}(x)]^{\mathrm{T}}$, $x \in \mathbb{R}^n$, $V_{\mathrm{s}}(0) = 0$, and (3.65)–(3.67) hold for all $i = 1, \ldots, q$ with $\mathcal{F}(x) = Ax$, $G(x) = B$, $h(x) = Cx$, $J(x) = D$, and $w(r) = Wr$.

Since $v_{\mathrm{s}i}(\cdot)$ is nonnegative-definite and $v_{\mathrm{s}i}(0) = 0$, $i = 1,\ldots,q$, it follows that there exists $P_i \in \mathbb{N}^n$, $i = 1,\ldots,q$, such that

$$v_{\mathrm{s}i}(x) = x^{\mathrm{T}} P_i x + v_{\mathrm{sr}i}(x), \quad x \in \mathbb{R}^n, \quad i = 1,\ldots,q, \tag{3.89}$$

where $v_{\mathrm{sr}i} : \mathbb{R}^n \to \mathbb{R}$ contains the higher-order terms of $v_{\mathrm{s}i}(x)$. Next, note that it follows from (3.65) that $\ell_i(0) = 0$, and hence, there exists $L_i \in \mathbb{R}^{s_i \times n}$ such that $\ell_i(x) = L_i x + \ell_{\mathrm{r}i}(x)$, $x \in \mathbb{R}^n$, where $\ell_{\mathrm{r}i}(\cdot)$ contains higher-order terms. Furthermore, it follows from (3.67) that $\mathcal{Z}_i = Z_i$, $Z_i \in \mathbb{R}^{s_i \times m}$, $i = 1,\ldots,q$, which implies (3.88).

Using the above expressions, (3.65) and (3.66) can be written as

$$0 = x^{\mathrm{T}} \left(A^{\mathrm{T}} P_i + P_i A - C^{\mathrm{T}} \hat{Q}_i C - \sum_{j=1}^{q} W_{(i,j)} P_j + L_i^{\mathrm{T}} L_i \right) x + \gamma_i(x), \tag{3.90}$$

$$0 = x^{\mathrm{T}} (P_i B - C^{\mathrm{T}}(\hat{S}_i + \hat{Q}_i D) + L_i^{\mathrm{T}} Z_i) + \Gamma_i(x), \tag{3.91}$$

where

$$\gamma_i(x) = v'_{\mathrm{sr}i}(x) A x - \sum_{j=1}^{q} W_{(i,j)} v_{\mathrm{sr}j}(x) + 2x^{\mathrm{T}} L_i^{\mathrm{T}} \ell_{\mathrm{r}i}(x) + \ell_{\mathrm{r}i}^{\mathrm{T}}(x) \ell_{\mathrm{r}i}(x), \tag{3.92}$$

$$\Gamma_i(x) = \tfrac{1}{2} v'_{\mathrm{sr}i}(x) B + \ell_{\mathrm{r}i}^{\mathrm{T}}(x) Z_i. \tag{3.93}$$

Now, viewing (3.90) and (3.91) as the Taylor's series expansion of (3.65) and (3.67), respectively, about $x = 0$ and noting that $\lim_{\|x\| \to 0} \frac{|\gamma_i(x)|}{\|x\|^2} = 0$ and $\lim_{\|x\| \to 0} \frac{|\Gamma_i(x)|}{\|x\|} = 0$, $i = 1,\ldots,q$, it follows that P_i, $i = 1,\ldots,q$, satisfy (3.86) and (3.87). $\qquad\square$

Note that (3.86)–(3.88) are equivalent to

$$\begin{bmatrix} \mathcal{A}_i & \mathcal{B}_i \\ \mathcal{B}_i^{\mathrm{T}} & \mathcal{C}_i \end{bmatrix} = - \begin{bmatrix} L_i^{\mathrm{T}} \\ Z_i^{\mathrm{T}} \end{bmatrix} \begin{bmatrix} L_i & Z_i \end{bmatrix} \leq 0, \quad i = 1,\ldots,q, \tag{3.94}$$

where, for all $i = 1,\ldots,q$,

$$\mathcal{A}_i = A^{\mathrm{T}} P_i + P_i A - C^{\mathrm{T}} \hat{Q}_i C - \sum_{j=1}^{q} W_{(i,j)} P_j, \tag{3.95}$$

$$\mathcal{B}_i = P_i B - C^{\mathrm{T}}(\hat{S}_i + \hat{Q}_i D), \tag{3.96}$$

$$\mathcal{C}_i = -(\hat{R}_i + D^{\mathrm{T}} \hat{S}_i + \hat{S}_i^{\mathrm{T}} D + D^{\mathrm{T}} \hat{Q}_i D). \tag{3.97}$$

Hence, vector dissipativity of large-scale linear dynamical systems with respect to vector quadratic supply rates can be characterized via (cascade) linear matrix inequalities (LMIs) [26]. A similar remark holds for Theorem 3.10 below.

The next result presents sufficient conditions guaranteeing vector dissipativity of $\mathcal{G}$ with respect to a vector quadratic supply rate in the case where the vector storage function is component decoupled.

Theorem 3.10. Consider the large-scale linear dynamical system $\mathcal{G}$ given by (3.84) and (3.85). Let $R_i \in \mathbb{S}^{m_i}$, $S_i \in \mathbb{R}^{l_i \times m_i}$, and $Q_i \in \mathbb{S}^{l_i}$, $i = 1, \ldots, q$, be given. Assume there exist matrices $W \in \mathbb{R}^{q \times q}$, $P_i \in \mathbb{N}^{n_i}$, $L_{ii} \in \mathbb{R}^{s_{ii} \times n_i}$, $Z_{ii} \in \mathbb{R}^{s_{ii} \times m_i}$, $i = 1, \ldots, q$, $L_{ij} \in \mathbb{R}^{s_{ij} \times n_i}$, and $Z_{ij} \in \mathbb{R}^{s_{ij} \times n_j}$, $i, j = 1, \ldots, q$, $i \neq j$, such that W is essentially nonnegative and semistable (respectively, asymptotically stable), and, for all $i = 1, \ldots, q$,

$$0 \geq A_{ii}^{\mathrm{T}} P_i + P_i A_{ii} - C_i^{\mathrm{T}} Q_i C_i - W_{(i,i)} P_i + L_{ii}^{\mathrm{T}} L_{ii} + \sum_{j=1, j \neq i}^{q} L_{ij}^{\mathrm{T}} L_{ij}, \quad (3.98)$$

$$0 = P_i B_i - C_i^{\mathrm{T}} S_i - C_i^{\mathrm{T}} Q_i D_i + L_{ii}^{\mathrm{T}} Z_{ii}, \quad (3.99)$$

$$0 \leq R_i + D_i^{\mathrm{T}} S_i + S_i^{\mathrm{T}} D_i + D_i^{\mathrm{T}} Q_i D_i - Z_{ii}^{\mathrm{T}} Z_{ii}, \quad (3.100)$$

and, for $j = 1, \ldots, q$, $j \neq i$,

$$0 = P_i A_{ij} + L_{ij}^{\mathrm{T}} Z_{ij}, \quad (3.101)$$

$$0 \leq W_{(i,j)} P_j - Z_{ij}^{\mathrm{T}} Z_{ij}. \quad (3.102)$$

Then $\mathcal{G}$ is vector dissipative (respectively, exponentially vector dissipative) with respect to the vector supply rate $S(u, y) \triangleq [s_1(u_1, y_1), \ldots, s_q(u_q, y_q)]^{\mathrm{T}}$, where $s_i(u_i, y_i) = u_i^{\mathrm{T}} R_i u_i + 2 y_i^{\mathrm{T}} S_i u_i + y_i^{\mathrm{T}} Q_i y_i$, $i = 1, \ldots, q$.

Proof. Since $P_i \in \mathbb{N}^{n_i}$, the function $v_{\mathrm{s}i}(x_i) \triangleq x_i^{\mathrm{T}} P_i x_i$, $x_i \in \mathbb{R}^{n_i}$, is nonnegative definite and $v_{\mathrm{s}i}(0) = 0$. Moreover, since $v_{\mathrm{s}i}(\cdot)$ is continuously differentiable it follows from (3.98)–(3.102) that for all $u_i \in \mathbb{R}^{m_i}$, $i = 1, \ldots, q$, and $t \geq t_0$,

$$\dot{v}_{\mathrm{s}i}(x_i(t)) = 2 x_i^{\mathrm{T}}(t) P_i \left[\sum_{j=1}^{q} A_{ij} x_j(t) + B_i u_i(t) \right]$$

$$\leq x_i^{\mathrm{T}}(t) \left[W_{(i,i)} P_i + C_i^{\mathrm{T}} Q_i C_i - L_{ii}^{\mathrm{T}} L_{ii} - \sum_{j=1, j \neq i}^{q} L_{ij}^{\mathrm{T}} L_{ij} \right] x_i(t)$$

$$- \sum_{j=1, j \neq i}^{q} 2 x_i^{\mathrm{T}}(t) L_{ij}^{\mathrm{T}} Z_{ij} x_j(t) + 2 x_i^{\mathrm{T}}(t) C_i^{\mathrm{T}} S_i u_i(t)$$

$$+ 2 x_i^{\mathrm{T}}(t) C_i^{\mathrm{T}} Q_i D_i u_i(t) - 2 x_i^{\mathrm{T}}(t) L_{ii}^{\mathrm{T}} Z_{ii} u_i(t)$$

$$+ \sum_{j=1, j \neq i}^{q} x_j^{\mathrm{T}}(t) [W_{(i,j)} P_j - Z_{ij}^{\mathrm{T}} Z_{ij}] x_j(t)$$

$$+ u_i^{\mathrm{T}}(t) R_i u_i(t) + 2 u_i^{\mathrm{T}}(t) D_i^{\mathrm{T}} S_i u_i(t)$$

$$+u_i^{\mathrm{T}}(t)D_i^{\mathrm{T}}Q_iD_iu_i(t) - u_i^{\mathrm{T}}(t)Z_{ii}^{\mathrm{T}}Z_{ii}u_i(t)$$

$$= \sum_{j=1}^{q} W_{(i,j)}v_{\mathrm{s}j}(x_j(t)) + u_i^{\mathrm{T}}(t)R_iu_i(t)$$

$$+2y_i^{\mathrm{T}}(t)S_iu_i(t) + y_i^{\mathrm{T}}(t)Q_iy_i(t)$$
$$-[L_{ii}x_i(t) + Z_{ii}u_i(t)]^{\mathrm{T}}[L_{ii}x_i(t) + Z_{ii}u_i(t)]$$

$$- \sum_{j=1,j\neq i}^{q} (L_{ij}x_i(t) + Z_{ij}x_j(t))^{\mathrm{T}}(L_{ij}x_i(t) + Z_{ij}x_j(t))$$

$$\leq s_i(u_i(t),y_i(t)) + \sum_{j=1}^{q} W_{(i,j)}v_{\mathrm{s}j}(x_j(t)), \tag{3.103}$$

or, equivalently, in vector form

$$\dot{V}_{\mathrm{s}}(x(t)) \leq\leq WV_{\mathrm{s}}(x(t)) + S(u,y), \quad u \in \mathcal{U}, \quad t \geq t_0, \tag{3.104}$$

where $V_{\mathrm{s}}(x) \triangleq [v_{\mathrm{s}1}(x_1),\ldots,v_{\mathrm{s}q}(x_q)]^{\mathrm{T}}$, $x \in \mathbb{R}^n$. Now, it follows from Proposition 3.2 that $\mathcal{G}$ is vector dissipative (respectively, exponentially vector dissipative) with respect to the vector supply rate $S(u,y)$ and with vector storage function $V_{\mathrm{s}}(x)$, $x \in \mathbb{R}^n$. $\qquad\square$

3.5 Stability of Feedback Interconnections of Large-Scale Nonlinear Dynamical Systems

In this section, we consider stability of feedback interconnections of large-scale nonlinear dynamical systems. Specifically, for the large-scale dynamical system $\mathcal{G}$ given by (3.5) and (3.6) we consider either a dynamic or static large-scale feedback system $\mathcal{G}_{\mathrm{c}}$. Then by appropriately combining vector storage functions for each system we show stability of the feedback interconnection. We begin by considering the large-scale nonlinear dynamical system (3.5) and (3.6) with the large-scale feedback system $\mathcal{G}_{\mathrm{c}}$ given by

$$\dot{x}_{\mathrm{c}}(t) = F_{\mathrm{c}}(x_{\mathrm{c}}(t),u_{\mathrm{c}}(t)), \quad x_{\mathrm{c}}(t_0) = x_{\mathrm{c}0}, \quad t \geq t_0, \tag{3.105}$$
$$y_{\mathrm{c}}(t) = H_{\mathrm{c}}(x_{\mathrm{c}}(t),u_{\mathrm{c}}(t)), \tag{3.106}$$

where $F_{\mathrm{c}} : \mathbb{R}^{n_{\mathrm{c}}} \times U_{\mathrm{c}} \to \mathbb{R}^{n_{\mathrm{c}}}$, $H_{\mathrm{c}} : \mathbb{R}^{n_{\mathrm{c}}} \times U_{\mathrm{c}} \to Y_{\mathrm{c}}$, $F_{\mathrm{c}} \triangleq [F_{\mathrm{c}1}^{\mathrm{T}},\ldots,F_{\mathrm{c}q}^{\mathrm{T}}]^{\mathrm{T}}$, $H_{\mathrm{c}} \triangleq [H_{\mathrm{c}1}^{\mathrm{T}},\ldots,H_{\mathrm{c}q}^{\mathrm{T}}]^{\mathrm{T}}$, $U_{\mathrm{c}} \subseteq \mathbb{R}^l$, $Y_{\mathrm{c}} \subseteq \mathbb{R}^m$. Moreover, for all $i = 1,\ldots,q$, we assume that

$$F_{\mathrm{c}i}(x_{\mathrm{c}},u_{\mathrm{c}i}) = f_{\mathrm{c}i}(x_{\mathrm{c}i}) + \mathcal{I}_{\mathrm{c}i}(x_{\mathrm{c}}) + G_{\mathrm{c}i}(x_{\mathrm{c}i})u_{\mathrm{c}i}, \tag{3.107}$$
$$H_{\mathrm{c}i}(x_{\mathrm{c}i},u_{\mathrm{c}i}) = h_{\mathrm{c}i}(x_{\mathrm{c}i}) + J_{\mathrm{c}i}(x_{\mathrm{c}i})u_{\mathrm{c}i}, \tag{3.108}$$

where $u_{\mathrm{c}i} \in U_{\mathrm{c}i} \subseteq \mathbb{R}^{l_i}$, $y_{\mathrm{c}i} \triangleq H_{\mathrm{c}i}(x_{\mathrm{c}i},u_{\mathrm{c}i}) \in Y_i \subseteq \mathbb{R}^{m_i}$, $(u_{\mathrm{c}i},y_{\mathrm{c}i})$ is the input-output pair for the ith subsystem of $\mathcal{G}_{\mathrm{c}}$, $f_{\mathrm{c}i} : \mathbb{R}^{n_{\mathrm{c}i}} \to \mathbb{R}^{n_{\mathrm{c}i}}$ and $\mathcal{I}_{\mathrm{c}i} : \mathbb{R}^{n_{\mathrm{c}}} \to$

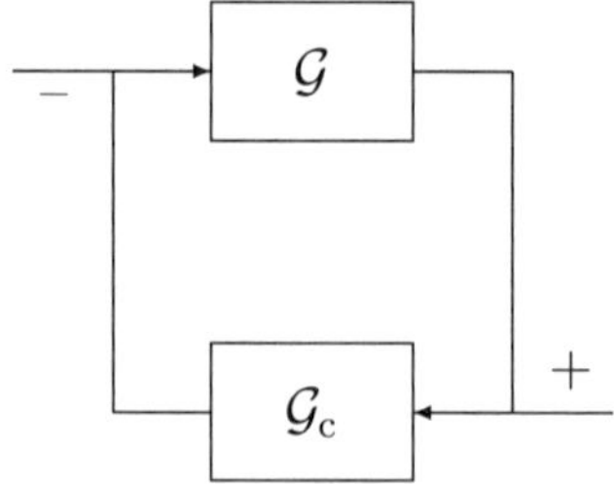

Figure 3.1 Feedback interconnection of large-scale systems $\mathcal{G}$ and $\mathcal{G}_{\mathrm{c}}$.

$\mathbb{R}^{n_{ci}}$ satisfy $f_{ci}(0) = 0$ and $\mathcal{I}_{ci}(0) = 0$, $G_{ci} : \mathbb{R}^{n_{ci}} \to \mathbb{R}^{n_{ci} \times l_i}$, $h_{ci} : \mathbb{R}^{n_{ci}} \to \mathbb{R}^{m_i}$ and satisfies $h_{ci}(0) = 0$, $J_{ci} : \mathbb{R}^{n_{ci}} \to \mathbb{R}^{m_i \times l_i}$, and $\sum_{i=1}^{q} n_{ci} = n_{\mathrm{c}}$.

Furthermore, we define the composite input and composite output for the system $\mathcal{G}_{\mathrm{c}}$ as $u_{\mathrm{c}} \triangleq [u_{c1}^{\mathrm{T}}, \ldots, u_{cq}^{\mathrm{T}}]^{\mathrm{T}}$ and $y_{\mathrm{c}} \triangleq [y_{c1}^{\mathrm{T}}, \ldots, y_{cq}^{\mathrm{T}}]^{\mathrm{T}}$, respectively. In this case, $U_{\mathrm{c}} = U_{c1} \times \cdots \times U_{cq}$ and $Y_{\mathrm{c}} = Y_{c1} \times \cdots \times Y_{cq}$. Note that with the feedback interconnection given by Figure 3.1, $u_{\mathrm{c}} = y$ and $y_{\mathrm{c}} = -u$. We assume that the negative feedback interconnection of $\mathcal{G}$ and $\mathcal{G}_{\mathrm{c}}$ is well posed, that is, $\det(I_{m_i} + J_{ci}(x_{ci}) J_i(x_i)) \neq 0$ for all $x_i \in \mathbb{R}^{n_i}$, $x_{ci} \in \mathbb{R}^{n_{ci}}$, and $i = 1, \ldots, q$. Furthermore, we assume that for the large-scale systems $\mathcal{G}$ and $\mathcal{G}_{\mathrm{c}}$, the conditions of Theorem 3.3 are satisfied, that is, if $V_{\mathrm{s}}(x)$, $x \in \mathbb{R}^n$, and $V_{\mathrm{cs}}(x_{\mathrm{c}})$, $x_{\mathrm{c}} \in \mathbb{R}^{n_{\mathrm{c}}}$, are vector storage functions for $\mathcal{G}$ and $\mathcal{G}_{\mathrm{c}}$, respectively, then there exist $p \in \mathbb{R}^q_+$ and $p_{\mathrm{c}} \in \mathbb{R}^q_+$ such that the functions $v_{\mathrm{s}}(x) = p^{\mathrm{T}} V_{\mathrm{s}}(x)$, $x \in \mathbb{R}^n$, and $v_{\mathrm{cs}}(x_{\mathrm{c}}) = p_{\mathrm{c}}^{\mathrm{T}} V_{\mathrm{cs}}(x_{\mathrm{c}})$, $x_{\mathrm{c}} \in \mathbb{R}^{n_{\mathrm{c}}}$, are positive definite.

The following result gives sufficient conditions for Lyapunov and asymptotic stability of the feedback interconnection given by Figure 3.1.

Theorem 3.11. Consider the large-scale nonlinear dynamical systems $\mathcal{G}$ and $\mathcal{G}_{\mathrm{c}}$ given by (3.5) and (3.6), and (3.105) and (3.106), respectively. Assume that $\mathcal{G}$ and $\mathcal{G}_{\mathrm{c}}$ are vector dissipative with respect to the vector supply rates $S(u, y)$ and $S_{\mathrm{c}}(u_{\mathrm{c}}, y_{\mathrm{c}})$, and with continuously differentiable vector storage functions $V_{\mathrm{s}}(\cdot)$ and $V_{\mathrm{cs}}(\cdot)$ and dissipation matrices $W \in \mathbb{R}^{q \times q}$ and $W_{\mathrm{c}} \in \mathbb{R}^{q \times q}$, respectively.

i) If there exists $\Sigma \triangleq \mathrm{diag}[\sigma_1, \ldots, \sigma_q] > 0$ such that $S(u, y) + \Sigma S_{\mathrm{c}}(u_{\mathrm{c}}, y_{\mathrm{c}})$ $\leq\leq 0$ and $\tilde{W} \in \mathbb{R}^{q \times q}$ is semistable (respectively, asymptotically stable), where

$$\tilde{W}_{(i,j)} \triangleq \max\left\{ W_{(i,j)}, (\Sigma W_{\mathrm{c}} \Sigma^{-1})_{(i,j)} \right\}$$

$$= \max\left\{ W_{(i,j)}, \frac{\sigma_i}{\sigma_j} W_{\mathrm{c}(i,j)} \right\}, \quad i, j = 1, \ldots, q, \quad (3.109)$$

then the negative feedback interconnection of $\mathcal{G}$ and $\mathcal{G}_{\mathrm{c}}$ is Lyapunov (respectively, asymptotically) stable.

ii) Let $Q_i \in \mathbb{S}^{l_i}$, $S_i \in \mathbb{R}^{l_i \times m_i}$, $R_i \in \mathbb{S}^{m_i}$, $Q_{ci} \in \mathbb{S}^{m_i}$, $S_{ci} \in \mathbb{R}^{m_i \times l_i}$, and $R_{ci} \in \mathbb{S}^{l_i}$, and suppose $S(u, y) = [s_1(u_1, y_1), \ldots, s_q(u_q, y_q)]^{\mathrm{T}}$ and $S_c(u_c, y_c) = [s_{c1}(u_{c1}, y_{c1}), \ldots, s_{cq}(u_{cq}, y_{cq})]^{\mathrm{T}}$, where $s_i(u_i, y_i) = u_i^{\mathrm{T}} R_i u_i + 2 y_i^{\mathrm{T}} S_i u_i + y_i^{\mathrm{T}} Q_i y_i$ and $s_{ci}(u_{ci}, y_{ci}) = u_{ci}^{\mathrm{T}} R_{ci} u_{ci} + 2 y_{ci}^{\mathrm{T}} S_{ci} u_{ci} + y_{ci}^{\mathrm{T}} Q_{ci} y_{ci}$, $i = 1, \ldots, q$. If there exists $\Sigma \triangleq \mathrm{diag}[\sigma_1, \ldots, \sigma_q] > 0$ such that for all $i = 1, \ldots, q$,

$$\tilde{Q}_i \triangleq \begin{bmatrix} Q_i + \sigma_i R_{ci} & -S_i + \sigma_i S_{ci}^{\mathrm{T}} \\ -S_i^{\mathrm{T}} + \sigma_i S_{ci} & R_i + \sigma_i Q_{ci} \end{bmatrix} \leq 0 \qquad (3.110)$$

and $\tilde{W} \in \mathbb{R}^{q \times q}$ is semistable (respectively, asymptotically stable), where $\tilde{W}_{(i,j)} \triangleq \max\{W_{(i,j)}, (\Sigma W_c \Sigma^{-1})_{(i,j)}\} = \max\{W_{(i,j)}, \frac{\sigma_i}{\sigma_j} W_{c(i,j)}\}$, $i, j = 1, \ldots, q$, then the negative feedback interconnection of $\mathcal{G}$ and $\mathcal{G}_c$ is Lyapunov (respectively, asymptotically) stable.

Proof. i) Consider the vector Lyapunov function candidate $V(x, x_c) = V_s(x) + \Sigma V_{cs}(x_c)$, $(x, x_c) \in \mathbb{R}^n \times \mathbb{R}^{n_c}$, and note that the corresponding vector Lyapunov derivative satisfies

$$\begin{aligned} \dot{V}(x, x_c) &= \dot{V}_s(x) + \Sigma \dot{V}_{cs}(x_c) \\ &\leq\leq S(u, y) + \Sigma S_c(u_c, y_c) + W V_s(x) + \Sigma W_c V_{cs}(x_c) \\ &\leq\leq W V_s(x) + \Sigma W_c \Sigma^{-1} \Sigma V_{cs}(x_c) \\ &\leq\leq \tilde{W}(V_s(x) + \Sigma V_{cs}(x_c)) \\ &= \tilde{W} V(x, x_c), \quad (x, x_c) \in \mathbb{R}^n \times \mathbb{R}^{n_c}. \end{aligned} \qquad (3.111)$$

Now, since for $V_s(x)$, $x \in \mathbb{R}^n$, and $V_{cs}(x_c)$, $x_c \in \mathbb{R}^{n_c}$, there exist, by assumption, $p \in \mathbb{R}_+^q$ and $p_c \in \mathbb{R}_+^q$ such that the functions $v_s(x) = p^{\mathrm{T}} V_s(x)$, $x \in \mathbb{R}^n$, and $v_{cs}(x_c) = p_c^{\mathrm{T}} V_{cs}(x_c)$, $x_c \in \mathbb{R}^{n_c}$, are positive definite and noting that $v_{cs}(x_c) \leq \max_{i=1,\ldots,q}\{p_{ci}\} \mathbf{e}^{\mathrm{T}} V_{cs}(x_c)$, where p_{ci} is the ith component of p_c and $\mathbf{e} \triangleq [1, \ldots, 1]^{\mathrm{T}}$, it follows that $\mathbf{e}^{\mathrm{T}} V_{cs}(x_c)$, $x_c \in \mathbb{R}^{n_c}$, is positive definite. Next, since $\min_{i=1,\ldots,q}\{p_i \sigma_i\} \mathbf{e}^{\mathrm{T}} V_{cs}(x_c) \leq p^{\mathrm{T}} \Sigma V_{cs}(x_c)$, it follows that $p^{\mathrm{T}} \Sigma V_{cs}(x_c)$, $x_c \in \mathbb{R}^{n_c}$, is positive definite. Hence, the function $v(x, x_c) = p^{\mathrm{T}} V(x, x_c)$, $(x, x_c) \in \mathbb{R}^n \times \mathbb{R}^{n_c}$, is positive definite. Now, the result is a direct consequence of Corollary 2.3.

ii) The proof follows from i) by noting that, for all $i = 1, \ldots, q$,

$$s_i(u_i, y_i) + \sigma_i s_{ci}(u_{ci}, y_{ci}) = \begin{bmatrix} y \\ y_c \end{bmatrix}^{\mathrm{T}} \tilde{Q}_i \begin{bmatrix} y \\ y_c \end{bmatrix}, \qquad (3.112)$$

and hence, $S(u, y) + \Sigma S_c(u_c, y_c) \leq\leq 0$. $\square$

For the next result note that if the large-scale nonlinear dynamical system $\mathcal{G}$ is vector dissipative with respect to the vector supply rate $S(u, y)$, where $s_i(u_i, y_i) = 2y_i^{\mathrm{T}} u_i$, $i = 1, \ldots, q$, then with $\kappa_i(y_i) = -\kappa_i y_i$, where $\kappa_i > 0$, $i = 1, \ldots, q$, it follows that $s_i(\kappa_i(y_i), y_i) = -2\kappa_i y_i^{\mathrm{T}} y_i < 0$, $y_i \neq 0$, $i = 1, \ldots, q$. Alternatively, if $\mathcal{G}$ is vector dissipative with respect to the vector supply rate $S(u, y)$, where $s_i(u_i, y_i) = \gamma_i^2 u_i^{\mathrm{T}} u_i - y_i^{\mathrm{T}} y_i$, where $\gamma_i > 0$, $i = 1, \ldots, q$, then with $\kappa_i(y_i) = 0$, it follows that $s_i(\kappa_i(y_i), y_i) = -y_i^{\mathrm{T}} y_i < 0$, $y_i \neq 0$, $i = 1, \ldots, q$. Hence, if $\mathcal{G}$ is zero-state observable and the dissipation matrix W is such that there exist $\alpha \geq 0$ and $p \in \mathbb{R}_+^q$ such that (2.4) holds, then it follows from Theorem 3.3 that (scalar) storage functions of the form $v_{\mathrm{s}}(x) = p^{\mathrm{T}} V_{\mathrm{s}}(x)$, $x \in \mathbb{R}^n$, where $V_{\mathrm{s}}(\cdot)$ is a vector storage function for $\mathcal{G}$, are positive definite. If $\mathcal{G}$ is exponentially vector dissipative, then p is positive.

Corollary 3.2. Consider the large-scale nonlinear dynamical systems $\mathcal{G}$ and $\mathcal{G}_{\mathrm{c}}$ given by (3.5) and (3.6), and (3.105) and (3.106), respectively. Assume that $\mathcal{G}$ and $\mathcal{G}_{\mathrm{c}}$ are zero-state observable and the dissipation matrices $W \in \mathbb{R}^{q \times q}$ and $W_{\mathrm{c}} \in \mathbb{R}^{q \times q}$ are such that there exist, respectively, $\alpha \geq 0$, $p \in \mathbb{R}_+^q$, $\alpha_{\mathrm{c}} \geq 0$, and $p_{\mathrm{c}} \in \mathbb{R}_+^q$ such that (2.4) is satisfied. Then the following statements hold:

i) If $\mathcal{G}$ and $\mathcal{G}_{\mathrm{c}}$ are vector passive and $\tilde{W} \in \mathbb{R}^{q \times q}$ is asymptotically stable, where $\tilde{W}_{(i,j)} \triangleq \max\{W_{(i,j)}, W_{\mathrm{c}(i,j)}\}$, $i, j = 1, \ldots, q$, then the negative feedback interconnection of $\mathcal{G}$ and $\mathcal{G}_{\mathrm{c}}$ is asymptotically stable.

ii) If $\mathcal{G}$ and $\mathcal{G}_{\mathrm{c}}$ are vector nonexpansive and $\tilde{W} \in \mathbb{R}^{q \times q}$ is asymptotically stable, where $\tilde{W}_{(i,j)} \triangleq \max\{W_{(i,j)}, W_{\mathrm{c}(i,j)}\}$, $i, j = 1, \ldots, q$, then the negative feedback interconnection of $\mathcal{G}$ and $\mathcal{G}_{\mathrm{c}}$ is asymptotically stable.

Proof. The proof is a direct consequence of Theorem 3.11. Specifically, $i)$ follows from Theorem 3.11 with $R_i = 0$, $S_i = I_{m_i}$, $Q_i = 0$, $R_{\mathrm{c}i} = 0$, $S_{\mathrm{c}i} = I_{m_i}$, $Q_{\mathrm{c}i} = 0$, $i = 1, \ldots, q$, and $\Sigma = I_q$, while $ii)$ follows from Theorem 3.11 with $R_i = \gamma_i^2 I_{m_i}$, $S_i = 0$, $Q_i = -I_{l_i}$, $R_{\mathrm{c}i} = \gamma_{\mathrm{c}i}^2 I_{l_i}$, $S_{\mathrm{c}i} = 0$, $Q_{\mathrm{c}i} = -I_{m_i}$, $i = 1, \ldots, q$, and $\Sigma = I_q$. $\qquad\square$

Thermodynamic Modeling of Large-Scale Interconnected Systems

4.1 Introduction

In this chapter, we use vector dissipativity theory to provide connections between large-scale dynamical systems and thermodynamics. Specifically, using a large-scale dynamical systems theory prospective for thermodynamics, we show that vector dissipativity notions lead to a precise formulation of the equivalence between dissipated energy (heat) and work in a large-scale dynamical system. Next, we give a deterministic definition of entropy for a large-scale dynamical system that is consistent with the classical thermodynamic definition of entropy and show that it satisfies a Clausius-type inequality leading to the law of entropy nonconservation. Furthermore, we introduce a dual notion to entropy, namely, *ectropy*, as a measure of the tendency of a large-scale dynamical system to do useful work and show that conservation of energy in an isolated system necessarily leads to nonconservation of ectropy and entropy. Then, we show that our thermodynamically consistent large-scale nonlinear dynamical system model is *semistable*, that is, it has convergent subsystem energies to Lyapunov stable energy equilibria. In addition, we show that the steady-state distribution of the large-scale system energies is uniform, leading to system energy equipartitioning corresponding to a minimum ectropy and a maximum entropy equilibrium state.

4.2 Conservation of Energy and the First Law of Thermodynamics

The fundamental and unifying concept in the analysis and control design of complex large-scale dynamical systems is the concept of energy. The energy of a state of a dynamical system is the measure of its ability to produce changes (motion) in its own system state as well as changes in the system states of its surroundings. These changes occur as a direct consequence of the energy flow between different subsystems within the dynamical system. Since heat (energy) is a fundamental concept of thermodynamics involving the capacity of hot bodies (more energetic subsystems) to produce work, thermodynamics is a theory of large-scale dynamical systems. As in thermodynamic systems, dynamical systems exhibit energy that becomes unavailable to do useful work. This in turn contributes to an increase in system

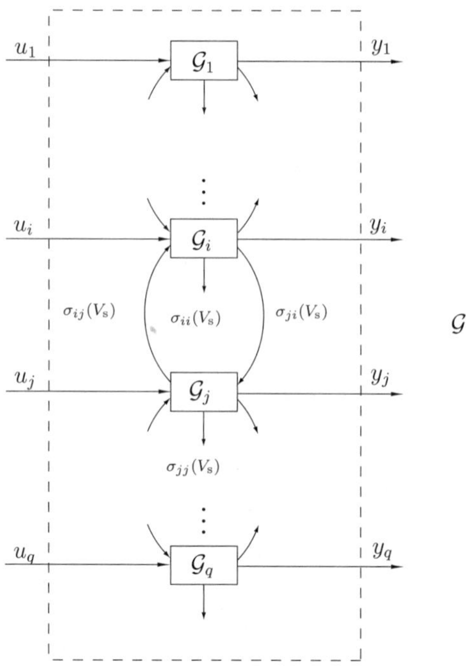

Figure 4.1 Large-scale dynamical system $\mathcal{G}$.

entropy; a measure of the tendency of a system to lose the ability to do useful work. Connections between thermodynamics and system theory as well as information theory are well known [20, 21, 27, 30, 69, 81, 113, 114, 147, 179].

To develop connections between vector dissipativity, large-scale nonlinear dynamical systems, and thermodynamics, consider the large-scale dynamical system $\mathcal{G}$ shown in Figure 4.1 involving q interconnected subsystems. Let $v_{si}(x)$, $x \in \mathcal{D}$, denote the energy of the ith subsystem and assume that, without loss of generality, $v_{si}(0) = 0$, $i = 1, \ldots, q$. Otherwise, assuming $v_{si}(\cdot)$ is lower bounded, we can consider the shifted subsystem energy $\hat{v}_{si}(x) = v_{si}(x) - v_{si}(0)$, $x \in \mathcal{D}$. Furthermore, let $s_i(u_i, y_i)$, $u_i \in \mathcal{U}_i$, $y_i \in \mathcal{Y}_i$, denote the external power supplied to the ith subsystem, let $\sigma_{ij} : \overline{\mathbb{R}}_+^q \to \overline{\mathbb{R}}_+$, $i \neq j$, $i, j = 1, \ldots, q$, denote the instantaneous rate of energy flow from the jth subsystem to the ith subsystem, and let $\sigma_{ii} : \overline{\mathbb{R}}_+^q \to \overline{\mathbb{R}}_+$, $i = 1, \ldots, q$, denote the instantaneous rate of energy loss from the ith subsystem.

An energy balance equation for the ith subsystem yields

$$v_{si}(x(T)) = v_{si}(x(t_0)) + \sum_{j=1, j \neq i}^{q} \int_{t_0}^{T} [\sigma_{ij}(V_\mathrm{s}(x(t))) - \sigma_{ji}(V_\mathrm{s}(x(t)))] \mathrm{d}t$$

$$- \int_{t_0}^{T} \sigma_{ii}(V_\mathrm{s}(x(t))) \mathrm{d}t + \int_{t_0}^{T} s_i(u_i(t), y_i(t)) \mathrm{d}t, \quad T \geq t_0, \quad (4.1)$$

or, equivalently, in vector form,

$$V_{\rm s}(x(T)) = V_{\rm s}(x(t_0)) + \int_{t_0}^{T} w(V_{\rm s}(x(t)))\mathrm{d}t - \int_{t_0}^{T} d(V_{\rm s}(x(t)))\mathrm{d}t$$

$$+ \int_{t_0}^{T} S(u(t), y(t))\mathrm{d}t, \quad T \geq t_0, \tag{4.2}$$

where $V_{\rm s}(x) = [v_{{\rm s}1}(x), \ldots, v_{{\rm s}q}(x)]^{\rm T}$, $d(V_{\rm s}(x)) = [\sigma_{11}(V_{\rm s}(x)), \ldots, \sigma_{qq}(V_{\rm s}(x))]^{\rm T}$, $S(u, y) = [s_1(u_1, y_1), \ldots, s_q(u_q, y_q)]^{\rm T}$, and $w = [w_1, \ldots, w_q]^{\rm T} : \overline{\mathbb{R}}_+^q \to \mathbb{R}^q$ is such that

$$w_i(r) = \sum_{j=1, j \neq i}^{q} [\sigma_{ij}(r) - \sigma_{ji}(r)], \quad r \in \overline{\mathbb{R}}_+^q. \tag{4.3}$$

Note that (4.1) yields a conservation of energy equation and implies that the energy stored in the ith subsystem is equal to the external energy supplied to the ith subsystem plus the energy gained by the ith subsystem from all other subsystems due to subsystem coupling minus the energy dissipated from the ith subsystem. Equivalently, (4.1) can be rewritten as

$$\dot{v}_{{\rm s}i}(x(t)) = \sum_{j=1, j \neq i}^{q} [\sigma_{ij}(V_{\rm s}(x(t))) - \sigma_{ji}(V_{\rm s}(x(t)))] - \sigma_{ii}(V_{\rm s}(x(t)))$$

$$+ s_i(u_i(t), y_i(t)), \quad t \geq t_0, \tag{4.4}$$

or, in vector form,

$$\dot{V}_{\rm s}(x(t)) = w(V_{\rm s}(x(t))) - d(V_{\rm s}(x(t))) + S(u(t), y(t)), \quad t \geq t_0, \quad u(\cdot) \in \mathcal{U}, \tag{4.5}$$

yielding a *power balance* equation.

Specifically, (4.4) shows that the rate of change of energy, or power, in the ith subsystem is equal to the power input to the ith subsystem plus the energy flow to the ith subsystem from all other subsystems minus the power dissipated by the ith subsystem. Note that (4.2) or, equivalently, (4.5) is a statement of the *first law of thermodynamics* for each of the subsystems with $v_{{\rm s}i}(x)$, $s_i(u_i, y_i)$, $\sigma_{ij}(\cdot)$, $i \neq j$, and $\sigma_{ii}(\cdot)$, $i = 1, \ldots, q$, playing the role of the ith subsystem internal energy, rate of work done on the ith subsystem, heat flow between subsystems due to coupling, and the rate of heat dissipated to the environment, respectively. In other words, (4.2) is a precise formulation of the equivalence between dissipated energy (heat) and work.

Next, since the instantaneous rate of energy loss $\sigma_{ii}(r)$, $r \in \overline{\mathbb{R}}_+^q$, $i = 1, \ldots, q$, is a nonnegative function, it follows from (4.1) that

$$v_{{\rm s}i}(x(T)) \leq v_{{\rm s}i}(x(t_0)) + \sum_{j=1, j \neq i}^{q} \int_{t_0}^{T} [\sigma_{ij}(V_{\rm s}(x(t))) - \sigma_{ji}(V_{\rm s}(x(t)))]\mathrm{d}t$$

$$+ \int_{t_0}^{T} s_i(u_i(t), y_i(t))\mathrm{d}t, \quad T \geq t_0. \tag{4.6}$$

In vector form, (4.6) can be equivalently written as

$$V_{\mathrm{s}}(x(T)) \leq\leq V_{\mathrm{s}}(x(t_0)) + \int_{t_0}^{T} w(V_{\mathrm{s}}(x(t)))\mathrm{d}t + \int_{t_0}^{T} S(u(t), y(t))\mathrm{d}t,$$
$$T \geq t_0. \tag{4.7}$$

For every $i = 1, \ldots, q$ and every $r', r'' \in \overline{\mathbb{R}}_+^q$ such that $r_i' = r_i''$ and $r_j' \leq r_j''$, $j \neq i$, $j = 1, \ldots, q$, we assume that $\sigma_{ij}(r') \leq \sigma_{ij}(r'')$ and $\sigma_{ji}(r') = \sigma_{ji}(r'')$. In this case, $w(\cdot) \in \mathcal{W}$. The above mathematical constraints physically imply that the more energy the jth subsystem has, the faster it can transfer this energy to the ith subsystem regardless of the energy distribution in the large-scale system $\mathcal{G}$. Furthermore, for any two energy distributions in $\mathcal{G}$ with the energy of the jth subsystem being the same for both distributions, the jth subsystem has the same energy transfer rate to the ith subsystem.

In the special case where the energy transfer rate from the jth subsystem to the ith subsystem is only dependent on the energy of the jth subsystem, that is, $\sigma_{ij}(r) = \sigma_{ij}(r_j)$, $i \neq j$, $i, j = 1, \ldots, q$, it is sufficient to assume that $\sigma_{ij}(r_j)$ is a nondecreasing function to ensure that $w(\cdot) \in \mathcal{W}$. This implies that the more energy the jth subsystem has, the faster it can transfer this energy to the ith subsystem. Next, assume that $\sigma_{ij}(0) = 0$, $i \neq j$, $i, j = 1, \ldots, q$, which implies that if the energy of each subsystem of $\mathcal{G}$ is zero, then the energy exchange between subsystems is not possible. Moreover, since $w(\cdot) \in \mathcal{W}$ and $w(0) = 0$ it follows that $w : \overline{\mathbb{R}}_+^q \to \mathbb{R}^q$ is essentially nonnegative, and hence, by Proposition 2.1 the solution $r(t)$, $t \geq t_0$, of the comparison system

$$\dot{r}(t) = w(r(t)), \quad r(t_0) = r_0, \quad t \geq t_0, \tag{4.8}$$

remains in the nonnegative orthant $\overline{\mathbb{R}}_+^q$ for all $r_0 \in \overline{\mathbb{R}}_+^q$.

Next, we show that the zero solution $r(t) \equiv 0$ to (4.8) is Lyapunov stable. To see this, consider $v(r) = \sum_{i=1}^{q} r_i$, $r \in \overline{\mathbb{R}}_+^q$, representing the sum of all subsystem energies, as a Lyapunov function candidate for (4.8). Clearly, $v(0) = 0$ and $v(r) > 0$, $r \in \overline{\mathbb{R}}_+^q$, $r \neq 0$. Furthermore,

$$\dot{v}(r(t)) = \sum_{i=1}^{q} \dot{r}_i(t) = \sum_{i=1}^{q} \sum_{j=1, j \neq i}^{q} [\sigma_{ij}(r(t)) - \sigma_{ji}(r(t))] = 0, \quad t \geq t_0, \tag{4.9}$$

which implies Lyapunov stability of the zero solution $r(t) \equiv 0$ to (4.8). Hence, by Definition 3.5 the large-scale dynamical system $\mathcal{G}$ is vector dissipative with respect to the vector supply rate $S(u, y)$.

Finally, in the case where $\sigma_{ij}(r) = \sigma_{ij} r_j$, $\sigma_{ij} \geq 0$, $i \neq j$, $i, j = 1, \ldots, q$, it follows that $w(r) = Wr$, where

$$W_{(i,j)} = \begin{cases} -\sum_{k=1,\,k\neq j}^{q} \sigma_{kj}, & i = j, \\ \sigma_{ij}, & i \neq j. \end{cases} \tag{4.10}$$

In this case, it follows from (4.10) that $\sum_{i=1}^{q} W_{(i,j)} = 0$, $j = 1,\ldots,q$, and hence, W is a semistable compartmental matrix. Furthermore, it follows from Corollary 3.1 with $p = \mathbf{e}$, where $\mathbf{e} \triangleq [1,\ldots,1]^{\mathrm{T}}$, that $0 \leq v_{\mathrm{a}}(x) \leq \sum_{i=1}^{q} v_{\mathrm{si}}(x) \leq v_{\mathrm{r}}(x) < \infty$, which implies that the large-scale nonlinear dynamical system $\mathcal{G}$ can deliver to its surroundings only a fraction of all of its stored subsystem energies and can store only a fraction of the work done to all of its subsystems.

4.3 Nonconservation of Entropy and the Second Law of Thermodynamics

The nonlinear power balance equation (4.5) can exhibit a full range of nonlinear behavior including bifurcations, limit cycles, and even chaos. However, a thermodynamically consistent energy flow model should ensure that the evolution of the system energy is diffusive (parabolic) in character with convergent subsystem energies. Hence, to ensure a thermodynamically consistent energy flow model we require the following assumptions. For the statement of these assumptions we first recall the following graph theoretic notions.

Definition 4.1 ([15]). A *directed graph* $\mathfrak{G}(\mathcal{C})$ associated with the *connectivity matrix* $\mathcal{C} \in \mathbb{R}^{q \times q}$ has *vertices* $\{1, 2, \ldots, q\}$ and an *arc* from vertex i to vertex j, $i \neq j$, if and only if $\mathcal{C}_{(j,i)} \neq 0$. A *graph* $\mathfrak{G}(\mathcal{C})$ associated with the connectivity matrix $\mathcal{C} \in \mathbb{R}^{q \times q}$ is a directed graph for which the *arc set* is symmetric, that is, $\mathcal{C} = \mathcal{C}^{\mathrm{T}}$. We say that $\mathfrak{G}(\mathcal{C})$ is *strongly connected* if for every ordered pair of vertices (i, j), $i \neq j$, there exists a *path* (i.e., sequence of arcs) leading from i to j.

Recall that $\mathcal{C} \in \mathbb{R}^{q \times q}$ is *irreducible*, that is, there does not exist a permutation matrix such that $\mathcal{C}$ is cogredient to a lower block triangular matrix, if and only if $\mathfrak{G}(\mathcal{C})$ is strongly connected (see Theorem 2.7 of [15]). Let $\phi_{ij}(V_{\mathrm{s}}) \triangleq \sigma_{ij}(V_{\mathrm{s}}) - \sigma_{ji}(V_{\mathrm{s}})$, $V_{\mathrm{s}} \in \overline{\mathbb{R}}_{+}^{q}$, define the energy flow between subsystems $\mathcal{G}_i$ and $\mathcal{G}_j$ of a large-scale dynamical system $\mathcal{G}$.

Assumption 4.1. The connectivity matrix $\mathcal{C} \in \mathbb{R}^{q \times q}$ associated with the large-scale dynamical system $\mathcal{G}$ is defined by

$$\mathcal{C}_{(i,j)} \triangleq \begin{cases} 0, & \text{if } \phi_{ij}(V_{\mathrm{s}}) \equiv 0, \\ 1, & \text{otherwise,} \end{cases} \quad i \neq j, \quad i,j = 1,\ldots,q, \tag{4.11}$$

and

$$\mathcal{C}_{(i,i)} \triangleq - \sum_{k=1,\, k\neq i}^{q} \mathcal{C}_{(k,i)}, \quad i = j, \quad i = 1,\ldots,q, \tag{4.12}$$

and satisfies rank $\mathcal{C} = q - 1$. Moreover, for every $i \neq j$ such that $\mathcal{C}_{(i,j)} = 1$, $\phi_{ij}(V_{\mathrm{s}}) = 0$ if and only if $v_{\mathrm{s}i} = v_{\mathrm{s}j}$.

Assumption 4.2. For $i,j = 1,\ldots,q$, $(v_{\mathrm{s}i} - v_{\mathrm{s}j})\phi_{ij}(V_{\mathrm{s}}) \leq 0$, $V_{\mathrm{s}} \in \overline{\mathbb{R}}_{+}^{q}$.

The fact that $\phi_{ij}(V_{\mathrm{s}}) = 0$ if and only if $v_{\mathrm{s}i} = v_{\mathrm{s}j}$, $i \neq j$, implies that subsystems $\mathcal{G}_i$ and $\mathcal{G}_j$ of $\mathcal{G}$ are *connected*; alternatively, $\phi_{ij}(V_{\mathrm{s}}) \equiv 0$ implies that $\mathcal{G}_i$ and $\mathcal{G}_j$ are *disconnected*. Assumption 4.1 implies that if the energies in the connected subsystems $\mathcal{G}_i$ and $\mathcal{G}_j$ are equal, then energy exchange between these subsystems is not possible. This statement is consistent with the *zeroth law of thermodynamics*, which postulates that temperature equality is a necessary and sufficient condition for thermal equilibrium. Furthermore, it follows from the fact that $\mathcal{C} = \mathcal{C}^{\mathrm{T}}$ and rank $\mathcal{C} = q-1$ that the connectivity matrix $\mathcal{C}$ is irreducible, which implies that for every pair of subsystems $\mathcal{G}_i$ and $\mathcal{G}_j$, $i \neq j$, of $\mathcal{G}$ there exists a sequence of connectors (arcs) of $\mathcal{G}$ that connect $\mathcal{G}_i$ and $\mathcal{G}_j$.

Assumption 4.2 implies that energy flows from more energetic subsystems to less energetic subsystems and is consistent with the *second law of thermodynamics*, which states that heat (energy) must flow in the direction of lower temperatures. Furthermore, note that $\phi_{ij}(V_{\mathrm{s}}) = -\phi_{ji}(V_{\mathrm{s}})$, $V_{\mathrm{s}} \in \overline{\mathbb{R}}_{+}^{q}$, $i \neq j$, $i,j = 1,\ldots,q$, which implies conservation of energy between lossless subsystems. With $S(t) \triangleq S(u(t), y(t)) \equiv 0$, Assumptions 4.1 and 4.2, along with the fact that $\phi_{ij}(V_{\mathrm{s}}) = -\phi_{ji}(V_{\mathrm{s}})$, $V_{\mathrm{s}} \in \overline{\mathbb{R}}_{+}^{q}$, $i \neq j$, $i,j = 1,\ldots,q$, imply that at a given instant of time energy can only be transported, stored, or dissipated but not created and the maximum amount of energy that can be transported and/or dissipated from a subsystem cannot exceed the energy in the subsystem.

Next, we give a deterministic definition of entropy for a large-scale dynamical system that is consistent with the classical thermodynamic definition of entropy.

Definition 4.2. For the large-scale dynamical system $\mathcal{G}$ with the power balance equation (4.5), the function $\mathcal{S} : \overline{\mathbb{R}}_{+}^{q} \to \mathbb{R}$ given by $\mathcal{S}(V_{\mathrm{s}}) = \mathbf{e}^{\mathrm{T}}\mathbf{log}_e(c\mathbf{e} + V_{\mathrm{s}}) - q\log_e c$, where $\mathbf{log}_e(c\mathbf{e} + V_{\mathrm{s}})$ denotes the vector natural logarithm given by $[\log_e(c + v_{\mathrm{s}1}),\ldots,\log_e(c + v_{\mathrm{s}q})]^{\mathrm{T}}$ and $c > 0$, is called the *entropy* of $\mathcal{G}$.

The entropy of $\mathcal{G}$ can be thought of as a measure of the tendency of a large-scale dynamical system to lose the ability to do useful work, lose order, and to settle to a more homogenous state. Note that the change in entropy of a large-scale dynamical system $\mathcal{G}$ is given by

$$\mathcal{S}(V_{\mathrm{s}}(t_2)) - \mathcal{S}(V_{\mathrm{s}}(t_1)) = \int_{t_1}^{t_2} \sum_{i=1}^{q} \frac{\dot{v}_{si}(t)}{c + v_{si}(t)} \mathrm{d}t, \tag{4.13}$$

or, equivalently,

$$\mathrm{d}\mathcal{S}(V_{\mathrm{s}}(t)) = \sum_{i=1}^{q} \frac{\mathrm{d}v_{si}(t)}{c + v_{si}(t)}. \tag{4.14}$$

Theorem 4.1. Consider the large-scale dynamical system $\mathcal{G}$ with power balance equation (4.5) and assume that Assumption 4.2 holds. Then, for every $V_{\mathrm{s}}(t_0) \in \overline{\mathbb{R}}_+^q$, the entropy of $\mathcal{G}$ satisfies

$$\mathcal{S}(V_{\mathrm{s}}(t_2)) \geq \mathcal{S}(V_{\mathrm{s}}(t_1)) + \int_{t_1}^{t_2} \mathbf{e}^{\mathrm{T}} Q(t) \mathrm{d}t, \quad t_2 \geq t_1 \geq t_0, \tag{4.15}$$

where $Q \triangleq \left[\frac{s_1 - \sigma_{11}(V_{\mathrm{s}})}{c + v_{s1}}, \ldots, \frac{s_q - \sigma_{qq}(V_{\mathrm{s}})}{c + v_{sq}}\right]^{\mathrm{T}}$.

Proof. Since $V_{\mathrm{s}}(t) \geq\geq 0$, $t \geq t_0$, and $\phi_{ij}(V_{\mathrm{s}}) = -\phi_{ji}(V_{\mathrm{s}})$, $V_{\mathrm{s}} \in \overline{\mathbb{R}}_+^q$, $i \neq j$, $i, j = 1, \ldots, q$, it follows that

$$\begin{aligned}
\dot{\mathcal{S}}(V_{\mathrm{s}}(t)) &= \sum_{i=1}^{q} \frac{\dot{v}_{si}(t)}{c + v_{si}(t)} \\
&= \sum_{i=1}^{q} \left[\frac{s_i(t) - \sigma_{ii}(V_{\mathrm{s}}(t))}{c + v_{si}(t)} + \sum_{j=1, j \neq i}^{q} \frac{\phi_{ij}(V_{\mathrm{s}}(t))}{c + v_{si}(t)} \right] \\
&= \sum_{i=1}^{q} \left[Q_i(t) + \sum_{j=i+1}^{q} \left(\frac{\phi_{ij}(V_{\mathrm{s}}(t))}{c + v_{si}(t)} - \frac{\phi_{ij}(V_{\mathrm{s}}(t))}{c + v_{sj}(t)} \right) \right] \\
&= \mathbf{e}^{\mathrm{T}} Q(t) + \sum_{i=1}^{q} \sum_{j=i+1}^{q} \frac{\phi_{ij}(V_{\mathrm{s}}(t))(v_{sj}(t) - v_{si}(t))}{(c + v_{si}(t))(c + v_{sj}(t))} \\
&\geq \mathbf{e}^{\mathrm{T}} Q(t). \tag{4.16}
\end{aligned}$$

Now, integrating (4.16) over $[t_1, t_2]$ yields (4.15). $\qquad\square$

Definition 4.3. For the large-scale dynamical system $\mathcal{G}$ with the power balance equation (4.5), the function $\mathcal{E} : \overline{\mathbb{R}}_+^q \to \mathbb{R}$ given by $\mathcal{E}(V_{\mathrm{s}}) = \frac{1}{2} V_{\mathrm{s}}^{\mathrm{T}} V_{\mathrm{s}}$ is called the *ectropy* of $\mathcal{G}$.

Ectropy is a measure of the extent to which the system energy deviates from a homogeneous state. Thus, ectropy is the dual of entropy and is a measure of the tendency of a large-scale dynamical system to do useful work and grow more organized.

Theorem 4.2. Consider the large-scale dynamical system $\mathcal{G}$ with power balance equation (4.5) and assume that Assumption 4.2 holds. Then, for every $V_{\mathrm{s}}(t_0) \in \overline{\mathbb{R}}_+^q$, the ectropy of $\mathcal{G}$ satisfies

$$\mathcal{E}(V_{\mathrm{s}}(t_2)) \le \mathcal{E}(V_{\mathrm{s}}(t_1)) + \int_{t_1}^{t_2} V_{\mathrm{s}}^{\mathrm{T}}(t) S(t) \mathrm{d}t, \quad t_2 \ge t_1 \ge t_0. \qquad (4.17)$$

Proof. Since $\phi_{ij}(V_{\mathrm{s}}) = -\phi_{ji}(V_{\mathrm{s}})$, $V_{\mathrm{s}} \in \overline{\mathbb{R}}_+^q$, $i \ne j$, $i = 1, \ldots, q$, it follows that

$$
\begin{aligned}
\dot{\mathcal{E}}(V_{\mathrm{s}}(t)) &= V_{\mathrm{s}}^{\mathrm{T}}(t) \dot{V}_{\mathrm{s}}(t) \\
&= V_{\mathrm{s}}^{\mathrm{T}}(t)[w(V_{\mathrm{s}}(t)) - d(V_{\mathrm{s}}(t)) + S(t)] \\
&= \sum_{i=1}^{q} v_{\mathrm{s}i}(t) \left[\sum_{j=1, j \ne i}^{q} \phi_{ij}(V_{\mathrm{s}}(t)) \right] - \sum_{i=1}^{q} v_{\mathrm{s}i}(t) \sigma_{ii}(V_{\mathrm{s}}(t)) \\
&\quad + V_{\mathrm{s}}^{\mathrm{T}}(t) S(t) \\
&= \sum_{i=1}^{q} \sum_{j=i+1}^{q} (v_{\mathrm{s}i}(t) - v_{\mathrm{s}j}(t)) \phi_{ij}(V_{\mathrm{s}}(t)) \\
&\quad - \sum_{i=1}^{q} v_{\mathrm{s}i}(t) \sigma_{ii}(V_{\mathrm{s}}(t)) + V_{\mathrm{s}}^{\mathrm{T}}(t) S(t) \\
&\le V_{\mathrm{s}}^{\mathrm{T}}(t) S(t). \qquad (4.18)
\end{aligned}
$$

Now, integrating (4.18) over $[t_1, t_2]$ yields (4.17). $\qquad \square$

4.4 Semistability and Large-Scale Systems

Inequality (4.15) is precisely Clausius' inequality for reversible and irreversible thermodynamics as applied to large-scale dynamical systems, whereas inequality (4.17) is an anti-Clausius inequality that shows that a thermodynamically consistent large-scale dynamical system is dissipative with respect to the supply rate $V_{\mathrm{s}}^{\mathrm{T}} S$ and with storage function corresponding to the system ectropy. Note that $\mathcal{S}(0) = 0$ or, equivalently, $\lim_{V_{\mathrm{s}} \to 0} \mathcal{S}(V_{\mathrm{s}}) = 0$, which is consistent with the *third law of thermodynamics* (Nernst's theorem), which states that the entropy of every system at absolute zero can always be taken to be equal to zero. Furthermore, note that since $\frac{\mathrm{d}\mathcal{S}_i}{\mathrm{d}v_{\mathrm{s}i}} = \frac{1}{c+v_{\mathrm{s}i}}$, where $\mathcal{S}_i \triangleq \log_e(c + v_{\mathrm{s}i}) - \log_e c$ denotes the ith subsystem entropy, it follows that the subsystem energies play the role of subsystem thermodynamic temperatures.

For an *isolated* large-scale dynamical system, that is, an *input-closed* (i.e., $S(t) \equiv 0$) and *output-closed* (i.e., $d(V_{\mathrm{s}}) \equiv 0$) dynamical system, (4.15) yields the fundamental (universal) inequality

$$\mathcal{S}(V_{\mathrm{s}}(t_2)) \ge \mathcal{S}(V_{\mathrm{s}}(t_1)), \quad t_2 \ge t_1. \qquad (4.19)$$

Inequality (4.19) implies that, for any dynamical change in an *isolated* large-scale system, the entropy of the final state can never be less than the entropy of the initial state. It is important to stress that this result holds for an isolated dynamical system. It is, however, possible with power supplied from an external dynamical system (e.g., a controller) to reduce the entropy of the large-scale dynamical system. The entropy of the closed-loop system, however, cannot decrease. The above observations imply that when an isolated large-scale dynamical system with thermodynamically consistent energy flow characteristics (i.e., Assumptions 4.1 and 4.2 hold) is at a state of maximum entropy consistent with its energy, it cannot be subject to any further dynamical change since any such change would result in a decrease of entropy. This of course implies that the state of *maximum entropy* is the stable state of an isolated system and this equilibrium state has to be semistable.

Analogously, it follows from (4.17) that an isolated large-scale dynamical system $\mathcal{G}$ satisfies the fundamental inequality

$$\mathcal{E}(V_{\mathrm{s}}(t_2)) \leq \mathcal{E}(V_{\mathrm{s}}(t_1)), \quad t_2 \geq t_1, \tag{4.20}$$

which implies that the ectropy of the final state of $\mathcal{G}$ is always less than the ectropy of the initial state of $\mathcal{G}$. Hence, for an isolated large-scale dynamical system the entropy increases if and only if the ectropy decreases. Thus, the state of *minimum ectropy* is the stable state of an isolated system and this equilibrium state has to be semistable. It is important to note, however, since (4.20) also holds in the case where $d(V_{\mathrm{s}}) \neq 0$, the system ectropy is a more fundamental concept as compared to the system entropy since (4.19) does not necessarily hold in the case where $d(V_{\mathrm{s}}) \neq 0$. The next theorem concretizes the above observations.

Theorem 4.3. Consider the large-scale dynamical system $\mathcal{G}$ with power balance equation (4.5) with $S(t) \equiv 0$ and $d(V_{\mathrm{s}}) \equiv 0$, and assume that Assumptions 4.1 and 4.2 hold. Then, for every $\alpha \geq 0$, $\alpha \mathbf{e}$ is a semistable equilibrium state of (4.5). Furthermore, $V_{\mathrm{s}}(t) \to \frac{1}{q}\mathbf{e}\mathbf{e}^{\mathrm{T}}V_{\mathrm{s}}(t_0)$ as $t \to \infty$ and $\frac{1}{q}\mathbf{e}\mathbf{e}^{\mathrm{T}}V_{\mathrm{s}}(t_0)$ is a semistable equilibrium state. Finally, if for some $k \in \{1,\ldots,q\}$, $\sigma_{kk}(V_{\mathrm{s}}) \not\equiv 0$ and $\sigma_{kk}(V_{\mathrm{s}}) = 0$ if and only if $v_{\mathrm{s}k} = 0$,[1] then the zero solution $V_{\mathrm{s}}(t) \equiv 0$ to (4.5) is a globally asymptotically stable equilibrium state of (4.5).

Proof. First, we show that $\alpha \mathbf{e} \in \overline{\mathbb{R}}_+^q$, $\alpha \geq 0$, is an equilibrium state for (4.5). If $v_{\mathrm{s}i} = v_{\mathrm{s}j}$ for all $i, j = 1, \ldots, q$, then $w_i(V_{\mathrm{s}}) = 0$ for all $i = 1, \ldots, q$, is immediate from Assumption 4.1. Next, if $w_i(V_{\mathrm{s}}) = 0$ for all $i = 1, \ldots, q$,

---

[1] The assumption $\sigma_{kk}(V_{\mathrm{s}}) \not\equiv 0$ and $\sigma_{kk}(V_{\mathrm{s}}) = 0$ if and only if $v_{\mathrm{s}k} = 0$ for some $k \in \{1,\ldots,q\}$ implies that if the kth subsystem possesses no energy, then nothing can dissipate from it. Conversely, if $\sigma_{kk}(V_{\mathrm{s}}) \not\equiv 0$ and there is no dissipation from the kth subsystem, then this subsystem has no energy.

then it follows from Assumption 4.2 that

$$
\begin{aligned}
0 &= \sum_{i=1}^{q} v_{si} w_i(V_s) \\
&= \sum_{i=1}^{q} \sum_{j=1}^{q} v_{si} \phi_{ij}(V_s) \\
&= \sum_{i=1}^{q} \sum_{j=i+1}^{q} (v_{si} - v_{sj}) \phi_{ij}(V_s) \\
&\leq 0,
\end{aligned}
\tag{4.21}
$$

where we have used the fact that $\phi_{ij}(V_s) = -\phi_{ji}(V_s)$ for all $i, j = 1, \ldots, q$. Hence, $(v_{si} - v_{sj})\phi_{ij}(V_s) = 0$ for all $i, j = 1, \ldots, q$. Then, $w_i(V_s) = 0$ for all $i = 1, \ldots, q$ if and only if $v_{s1} = \cdots = v_{sq}$, which shows that $\alpha \mathbf{e} \in \overline{\mathbb{R}}_+^q$, $\alpha \geq 0$, is an equilibrium state of (4.5).

To show Lyapunov stability of the equilibrium state $\alpha \mathbf{e}$ consider the system shifted ectropy $\mathcal{E}_s(V_s) = \frac{1}{2}(V_s - \alpha \mathbf{e})^{\mathrm{T}}(V_s - \alpha \mathbf{e})$ as a Lyapunov function candidate. Now, since $\phi_{ij}(V_s) = -\phi_{ji}(V_s)$, $V_s \in \overline{\mathbb{R}}_+^q$, $i \neq j$, $i, j = 1, \ldots, q$, and $\mathbf{e}^{\mathrm{T}} w(V_s) = 0$, $V_s \in \overline{\mathbb{R}}_+^q$, it follows from Assumption 4.2 that

$$
\begin{aligned}
\dot{\mathcal{E}}_s(V_s) &= (V_s - \alpha \mathbf{e})^{\mathrm{T}} \dot{V}_s \\
&= (V_s - \alpha \mathbf{e})^{\mathrm{T}} w(V_s) \\
&= V_s^{\mathrm{T}} w(V_s) \\
&= \sum_{i=1}^{q} v_{si} \left[\sum_{j=1, j\neq i}^{q} \phi_{ij}(V_s) \right] \\
&= \sum_{i=1}^{q} \sum_{j=i+1}^{q} (v_{si} - v_{sj}) \phi_{ij}(V_s) \\
&= \sum_{i=1}^{q} \sum_{j \in \mathcal{K}_i} (v_{si} - v_{sj}) \phi_{ij}(V_s) \\
&\leq 0, \quad V_s \in \overline{\mathbb{R}}_+^q,
\end{aligned}
\tag{4.22}
$$

where $\mathcal{K}_i \triangleq \mathcal{N}_i \setminus \cup_{l=1}^{i-1}\{l\}$ and $\mathcal{N}_i \triangleq \{j \in \{1, \ldots, q\} : \phi_{ij}(V_s) = 0$ if and only if $v_{si} = v_{sj}\}$, $i = 1, \ldots, q$, which establishes Lyapunov stability of the equilibrium state $\alpha \mathbf{e}$.

To show that $\alpha \mathbf{e}$ is semistable, let $\mathcal{R} \triangleq \{V_s \in \overline{\mathbb{R}}_+^q : \dot{\mathcal{E}}_s(V_s) = 0\} = \{V_s \in \overline{\mathbb{R}}_+^q : (v_{si} - v_{sj})\phi_{ij}(V_s) = 0, i = 1, \ldots, q, j \in \mathcal{K}_i\}$. Now, by Assumption 4.1 the directed graph associated with the connectivity matrix $\mathcal{C}$ for the large-scale dynamical system $\mathcal{G}$ is strongly connected, which implies that $\mathcal{R} = \{V_s \in \overline{\mathbb{R}}_+^q : v_{s1} = \cdots = v_{sq}\}$. Since the set $\mathcal{R}$ consists of the equilibrium states of (4.5), it follows that the largest invariant set $\mathcal{M}$ contained in $\mathcal{R}$ is given by $\mathcal{M} = \mathcal{R}$. Hence, it follows from the Krasovskii-LaSalle invariant set theorem [70] that for every initial condition $V_\mathrm{s}(t_0) \in \overline{\mathbb{R}}_+^q$, $V_\mathrm{s}(t) \to \mathcal{M}$ as $t \to \infty$, and hence, $\alpha \mathbf{e}$ is a semistable equilibrium state of (4.5). Next, note that since $\mathbf{e}^\mathrm{T} V_\mathrm{s}(t) = \mathbf{e}^\mathrm{T} V_\mathrm{s}(t_0)$ and $V_\mathrm{s}(t) \to \mathcal{M}$ as $t \to \infty$, it follows that $V_\mathrm{s}(t) \to \frac{1}{q}\mathbf{e}\mathbf{e}^\mathrm{T} V_\mathrm{s}(t_0)$ as $t \to \infty$. Hence, with $\alpha = \frac{1}{q}\mathbf{e}^\mathrm{T} V_\mathrm{s}(t_0)$, $\alpha \mathbf{e} = \frac{1}{q}\mathbf{e}\mathbf{e}^\mathrm{T} V_\mathrm{s}(t_0)$ is a semistable equilibrium state of (4.5).

Finally, to show that in the case where for some $k \in \{1,\ldots,q\}$, $\sigma_{kk}(V_\mathrm{s}) \not\equiv 0$ and $\sigma_{kk}(V_\mathrm{s}) = 0$ if and only if $v_{sk} = 0$, the zero solution $V_\mathrm{s}(t) \equiv 0$ to (4.5) is globally asymptotically stable, consider the system ectropy $\mathcal{E}(V_\mathrm{s}) = \frac{1}{2} V_\mathrm{s}^\mathrm{T} V_\mathrm{s}$ as a candidate Lyapunov function. Note that $\mathcal{E}(0) = 0$, $\mathcal{E}(V_\mathrm{s}) > 0$, $V_\mathrm{s} \in \overline{\mathbb{R}}_+^q$, $V_\mathrm{s} \neq 0$, and $\mathcal{E}(V_\mathrm{s})$ is radially unbounded. Now, the Lyapunov derivative along the system energy trajectories of (4.5) is given by

$$
\begin{aligned}
\dot{\mathcal{E}}(V_\mathrm{s}) &= V_\mathrm{s}^\mathrm{T}[w(V_\mathrm{s}) - d(V_\mathrm{s})] \\
&= V_\mathrm{s}^\mathrm{T} w(V_\mathrm{s}) - v_{sk}\sigma_{kk}(V_\mathrm{s}) \\
&= \sum_{i=1}^{q} v_{si}\left[\sum_{j=1,j\neq i}^{q} \phi_{ij}(V_\mathrm{s})\right] - v_{sk}\sigma_{kk}(V_\mathrm{s}) \\
&= \sum_{i=1}^{q}\sum_{j=i+1}^{q} (v_{si} - v_{sj})\phi_{ij}(V_\mathrm{s}) - v_{sk}\sigma_{kk}(V_\mathrm{s}) \\
&= \sum_{i=1}^{q}\sum_{j\in\mathcal{K}_i} (v_{si} - v_{sj})\phi_{ij}(V_\mathrm{s}) - v_{sk}\sigma_{kk}(V_\mathrm{s}) \\
&\leq 0, \quad V_\mathrm{s} \in \overline{\mathbb{R}}_+^q,
\end{aligned}
\tag{4.23}
$$

which shows that the zero solution $V_\mathrm{s}(t) \equiv 0$ to (4.5) is Lyapunov stable.

To show global asymptotic stability of the zero equilibrium state, let $\mathcal{R} \triangleq \{V_\mathrm{s} \in \overline{\mathbb{R}}_+^q : \dot{\mathcal{E}}(V_\mathrm{s}) = 0\} = \{V_\mathrm{s} \in \overline{\mathbb{R}}_+^q : v_{sk}\sigma_{kk}(V_\mathrm{s}) = 0, k \in \{1,\ldots,q\}\} \cap \{V_\mathrm{s} \in \overline{\mathbb{R}}_+^q : (v_{si} - v_{sj})\phi_{ij}(V_\mathrm{s}) = 0, i = 1,\ldots,q, j \in \mathcal{K}_i\}$. Now, since Assumption 4.1 holds and $\sigma_{kk}(V_\mathrm{s}) = 0$ if and only if $v_{sk} = 0$, it follows that $\mathcal{R} = \{V_\mathrm{s} \in \overline{\mathbb{R}}_+^q : v_{sk} = 0, k \in \{1,\ldots,q\}\} \cap \{V_\mathrm{s} \in \overline{\mathbb{R}}_+^q : v_{s1} = v_{s2} = \cdots = v_{sq}\} = \{0\}$ and the largest invariant set $\mathcal{M}$ contained in $\mathcal{R}$ is given by $\mathcal{M} = \{0\}$. Hence, it follows from the Krasovskii-LaSalle invariant set theorem that for every initial condition $V_\mathrm{s}(t_0) \in \overline{\mathbb{R}}_+^q$, $V_\mathrm{s}(t) \to \mathcal{M} = \{0\}$ as $t \to \infty$, which proves global asymptotic stability of the zero equilibrium state of (4.5). $\square$

In Theorem 4.3 we used the shifted ectropy function to show that for an isolated large-scale dynamical system $\mathcal{G}$, $V_\mathrm{s}(t) \to \frac{1}{q}\mathbf{e}\mathbf{e}^\mathrm{T} V_\mathrm{s}(t_0)$ as $t \to \infty$ and $\frac{1}{q}\mathbf{e}\mathbf{e}^\mathrm{T} V_\mathrm{s}(t_0)$ is a semistable equilibrium state. This result can also be arrived at using the system entropy. To see this, note that since $\mathbf{e}^\mathrm{T} w(V_\mathrm{s}) = 0$, $V_\mathrm{s} \in \overline{\mathbb{R}}_+^q$, it follows that $\mathbf{e}^\mathrm{T} \dot{V}_\mathrm{s}(t) = 0$, $t \geq t_0$. Hence, $\mathbf{e}^\mathrm{T} V_\mathrm{s}(t) = \mathbf{e}^\mathrm{T} V_\mathrm{s}(t_0)$, $t \geq t_0$.

Furthermore, since $V_{\mathrm{s}}(t) \geq\geq 0$, $t \geq t_0$, it follows that $0 \leq\leq V_{\mathrm{s}}(t) \leq\leq$ $\mathbf{e}\mathbf{e}^{\mathrm{T}}V_{\mathrm{s}}(t_0)$, $t \geq t_0$, which implies that all solutions to (4.5) are bounded. Next, since, by (4.19), the entropy $\mathcal{S}(V_{\mathrm{s}}(t))$, $t \geq t_0$, of $\mathcal{G}$ is monotonically increasing and $V_{\mathrm{s}}(t)$, $t \geq t_0$, is bounded, it follows from the Krasovskii-LaSalle invariant set theorem that for every initial condition $V_{\mathrm{s}}(t_0) \in \overline{\mathbb{R}}_+^q$, $V_{\mathrm{s}}(t) \to \mathcal{M}$ as $t \to \infty$, where $\mathcal{M}$ is the largest invariant set contained in $\mathcal{R} \triangleq \{V_{\mathrm{s}} \in \overline{\mathbb{R}}_+^q : -\dot{\mathcal{S}}(V_{\mathrm{s}}) = 0\}$. It now follows from the last inequality of (4.16) that $\mathcal{R} = \{V_{\mathrm{s}} \in \overline{\mathbb{R}}_+^q : (v_{\mathrm{s}i} - v_{\mathrm{s}j})\phi_{ij}(V_{\mathrm{s}}) = 0, i = 1,\ldots,q, j \in \mathcal{K}_i\}$, which, since the directed graph associated with the connectivity matrix $\mathcal{C}$ for the large-scale dynamical system $\mathcal{G}$ is strongly connected, implies that $\mathcal{R} = \{V_{\mathrm{s}} \in \overline{\mathbb{R}}_+^q : v_{\mathrm{s}i} = \cdots = v_{\mathrm{s}q}\}$. Since the set $\mathcal{R}$ consists of equilibrium states of (4.5), it follows that $\mathcal{M} = \mathcal{R}$, which along with (4.22), establishes semistability of the equilibrium states $\alpha\mathbf{e}$, $\alpha \geq 0$.

4.5 Energy Equipartition

Theorem 4.3 implies that the steady-state value of the energy in each subsystem $\mathcal{G}_i$ of an isolated system $\mathcal{G}$ is equal; that is, the steady-state energy of the isolated large-scale dynamical system $\mathcal{G}$ given by

$$V_{\mathrm{s}\infty} = \frac{1}{q}\mathbf{e}\mathbf{e}^{\mathrm{T}}V_{\mathrm{s}}(t_0) = \left[\frac{1}{q}\sum_{i=1}^{q} v_{\mathrm{s}i}(t_0)\right]\mathbf{e} \tag{4.24}$$

is uniformly distributed over all subsystems of $\mathcal{G}$. This phenomenon is known as *equipartition of energy*[2] [20, 21, 81, 88, 129, 148] and is an emergent behavior in thermodynamic systems. The next proposition shows that among all possible energy distributions in an isolated large-scale dynamical system $\mathcal{G}$, energy equipartition corresponds to the minimum value of the system's ectropy and the maximum value of the system's entropy (see Figure 4.2).

Proposition 4.1. Consider the large-scale dynamical system $\mathcal{G}$ with power balance equation (4.5), let $\mathcal{E} : \overline{\mathbb{R}}_+^q \to \mathbb{R}$ and $\mathcal{S} : \overline{\mathbb{R}}_+^q \to \mathbb{R}$ denote the ectropy and entropy of $\mathcal{G}$, respectively, and define $\mathcal{D}_{\mathrm{c}} \triangleq \{V_{\mathrm{s}} \in \overline{\mathbb{R}}_+^q : \mathbf{e}^{\mathrm{T}}V_{\mathrm{s}} = \beta\}$, where $\beta \geq 0$. Then,

$$\arg\min_{V_{\mathrm{s}}\in\mathcal{D}_{\mathrm{c}}}(\mathcal{E}(V_{\mathrm{s}})) = \arg\max_{V_{\mathrm{s}}\in\mathcal{D}_{\mathrm{c}}}(\mathcal{S}(V_{\mathrm{s}})) = V_{\mathrm{s}}^* = \frac{\beta}{q}\mathbf{e}. \tag{4.25}$$

Furthermore, $\mathcal{E}_{\min} \triangleq \mathcal{E}(V_{\mathrm{s}}^*) = \frac{1}{2}\frac{\beta^2}{q}$ and $\mathcal{S}_{\max} \triangleq \mathcal{S}(V_{\mathrm{s}}^*) = q\log_e(c + \frac{\beta}{q}) - q\log_e c$.

Proof. The existence and uniqueness of V_{s}^* follows from the fact that $\mathcal{E}(V_{\mathrm{s}})$ and $-\mathcal{S}(V_{\mathrm{s}})$ are strictly convex continuous functions on the compact

---

[2]The phenomenon of equipartition of energy is closely related to the notion of a *monotemperaturic* system discussed in [27, 81].

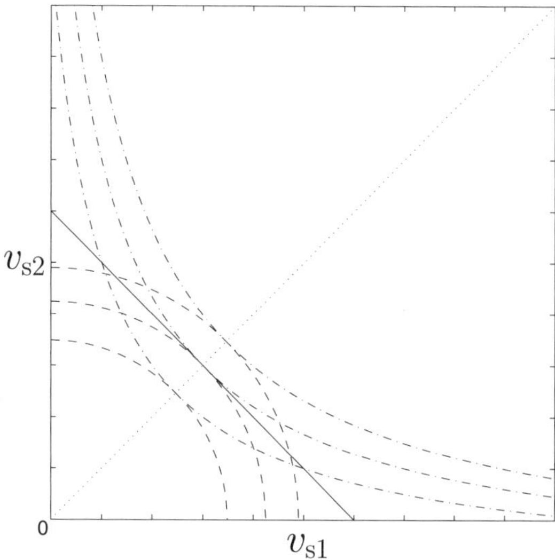

Figure 4.2 Thermodynamic equilibria $(\cdots)$, constant energy surfaces (———), constant ectropy surfaces $(---)$, and constant entropy surfaces $(-\cdot-\cdot-)$.

set $\mathcal{D}_c$. To minimize $\mathcal{E}(V_s) = \frac{1}{2}V_s^{\mathrm{T}}V_s$, $V_s \in \overline{\mathbb{R}}_+^q$, subject to $V_s \in \mathcal{D}_c$ form the Lagrangian $\mathcal{L}(V_s,\lambda) = \frac{1}{2}V_s^{\mathrm{T}}V_s + \lambda(\mathbf{e}^{\mathrm{T}}V_s - \beta)$, where $\lambda \in \mathbb{R}$ is the Lagrange multiplier. If V_s^* solves this minimization problem, then

$$0 = \left.\frac{\partial \mathcal{L}}{\partial V_s}\right|_{V_s=V_s^*} = V_s^{*\mathrm{T}} + \lambda\mathbf{e}^{\mathrm{T}} = 0 \qquad (4.26)$$

and hence $V_s^* = -\lambda\mathbf{e}$. Now, it follows from $\mathbf{e}^{\mathrm{T}}V_s = \beta$ that $\lambda = -\frac{\beta}{q}$, which implies that $V_s^* = \frac{\beta}{q}\mathbf{e} \in \overline{\mathbb{R}}_+^q$. The fact that V_s^* minimizes the ectropy on the compact set $\mathcal{D}_c$ can be shown by computing the Hessian of the ectropy for the constrained parameter optimization problem and showing that the Hessian is positive definite at V_s^*. $\mathcal{E}_{\min} = \frac{1}{2}\frac{\beta^2}{q}$ is now immediate.

Analogously, to maximize $\mathcal{S}(V_s) = \mathbf{e}^{\mathrm{T}}\mathbf{log}_e(c\mathbf{e} + V_s) - q\log_e c$ on the compact set $\mathcal{D}_c$, form the Lagrangian $\mathcal{L}(V_s,\lambda) \triangleq \sum_{i=1}^q \log_e(c+v_{si}) + \lambda(\mathbf{e}^{\mathrm{T}}V_s - \beta)$, where $\lambda \in \mathbb{R}$ is a Lagrange multiplier. If V_s^* solves this maximization problem, then

$$0 = \left.\frac{\partial \mathcal{L}}{\partial V_s}\right|_{V_s=V_s^*} = \left[\frac{1}{c+v_{s1}^*} + \lambda, \ldots, \frac{1}{c+v_{sq}^*} + \lambda\right] = 0. \qquad (4.27)$$

Thus, $\lambda = -\frac{1}{c+v_{si}^*}$, $i = 1, \ldots, q$. If $\lambda = 0$, then the only value of V_s^* that satisfies (4.27) is $V_s^* = \infty$, which does not satisfy the constraint equation

$\mathbf{e}^{\mathrm{T}}V_{\mathrm{s}} = \beta$ for finite $\beta \geq 0$. Hence, $\lambda \neq 0$ and $v_{si}^{*} = -(\frac{1}{\lambda} + c)$, $i = 1,\ldots,q$, which implies $V_{\mathrm{s}}^{*} = -(\frac{1}{\lambda}+c)\mathbf{e}$. Now, it follows from $\mathbf{e}^{\mathrm{T}}V_{\mathrm{s}} = \beta$ that $-(\frac{1}{\lambda}+c) = \frac{\beta}{q}$ and hence $V_{\mathrm{s}}^{*} = \frac{\beta}{q}\mathbf{e} \in \overline{\mathbb{R}}_{+}^{q}$. The fact that V_{s}^{*} maximizes the entropy on the compact set $\mathcal{D}_{\mathrm{c}}$ can be shown by computing the Hessian and showing that it is negative definite at V_{s}^{*}. $\mathcal{S}_{\max} = q\log_{e}(c + \frac{\beta}{q}) - q\log_{e} c$ is now immediate. $\square$

It follows from (4.19), (4.20), and Proposition 4.1 that conservation of energy necessarily implies nonconservation of ectropy and entropy. Hence, in an isolated large-scale dynamical system $\mathcal{G}$ all the energy, though always conserved, will eventually be degraded (diluted) to the point where it cannot produce any useful work. Hence, all motion would cease and the dynamical system would be fated to a state of eternal rest (semistability) wherein all subsystems will possess identical energies (energy equipartition). Ectropy would be a minimum and entropy would be a maximum, giving rise to a state of absolute disorder. This is precisely what is known in theoretical physics as the *heat death of the universe* [81].

4.6 Entropy Increase and the Second Law of Thermodynamics

In the preceding discussion it was assumed that our large-scale nonlinear dynamical system model is such that energy flows from more energetic subsystems to less energetic subsystems, that is, heat (energy) flows in the direction of lower temperatures. Although this universal phenomenon can be predicted with virtual certainty, it follows as a manifestation of entropy and ectropy nonconservation for the case of two subsystems. To see this, consider the isolated large-scale dynamical system $\mathcal{G}$ with power balance equation (4.5) (with $S(t) \equiv 0$ and $d(V_{\mathrm{s}}) \equiv 0$) and assume that the system entropy is monotonically increasing and hence $\dot{\mathcal{S}}(V_{\mathrm{s}}(t)) \geq 0$, $t \geq t_0$. Now, since

$$\dot{\mathcal{S}}(V_{\mathrm{s}}(t)) = \sum_{i=1}^{q} \frac{\dot{v}_{si}(t)}{c + v_{si}(t)}$$

$$= \sum_{i=1}^{q} \sum_{j=1,\, j\neq i}^{q} \frac{\phi_{ij}(V_{\mathrm{s}}(t))}{c + v_{si}(t)}$$

$$= \sum_{i=1}^{q} \sum_{j=i+1}^{q} \left(\frac{\phi_{ij}(V_{\mathrm{s}}(t))}{c + v_{si}(t)} - \frac{\phi_{ij}(V_{\mathrm{s}}(t))}{c + v_{sj}(t)} \right)$$

$$= \sum_{i=1}^{q} \sum_{j\in\mathcal{K}_i} \frac{\phi_{ij}(V_{\mathrm{s}}(t))(v_{sj}(t) - v_{si}(t))}{(c + v_{si}(t))(c + v_{sj}(t))}$$

$$\geq 0, \quad t \geq t_0, \tag{4.28}$$

it follows that for $q = 2$, $(v_{s1} - v_{s2})\phi_{12}(V_s) \leq 0$, $V_s \in \overline{\mathbb{R}}_+^2$, which implies that energy (heat) flows naturally from a more energetic subsystem (hot object) to a less energetic subsystem (cooler object). The universality of this emergent behavior thus follows from the fact that entropy (respectively, ectropy), accompanying energy transfer, always increases (respectively, decreases).

In the case where we have multiple subsystems, it is clear from (4.28) that entropy and ectropy nonconservation does not necessarily imply Assumption 4.2. However, if we invoke the additional condition (Assumption 4.3) that if for any pair of connected subsystems $\mathcal{G}_k$ and $\mathcal{G}_l$, $k \neq l$, with $v_{sk} \geq v_{sl}$ (respectively, $v_{sk} \leq v_{sl}$) and for any other pair of connected subsystems $\mathcal{G}_m$ and $\mathcal{G}_n$, $m \neq n$, with $v_{sm} \geq v_{sn}$ (respectively, $v_{sm} \leq v_{sn}$) the inequality $\phi_{kl}(V_s)\phi_{mn}(V_s) \geq 0$, $V_s \in \overline{\mathbb{R}}_+^q$, holds, then nonconservation of entropy and ectropy in an isolated large-scale dynamical system implies Assumption 4.2. The above inequality postulates that the direction of energy flow for any pair of *energy similar* subsystems is consistent, that is, if for a given pair of connected subsystems at a given energy level the energy flows in a certain direction, then for any other pair of connected subsystems with the same energy level, the energy flow direction is consistent with the original pair of subsystems. Note that this assumption does *not* specify the direction of energy flow between subsystems.

To see that $\dot{\mathcal{S}}(V_s(t)) \geq 0$, $t \geq t_0$, along with Assumption 4.3 implies Assumption 4.2, note that since (4.28) holds for all $t \geq t_0$ and $V_s(t_0) \in \overline{\mathbb{R}}_+^q$ is arbitrary, (4.28) implies

$$\sum_{i=1}^{q} \sum_{j \in \mathcal{K}_i} \frac{\phi_{ij}(V_s)(v_{sj} - v_{si})}{(c + v_{si})(c + v_{sj})} \geq 0, \quad V_s \in \overline{\mathbb{R}}_+^q. \tag{4.29}$$

Now, it follows from (4.29) that for any fixed system energy level $V_s \in \overline{\mathbb{R}}_+^q$ there exists at least one pair of connected subsystems $\mathcal{G}_k$ and $\mathcal{G}_l$, $k \neq l$, such that $\phi_{kl}(V_s)(v_{sl} - v_{sk}) \geq 0$. Thus, if $v_{sk} \geq v_{sl}$ (respectively, $v_{sk} \leq v_{sl}$), then $\phi_{kl}(V_s) \leq 0$ (respectively, $\phi_{kl}(V_s) \geq 0$). Furthermore, it follows from Assumption 4.3 that for any other pair of connected subsystems $\mathcal{G}_m$ and $\mathcal{G}_n$, $m \neq n$, with $v_{sm} \geq v_{sn}$ (respectively, $v_{sm} \leq v_{sn}$) the inequality $\phi_{mn}(V_s) \leq 0$ (respectively, $\phi_{mn}(V_s) \geq 0$) holds, which implies that

$$\phi_{mn}(V_s)(v_{sn} - v_{sm}) \geq 0, \quad m \neq n. \tag{4.30}$$

Thus, it follows from (4.30) that energy (heat) flows naturally from more energetic subsystems (hot objects) to less energetic subsystems (cooler objects). Of course, since in an isolated large-scale dynamical system $\mathcal{G}$ ectropy decreases if and only if entropy increases, the same result can be arrived at by considering the ectropy of $\mathcal{G}$. Finally, since Assumption 4.2 holds, it follows from the conservation of energy and the fact that the large-scale dynamical system $\mathcal{G}$ is strongly connected that nonconservation of entropy and ectropy necessarily implies energy equipartition.

4.7 Thermodynamic Models with Linear Energy Exchange

In this section, we assume a linear energy exchange between subsystems, that is, $\sigma_{ij}(r) = \sigma_{ij}r_j$, $\sigma_{ij} \geq 0$, $i, j = 1, \ldots, q$, and we let $s_i(t) = s_i(u_i(t), y_i(t))$, $t \geq t_0$, $i = 1, \ldots, q$, so that the vector form of the energy balance equation (4.1), with $t_0 = 0$, is given by

$$V_{\mathrm{s}}(x(T)) = V_{\mathrm{s}}(x(0)) + \int_0^T WV_{\mathrm{s}}(x(t))\mathrm{d}t + \int_0^T S(t)\mathrm{d}t, \quad T \geq 0, \quad (4.31)$$

or, in power balance form,

$$\dot{V}_{\mathrm{s}}(x(t)) = WV_{\mathrm{s}}(x(t)) + S(t), \quad V_{\mathrm{s}}(x(0)) = V_{\mathrm{s}}(x_0), \quad t \geq 0, \quad (4.32)$$

where $x_0 \in \mathcal{D}$ and $W \in \mathbb{R}^{q \times q}$ is such that

$$W_{(i,j)} = \left\{ \begin{array}{ll} -\sum_{k=1}^q \sigma_{kj}, & i = j, \\ \sigma_{ij}, & i \neq j. \end{array} \right. \quad (4.33)$$

Note that (4.33) implies $\sum_{i=1}^q W_{(i,j)} \leq 0$, $j = 1, \ldots, q$, and hence, W is a semistable compartmental matrix. If $\sigma_{ii} > 0$, $i = 1, \ldots, q$, then W is an asymptotically stable compartmental matrix.

An important special case of (4.32) is the case where W is symmetric or, equivalently, $\sigma_{ij} = \sigma_{ji}$, $i \neq j$, $i, j = 1, \ldots, q$. In this case, it follows from (4.32) that for each subsystem the power balance equation satisfies

$$\dot{v}_{\mathrm{s}i}(x(t)) + \sigma_{ii}v_{\mathrm{s}i}(x(t)) + \sum_{j=1, j \neq i}^q \sigma_{ij}[v_{\mathrm{s}i}(x(t)) - v_{\mathrm{s}j}(x(t))] = s_i(t), \quad t \geq 0.$$

$$(4.34)$$

Note that $\phi_i(x) \triangleq \sum_{j=1, j \neq i}^q \sigma_{ij}[v_{\mathrm{s}i}(x) - v_{\mathrm{s}j}(x)]$, $i = 1, \ldots, q$, represents the energy flow from the ith subsystem to all other subsystems and is given by the sum of the individual energy flows from the ith subsystem to the jth subsystem. Furthermore, these energy flows are proportional to the energy differences of the subsystems, that is, $v_{\mathrm{s}i}(x) - v_{\mathrm{s}j}(x)$. Hence, (4.34) is a power balance equation that governs the energy exchange among coupled subsystems and is completely analogous to the equations of conduction and convection heat transfer with subsystem energies playing the role of temperatures. Furthermore, note that since $\sigma_{ij} \geq 0$, $i, j = 1, \ldots, q$, energy flows from more energetic subsystems to less energetic subsystems, which is consistent with the second law of thermodynamics requiring that heat (energy) *must* flow in the direction of lower temperatures.

The next proposition is needed for developing expressions for steady-state energy distributions of large-scale dynamical systems.

Proposition 4.2 ([81]). Consider the large-scale dynamical system $\mathcal{G}$ with power balance equation given by (4.32). Suppose $V_{\mathrm{s}}(x_0) \geq\geq 0$, $x_0 \in \mathcal{D}$, and $S(t) \geq\geq 0$, $t \geq 0$. Then the solution $V_{\mathrm{s}}(x(t))$, $t \geq 0$, to (4.32) is nonnegative for all $t \geq 0$ if and only if W is essentially nonnegative.

Next, we develop expressions for the steady-state energy distribution for a large-scale nonlinear dynamical system $\mathcal{G}$ for the cases where supplied system power $S(t)$ is a periodic function with period $\tau > 0$, that is, $S(t + \tau) = S(t)$, $t \geq 0$, and $S(t)$ is constant, that is, $S(t) \equiv S$. Define $e(t) \triangleq V_{\mathrm{s}}(x(t)) - V_{\mathrm{s}}(x(t+\tau))$, $t \geq 0$, and note that

$$\dot{e}(t) = We(t), \quad e(0) = V_{\mathrm{s}}(x_0) - V_{\mathrm{s}}(x(\tau)), \quad t \geq 0, \tag{4.35}$$

where $x_0 \in \mathcal{D}$. Hence, since

$$e(t) = e^{Wt}[V_{\mathrm{s}}(x_0) - V_{\mathrm{s}}(x(\tau))], \quad t \geq 0, \tag{4.36}$$

and W is semistable, it follows from $iv)$ of Lemma 2.2 of [21] that

$$\lim_{t\to\infty} e(t) = \lim_{t\to\infty} [V_{\mathrm{s}}(x(t)) - V_{\mathrm{s}}(x(t+\tau))]$$
$$= (I_q - WW^{\#})[V_{\mathrm{s}}(x_0) - V_{\mathrm{s}}(x(\tau))], \tag{4.37}$$

which represents a constant offset to the steady-state error energy distribution in the large-scale nonlinear dynamical system $\mathcal{G}$. For the case where $S(t) \equiv S$, $\tau \to \infty$, and hence, the following result is immediate. This result first appeared in [21].

Proposition 4.3. Consider the large-scale dynamical system $\mathcal{G}$ with power balance equation given by (4.32). Suppose that $V_{\mathrm{s}}(x_0) \geq\geq 0$, $x_0 \in \mathcal{D}$, and $S(t) \equiv S \geq\geq 0$. Then, $V_{\mathrm{s}\infty} \triangleq \lim_{t\to\infty} V_{\mathrm{s}}(x(t))$ exists if and only if $S \in \mathcal{R}(W)$. In this case,

$$V_{\mathrm{s}\infty} = (I_q - WW^{\#})V_{\mathrm{s}}(x_0) - W^{\#}S, \quad x_0 \in \mathcal{D}, \tag{4.38}$$

and $V_{\mathrm{s}\infty} \geq\geq 0$. If, in addition, W is nonsingular, then $V_{\mathrm{s}\infty}$ exists for all $S \geq\geq 0$ and is given by

$$V_{\mathrm{s}\infty} = -W^{-1}S. \tag{4.39}$$

Proof. Note that the solution $V_{\mathrm{s}}(x(t))$, $t \geq 0$, to (4.32) is given by

$$V_{\mathrm{s}}(x(t)) = e^{Wt}V_{\mathrm{s}}(x_0) + \int_0^t e^{W(t-s)}S(s)\mathrm{d}s, \quad t \geq 0. \tag{4.40}$$

Now, the result is a direct consequence of Proposition 4.2 and $iv)$, $vii)$, $viii)$, and $ix)$ of Lemma 2.2 of [21]. $\qquad\square$

Next, we specialize the result of Proposition 4.3 to the case where there is no energy dissipation from each subsystem $\mathcal{G}_i$ of $\mathcal{G}$, that is, $\sigma_{ii} = 0$, $i = 1,\ldots,q$. Note that in this case $\mathbf{e}^{\mathrm{T}}W = 0$, and hence, rank $W \leq q - 1$. Furthermore, if $S = 0$ it follows from (4.32) that $\mathbf{e}^{\mathrm{T}}\dot{V}_{\mathrm{s}}(x(t)) = \mathbf{e}^{\mathrm{T}}WV_{\mathrm{s}}(x(t)) =$

0, $t \geq 0$, and hence, the total energy of the isolated large-scale nonlinear dynamical system $\mathcal{G}$ is conserved.

Proposition 4.4. Consider the large-scale dynamical system $\mathcal{G}$ with power balance equation given by (4.32). Let $W \in \mathbb{R}^{q \times q}$ be compartmental and assume rank $W = q-1$, $\sigma_{ii} = 0$, $i = 1, \ldots, q$, and $W\mathbf{e} = 0$. If $V_{\mathrm{s}}(x_0) \geq\geq 0$, $x_0 \in \mathcal{D}$, and $S = 0$, then the steady-state energy distribution $V_{\mathrm{s}\infty}$ of the isolated large-scale dynamical system $\mathcal{G}$ is given by

$$V_{\mathrm{s}\infty} = \left[\frac{1}{q} \sum_{i=1}^{q} v_{\mathrm{s}i}(x_0) \right] \mathbf{e}, \quad x_0 \in \mathcal{D}. \tag{4.41}$$

Proof. The proof is similar to the proof of Theorem 4.3 with $w(V_{\mathrm{s}}) = WV_{\mathrm{s}}$. $\qquad\square$

Finally, we examine the steady-state energy distribution for large-scale nonlinear dynamical systems $\mathcal{G}$ in the case of strong coupling between subsystems; that is, $\sigma_{ij} \to \infty$, $i \neq j$. For this analysis we assume that W given by (4.33) is symmetric, that is, $\sigma_{ij} = \sigma_{ji}$, $i \neq j$, $i,j = 1, \ldots, q$, and $\sigma_{ii} > 0$, $i = 1, \ldots, q$. Thus, $-W$ is a nonsingular M-matrix for all values of σ_{ij}, $i \neq j$, $i,j = 1, \ldots, q$. Moreover, in this case it can be shown that if $\frac{\sigma_{ij}}{\sigma_{kl}} \to 1$ as $\sigma_{ij} \to \infty$, $i \neq j$, and $\sigma_{kl} \to \infty$, $k \neq l$, then

$$\lim_{\sigma_{ij} \to \infty, i \neq j} W^{-1} = -\frac{1}{\sum_{i=1}^{q} \sigma_{ii}} \mathbf{e}\mathbf{e}^{\mathrm{T}}. \tag{4.42}$$

Hence, in the limit of strong coupling the steady-state energy distribution $V_{\mathrm{s}\infty}$ given by (4.39) becomes

$$V_{\mathrm{s}\infty} = \lim_{\sigma_{ij} \to \infty, i \neq j} (-W^{-1}S) = \left[\frac{\mathbf{e}^{\mathrm{T}}S}{\sum_{i=1}^{q} \sigma_{ii}} \right] \mathbf{e}, \tag{4.43}$$

which implies energy equipartition. This result first appeared in [21].

Control of Large-Scale Dynamical Systems via Vector Lyapunov Functions

5.1 Introduction

One of the most basic issues in system theory is the stability of dynamical systems. The most complete contribution to the stability analysis of nonlinear dynamical systems is due to Lyapunov [128]. Lyapunov's results, along with the Krasovskii-LaSalle invariance principle [116, 120, 121], provide a powerful framework for analyzing the stability of nonlinear dynamical systems. Lyapunov methods have also been used by control system designers to obtain stabilizing feedback controllers for nonlinear systems. In particular, for smooth feedback, Lyapunov-based methods were inspired by Jurdjevic and Quinn [105] who give sufficient conditions for smooth stabilization based on the ability of constructing a Lyapunov function for the closed-loop system.

More recently, Artstein [6] introduced the notion of a control Lyapunov function whose existence guarantees a feedback control law that globally stabilizes a nonlinear dynamical system. In general, the feedback control law is not necessarily smooth but can be guaranteed to be at least continuous at the origin in addition to being smooth everywhere else. Even though for certain classes of nonlinear dynamical systems a universal construction of a feedback stabilizer can be obtained using control Lyapunov functions [165, 167], there does not exist a unified procedure for finding a Lyapunov function candidate that will stabilize the closed-loop system for general nonlinear systems.

In an attempt to simplify the construction of Lyapunov functions for the analysis and control design of nonlinear dynamical systems, several researchers have resorted to vector Lyapunov functions as an alternative to scalar Lyapunov functions. As discussed in Chapter 1, vector Lyapunov functions were first introduced by Bellman [14] and Matrosov [133], and further developed in [65, 118, 127, 136], with [65, 127, 131, 132, 136, 159, 160] exploiting their utility for analyzing large-scale systems. The use of vector Lyapunov functions in dynamical system theory offers a very flexible framework since each component of the vector Lyapunov function can satisfy less rigid requirements as compared to a single scalar Lyapunov function. Weakening the hypothesis on the Lyapunov function enlarges the class of Lyapunov functions that can be used for analyzing system stability. In particular, each component of a vector Lyapunov function need not be positive definite with a negative or even negative-semidefinite derivative. Alternatively, the time derivative of the vector Lyapunov function need only satisfy an element-by-element inequality involving a vector field of a certain comparison system. Since in this case the stability properties of the comparison system imply the stability properties of the dynamical system, the use of vector Lyapunov theory can significantly reduce the complexity (i.e., dimensionality) of the dynamical system being analyzed. Extensions of vector Lyapunov function theory that include relaxed conditions on standard vector Lyapunov functions as well as matrix Lyapunov functions appear in [52, 131, 132].

In this chapter, we introduce the notion of a *control vector Lyapunov function* as a generalization of control Lyapunov functions and show that asymptotic stabilizability of a nonlinear dynamical system is equivalent to the existence of a control vector Lyapunov function. In addition, using control vector Lyapunov functions, we present a universal decentralized feedback stabilizer for a decentralized affine in the control nonlinear dynamical system with guaranteed gain and sector margins. Furthermore, we establish connections between vector dissipativity notions [80] and inverse optimality of decentralized nonlinear regulators. These results are then used to develop decentralized controllers for large-scale dynamical systems with robustness guarantees against full modeling and input uncertainty.

5.2 Control Vector Lyapunov Functions

In this section, we consider a feedback control problem and introduce the notion of a control vector Lyapunov function as a generalization of control Lyapunov functions. Specifically, consider the nonlinear controlled dynamical system given by

$$\dot{x}(t) = F(x(t), u(t)), \quad x(t_0) = x_0, \quad t \geq t_0, \tag{5.1}$$

where $x_0 \in \mathcal{D}$, $\mathcal{D} \subseteq \mathbb{R}^n$ is an open set with $0 \in \mathcal{D}$, $u(t) \in U \subseteq \mathbb{R}^m$, $t \geq t_0$, is the control input, and $F : \mathcal{D} \times U \to \mathbb{R}^n$ is Lipschitz continuous for all $(x, u) \in \mathcal{D} \times U$ and satisfies $F(0, 0) = 0$. We assume that the control input $u(\cdot)$ in (5.1) is restricted to the class of *admissible controls* consisting of measurable functions $u(\cdot) \in \mathcal{U}$ such that $u(t) \in U$ for all $t \geq t_0$, where the constraint set $\mathcal{U}$ is given with $0 \in U$. Furthermore, we assume that $u(\cdot)$ satisfies sufficient regularity conditions such that the nonlinear dynamical system (5.1) has a unique solution forward in time. A measurable mapping $\phi : \mathcal{D} \to U$ satisfying $\phi(0) = 0$ is called a *control law*. Furthermore, if $u(t) = \phi(x(t))$, where ϕ is a control law and $x(t)$, $t \geq t_0$, satisfies (5.1), then $u(\cdot)$ is called a *feedback control law*.

Definition 5.1. If there exist a continuously differentiable vector function $V = [v_1, \ldots, v_q]^{\mathrm{T}} : \mathcal{D} \to \mathcal{Q} \cap \overline{\mathbb{R}}_+^q$, a continuous function $w = [w_1, \ldots, w_q]^{\mathrm{T}} : \mathcal{Q} \times \mathcal{D} \to \mathbb{R}^q$, and a positive vector $p \in \mathbb{R}_+^q$ such that $V(0) = 0$, $v(x) \triangleq p^{\mathrm{T}} V(x)$, $x \in \mathcal{D}$, is positive definite, $w(\cdot, x) \in \mathcal{W}$, $w(0,0) = 0$, $\mathcal{F}(x) \triangleq \cap_{i=1}^q \mathcal{F}_i(x) \neq \emptyset$, $x \in \mathcal{D}$, $x \neq 0$, where $\mathcal{F}_i(x) \triangleq \{u \in U : v_i'(x)F(x, u) < w_i(V(x), x)\}$, $x \in \mathcal{D}$, $x \neq 0$, $i = 1, \ldots, q$, then the vector function $V : \mathcal{D} \to \mathcal{Q} \cap \overline{\mathbb{R}}_+^q$ is called a *control vector Lyapunov function candidate*.

It follows from Definition 5.1 that if there exists a control vector Lyapunov function candidate, then there exists a feedback control law $\phi : \mathcal{D} \to U$ such that

$$V'(x)F(x, \phi(x)) << w(V(x), x), \quad x \in \mathcal{D}, \quad x \neq 0. \tag{5.2}$$

Moreover, if the nonlinear dynamical system

$$\dot{z}(t) = w(z(t), x(t)), \quad z(t_0) = z_0, \quad t \geq t_0, \tag{5.3}$$

$$\dot{x}(t) = F(x(t), \phi(x(t))), \quad x(t_0) = x_0, \tag{5.4}$$

where $z_0 \in \mathcal{Q}$ and $x_0 \in \mathcal{D}$, is asymptotically stable with respect to z uniformly in x_0, then it follows from Theorem 2.3 that the zero solution $x(t) \equiv 0$ to (5.4) is asymptotically stable. In this case, the vector function $V : \mathcal{D} \to \overline{\mathbb{R}}_+^q$ given in Definition 5.1 is called a *control vector Lyapunov function*. Furthermore, if $\mathcal{D} = \mathbb{R}^n$, $\mathcal{Q} = \mathbb{R}^q$, $U = \mathbb{R}^m$, $v : \mathbb{R}^n \to \overline{\mathbb{R}}_+$ is radially unbounded, and the system (5.3) and (5.4) is globally asymptotically stable with respect to z uniformly in x_0, then the zero solution $x(t) \equiv 0$ to (5.1) is globally asymptotically stabilizable.

If in Definition 5.1 $w(z, x) = w(z)$ and the zero solution $z(t) \equiv 0$ to

$$\dot{z}(t) = w(z(t)), \quad z(t_0) = z_0, \quad t \geq t_0, \tag{5.5}$$

where $z_0 \in \mathcal{Q}$, is asymptotically stable, then it follows from Corollary 2.3 that $V : \mathcal{D} \to \mathcal{Q} \cap \overline{\mathbb{R}}_+^q$ is a control vector Lyapunov function.

Conversely, suppose that there exists a stabilizing feedback control law $\phi : \mathcal{D} \to U$ such that the zero solution $x(t) \equiv 0$ to (5.4) is asymptotically stable. Then it follows from Theorem 2.6 that there exist a continuously differentiable vector function $V = [v_1, \ldots, v_q]^{\mathrm{T}} : \mathcal{D}_0 \to \overline{\mathbb{R}}_+^q$, a continuous function $w = [w_1, \ldots, w_q]^{\mathrm{T}} : \overline{\mathbb{R}}_+^q \to \mathbb{R}^q$, and a positive vector $p \in \mathbb{R}_+^q$ such that $V(0) = 0$, the scalar function $v : \mathcal{D}_0 \to \overline{\mathbb{R}}_+$ defined by $v(x) \triangleq p^{\mathrm{T}} V(x)$, $x \in \mathcal{D}_0$, is positive definite, $w(\cdot) \in \mathcal{W}$, $w(0) = 0$, and $V'(x)F(x, \phi(x)) << w(V(x))$, $x \in \mathcal{D}_0$, $x \neq 0$. Thus, $\mathcal{F}(x) \neq \emptyset$, $x \in \mathcal{D}_0$, $x \neq 0$. Moreover, since, by Theorem 2.6, the zero solution $z(t) \equiv 0$ to (5.5) is asymptotically stable, it follows from the discussion above that $V : \mathcal{D}_0 \to \overline{\mathbb{R}}_+^q$ is a control vector Lyapunov function. Hence, a given nonlinear dynamical system of the form (5.1) is feedback asymptotically stabilizable if and only if there exists a control vector Lyapunov function.

In the case where $q = 1$ and $w(z,x) \equiv w(z)$, Definition 5.1 implies the existence of a positive-definite continuously differentiable function $v : \mathcal{D} \to \mathcal{Q} \cap \overline{\mathbb{R}}_+$ and a continuous function $w : \mathcal{Q} \to \mathbb{R}$, where $\mathcal{Q} \subseteq \mathbb{R}$, such that $w(0) = 0$ and $\mathcal{F}(x) = \{u \in U : v'(x)F(x,u) < w(v(x))\} \neq \varnothing$, $x \in \mathcal{D}$, $x \neq 0$, which is equivalent to

$$\inf_{u \in U} v'(x)F(x,u) < w(v(x)), \quad x \in \mathcal{D}, \quad x \neq 0. \tag{5.6}$$

Now, (5.6) implies the existence of a feedback control law $\phi : \mathcal{D} \to U$ such that $v'(x)F(x,\phi(x)) < w(v(x))$, $x \in \mathcal{D}$, $x \neq 0$. Moreover, if $v : \mathcal{D} \to \overline{\mathbb{R}}_+$ is a control vector Lyapunov function (with $q = 1$), then it follows from the discussion above that the zero solution $z(t) \equiv 0$ to the system (5.5) is asymptotically stable and, since $q = 1$, this implies that $w(z) < 0$, $z \in \mathcal{Q} \cap \overline{\mathbb{R}}_+$, $z \neq 0$. Thus, since $v(\cdot)$ is positive definite, (5.6) can be rewritten as

$$\inf_{u \in U} v'(x)F(x,u) < 0, \quad x \in \mathcal{D}, \quad x \neq 0, \tag{5.7}$$

which is equivalent to the standard definition of a control Lyapunov function [6].

Next, consider the case where the control input to (5.1) possesses a decentralized control architecture so that the dynamics of (5.1) are given by

$$\dot{x}_i(t) = F_i(x(t), u_i(t)), \quad t \geq t_0, \quad i = 1, \ldots, q, \tag{5.8}$$

where $x_i(t) \in \mathbb{R}^{n_i}$, $x(t) = [x_1^{\mathrm{T}}(t), \ldots, x_q^{\mathrm{T}}(t)]^{\mathrm{T}}$, $u_i(t) \in U_i \subseteq \mathbb{R}^{m_i}$, $t \geq t_0$, $\sum_{i=1}^{q} n_i = n$, and $\sum_{i=1}^{q} m_i = m$. Note that $x_i(t) \in \mathbb{R}^{n_i}$, $t \geq t_0$, $i = 1, \ldots, q$, as long as $x(t) \in \mathcal{D}$, $t \geq t_0$, and the set of control inputs is given by $U = U_1 \times \cdots \times U_q \subseteq \mathbb{R}^m$. In the case of a component decoupled control vector Lyapunov function candidate, that is, $V(x) = [v_1(x_1), \ldots, v_q(x_q)]^{\mathrm{T}}$, $x \in \mathcal{D}$, it suffices to require in Definition 5.1 that

$$\mathcal{F}_i(x) = \{u \in U : v_i'(x_i)F_i(x,u_i) < w_i(V(x),x)\} \neq \varnothing,$$
$$x \in \mathcal{D}, \quad x \neq 0, \quad i = 1, \ldots, q, \tag{5.9}$$

to ensure that $\mathcal{F}(x) = \cap_{i=1}^{q}\mathcal{F}_i(x) \neq \varnothing$, $x \in \mathcal{D}$, $x \neq 0$. Note that for a component decoupled control vector Lyapunov function $V : \mathcal{D} \to \mathcal{Q} \cap \overline{\mathbb{R}}_+^q$, (5.9) holds if and only if

$$\inf_{u \in U} V'(x)F(x,u) << w(V(x),x), \quad x \in \mathcal{D}, \quad x \neq 0, \tag{5.10}$$

where the infimum in (5.10) is taken componentwise, that is, for each component of (5.10) the infimum is calculated separately. It follows from (5.10) that there exists a feedback control law $\phi : \mathcal{D} \to U$ such that $\phi(x) = [\phi_1^{\mathrm{T}}(x), \ldots, \phi_q^{\mathrm{T}}(x)]^{\mathrm{T}}$, $x \in \mathcal{D}$, where $\phi_i : \mathcal{D} \to U_i$, and $v_i'(x_i)F_i(x,\phi_i(x)) <$

$w_i(V(x), x)$, $x \in \mathcal{D}$, $x \neq 0$, $i = 1, \ldots, q$. Note that if $w_i(V(x), x) = 0$ for $x \in \mathcal{D}$ with $x_i = 0$, then condition (5.9) holds for all $x \in \mathcal{D}$ such that $x_i \neq 0$.

Next, we consider the special case of a nonlinear dynamical system of the form (5.8) with affine control inputs given by

$$\dot{x}_i(t) = f_i(x(t)) + G_i(x(t))u_i(t), \quad t \geq t_0, \quad i = 1, \ldots, q, \tag{5.11}$$

where $f_i : \mathbb{R}^n \to \mathbb{R}^{n_i}$ satisfying $f_i(0) = 0$ and $G_i : \mathbb{R}^n \to \mathbb{R}^{n_i \times m_i}$ are smooth functions (at least continuously differentiable mappings) for all $i = 1, \ldots, q$, and $u_i(t) \in \mathbb{R}^{m_i}$, $t \geq t_0$, $i = 1, \ldots, q$.

Theorem 5.1. Consider the controlled nonlinear dynamical system given by (5.11). If there exist a continuously differentiable, component decoupled vector function $V : \mathbb{R}^n \to \overline{\mathbb{R}}_+^q$, a continuous function $w = [w_1, \ldots, w_q]^{\mathrm{T}} : \overline{\mathbb{R}}_+^q \times \mathbb{R}^n \to \mathbb{R}^q$, and a positive vector $p \in \mathbb{R}_+^q$ such that $V(0) = 0$, the scalar function $v : \mathbb{R}^n \to \overline{\mathbb{R}}_+$ defined by $v(x) \triangleq p^{\mathrm{T}}V(x)$, $x \in \mathbb{R}^n$, is positive definite and radially unbounded, $w(\cdot, x) \in \mathcal{W}$, $w(0, 0) = 0$, and

$$v_i'(x_i)f_i(x) < w_i(V(x), x), \quad x \in \mathcal{R}_i, \quad i = 1, \ldots, q, \tag{5.12}$$

where $\mathcal{R}_i \triangleq \{x \in \mathbb{R}^n, x \neq 0 : v_i'(x_i)G_i(x) = 0\}$, $i = 1, \ldots, q$, then $V : \mathbb{R}^n \to \overline{\mathbb{R}}_+^q$ is a control vector Lyapunov function candidate. If, in addition, there exists $\phi : \mathbb{R}^n \to U$ such that $\phi(x) = [\phi_1^{\mathrm{T}}(x), \ldots, \phi_q^{\mathrm{T}}(x)]^{\mathrm{T}}$, $x \in \mathbb{R}^n$, and the system (5.3) and (5.4) is globally asymptotically stable with respect to z uniformly in x_0, then the zero solution $x(t) \equiv 0$ to (5.4) is globally asymptotically stable and $V : \mathbb{R}^n \to \overline{\mathbb{R}}_+^q$ is a control vector Lyapunov function.

Proof. Note that for all $i = 1, \ldots, q$,

$$\inf_{u_i \in \mathbb{R}^{m_i}} v_i'(x_i)(f_i(x) + G_i(x)u_i) = \begin{cases} -\infty, & x \notin \mathcal{R}_i, \\ v_i'(x_i)f_i(x), & x \in \mathcal{R}_i, \end{cases}$$
$$< w_i(V(x), x), \quad x \in \mathbb{R}^n, \tag{5.13}$$

which implies (5.10). Now, the proof is a direct consequence of the definition of a control vector Lyapunov function by noting the equivalence between (5.9) and (5.10) for component decoupled vector Lyapunov functions. $\square$

Using Theorem 5.1 we can construct an explicit feedback control law that is a function of the control vector Lyapunov function $V(\cdot)$. Specifically, consider the feedback control law $\phi(x) = [\phi_1^{\mathrm{T}}(x), \ldots, \phi_q^{\mathrm{T}}(x)]^{\mathrm{T}}$, $x \in \mathbb{R}^n$, given by

$$\phi_i(x) = \begin{cases} -\left(c_{0i} + \eta_i(x) + \sqrt{1 + \eta_i^2(x)}\right)\beta_i(x), & \beta_i(x) \neq 0, \\ 0, & \beta_i(x) = 0, \end{cases} \tag{5.14}$$

where $\alpha_i(x) \triangleq v_i'(x_i)f_i(x)$, $x \in \mathbb{R}^n$, $\beta_i(x) \triangleq G_i^{\mathrm{T}}(x)v_i'^{\mathrm{T}}(x_i)$, $\eta_i(x) \triangleq (\alpha_i(x) - w_i(V(x), x))/(\beta_i^{\mathrm{T}}(x)\beta_i(x))$, $x \in \mathbb{R}^n$, and $c_{0i} > 0$, $i = 1, \ldots, q$. The derivative $\dot{V}(\cdot)$ along the trajectories of the dynamical system (5.11), with $u = \phi(x)$, $x \in \mathbb{R}^n$, given by (5.14), is given by

$$\begin{aligned}
\dot{v}_i(x_i) &= v_i'(x_i)(f_i(x) + G_i(x)\phi_i(x)) \\
&= \alpha_i(x) + \beta_i^{\mathrm{T}}(x)\phi_i(x) \\
&= \begin{cases} -\left(c_{0i} + \sqrt{1 + \eta_i^2(x)}\right)\beta_i^{\mathrm{T}}(x)\beta_i(x) + w_i(V(x), x), & \beta_i(x) \neq 0, \\ \alpha_i(x), & \beta_i(x) = 0, \end{cases} \\
&< w_i(V(x), x), \quad x \in \mathbb{R}^n.
\end{aligned} \tag{5.15}$$

Thus, if the zero solution $z(t) \equiv 0$ to (5.3) and (5.4) is globally asymptotically stable with respect to z uniformly in x_0, then it follows from Theorem 2.3 that the zero solution $x(t) \equiv 0$ to (5.11) with $u = \phi(x) = [\phi_1^{\mathrm{T}}(x), \ldots, \phi_q^{\mathrm{T}}(x)]^{\mathrm{T}}$, $x \in \mathbb{R}^n$, given by (5.14) is globally asymptotically stable.

If in Theorem 5.1 $w(z, x) = w(z)$ and the zero solution $z(t) \equiv 0$ to (5.5) is globally asymptotically stable, then it follows from Corollary 2.3 that the feedback control law given by (5.14) is a globally asymptotically stabilizing controller for the nonlinear dynamical system (5.11).

In the case where $q = 1$, the function $w(\cdot, \cdot)$ in Theorem 5.1 can be set to be identically zero, that is, $w(z, x) \equiv 0$. In this case, the feedback control law (5.14) specializes to Sontag's universal formula [165] and is a global stabilizer for (5.11).

Since $f_i(\cdot)$ and $G_i(\cdot)$ are smooth and $v_i(\cdot)$ is continuously differentiable for all $i = 1, \ldots, q$, it follows that $\alpha_i(x)$ and $\beta_i(x)$, $x \in \mathbb{R}^n$, $i = 1, \ldots, q$, are continuous functions, and hence, $\phi_i(x)$ given by (5.14) is continuous for all $x \in \mathbb{R}^n$ if either $\beta_i(x) \neq 0$ or $\alpha_i(x) - w_i(V(x), x) < 0$ for all $i = 1, \ldots, q$. Hence, the feedback control law given by (5.14) is continuous everywhere except for the origin. The following result provides necessary and sufficient conditions under which the feedback control law given by (5.14) is guaranteed to be continuous at the origin in addition to being continuous everywhere else.

Proposition 5.1. The feedback control law $\phi(x)$ given by (5.14) is continuous on $\mathbb{R}^n$ if and only if for every $\varepsilon > 0$, there exists $\delta > 0$ such that for all $0 < \|x\| < \delta$ there exists $u_i \in \mathbb{R}^{m_i}$ such that $\|u_i\| < \varepsilon$ and $\alpha_i(x) + \beta_i^{\mathrm{T}}(x)u_i < w_i(V(x), x)$, $i = 1, \ldots, q$.

Proof. First note that since $v_i(x_i)$, $x_i \in \mathbb{R}^{n_i}$, is a nonnegative function and $v_i(0) = 0$, it follows from a Taylor series expansion about $x_i = 0$ that $v_i'(0) = 0$, $i = 1, \ldots, q$, and hence, $\phi(0) = 0$. To show necessity assume that the feedback control law given by (5.14) is continuous on $\mathbb{R}^n$, that is, $\phi_i(x)$ is continuous on $\mathbb{R}^n$ for all $i = 1, \ldots, q$. Then for every $\varepsilon > 0$, there exists $\delta > 0$ such that $\|\phi_i(x)\| < \varepsilon$ for all $0 < \|x\| < \delta$ and, by (5.14), $\alpha_i(x) + \beta_i^{\mathrm{T}}(x)\phi_i(x) < w_i(V(x), x)$, $i = 1, \ldots, q$. Thus, necessity follows with

$u_i = \phi_i(x)$, $i = 1, \ldots, q$.

To show sufficiency, assume that for $\varepsilon > 0$, there exists $\delta > 0$ such that for all $0 < \|x\| < \delta$ there exists $u_i \in \mathbb{R}^{m_i}$ such that $\|u_i\| < \varepsilon$ and $\alpha_i(x) + \beta_i(x)u_i < w_i(V(x), x)$, $i = 1, \ldots, q$. In this case, since $\|u_i\| < \varepsilon$ it follows from the Cauchy-Schwarz inequality that $\alpha_i(x) - w_i(V(x), x) < \varepsilon\|\beta_i(x)\|$, $i = 1, \ldots, q$. Furthermore, since $v_i(\cdot)$, $i = 1, \ldots, q$, is continuously differentiable and $G_i(\cdot)$, $i = 1, \ldots, q$, is continuous, it follows that there exists $\hat{\delta} > 0$ such that for all $0 < \|x\| < \hat{\delta}$, $\|\beta_i(x)\| < \varepsilon$, $i = 1, \ldots, q$. Hence, for all $0 < \|x\| < \delta_{\min}$, where $\delta_{\min} \triangleq \min\{\delta, \hat{\delta}\}$, it follows that $\alpha_i(x) - w_i(V(x), x) < \varepsilon\|\beta_i(x)\|$ and $\|\beta_i(x)\| < \varepsilon$, $i = 1, \ldots, q$.

Furthermore, if $\beta_i(x) = 0$, then $\|\phi_i(x)\| = 0$, and if $\beta_i(x) \neq 0$, then it follows from (5.14) that

$$\|\phi_i(x)\| \leq c_{0i}\|\beta_i(x)\| + \left(\eta_i(x) + \sqrt{1 + \eta_i^2(x)}\right)\|\beta_i(x)\|$$

$$\leq \frac{2(\alpha_i(x) - w_i(V(x), x)) + (c_{0i} + 1)\|\beta_i(x)\|^2}{\|\beta_i(x)\|}$$

$$\leq (c_{0i} + 3)\varepsilon, \quad 0 < \|x\| < \delta_{\min}, \quad \alpha_i(x) > w_i(V(x), x), \quad i = 1, \ldots, q,$$

$$(5.16)$$

and

$$\|\phi_i(x)\| \leq c_{0i}\|\beta_i(x)\| + \left(\eta_i(x) + \sqrt{1 + \eta_i^2(x)}\right)\|\beta_i(x)\|$$

$$\leq c_{0i}\|\beta_i(x)\| + \frac{\beta_i^{\mathrm{T}}(x)\beta_i(x)}{\|\beta_i(x)\|}$$

$$= (c_{0i} + 1)\|\beta_i(x)\|$$

$$\leq (c_{0i} + 1)\varepsilon, \quad 0 < \|x\| < \delta_{\min}, \quad \alpha_i(x) \leq w_i(V(x), x), \quad i = 1, \ldots, q.$$

$$(5.17)$$

Hence, it follows that for every $\hat{\varepsilon} \triangleq (c_{0i} + 3)\varepsilon > 0$ there exists $\delta_{\min} > 0$ such that, for all $\|x\| < \delta_{\min}$, $\|\phi_i(x)\| < \hat{\varepsilon}$, which implies that $\phi_i(\cdot)$, $i = 1, \ldots, q$, is continuous at the origin, and hence, $\phi(\cdot) = [\phi_1^{\mathrm{T}}(\cdot), \ldots, \phi_q^{\mathrm{T}}(\cdot)]^{\mathrm{T}}$ is continuous at the origin. $\square$

5.3 Stability Margins, Inverse Optimality, and Vector Dissipativity

In this section, we show that the feedback control law given by (5.14) is robust to sector bounded input nonlinearities. Specifically, we consider the nonlinear dynamical system (5.11) with nonlinear uncertainties in the input so that the dynamics of the system are given by

$$\dot{x}_i(t) = f_i(x(t)) + G_i(x(t))\sigma_i(u_i(t)), \quad t \geq t_0, \quad i = 1, \ldots, q, \quad (5.18)$$

where $\sigma_i(\cdot) \in \Phi_i \triangleq \{\sigma_i : \mathbb{R}^{m_i} \to \mathbb{R}^{m_i} : \sigma_i(0) = 0$ and $\frac{1}{2}u_i^{\mathrm{T}}u_i \leq \sigma_i^{\mathrm{T}}(u_i)u_i <$
$\infty, u_i \in \mathbb{R}^{m_i}\}$, $i = 1, \ldots, q$. In addition, we show that for the dynamical
system (5.11) the feedback control law given by (5.14) is inverse optimal
in the sense that it minimizes a derived performance functional over the
set of stabilizing controllers $\mathcal{S}(x_0) \triangleq \{u(\cdot) : u(\cdot)$ is admissible and $x(t) \to$
0 as $t \to \infty\}$.

Theorem 5.2. Consider the nonlinear dynamical system (5.18) and
assume that the conditions of Theorem 5.1 hold with $w(z,x) \equiv w(z)$, and
with the zero solution $z(t) \equiv 0$ to (5.5) being globally asymptotically stable.
Then with the feedback control law given by (5.14) the nonlinear dynamical
system (5.18) is globally asymptotically stable for all $\sigma_i(\cdot) \in \Phi_i$, $i = 1, \ldots, q$.
Moreover, for the dynamical system (5.11) the feedback control law (5.14)
minimizes the performance functional given by

$$J(x_0, u(\cdot)) = \int_{t_0}^{\infty} \sum_{i=1}^{q} [L_{1i}(x(t)) + u_i^{\mathrm{T}}(t)R_{2i}(x(t))u_i(t)]\mathrm{d}t \qquad (5.19)$$

in the sense that

$$J(x_0, \phi(x(\cdot))) = \min_{u(\cdot) \in \mathcal{S}(x_0)} J(x_0, u(\cdot)), \quad x_0 \in \mathbb{R}^n, \qquad (5.20)$$

where $L_{1i}(x) \triangleq -\alpha_i(x) + \frac{\gamma_i(x)}{2}\beta_i^{\mathrm{T}}(x)\beta_i(x)$,

$$R_{2i}(x) \triangleq \begin{cases} \frac{1}{2\gamma_i(x)}I_{m_i}, & \beta_i(x) \neq 0, \\ 0, & \beta_i(x) = 0, \end{cases} \qquad (5.21)$$

and

$$\gamma_i(x) \triangleq \begin{cases} c_{0i} + \eta_i(x) + \sqrt{1 + \eta_i^2(x)} > 0, & \beta_i(x) \neq 0, \\ 0, & \beta_i(x) = 0, \end{cases} \qquad (5.22)$$

for all $i = 1, \ldots, q$. Finally, $J(x_0, \phi(x(\cdot))) = \mathbf{e}^{\mathrm{T}}V(x_0)$, $x_0 \in \mathbb{R}^n$, where
$V : \mathbb{R}^n \to \overline{\mathbb{R}}_+^q$ is a control vector Lyapunov function for the dynamical
system (5.11).

Proof. It follows from Theorem 5.1 that the feedback control law
(5.14) globally asymptotically stabilizes the dynamical system (5.11) and
the vector function $V : \mathbb{R}^n \to \overline{\mathbb{R}}_+^q$ is a control vector Lyapunov function for
the dynamical system (5.11). Note that with (5.22) the feedback control
law (5.14) can be rewritten as $\phi_i(x) = -\gamma_i(x)\beta_i(x)$, $x \in \mathbb{R}^n$, $i = 1, \ldots, q$.
Let the control vector Lyapunov function $V : \mathbb{R}^n \to \overline{\mathbb{R}}_+^q$ for (5.11) be a
vector Lyapunov function candidate for (5.18). Then the vector Lyapunov
derivative components are given by

$$\dot{v}_i(x_i) = v_i'(x_i)(f_i(x) + G_i(x)\sigma_i(\phi_i(x))) = \alpha_i(x) + \beta_i^{\mathrm{T}}(x)\sigma_i(\phi_i(x)),$$
$$x \in \mathbb{R}^n, \quad i = 1, \ldots, q. \qquad (5.23)$$

Note that $\phi_i(x) = 0$, and hence, $\sigma_i(\phi_i(x)) = 0$ whenever $\beta_i(x) = 0$ for all $i = 1, \ldots, q$. In this case, it follows from (5.12) that $\dot{v}_i(x_i) < w_i(V(x))$, $x \in \mathbb{R}^n$, $\beta_i(x) = 0$, $x \neq 0$, $i = 1, \ldots, q$.

Next, consider the case where $\beta_i(x) \neq 0$, $i = 1, \ldots, q$. In this case, note that

$$\alpha_i(x) - w_i(V(x)) - \frac{\gamma_i(x)}{2}\beta_i^{\mathrm{T}}(x)\beta_i(x)$$

$$= \frac{-c_{0i}\beta_i^{\mathrm{T}}(x)\beta_i(x)}{2} + \frac{1}{2}\left(\eta_i(x) - \sqrt{1 + \eta_i^2(x)}\right)\beta_i^{\mathrm{T}}(x)\beta_i(x)$$

$$< 0, \quad x \in \mathbb{R}^n, \quad \beta_i(x) \neq 0, \tag{5.24}$$

for all $i = 1, \ldots, q$. Thus, the vector Lyapunov derivative components given by (5.23) satisfy

$$\dot{v}_i(x_i) < w_i(V(x)) + \frac{\gamma_i(x)}{2}\beta_i^{\mathrm{T}}(x)\beta_i(x) + \beta_i^{\mathrm{T}}(x)\sigma_i(\phi_i(x))$$

$$= w_i(V(x)) + \frac{1}{2\gamma_i(x)}\phi_i^{\mathrm{T}}(x)\phi_i(x) - \frac{1}{\gamma_i(x)}\phi_i^{\mathrm{T}}(x)\sigma_i(\phi_i(x))$$

$$= w_i(V(x)) + \frac{1}{\gamma_i(x)}\left[\frac{\phi_i^{\mathrm{T}}(x)\phi_i(x)}{2} - \phi_i^{\mathrm{T}}(x)\sigma_i(\phi_i(x))\right]$$

$$\leq w_i(V(x)), \quad x \in \mathbb{R}^n, \quad \beta_i(x) \neq 0, \tag{5.25}$$

for all $\sigma_i(\cdot) \in \Phi_i$ and $i = 1, \ldots, q$. Since the dynamical system (5.5) is globally asymptotically stable it follows from Corollary 2.3 that the nonlinear dynamical system (5.18) is globally asymptotically stable for all $\sigma_i(\cdot) \in \Phi_i$, $i = 1, \ldots, q$.

To show that the feedback control law (5.14) minimizes (5.19) in the sense of (5.20), define the Hamiltonian

$$\mathcal{H}(x, u) \triangleq \sum_{i=1}^{q}[L_{1i}(x) + u_i^{\mathrm{T}}R_{2i}(x)u_i + v_i'(f_i(x) + G_i(x)u_i)], \tag{5.26}$$

and note that $\mathcal{H}(x, \phi(x)) = 0$ and $\mathcal{H}(x, u) \geq 0$, $x \in \mathbb{R}^n$, $u \in \mathbb{R}^m$, since $\mathcal{H}(x, u) = \sum_{i=1}^{q}(u_i - \phi_i(x))^{\mathrm{T}}R_{2i}(x)(u_i - \phi_i(x))$, $x \in \mathbb{R}^n$, $u \in \mathbb{R}^m$. Thus,

$$J(x_0, u(\cdot)) = \int_{t_0}^{\infty}\left[\mathcal{H}(x(t), u(t)) - \sum_{i=1}^{q}v_i'(x(t))(f_i(x(t)) + G_i(x(t))u_i(t))\right]\mathrm{d}t$$

$$= -\int_{t_0}^{\infty}\mathbf{e}^{\mathrm{T}}\dot{V}(x(t))\mathrm{d}t + \int_{t_0}^{\infty}H(x(t), u(t))\mathrm{d}t$$

$$= -\lim_{t \to \infty}\mathbf{e}^{\mathrm{T}}V(x(t)) + \mathbf{e}^{\mathrm{T}}V(x_0) + \int_{t_0}^{\infty}H(x(t), u(t))\mathrm{d}t$$

$$\geq \mathbf{e}^{\mathrm{T}}V(x_0)$$

$$= J(x_0, \phi(x(\cdot))), \tag{5.27}$$

which yields (5.20).   □

It follows from Theorem 5.2 that with the feedback stabilizing control law (5.14) the nonlinear dynamical system (5.11) has a sector (and hence gain) margin $(\frac{1}{2}, \infty)$ in each decentralized input channel. For details on stability margins for nonlinear dynamical systems, see [39, 70, 157].

Finally, note that Theorem 5.2 implies that

$$\alpha_i(x) - w_i(V(x)) - \theta_i \gamma_i(x)\beta_i^{\mathrm{T}}(x)\beta_i(x) \le 0, \quad x \in \mathbb{R}^n, \tag{5.28}$$

for all $\theta_i \in [\frac{1}{2}, \infty)$ and $i = 1, \ldots, q$. Thus, if

$$\left| \frac{\alpha_i(x) - w_i(V(x))}{\beta_i^{\mathrm{T}}(x)\beta_i(x)} \right| \le \mu_i, \quad x \in \mathbb{R}^n, \quad \beta_i(x) \ne 0, \quad i = 1, \ldots, q, \tag{5.29}$$

then $|\gamma_i(x)| \le c_{0i} + \mu_i + \mu_i\sqrt{1 + \mu_i^2} \triangleq \lambda_i$, $x \in \mathbb{R}^n$, $i = 1, \ldots, q$. In this case, the vector Lyapunov derivative components for the dynamical system (5.11) with the output $y = [y_1^{\mathrm{T}}, \ldots, y_q^{\mathrm{T}}]^{\mathrm{T}}$, where $y_i(x) \triangleq \beta_i(x)$, $x \in \mathbb{R}^n$, $i = 1, \ldots, q$, satisfy

$$\begin{aligned}
\dot{v}_i(x_i) &= \alpha_i(x) + \beta_i^{\mathrm{T}}(x)u_i \\
&\le w_i(V(x)) + \theta_i \gamma_i(x)\beta_i^{\mathrm{T}}(x)\beta_i(x) + \beta_i^{\mathrm{T}}(x)u_i \\
&\le w_i(V(x)) + \theta_i \lambda_i y_i^{\mathrm{T}} y_i + y_i^{\mathrm{T}} u_i, \quad x \in \mathbb{R}^n, \quad u_i \in \mathbb{R}^{m_i}, \quad i = 1, \ldots, q.
\end{aligned} \tag{5.30}$$

Inequlity (5.30) imples that (5.11) is *exponentially vector dissipative* with respect to the *vector supply rate* $S(u, y) \triangleq [s_1(u_1, y_1), \ldots, s_q(u_q, y_q)]^{\mathrm{T}}$, where $s_i(u_i, y_i) = \theta_i \lambda_i y_i^{\mathrm{T}} y_i + y_i^{\mathrm{T}} u_i$, $i = 1, \ldots, q$, and with the control vector Lyapunov function $V : \mathbb{R}^n \to \overline{\mathbb{R}}_+^q$ being a *vector storage function*.

5.4 Decentralized Control for Large-Scale Nonlinear Dynamical Systems

In this section, we apply the proposed control framework to decentralized control of large-scale nonlinear dynamical systems. Specifically, we consider the large-scale dynamical system $\mathcal{G}$ shown in Figure 5.1 involving energy exchange between n interconnected subsystems. Let $x_i : [0, \infty) \to \overline{\mathbb{R}}_+$ denote the energy (and hence a nonnegative quantity) of the ith subsystem, let $u_i : [0, \infty) \to \mathbb{R}$ denote the control input to the ith subsystem, and let $\sigma_{ij} : \overline{\mathbb{R}}_+^n \to \overline{\mathbb{R}}_+$, $i \ne j$, $i, j = 1, \ldots, n$, denote the instantaneous rate of energy flow from the jth subsystem to the ith subsystem.

An energy balance yields the large-scale dynamical system [81]

$$\dot{x}(t) = f(x(t)) + G(x(t))u(t), \quad x(t_0) = x_0, \quad t \ge t_0, \tag{5.31}$$

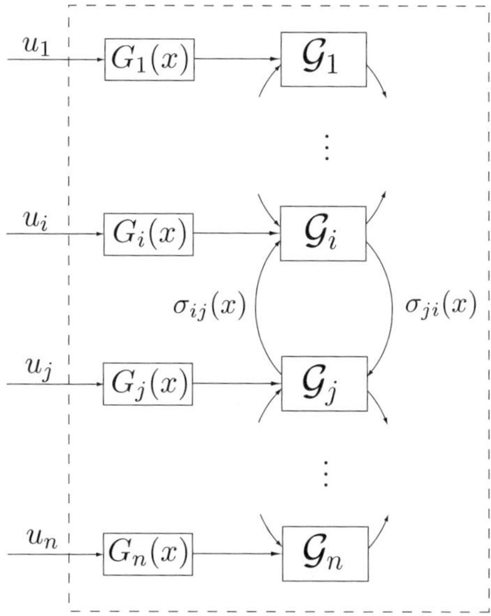

Figure 5.1 Large-scale dynamical system $\mathcal{G}$.

where $x(t) = [x_1(t), \ldots, x_n(t)]^{\mathrm{T}}$, $t \geq t_0$, $f_i(x) = \sum_{j=1, j \neq i}^{n} \phi_{ij}(x)$, where $\phi_{ij}(x) \triangleq \sigma_{ij}(x) - \sigma_{ji}(x)$, $x \in \overline{\mathbb{R}}_+^n$, $i \neq j$, $i,j = 1, \ldots, n$, denotes the net energy flow from the jth subsystem to the ith subsystem, $G(x) = \mathrm{diag}[G_1(x), \ldots, G_n(x)] = \mathrm{diag}[x_1, \ldots, x_n]$, $x \in \overline{\mathbb{R}}_+^n$, and $u(t) \in \mathbb{R}^n$, $t \geq t_0$. Here, we assume that $\sigma_{ij} : \overline{\mathbb{R}}_+^n \to \overline{\mathbb{R}}_+$, $i \neq j$, $i,j = 1, \ldots, n$, are locally Lipschitz continuous on $\overline{\mathbb{R}}_+^n$, $\sigma_{ij}(0) = 0$, $i \neq j$, $i,j = 1, \ldots, n$, and $u = [u_1, \ldots, u_n]^{\mathrm{T}} : \mathbb{R} \to \mathbb{R}^n$ is such that $u_i : \mathbb{R} \to \mathbb{R}$, $i = 1, \ldots, n$, are bounded piecewise continuous functions of time.

Furthermore, we assume that $\sigma_{ij}(x) = 0$, $x \in \overline{\mathbb{R}}_+^n$, whenever $x_j = 0$, $i \neq j$, $i,j = 1, \ldots, n$. In this case, $f(\cdot)$ is *essentially nonnegative* (i.e., $f_i(x) \geq 0$ for all $x \in \overline{\mathbb{R}}_+^n$ such that $x_i = 0$, $i = 1, \ldots, n$). The above constraint implies that if the energy of the jth subsystem of $\mathcal{G}$ is zero, then this subsystem cannot supply any energy to its surroundings. Finally, to ensure that the trajectories of the closed-loop system remain in the nonnegative orthant of the state space for all nonnegative initial conditions, we seek a feedback control law $u(\cdot)$ that guarantees the closed-loop system dynamics are essentially nonnegative [72].

For the dynamical system $\mathcal{G}$, consider the control vector Lyapunov function candidate $V(x) = [v_1(x_1), \ldots, v_n(x_n)]^{\mathrm{T}}$, $x \in \overline{\mathbb{R}}_+^n$, given by

$$V(x) = [x_1, \ldots, x_n]^{\mathrm{T}}, \quad x \in \overline{\mathbb{R}}_+^n. \tag{5.32}$$

Note that $V(0) = 0$ and $v(x) \triangleq \mathbf{e}^{\mathrm{T}} V(x)$, $x \in \overline{\mathbb{R}}_+^n$, is positive definite and radially unbounded. Furthermore, consider the function

$$w(V(x), x) = \begin{bmatrix} -\sigma_{11}(v_1(x_1)) + \sum_{j=1, j\neq 1}^{n} \phi_{1j}(x) \\ \vdots \\ -\sigma_{nn}(v_n(x_n)) + \sum_{j=1, j\neq n}^{n} \phi_{nj}(x) \end{bmatrix}, \quad x \in \overline{\mathbb{R}}_+^n, \quad (5.33)$$

where $\sigma_{ii} : \overline{\mathbb{R}}_+ \to \overline{\mathbb{R}}_+$, $i = 1, \ldots, n$, are positive definite functions, and note that $w(\cdot, x) \in \mathcal{W}$, $x \in \overline{\mathbb{R}}_+^n$, and $w(0, 0) = 0$. Also note that $\mathcal{R}_i \triangleq \{x \in \overline{\mathbb{R}}_+^n, x_i \neq 0 : V_i'(x_i)G_i(x) = 0\} = \{x \in \overline{\mathbb{R}}_+^n, x_i \neq 0 : x_i = 0\} = \emptyset$, and hence, condition (5.12) is satisfied for $V(\cdot)$ and $w(\cdot, \cdot)$ given by (5.32) and (5.33), respectively.

To show that the dynamical system

$$\dot{z}(t) = w(z(t), x(t)), \quad z(t_0) = z_0, \quad t \geq t_0, \quad (5.34)$$

where $z(t) \in \overline{\mathbb{R}}_+^n$, $t \geq t_0$, $x(t)$, $t \geq t_0$, is the solution to (5.31), and the ith component of $w(z, x)$ is given by $w_i(z, x) = -\sigma_{ii}(z_i) + \sum_{j=1, j\neq i}^{n} \phi_{ij}(x)$, $z \in \overline{\mathbb{R}}_+^n$, $x \in \overline{\mathbb{R}}_+^n$, is globally asymptotically stable with respect to z uniformly in x_0, consider the partial Lyapunov function candidate $\tilde{v}(z) = \mathbf{e}^{\mathrm{T}}z$, $z \in \overline{\mathbb{R}}_+^n$. Note that $\tilde{v}(\cdot)$ is radially unbounded, $\tilde{v}(0) = 0$, $\tilde{v}(z) > 0$, $z \in \overline{\mathbb{R}}_+^n$, $z \neq 0$, and $\dot{\tilde{v}}(z) = -\sum_{i=1}^{n} \sigma_{ii}(z_i) + \sum_{i=1}^{n} \sum_{j=1, j\neq i}^{n} \phi_{ij}(x) = -\sum_{i=1}^{n} \sigma_{ii}(z_i) < 0$, $z \in \overline{\mathbb{R}}_+^n$, $z \neq 0$. Thus, it follows from Corollary 1 of [41] that the dynamical system (5.34) and (5.31) is globally asymptotically stable with respect to z uniformly in x_0. Hence, it follows from Theorem 5.1 that $V(x)$, $x \in \overline{\mathbb{R}}_+^n$, given by (5.32) is a control vector Lyapunov function for the dynamical system (5.31).

Next, using (5.14) with $\alpha_i(x) = V_i'(x_i)f_i(x) = \sum_{j=1, j\neq i}^{n} \phi_{ij}(x)$, $\beta_i(x) = x_i$, $x \in \overline{\mathbb{R}}_+^n$, and $c_{0i} > 0$, $i = 1, \ldots, n$, we construct a globally stabilizing decentralized feedback controller for (5.31). It can be seen from the structure of the feedback control law that the closed-loop system dynamics are essentially nonnegative. Furthermore, since $\alpha_i(x) - w_i(V(x), x) = \sigma_{ii}(v_i(x_i))$, $x \in \overline{\mathbb{R}}_+^n$, $i = 1, \ldots, n$, this feedback controller is fully independent from $f(x)$, which represents the internal interconnections of the large-scale system dynamics and hence is robust against full modeling uncertainty in $f(x)$. Moreover, it follows from Theorem 5.2 that the dynamical system (5.31) with the feedback stabilizing control law (5.14) has a sector (and hence gain) margin $(\frac{1}{2}, \infty)$ in each decentralized input channel, and hence, additionally guarantees robustness to multiplicative input uncertainty. Finally, the feedback controller minimizes the derived cost functional given by (5.19).

For the following simulation we consider (5.31) with $\sigma_{ij}(x) = \sigma_{ij}x_ix_j$ and $\sigma_{ii}(x) = \sigma_{ii}x_i^2$, where $\sigma_{ij} \geq 0$, $i \neq j$, $i, j = 1, \ldots, n$, and $\sigma_{ii} > 0$, $i = 1, \ldots, n$. Note that in this case the conditions of Proposition 5.1 are satisfied, and hence, the feedback control law (5.14) is continuous on $\overline{\mathbb{R}}_+^n$. For our simulation we set $n = 3$, $\sigma_{11} = 0.1$, $\sigma_{22} = 0.2$, $\sigma_{33} = 0.01$, $\sigma_{12} = 2$,

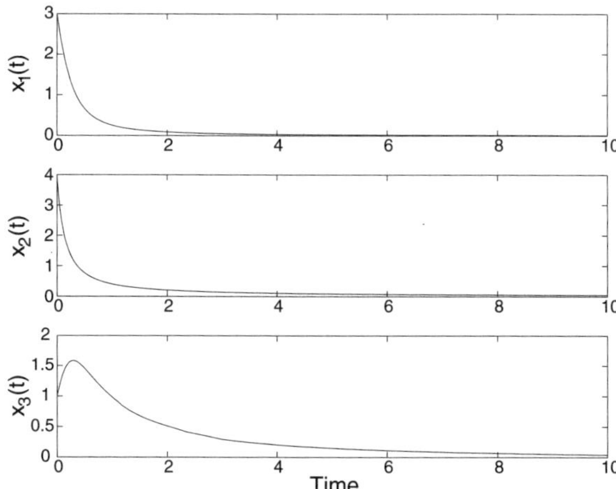

Figure 5.2 Controlled system states versus time.

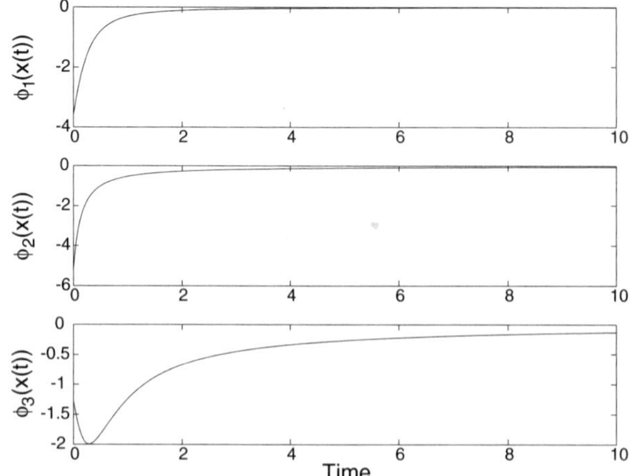

Figure 5.3 Control signals in each decentralized control channel versus time.

$\sigma_{13} = 3$, $\sigma_{21} = 1.5$, $\sigma_{23} = 0.3$, $\sigma_{31} = 4.4$, $\sigma_{32} = 0.6$, $c_{01} = 1$, $c_{02} = 1$, and $c_{03} = 0.25$, with initial condition $x_0 = [3, 4, 1]^{\mathrm{T}}$. Figure 5.2 shows the states of the closed-loop system versus time and Figure 5.3 shows control signals for each decentralized control channel as a function of time.

Finite-Time Stabilization of Large-Scale Systems via Control Vector Lyapunov Functions

6.1 Introduction

The notions of asymptotic and exponential stability in dynamical systems theory imply convergence of the system trajectories to an equilibrium state over the infinite horizon. In many applications, however, it is desirable that a dynamical system possesses the property that trajectories that converge to a Lyapunov stable equilibrium state must do so in finite time rather than merely asymptotically. Most of the existing control techniques in the literature ensure that the closed-loop system dynamics of a controlled system are Lipschitz continuous, which implies uniqueness of system solutions in forward and backward times. Hence, convergence to an equilibrium state is achieved over an infinite time interval.

In order to achieve convergence in finite time, the closed-loop system dynamics need to be non-Lipschitzian giving rise to non-uniqueness of solutions in backward time. Uniqueness of solutions in forward time, however, can be preserved in the case of finite-time convergence. Sufficient conditions that ensure uniqueness of solutions in forward time in the absence of Lipschitz continuity are given in [1, 58, 108, 176]. In addition, it is shown in [43, Theorem 4.3, p. 59] that uniqueness of solutions in forward time along with continuity of the system dynamics ensure that the system solutions are continuous functions of the system initial conditions even when the dynamics are not Lipschitz continuous.

Finite-time convergence to a Lyapunov stable equilibrium, that is, finite-time stability, was rigorously studied in [24, 25] using Hölder continuous Lyapunov functions. Finite-time stabilization of second-order systems was considered in [23, 86]. More recently, researchers have considered finite-time stabilization of higher-order systems [90] as well as finite-time stabilization using output feedback [91]. Design of globally strongly stabilizing continuous controllers for nonlinear systems using the theory of homogeneous systems was studied in [149]. Alternatively, discontinuous finite-time stabilizing feedback controllers have also been developed in the literature [59, 152, 153]. However, for practical implementations, discontinuous feedback controllers can lead to chattering due to system uncertainty or measurement noise, and hence, may excite unmodeled high-frequency system dynamics.

In this chapter, we develop a general framework for finite-time stability analysis of nonlinear dynamical systems using vector Lyapunov functions. Specifically, we construct a vector comparison system that is finite-time stable and, using the vector comparison principle [14, 118, 133, 141, 159], relate this finite-time stability property to the stability properties of the nonlinear dynamical system. We show that in the case of a scalar comparison system this result specializes to the result in [24]. Furthermore, we design universal finite-time stabilizing decentralized controllers for large-scale dynamical systems based on the newly proposed notion of a *control vector Lyapunov function* introduced in Chapter 5. In addition, we present necessary and sufficient conditions for continuity of such controllers. Moreover, we specialize these results to the case of a scalar Lyapunov function to obtain universal finite-time stabilizers for nonlinear systems that are affine in the control. Finally, we demonstrate the utility of the proposed framework on two numerical examples.

6.2 Finite-Time Stability via Vector Lyapunov Functions

Consider the nonlinear dynamical system given by

$$\dot{x}(t) = f(x(t)), \quad x(t_0) = x_0, \quad t \in \mathcal{I}_{x_0}, \tag{6.1}$$

where $x(t) \in \mathcal{D} \subseteq \mathbb{R}^n$, $t \in \mathcal{I}_{x_0}$, is the system state vector, $\mathcal{I}_{x_0}$ is the maximal interval of existence of a solution $x(t)$ of (6.1), $\mathcal{D}$ is an open set, $0 \in \mathcal{D}$, $f(0) = 0$, and $f(\cdot)$ is continuous on $\mathcal{D}$. A continuously differentiable function $x : \mathcal{I}_{x_0} \to \mathcal{D}$ is said to be a *solution* of (6.1) on the interval $\mathcal{I}_{x_0} \subset \mathbb{R}$ if $x(\cdot)$ satisfies (6.1) for all $t \in \mathcal{I}_{x_0}$. Recall that every bounded solution to (6.1) can be extended on a semi-infinite time interval $[0, \infty)$ [87]. We assume that (6.1) possesses unique solutions in forward time for all initial conditions except possibly the origin in the following sense. For every $x \in \mathcal{D} \backslash \{0\}$ there exists $\tau_x > 0$ such that, if $y_1 : [0, \tau_1) \to \mathcal{D}$ and $y_2 : [0, \tau_2) \to \mathcal{D}$ are two solutions of (6.1) with $y_1(0) = y_2(0) = x$, then $\tau_x \leq \min\{\tau_1, \tau_2\}$ and $y_1(t) = y_2(t)$ for all $t \in [0, \tau_x)$. Without loss of generality, we assume that for each x, τ_x is chosen to be the largest such number in $\overline{\mathbb{R}}_+$. In this case, we denote the *trajectory* or *solution curve* of (6.1) on $[0, \tau_x)$ satisfying the consistency property $s(0, x) = x$ and the semi-group property $s(t, s(\tau, x)) = s(t + \tau, x)$ for every $x \in \mathcal{D}$ and $t, \tau \in [0, \tau_x)$ by $s(\cdot, x)$ or $s^x(\cdot)$. Sufficient conditions for forward uniqueness in the absence of Lipschitz continuity can be found in [1], [58, Section 10], [108], and [176, Section 1].

The next definition introduces the notion of finite-time stability [24].

Definition 6.1. Consider the nonlinear dynamical system (6.1). The zero solution $x(t) \equiv 0$ to (6.1) is *finite-time stable* if there exist an open neighborhood $\mathcal{N} \subseteq \mathcal{D}$ of the origin and a function $T : \mathcal{N}\backslash\{0\} \to (0, \infty)$, called the *settling-time function*, such that the following statements hold:

i) *Finite-time convergence.* For every $x \in \mathcal{N}\backslash\{0\}$, $s^x(t)$ is defined on $[0, T(x))$, $s^x(t) \in \mathcal{N}\backslash\{0\}$ for all $t \in [0, T(x))$, and $\lim_{t \to T(x)} s(x, t) = 0$.

ii) *Lyapunov stability.* For every $\varepsilon > 0$ there exists $\delta > 0$ such that $\mathcal{B}_\delta(0) \subset \mathcal{N}$ and for every $x \in \mathcal{B}_\delta(0)\backslash\{0\}$, $s(t, x) \in \mathcal{B}_\varepsilon(0)$ for all $t \in [0, T(x))$.

The zero solution $x(t) \equiv 0$ of (6.1) is *globally finite-time stable* if it is finite-time stable with $\mathcal{N} = \mathcal{D} = \mathbb{R}^n$.

Note that if the zero solution $x(t) \equiv 0$ to (6.1) is finite-time stable, then it is asymptotically stable, and hence, finite-time stability is a stronger notion than asymptotic stability. Next, consider an example of a finite-time stable system with a continuous but non-Lipschitzian vector field.

Example 6.1 ([24]). Consider the scalar system

$$\dot{x}(t) = -k\operatorname{sign}(x(t))|x(t)|^\alpha, \quad x(0) = x_0, \quad t \geq 0, \tag{6.2}$$

where $x_0 \in \mathbb{R}$, $\operatorname{sign}(x) \triangleq \frac{x}{|x|}$, $x \neq 0$, $\operatorname{sign}(0) \triangleq 0$, $k > 0$, and $\alpha \in (0, 1)$. The right-hand side of (6.2) is continuous everywhere and locally Lipschitz everywhere except the origin. Hence, every initial condition in $\mathbb{R}\backslash\{0\}$ has a unique solution in forward time on a sufficiently small time interval. The solution to (6.2) is obtained by direct integration and is given by

$$s(t, x_0) = \begin{cases} \operatorname{sign}(x_0)\left[|x_0|^{1-\alpha} - k(1-\alpha)t\right]^{\frac{1}{1-\alpha}}, \\ \qquad\qquad t < \frac{1}{k(1-\alpha)}|x_0|^{1-\alpha}, \quad x_0 \neq 0, \\ 0, \qquad\quad t \geq \frac{1}{k(1-\alpha)}|x_0|^{1-\alpha}, \quad x_0 \neq 0, \\ 0, \qquad\quad t \geq 0, \quad x_0 = 0. \end{cases} \tag{6.3}$$

It is clear from (6.3) that $i)$ in Definition 6.1 is satisfied with $\mathcal{N} = \mathcal{D} = \mathbb{R}$ and with the settling-time function $T : \mathbb{R} \to \mathbb{R}_+$ given by

$$T(x_0) = \frac{1}{k(1-\alpha)}|x_0|^{1-\alpha}, \quad x_0 \in \mathbb{R}. \tag{6.4}$$

Lyapunov stability follows by considering the Lyapunov function $V(x) = x^2$, $x \in \mathbb{R}$. Thus, the zero solution $x(t) \equiv 0$ to (6.2) is globally finite-time stable. $\triangle$

Next, we present sufficient conditions for finite-time stability using a vector Lyapunov function involving a vector differential inequality.

Theorem 6.1. Consider the nonlinear dynamical system (6.1). Assume there exist a continuously differentiable vector function $V : \mathcal{D} \to$

$\mathcal{Q} \cap \overline{\mathbb{R}}_+^q$, where $\mathcal{Q} \subset \mathbb{R}^q$ and $0 \in \mathcal{Q}$, and a positive vector $p \in \mathbb{R}_+^q$ such that $V(0) = 0$, the scalar function $p^{\mathrm{T}} V(x)$, $x \in \mathcal{D}$, is positive definite, and

$$V'(x)f(x) \leqq w(V(x)), \quad x \in \mathcal{D}, \tag{6.5}$$

where $w : \mathcal{Q} \to \mathbb{R}^q$ is continuous, $w(\cdot) \in \mathcal{W}$, and $w(0) = 0$. In addition, assume that the vector comparison system

$$\dot{z}(t) = w(z(t)), \quad z(0) = z_0, \quad t \in \mathcal{I}_{z_0}, \tag{6.6}$$

has a unique solution in forward time $z(t)$, $t \in \mathcal{I}_{z_0}$, and there exist a continuously differentiable function $v : \mathcal{Q} \to \mathbb{R}$, real numbers $c > 0$ and $\alpha \in (0,1)$, and a neighborhood $\mathcal{M} \subseteq \mathcal{Q}$ of the origin such that $v(\cdot)$ is positive definite and

$$v'(z)w(z) \leq -c(v(z))^\alpha, \quad z \in \mathcal{M}. \tag{6.7}$$

Then the zero solution $x(t) \equiv 0$ to (6.1) is finite-time stable. Moreover, if $\mathcal{N}$ is as in Definition 6.1 and $T : \mathcal{N} \to [0,\infty)$ is the settling-time function, then

$$T(x_0) \leq \frac{1}{c(1-\alpha)}(v(V(x_0)))^{1-\alpha}, \quad x_0 \in \mathcal{N}, \tag{6.8}$$

and $T(\cdot)$ is continuous on $\mathcal{N}$. If, in addition, $\mathcal{D} = \mathbb{R}^n$, $v(\cdot)$ is radially unbounded, and (6.7) holds on $\mathbb{R}^q$, then the zero solution $x(t) \equiv 0$ to (6.1) is globally finite-time stable.

Proof. Note that $p^{\mathrm{T}} V(x) \leq \max_{i=1,\ldots,q}\{p_i\} \mathbf{e}^{\mathrm{T}} V(x)$, $x \in \mathcal{D}$. Hence, since $p^{\mathrm{T}} V(x)$, $x \in \mathcal{D}$, is positive definite, that is, $p^{\mathrm{T}} V(0) = 0$ and $p^{\mathrm{T}} V(x) > 0$, $x \neq 0$, it follows that the function $\mathbf{e}^{\mathrm{T}} V(x)$, $x \in \mathcal{D}$, is also positive definite.

Let $\mathcal{V} \subseteq \mathcal{M}$ be a bounded open set such that $0 \in \mathcal{V}$ and $\overline{\mathcal{V}} \subset \mathcal{Q}$. Then $\partial\mathcal{V}$ is compact and $0 \notin \partial\mathcal{V}$. Now, it follows from Weierstrass' theorem [70] that the continuous function $v(\cdot)$ attains a minimum on $\partial\mathcal{V}$ and since $v(\cdot)$ is positive definite, $\min_{z \in \partial\mathcal{V}} v(z) > 0$. Let $0 < \beta < \min_{z \in \partial\mathcal{V}} v(z)$ and $\mathcal{D}_\beta \triangleq \{z \in \mathcal{V} : v(z) \leq \beta\}$. It follows from (6.7) that $\mathcal{D}_\beta \subset \mathcal{M}$ is invariant with respect to (6.6). Furthermore, it follows from (6.7), the positive definiteness of $v(\cdot)$, and standard Lyapunov arguments that for every $\hat{\varepsilon} > 0$, there exists $\hat{\delta} > 0$ such that $\mathcal{B}_{\hat{\delta}}(0) \subset \mathcal{D}_\beta \subset \mathcal{M}$ and

$$\|z(t)\|_1 \leq \hat{\varepsilon}, \quad \|z_0\|_1 < \hat{\delta}, \tag{6.9}$$

where $\|\cdot\|_1$ denotes the absolute sum norm, $\mathcal{B}_{\hat{\delta}}(0)$ is defined in terms of the absolute sum norm $\|\cdot\|_1$, and $t \in \mathcal{I}_{z_0}$. Moreover, since the solution $z(t)$ to (6.6) is bounded for all $t \in \mathcal{I}_{z_0}$, it can be extended on the semi-infinite interval $[0,\infty)$ [87], and hence, $z(t)$ is defined for all $t \geq 0$.

Furthermore, it follows from Corollary 2.2 with $q = 1$, $w(y,x) = w(y) = -cy^\alpha$, and $z(t) = s(t, v(z_0))$, where $\alpha \in (0,1)$, that

$$v(z(t)) \leq s(t, v(z_0)), \quad z_0 \in \mathcal{B}_{\hat{\delta}}(0), \quad t \in [0,\infty), \tag{6.10}$$

where $s(\cdot,\cdot)$ is given by (6.3) with $k = c$. Now, it follows from (6.3), (6.10), and the positive definiteness of $v(\cdot)$ that

$$z(t) = 0, \quad t \geq \frac{1}{c(1-\alpha)}(v(z_0))^{1-\alpha}, \quad z_0 \in \mathcal{B}_{\hat{\delta}}(0), \tag{6.11}$$

which implies finite-time convergence of the trajectories of (6.6) for all $z_0 \in \mathcal{B}_{\hat{\delta}}(0)$. This along with (6.9) implies finite-time stability of the zero solution $z(t) \equiv 0$ to (6.6).

Next, it follows from the continuity of $V(\cdot)$ that there exists $\delta_1 > 0$ such that $\|V(x_0)\|_1 < \hat{\delta}$ for all $\|x_0\| < \delta_1$, where $\|\cdot\|$ is the Euclidian norm on $\mathbb{R}^n$. Now, choose $z_0 = V(x_0) \in \mathcal{B}_{\hat{\delta}}(0)$ for all $\|x_0\| < \delta_1$. In this case, it follows from (6.5) and Corollary 2.2 that $V(x(t)) \leq\leq z(t)$ on a compact interval $[0, \tau_{x_0}]$, where $[0, \tau_{x_0})$ is the maximal interval of existence of the solution $x(t)$ to (6.1). Since $z(t)$, $t \geq 0$, is bounded and $\mathbf{e}^{\mathrm{T}} V(\cdot)$ is positive definite it follows that $x(t)$, $t \in [0, \tau_{x_0}]$, is bounded, and hence, $x(t)$ can be extended to the semi-infinite interval $[0, \infty)$. Using (6.11) it follows that

$$\mathbf{e}^{\mathrm{T}} V(x(t)) = \mathbf{e}^{\mathrm{T}} z(t) = 0, \quad t \geq \frac{1}{c(1-\alpha)}(v(z_0))^{1-\alpha}, \quad z_0 = V(x_0) \in \mathcal{B}_{\hat{\delta}}(0). \tag{6.12}$$

Since $\mathbf{e}^{\mathrm{T}} V(\cdot)$ is positive definite, it follows that

$$x(t) = 0, \quad t \geq \frac{1}{c(1-\alpha)}(v(V(x_0)))^{1-\alpha}, \quad \|x_0\| < \delta_1, \tag{6.13}$$

which implies finite-time convergence of the trajectories of (6.1) for all $\|x_0\| < \delta_1$. Furthermore, it follows from (6.13) that the settling-time function satisfies

$$T(x_0) \leq \frac{1}{c(1-\alpha)}(v(V(x_0)))^{1-\alpha}, \quad \|x_0\| < \delta_1. \tag{6.14}$$

Next, note that since $\mathbf{e}^{\mathrm{T}} V(x)$, $x \in \mathcal{D}$, is positive definite, there exist $r > 0$ and class $\mathcal{K}$ functions [85] $\alpha, \beta : [0, r] \to \overline{\mathbb{R}}_+$ such that $\mathcal{B}_r(0) \subset \mathcal{D}$, where $\mathcal{B}_r(0)$ is defined in terms of the Euclidean norm $\|\cdot\|$, and

$$\alpha(\|x\|) \leq \mathbf{e}^{\mathrm{T}} V(x) \leq \beta(\|x\|), \quad x \in \mathcal{B}_r(0). \tag{6.15}$$

Let $\varepsilon > 0$ and choose $0 < \hat{\varepsilon} < \min\{\varepsilon, r\}$. In this case, it follows from the Lyapunov stability of the nonlinear vector comparison system (6.6) that there exists $\mu = \mu(\hat{\varepsilon}) = \mu(\varepsilon) > 0$ such that if $\|z_0\|_1 < \mu$, then $\|z(t)\|_1 < \alpha(\hat{\varepsilon})$, $t \geq 0$. Now, choose $z_0 = V(x_0) \geq\geq 0$, $x_0 \in \mathcal{D}$. Since $V(x)$, $x \in \mathcal{D}$, is continuous, $\mathbf{e}^{\mathrm{T}} V(x)$, $x \in \mathcal{D}$, is also continuous. Hence, for $\mu = \mu(\hat{\varepsilon}) > 0$ there exists $\delta = \delta(\mu(\hat{\varepsilon})) = \delta(\varepsilon) > 0$ such that $\delta < \min\{\delta_1, \hat{\varepsilon}\}$, and if $\|x_0\| < \delta$, then $\mathbf{e}^{\mathrm{T}} V(x_0) = \mathbf{e}^{\mathrm{T}} z_0 = \|z_0\|_1 < \mu$, which implies that $\|z(t)\|_1 < \alpha(\hat{\varepsilon})$, $t \geq 0$. Now, with $z_0 = V(x_0) \geq\geq 0$, $x_0 \in \mathcal{D}$, and the assumption that $w(\cdot) \in \mathcal{W}$, it follows from (6.5) and Corollary 2.2 that $0 \leq\leq V(x(t)) \leq\leq z(t)$ on any compact interval $[0, \tau]$, and hence, $\mathbf{e}^{\mathrm{T}} z(t) = \|z(t)\|_1$, $t \in [0, \tau]$.

Let $\tau > 0$ be such that $x(t) \in \mathcal{B}_r(0)$, $t \in [0, \tau]$, for all $x_0 \in \mathcal{B}_\delta(0)$. Thus, using (6.15), if $\|x_0\| < \delta$, then

$$\alpha(\|x(t)\|) \leq \mathbf{e}^{\mathrm{T}} V(x(t)) \leq \mathbf{e}^{\mathrm{T}} z(t) < \alpha(\hat{\varepsilon}), \quad t \in [0, \tau], \qquad (6.16)$$

which implies $\|x(t)\| < \hat{\varepsilon} < \varepsilon$, $t \in [0, \tau]$. Now, suppose, *ad absurdum*, that for some $x_0 \in \mathcal{B}_\delta(0)$ there exists $\hat{t} > \tau$ such that $\|x(\hat{t})\| = \hat{\varepsilon}$. Then, for $z_0 = V(x_0)$ and the compact interval $[0, \hat{t}]$ it follows from (6.5) and Corollary 2.2 that $V(x(\hat{t})) \leq\leq z(\hat{t})$, which implies that $\alpha(\hat{\varepsilon}) = \alpha(\|x(\hat{t})\|) \leq \mathbf{e}^{\mathrm{T}} V(x(\hat{t})) \leq \mathbf{e}^{\mathrm{T}} z(\hat{t}) < \alpha(\hat{\varepsilon})$, leading to a contradiction. Hence, for a given $\varepsilon > 0$ there exists $\delta = \delta(\varepsilon) > 0$ such that for all $x_0 \in \mathcal{B}_\delta(0)$, $\|x(t)\| < \varepsilon$, $t \geq t_0$, which implies Lyapunov stability of the zero solution $x(t) \equiv 0$ to (6.1). This, along with (6.13), implies finite-time stability of the zero solution $x(t) \equiv 0$ to (6.1) with $\mathcal{N} \triangleq \mathcal{B}_\delta(0)$. Equation (6.8) implies that $T(\cdot)$ is continuous at the origin, and hence, by Proposition 2.4 of [24], continuous on $\mathcal{N}$.

Finally, if $\mathcal{D} = \mathbb{R}^n$ and $v(\cdot)$ is radially unbounded, then global finite-time stability follows using standard arguments. $\square$

Assume the conditions of Theorem 6.1 are satisfied with $q = 1$. In this case, there exists a continuously differentiable, positive definite function $V : \mathcal{D} \to \mathcal{Q} \cap \overline{\mathbb{R}}_+$ such that (6.5) holds, and there exists a continuously differentiable, positive definite function $v : \mathcal{Q} \to \overline{\mathbb{R}}_+$ such that (6.7) holds. Since $q = 1$ and $\mathcal{M}$ is a neighborhood of the origin, it follows that there exists $\gamma > 0$ such that $[0, \gamma] \subset \mathcal{M}$. Furthermore, since $v(\cdot)$ is positive definite, there exists $\beta > 0$ such that $v'(z) \geq 0$ for all $z \in [0, \beta]$. Next, consider the function $\tilde{v}(x) \triangleq v(V(x))$, $x \in \mathcal{D}$, and note that $\tilde{v}(\cdot)$ is positive definite. Define $\mathcal{V} \triangleq \{x \in \mathcal{D} : V(x) \leq \min\{\beta, \gamma\}\}$. Then it follows from (6.5) and (6.7) that

$$\begin{aligned}
\dot{\tilde{v}}(x) &= v'(V(x)) V'(x) f(x) \\
&\leq v'(V(x)) w(V(x)) \\
&\leq -c(v(V(x)))^\alpha \\
&= -c(\tilde{v}(x))^\alpha, \quad x \in \mathcal{V}, \qquad (6.17)
\end{aligned}$$

which implies condition (4.7) in Theorem 4.2 of [24]. Thus, in the case where $q = 1$, Theorem 6.1 specializes to Theorem 4.2 of [24].

The next result is a specialization of Theorem 6.1 to the case where the structure of the comparison dynamics directly guarantees finite-time stability of the comparison system. That is, there is *no* need to require the existence of a scalar function $v(\cdot)$ such that (6.7) holds to guarantee finite-time stability of the nonlinear dynamical system (6.1).

Corollary 6.1. Consider the nonlinear dynamical system (6.1). Assume there exist a continuously differentiable vector function $V : \mathcal{D} \to$

$\mathcal{Q} \cap \overline{\mathbb{R}}_+^q$, where $\mathcal{Q} \subset \mathbb{R}^q$ and $0 \in \mathcal{Q}$, and a positive vector $p \in \mathbb{R}_+^q$ such that $V(0) = 0$, the scalar function $p^{\mathrm{T}} V(x)$, $x \in \mathcal{D}$, is positive definite, and

$$V'(x) f(x) \leq\leq W(V(x))^{[\alpha]}, \quad x \in \mathcal{D}, \tag{6.18}$$

where $\alpha \in (0,1)$, $W \in \mathbb{R}^{q \times q}$ is essentially nonnegative and Hurwitz, and $(V(x))^{[\alpha]} \triangleq [(V_1(x))^\alpha, \ldots, (V_q(x))^\alpha]^{\mathrm{T}}$. Then the zero solution $x(t) \equiv 0$ to (6.1) is finite-time stable. If, in addition, $\mathcal{D} = \mathbb{R}^n$, then the zero solution $x(t) \equiv 0$ to (6.1) is globally finite-time stable.

Proof. Consider the comparison system given by

$$\dot{z}(t) = W(z(t))^{[\alpha]}, \quad z(0) = z_0, \quad t \geq 0, \tag{6.19}$$

where $z_0 \in \overline{\mathbb{R}}_+^q$. Note that the right-hand side in (6.19) is of class $\mathcal{W}$ and is essentially nonnegative and, hence, the solutions to (6.19) are nonnegative for all nonnegative initial conditions [69]. Since $W \in \mathbb{R}^{q \times q}$ is essentially nonnegative and Hurwitz, it follows from Lemma 2.1 that there exist positive vectors $\hat{p} \in \mathbb{R}_+^q$ and $r \in \mathbb{R}_+^q$ such that

$$0 = W^{\mathrm{T}} \hat{p} + r. \tag{6.20}$$

Now, consider the Lyapunov function $v(z) = \hat{p}^{\mathrm{T}} z$, $z \in \overline{\mathbb{R}}_+^q$. Note that $v(0) = 0$, $v(z) > 0$, $z \in \overline{\mathbb{R}}_+^q$, $z \neq 0$, and $v(\cdot)$ is radially unbounded. Let $\beta \triangleq \min_{i=1,\ldots,q} r_i$, $\gamma \triangleq \max_{i=1,\ldots,q} \hat{p}_i^\alpha$, where r_i and $\hat{p}_i$ are the ith components of $r \in \mathbb{R}_+^q$ and $\hat{p} \in \mathbb{R}_+^q$, respectively. Then

$$\begin{aligned}
\dot{v}(z) &= \hat{p}^{\mathrm{T}} W z^{[\alpha]} \\
&= -r^{\mathrm{T}} z^{[\alpha]} \\
&\leq -\frac{\beta}{\gamma} \gamma \left(\sum_{i=1}^q z_i^\alpha \right) \\
&\leq -\frac{\beta}{\gamma} \left(\sum_{i=1}^q \hat{p}_i^\alpha z_i^\alpha \right) \\
&\leq -\frac{\beta}{\gamma} \left(\sum_{i=1}^q \hat{p}_i z_i \right)^\alpha \\
&\leq -\frac{\beta}{\gamma} (v(z))^\alpha \\
&= -c(v(z))^\alpha, \quad z \in \overline{\mathbb{R}}_+^q, \tag{6.21}
\end{aligned}$$

where $c \triangleq \frac{\beta}{\gamma}$. Thus, it follows from Theorem 4.2 of [24] that the comparison system (6.19) is finite-time stable with the settling-time function $T(z_0) \leq \frac{1}{c(1-\alpha)} (v(z_0))^{1-\alpha}$, $z_0 \in \overline{\mathbb{R}}_+^q$. Next, it follows from Corollary 4.1 of [141] that the nonlinear dynamical system (6.1) is asymptotically stable with the domain of attraction $\mathcal{N} \subset \mathcal{D}$. Now, the result is a direct consequence of Theorem 6.1. $\qquad\square$

It follows from Corollary 6.1 that the nonlinear dynamical system (6.1) has a settling-time function $T(x_0) \le \frac{1}{c(1-\alpha)}(v(V(x_0)))^{1-\alpha}$, $x_0 \in \mathcal{N}$, where $v(z) = \hat{p}^{\mathrm{T}}z$, $z \in \overline{\mathbb{R}}_+^q$, and $\hat{p} \in \mathbb{R}_+^q$ is as in the proof of Corollary 6.1.

6.3 Finite-Time Stabilization of Large-Scale Dynamical Systems

In Chapter 5, the notion of a *control vector Lyapunov function* was introduced as a generalization of the classical notion of a control Lyapunov function. Furthermore, a universal stabilizing feedback control law was constructed based on a control vector Lyapunov function. In this section, we show that this control law can be used to stabilize large-scale dynamical systems in finite time provided that the comparison system possesses non-Lipschitzian dynamics.

Specifically, consider the large-scale dynamical system composed of q interconnected subsystems given by

$$\dot{x}_i(t) = f_i(x(t)) + G_i(x(t))u_i(t), \quad t \ge t_0, \quad i = 1, \ldots, q, \qquad (6.22)$$

where $f_i : \mathbb{R}^n \to \mathbb{R}^{n_i}$ satisfying $f_i(0) = 0$ and $G_i : \mathbb{R}^n \to \mathbb{R}^{n_i \times m_i}$ are continuous functions for all $i = 1, \ldots, q$, and $u_i(\cdot)$, $i = 1, \ldots, q$, satisfy sufficient regularity conditions such that the nonlinear dynamical system (6.22) has a unique solution forward in time. Let $V = [V_1, \ldots, V_q]^{\mathrm{T}} : \mathbb{R}^n \to \overline{\mathbb{R}}_+^q$ be a component decoupled continuously differentiable vector function, that is, $V(x) = [V_1(x_1), \ldots, V_q(x_q)]^{\mathrm{T}}$, $x \in \mathbb{R}^n$, $p \in \mathbb{R}_+^q$ be a positive vector, and $w : \overline{\mathbb{R}}_+^q \to \mathbb{R}^q$ be a continuous function such that $V(0) = 0$, $p^{\mathrm{T}}V(x)$, $x \in \mathbb{R}^n$, is positive definite, and $w(\cdot) \in \mathcal{W}$ with $w(0) = 0$.

Define $\alpha_i(x) \triangleq V_i'(x_i)f_i(x)$, $x \in \mathbb{R}^n$, and $\beta_i(x) \triangleq G_i^{\mathrm{T}}(x)V_i'^{\mathrm{T}}(x_i)$, $x \in \mathbb{R}^n$, and assume that

$$V_i'(x_i)f_i(x) < w_i(V(x)), \quad x \in \mathcal{R}_i, \quad i = 1, \ldots, q, \qquad (6.23)$$

where $\mathcal{R}_i \triangleq \{x \in \mathbb{R}^n, x \ne 0 : \beta_i(x) = 0\}$, $i = 1, \ldots, q$. Construct the feedback control law $\phi(x) = [\phi_1^{\mathrm{T}}(x), \ldots, \phi_q^{\mathrm{T}}(x)]^{\mathrm{T}}$, $x \in \mathbb{R}^n$, given by

$$\phi_i(x) = \begin{cases} -\left(c_{0i} + \dfrac{(\alpha_i(x) - w_i(V(x))) + \sqrt{(\alpha_i(x) - w_i(V(x)))^2 + (\beta_i^{\mathrm{T}}(x)\beta_i(x))^2}}{\beta_i^{\mathrm{T}}(x)\beta_i(x)}\right)\beta_i(x), \\ \qquad\qquad\qquad\qquad\qquad\qquad\qquad\qquad\quad \text{if } \beta_i(x) \ne 0, \\ \qquad\qquad\qquad 0, \qquad\qquad\qquad\qquad\qquad \text{if } \beta_i(x) = 0, \end{cases}$$

$$(6.24)$$

where $c_{0i} > 0$, $i = 1, \ldots, q$.

The vector Lyapunov derivative components $\dot{V}_i(\cdot)$, $i = 1, \ldots, q$, along the trajectories of the closed-loop dynamical system (6.22), with $u = \phi(x)$,

$x \in \mathbb{R}^n$, given by (6.24), satisfy

$$\begin{aligned}
\dot{V}_i(x_i) &= V_i'(x_i)[f_i(x) + G_i(x)\phi_i(x)] \\
&= \alpha_i(x) + \beta_i^{\mathrm{T}}(x)\phi_i(x) \\
&= \begin{cases}
-c_{0i}\beta_i^{\mathrm{T}}(x)\beta_i(x) - \sqrt{(\alpha_i(x) - w_i(V(x)))^2 + (\beta_i^{\mathrm{T}}(x)\beta_i(x))^2} \\
\quad +w_i(V(x)), & \text{if } \beta_i(x) \neq 0, \\
\alpha_i(x), & \text{if } \beta_i(x) = 0,
\end{cases} \\
&< w_i(V(x)), \quad x \in \mathbb{R}^n.
\end{aligned}$$

It follows from Theorem 6.1 that if there exist $v : \overline{\mathbb{R}}_+^q \to \mathbb{R}$, $c > 0$, and $\alpha \in (0, 1)$ such that $v(\cdot)$ is positive definite and

$$v'(z)w(z) \leq -c(v(z))^\alpha, \quad z \in \mathcal{M}, \tag{6.25}$$

where $\mathcal{M}$ is a neighborhood of $\overline{\mathbb{R}}_+^q$ containing the origin, then the zero solution $x(t) \equiv 0$ to (6.22) is finite-time stable with the settling time $T(x_0) \leq \frac{1}{c(1-\alpha)}(v(V(x_0)))^{1-\alpha}$, $x_0 \in \mathbb{R}^n$. In this case, it follows from Theorem 5.1 that $V(x)$, $x \in \mathbb{R}^n$, is a control vector Lyapunov function.

If $\mathcal{R}_i = \varnothing$, $i = 1, \ldots, q$, then the function $w(\cdot)$ in (6.24) can be chosen to be

$$w(z) = W z^{[\alpha]}, \quad z \in \overline{\mathbb{R}}_+^q, \tag{6.26}$$

where $W \in \mathbb{R}^{q \times q}$ is essentially nonnegative and Hurwitz, $\alpha \in (0, 1)$, and $z^{[\alpha]} \triangleq [z_1^\alpha, \ldots, z_q^\alpha]^{\mathrm{T}}$. In this case, condition (6.25) need *not* be verified and it follows from Corollary 6.1 that the closed-loop system (6.22) and (6.24) with $w(\cdot)$ given by (6.26) is finite-time stable and, hence, the controller (6.24) is finite-time stabilizing controller for (6.22).

Since $f_i(\cdot)$ and $G_i(\cdot)$ are continuous and $V_i(\cdot)$ is continuously differentiable for all $i = 1, \ldots, q$, it follows that $\alpha_i(x)$ and $\beta_i(x)$, $x \in \mathbb{R}^n$, $i = 1, \ldots, q$, are continuous functions, and hence, $\phi_i(x)$ given by (6.24) is continuous for all $x \in \mathbb{R}^n$ if either $\beta_i(x) \neq 0$ or $\alpha_i(x) - w_i(V(x)) < 0$ for all $i = 1, \ldots, q$. Hence, the feedback control law given by (6.24) is continuous everywhere except for the origin. However, as shown in Proposition 5.1, the feedback control law $\phi(x)$ given by (6.24) is continuous on $\mathbb{R}^n$ if and only if for every $\varepsilon > 0$, there exists $\delta > 0$ such that for all $0 < \|x\| < \delta$ there exists $u_i \in \mathbb{R}^{m_i}$ such that $\|u_i\| < \varepsilon$ and $\alpha_i(x) + \beta_i^{\mathrm{T}}(x)u_i < w_i(V(x))$, $i = 1, \ldots, q$.

The following corollary addressing the case where $q = 1$ is immediate from the above arguments. In this case, the nonlinear dynamical system (6.22) specializes to

$$\dot{x}(t) = f(x(t)) + G(x(t))u(t), \quad x(t_0) = x_0, \quad t \geq t_0, \tag{6.27}$$

where $x_0 \in \mathbb{R}^n$ and $f : \mathbb{R}^n \to \mathbb{R}^n$ satisfying $f(0) = 0$ and $G : \mathbb{R}^n \to \mathbb{R}^{n \times m}$ are continuous functions.

Corollary 6.2. Consider the nonlinear dynamical system (6.27). Assume there exists a continuously differentiable function $V : \mathcal{D} \to \overline{\mathbb{R}}_+$ such that $V(\cdot)$ is positive definite, $w(V(x)) \triangleq -c(V(x))^\alpha$, $x \in \mathbb{R}^n$, and

$$V'(x)f(x) \leq w(V(x)) = -c(V(x))^\alpha, \quad x \in \mathcal{R}, \tag{6.28}$$

where $c > 0$, $\alpha \in (0,1)$, $\mathcal{R} \triangleq \{x \in \mathbb{R}^n, x \neq 0 : V'(x)G(x) = 0\}$. Then the nonlinear dynamical system (6.27) with the feedback controller $u = \phi(x)$, $x \in \mathbb{R}^n$, given by

$$\phi(x) = \begin{cases} -\left(c_0 + \dfrac{(\alpha(x)-w(V(x)))+\sqrt{(\alpha(x)-w(V(x)))^2+(\beta^{\mathrm{T}}(x)\beta(x))^2}}{\beta^{\mathrm{T}}(x)\beta(x)}\right)\beta(x), \\ \qquad\qquad\qquad\qquad\qquad\qquad\qquad \text{if } \beta(x) \neq 0, \\ \qquad\qquad 0, \qquad\qquad\qquad\qquad\quad \text{if } \beta(x) = 0, \end{cases} \tag{6.29}$$

where $c_0 > 0$, $\alpha(x) \triangleq V'(x)f(x)$, $x \in \mathbb{R}^n$, and $\beta(x) \triangleq G^{\mathrm{T}}(x)V'^{\mathrm{T}}(x)$, $x \in \mathbb{R}^n$, is finite-time stable with the settling time $T(x_0) \leq \frac{1}{c(1-\alpha)}(V(x_0))^{1-\alpha}$, $x_0 \in \mathbb{R}^n$. Furthermore, $V(\cdot)$ is a control Lyapunov function.

Next, we show that the control law (6.29) ensures finite-time stability for a perturbed version of (6.27) with bounded perturbations. Specifically, consider the more accurate description of the system (6.27) given by the perturbed model

$$\dot{x}(t) = f(x(t)) + G(x(t))u(t) + g(t, x(t)), \quad x(t_0) = x_0, \quad t \geq t_0, \tag{6.30}$$

where $g : [t_0, \infty) \times \mathbb{R}^n \to \mathbb{R}^n$ is a continuous function that captures disturbances, uncertainties, parameter variations, or modeling errors. Assume that there exists a continuously differentiable function $V : \mathbb{R}^n \to \overline{\mathbb{R}}_+$ such that the conditions of Corollary 6.2 are satisfied. Then it follows from Theorem 5.2 of [24] that there exist $\delta_0 > 0$, $\ell > 0$, $\tau > 0$, and an open neighborhood $\mathcal{V}$ of the origin such that for every continuous function $g(\cdot, \cdot)$ with

$$\delta = \sup_{[t_0,\infty) \times \mathbb{R}^n} \|g(t, x)\| < \delta_0, \tag{6.31}$$

the solutions $x(t)$, $t \geq t_0$, to the closed-loop system (6.30) with $u(t)$ given by (6.29) and $x_0 \in \mathcal{V}$ are such that $x(t) \in \mathcal{V}$, $t \geq t_0$, and

$$\|x(t)\| \leq \ell\delta^\gamma, \quad t \geq \tau, \tag{6.32}$$

where $\gamma = \frac{1-\alpha}{\alpha}$.

Note that, if in Corollary 6.2 $\alpha \in (0, \frac{1}{2})$, then $\gamma > 1$, which makes the bound in (6.32) smaller for sufficiently small δ compared to the case when $0 < \gamma < 1$. In addition, if $g(\cdot, \cdot)$ is such that

$$\|g(t, x)\| \leq L\|x\|, \quad (t, x) \in [t_0, \infty) \times \mathbb{R}^n, \tag{6.33}$$

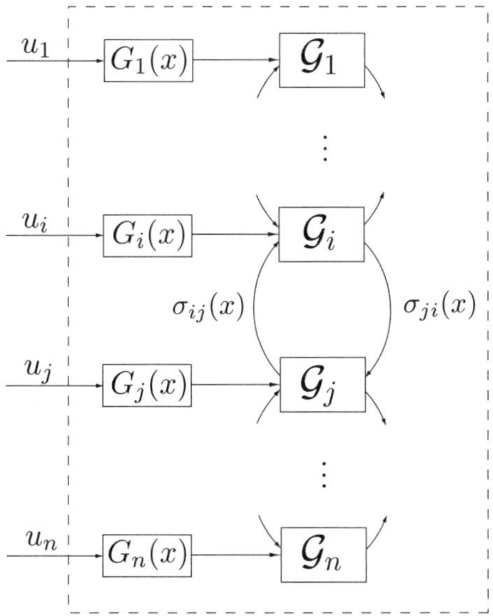

Figure 6.1 Large-scale dynamical system $\mathcal{G}$.

where $L \geq 0$, then it follows from Theorem 5.3 of [24] that $x(t) = 0$, $t \geq \tau$, for all $x_0 \in \mathcal{V}$. Finally, if $g : \mathbb{R}^n \to \mathbb{R}^n$ is only a function of the dynamical system state and

$$\|g(x)\| \leq L\|x\|, \quad x \in \mathbb{R}^n, \tag{6.34}$$

where $L \geq 0$, then it follows from Theorem 5.4 of [24] that the zero solution $x(t) \equiv 0$ to the closed-loop system (6.30) with $u(t)$ given by (6.29) is finite-time stable.

Next, consider the large-scale dynamical system $\mathcal{G}$ shown in Figure 6.1 involving energy exchange between n interconnected subsystems. Let $x_i : [0, \infty) \to \overline{\mathbb{R}}_+$ denote the energy (and hence a nonnegative quantity) of the ith subsystem, let $u_i : [0, \infty) \to \mathbb{R}$ denote the control input to the ith subsystem, and let $\sigma_{ij} : \overline{\mathbb{R}}_+^n \to \overline{\mathbb{R}}_+$, $i \neq j$, $i, j = 1, \ldots, n$, denote the instantaneous rate of energy flow from the jth subsystem to the ith subsystem.

An energy balance yields the large-scale dynamical system [81]

$$\dot{x}(t) = f(x(t)) + G(x(t))u(t), \quad x(t_0) = x_0, \quad t \geq t_0, \tag{6.35}$$

where $x(t) = [x_1(t), \ldots, x_n(t)]^{\mathrm{T}}$, $t \geq t_0$, $f_i(x) = \sum_{j=1, j \neq i}^{n} \phi_{ij}(x)$, where $\phi_{ij}(x) \triangleq \sigma_{ij}(x) - \sigma_{ji}(x)$, $x \in \overline{\mathbb{R}}_+^n$, $i \neq j$, $i, j = 1, \ldots, q$, denotes the net energy flow from the jth subsystem to the ith subsystem, $G(x) = \mathrm{diag}[G_1(x_1), \ldots, G_n(x_n)]$, $x \in \overline{\mathbb{R}}_+^n$, $G_i(x_i) = 0$ if and only if $x_i = 0$ for all $i = 1, \ldots, n$, and $u(t) \in \mathbb{R}^n$, $t \geq t_0$. Here, we assume that $\sigma_{ij}(x) = 0$, $x \in \overline{\mathbb{R}}_+^n$, whenever $x_j = 0$, $i \neq j$, $i, j = 1, \ldots, n$. In this case, $f(\cdot)$ is *essentially nonnegative* (i.e., $f_i(x) \geq 0$ for all $x \in \overline{\mathbb{R}}_+^n$ such that $x_i = 0$,

$i = 1, \ldots, n$). The above constraint implies that if the energy of the jth subsystem of $\mathcal{G}$ is zero, then this subsystem cannot supply any energy to its surroundings. In addition, we assume that $\phi_{ij}(x') \leq \phi_{ij}(x'')$, $i \neq j$, $i, j = 1, \ldots, n$, for all $x', x'' \in \mathbb{R}^n$ such that $x_i' = x_i''$ and $x_k' \leq x_k''$, $k \neq i$, where x_i is the ith component of x. The above assumption implies that the more energy the surroundings of the ith subsystem possess, the more energy is gained by the ith subsystem from the energy exchange due to subsystem interconnections. Finally, to ensure that the trajectories of the closed-loop system remain in the nonnegative orthant of the state space for all nonnegative initial conditions, we seek a feedback control law $u(\cdot)$ that guarantees the closed-loop system dynamics are essentially nonnegative [69].

For the dynamical system $\mathcal{G}$, consider the control vector Lyapunov function candidate $V(x) = [V_1(x_1), \ldots, V_n(x_n)]^{\mathrm{T}}$, $x \in \overline{\mathbb{R}}_+^n$, given by

$$V(x) = [x_1, \ldots, x_n]^{\mathrm{T}}, \quad x \in \overline{\mathbb{R}}_+^n. \tag{6.36}$$

Note that $V(0) = 0$ and $\mathbf{e}^{\mathrm{T}}V(x)$, $x \in \overline{\mathbb{R}}_+^n$, is positive definite and radially unbounded. Furthermore, consider the function

$$w(V(x)) = \begin{bmatrix} -V_1^{1/2}(x_1) + \sum_{j=1, j \neq 1}^n \phi_{1j}(V(x)) \\ \vdots \\ -V_n^{1/2}(x_n) + \sum_{j=1, j \neq n}^n \phi_{nj}(V(x)) \end{bmatrix}, \quad x \in \overline{\mathbb{R}}_+^n, \tag{6.37}$$

and note that it follows from the above constraints that $w(\cdot) \in \mathcal{W}$ and $w(0) = 0$. Furthermore, note that $\mathcal{R}_i \triangleq \{x \in \overline{\mathbb{R}}_+^n, x_i \neq 0 : V_i'(x_i)G_i(x_i) = 0\} = \{x \in \overline{\mathbb{R}}_+^n, x_i \neq 0 : x_i = 0\} = \varnothing$, and hence, condition (6.23) is satisfied for $V(\cdot)$ and $w(\cdot)$ given by (6.36) and (6.37), respectively.

Next, consider the vector comparison system

$$\dot{z}(t) = w(z(t)), \quad z(t_0) = z_0, \quad t \geq t_0, \tag{6.38}$$

where $z_0 \in \overline{\mathbb{R}}_+^n$ and the ith component of $w(z)$ is given by $w_i(z) = -z_i^{1/2} + \sum_{j=1, j \neq i}^n \phi_{ij}(z)$, $z \in \overline{\mathbb{R}}_+^n$. In addition, consider the Lyapunov function candidate $v(z) = \mathbf{e}^{\mathrm{T}}z$, $z \in \overline{\mathbb{R}}_+^n$, and note that $v(\cdot)$ is radially unbounded, $v(0) = 0$, $v(z) > 0$, $z \in \overline{\mathbb{R}}_+^n$, $z \neq 0$, and

$$v'(z)w(z) = -\sum_{i=1}^n z_i^{1/2} + \sum_{i=1}^n \sum_{j=1, j \neq i}^n \phi_{ij}(z)$$

$$= -\sum_{i=1}^n z_i^{1/2}$$

$$\leq -\left(\sum_{i=1}^n z_i\right)^{1/2}$$

$$= -(v(z))^{1/2}, \quad z \in \overline{\mathbb{R}}_+^n. \tag{6.39}$$

Thus, it follows from Theorem 6.1 with $c = 1$, $\alpha = \frac{1}{2}$, and $\mathcal{M} = \overline{\mathbb{R}}_+^n$ that the large-scale dynamical system (6.35) is finite-time stable with a settling time $T(x_0) \le 2(\mathbf{e}^{\mathrm{T}} x_0)^{1/2}$, $x_0 \in \overline{\mathbb{R}}_+^n$, and $V(x)$, $x \in \overline{\mathbb{R}}_+^n$, given by (6.36) is a control vector Lyapunov function for (6.35).

Finally, the feedback control law $\phi(x) = [\phi_1^{\mathrm{T}}(x), \ldots, \phi_n^{\mathrm{T}}(x)]^{\mathrm{T}}$, where $\phi_i(x)$, $i = 1, \ldots, n$, is given by (6.24) with

$$\alpha_i(x) = V_i'(x_i) f_i(x) = \sum_{j=1, j \neq i}^{n} \phi_{ij}(x), \quad x \in \overline{\mathbb{R}}_+^n,$$

$$\beta_i(x) = G_i(x_i), \quad x \in \overline{\mathbb{R}}_+^n,$$

and $c_{0i} > 0$, $i = 1, \ldots, n$, is a finite-time globally stabilizing decentralized feedback controller for (6.35). It can be seen from the structure of the feedback control law that the closed-loop system dynamics are essentially nonnegative. Furthermore, since $\alpha_i(x) - w_i(V(x)) = (V_i(x_i))^{1/2}$, $x \in \overline{\mathbb{R}}_+^n$, $i = 1, \ldots, n$, this feedback controller is fully independent from $f(x)$, which represents the internal interconnections of the large-scale system dynamics, and hence, is robust against full modeling uncertainty in $f(x)$.

6.4 Finite-Time Stabilization for Large-Scale Homogeneous Systems

In this section, we use geometric homogeneity developed in [10, 25] to construct finite-time controllers for large-scale homogeneous systems. First, we introduce the concept of homogeneity in relation to a scaling operation or dilation.

Definition 6.2 ([10, 25]). Let $x \triangleq [x_1, \ldots, x_n]^{\mathrm{T}} \in \mathbb{R}^n$. A *dilation* $\Delta_\lambda(x) : (\lambda, x_1, \ldots, x_n) \mapsto (\lambda^{r_1} x_1, \ldots, \lambda^{r_n} x_n)$ is a mapping that assigns to every $\lambda > 0$ a diffeomorphism $\Delta_\lambda(x) = (\lambda^{r_1} x_1, \ldots, \lambda^{r_n} x_n)$, where $(x_1, \ldots, x_n)$ is a suitable coordinate on $\mathbb{R}^n$ and $r_i > 0$, $i = 1, \ldots, n$, are constants. A function $V : \mathbb{R}^n \to \mathbb{R}$ is *homogeneous of degree $l \in \mathbb{R}$ with respect to the dilation* $\Delta_\lambda(x)$ if $V(\lambda^{r_1} x_1, \ldots, \lambda^{r_n} x_n) = \lambda^l V(x_1, \ldots, x_n)$. Finally, a vector field $f(x) \triangleq [f_1(x), \ldots, f_n(x)]^{\mathrm{T}} : \mathbb{R}^n \to \mathbb{R}$ is *homogeneous of degree $k \in \mathbb{R}$ with respect to the dilation* $\Delta_\lambda(x)$ if $f_i(\lambda^{r_1} x_1, \ldots, \lambda^{r_n} x_n) = \lambda^{k+r_i} f_i(x_1, \ldots, x_n)$, $\lambda > 0$, $i = 1, \ldots, n$.

Proposition 6.1 ([25]). Consider the nonlinear dynamical system given by (6.1). Assume $f(\cdot)$ is homogeneous of degree $k \in \mathbb{R}$ with respect to the dilation $\Delta_\lambda(x)$. Furthermore, assume $f(\cdot)$ is continuous on $\mathcal{D}$ and $x = 0$ is an asymptotically stable equilibrium point of (6.1). If $k < 0$, then $x = 0$ is a finite-time stable equilibrium point of (6.1). Alternatively, suppose

$f(x) = g_1(x) + \cdots + g_q(x)$, $x \in \mathcal{D}$, where for each $i = 1, \ldots, q$, the vector field $g_i(\cdot)$ is continuous on $\mathcal{D}$, homogeneous of degree $k_i \in \mathbb{R}$ with respect to the dilation $\Delta_\lambda(x)$, and $k_1 < \cdots < k_q$. If $x = 0$ is a finite-time stable equilibrium point of $g_1(\cdot)$, then $x = 0$ is a finite-time stable equilibrium point of $f(\cdot)$.

If in Theorem 6.1 the comparison function $w(\cdot)$ is homogeneous of degree $k < 0$ with respect to the dilation $\Delta_\lambda(z)$ and $z = 0$ is an asymptotically stable equilibrium point of (6.6), then the zero solution $x(t) \equiv 0$ to (6.1) is finite-time stable. In this case, there is *no* need to construct a scalar positive definite function $v(\cdot)$ such that (6.7) holds.

Now, consider the large-scale dynamical system $\mathcal{G}$ involving energy exchange between n interconnected subsystems given by (6.35). Furthermore, assume that there exists a constant $k \in \mathbb{R}$ such that

$$\phi_{ij}(\lambda^{r_1} x_1, \ldots, \lambda^{r_n} x_n) = \lambda^{r_i + k} \phi_{ij}(x_1, \ldots, x_n), \quad i, j = 1, \ldots, q, \quad i \neq j,$$

$$(6.40)$$

for every $\lambda > 0$ and for given $r_i > 0$, $i = 1, \ldots, n$. Next, consider the decentralized controller given by

$$u_i = \psi_i(x_i), \quad i = 1, \ldots, n, \tag{6.41}$$

with $\psi_i(x_i)$ satisfying

$$G_i(\lambda^{r_i} x_i) \psi_i(\lambda^{r_i} x_i) = \lambda^{r_i + l} G_i(x_i) \psi_i(x_i), \quad i = 1, \ldots, n, \quad x \in \mathbb{R}^n, \tag{6.42}$$

and

$$\sum_{i=1}^{n} G_i(x_i) \psi_i(x_i) < 0, \quad x \in \mathbb{R}^n, \tag{6.43}$$

for every $\lambda > 0$ and for given $r_i > 0$, $i = 1, \ldots, n$, where $l \in \mathbb{R}$, $G(x) = \mathrm{diag}[G_1(x_1), \ldots, G_n(x_n)]$, and $G_i(x_i) = 0$ if and only if $x_i = 0$, $i = 1, \ldots, n$. If $l = k < 0$, then it follows from Proposition 6.1 that the closed-loop system (6.35) with $u(t) = [\psi_1(x_1), \ldots, \psi_n(x_n)]^{\mathrm{T}}$ is globally finite-time stable. Alternatively, if $l < k$ and $l < 0$, then it follows from Proposition 6.1 that the closed-loop system (6.35) with $u(t) = [\psi_1(x_1), \ldots, \psi_n(x_n)]^{\mathrm{T}}$ is finite-time stable.

Note that if $l < k$ and $l < 0$, then stability is only local [25]. To obtain a global result in this case, we need to examine the control vector Lyapunov function of the large-scale homogeneous system. Specifically, for the dynamical system $\mathcal{G}$ given by (6.35), consider the control vector Lyapunov function candidate $V(\cdot)$ given by (6.36). Furthermore, consider the function

$$w(V(x)) = \begin{bmatrix} -\sigma_1(V_1(x_1)) + \sum_{j=1, j \neq 1}^{n} \phi_{1j}(V(x)) \\ \vdots \\ -\sigma_n(V_n(x_n)) + \sum_{j=1, j \neq n}^{n} \phi_{nj}(V(x)) \end{bmatrix}, \quad x \in \overline{\mathbb{R}}_+^n,$$

where $\sigma_i(\cdot)$ satisfies $\sigma_i(\lambda^{r_i} x_i) = \lambda^{r_i+l}\sigma_i(x_i)$ for each $\lambda > 0$ and given $r_i > 0$, $i = 1, \ldots, n$, $l < 0$, $x_i \in \overline{\mathbb{R}}_+$, $\sigma_i(0) = 0$, $\sigma_i(z) > 0$ for $z \neq 0$, $z \in \mathbb{R}$, and $\phi_{ij}(\cdot)$ satisfies (6.40) with $k > l$ and $i, j = 1, \ldots, n$, $i \neq j$.

Next, consider the comparison system given by (6.38) where the ith component of $w(z)$ is given by $w_i(z) = -\sigma_i(z_i) + \sum_{j=1, j\neq i}^{n} \phi_{ij}(z)$, $z \in \mathbb{R}_+^n$. Then it follows from Proposition 6.1 that (6.38) is finite-time stable. Furthermore, consider the Lyapunov function candidate $v(z) = \mathbf{e}^{\mathrm{T}} z$, $z \in \mathbb{R}_+^n$, and note that $v(\cdot)$ is radially unbounded, $v(0) = 0$, $v(z) > 0$, $z \in \overline{\mathbb{R}}_+^n$, $z \neq 0$, and

$$
\begin{aligned}
v'(z)w(z) &= -\sum_{i=1}^{n} \sigma_i(z_i) + \sum_{i=1}^{n} \sum_{j=1, j\neq i}^{n} \phi_{ij}(z) \\
&= -\sum_{i=1}^{n} \sigma_i(z_i) \\
&< 0, \quad z \neq 0, \quad z \in \overline{\mathbb{R}}_+^n,
\end{aligned} \tag{6.44}
$$

which implies that (6.38) is globally asymptotically stable. Hence, (6.38) is globally asymptotically stable, and thus, the large-scale homogeneous system (6.35) with $u_i = \psi_i(x_i)$, $i = 1, \ldots, n$, is globally finite-time stable and $V(\cdot)$ given by (6.36) is a control vector Lyapunov function for (6.35). Finally, (6.24) with $\alpha_i(x) = V_i'(x_i)f_i(x) = \sum_{j=1, j\neq i}^{n} \phi_{ij}(x)$, $\beta_i(x) = G_i(x_i)$, $x \in \mathbb{R}_+^n$, and $c_{0i} > 0$, $i = 1, \ldots, n$, is a finite-time globally stabilizing decentralized feedback controller for (6.35). It can be seen from the structure of the feedback control law that the closed-loop system dynamics are essentially nonnegative. Furthermore, since $\alpha_i(x) - w_i(V(x)) = \sigma_i(V_i(x_i))$, $x \in \mathbb{R}_+^n$, $i = 1, \ldots, n$, this feedback controller is fully independent from $f(x)$ which represents the internal interconnections of the large-scale system dynamics, and hence, is robust against full modeling uncertainty in $f(x)$.

6.5 Decentralized Control for Finite-Time Stabilization of Large-Scale Systems

In this section, we provide two numerical examples to illustrate the efficacy of the proposed approach. In our first example, we consider the large-scale dynamical system shown in Figure 6.1 with the power balance equation (6.35) where $\sigma_{ij}(x) = \sigma_{ij}x_j^2$, $\sigma_{ij} \geq 0$, $i \neq j$, $i, j = 1, \ldots, n$, and $G_i(x_i) = x_i^{1/4}$, $i = 1, \ldots, n$. Note that in this case $\phi_{ij}(x') \leq \phi_{ij}(x'')$, $i \neq j$, $i, j = 1, \ldots, n$, for all $x', x'' \in \mathbb{R}_+^n$ such that $x_i' = x_i''$ and $x_k' \leq x_k''$, $k \neq i$. Furthermore, with $u_i = -2x_i^{1/4}$, $i = 1, \ldots, n$, the conditions of Proposition 5.1 are satisfied, and hence, the feedback control law (6.24) is continuous on $\overline{\mathbb{R}}_+^n$. For our simulation we set $n = 3$, $\sigma_{12} = 2$, $\sigma_{13} = 3$, $\sigma_{21} = 1.5$, $\sigma_{23} = 0.3$, $\sigma_{31} = 4.4$, $\sigma_{32} = 0.6$, $c_{01} = 1$, $c_{02} = 1$, and $c_{03} = 0.25$, with initial condition

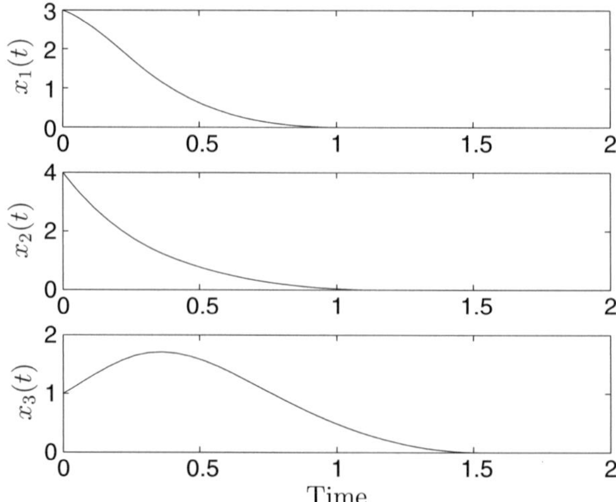

Figure 6.2 Controlled system states versus time.

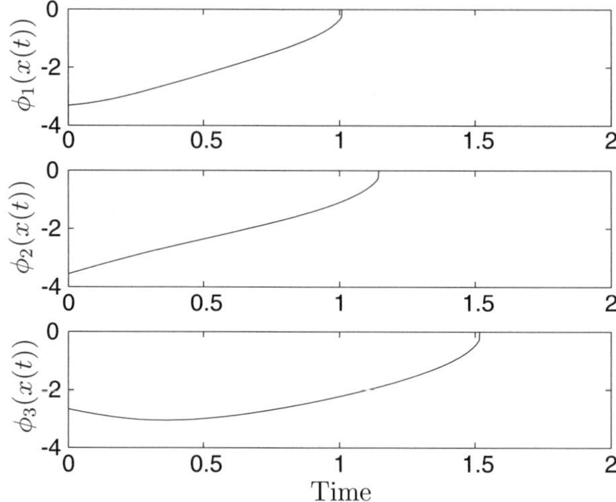

Figure 6.3 Control signals in each decentralized control channel versus time.

$x_0 = [3, 4, 1]^{\mathrm{T}}$. Figure 6.2 shows the states of the closed-loop system versus time and Figure 6.3 shows control signals for each decentralized control channel as a function of time.

For the next example, we consider control of thermoacoustic instabilities in combustion processes. Engineering applications involving steam and gas turbines and jet and ramjet engines for power generation and propulsion technology involve combustion processes. Due to the inherent coupling between several intricate physical phenomena in these processes involving acoustics, thermodynamics, fluid mechanics, and chemical kinetics, the dynamic behavior of combustion systems is characterized by highly complex nonlinear models [8, 9, 47, 103]. The unstable dynamic coupling between heat release in combustion processes generated by reacting mixtures releasing chemical energy and unsteady motions in the combustor develop acoustic pressure and velocity oscillations that can severely affect operating conditions and system performance. These pressure oscillations, known as *thermoacoustic instabilities*, often lead to high vibration levels, causing mechanical failures, high levels of acoustic noise, high burn rates, and even component melting. Hence, the need for active control to mitigate combustion-induced pressure instabilities is critical.

Next, we design a finite-time stabilizing controller for a two-mode, nonlinear time-averaged combustion model with nonlinearities present due to the second-order gas dynamics. This model is developed in [47] and is given by

$$\dot{x}_1(t) = \alpha_1 x_1(t) + \theta_1 x_2(t) - \beta(x_1(t)x_3(t) + x_2(t)x_4(t)) + u_1(t),$$
$$x_1(0) = x_{10}, \quad t \geq 0, \quad (6.45)$$
$$\dot{x}_2(t) = -\theta_1 x_1(t) + \alpha_1 x_2(t) + \beta(x_2(t)x_3(t) - x_1(t)x_4(t)) + u_2(t),$$
$$x_2(0) = x_{20}, \quad (6.46)$$
$$\dot{x}_3(t) = \alpha_2 x_3(t) + \theta_2 x_4(t) + \beta(x_1^2(t) - x_2^2(t)) + u_3(t), \quad x_3(0) = x_{30},$$
$$(6.47)$$
$$\dot{x}_4(t) = -\theta_2 x_3(t) + \alpha_2 x_4(t) + 2\beta x_1(t)x_2(t) + u_4(t), \quad x_4(0) = x_{40},$$
$$(6.48)$$

where $\alpha_1, \alpha_2 \in \mathbb{R}$ represent growth/decay constants, $\theta_1, \theta_2 \in \mathbb{R}$ represent frequency shift constants, $\beta = ((\gamma + 1)/8\gamma)\omega_1$, where γ denotes the ratio of specific heats, ω_1 is the frequency of the fundamental mode, and u_i, $i = 1, \ldots, 4$, are control input signals. For the data parameters $\alpha_1 = 5$, $\alpha_2 = -55$, $\theta_1 = 4$, $\theta_2 = 32$, $\gamma = 1.4$, $\omega_1 = 1$, and $x_0 = [2, 3, 1, 1]^{\mathrm{T}}$, the open-loop (i.e., $u_i(t) \equiv 0, i = 1, \ldots, 4$) dynamics (6.45)–(6.48) result in a limit cycle instability.

To stabilize this system in finite time we design a feedback control law given by (6.29), where $V(x) = \frac{1}{2}x^{\mathrm{T}}x$, $x \in \mathbb{R}^4$, $c = 1$, $c_0 = 1$, $\alpha = \frac{3}{4}$. In this case, $V'(x) = x^{\mathrm{T}}$, $G(x) = I_4$, and hence, $\mathcal{R} = \{x \in \mathbb{R}^4, x \neq 0 : x^{\mathrm{T}} = 0\} = \emptyset$. Thus, condition (6.28) is trivially satisfied and it follows from Corollary 6.2 that the closed-loop system (6.45)–(6.48) with the feedback control law (6.29) is finite-time stable. Furthermore, the hypotheses of Proposition 5.1 are satisfied for the case where $q = 1$, and hence, the control law (6.29) is continuous in $\mathbb{R}^4$. Specifically, with $u = -f(x) - 2^{-3/4}g(x)$, where

$$f(x) = \begin{bmatrix} \alpha_1 x_1 + \theta_1 x_2 - \beta(x_1 x_3 + x_2 x_4) \\ -\theta_1 x_1 + \alpha_1 x_2 + \beta(x_2 x_3 - x_1 x_4) \\ \alpha_2 x_3 + \theta_2 x_4 + \beta(x_1^2 - x_2^2) \\ -\theta_2 x_3 + \alpha_2 x_4 + 2\beta x_1 x_2 \end{bmatrix}, \quad g(x) = \begin{bmatrix} x_1^{1/3} \\ x_2^{1/3} \\ x_3^{1/3} \\ x_4^{1/3} \end{bmatrix},$$

the inequality

$$\alpha(x) + \beta^{\mathrm{T}}(x)u \le w(V(x)), \quad 0 < \|x\| < \delta, \tag{6.49}$$

is satisfied, where $\alpha(x) \triangleq V'(x)f(x)$, $\beta(x) \triangleq G^{\mathrm{T}}(x)V'^{\mathrm{T}}(x)$, $w(V(x)) = -(V(x))^{3/4}$, $x \in \mathbb{R}^4$, and $0 < \delta < 1$.

To see this, note that

$$\begin{aligned}
\alpha(x) + \beta^{\mathrm{T}}(x)u &= -2^{-3/4}x^{\mathrm{T}}g(x) \\
&= -2^{-3/4}\sum_{i=1}^{4} x_i^{4/3} \\
&\le -2^{-3/4}\left(\sum_{i=1}^{4} x_i^2\right)^{3/4} \\
&= -(V(x))^{3/4} \\
&= w(V(x)), \quad 0 < \|x\| < \delta < 1.
\end{aligned} \tag{6.50}$$

In addition, since $f(\cdot)$ and $g(\cdot)$ are continuous and $f(0) = g(0) = 0$, it follows from (6.50) that for every $\varepsilon > 0$, there exists $0 < \delta < 1$ such that for all $0 < \|x\| < \delta$ there exists $u \in \mathbb{R}^4$ such that $\|u\| < \varepsilon$ and inequality (6.49) holds. Thus, the feedback control law (6.29) is continuous in $\mathbb{R}^4$. Figure 6.4 shows the states of the closed-loop system versus time and Figure 6.5 shows the control signals versus time.

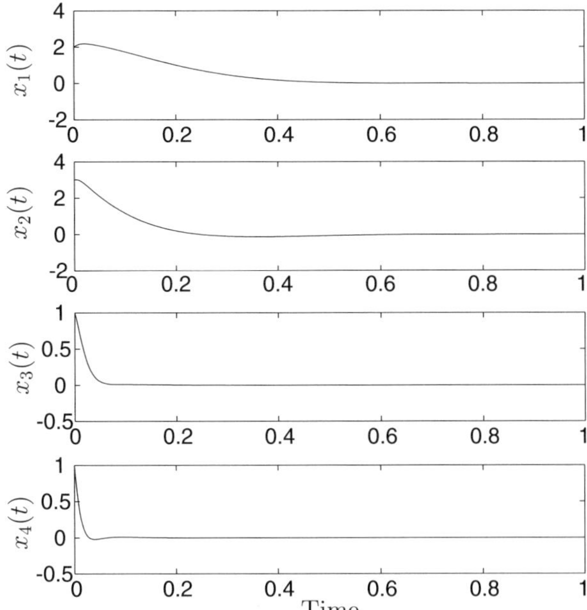

Figure 6.4 Controlled system states versus time.

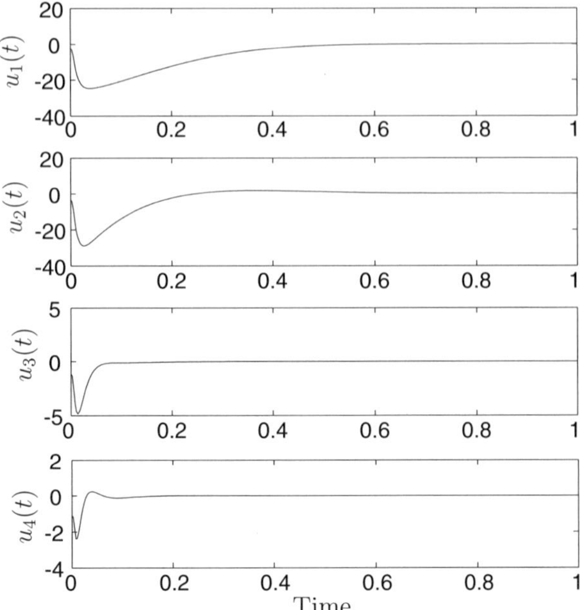

Figure 6.5 Control signals in each control channel versus time.

Coordination Control for Multiagent Interconnected Systems

7.1 Introduction

Modern complex multiagent dynamical systems are highly interconnected and mutually interdependent, both physically and through a multitude of information and communication networks. Distributed decision-making for coordination of networks of dynamic agents involving information flow can be naturally captured by graph-theoretic notions. These dynamical network systems cover a very broad spectrum of applications, including cooperative control of unmanned air vehicles (UAVs), autonomous underwater vehicles (AUVs), distributed sensor networks, air and ground transportation systems, swarms of air and space vehicle formations, and congestion control in communication networks, to cite but a few examples. Hence, it is not surprising that a considerable research effort has been devoted to control of networks and control over networks in recent years [15,27,29,37].

A key application area of multiagent network coordination within aerospace systems is cooperative control of vehicle formations using distributed and decentralized controller architectures. Distributed control refers to a control architecture wherein the control is distributed via multiple computational units that are interconnected through information and communication networks, whereas decentralized control refers to a control architecture wherein local decisions are based only on local information. Vehicle formations are typically dynamically decoupled, that is, the motion of a given agent or vehicle does not directly affect the motion of the other agents or vehicles. The multiagent system is coupled via the task which the agents or vehicles are required to perform.

The complexity of cooperative maneuvers that multiagent systems need to perform as well as environmental constraints often necessitate the design of control algorithms that use information on current position and velocity of each vehicle to steer them while maintaining a specified formation. In particular, for mobile agents operating in a foggy environment or located far from each other, open-loop visual control for coordinated motion becomes impractical. In this case, feedback control algorithms are required for individual vehicle steering which determine how a given vehicle maneuvers based on positions and velocities of nearby vehicles and/or on those of a formation leader. The leader could be real, that is, one of the vehicles in a formation leads the others, or the leader could be virtual, that is, vehicles synthesize a leader and the motions of the vehicles in a formation are defined with respect to a virtual agent whose positions and velocities are known at each instant of time.

Analysis and control design for networks of mobile agents has received considerable attention in the literature. Common formations of multiagent systems include flocking [142, 166], cyclic pursuit [130], (virtual) leader following [53, 164], and rendezvous [36, 151]. Graph-theoretic notions [51] are essential in the analysis and control design for a system of mobile agents performing a common task [56, 102]. Several researchers have proposed different techniques for analyzing network systems. Specifically, the authors in [45, 159] use graph theory to model interconnected systems, whereas [49, 56, 134, 135] invoke graph-theoretic notions for stability analysis of formations of large numbers of agents. Alternatively, the authors in [124, 142, 143] use potential functions to analyze flocking and [144] resorts to control Lyapunov functions to design feedback controllers for coordinated motion of multi-robot platforms. Distributed control of robotic networks has been extensively studied in [31, 151] where the authors develop a variety of control algorithms for network consensus. Furthermore, distributed nonlinear static and dynamic control architectures for multiagent coordination using thermodynamic principles are presented in [95]. A survey of recent research results in cooperative control of multivehicle systems is presented in [140].

In this chapter, we develop stability analysis and a control design framework for multiagent coordination predicated on vector Lyapunov functions. In multiagent systems, several Lyapunov functions arise naturally where each agent can be associated with a generalized energy function corresponding to a component of a vector Lyapunov function. Furthermore, since a specified formation of multiple vehicles can be characterized by a time-varying set in the state space, the problem of control design for multiagent coordinated motion is equivalent to design of stabilizing controllers for time-varying sets of nonlinear dynamical systems. Thus, using a stability and control design framework for time-varying sets, we design distributed control algorithms for stabilization of multi-vehicle formations. These distributed control algorithms use only local information of the individual vehicle relative position and velocity with respect to the leader to maintain a specified formation for a system of multiple vehicles. Finally, we specialize the results obtained for time-varying sets to address stabilization of time-invariant sets and develop stabilizing control algorithms for static formations (rendezvous) of multiple vehicles. The developed cooperative control algorithms are shown to globally exponentially stabilize both moving and static formations.

7.2 Stability and Stabilization of Time-Varying Sets

In this section, we present results on stability and stabilization of time-varying sets for time-varying nonlinear dynamical systems using vector Lyapunov functions [14,118,133,141,159]. Specifically, consider the time-varying nonlinear dynamical system given by

$$\dot{x}(t) = f(t, x(t)), \quad x(t_0) = x_0, \quad t \geq t_0, \tag{7.1}$$

where $x(t) \in \mathcal{D} \subseteq \mathbb{R}^n$, $t \geq t_0$, is the solution to (7.1), $t_0 \in [0, \infty)$, $\mathcal{D}$ is an open set, $0 \in \mathcal{D}$, $f(t, 0) = 0$, $t \geq t_0$, and $f : [t_0, \infty) \times \mathcal{D} \to \mathbb{R}^n$ is such that $f(\cdot, \cdot)$ is jointly continuous in t and x, for every $t \in [t_0, \infty)$, and $f(t, \cdot)$ is locally Lipschitz in x uniformly in t for all compact subsets of $[0, \infty)$.

The following definition introduces several types of stability for time-varying sets of nonlinear time-varying dynamical systems. For this definition, $\mathcal{D}_0^t \triangleq \mathcal{D}_0(t)$, $t \geq t_0$, is a time-varying set such that, for every $t \geq t_0$, $\mathcal{D}_0(t)$ is a compact set and $\mathcal{O}_\varepsilon(\mathcal{D}_0(t)) \triangleq \{x \in \mathcal{D} : \mathrm{dist}(x, \mathcal{D}_0(t)) < \varepsilon\}$, $t \geq t_0$, defines the ε-neighborhood of $\mathcal{D}_0(t)$ at each instant of time $t \geq t_0$, where $\mathrm{dist}(x, \mathcal{D}_0(t)) \triangleq \inf_{y \in \mathcal{D}_0(t)} \|y - x\|$, $t \geq t_0$.

Definition 7.1. Consider the nonlinear time-varying dynamical system (7.1). Let $\mathcal{D}_0^t$ be positively invariant with respect to (7.1) and, at each time instant $t \in [t_0, \infty)$, $\mathcal{D}_0(t)$ is a compact set.

i) $\mathcal{D}_0^t$ is *Lyapunov stable* if, for every $\varepsilon > 0$ and $t_0 \in [0, \infty)$, there exists $\delta = \delta(\varepsilon, t_0) > 0$ such that $x_0 \in \mathcal{O}_\delta(\mathcal{D}_0(t_0))$ implies that $x(t) \in \mathcal{O}_\varepsilon(\mathcal{D}_0(t))$ for all $t \geq t_0$.

ii) $\mathcal{D}_0^t$ is *uniformly Lyapunov stable* if, for every $\varepsilon > 0$, there exists $\delta = \delta(\varepsilon) > 0$ such that $x_0 \in \mathcal{O}_\delta(\mathcal{D}_0(t_0))$ implies that $x(t) \in \mathcal{O}_\varepsilon(\mathcal{D}_0(t))$ for all $t \geq t_0$ and for all $t_0 \in [0, \infty)$.

iii) $\mathcal{D}_0^t$ is *asymptotically stable* if $\mathcal{D}_0^t$ is *Lyapunov stable* and, for every $t_0 \in [0, \infty)$, there exists $\delta = \delta(t_0) > 0$ such that $x_0 \in \mathcal{O}_\delta(\mathcal{D}_0(t_0))$ implies that $\lim_{t \to \infty} \mathrm{dist}(x(t), \mathcal{D}_0(t)) = 0$.

iv) $\mathcal{D}_0^t$ is *uniformly asymptotically stable* if $\mathcal{D}_0^t$ is uniformly Lyapunov stable and there exists $\delta > 0$ such that $x_0 \in \mathcal{O}_\delta(\mathcal{D}_0(t_0))$ implies that $\lim_{t \to \infty} \mathrm{dist}(x(t), \mathcal{D}_0(t)) = 0$ uniformly in t_0 and x_0 for all $t_0 \in [0, \infty)$.

v) $\mathcal{D}_0^t$ is *globally asymptotically stable* if $\mathcal{D}_0^t$ is Lyapunov stable and $\lim_{t \to \infty} \mathrm{dist}(x(t), \mathcal{D}_0(t)) = 0$ for all $x_0 \in \mathbb{R}^n$ and $t_0 \in [0, \infty)$.

vi) $\mathcal{D}_0^t$ is *globally uniformly asymptotically stable* if $\mathcal{D}_0^t$ is uniformly Lyapunov stable and $\lim_{t \to \infty} \mathrm{dist}(x(t), \mathcal{D}_0(t)) = 0$ uniformly in t_0 and x_0 for all $x_0 \in \mathbb{R}^n$ and $t_0 \in [0, \infty)$.

vii) $\mathcal{D}_0^t$ is *uniformly exponentially stable* if there exist scalars $\alpha > 0$, $\beta > 0$, $\delta > 0$ such that $x_0 \in \mathcal{O}_\delta(\mathcal{D}_0(t_0))$ implies that $\mathrm{dist}(x(t), \mathcal{D}_0(t)) \leq \alpha \, \mathrm{dist}(x_0, \mathcal{D}_0(t_0))e^{-\beta(t-t_0)}$, $t \geq t_0$, for all $t_0 \in [0, \infty)$.

viii) $\mathcal{D}_0^t$ is *globally uniformly exponentially stable* if there exist scalars $\alpha > 0$, $\beta > 0$ such that $\mathrm{dist}(x(t), \mathcal{D}_0(t)) \leq \alpha \, \mathrm{dist}(x_0, \mathcal{D}_0(t_0))e^{-\beta(t-t_0)}$, $t \geq t_0$, for all $x_0 \in \mathbb{R}^n$ and $t_0 \in [0, \infty)$.

The following definition introduces the notion of class $\mathcal{W}$ functions involving *time-varying quasi-monotone increasing* functions.

Definition 7.2 ([141, 159]). A function $w = [w_1, ..., w_q]^{\mathrm{T}} : [0, \infty) \times \mathbb{R}^q \to \mathbb{R}^q$ is of *class* $\mathcal{W}$ if, for every fixed $t \in [0, \infty)$, each component $w_i(\cdot, \cdot)$, $i = 1, ..., q$, of $w(\cdot, \cdot)$ satisfies $w_i(t, z') \leq w_i(t, z'')$ for all $z', z'' \in \mathbb{R}^q$ such that $z'_j \leq z''_j$, $j = 1, ..., q$, $j \neq i$, and $z'_i = z''_i$, where z_i denotes the ith component of z.

Note that if $w(t, \cdot) \in \mathcal{W}$ and $w(t, 0) \equiv 0$, then $w(\cdot, \cdot)$ is essentially nonnegative [72], which implies that a time-varying dynamical system whose dynamics are represented by $w(\cdot, \cdot)$ exhibits solutions that belong to the nonnegative orthant for all nonnegative initial conditions [72]. The following result presents sufficient conditions for several types of stability of time-varying sets with respect to nonlinear time-varying dynamical systems using vector Lyapunov functions.

Theorem 7.1. Consider the nonlinear time-varying dynamical system (7.1). Assume that there exists a continuously differentiable vector function $V(t, x) = [V_1, \ldots, V_q]^{\mathrm{T}} : [0, \infty) \times \mathcal{D} \to \mathcal{Q} \cap \overline{\mathbb{R}}_+^q$, where $\mathcal{Q} \subset \mathbb{R}^q$ and $0 \in \mathcal{Q}$, a continuous function $w = [w_1, \ldots, w_q]^{\mathrm{T}} : [0, \infty) \times \mathcal{Q} \to \mathbb{R}^q$, and class $\mathcal{K}$ functions $\alpha(\cdot)$, $\beta(\cdot)$ such that $V_i(t, x) = 0$, $x \in \mathcal{D}_i(t)$, $t \geq t_0$, where $\mathcal{D}_i(t) \subset \mathcal{D}$, $t \geq t_0$, $V_i(t, x) > 0$, $x \in \mathcal{D}\backslash\mathcal{D}_i(t)$, $t \geq t_0$, $i = 1, \ldots, q$, $\mathcal{D}_0^t = \mathcal{D}_0(t) \triangleq \cap_{i=1}^q \mathcal{D}_i(t) \neq \emptyset$ is positively invariant with respect to (7.1) and $\mathcal{D}_0(t)$ is compact at each time instant $t \geq t_0$, $w(t, \cdot) \in \mathcal{W}$, $w(t, 0) = 0$, $t \geq 0$,

$$\alpha(\mathrm{dist}(x, \mathcal{D}_0(t))) \leq \mathbf{e}^{\mathrm{T}} V(t, x) \leq \beta(\mathrm{dist}(x, \mathcal{D}_0(t))), \quad (t, x) \in [0, \infty) \times \mathcal{D}, \tag{7.2}$$

and

$$\frac{\partial V_i(t, x)}{\partial t} + V_i'(t, x) f(t, x) \leq w_i(t, V(t, x)), \quad (t, x) \in [0, \infty) \times \mathcal{D},$$
$$i = 1, \ldots, q. \tag{7.3}$$

In addition, assume that the vector comparison system

$$\dot{z}(t) = w(t, z(t)), \quad z(0) = z_0, \quad t \geq t_0, \tag{7.4}$$

has a unique solution $z(t)$, $t \geq t_0$, forward in time. Then the following statements hold:

i) If the zero solution to (7.4) is uniformly Lyapunov stable, then $\mathcal{D}_0^t$ is uniformly Lyapunov stable with respect to (7.1).

ii) If the zero solution to (7.4) is uniformly asymptotically stable, then $\mathcal{D}_0^t$ is uniformly asymptotically stable with respect to (7.1).

iii) If $\mathcal{D} = \mathbb{R}^n$, $\mathcal{Q} = \mathbb{R}^q$, $\alpha(\cdot)$ and $\beta(\cdot)$ are class $\mathcal{K}_\infty$ functions, and the zero solution to (7.4) is globally uniformly asymptotically stable, then $\mathcal{D}_0^t$ is globally uniformly asymptotically stable with respect to (7.1).

iv) If there exist constants $\nu \geq 1$, $\alpha > 0$, $\beta > 0$ such that

$$\alpha \left[\mathrm{dist}(x, \mathcal{D}_0(t))\right]^\nu \leq \mathbf{e}^{\mathrm{T}} V(t, x) \leq \beta \left[\mathrm{dist}(x, \mathcal{D}_0(t))\right]^\nu,$$
$$(t, x) \in [0, \infty) \times \mathcal{D}, \quad (7.5)$$

and the zero solution to (7.4) is uniformly exponentially stable, then $\mathcal{D}_0^t$ is uniformly exponentially stable with respect to (7.1).

v) If $\mathcal{D} = \mathbb{R}^n$, $\mathcal{Q} = \mathbb{R}^q$ and there exist constants $\nu \geq 1$, $\alpha > 0$, $\beta > 0$ such that (7.5) holds and the zero solution to (7.4) is globally uniformly exponentially stable, then $\mathcal{D}_0^t$ is globally uniformly exponentially stable with respect to (7.1).

Proof. *i)* Let $\varepsilon > 0$. It follows from uniform Lyapunov stability of the nonlinear dynamical system (7.4) that there exists $\mu = \mu(\varepsilon) > 0$ such that if $\|z_0\|_1 < \mu$ and $z_0 \in \overline{\mathbb{R}}_+^q$, where $\| \cdot \|_1$ denotes the absolute sum norm, then $\|z(t)\|_1 < \alpha(\varepsilon)$, $t \geq t_0$, and $z(t) \in \overline{\mathbb{R}}_+^q$, $t \geq t_0$. Now, choose $z_0 = V(t_0, x_0) \geq\geq 0$, $x_0 \in \mathcal{D}$, $t_0 \in [0, \infty)$, and, for $\mu = \mu(\varepsilon) > 0$, choose $\delta = \delta(\mu(\varepsilon)) = \delta(\varepsilon) > 0$ such that $\beta(\delta) = \mu$. Then, for $x_0 \in \mathcal{O}_\delta(\mathcal{D}_0(t_0))$, it follows from (7.2) that

$$\|z_0\|_1 = \mathbf{e}^{\mathrm{T}} z_0 = \mathbf{e}^{\mathrm{T}} V(t_0, x_0) \leq \beta(\mathrm{dist}(x_0, \mathcal{D}_0(t_0))) < \beta(\delta) = \mu, \quad (7.6)$$

which implies that $\mathbf{e}^{\mathrm{T}} z(t) = \|z(t)\|_1 < \alpha(\varepsilon)$, $t \geq t_0$. Now, with $z_0 = V(t_0, x_0) \geq\geq 0$ and the assumption that $w(t, \cdot) \in \mathcal{W}$ it follows from (7.3) and the vector comparison principle [141] that $0 \leq\leq V(t, x(t)) \leq\leq z(t)$, $t \geq t_0$. Thus, using (7.2), it follows that if $x_0 \in \mathcal{O}_\delta(\mathcal{D}_0(t_0))$, then

$$\alpha(\mathrm{dist}(x(t), \mathcal{D}_0(t))) \leq \mathbf{e}^{\mathrm{T}} V(t, x(t)) \leq \mathbf{e}^{\mathrm{T}} z(t) < \alpha(\varepsilon), \quad t \geq t_0, \quad (7.7)$$

which implies that $x(t) \in \mathcal{O}_\varepsilon(\mathcal{D}_0(t))$, $t \geq t_0$. This proves uniform Lyapunov stability of the time-varying set $\mathcal{D}_0^t$ with respect to (7.1).

ii) Uniform Lyapunov stability of $\mathcal{D}_0^t$ with respect to (7.1) follows from *i)*. Furthermore, since the zero solution to (7.4) is uniformly asymptotically stable, there exists $\mu > 0$ such that if $\|z_0\|_1 < \mu$, then $\lim_{t \to \infty} z(t) = 0$. As in *i)*, let $z_0 = V(t_0, x_0) \geq\geq 0$, $x_0 \in \mathcal{D}$, $t_0 \in [0, \infty)$, and choose $\delta = \delta(\mu) > 0$

such that $\beta(\delta) = \mu$. In this case, if $x_0 \in \mathcal{O}_\delta(\mathcal{D}_0(t_0))$, it follows from (7.2) that

$$\|z_0\|_1 = \mathbf{e}^{\mathrm{T}} z_0 = \mathbf{e}^{\mathrm{T}} V(t_0, x_0) \le \beta(\mathrm{dist}(x_0, \mathcal{D}_0(t_0))) < \beta(\delta) = \mu, \quad (7.8)$$

which implies that $\lim_{t\to\infty} z(t) = 0$. Since $w(t, \cdot) \in \mathcal{W}$ and $z_0 = V(t_0, x_0)$, it follows from (7.3) and the vector comparison principle that $0 \le\le V(t, x(t))$ $\le\le z(t)$, $t \ge t_0$. Now, using (7.2), it follows that, for $x_0 \in \mathcal{O}_\delta(\mathcal{D}_0(t_0))$,

$$\alpha(\mathrm{dist}(x(t), \mathcal{D}_0(t))) \le \mathbf{e}^{\mathrm{T}} V(t, x(t)) \le \mathbf{e}^{\mathrm{T}} z(t), \quad (7.9)$$

for all $t \ge t_0$. Since $\lim_{t\to\infty} z(t) = 0$, it follows from (7.9) that $\lim_{t\to\infty} \mathrm{dist}$ $(x(t), \mathcal{D}_0(t)) = 0$, which proves uniform asymptotic stability of $\mathcal{D}_0^t$ with respect to (7.1).

iii) Uniform Lyapunov stability of $\mathcal{D}_0^t$ follows from *i*). Next, for every $x_0 \in \mathbb{R}^n$, $t_0 \in [0, \infty)$, and $z_0 = V(t_0, x_0)$, identical arguments as in *ii*) can be used to show that $\lim_{t\to\infty} \mathrm{dist}(x(t), \mathcal{D}_0(t)) = 0$, which, along with the uniform Lyapunov stability, implies global uniform asymptotic stability of $\mathcal{D}_0^t$ with respect to (7.1).

iv) It follows from the uniform exponential stability of the nonlinear dynamical system (7.4) that there exist positive constants γ, μ, and η such that if $\|z_0\|_1 < \mu$, then

$$\|z(t)\|_1 \le \gamma \|z_0\|_1 e^{-\eta(t-t_0)}, \quad t \ge t_0. \quad (7.10)$$

As in *i*), let $z_0 = V(t_0, x_0) \ge\ge 0, x_0 \in \mathcal{D}, t_0 \in [0, \infty)$ and choose $\delta = \delta(\mu) = \left(\frac{\mu}{\beta}\right)^{\frac{1}{\nu}} > 0$. In this case, if $x_0 \in \mathcal{O}_\delta(\mathcal{D}_0(t_0))$, it follows from (7.5) that

$$\|z_0\|_1 = \mathbf{e}^{\mathrm{T}} z_0 = \mathbf{e}^{\mathrm{T}} V(t_0, x_0) \le \beta \left[\mathrm{dist}(x_0, \mathcal{D}_0(t_0))\right]^\nu < \beta \delta^\nu = \mu. \quad (7.11)$$

Since $w(t, \cdot) \in \mathcal{W}$ and $z_0 = V(t_0, x_0)$, it follows from (7.3) and the vector comparison principle that $0 \le\le V(t, x(t)) \le\le z(t), t \ge t_0$. Now, using (7.5) and (7.10), it follows that, for $x_0 \in \mathcal{O}_\delta(\mathcal{D}_0(t_0))$,

$$\begin{aligned}
\alpha \left[\mathrm{dist}(x(t), \mathcal{D}_0(t))\right]^\nu &\le \mathbf{e}^{\mathrm{T}} V(t, x(t)) \\
&\le \mathbf{e}^{\mathrm{T}} z(t) \\
&\le \gamma \|z_0\|_1 e^{-\eta(t-t_0)} \\
&\le \gamma \beta \left[\mathrm{dist}(x_0, \mathcal{D}_0(t_0))\right]^\nu e^{-\eta(t-t_0)}, \quad t \ge t_0, \quad (7.12)
\end{aligned}$$

which implies that

$$\mathrm{dist}(x(t), \mathcal{D}_0(t)) \le \left(\frac{\gamma\beta}{\alpha}\right)^{\frac{1}{\nu}} \mathrm{dist}(x_0, \mathcal{D}_0(t_0)) e^{-\frac{\eta}{\nu}(t-t_0)}, \quad t \ge t_0, \quad (7.13)$$

establishing uniform exponential stability of $\mathcal{D}_0^t$ with respect to (7.1).

v) The proof is identical to the proof of *iv*). $\square$

If we take $w(t, z) \equiv w(z)$ in Theorem 7.1, then uniform stability of (7.4) is equivalent to the regular notion of stability for autonomous systems. In addition, if $\mathcal{D}_i(t) \triangleq \{x \in \mathbb{R}^n : \mathcal{X}_i(t, x) = 0\}$, $t \geq t_0$, where $\mathcal{X}_i : [0, \infty) \times \mathbb{R}^n \to \mathbb{R}^{s_i}$ are continuous functions for all $i = 1, \ldots, q$, then Theorem 7.1 still holds for the definition of a distance given by

$$\operatorname{dist}(x, \mathcal{D}_0(t)) \triangleq \left[\mathcal{X}^{\mathrm{T}}(t, x)\mathcal{X}(t, x)\right]^{\frac{1}{2}}, \tag{7.14}$$

where $\mathcal{X}(t, x) \triangleq [\mathcal{X}_1^{\mathrm{T}}(t, x), \ldots, \mathcal{X}_q^{\mathrm{T}}(t, x)]^{\mathrm{T}}$.

Next, we use the result of Theorem 7.1 to design stabilizing controllers for time-varying sets of multiagent dynamical systems composed of q agents whose dynamics are given by

$$\dot{x}_i(t) = f_i(t, x(t)) + G_i(t, x(t))u_i(t), \quad t \geq t_0, \quad i = 1, \ldots, q, \tag{7.15}$$

where $x(t) = [x_1^{\mathrm{T}}(t) \ldots, x_q^{\mathrm{T}}(t)]^{\mathrm{T}}$, $x_i(t) \in \mathbb{R}^{n_i}$, $t \geq 0$, and $f_i : [0, \infty) \times \mathbb{R}^n \to \mathbb{R}^{n_i}$ and $G_i : [0, \infty) \times \mathbb{R}^n \to \mathbb{R}^{n_i \times m_i}$ are continuous functions for all $i = 1, \ldots, q$. Consider the time-varying sets given by $\mathcal{D}_i(t) \triangleq \{x \in \mathbb{R}^n : \mathcal{X}_i(t, x_i) = 0\}$, $t \geq t_0$, where $\mathcal{X}_i : [0, \infty) \times \mathbb{R}^{n_i} \to \mathbb{R}^{s_i}$ are continuous functions for all $i = 1, \ldots, q$. Define the motion of the ith agent on the set $\mathcal{D}_i(\cdot)$ as $x_{\mathrm{e}i}(t)$, $t \geq t_0$, and note that $\mathcal{X}_i(t, x_{\mathrm{e}i}(t)) \equiv 0$. Furthermore, assume there exist vector functions $u_{\mathrm{e}i}(t)$, $t \geq t_0$, $i = 1, \ldots, q$, such that

$$G_i(t, x_{\mathrm{e}}(t))u_{\mathrm{e}i}(t) = \dot{x}_{\mathrm{e}i}(t) - f_i(t, x_{\mathrm{e}}(t)), \quad t \geq t_0, \quad i = 1, \ldots, q, \tag{7.16}$$

where $x_{\mathrm{e}}(t) \triangleq [x_{\mathrm{e}1}^{\mathrm{T}}(t), \ldots, x_{\mathrm{e}q}^{\mathrm{T}}(t)]^{\mathrm{T}}$, $t \geq t_0$.

The next result presents a controller design that guarantees stabilization of a time-varying set for the time-varying nonlinear dynamical system (7.15) using vector Lyapunov functions.

Theorem 7.2. Consider the multiagent dynamical system given by (7.15). Assume there exist a continuously differentiable, component decoupled vector function $V : [0, \infty) \times \mathbb{R}^n \to \overline{\mathbb{R}}_+^q$, that is, $V(t, x) = [V_1(t, x_1), \ldots, V_q(t, x_q)]^{\mathrm{T}}$, $(t, x) \in [0, \infty) \times \mathbb{R}^n$, a continuous function $w = [w_1, \ldots, w_q]^{\mathrm{T}} : [0, \infty) \times \overline{\mathbb{R}}_+^q \to \mathbb{R}^q$, and class $\mathcal{K}$ functions $\alpha : [0, \infty) \to [0, \infty)$ and $\beta : [0, \infty) \to [0, \infty)$ such that $V_i(t, x_i) = 0$, $x \in \mathcal{D}_i(t) \subset \mathbb{R}^n$, $t \geq t_0$, $i = 1, \ldots, q$, $V_i(t, x_i) > 0$, $x \in \mathbb{R}^n \backslash \mathcal{D}_i(t)$, $t \geq t_0$, $i = 1, \ldots, q$, $\mathcal{D}_0^t = \mathcal{D}_0(t) \triangleq \cap_{i=1}^q \mathcal{D}_i(t) \neq \emptyset$, $t \geq t_0$, is a compact set at each $t \geq t_0$, $w(t, \cdot) \in \mathcal{W}$, $w(t, 0) = 0$, $t \geq 0$,

$$\alpha(\operatorname{dist}(x, \mathcal{D}_0(t))) \leq \mathbf{e}^{\mathrm{T}}V(t, x) \leq \beta(\operatorname{dist}(x, \mathcal{D}_0(t))), \quad (t, x) \in [0, \infty) \times \mathbb{R}^n, \tag{7.17}$$

and, for all $i = 1, \ldots, q$,

$$\frac{\partial V_i(t, x_i)}{\partial t} + V_i'(t, x_i)f_i(t, x) \leq w_i(t, V(t, x)), \quad (t, x) \in \mathcal{R}_i, \tag{7.18}$$

where $\mathcal{R}_i \triangleq \{(t,x) \in [0,\infty) \times \mathbb{R}^n : V_i'(t,x_i)G_i(t,x) = 0\}$, $i = 1,\ldots,q$. In addition, assume that the zero solution $z(t) \equiv 0$ to

$$\dot{z}(t) = w(t, z(t)), \quad z(0) = z_0, \quad t \geq t_0, \tag{7.19}$$

is uniformly asymptotically stable. Then $\mathcal{D}_0^t$ is uniformly asymptotically stable with respect to the nonlinear dynamical system (7.15) with the feedback control law $u = \phi(t,x) = [\phi_1^{\mathrm{T}}(t,x), \ldots, \phi_q^{\mathrm{T}}(t,x)]^{\mathrm{T}}$, $x \in \mathbb{R}^n$, $t \in [0,\infty)$, given by

$$\phi_i(t,x) = \begin{cases} u_{ei}(t) - \left(c_{0i} + \dfrac{\mu_i(t,x) + \sqrt{\lambda_i(t,x)}}{\sigma_i^{\mathrm{T}}(t,x)\sigma_i(t,x)}\right)\sigma_i(t,x), & \text{if } \sigma_i(t,x) \neq 0, \\ u_{ei}(t), & \text{if } \sigma_i(t,x) = 0, \end{cases} \tag{7.20}$$

where $u_{ei}(t)$, $t \geq t_0$, satisfies (7.16), $\lambda_i(t,x) \triangleq \mu_i^2(t,x) + (\sigma_i^{\mathrm{T}}(t,x)\sigma_i(t,x))^2$, $\mu_i(t,x) \triangleq \rho_i(t,x) - w_i(t, V(t,x)) + \frac{\partial V_i(t,x_i)}{\partial t} + \sigma_i^{\mathrm{T}}(t,x)u_{ei}(t)$, $\rho_i(t,x) \triangleq V_i'(t,x_i) \cdot f_i(t,x)$, $\sigma_i(t,x) \triangleq G_i^{\mathrm{T}}(t,x)V_i'^{\mathrm{T}}(t,x_i)$, and $c_{0i} > 0$, $i = 1,\ldots,q$. If, in addition, $\alpha(\cdot)$ and $\beta(\cdot)$ are class $\mathcal{K}_\infty$ functions and the zero solution $z(t) \equiv 0$ to (7.19) is globally uniformly asymptotically stable, then $\mathcal{D}_0^t$ is globally uniformly asymptotically stable with respect to (7.15) with the feedback control law $u = \phi(t,x)$ given by (7.20). Furthermore, if there exist constants $\nu \geq 1$, $\alpha > 0$, and $\beta > 0$ such that

$$\alpha\left[\mathrm{dist}(x, \mathcal{D}_0(t))\right]^\nu \leq \mathbf{e}^{\mathrm{T}}V(t,x) \leq \beta\left[\mathrm{dist}(x, \mathcal{D}_0(t))\right]^\nu, \quad (t,x) \in [0,\infty) \times \mathbb{R}^n, \tag{7.21}$$

and the zero solution to (7.19) is uniformly exponentially stable, then $\mathcal{D}_0^t$ is uniformly exponentially stable with respect to (7.15) with the feedback control law (7.20). Finally, if (7.21) holds and the zero solution to (7.19) is globally uniformly exponentially stable, then $\mathcal{D}_0^t$ is globally uniformly exponentially stable with respect to (7.15) with the feedback control law (7.20).

Proof. The vector Lyapunov derivative components $\dot{V}_i(\cdot,\cdot)$, $i = 1,\ldots,q$, along the trajectories of the closed-loop system (7.15), with $u = \phi(t,x)$, $(t,x) \in [0,\infty) \times \mathbb{R}^n$, given by (7.20), satisfy

$$\dot{V}_i(t,x_i) = \frac{\partial V_i(t,x_i)}{\partial t} + \rho_i(t,x) + \sigma_i^{\mathrm{T}}(t,x)\phi_i(t,x)$$

$$= \begin{cases} \frac{\partial V_i(t,x_i)}{\partial t} + \rho_i(t,x) + \sigma_i^{\mathrm{T}}(t,x)u_{ei}(t) \\ \qquad - c_{0i}\sigma_i^{\mathrm{T}}(t,x)\sigma_i(t,x) - \mu_i(t,x) \\ \qquad - \sqrt{\lambda_i(t,x)}, & \text{if } \sigma_i(t,x) \neq 0, \\ \frac{\partial V_i(t,x_i)}{\partial t} + \rho_i(t,x), & \text{if } \sigma_i(t,x) = 0, \end{cases}$$

$$\leq w_i(t, V(t,x)), \quad (t,x) \in [0,\infty) \times \mathbb{R}^n.$$

Now, the result is a direct consequence of Theorem 7.1. $\qquad\square$

Note that if in Theorem 7.2 $q = 1$, $\mathcal{D}_0^t \equiv \{0\}$, and (7.15) is a time-invariant system, then we can set $x_{\mathrm{e}}(t) \equiv 0$, $u_{\mathrm{e}}(t) \equiv 0$, $w(t, z) \equiv 0$, and $V(t, x) \equiv V(x)$. In this case, the feedback control law (7.20) specializes to Sontag's universal formula [165]. Alternatively, if $\mathcal{R}_i = \varnothing$, $i = 1, \ldots, q$, then $w(\cdot, \cdot)$ in (7.18) can be chosen arbitrarily so that the comparison system (7.19) is (globally) uniformly asymptotically (respectively, exponentially) stable. In addition, since $\mathcal{D}_i(t) = \{x \in \mathbb{R}^n : \mathcal{X}_i(t, x_i) = 0\}$, $t \geq t_0$, then $V_i(\cdot, \cdot)$, $i = 1, \ldots, q$, can be chosen arbitrarily provided that $V_i(t, x_i) = 0$, $x \in \mathcal{D}_i(t)$, $t \geq t_0$, $V_i(t, x_i) > 0$, $x \in \mathbb{R}^n \backslash \mathcal{D}_i(t)$, $t \geq t_0$, $i = 1, \ldots, q$, and (7.17) (respectively, (7.21)) holds.

For example, $V_i(\cdot, \cdot)$ can be taken as $V_i(t, x_i) = \mathcal{X}_i^{\mathrm{T}}(t, x_i) P_i \mathcal{X}_i(t, x_i)$, $x_i \in \mathbb{R}^{n_i}$, where $P_i \in \mathbb{R}^{s_i \times s_i}$ is such that $P_i > 0$, $i = 1, \ldots, q$. In this case, $\alpha(\mathrm{dist}(x, \mathcal{D}_0(t))) = \alpha \left[\mathrm{dist}(x, \mathcal{D}_0(t))\right]^2$, where $0 < \alpha \leq \min_{i=1,\ldots,q}\{\lambda_{\min}(P_i)\}$, $\beta(\mathrm{dist}(x, \mathcal{D}_0(t))) = \beta \left[\mathrm{dist}(x, \mathcal{D}_0(t))\right]^2$, where $\beta \geq \max_{i=1,\ldots,q}\{\lambda_{\max}(P_i)\}$, and $\mathrm{dist}(x, \mathcal{D}_0(t)) \triangleq \left[\mathcal{X}^{\mathrm{T}}(t, x)\mathcal{X}(t, x)\right]^{\frac{1}{2}}$, where $\mathcal{X}(t, x) \triangleq [\mathcal{X}_1^{\mathrm{T}}(t, x_1), \ldots, \mathcal{X}_q^{\mathrm{T}}(t, x_q)]^{\mathrm{T}}$. In this case, it follows from Theorem 7.2 that for the closed-loop system (7.15) and (7.20) the time-varying set $\mathcal{D}_0^t$ is (globally) uniformly asymptotically (respectively, exponentially) stable.

7.3 Control Design for Multivehicle Coordinated Motion

In this section, we apply the results of Section 7.2 to a problem of coordinated motion of multiple vehicles in pursuit of a (virtual) leader. Specifically, we design a distributed feedback control law that drives individual vehicles to a configuration with specified distance and orientation with respect to a leader while maintaining this configuration throughout the motion of the leader. The leader can be either real or virtual. In the latter case, the agents synthesize a motion with respect to which they need to maintain a specified formation.

Consider the planar motion of q agents with the individual agent dynamics given by

$$\ddot{x}_i(t) = u_{xi}(t), \quad x_i(0) = x_{i0}, \quad \dot{x}_i(0) = \dot{x}_{i0}, \quad t \geq 0, \qquad (7.22)$$
$$\ddot{y}_i(t) = u_{yi}(t), \quad y_i(0) = y_{i0}, \quad \dot{y}_i(0) = \dot{y}_{i0}, \qquad (7.23)$$

where $x_i : [0, \infty) \to \mathbb{R}$ and $y_i : [0, \infty) \to \mathbb{R}$, $i = 1, \ldots, q$, are the displacements of the ith agent in the horizontal and vertical directions, respectively, and u_{xi} and u_{yi} are the control forces acting on the ith agent in the horizontal and vertical directions, respectively. Next, define $\eta_i \triangleq [x_i, y_i, \dot{x}_i, \dot{y}_i]^{\mathrm{T}}$, $i = 1, \ldots, q$, and $\eta \triangleq [\eta_1^{\mathrm{T}}, \ldots, \eta_q^{\mathrm{T}}]^{\mathrm{T}}$. Then the generalized dynamics (7.22) and (7.23) for q agents can be written in the state space form as

$$\dot{\eta}(t) = (I_q \otimes A)\eta(t) + (I_q \otimes B)u(t), \quad \eta(0) = \eta_0, \quad t \geq 0, \qquad (7.24)$$

where $\eta_0 = [\eta_{10}^{\mathrm{T}}, \ldots, \eta_{q0}^{\mathrm{T}}]^{\mathrm{T}}$, $\eta_{i0} = [x_{i0}, y_{i0}, \dot{x}_{i0}, \dot{y}_{i0}]^{\mathrm{T}}$, $u \triangleq [u_1^{\mathrm{T}}, \ldots, u_q^{\mathrm{T}}]^{\mathrm{T}}$, $u_i \triangleq$

$[u_{xi}, u_{yi}]^{\mathrm{T}}$, "$\otimes$" is the Kronecker product [18], $I_q \in \mathbb{R}^{q \times q}$ is the identity matrix, and A, B are given by

$$A = \begin{bmatrix} 0 & 0 & 1 & 0 \\ 0 & 0 & 0 & 1 \\ 0 & 0 & 0 & 0 \\ 0 & 0 & 0 & 0 \end{bmatrix}, \qquad B = \begin{bmatrix} 0 & 0 \\ 0 & 0 \\ 1 & 0 \\ 0 & 1 \end{bmatrix}. \tag{7.25}$$

Furthermore, define the time-varying sets

$$\mathcal{D}_i(t) \triangleq \{\eta \in \mathbb{R}^{4q} : \eta_i - p_i(t) = 0\}, \quad t \geq 0, \quad i = 1, \ldots, q, \tag{7.26}$$

where

$$p_i(t) \triangleq \begin{bmatrix} x_{\mathrm{L}}(t) + l_{xi\mathrm{L}} \\ y_{\mathrm{L}}(t) + l_{yi\mathrm{L}} \\ \dot{x}_{\mathrm{L}}(t) \\ \dot{y}_{\mathrm{L}}(t) \end{bmatrix}, \quad t \geq 0, \quad i = 1, \ldots, q, \tag{7.27}$$

$x_{\mathrm{L}}(t)$, $y_{\mathrm{L}}(t)$, $t \geq 0$, are, respectively, the horizontal and vertical positions of the leader, $\dot{x}_{\mathrm{L}}(t)$, $\dot{y}_{\mathrm{L}}(t)$, $t \geq 0$, are, respectively, the horizontal and vertical velocities of the leader, and $l_{xi\mathrm{L}}, l_{yi\mathrm{L}} \in \mathbb{R}$ are, respectively, the desired horizontal and vertical distances between the ith agent and the leader. Note that each set $\mathcal{D}_i(t)$, $t \geq 0$, $i = 1, \ldots, q$, defines relative position and velocity of the ith agent with respect to the leader. To construct the set $\mathcal{D}_i(t)$, $t \geq 0$, $i = 1, \ldots, q$, only local information of the ith agent position and velocity is needed. The position and velocity of the leader at each instant of time are assumed to be known. Furthermore, the intersection of the sets (7.26) given by

$$\mathcal{D}_0^t = \mathcal{D}_0(t) \triangleq \bigcap_{i=1,\ldots,q} \mathcal{D}_i(t), \quad t \geq 0, \tag{7.28}$$

characterizes the desired formation of the agents with respect to the leader where all agents maintain specified distances and velocities with respect to the leader.

Note that this approach of characterizing multivehicle formations via time-varying sets also captures formations where only neighbor-to-neighbor interactions are permitted [31, 95, 102, 142, 151]. In this case, as long as the connectivity graph describing the entire multivehicle formation is strongly connected [56, 95], the formation is uniquely defined by a time-varying set characterizing neighbor-to-neighbor relative positions and velocities.

Next, we define the component decoupled vector function $V : [0, \infty) \times \mathbb{R}^{4q} \to \mathbb{R}^q$ such that $V(t, \eta) = [V_1(t, \eta_1), \ldots, V_q(t, \eta_q)]^{\mathrm{T}}$, where

$$V_i(t, \eta_i) = (\eta_i - p_i(t))^{\mathrm{T}} P(\eta_i - p_i(t)), \quad \eta_i \in \mathbb{R}^4, \quad t \geq 0, \quad i = 1, \ldots, q, \tag{7.29}$$

and

$$P = \begin{bmatrix} 1 & 0 & 1 & 0 \\ 0 & 1 & 0 & 1 \\ 0 & 0 & 1 & 0 \\ 0 & 0 & 0 & 1 \end{bmatrix} > 0. \qquad (7.30)$$

Note that $V_i(t, \eta_i) = 0$, $\eta \in \mathcal{D}_i(t)$, $t \geq 0$, and $V_i(t, \eta_i) > 0$, $\eta \in \mathbb{R}^{4q} \setminus \mathcal{D}_i(t)$, $t \geq 0$, $i = 1, \ldots, q$. In addition, since $\lambda_{\min}(P) = \lambda_{\max}(P) = 1$, condition (7.21) is satisfied with $\alpha = \frac{1}{2}$, $\beta = 2$, $\nu = 2$, $\mathrm{dist}(\eta, \mathcal{D}_0(t)) \triangleq \left[(\eta - p(t))^{\mathrm{T}} (\eta - p(t)) \right]^{\frac{1}{2}}$, $\eta \in \mathbb{R}^{4q}$, $t \geq 0$, where $p(t) \triangleq [p_1^{\mathrm{T}}(t), \ldots, p_q^{\mathrm{T}}(t)]^{\mathrm{T}}$. Furthermore, it can be shown that, for $\mathcal{R}_i \triangleq \{ (\eta, t) \in \mathbb{R}^{4q} \times [0, \infty) : V_i'(t, \eta_i) B = 0 \}$, $i = 1, \ldots, q$, condition (7.18) is satisfied with

$$\frac{\partial V_i(t, \eta_i(t))}{\partial t} + V_i'(t, \eta_i(t)) A \eta_i(t) \leq -\gamma_i V_i(t, \eta_i(t)), \quad (t, \eta) \in \mathcal{R}_i,$$
$$i = 1, \ldots, q, \qquad (7.31)$$

for $\gamma_i \in (0, 1]$, $i = 1, \ldots, q$.

In this case, the zero solution to (7.19) is globally exponentially stable with

$$w(z) = [-\gamma_1 z_1, \ldots, -\gamma_q z_q]^{\mathrm{T}}. \qquad (7.32)$$

Hence, it follows from Theorem 7.2 that the time-varying set $\mathcal{D}_0^t$ defined by (7.28) is globally uniformly exponentially stable with respect to (7.24) with the feedback control law $u_i = \phi_i(t, \eta_i)$, $i = 1, \ldots, q$, given by (7.20) with $\mu_i(t, \eta_i) \triangleq \rho_i(t, \eta_i) - w_i(V_i(t, \eta_i)) + \frac{\partial V_i(t, \eta_i)}{\partial t} + \sigma_i^{\mathrm{T}}(t, \eta_i) u_{ei}(t)$, $\rho_i(t, \eta_i) \triangleq V_i'(t, \eta_i) A \eta_i$, $\sigma_i(t, \eta_i) \triangleq B^{\mathrm{T}} V_i'^{\mathrm{T}}(t, \eta_i)$, $u_{ei}(t) = [\ddot{x}_{\mathrm{L}}(t), \ddot{y}_{\mathrm{L}}(t)]^{\mathrm{T}}$, and $w(V(t, \eta))$ given by (7.32). Note that the feedback control law $u_i = \phi_i(t, \eta_i)$, $i = 1, \ldots, q$, is a distributed control algorithm [31, 151] that uses only local information of the relative position and velocity of the ith agent with respect to the leader. This allows us to use a fixed controller structure to steer individual agents while maintaining a specified formation with respect to the leader.

In the following simulation, we consider two agents pursuing a leader in a triangular formation. For this design, we set $l_{x1\mathrm{L}} = -2$, $l_{y1\mathrm{L}} = 2\sqrt{3}$, $l_{x2\mathrm{L}} = -4$, $l_{y2\mathrm{L}} = 0$, $c_{0i} = 0.5$, $i = 1, 2$, $\gamma_i = \frac{1}{5}$, $i = 1, 2$, $\eta_{10} = [8, 3, -1, -3]^{\mathrm{T}}$, and $\eta_{20} = [-3, -5, 3, -1]^{\mathrm{T}}$. With this choice of the parameters $l_{xi\mathrm{L}}$ and $l_{yi\mathrm{L}}$, $i = 1, 2$, the agents will form a configuration of an equilateral triangle with respect to the leader. Furthermore, the leader is set to be moving counterclockwise around a circle of radius 1 characterized by $x_{\mathrm{L}}(t) = \cos t$, $y_{\mathrm{L}}(t) = \sin t$, $t \geq 0$. For the feedback controller (7.20), Figure 7.1 shows the position phase portrait of two agents following the leader and Figure 7.2 shows the time history of the control forces acting on each agent.

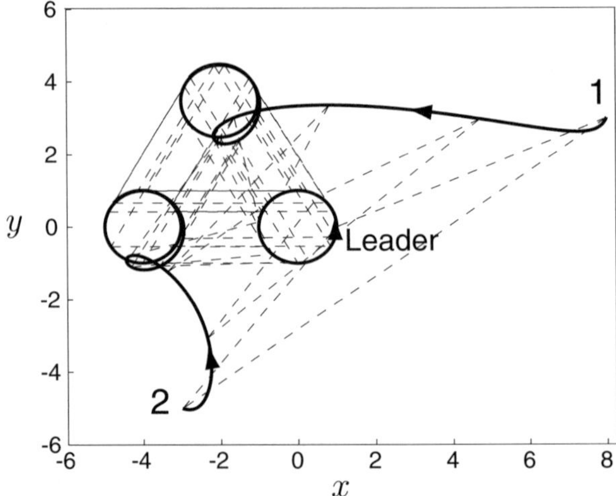

Figure 7.1 Position phase portrait of two agents following the leader.

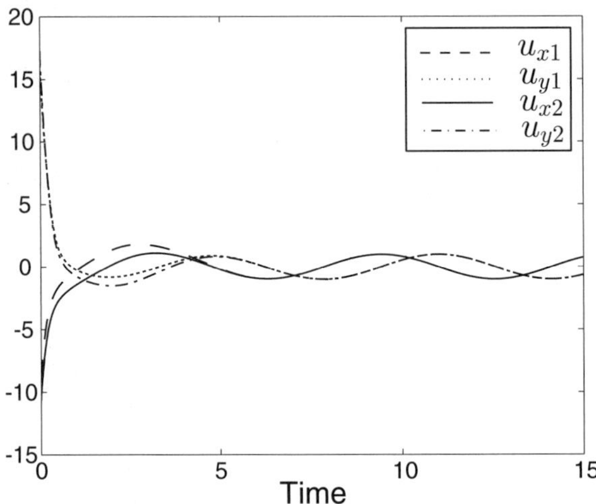

Figure 7.2 Control forces in horizontal and vertical directions versus time.

Next, we compare the performance of the control law (7.20) with the performance of two other cooperative control algorithms developed in [151] for the same formation control problem as above with the same data. Specifically, the first control law that we consider for the system (7.22) and (7.23) is given by

$$[u_{xi}(t), u_{yi}(t)]^{\mathrm{T}} = -K_g \tilde{h}_i(t) - D_g \dot{\tilde{h}}_i(t)$$
$$-K_f(\tilde{h}_i(t) - \tilde{h}_{i-1}(t))$$
$$-D_f(\dot{\tilde{h}}_i(t) - \dot{\tilde{h}}_{i-1}(t))$$

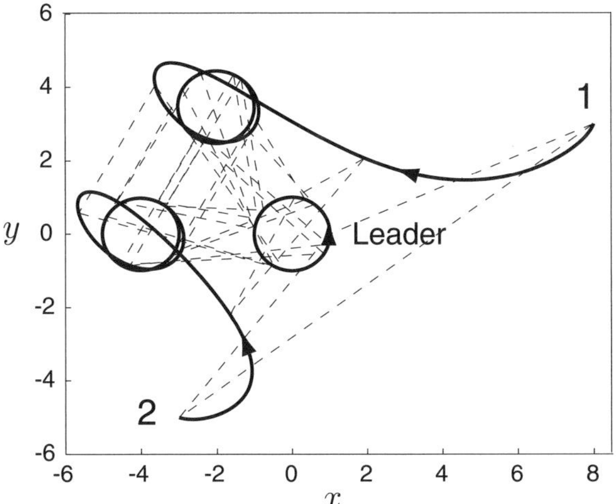

Figure 7.3 Position phase portrait of two agents following the leader.

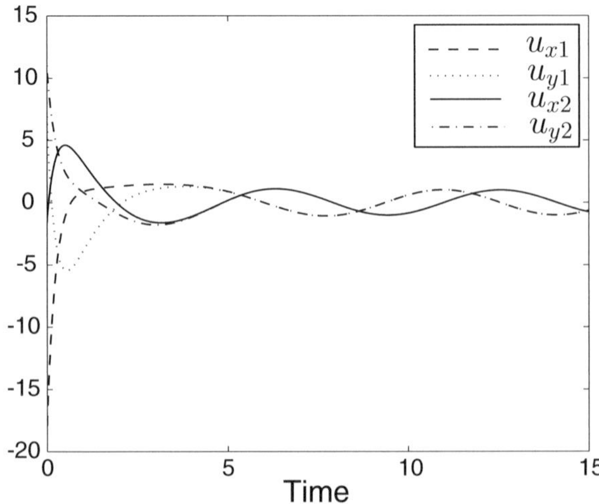

Figure 7.4 Control forces in horizontal and vertical directions versus time.

$$-K_f(\tilde{h}_i(t) - \tilde{h}_{i+1}(t))$$
$$-D_f(\dot{\tilde{h}}_i(t) - \dot{\tilde{h}}_{i+1}(t)),\qquad(7.33)$$

where $i = 1, 2$, $\tilde{h}_i(t) \triangleq h_i(t) - h_{id}(t)$, $h_i(t) \triangleq [x_i(t), y_i(t)]^{\mathrm{T}}$, $h_{id}(t) \triangleq [x_{\mathrm{L}}(t) + l_{xi\mathrm{L}}, y_{\mathrm{L}}(t) + l_{yi\mathrm{L}}]^{\mathrm{T}}$, $\tilde{h}_3(t) \triangleq \tilde{h}_1(t)$, $\tilde{h}_0(t) \triangleq \tilde{h}_2(t)$, $h_3(t) \triangleq h_1(t)$, and $h_0(t) \triangleq h_2(t)$. The control gains $K_g \in \mathbb{R}^{2\times2}$ and $D_g \in \mathbb{R}^{2\times2}$ are positive-definite matrices and control gains $K_f \in \mathbb{R}^{2\times2}$ and $D_f \in \mathbb{R}^{2\times2}$ are nonnegative-definite matrices. The second control law for the system (7.22) and (7.23)

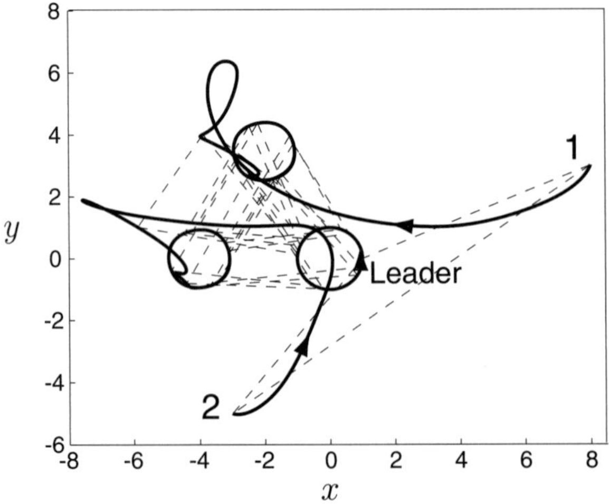

Figure 7.5 Position phase portrait of two agents following the leader.

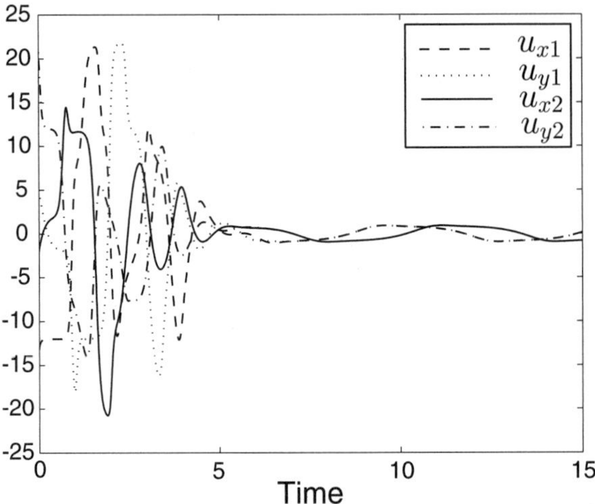

Figure 7.6 Control forces in horizontal and vertical directions versus time.

accounts for the actuator saturation and is given by

$$
\begin{aligned}
[u_{xi}(t), u_{yi}(t)]^{\mathrm{T}} = {}&-k_g \tanh(k\tilde{h}_i(t)) - d_g \tanh(k\dot{h}_i(t)) \\
&-k_f \tanh[k(\tilde{h}_i(t) - \tilde{h}_{i-1}(t))] \\
&-k_f \tanh[k(\tilde{h}_i(t) - \tilde{h}_{i+1}(t))],
\end{aligned} \tag{7.34}
$$

where $k_g > 0$, $k > 0$, $d_g > 0$, $k_f \geq 0$, and $\tanh(\cdot)$ is the hyperbolic tangent defined componentwise.

Figures 7.3 and 7.4 show the performance of the controller (7.33) with

$K_g = D_g = K_f = D_f = I_2$. Furthermore, Figures 7.5 and 7.6 show the performance of the controller (7.34) with $k_g = 7$, $d_g = 5$, $k_f = 5$, and $k = 1$. In both cases, the values of the control gains were slightly altered from the ones in [151] to yield the best compromise between the convergence time and the control effort. It was observed that both controllers, (7.33) and (7.34), retain a steady-state error between the desired and actual positions of each agent. This corresponds to a triangular steady-state formation of two agents with respect to the leader that oscillates around a desired equilateral triangle but never converges to it. Alternatively, controller (7.20) ensures exponential stabilization of the desired formation. In addition, the rate of change for the controller (7.34) is significantly higher than that of (7.20) and (7.33).

7.4 Stability and Stabilization of Time-Invariant Sets

In this section, we present results on stabilization of time-invariant sets for time-invariant nonlinear dynamical systems using vector Lyapunov functions. Consider the time-invariant nonlinear dynamical system given by

$$\dot{x}(t) = f(x(t)), \quad x(0) = x_0, \quad t \geq 0, \tag{7.35}$$

where $x(t) \in \mathcal{D} \subseteq \mathbb{R}^n$, $t \geq 0$, is the system state vector, $\mathcal{D}$ is an open set, $0 \in \mathcal{D}$, $f(0) = 0$, and $f(\cdot)$ is Lipschitz continuous on $\mathcal{D}$.

Definition 7.3. For the nonlinear dynamical system (7.35), let $\mathcal{D}_0 \subset \mathcal{D}$ be a compact positively invariant set with respect to (7.35). $\mathcal{D}_0$ is *Lyapunov stable* if, for every open neighborhood $\mathcal{O}_1 \subseteq \mathcal{D}$ of $\mathcal{D}_0$, there exists an open neighborhood $\mathcal{O}_2 \subseteq \mathcal{O}_1$ of $\mathcal{D}_0$ such that $x(t) \in \mathcal{O}_1$, $t \geq 0$, for all $x_0 \in \mathcal{O}_2$. $\mathcal{D}_0$ is *asymptotically stable* if $\mathcal{D}_0$ is Lyapunov stable and there exists a neighborhood $\mathcal{O}_1$ of $\mathcal{D}_0$ such that $\mathrm{dist}(x(t), \mathcal{D}_0) \to 0$ as $t \to \infty$ for all $x_0 \in \mathcal{O}_1$. $\mathcal{D}_0$ is *globally asymptotically stable* if $\mathcal{D}_0$ is Lyapunov stable and $\mathrm{dist}(x(t), \mathcal{D}_0) \to 0$ as $t \to \infty$ for all $x_0 \in \mathbb{R}^n$. $\mathcal{D}_0$ is *exponentially stable* if there exist $\alpha > 0$, $\beta > 0$, and a neighborhood $\mathcal{O}_1$ of $\mathcal{D}_0$ such that $\mathrm{dist}(x(t), \mathcal{D}_0) \leq \alpha \, \mathrm{dist}(x_0, \mathcal{D}_0) e^{-\beta t}$, $t \geq 0$, for all $x_0 \in \mathcal{O}_1$. Finally, $\mathcal{D}_0$ is *globally exponentially stable* if there exist $\alpha > 0$ and $\beta > 0$ such that $\mathrm{dist}(x(t), \mathcal{D}_0) \leq \alpha \, \mathrm{dist}(x_0, \mathcal{D}_0) e^{-\beta t}$, $t \geq 0$, for all $x_0 \in \mathbb{R}^n$.

Theorem 7.3. Consider the nonlinear dynamical system (7.35). Assume there exists a continuously differentiable vector function $V = [V_1, \ldots, V_q]^{\mathrm{T}} : \mathcal{D} \to \mathcal{Q} \cap \overline{\mathbb{R}}_+^q$, where $\mathcal{Q} \subset \mathbb{R}^q$ and $0 \in \mathcal{Q}$, such that $V_i(x) = 0$, $x \in \mathcal{D}_i$, where $\mathcal{D}_i \subset \mathcal{D}$, $i = 1, \ldots, q$, $V_i(x) > 0$, $x \in \mathcal{D} \backslash \mathcal{D}_i$, $i = 1, \ldots, q$, $\mathcal{D}_0 \triangleq \cap_{i=1}^q \mathcal{D}_i \neq \varnothing$ is a compact positively invariant set with respect to (7.35), and

$$V'(x)f(x) \leq\leq w(V(x)), \quad x \in \mathcal{D}, \tag{7.36}$$

where $w : \mathcal{Q} \to \mathbb{R}^q$ is continuous, $w(\cdot) \in \mathcal{W}$, and $w(0) = 0$. In addition, assume that the vector comparison system

$$\dot{z}(t) = w(z(t)), \quad z(0) = z_0, \quad t \geq 0, \tag{7.37}$$

has a unique solution in forward time $z(t)$, $t \geq 0$. Then the following statements hold:

   $i)$ If the zero solution $z(t) \equiv 0$ to (7.37) is Lyapunov stable, then $\mathcal{D}_0$ is Lyapunov stable with respect to (7.35).

  $ii)$ If the zero solution $z(t) \equiv 0$ to (7.37) is asymptotically stable, then $\mathcal{D}_0$ is asymptotically stable with respect to (7.35).

 $iii)$ If $\mathcal{D} = \mathbb{R}^n$, $\mathcal{Q} = \mathbb{R}^q$, and $v(x) \triangleq \mathbf{e}^{\mathrm{T}} V(x)$, $x \in \mathbb{R}^n$, is such that $v(x) \to \infty$ as $\mathrm{dist}(x, \mathcal{D}_0) \to \infty$, and the zero solution $z(t) \equiv 0$ to (7.37) is globally asymptotically stable, then $\mathcal{D}_0$ is globally asymptotically stable with respect to (7.35).

  $iv)$ If there exist constants $\nu \geq 1$, $\alpha > 0$, and $\beta > 0$ such that

$$\alpha[\mathrm{dist}(x, \mathcal{D}_0)]^{\nu} \leq \mathbf{e}^{\mathrm{T}} V(x) \leq \beta[\mathrm{dist}(x, \mathcal{D}_0)]^{\nu}, \quad x \in \mathcal{D}, \tag{7.38}$$

and the zero solution $z(t) \equiv 0$ to (7.37) is exponentially stable, then $\mathcal{D}_0$ is exponentially stable with respect to (7.35).

   $v)$ If $\mathcal{D} = \mathbb{R}^n$, $\mathcal{Q} = \mathbb{R}^q$, (7.38) holds for all $x_0 \in \mathbb{R}^n$, and the zero solution $z(t) \equiv 0$ to (7.37) is globally exponentially stable, then $\mathcal{D}_0$ is globally exponentially stable with respect to (7.35).

Proof. The proof is similar to the proof of Theorem 7.1 and, hence, is omitted. $\square$

Next, we use the result of Theorem 7.3 to design controllers to stabilize time-invariant sets for multiagent dynamical systems composed of q agents whose dynamics are given by

$$\dot{x}_i(t) = f_i(x(t)) + G_i(x(t))u_i(t), \quad t \geq t_0, \quad i = 1, \ldots, q, \tag{7.39}$$

where $x(t) = [x_1^{\mathrm{T}}(t) \ldots, x_q^{\mathrm{T}}(t)]^{\mathrm{T}}$, $x_i(t) \in \mathbb{R}^{n_i}$, $t \geq 0$, $f_i : \mathbb{R}^n \to \mathbb{R}^{n_i}$ satisfying $f_i(0) = 0$ and $G_i : \mathbb{R}^n \to \mathbb{R}^{n_i \times m_i}$ are continuous functions for all $i = 1, \ldots, q$, and $u_i(\cdot)$, $i = 1, \ldots, q$, satisfy sufficient regularity conditions such that the nonlinear dynamical system (7.39) has a unique solution forward in time.

Theorem 7.4. Consider the multiagent dynamical system given by (7.39). Assume there exist a continuously differentiable, component decoupled vector function $V : \mathbb{R}^n \to \overline{\mathbb{R}}_+^q$, that is, $V(x) = [V_1(x_1), \ldots, V_q(x_q)]^{\mathrm{T}}$, $x \in \mathbb{R}^n$, and continuous function $w = [w_1, \ldots, w_q]^{\mathrm{T}} : \overline{\mathbb{R}}_+^q \to \mathbb{R}^q$, such that

$V_i(x_i) = 0$, $x \in \mathcal{D}_i \subset \mathbb{R}^n$, $i = 1, \ldots, q$, $V_i(x_i) > 0$, $x \in \mathbb{R}^n \backslash \mathcal{D}_i$, $i = 1, \ldots, q$, $\mathcal{D}_0 \triangleq \cap_{i=1}^q \mathcal{D}_i \neq \varnothing$ is a compact set, $w(\cdot) \in \mathcal{W}$, $w(0) = 0$, and, for all $i = 1, \ldots, q$,

$$V_i'(x_i) f_i(x) \leq w_i(V(x)), \quad x \in \mathcal{R}_i, \tag{7.40}$$

where $\mathcal{R}_i \triangleq \{x \in \mathbb{R}^n : V_i'(x_i) G_i(x) = 0\}$, $i = 1, \ldots, q$. In addition, assume that the zero solution $z(t) \equiv 0$ to (7.37) is asymptotically stable. Then $\mathcal{D}_0$ is asymptotically stable with respect to the nonlinear dynamical system (7.39) with the feedback control law $u = \phi(x) = [\phi_1^{\mathrm{T}}(x), \ldots, \phi_q^{\mathrm{T}}(x)]^{\mathrm{T}}$, $x \in \mathbb{R}^n$, given by

$$\phi_i(x) = \begin{cases} -\left(c_{0i} + \dfrac{(\rho_i(x) - w_i(V(x))) + \sqrt{\lambda_i(x)}}{\sigma_i^{\mathrm{T}}(x) \sigma_i(x)} \right) \sigma_i(x), & \text{if } \sigma_i(x) \neq 0, \\ 0, & \text{if } \sigma_i(x) = 0, \end{cases} \tag{7.41}$$

where $\lambda_i(x) \triangleq (\rho_i(x) - w_i(V(x)))^2 + (\sigma_i^{\mathrm{T}}(x) \sigma_i(x))^2$, $\rho_i(x) \triangleq V_i'(x_i) f_i(x)$, $x \in \mathbb{R}^n$, $\sigma_i(x) \triangleq G_i^{\mathrm{T}}(x) V_i'^{\mathrm{T}}(x_i)$, $x \in \mathbb{R}^n$, and $c_{0i} > 0$, $i = 1, \ldots, q$. If, in addition, $v(x) \triangleq \mathbf{e}^{\mathrm{T}} V(x)$, $x \in \mathbb{R}^n$, is such that $v(x) \to \infty$ as $\mathrm{dist}(x, \mathcal{D}_0) \to \infty$, and the zero solution $z(t) \equiv 0$ to (7.37) is globally asymptotically stable, then $\mathcal{D}_0$ is globally asymptotically stable with respect to (7.39) with the feedback control law (7.41). Furthermore, if there exist constants $\nu \geq 1$, $\alpha > 0$, and $\beta > 0$ such that

$$\alpha \left[\mathrm{dist}(x, \mathcal{D}_0) \right]^\nu \leq \mathbf{e}^{\mathrm{T}} V(x) \leq \beta \left[\mathrm{dist}(x, \mathcal{D}_0) \right]^\nu, \quad x \in \mathbb{R}^n, \tag{7.42}$$

and the zero solution to (7.37) is exponentially stable, then $\mathcal{D}_0$ is exponentially stable with respect to (7.39) with the feedback control law (7.41). Finally, if (7.42) holds and the zero solution to (7.37) is globally exponentially stable, then $\mathcal{D}_0$ is globally exponentially stable with respect to (7.39) with the feedback control law (7.41).

Proof. The vector Lyapunov derivative components $\dot{V}_i(\cdot)$, $i = 1, \ldots, q$, along the trajectories of the closed-loop dynamical system (7.39), with $u = \phi(x)$, $x \in \mathbb{R}^n$, given by (7.41), are given by

$$\begin{aligned} \dot{V}_i(x_i) &= V_i'(x_i)[f_i(x) + G_i(x)\phi_i(x)] \\ &= \rho_i(x) + \sigma_i^{\mathrm{T}}(x)\phi_i(x) \\ &= \begin{cases} -c_{0i}\sigma_i^{\mathrm{T}}(x)\sigma_i(x) - \sqrt{\lambda_i(x)} + w_i(V(x)), & \text{if } \sigma_i(x) \neq 0, \\ \rho_i(x), & \text{if } \sigma_i(x) = 0, \end{cases} \\ &\leq w_i(V(x)), \quad x \in \mathbb{R}^n. \end{aligned} \tag{7.43}$$

Now, the result is a direct consequence of Theorem 7.3. $\qquad \square$

If $\mathcal{R}_i = \varnothing$, $i = 1, \ldots, q$, then $w(\cdot)$ in (7.40) and (7.41) can be chosen arbitrarily so that the comparison system (7.37) is (globally) asymptotically (respectively, exponentially) stable. In addition, if $\mathcal{D}_i = \{x \in \mathbb{R}^n : \mathcal{X}_i(x_i) = $

$0\}$, where $\mathcal{X}_i : \mathbb{R}^{n_i} \to \mathbb{R}^{s_i}$ are continuous functions for all $i = 1, \ldots, q$, then $V_i(\cdot)$, $i = 1, \ldots, q$, can be chosen arbitrarily provided that $V_i(x_i) = 0$, $x \in \mathcal{D}_i$, and $V_i(x_i) > 0$, $x \in \mathbb{R}^n \backslash \mathcal{D}_i$, $i = 1, \ldots, q$. For example, $V_i(\cdot)$ can be taken as $V_i(x_i) = \mathcal{X}_i^{\mathrm{T}}(x_i) P_i \mathcal{X}_i(x_i)$, $x_i \in \mathbb{R}^{n_i}$, where $P_i \in \mathbb{R}^{s_i \times s_i}$ is such that $P_i > 0$, $i = 1, \ldots, q$.

7.5 Control Design for Static Formations

In this section, we apply the results of Section 7.4 to stabilize static formations of multiple vehicles. Specifically, we design a feedback control law that drives two agents to a configuration with specified distance between the agents and orientation with respect to the horizontal.

In particular, consider the dynamics of the two agents given by (7.22) and (7.23) and note that we can rewrite them in the state space form as

$$\dot{\xi}_1(t) = A\xi_1(t) + B\tilde{u}_1(t), \quad \xi_1(0) = \xi_{10}, \quad t \geq 0, \tag{7.44}$$

$$\dot{\xi}_2(t) = A\xi_2(t) + B\tilde{u}_2(t), \quad \xi_2(0) = \xi_{20}, \tag{7.45}$$

where $\xi_1 \triangleq [x_1, x_2, \dot{x}_1, \dot{x}_2]^{\mathrm{T}}$, $\xi_2 \triangleq [y_1, y_2, \dot{y}_1, \dot{y}_2]^{\mathrm{T}}$, $\xi \triangleq [\xi_1^{\mathrm{T}}, \xi_2^{\mathrm{T}}]^{\mathrm{T}}$, $\tilde{u}_1 \triangleq [u_{x1}, u_{x2}]^{\mathrm{T}}$, $\tilde{u}_2 \triangleq [u_{y1}, u_{y2}]^{\mathrm{T}}$, and A, B are given by (7.25).

Next, define the sets $\mathcal{D}_1 \triangleq \{\xi \in \mathbb{R}^8 : E\xi_1 - p_x = 0\}$ and $\mathcal{D}_2 \triangleq \{\xi \in \mathbb{R}^8 : E\xi_2 - p_y = 0\}$, where

$$E = \begin{bmatrix} 1 & -1 & 0 & 0 \\ 0 & 0 & 1 & 0 \\ 0 & 0 & 0 & 1 \end{bmatrix}, \quad p_x \triangleq \begin{bmatrix} l_x \\ 0 \\ 0 \end{bmatrix}, \quad p_y \triangleq \begin{bmatrix} l_y \\ 0 \\ 0 \end{bmatrix}, \tag{7.46}$$

and $l_x, l_y \in \mathbb{R}$. Note that $\mathcal{D}_0 \triangleq \mathcal{D}_1 \cap \mathcal{D}_2$ determines a family of formations for the two agents where both agents are at the equilibrium and the distance between the agents and the angle with respect to the horizontal, respectively, is given by

$$L = (l_x^2 + l_y^2)^{\frac{1}{2}}, \quad \theta = \begin{cases} \arctan\left(\frac{l_y}{l_x}\right), & l_x \geq 0, \\ \pi + \arctan\left(\frac{l_y}{l_x}\right), & l_x < 0. \end{cases} \tag{7.47}$$

Furthermore, note that for every pair of $L > 0$ and $\theta \in [-\frac{\pi}{2}, \frac{3}{2}\pi]$, there exist unique $l_x \in \mathbb{R}$ and $l_y \in \mathbb{R}$ such that (7.47) is satisfied.

Next, define a component decoupled vector function $V : \mathbb{R}^8 \to \mathbb{R}^2$ such that $V(\xi) = [V_1(\xi_1), V_2(\xi_2)]^{\mathrm{T}}$, $\xi = [\xi_1^{\mathrm{T}}, \xi_2^{\mathrm{T}}]^{\mathrm{T}} \in \mathbb{R}^8$, where

$$V_1(\xi_1) = \frac{1}{2}(E\xi_1 - p_x)^{\mathrm{T}} P(E\xi_1 - p_x), \quad \xi_1 \in \mathbb{R}^4, \tag{7.48}$$

$$V_2(\xi_2) = \frac{1}{2}(E\xi_2 - p_y)^{\mathrm{T}} P(E\xi_2 - p_y), \quad \xi_2 \in \mathbb{R}^4, \tag{7.49}$$

and

$$P = \begin{bmatrix} 2 & 1 & 0 \\ 1 & 2 & 0 \\ 0 & 0 & 1 \end{bmatrix} > 0. \tag{7.50}$$

Note that $V_i(\xi_i) = 0$, $\xi \in \mathcal{D}_i$, and $V_i(\xi_i) > 0$, $\xi \in \mathbb{R}^8 \setminus \mathcal{D}_i$, $i = 1, 2$. It can be shown that, for $\mathcal{R}_i \triangleq \{\xi \in \mathbb{R}^8 : V_i'(\xi_i)B = 0\}$, condition (7.40) is satisfied with

$$V_i'(\xi_i)A\xi_i \leq -\gamma_i V_i(\xi_i), \quad \xi \in \mathcal{R}_i, \quad i = 1, 2, \tag{7.51}$$

for $\gamma_i \in (0, 1]$, $i = 1, 2$.

In this case, the zero solution to (7.37) is globally exponentially stable with

$$w(V) = \begin{bmatrix} -\gamma_1 V_1 \\ -\gamma_2 V_2 \end{bmatrix}. \tag{7.52}$$

Furthermore, since $\lambda_{\min}(P) = 1$ and $\lambda_{\max}(P) = 3$, condition (7.42) is satisfied with $\alpha = \frac{1}{2}$, $\beta = \frac{3}{2}$, $\nu = 2$, and

$$\text{dist}(\xi, \mathcal{D}_0) \triangleq \left([(E\xi_1 - p_x)^{\mathrm{T}}, (E\xi_2 - p_y)^{\mathrm{T}}] \begin{bmatrix} E\xi_1 - p_x \\ E\xi_2 - p_y \end{bmatrix} \right)^{\frac{1}{2}}, \quad \xi \in \mathbb{R}^8. \tag{7.53}$$

Thus, it follows from Theorem 7.4 that $\mathcal{D}_0$ is globally exponentially stable with respect to (7.44), (7.45) with the feedback control law $\tilde{u}_i = \phi_i(\xi_i)$ given by (7.41), where $\rho_i(\xi_i) = V_i'(\xi_i)A\xi_i$, $\sigma_i(\xi_i) = B^{\mathrm{T}}V_i'^{\mathrm{T}}(\xi_i)$, $i = 1, 2$, and $w(V)$ is given by (7.52).

In the following simulation, we set $l_x = \frac{1}{\sqrt{2}}$, $l_y = \frac{1}{\sqrt{2}}$, $c_{0i} = 0.2$, $\gamma_i = \frac{1}{2}$, $i = 1, 2$, $\xi_{10} = [2, 5, -3, 2]^{\mathrm{T}}$, and $\xi_{20} = [3, 4, 4, 1]^{\mathrm{T}}$. With this choice of the parameters l_x and l_y, the steady-state distance between agents is 1 with the angle with respect to the horizontal being $\frac{\pi}{4}$. Figure 7.7 shows the position phase portrait of the two agents and Figure 7.8 shows the time history of the control forces acting on each agent.

7.6 Obstacle Avoidance in Multivehicle Coordination

Obstacle avoidance strategies for multivehicle problems include decentralized control approaches where local control laws are defined for each agent based on local information [2, 174] and behavior-based methods [50, 177]. Perhaps the most promising approach to obstacle avoidance is the potential field method, which has been extensively utilized for mobile robots with static and dynamic obstacles implemented in real time experiments [46, 57, 60, 146] using robust sliding mode controllers [55].

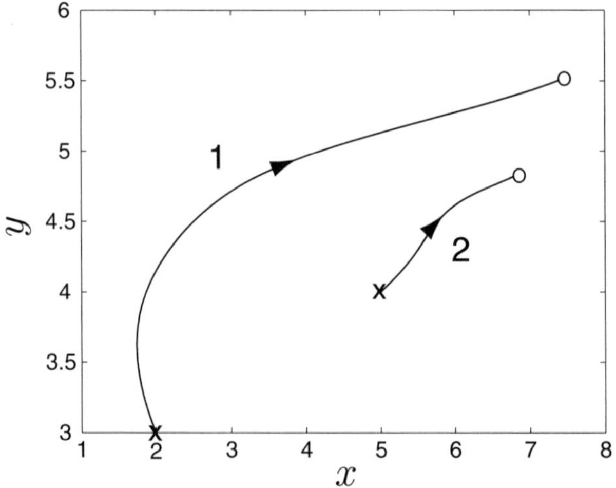

Figure 7.7 Position phase portrait of two agents.

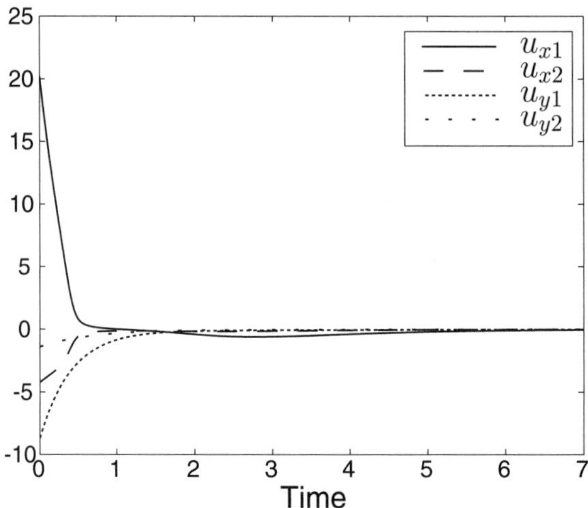

Figure 7.8 Control forces in horizontal and vertical directions versus time.

A more recent approach to obstacle avoidance is the limit cycle-based method introduced in [54]. Specifically, the authors in [54] use limit cycles to generate trajectories for robot manipulators by defining unstable limit cycles as objects of finite size and shape in order to avoid complex obstacles. The use of stable limit cycles as a navigation method has been introduced for obstacle avoidance of mobile robots in [111, 112]. The approach only considers circular limit cycles for mobile robots which are suitable for shapes with approximately the same length and width.

In this section, we use the results in this chapter to present an obstacle avoidance strategy that involves transitional trajectories defined as solutions to a set of ordinary differential equations possessing a stable limit cycle of a given elliptical shape. Specifically, the obstacle is encircled by an ellipse and once detected, the trajectory of an agent is replanned in such a way so that the new trajectory follows a solution to the aforementioned system of ordinary differential equations. As soon as the obstacle is cleared, the trajectory of the agent is set back to the original trajectory the agent was following before encountering the obstacle.

For planar agent motion, our obstacle avoidance strategy is based on approximating obstacles as continuously differentiable shapes that can be represented as a limit cycle solution of a planar system of ordinary differential equations. Specifically, we encircle an obstacle by an ellipse which serves as a limit cycle orbit of a certain two-dimensional dynamical system. As soon as the obstacle is detected, the trajectory of an agent is replanned so as to follow a new solution that clears the obstacle.

To elucidate this approach, let the state variables of the transitional trajectory be given by

$$\tilde{x} = x - x_{\mathrm{c}}, \tag{7.54}$$

$$\tilde{y} = y - y_{\mathrm{c}}, \tag{7.55}$$

where x and y denote horizontal and vertical displacements of an agent, respectively, and $(x_{\mathrm{c}}, y_{\mathrm{c}})$ denotes the location of the limit cycle origin. We consider limit cycles of elliptical form given by the zero level set of the function

$$l(\tilde{x}, \tilde{y}) = \left[\frac{\tilde{x} \cos\phi + \tilde{y} \sin\phi}{a} \right]^2 + \left[\frac{-\tilde{x} \sin\phi + \tilde{y} \cos\phi}{b} \right]^2 - 1, \tag{7.56}$$

where $\tilde{x}$ and $\tilde{y}$ are defined in (7.54) and (7.55), a and b are the semi-major and semi-minor axes, respectively, and ϕ is the angle representing the orientation of the ellipse's semi-major axis relative to the horizontal axis. Thus,

$$l(\tilde{x}, \tilde{y}) \equiv 0 \tag{7.57}$$

defines an ellipse centered at $(x_{\mathrm{c}}, y_{\mathrm{c}})$ with semi-major and semi-minor axes a and b, respectively, and with the semi-major axis forming angle ϕ relative to the horizontal axis.

Next, consider a planar dynamical system that exhibits a limit cycle of the form (7.57) given by

$$\dot{\tilde{x}}(t) = h_1(\tilde{x}(t), \tilde{y}(t)) - \tilde{x}(t)l(\tilde{x}(t), \tilde{y}(t)), \quad \tilde{x}(0) = \tilde{x}_0, \quad t \geq 0, \tag{7.58}$$

$$\dot{\tilde{y}}(t) = h_2(\tilde{x}(t), \tilde{y}(t)) - \tilde{y}(t)l(\tilde{x}(t), \tilde{y}(t)), \quad \tilde{y}(0) = \tilde{y}_0, \tag{7.59}$$

where $h_1(\tilde{x}, \tilde{y})$ and $h_2(\tilde{x}, \tilde{y})$ represent the agent dynamics on the limit cycle, that is, the dynamics on $l(\tilde{x}, \tilde{y}) = 0$. The dynamics of (7.58) and (7.59) must ensure that a trajectory starting from every point outside of the limit cycle, that is, every point $(\tilde{x}, \tilde{y}) \in \mathbb{R}^2$ such that $l(\tilde{x}, \tilde{y}) > 0$, will converge to the limit cycle without crossing it.

The motion of a particle along the ellipse given by (7.57) with the angular speed ω is given by

$$\tilde{x}(t) = a \cos \phi \cos \omega t - b \sin \phi \sin \omega t, \tag{7.60}$$

$$\tilde{y}(t) = a \sin \phi \cos \omega t + b \cos \phi \sin \omega t. \tag{7.61}$$

Thus, the time derivative of (7.60) and (7.61) is given by

$$\dot{\tilde{x}}(t) = -\omega(a \cos \phi \sin \omega t - b \sin \phi \cos \omega t), \tag{7.62}$$

$$\dot{\tilde{y}}(t) = \omega(-a \sin \phi \sin \omega t + b \cos \phi \cos \omega t). \tag{7.63}$$

Note that $\omega > 0$ and $\omega < 0$ represent counterclockwise and clockwise rotation of the particle, respectively. Eliminating $\cos \omega t$ and $\sin \omega t$ from (7.60)–(7.63) yields

$$\dot{\tilde{x}}(t) = \frac{\omega}{ab}(h_{e11}\tilde{x}(t) - h_{e12}\tilde{y}(t)), \tag{7.64}$$

$$\dot{\tilde{y}}(t) = \frac{\omega}{ab}(h_{e21}\tilde{x}(t) - h_{e11}\tilde{y}(t)), \tag{7.65}$$

where

$$h_{e11} = (a^2 - b^2) \sin \phi \cos \phi, \tag{7.66}$$

$$h_{e12} = a^2 \cos^2 \phi + b^2 \sin^2 \phi, \tag{7.67}$$

$$h_{e21} = b^2 \cos^2 \phi + a^2 \sin^2 \phi, \tag{7.68}$$

which represents the particle dynamics on the limit cycle. Thus, $h_1(\tilde{x}, \tilde{y})$ and $h_2(\tilde{x}, \tilde{y})$ in (7.58) and (7.59) are given by

$$h_1(\tilde{x}, \tilde{y}) = \frac{\omega}{ab}(h_{e11}\tilde{x} - h_{e12}\tilde{y}), \tag{7.69}$$

$$h_2(\tilde{x}, \tilde{y}) = \frac{\omega}{ab}(h_{e21}\tilde{x} - h_{e11}\tilde{y}). \tag{7.70}$$

Proposition 7.1. The system trajectories defined by (7.58) and (7.59) asymptotically converge to the elliptical limit cycle given by (7.57) for all initial conditions in $\mathcal{N} \triangleq \{(\tilde{x}, \tilde{y}) \in \mathbb{R}^2 : l(\tilde{x}, \tilde{y}) > 0\}$.

Proof. Consider the Lyapunov function candidate given by

$$V(\tilde{x}, \tilde{y}) = \frac{a^2 b^2}{2} l(\tilde{x}, \tilde{y}), \quad (\tilde{x}, \tilde{y}) \in \mathcal{N}. \tag{7.71}$$

Note that $V(\tilde{x}, \tilde{y}) > 0$ for all $(\tilde{x}, \tilde{y}) \in \mathcal{N}$. Now, the Lyapunov derivative along the trajectories of (7.58) and (7.59) is given by

$$\dot{V}(\tilde{x}, \tilde{y}) = -l(\tilde{x}, \tilde{y})[a^2(-\tilde{x} \sin \phi + \tilde{y} \cos \phi)^2 + b^2(\tilde{x} \cos \phi + \tilde{y} \sin \phi)^2] < 0,$$
$$(\tilde{x}, \tilde{y}) \in \mathcal{N}, \tag{7.72}$$

which implies asymptotic convergence of the trajectories of (7.58) and (7.59) to the ellipse (7.57) for all initial conditions in $\mathcal{N}$. $\qquad\square$

Note that an agent is only required to remain on the trajectory converging to a limit cycle for a finite time. As soon as the obstacle is cleared the agent returns back to its original trajectory. Next, we consider the multivehicle coordinated motion discussed in Section 7.3 with obstacles present in the path of the agents as well as the leader. Specifically, before an obstacle is encountered, the decentralized feedback controllers given by (7.20) drive each vehicle in the formation. Furthermore, in accordance with our obstacle avoidance strategy, when obstacles are detected they are approximated by ellipses. In this case, the dynamics of the ith agent is driven by

$$\dot{\tilde{x}}_i(t) = h_1(\tilde{x}_i(t), \tilde{y}_i(t)) - \tilde{x}_i(t)l(\tilde{x}_i(t), \tilde{y}_i(t)), \quad \tilde{x}_i(0) = \tilde{x}_{i0}, \quad t \geq 0, \quad (7.73)$$

$$\dot{\tilde{y}}_i(t) = h_2(\tilde{x}_i(t), \tilde{y}_i(t)) - \tilde{y}_i(t)l(\tilde{x}_i(t), \tilde{y}_i(t)), \quad \tilde{y}_i(0) = \tilde{y}_{i0}, \quad (7.74)$$

where $\tilde{x}_i \triangleq x_i - x_{ci}$, $\tilde{y}_i \triangleq y_i - y_{ci}$, (x_{ci}, y_{ci}) is the position of the center of an ellipse, and $h_1(\cdot, \cdot)$, $h_2(\cdot, \cdot)$, and $l(\cdot, \cdot)$ are given by (7.69), (7.70), and (7.56), respectively.

First, we consider the case when an obstacle is detected in the path of a formation leader. In this case, in addition to the elliptical limit cycle that encircles an obstacle, we define an elliptical region surrounding the obstacle. This region includes all points in $\mathbb{R}^2$ satisfying $l(\tilde{x}_L, \tilde{y}_L) > k$, where k is a safety factor that specifies the size of the elliptical region. Introducing such an elliptical region as a safety zone ensures that the leader will not collide with the obstacle and will have enough time to change its trajectory and converge to the elliptical limit cycle. When an obstacle is detected in the elliptical region, the leader changes its path from its original path to the path characterized by the solution to (7.73) and (7.74).

In order to ensure a smooth transition between the original and modified paths, we define an intermediate path for the leader given by a fifth-order polynomial. Specifically, as soon as the leader reaches the boundary of the elliptical region given by $l(\tilde{x}_L, \tilde{y}_L) = k$ at time $t = t_0$, the trajectory of the leader is driven to follow

$$x_L(t) = a_5\Delta t^5 + a_4\Delta t^4 + a_3\Delta t^3 + a_2\Delta t^2 + a_1\Delta t + a_0, \quad (7.75)$$

$$y_L(t) = b_5\Delta t^5 + b_4\Delta t^4 + b_3\Delta t^3 + b_2\Delta t^2 + b_1\Delta t + b_0, \quad (7.76)$$

where $\Delta t \triangleq t - t_0$ and coefficients a_i, b_i, $i = 1, \ldots, 5$, are determined from the boundary conditions for the position, velocity, and acceleration of the leader at times t_0 and t_1, where t_1 denotes the end time of the transitional path (7.75) and (7.76). After the transitional phase given by (7.75) and (7.76) is executed, the motion of the leader switches to the dynamics given by (7.73) and (7.74).

The decentralized control algorithm developed in Section 7.2 guarantees that when the leader bypasses the obstacle, the agents will follow the leader in a specified formation enforced by the exponential stability of the time-varying set describing the formation. As soon as the obstacle is cleared, the formation leader will switch its trajectory back to the original trajectory. In order to find the proper point of departure from the motion given by (7.73) and (7.74) to the motion on the original path, a line-drawing method has been used. In particular, a line is drawn between the current leader's position and its original position corresponding to the case where no obstacle is present. As long as there is an intersection between this line and the ellipse $l(\tilde{x}_{\mathrm{L}}, \tilde{y}_{\mathrm{L}}) = 0$, the leader's dynamics will remain driven by (7.73) and (7.74), and as soon as there is no intersection between the line and the ellipse, the leader will switch back to its original path and obstacle avoidance is guaranteed. The transition phase when the leader is departing from the motion characterized by (7.73) and (7.74) to its original trajectory is again described by the fifth-order polynomials given by (7.75) and (7.76).

Next, we consider the case when the kth agent in the formation encounters an obstacle. Recall that we design a stabilizing controller for the time-varying set (7.26) for each agent to ensure that the agent will be on a specified formation while following the formation leader. Now, as soon as the kth agent detects an obstacle, the time-varying set (7.26) for this agent is switched to

$$\tilde{\mathcal{D}}_k(t) \triangleq \{\eta \in \mathbb{R}^{4q} : \eta_k - \tilde{p}_k(t) = 0\}, \quad t \geq 0, \tag{7.77}$$

where

$$\tilde{p}_k(t) \triangleq \begin{bmatrix} \tilde{x}_k(t) + x_{\mathrm{c}k} \\ \tilde{y}_k(t) + y_{\mathrm{c}k} \\ \dot{\tilde{x}}_k(t) \\ \dot{\tilde{y}}_k(t) \end{bmatrix}, \quad t \geq 0, \tag{7.78}$$

and $\tilde{x}_k(t)$, $t \geq 0$, and $\tilde{y}_k(t)$, $t \geq 0$, are solutions to (7.73) and (7.74). The intersection of the sets (7.26) and (7.77) given by

$$\tilde{\mathcal{D}}_0^t = \tilde{\mathcal{D}}_0(t) \triangleq \bigcap_{i=1,\dots,q,\, i \neq k} \mathcal{D}_i(t) \bigcap \tilde{\mathcal{D}}_k(t), \quad t \geq 0, \tag{7.79}$$

characterizes a temporary desired formation of the agents with respect to the leader over the time interval until the obstacle is cleared by the kth agent. Specifically, when the kth agent encounters an obstacle, the time-varying set describing the desired formation will switch from (7.26) to (7.77). Then, after the obstacle is cleared, the time-varying set describing the desired formation will switch back to the original set (7.26). By switching between the above time-varying sets, we guarantee obstacle avoidance for each agent while maintaining the desired formation at steady state.

In the following simulation, we consider two agents pursuing a leader in a triangular formation with an obstacle in the path of each agent as

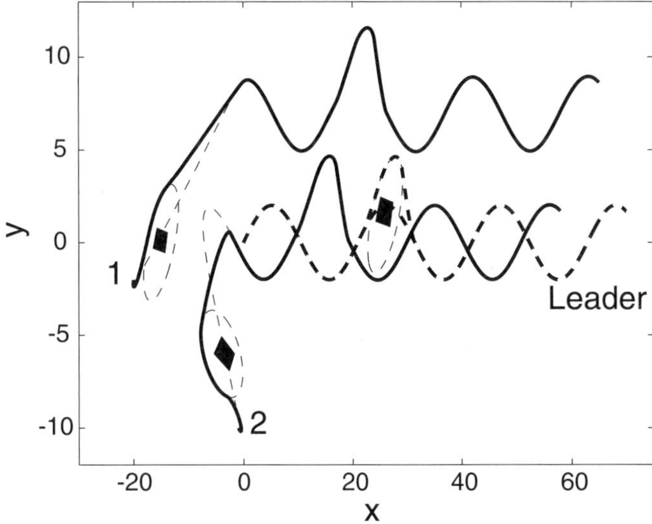

Figure 7.9 Position phase portrait of two agents following the leader. Thick dashed line represents trajectory of the leader.

well as the leader. We set $l_{x1L} = -5$, $l_{y1L} = 7$, $l_{x2L} = -12$, $l_{y2L} = 0$, $c_{0i} = 0.2$, $i = 1, 2$, $\gamma_i = \frac{1}{5}$, $i = 1, 2$, $\eta_{10} = [-20, -2, -1, -2]^T$, and $\eta_{20} = [-1, -10, 3, -2]^T$. With this choice of the parameters the agents will form an equilateral triangle configuration with respect to the leader. Furthermore, the leader is set to be moving on a sinusoidal path given by $x_L(t) = t$ and $y_L(t) = 2\sin(0.3t)$, $t \geq 0$. We define an obstacle for the first agent encircled by an ellipse with the parameters $x_c = -15$, $y_c = 0$, $a = 2$, $b = 4$, $\phi = 0.77$ rad, and $\omega = -0.75$ rad/s, and we define an obstacle for the second agent encircled by an ellipse with the parameters $x_c = -4$, $y_c = -6$, $a = 2$, $b = 4$, $\phi = -0.37$ rad, and $\omega = -0.32$ rad/s. The obstacle for the leader is encircled by the ellipse with parameters $x_c = 26$, $y_c = 1.5$, $a = 2$, $b = 4$, $\phi = 0.77$ rad, and $\omega = -0.22$ rad/s. As soon as the leader detects the obstacle, it switches its path to the path characterized by the solution of (7.73) and (7.74), and after the obstacle is cleared, the leader switches its trajectory back to the original sinusoidal path.

With feedback controller (7.20), Figure 7.9 shows the position phase portrait of the two agents following the leader while each agent and the leader avoid obstacles. Figure 7.10 shows that the agents eventually converge to the desired triangular formation after all obstacles are cleared. Finally, Figure 7.11 shows the time history of the control forces acting on each agent.

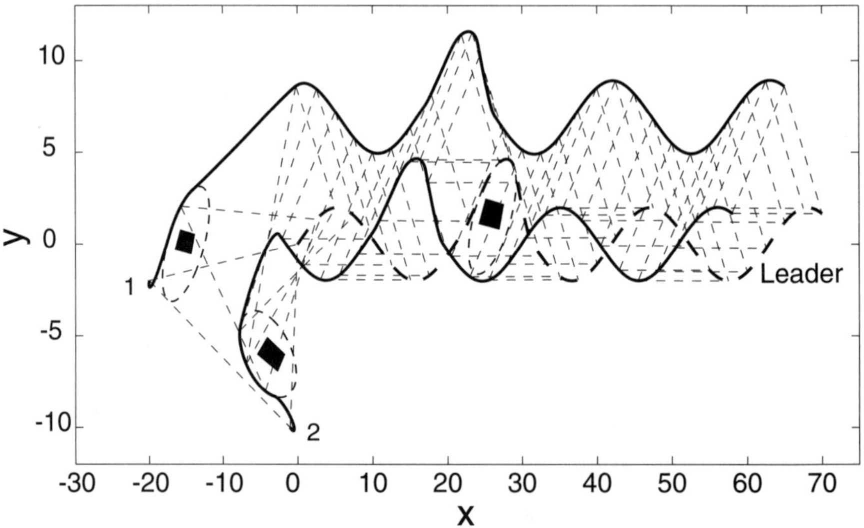

Figure 7.10 Position phase portrait of two agents following the leader in a triangular formation. Thick dashed line represents trajectory of the leader.

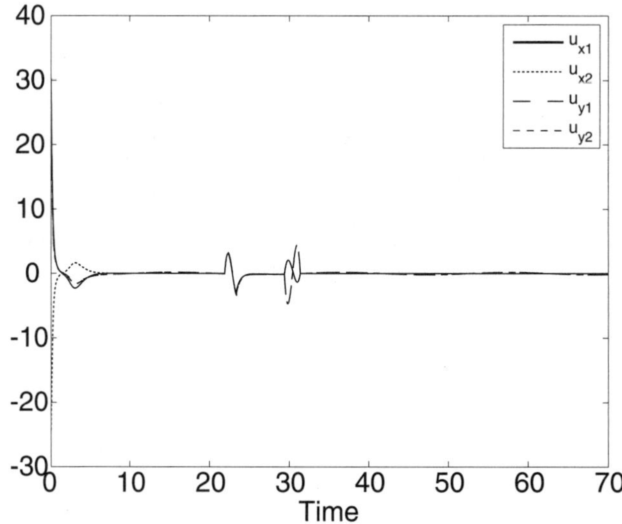

Figure 7.11 Control forces in horizontal and vertical directions versus time.

Large-Scale Discrete-Time Interconnected Dynamical Systems

8.1 Introduction

Since most physical processes evolve naturally in continuous-time, it is not surprising that the bulk of large-scale dynamical system theory has been developed for continuous-time systems. Nevertheless, it is the overwhelming trend to implement controllers digitally. Hence, in this chapter we extend the notions of dissipativity theory [70,170,171] to develop *vector dissipativity* notions for large-scale nonlinear discrete-time dynamical systems. In particular, we introduce a generalized definition of dissipativity for large-scale nonlinear discrete-time dynamical systems in terms of a *vector dissipation inequality* involving a *vector supply rate*, a *vector storage function*, and a nonnegative, semistable dissipation matrix. Generalized notions of vector available storage and vector required supply are also defined and shown to be element-by-element ordered, nonnegative, and finite. On the subsystem level, the proposed approach provides a discrete energy flow balance in terms of the stored subsystem energy, the supplied subsystem energy, the subsystem energy gained from all other subsystems independent of the subsystem coupling strengths, and the subsystem energy dissipated.

Furthermore, for large-scale discrete-time dynamical systems decomposed into interconnected subsystems, dissipativity of the composite system is shown to be determined from the dissipativity properties of the individual subsystems and the nature of the interconnections. In particular, we develop extended Kalman-Yakubovich-Popov conditions, in terms of the local subsystem dynamics and the interconnection constraints, for characterizing vector dissipativeness via vector storage functions for large-scale discrete-time dynamical systems. Finally, using the concepts of vector dissipativity and vector storage functions as candidate vector Lyapunov functions, we develop feedback interconnection stability results of large-scale discrete-time nonlinear dynamical systems. General stability criteria are given for Lyapunov and asymptotic stability of feedback interconnections of large-scale discrete-time dynamical systems. In the case of vector quadratic supply rates involving net subsystem powers and input-output subsystem energies, these results provide a positivity and small gain theorem for large-scale discrete-time systems predicated on vector Lyapunov functions.

8.2 Vector Dissipativity Theory for Discrete-Time Large-Scale Nonlinear Dynamical Systems

In this section, we extend the notion of dissipative dynamical systems to develop the generalized notion of vector dissipativity for discrete-time large-scale nonlinear dynamical systems. First, however, we recall the regular notions of dissipativity [40, 70] and geometric dissipativity [70] for discrete-time nonlinear dynamical systems $\mathcal{G}$ of the form

$$x(k+1) = f(x(k)) + G(x(k))u(k), \quad x(k_0) = x_0, \quad k \geq k_0, \qquad (8.1)$$
$$y(k) = h(x(k)) + J(x(k))u(k), \qquad (8.2)$$

where $x \in \mathcal{D} \subseteq \mathbb{R}^n$, $u \in U \subseteq \mathbb{R}^m$, $y \in Y \subseteq \mathbb{R}^l$, $f : \mathcal{D} \to \mathbb{R}^n$ and satisfies $f(0) = 0$, $G : \mathcal{D} \to \mathbb{R}^{n \times m}$, $h : \mathcal{D} \to \mathbb{R}^l$ and satisfies $h(0) = 0$, and $J : \mathcal{D} \to \mathbb{R}^{l \times m}$. For the discrete-time nonlinear dynamical system $\mathcal{G}$ we assume that the required properties for the existence and uniqueness of solutions are satisfied, that is, $u(\cdot)$ satisfies sufficient regularity conditions such that (8.1) has a unique solution forward in time. Note that since all input-output pairs $u(\cdot) \in \mathcal{U}$ and $y(\cdot) \in \mathcal{Y}$ of the discrete-time nonlinear dynamical system $\mathcal{G}$ are defined on $\overline{\mathbb{Z}}_+$, the *supply rate* [170] satisfying $s(0,0) = 0$ is locally summable for all input-output pairs satisfying (8.1) and (8.2), that is, for all input-output pairs $u(\cdot) \in \mathcal{U}$ and $y(\cdot) \in \mathcal{Y}$ satisfying (8.1) and (8.2), $s(\cdot, \cdot)$ satisfies $\sum_{k=k_1}^{k_2} |s(u(k), y(k))| < \infty$, $k_1, k_2 \in \overline{\mathbb{Z}}_+$.

Definition 8.1 ([70])**.** The discrete-time nonlinear dynamical system $\mathcal{G}$ given by (8.1) and (8.2) is *geometrically dissipative* (respectively, *dissipative*) with respect to the supply rate $s(u,y)$ if there exist a continuous nonnegative-definite function $v_{\mathrm{s}} : \mathbb{R}^n \to \overline{\mathbb{R}}_+$, called a *storage function*, and a scalar $\rho > 1$ (respectively, $\rho = 1$) such that $v_{\mathrm{s}}(0) = 0$ and the *dissipation inequality*

$$\rho^{k_2} v_{\mathrm{s}}(x(k_2)) \leq \rho^{k_1} v_{\mathrm{s}}(x(k_1)) + \sum_{i=k_1}^{k_2-1} \rho^{i+1} s(u(i), y(i)), \quad k_2 \geq k_1, \qquad (8.3)$$

is satisfied for all $k_2 \geq k_1 \geq k_0$, where $x(k), k \geq k_0$, is the solution to (8.1) with $u(\cdot) \in \mathcal{U}$. The discrete-time nonlinear dynamical system $\mathcal{G}$ given by (8.1) and (8.2) is *lossless with respect to the supply rate $s(u,y)$* if the dissipation inequality is satisfied as an equality with $\rho = 1$ for all $k_2 \geq k_1 \geq k_0$.

An equivalent statement for dissipativity of the dynamical system (8.1) and (8.2) is

$$\Delta v_{\mathrm{s}}(x(k)) \leq s(u(k), y(k)), \quad k \geq k_0, \quad u(\cdot) \in \mathcal{U}, \quad y(\cdot) \in \mathcal{Y}. \qquad (8.4)$$

Alternatively, an equivalent statement for geometric dissipativity of the dynamical system (8.1) and (8.2) is

$$\rho v_{\mathrm{s}}(x(k+1)) - v_{\mathrm{s}}(x(k)) \leq \rho s(u(k), y(k)), \quad k \geq k_0, \quad u(\cdot) \in \mathcal{U}, \quad y(\cdot) \in \mathcal{Y}. \tag{8.5}$$

Next, to develop vector dissipativity theory for discrete-time large-scale nonlinear dynamical systems, consider the discrete-time nonlinear dynamical systems $\mathcal{G}$ of the form

$$x(k+1) = F(x(k), u(k)), \quad x(k_0) = x_0, \quad k \geq k_0, \tag{8.6}$$
$$y(k) = H(x(k), u(k)), \tag{8.7}$$

where $x \in \mathcal{D} \subseteq \mathbb{R}^n$, $u \in U \subseteq \mathbb{R}^m$, $y \in Y \subseteq \mathbb{R}^l$, $F : \mathcal{D} \times U \to \mathbb{R}^n$, $H : \mathcal{D} \times U \to Y$, $\mathcal{D}$ is an open set with $0 \in \mathcal{D}$, and $F(0,0) = 0$. Here, we assume that $\mathcal{G}$ represents a discrete-time large-scale dynamical system composed of q interconnected controlled subsystems $\mathcal{G}_i$ such that, for all $i = 1, \ldots, q$,

$$F_i(x, u_i) = f_i(x_i) + \mathcal{I}_i(x) + G_i(x_i)u_i, \tag{8.8}$$
$$H_i(x_i, u_i) = h_i(x_i) + J_i(x_i)u_i, \tag{8.9}$$

where $x_i \in \mathbb{R}^{n_i}$, $u_i \in U_i \subseteq \mathbb{R}^{m_i}$, $y_i \triangleq H_i(x_i, u_i) \in Y_i \subseteq \mathbb{R}^{l_i}$, (u_i, y_i) is the input-output pair for the ith subsystem such that $u_i(\cdot) \in \mathcal{U}_i$, $y_i(\cdot) \in \mathcal{Y}_i$, where $\mathcal{U}_i$ and $\mathcal{Y}_i$ denote the ith subsystem input and output spaces, $f_i : \mathbb{R}^{n_i} \to \mathbb{R}^{n_i}$ and $\mathcal{I}_i : \mathcal{D} \to \mathbb{R}^{n_i}$ are continuous and satisfy $f_i(0) = 0$ and $\mathcal{I}_i(0) = 0$, $G_i : \mathbb{R}^{n_i} \to \mathbb{R}^{n_i \times m_i}$ is continuous, $h_i : \mathbb{R}^{n_i} \to \mathbb{R}^{l_i}$ and satisfies $h_i(0) = 0$, $J_i : \mathbb{R}^{n_i} \to \mathbb{R}^{l_i \times m_i}$, $\sum_{i=1}^{q} n_i = n$, $\sum_{i=1}^{q} m_i = m$, and $\sum_{i=1}^{q} l_i = l$. Furthermore, for the system $\mathcal{G}$ we assume that the required properties for the existence and uniqueness of solutions are satisfied. We define the composite input and composite output for the discrete-time large-scale system $\mathcal{G}$ as $u \triangleq [u_1^{\mathrm{T}}, \ldots, u_q^{\mathrm{T}}]^{\mathrm{T}}$ and $y \triangleq [y_1^{\mathrm{T}}, \ldots, y_q^{\mathrm{T}}]^{\mathrm{T}}$, respectively. Note that in this case the set $U = U_1 \times \cdots \times U_q$ contains the set of input values and $Y = Y_1 \times \cdots \times Y_q$ contains the set of output values whereas $\mathcal{U} = \mathcal{U}_1 \times \cdots \times \mathcal{U}_q$ and $\mathcal{Y} = \mathcal{Y}_1 \times \cdots \times \mathcal{Y}_q$ define the input and output spaces for (8.6) and (8.7).

Definition 8.2. For the discrete-time large-scale nonlinear dynamical system $\mathcal{G}$ given by (8.6) and (8.7) a vector function $S = [s_1, \ldots, s_q]^{\mathrm{T}} : U \times Y \to \mathbb{R}^q$ such that $S(u, y) \triangleq [s_1(u_1, y_1), \ldots, s_q(u_q, y_q)]^{\mathrm{T}}$ and $S(0, 0) = 0$ is called a *vector supply rate*.

Note that since all input-output pairs $(u_i, y_i) \in \mathcal{U}_i \times \mathcal{Y}_i$, $i = 1, \ldots, q$, satisfying (8.6) and (8.7) are defined on $\overline{\mathbb{Z}}_+$, $s_i(\cdot, \cdot)$ in Definition 8.2 satisfies $\sum_{k=k_1}^{k_2} |s_i(u_i(k), y_i(k))| < \infty$, $k_1, k_2 \in \overline{\mathbb{Z}}_+$.

Definition 8.3. The discrete-time large-scale nonlinear dynamical system $\mathcal{G}$ given by (8.6) and (8.7) is *vector dissipative* (respectively, *geometrically vector dissipative*) *with respect to the vector supply rate* $S(u, y)$ if there exist a continuous, nonnegative definite vector function $V_{\mathrm{s}} = [v_{\mathrm{s}1}, \ldots, v_{\mathrm{s}q}]^{\mathrm{T}} :$ $\mathcal{D} \to \overline{\mathbb{R}}_+^q$, called a *vector storage function*, and a nonsingular nonnegative *dissipation matrix* $W \in \mathbb{R}^{q \times q}$ such that $V_{\mathrm{s}}(0) = 0$, W is semistable (respectively, asymptotically stable), and the *vector dissipation inequality*

$$V_{\mathrm{s}}(x(k)) \leq\leq W^{k-k_0} V_{\mathrm{s}}(x(k_0)) + \sum_{i=k_0}^{k-1} W^{k-1-i} S(u(i), y(i)), \quad k \geq k_0,$$

(8.10)

is satisfied, where $x(k)$, $k \geq k_0$, is the solution to (8.6) with $u(\cdot) \in \mathcal{U}$. The discrete-time large-scale nonlinear dynamical system $\mathcal{G}$ given by (8.6) and (8.7) is *vector lossless with respect to the vector supply rate* $S(u, y)$ if the vector dissipation inequality is satisfied as an equality with W semistable.

Note that if the subsystems $\mathcal{G}_i$ of $\mathcal{G}$ are *disconnected*, that is, $\mathcal{I}_i(x) \equiv 0$ for all $i = 1, \ldots, q$, and $W \in \mathbb{R}^{q \times q}$ is diagonal, positive definite, and semistable, then it follows from Definition 8.3 that each of the isolated subsystems $\mathcal{G}_i$ is dissipative or geometrically dissipative in the sense of Definition 8.1. A similar remark holds in the case where $q = 1$.

Next, define the *vector available storage* of the discrete-time large-scale nonlinear dynamical system $\mathcal{G}$ by

$$V_{\mathrm{a}}(x_0) \triangleq \sup_{K \geq k_0, \, u(\cdot)} \left[-\sum_{k=k_0}^{K-1} W^{-(k+1-k_0)} S(u(k), y(k)) \right], \quad (8.11)$$

where $x(k)$, $k \geq k_0$, is the solution to (8.6) with $x(k_0) = x_0$ and admissible inputs $u(\cdot) \in \mathcal{U}$. The supremum in (8.11) is taken componentwise, which implies that for each component of $V_{\mathrm{a}}(\cdot)$ the supremum is calculated separately. Note, that $V_{\mathrm{a}}(x_0) \geq\geq 0$, $x_0 \in \mathcal{D}$, since $V_{\mathrm{a}}(x_0)$ is the supremum over a set of vectors containing the zero vector ($K = k_0$). To state the main results of this section the following definition is required.

Definition 8.4 ([70]). The discrete-time large-scale nonlinear dynamical system $\mathcal{G}$ given by (8.6) and (8.7) is *completely reachable* if for all $x_0 \in \mathcal{D} \subseteq \mathbb{R}^n$, there exist a $k_{\mathrm{i}} < k_0$ and a square summable input $u(\cdot)$ defined on $[k_{\mathrm{i}}, k_0]$ such that the state $x(k)$, $k \geq k_{\mathrm{i}}$, can be driven from $x(k_{\mathrm{i}}) = 0$ to $x(k_0) = x_0$. A discrete-time large-scale nonlinear dynamical system $\mathcal{G}$ is *zero-state observable* if $u(k) \equiv 0$ and $y(k) \equiv 0$ imply $x(k) \equiv 0$.

Theorem 8.1. Consider the discrete-time large-scale nonlinear dynamical system $\mathcal{G}$ given by (8.6) and (8.7), and assume that $\mathcal{G}$ is completely reachable. Let $W \in \mathbb{R}^{q \times q}$ be nonsingular, nonnegative, and semistable (respectively, asymptotically stable). Then

$$\sum_{k=k_0}^{K-1} W^{-(k+1-k_0)} S(u(k), y(k)) \geq\geq 0, \quad K \geq k_0, \quad u(\cdot) \in \mathcal{U}, \qquad (8.12)$$

for $x(k_0) = 0$ if and only if $V_a(0) = 0$ and $V_a(x)$ is finite for all $x \in \mathcal{D}$. Moreover, if (8.12) holds, then $V_a(x)$, $x \in \mathcal{D}$, is a vector storage function for $\mathcal{G}$, and hence, $\mathcal{G}$ is vector dissipative (respectively, geometrically vector dissipative) with respect to the vector supply rate $S(u, y)$.

Proof. Suppose $V_a(0) = 0$ and $V_a(x)$, $x \in \mathcal{D}$, is finite. Then

$$0 = V_a(0) = \sup_{K \geq k_0, u(\cdot)} \left[-\sum_{k=k_0}^{K-1} W^{-(k+1-k_0)} S(u(k), y(k)) \right], \qquad (8.13)$$

which implies (8.12).

Next, suppose (8.12) holds. Then, for $x(k_0) = 0$,

$$\sup_{K \geq k_0, u(\cdot)} \left[-\sum_{k=k_0}^{K-1} W^{-(k+1-k_0)} S(u(k), y(k)) \right] \leq\leq 0, \qquad (8.14)$$

which implies that $V_a(0) \leq\leq 0$. However, since $V_a(x_0) \geq\geq 0$, $x_0 \in \mathcal{D}$, it follows that $V_a(0) = 0$. Moreover, since $\mathcal{G}$ is completely reachable it follows that for every $x_0 \in \mathcal{D}$ there exists $\hat{k} > k_0$ and an admissible input $u(\cdot)$ defined on $[k_0, \hat{k}]$ such that $x(\hat{k}) = x_0$. Now, since (8.12) holds for $x(k_0) = 0$ it follows that for all admissible $u(\cdot) \in \mathcal{U}$,

$$\sum_{k=k_0}^{K-1} W^{-(k+1-k_0)} S(u(k), y(k)) \geq\geq 0, \quad K \geq \hat{k}, \qquad (8.15)$$

or, equivalently, multiplying (8.15) by the nonnegative matrix $W^{\hat{k}-k_0}$, $\hat{k} > k_0$, yields

$$-\sum_{k=\hat{k}}^{K-1} W^{-(k+1-\hat{k})} S(u(k), y(k)) \leq\leq \sum_{k=k_0}^{\hat{k}-1} W^{-(k+1-\hat{k})} S(u(k), y(k))$$

$$\leq\leq Q(x_0)$$

$$<<\infty, \quad K \geq \hat{k}, \quad u(\cdot) \in \mathcal{U}, \qquad (8.16)$$

where $Q : \mathcal{D} \to \mathbb{R}^q$. Hence,

$$V_a(x_0) = \sup_{K \geq \hat{k}, u(\cdot)} \left[-\sum_{k=\hat{k}}^{K-1} W^{-(k+1-\hat{k})} S(u(k), y(k)) \right]$$

$$\leq\leq Q(x_0)$$
$$<<\infty, \quad x_0 \in \mathcal{D}, \tag{8.17}$$

which implies that $V_{\mathrm{a}}(x_0)$, $x_0 \in \mathcal{D}$, is finite.

Finally, since (8.12) implies that $V_{\mathrm{a}}(0) = 0$ and $V_{\mathrm{a}}(x)$, $x \in \mathcal{D}$, is finite it follows from the definition of the vector available storage that

$$-V_{\mathrm{a}}(x_0) \leq\leq \sum_{k=k_0}^{K-1} W^{-(k+1-k_0)} S(u(k), y(k))$$

$$= \sum_{k=k_0}^{k_{\mathrm{f}}-1} W^{-(k+1-k_0)} S(u(k), y(k))$$

$$+ \sum_{k=k_{\mathrm{f}}}^{K-1} W^{-(k+1-k_0)} S(u(k), y(k)), \quad K \geq k_0. \tag{8.18}$$

Now, multiplying (8.18) by the nonnegative matrix $W^{k_{\mathrm{f}}-k_0}$, $k_{\mathrm{f}} > k_0$, it follows that

$$W^{k_{\mathrm{f}}-k_0} V_{\mathrm{a}}(x_0) + \sum_{k=k_0}^{k_{\mathrm{f}}-1} W^{-(k+1-k_{\mathrm{f}})} S(u(k), y(k))$$

$$\geq\geq \sup_{K \geq k_{\mathrm{f}}, u(\cdot)} \left[-\sum_{k=k_{\mathrm{f}}}^{K-1} W^{-(k+1-k_{\mathrm{f}})} S(u(k), y(k)) \right]$$

$$= V_{\mathrm{a}}(x(k_{\mathrm{f}})), \tag{8.19}$$

which implies that $V_{\mathrm{a}}(x)$, $x \in \mathcal{D}$, is a vector storage function, and hence, $\mathcal{G}$ is vector dissipative (respectively, geometrically vector dissipative) with respect to the vector supply rate $S(u, y)$. $\qquad\square$

It follows from Lemma 2.2 that if $W \in \mathbb{R}^{q \times q}$ is nonsingular, nonnegative, and semistable (respectively, asymptotically stable), then there exist a scalar $\alpha \geq 1$ (respectively, $\alpha > 1$) and a nonnegative vector $p \in \overline{\mathbb{R}}_+^q$, $p \neq 0$, (respectively, $p \in \mathbb{R}_+^q$) such that (2.97) holds. In this case,

$$p^{\mathrm{T}} W^{-k} = \alpha p^{\mathrm{T}} W^{-(k-1)} = \cdots = \alpha^k p^{\mathrm{T}}, \quad k \in \overline{\mathbb{Z}}_+. \tag{8.20}$$

Using (8.20), we define the (scalar) *available storage* for the discrete-time large-scale nonlinear dynamical system $\mathcal{G}$ by

$$v_{\mathrm{a}}(x_0) \triangleq \sup_{K \geq k_0, u(\cdot)} \left[-\sum_{k=k_0}^{K-1} p^{\mathrm{T}} W^{-(k+1-k_0)} S(u(k), y(k)) \right]$$

$$= \sup_{K \geq k_0, u(\cdot)} \left[-\sum_{k=k_0}^{K-1} \alpha^{k+1-k_0} s(u(k), y(k)) \right], \tag{8.21}$$

where $s : U \times Y \to \mathbb{R}$ defined as $s(u,y) \triangleq p^{\mathrm{T}}S(u,y)$ is the (scalar) supply rate for the discrete-time large-scale nonlinear dynamical system $\mathcal{G}$. Clearly, $v_{\mathrm{a}}(x) \geq 0$ for all $x \in \mathcal{D}$. As in standard discrete-time dissipativity theory, the available storage $v_{\mathrm{a}}(x)$, $x \in \mathcal{D}$, denotes the maximum amount of (scaled) energy that can be extracted from the discrete-time large-scale nonlinear dynamical system $\mathcal{G}$ at any instant K.

The following theorem relates vector storage functions and vector supply rates to scalar storage functions and scalar supply rates of discrete-time large-scale dynamical systems.

Theorem 8.2. Consider the discrete-time large-scale nonlinear dynamical system $\mathcal{G}$ given by (8.6) and (8.7). Suppose $\mathcal{G}$ is vector dissipative (respectively, geometrically vector dissipative) with respect to the vector supply rate $S : U \times Y \to \mathbb{R}^q$ and with vector storage function $V_{\mathrm{s}} : \mathcal{D} \to \overline{\mathbb{R}}^q_+$. Then there exists $p \in \overline{\mathbb{R}}^q_+$, $p \neq 0$, (respectively, $p \in \mathbb{R}^q_+$) such that $\mathcal{G}$ is dissipative (respectively, geometrically dissipative) with respect to the scalar supply rate $s(u,y) = p^{\mathrm{T}}S(u,y)$ and with storage function $v_{\mathrm{s}}(x) \triangleq p^{\mathrm{T}}V_{\mathrm{s}}(x)$, $x \in \mathcal{D}$. Moreover, in this case $v_{\mathrm{a}}(x)$, $x \in \mathcal{D}$, is a storage function for $\mathcal{G}$ and

$$0 \leq v_{\mathrm{a}}(x) \leq v_{\mathrm{s}}(x), \quad x \in \mathcal{D}. \tag{8.22}$$

Proof. Suppose $\mathcal{G}$ is vector dissipative (respectively, geometrically vector dissipative) with respect to the vector supply rate $S(u,y)$. Then there exist a nonsingular, nonnegative, and semistable (respectively, asymptotically stable) dissipation matrix W and a vector storage function $V_{\mathrm{s}} : \mathcal{D} \to \overline{\mathbb{R}}^q_+$ such that the dissipation inequality (8.10) holds. Furthermore, it follows from Lemma 2.2 that there exist $\alpha \geq 1$ (respectively, $\alpha > 1$) and a nonzero vector $p \in \overline{\mathbb{R}}^q_+$ (respectively, $p \in \mathbb{R}^q_+$) satisfying (2.97). Hence, premultiplying (8.10) by p^{T} and using (8.20) it follows that

$$v_{\mathrm{s}}(x(k)) \leq \alpha^{-(k-k_0)}v_{\mathrm{s}}(x(k_0)) + \sum_{i=k_0}^{k-1} \alpha^{-(k-1-i)}s(u(i),y(i)),$$

$$k \geq k_0, \quad u(\cdot) \in \mathcal{U}, \tag{8.23}$$

where $v_{\mathrm{s}}(x) = p^{\mathrm{T}}V_{\mathrm{s}}(x)$, $x \in \mathcal{D}$, which implies dissipativity (respectively, geometric dissipativity) of $\mathcal{G}$ with respect to the supply rate $s(u,y)$ and with storage function $v_{\mathrm{s}}(x)$, $x \in \mathcal{D}$.

Moreover, since $v_{\mathrm{s}}(0) = 0$, it follows from (8.23) that for $x(k_0) = 0$,

$$\sum_{i=k_0}^{k-1} \alpha^{i+1-k_0}s(u(i),y(i)) \geq 0, \quad k \geq k_0, \quad u(\cdot) \in \mathcal{U}, \tag{8.24}$$

which, using (8.21), implies that $v_{\mathrm{a}}(0) = 0$. Now, it can be easily shown that $v_{\mathrm{a}}(x)$, $x \in \mathcal{D}$, satisfies (8.23), and hence, the available storage defined by (8.21) is a storage function for $\mathcal{G}$.

Finally, it follows from (8.23) that

$$v_{\mathrm{s}}(x(k_0)) \geq \alpha^{k-k_0} v_{\mathrm{s}}(x(k)) - \sum_{i=k_0}^{k-1} \alpha^{i+1-k_0} s(u(i), y(i))$$

$$\geq - \sum_{i=k_0}^{k-1} \alpha^{i+1-k_0} s(u(i), y(i)), \quad k \geq k_0, \quad u(\cdot) \in \mathcal{U}, \qquad (8.25)$$

which implies

$$v_{\mathrm{s}}(x(k_0)) \geq \sup_{k \geq k_0,\, u(\cdot)} \left[- \sum_{i=k_0}^{k-1} \alpha^{i+1-k_0} s(u(i), y(i)) \right] = v_{\mathrm{a}}(x(k_0)), \qquad (8.26)$$

and hence, (8.22) holds. $\square$

It follows from Theorem 8.1 that if (8.12) holds for $x(k_0) = 0$, then the vector available storage $V_{\mathrm{a}}(x)$, $x \in \mathcal{D}$, is a vector storage function for $\mathcal{G}$. In this case, it follows from Theorem 8.2 that there exists $p \in \overline{\mathbb{R}}_+^q$, $p \neq 0$, such that $v_{\mathrm{s}}(x) \triangleq p^{\mathrm{T}} V_{\mathrm{a}}(x)$ is a storage function for $\mathcal{G}$ that satisfies (8.23), and hence, by (8.22), $v_{\mathrm{a}}(x) \leq p^{\mathrm{T}} V_{\mathrm{a}}(x)$, $x \in \mathcal{D}$. It is important to note that it follows from Theorem 8.2 that if $\mathcal{G}$ is vector dissipative, then $\mathcal{G}$ can either be (scalar) dissipative or (scalar) geometrically dissipative.

The following theorem provides sufficient conditions guaranteeing that all scalar storage functions defined in terms of vector storage functions, that is, $v_{\mathrm{s}}(x) = p^{\mathrm{T}} V_{\mathrm{s}}(x)$, of a given vector dissipative discrete-time large-scale nonlinear dynamical system are positive definite.

Theorem 8.3. Consider the discrete-time large-scale nonlinear dynamical system $\mathcal{G}$ given by (8.6) and (8.7), and assume that $\mathcal{G}$ is zero-state observable. Furthermore, assume that $\mathcal{G}$ is vector dissipative (respectively, geometrically vector dissipative) with respect to the vector supply rate $S(u, y)$ and there exist $\alpha \geq 1$ and $p \in \mathbb{R}_+^q$ such that (2.97) holds. In addition, assume that there exist functions $\kappa_i : Y_i \to U_i$ such that $\kappa_i(0) = 0$ and $s_i(\kappa_i(y_i), y_i) < 0$, $y_i \neq 0$, for all $i = 1, \ldots, q$. Then for all vector storage functions $V_{\mathrm{s}} : \mathcal{D} \to \overline{\mathbb{R}}_+^q$ the storage function $v_{\mathrm{s}}(x) \triangleq p^{\mathrm{T}} V_{\mathrm{s}}(x)$, $x \in \mathcal{D}$, is positive definite, that is, $v_{\mathrm{s}}(0) = 0$ and $v_{\mathrm{s}}(x) > 0$, $x \in \mathcal{D}$, $x \neq 0$.

Proof. It follows from Theorem 8.2 that $v_{\mathrm{a}}(x)$, $x \in \mathcal{D}$, is a storage function for $\mathcal{G}$ that satisfies (8.23). Next, suppose, *ad absurdum*, that there exists $x \in \mathcal{D}$ such that $v_{\mathrm{a}}(x) = 0$, $x \neq 0$. Then it follows from the definition of $v_{\mathrm{a}}(x)$, $x \in \mathcal{D}$, that for $x(k_0) = x$,

$$\sum_{k=k_0}^{K-1} \alpha^{k+1-k_0} s(u(k), y(k)) \geq 0, \quad K \geq k_0, \quad u(\cdot) \in \mathcal{U}. \qquad (8.27)$$

However, for $u_i = k_i(y_i)$ we have $s_i(\kappa_i(y_i), y_i) < 0$, $y_i \neq 0$, for all $i = 1, \ldots, q$ and since $p >> 0$ it follows that $y_i(k) = 0$, $k \geq k_0$, $i = 1, \ldots, q$, which further implies that $u_i(k) = 0$, $k \geq k_0$, $i = 1, \ldots, q$. Since $\mathcal{G}$ is zero-state observable it follows that $x = 0$, and hence, $v_a(x) = 0$ if and only if $x = 0$. The result now follows from (8.22). Finally, for the geometrically vector dissipative case it follows from Lemma 2.2 that $p >> 0$ with the rest of the proof identical, as above. $\square$

Next, we introduce the concept of *vector required supply* of a discrete-time large-scale nonlinear dynamical system. Specifically, define the vector required supply of the discrete-time large-scale dynamical system $\mathcal{G}$ by

$$V_r(x_0) \triangleq \inf_{K \geq -k_0+1,\, u(\cdot)} \sum_{k=-K}^{k_0-1} W^{-(k+1-k_0)} S(u(k), y(k)), \qquad (8.28)$$

where $x(k)$, $k \geq -K$, is the solution to (8.6) with $x(-K) = 0$ and $x(k_0) = x_0$. Note that since, with $x(k_0) = 0$, the infimum in (8.28) is the zero vector it follows that $V_r(0) = 0$. Moreover, since $\mathcal{G}$ is completely reachable it follows that $V_r(x) << \infty$, $x \in \mathcal{D}$. Using the notion of the vector required supply we present necessary and sufficient conditions for dissipativity of a large-scale dynamical system with respect to a vector supply rate.

Theorem 8.4. Consider the discrete-time large-scale nonlinear dynamical system $\mathcal{G}$ given by (8.6) and (8.7), and assume that $\mathcal{G}$ is completely reachable. Then $\mathcal{G}$ is vector dissipative (respectively, geometrically vector dissipative) with respect to the vector supply rate $S(u, y)$ if and only if

$$0 \leq\leq V_r(x) << \infty, \quad x \in \mathcal{D}. \qquad (8.29)$$

Moreover, if (8.29) holds, then $V_r(x)$, $x \in \mathcal{D}$, is a vector storage function for $\mathcal{G}$. Finally, if the vector available storage $V_a(x)$, $x \in \mathcal{D}$, is a vector storage function for $\mathcal{G}$, then

$$0 \leq\leq V_a(x) \leq\leq V_r(x) << \infty, \quad x \in \mathcal{D}. \qquad (8.30)$$

Proof. Suppose (8.29) holds and let $x(k)$, $k \in \overline{\mathbb{Z}}_+$, satisfy (8.6) with admissible inputs $u(\cdot) \in \mathcal{U}$, $k \in \overline{\mathbb{Z}}_+$, and $x(k_0) = x_0$. Then it follows from the definition of $V_r(\cdot)$ that for $-K \leq k_f \leq k_0 - 1$ and $u(\cdot) \in \mathcal{U}$,

$$V_r(x_0) \leq\leq \sum_{k=-K}^{k_0-1} W^{-(k+1-k_0)} S(u(k), y(k))$$

$$= \sum_{k=-K}^{k_f-1} W^{-(k+1-k_0)} S(u(k), y(k))$$

$$+ \sum_{k=k_{\mathrm{f}}}^{k_0-1} W^{-(k+1-k_0)} S(u(k), y(k)), \qquad (8.31)$$

and hence,

$$V_{\mathrm{r}}(x_0) \leq\leq W^{k_0-k_{\mathrm{f}}} \inf_{K \geq -k_{\mathrm{f}}+1,\, u(\cdot)} \left[\sum_{k=-K}^{k_{\mathrm{f}}-1} W^{-(k+1-k_{\mathrm{f}})} S(u(k), y(k)) \right]$$

$$+ \sum_{k=k_{\mathrm{f}}}^{k_0-1} W^{-(k+1-k_0)} S(u(k), y(k))$$

$$= W^{k_0-k_{\mathrm{f}}} V_{\mathrm{r}}(x(k_{\mathrm{f}})) + \sum_{k=k_{\mathrm{f}}}^{k_0-1} W^{k_0-1-k} S(u(k), y(k)), \qquad (8.32)$$

which shows that $V_{\mathrm{r}}(x)$, $x \in \mathcal{D}$, is a vector storage function for $\mathcal{G}$, and hence, $\mathcal{G}$ is vector dissipative with respect to the vector supply rate $S(u, y)$.

Conversely, suppose that $\mathcal{G}$ is vector dissipative with respect to the vector supply rate $S(u, y)$. Then there exists a nonnegative vector storage function $V_{\mathrm{s}}(x)$, $x \in \mathcal{D}$, such that $V_{\mathrm{s}}(0) = 0$. Since $\mathcal{G}$ is completely reachable it follows that for $x(k_0) = x_0$ there exist $K > -k_0$ and $u(k)$, $k \in [-K, k_0]$, such that $x(-K) = 0$. Hence, it follows from the vector dissipation inequality (8.10) that

$$0 \leq\leq V_{\mathrm{s}}(x(k_0)) \leq\leq W^{k_0+K} V_{\mathrm{s}}(x(-K)) + \sum_{k=-K}^{k_0-1} W^{k_0-1-k} S(u(k), y(k)),$$

$$(8.33)$$

which implies that for all $K \geq -k_0 + 1$ and $u(\cdot) \in \mathcal{U}$,

$$0 \leq\leq \sum_{k=-K}^{k_0-1} W^{-(k+1-k_0)} S(u(k), y(k)) \qquad (8.34)$$

or, equivalently,

$$0 \leq\leq \inf_{K \geq -k_0+1,\, u(\cdot)} \sum_{k=-K}^{k_0-1} W^{-(k+1-k_0)} S(u(k), y(k)) = V_{\mathrm{r}}(x_0). \qquad (8.35)$$

Since, by complete reachability $V_{\mathrm{r}}(x) << \infty$, $x \in \mathcal{D}$, it follows that (8.29) holds.

Finally, suppose that $V_{\mathrm{a}}(x)$, $x \in \mathcal{D}$, is a vector storage function. Then for $x(-K) = 0$, $x(k_0) = x_0$, and $u(\cdot) \in \mathcal{U}$, it follows that

$$V_{\mathrm{a}}(x(k_0)) \leq\leq W^{k_0+K} V_{\mathrm{a}}(x(-K)) + \sum_{k=-K}^{k_0-1} W^{k_0-1-k} S(u(k), y(k)), \quad (8.36)$$

which implies that

$$0 \leq\leq V_{\mathrm{a}}(x(k_0))$$

$$\leq\leq \inf_{K \geq -k_0+1,\, u(\cdot)} \sum_{k=-K}^{k_0-1} W^{-(k+1-k_0)} S(u(k), y(k))$$

$$= V_{\mathrm{r}}(x(k_0)), \quad x \in \mathcal{D}. \tag{8.37}$$

Since $x(k_0) = x_0 \in \mathcal{D}$ is arbitrary and, by complete reachability, $V_{\mathrm{r}}(x) << \infty$, $x \in \mathcal{D}$, (8.37) implies (8.30). $\qquad\square$

The next result is a direct consequence of Theorems 8.1 and 8.4.

Proposition 8.1. Consider the discrete-time large-scale nonlinear dynamical system $\mathcal{G}$ given by (8.6) and (8.7). Let $M = \mathrm{diag}\,[\mu_1, \ldots, \mu_q]$ be such that $0 \leq \mu_i \leq 1$, $i = 1, \ldots, q$. If $V_{\mathrm{a}}(x)$, $x \in \mathcal{D}$, and $V_{\mathrm{r}}(x)$, $x \in \mathcal{D}$, are vector storage functions for $\mathcal{G}$, then

$$V_{\mathrm{s}}(x) = M V_{\mathrm{a}}(x) + (I_q - M) V_{\mathrm{r}}(x), \quad x \in \mathcal{D}, \tag{8.38}$$

is a vector storage function for $\mathcal{G}$.

Proof. Note that $M \geq\geq 0$ and $I_q - M \geq\geq 0$ if and only if $M = \mathrm{diag}\,[\mu_1, \ldots, \mu_q]$ and $\mu_i \in [0, 1]$, $i = 1, \ldots, q$. Now, the result is a direct consequence of the vector dissipation inequality (8.10) by noting that if $V_{\mathrm{a}}(x)$ and $V_{\mathrm{r}}(x)$ satisfy (8.10), then $V_{\mathrm{s}}(x)$ satisfies (8.10). $\qquad\square$

Next, recall that if $\mathcal{G}$ is vector dissipative (respectively, geometrically vector dissipative), then there exist $p \in \overline{\mathbb{R}}_+^q$, $p \neq 0$, and $\alpha \geq 1$ (respectively, $p \in \mathbb{R}_+^q$ and $\alpha > 1$) such that (2.97) and (8.20) hold. Now, define the (scalar) *required supply* for the large-scale nonlinear dynamical system $\mathcal{G}$ by

$$v_{\mathrm{r}}(x_0) \triangleq \inf_{K \geq -k_0+1,\, u(\cdot)} \sum_{k=-K}^{k_0-1} p^{\mathrm{T}} W^{-(k+1-k_0)} S(u(k), y(k))$$

$$= \inf_{K \geq -k_0+1,\, u(\cdot)} \sum_{k=-K}^{k_0-1} \alpha^{k+1-k_0} s(u(k), y(k)), \quad x_0 \in \mathcal{D}, \tag{8.39}$$

where $s(u, y) = p^{\mathrm{T}} S(u, y)$ and $x(k)$, $k \geq -K$, is the solution to (8.6) with $x(-K) = 0$ and $x(k_0) = x_0$. It follows from (8.39) that the required supply of a discrete-time large-scale nonlinear dynamical system is the minimum amount of generalized energy that can be delivered to the discrete-time large-scale system in order to transfer it from an initial state $x(-K) = 0$ to a given state $x(k_0) = x_0$. Using the same arguments as in the case of the vector required supply, it follows that $v_{\mathrm{r}}(0) = 0$ and $v_{\mathrm{r}}(x) < \infty$, $x \in \mathcal{D}$.

Next, using the notion of required supply, we show that all storage functions of the form $v_{\mathrm{s}}(x) = p^{\mathrm{T}} V_{\mathrm{s}}(x)$, where $p \in \overline{\mathbb{R}}_+^q$, $p \neq 0$, are bounded from above by the required supply and bounded from below by the available storage. Hence, a dissipative discrete-time large-scale nonlinear dynamical system can deliver to its surroundings only a fraction of all of its stored subsystem energies and can store only a fraction of the work done to all of its subsystems.

Corollary 8.1. Consider the discrete-time large-scale nonlinear dynamical system $\mathcal{G}$ given by (8.6), (8.7). Assume that $\mathcal{G}$ is vector dissipative with respect to a vector supply rate $S(u,y)$ and with vector storage function $V_{\mathrm{s}} : \mathcal{D} \to \overline{\mathbb{R}}_+^q$. Then $v_{\mathrm{r}}(x)$, $x \in \mathcal{D}$, is a storage function for $\mathcal{G}$. Moreover, if $v_{\mathrm{s}}(x) \triangleq p^{\mathrm{T}} V_{\mathrm{s}}(x)$, $x \in \mathcal{D}$, where $p \in \overline{\mathbb{R}}_+^q$, $p \neq 0$, then

$$0 \leq v_{\mathrm{a}}(x) \leq v_{\mathrm{s}}(x) \leq v_{\mathrm{r}}(x) < \infty, \quad x \in \mathcal{D}. \tag{8.40}$$

Proof. It follows from Theorem 8.2 that if $\mathcal{G}$ is vector dissipative with respect to the vector supply rate $S(u,y)$ and with a vector storage function $V_{\mathrm{s}} : \mathcal{D} \to \overline{\mathbb{R}}_+^q$, then there exists $p \in \overline{\mathbb{R}}_+^q$, $p \neq 0$, such that $\mathcal{G}$ is dissipative with respect to the supply rate $s(u,y) = p^{\mathrm{T}} S(u,y)$ and with storage function $v_{\mathrm{s}}(x) = p^{\mathrm{T}} V_{\mathrm{s}}(x)$, $x \in \mathcal{D}$. Hence, it follows from (8.23), with $x(-K) = 0$ and $x(k_0) = x_0$, that

$$\sum_{k=-K}^{k_0-1} \alpha^{k+1-k_0} s(u(k), y(k)) \geq 0, \quad K \geq -k_0, \quad u(\cdot) \in \mathcal{U}, \tag{8.41}$$

which implies that $v_{\mathrm{r}}(x_0) \geq 0$, $x_0 \in \mathcal{D}$. Furthermore, it is easy to see from the definition of a required supply that $v_{\mathrm{r}}(x)$, $x \in \mathcal{D}$, satisfies the dissipation inequality (8.23). Hence, $v_{\mathrm{r}}(x)$, $x \in \mathcal{D}$, is a storage function for $\mathcal{G}$.

Moreover, it follows from the dissipation inequality (8.23), with $x(-K) = 0$, $x(k_0) = x_0$, and $u \in \mathcal{U}$, that

$$\alpha^{k_0} v_{\mathrm{s}}(x(k_0)) \leq \alpha^{-K} v_{\mathrm{s}}(x(-K)) + \sum_{k=-K}^{k_0-1} \alpha^{k+1} s(u(k), y(k))$$

$$= \sum_{k=-K}^{k_0-1} \alpha^{k+1} s(u(k), y(k)), \tag{8.42}$$

which implies that

$$v_{\mathrm{s}}(x(k_0)) \leq \inf_{K \geq -k_0+1, \, u(\cdot)} \sum_{k=-K}^{k_0-1} \alpha^{k+1-k_0} s(u(k), y(k)) = v_{\mathrm{r}}(x(k_0)). \tag{8.43}$$

Finally, it follows from Theorem 8.2 that $v_{\mathrm{a}}(x)$, $x \in \mathcal{D}$, is a storage function for $\mathcal{G}$, and hence, using (8.22) and (8.43), (8.40) holds. $\square$

It follows from Theorem 8.4 that if $\mathcal{G}$ is vector dissipative with respect to the vector supply rate $S(u,y)$, then $V_{\mathrm{r}}(x)$, $x \in \mathcal{D}$, is a vector storage function for $\mathcal{G}$ and, by Theorem 8.2, there exists $p \in \overline{\mathbb{R}}_+^q$, $p \neq 0$, such that $v_{\mathrm{s}}(x) \triangleq p^{\mathrm{T}} V_{\mathrm{r}}(x)$, $x \in \mathcal{D}$, is a storage function for $\mathcal{G}$ satisfying (8.23). Hence, it follows from Corollary 8.1 that $p^{\mathrm{T}} V_{\mathrm{r}}(x) \leq v_{\mathrm{r}}(x)$, $x \in \mathcal{D}$.

The next result relates vector (respectively, scalar) available storage and vector (respectively, scalar) required supply for vector lossless discrete-time large-scale dynamical systems.

Theorem 8.5. Consider the discrete-time large-scale nonlinear dynamical system $\mathcal{G}$ given by (8.6) and (8.7). Assume that $\mathcal{G}$ is completely reachable to and from the origin. If $\mathcal{G}$ is vector lossless with respect to the vector supply rate $S(u, y)$ and $V_{\mathrm{a}}(x)$, $x \in \mathcal{D}$, is a vector storage function, then $V_{\mathrm{a}}(x) = V_{\mathrm{r}}(x)$, $x \in \mathcal{D}$. Moreover, if $V_{\mathrm{s}}(x)$, $x \in \mathcal{D}$, is a vector storage function, then all (scalar) storage functions of the form $v_{\mathrm{s}}(x) = p^{\mathrm{T}} V_{\mathrm{s}}(x)$, $x \in \mathcal{D}$, where $p \in \overline{\mathbb{R}}_+^q$, $p \neq 0$, are given by

$$v_{\mathrm{s}}(x_0) = v_{\mathrm{a}}(x_0) = v_{\mathrm{r}}(x_0) = - \sum_{k=k_0}^{K_+ - 1} \alpha^{k+1-k_0} s(u(k), y(k))$$

$$= \sum_{k=-K_-}^{k_0 - 1} \alpha^{k+1-k_0} s(u(k), y(k)), \qquad (8.44)$$

where $x(k)$, $k \geq k_0$, is the solution to (8.6) with $u(\cdot) \in \mathcal{U}$, $x(k_0) = x_0 \in \mathcal{D}$, and $s(u, y) = p^{\mathrm{T}} S(u, y)$ for every K_-, K_+ such that $x(-K_-) = 0$ and $x(K_+) = 0$.

Proof. Suppose $\mathcal{G}$ is vector lossless with respect to the vector supply rate $S(u, y)$. Since $\mathcal{G}$ is completely reachable to and from the origin it follows that for every $x_0 = x(k_0) \in \mathcal{D}$ there exist $K_+ > k_0$, $-K_- < k_0$, and $u(k) \in U$, $k \in [-K_-, K_+]$, such that $x(-K_-) = 0$, $x(K_+) = 0$, and $x(k_0) = x_0$. Now, it follows from the dissipation inequality (8.10), which is satisfied as an equality, that

$$0 = \sum_{k=-K_-}^{K_+ - 1} W^{K_+ - 1 - k} S(u(k), y(k)), \qquad (8.45)$$

or, equivalently,

$$0 = \sum_{k=-K_-}^{K_+ - 1} W^{-(k+1-k_0)} S(u(k), y(k))$$

$$= \sum_{k=-K_-}^{k_0 - 1} W^{-(k+1-k_0)} S(u(k), y(k))$$

$$+ \sum_{k=k_0}^{K_+-1} W^{-(k+1-k_0)} S(u(k), y(k))$$

$$\geq\geq \inf_{K \geq -k_0+1,\, u(\cdot)} \sum_{k=-K}^{k_0-1} W^{-(k+1-k_0)} S(u(k), y(k))$$

$$+ \inf_{K \geq k_0,\, u(\cdot)} \sum_{k=k_0}^{K-1} W^{-(k+1-k_0)} S(u(k), y(k))$$

$$= V_{\mathrm{r}}(x_0) - V_{\mathrm{a}}(x_0), \tag{8.46}$$

which implies that $V_{\mathrm{r}}(x_0) \leq\leq V_{\mathrm{a}}(x_0)$, $x_0 \in \mathcal{D}$. However, it follows from Theorem 8.4 that if $\mathcal{G}$ is vector dissipative and $V_{\mathrm{a}}(x)$, $x \in \mathcal{D}$, is a vector storage function, then $V_{\mathrm{a}}(x) \leq\leq V_{\mathrm{r}}(x)$, $x \in \mathcal{D}$, which along with (8.46) implies that $V_{\mathrm{a}}(x) = V_{\mathrm{r}}(x)$, $x \in \mathcal{D}$. Furthermore, since $\mathcal{G}$ is vector lossless there exist a nonzero vector $p \in \overline{\mathbb{R}}_+^q$ and a scalar $\alpha \geq 0$ satisfying (2.97).

Next, it follows from (8.45) that

$$0 = \sum_{k=-K_-}^{K_+-1} p^{\mathrm{T}} W^{-(k+1-k_0)} S(u(k), y(k))$$

$$= \sum_{k=-K_-}^{K_+-1} \alpha^{k+1-k_0} s(u(k), y(k))$$

$$= \sum_{k=-K_-}^{k_0-1} \alpha^{k+1-k_0} s(u(k), y(k)) + \sum_{k=k_0}^{K_+-1} \alpha^{k+1-k_0} s(u(k), y(k))$$

$$\geq \inf_{K \geq -k_0+1,\, u(\cdot)} \sum_{k=-K}^{k_0-1} \alpha^{k+1-k_0} s(u(k), y(k))$$

$$+ \inf_{K \geq k_0,\, u(\cdot)} \sum_{k=k_0}^{K-1} \alpha^{k+1-k_0} s(u(k), y(k))$$

$$= v_{\mathrm{r}}(x_0) - v_{\mathrm{a}}(x_0), \quad x_0 \in \mathcal{D}, \tag{8.47}$$

which along with (8.40) implies that for any (scalar) storage function of the form $v_{\mathrm{s}}(x) = p^{\mathrm{T}} V_{\mathrm{s}}(x)$, $x \in \mathcal{D}$, the equality $v_{\mathrm{a}}(x) = v_{\mathrm{s}}(x) = v_{\mathrm{r}}(x)$, $x \in \mathcal{D}$, holds. Moreover, since $\mathcal{G}$ is vector lossless the inequalities (8.23) and (8.42) are satisfied as equalities and

$$v_{\mathrm{s}}(x_0) = - \sum_{k=k_0}^{K_+-1} \alpha^{k+1-k_0} s(u(k), y(k)) = \sum_{k=-K_-}^{k_0-1} \alpha^{k+1-k_0} s(u(k), y(k)),$$

$$\tag{8.48}$$

where $x(k)$, $k \geq k_0$, is the solution to (8.6) with $u(\cdot) \in \mathcal{U}$, $x(-K_-) = 0$, $x(K_+) = 0$, and $x(k_0) = x_0 \in \mathcal{D}$. $\qquad\square$

The next proposition presents a characterization for vector dissipativity of discrete-time large-scale nonlinear dynamical systems.

Proposition 8.2. Consider the discrete-time large-scale nonlinear dynamical system $\mathcal{G}$ given by (8.6) and (8.7), and assume $V_{\mathrm{s}} = [v_{\mathrm{s}1}, \ldots, v_{\mathrm{s}q}]^{\mathrm{T}} : \mathcal{D} \to \overline{\mathbb{R}}_+^q$ is a continuous vector storage function for $\mathcal{G}$. Then $\mathcal{G}$ is vector dissipative with respect to the vector supply rate $S(u, y)$ if and only if

$$V_{\mathrm{s}}(x(k+1)) \leq\leq W V_{\mathrm{s}}(x(k)) + S(u(k), y(k)), \quad k \geq k_0, \quad u(k) \in U. \quad (8.49)$$

Proof. The proof is immediate from (8.10) and, hence, is omitted. $\square$

As a special case of vector dissipativity theory we can analyze the stability of discrete-time large-scale nonlinear dynamical systems. Specifically, assume that the discrete-time large-scale dynamical system $\mathcal{G}$ is vector dissipative (respectively, geometrically vector dissipative) with respect to the vector supply rate $S(u, y)$ and with a continuous vector storage function $V_{\mathrm{s}} : \mathcal{D} \to \overline{\mathbb{R}}_+^q$. Moreover, assume that the conditions of Theorem 8.3 are satisfied. Then it follows from Proposition 8.2, with $u(k) \equiv 0$ and $y(k) \equiv 0$, that

$$V_{\mathrm{s}}(x(k+1)) \leq\leq W V_{\mathrm{s}}(x(k)), \quad k \geq k_0, \qquad (8.50)$$

where $x(k)$, $k \geq k_0$, is a solution to (8.6) with $x(k_0) = x_0$ and $u(k) \equiv 0$. Now, it follows from Corollary 2.6, with $w(r) = Wr$, that the zero solution $x(k) \equiv 0$ to (8.6), with $u(k) \equiv 0$, is Lyapunov (respectively, asymptotically) stable.

More generally, the problem of control system design for discrete-time large-scale nonlinear dynamical systems can be addressed within the framework of vector dissipativity theory. In particular, suppose that there exists a continuous vector function $V_{\mathrm{s}} : \mathcal{D} \to \overline{\mathbb{R}}_+^q$ such that $V_{\mathrm{s}}(0) = 0$ and

$$V_{\mathrm{s}}(x(k+1)) \leq\leq \mathcal{F}(V_{\mathrm{s}}(x(k)), u(k)), \quad k \geq k_0, \quad u(k) \in U, \qquad (8.51)$$

where $\mathcal{F} : \overline{\mathbb{R}}_+^q \times \mathbb{R}^m \to \mathbb{R}^q$ and $\mathcal{F}(0, 0) = 0$. Then the control system design problem for a discrete-time large-scale dynamical system reduces to constructing an *energy* feedback control law $\phi : \overline{\mathbb{R}}_+^q \to U$ of the form

$$u = \phi(V_{\mathrm{s}}(x)) \triangleq [\phi_1^{\mathrm{T}}(V_{\mathrm{s}}(x)), \ldots, \phi_q^{\mathrm{T}}(V_{\mathrm{s}}(x))]^{\mathrm{T}}, \quad x \in \mathcal{D}, \qquad (8.52)$$

where $\phi_i : \overline{\mathbb{R}}_+^q \to U_i$, $\phi_i(0) = 0$, $i = 1, \ldots, q$, such that the zero solution $r(k) \equiv 0$ to the comparison system

$$r(k+1) = w(r(k)), \quad r(k_0) = V_{\mathrm{s}}(x(k_0)), \quad k \geq k_0, \qquad (8.53)$$

is rendered asymptotically stable, where $w(r) \triangleq \mathcal{F}(r, \phi(r))$ is of class $\mathcal{W}_{\mathrm{d}}$. In this case, if there exists $p \in \overline{\mathbb{R}}_+^q$ such that $v_{\mathrm{s}}(x) \triangleq p^{\mathrm{T}} V_{\mathrm{s}}(x)$, $x \in \mathcal{D}$, is positive definite, then it follows from Corollary 2.6 that the zero solution $x(k) \equiv 0$ to (8.6), with u given by (8.52), is asymptotically stable.

As in the continuous-time case, using an energy feedback control architecture and exploiting the comparison system within the control design for discrete-time large-scale nonlinear dynamical systems can significantly reduce the dimensionality of a control synthesis problem in terms of a number of states that need to be stabilized. It should be noted, however, that for stability analysis of discrete-time large-scale dynamical systems the comparison system need not be linear as implied by (8.50). A discrete-time nonlinear comparison system would still guarantee stability of a discrete-time large-scale dynamical system provided that the conditions of Corollary 2.6 are satisfied.

8.3 Extended Kalman-Yakubovich-Popov Conditions for Discrete-Time Large-Scale Nonlinear Dynamical Systems

In this section, we show that vector dissipativeness (respectively, geometric vector dissipativeness) of a discrete-time large-scale nonlinear dynamical system $\mathcal{G}$ of the form (8.6) and (8.7) can be characterized in terms of the local subsystem functions $f_i(\cdot)$, $G_i(\cdot)$, $h_i(\cdot)$, and $J_i(\cdot)$, along with the interconnection structures $\mathcal{I}_i(\cdot)$ for $i = 1, \ldots, q$. For the results in this section we consider the special case of dissipative systems with quadratic vector supply rates and set $\mathcal{D} = \mathbb{R}^n$, $U_i = \mathbb{R}^{m_i}$, and $Y_i = \mathbb{R}^{l_i}$. Specifically, let $R_i \in \mathbb{S}^{m_i}$, $S_i \in \mathbb{R}^{l_i \times m_i}$, and $Q_i \in \mathbb{S}^{l_i}$ be given and assume $S(u, y)$ is such that $s_i(u_i, y_i) = y_i^{\mathrm{T}} Q_i y_i + 2 y_i^{\mathrm{T}} S_i u_i + u_i^{\mathrm{T}} R_i u_i$, $i = 1, \ldots, q$.

For the statement of the next result recall that $x = [x_1^{\mathrm{T}}, \ldots, x_q^{\mathrm{T}}]^{\mathrm{T}}$, $u = [u_1^{\mathrm{T}}, \ldots, u_q^{\mathrm{T}}]^{\mathrm{T}}$, $y = [y_1^{\mathrm{T}}, \ldots, y_q^{\mathrm{T}}]^{\mathrm{T}}$, $x_i \in \mathbb{R}^{n_i}$, $u_i \in \mathbb{R}^{m_i}$, $y_i \in \mathbb{R}^{l_i}$, $i = 1, \ldots, q$, $\sum_{i=1}^q n_i = n$, $\sum_{i=1}^q m_i = m$, and $\sum_{i=1}^q l_i = l$. Furthermore, for (8.6) and (8.7) define $\mathcal{F} : \mathbb{R}^n \to \mathbb{R}^n$, $G : \mathbb{R}^n \to \mathbb{R}^{n \times m}$, $h : \mathbb{R}^n \to \mathbb{R}^l$, and $J : \mathbb{R}^n \to \mathbb{R}^{l \times m}$ by $\mathcal{F}(x) \triangleq [\mathcal{F}_1^{\mathrm{T}}(x), \ldots, \mathcal{F}_q^{\mathrm{T}}(x)]^{\mathrm{T}}$, where $\mathcal{F}_i(x) \triangleq f_i(x_i) + \mathcal{I}_i(x)$, $i = 1, \ldots, q$, $G(x) \triangleq \mathrm{block\text{-}diag}[G_1(x_1), \ldots, G_q(x_q)]$, $h(x) \triangleq [h_1^{\mathrm{T}}(x_1), \ldots, h_q^{\mathrm{T}}(x_q)]^{\mathrm{T}}$, and $J(x) \triangleq \mathrm{block\text{-}diag}[J_1(x_1), \ldots, J_q(x_q)]$. In addition, for all $i = 1, \ldots, q$, define $\hat{R}_i \in \mathbb{S}^m$, $\hat{S}_i \in \mathbb{R}^{l \times m}$, and $\hat{Q}_i \in \mathbb{S}^l$ such that each of these matrices consists of zero blocks except, respectively, for the matrix blocks $R_i \in \mathbb{S}^{m_i}$, $S_i \in \mathbb{R}^{l_i \times m_i}$, and $Q_i \in \mathbb{S}^{l_i}$ on (i, i) position. Finally, we introduce a more general definition of vector dissipativity involving an underlying nonlinear comparsion system.

Definition 8.5. The discrete-time large-scale nonlinear dynamical system $\mathcal{G}$ given by (8.6) and (8.7) is *vector dissipative* (respectively, *geometrically vector dissipative*) *with respect to the vector supply rate* $S(u, y)$ *if there* exist a continuous, nonnegative definite vector function $V_{\mathrm{s}} = [v_{\mathrm{s}1}, \ldots, v_{\mathrm{s}q}]^{\mathrm{T}}$: $\mathcal{D} \to \overline{\mathbb{R}}_+^q$, called a *vector storage function*, and a class $\mathcal{W}_{\mathrm{d}}$ function w : $\overline{\mathbb{R}}_+^q \to \mathbb{R}^q$ such that $V_{\mathrm{s}}(0) = 0$, $w(0) = 0$, the zero solution $r(k) \equiv 0$ to the comparison system

$$r(k+1) = w(r(k)), \quad r(k_0) = r_0, \quad k \geq k_0, \tag{8.54}$$

is Lyapunov (respectively, asymptotically) stable, and the *vector dissipation inequality*

$$V_{\mathrm{s}}(x(k+1)) \leq\leq w(V_{\mathrm{s}}(x(k))) + S(u(k), y(k)), \quad k \geq k_0, \tag{8.55}$$

is satisfied, where $x(k)$, $k \geq k_0$, is the solution to (8.6) with $u(\cdot) \in \mathcal{U}$. The discrete-time large-scale nonlinear dynamical system $\mathcal{G}$ given by (8.6) and (8.7) is *vector lossless with respect to the vector supply rate* $S(u, y)$ if the vector dissipation inequality is satisfied as an equality with the zero solution $r(k) \equiv 0$ to (8.54) being Lyapunov stable.

If in Definition 8.5 the function $w : \overline{\mathbb{R}}_+^q \to \mathbb{R}^q$ is such that $w(r) = Wr$, where $W \in \mathbb{R}^{q \times q}$, then W is nonnegative and Definition 8.5 collapses to Definition 8.3.

Theorem 8.6. Consider the discrete-time large-scale nonlinear dynamical system $\mathcal{G}$ given by (8.6) and (8.7). Let $R_i \in \mathbb{S}^{m_i}$, $S_i \in \mathbb{R}^{l_i \times m_i}$, and $Q_i \in \mathbb{S}^{l_i}$, $i = 1, \ldots, q$. If there exist functions $V_{\mathrm{s}} = [v_{\mathrm{s}1}, \ldots, v_{\mathrm{s}q}]^{\mathrm{T}} : \mathbb{R}^n \to \overline{\mathbb{R}}_+^q$, $P_{1i} : \mathbb{R}^n \to \mathbb{R}^{1 \times m}$, $P_{2i} : \mathbb{R}^n \to \mathbb{N}^m$, $w = [w_1, \ldots, w_q]^{\mathrm{T}} : \overline{\mathbb{R}}_+^q \to \mathbb{R}^q$, $\ell_i : \mathbb{R}^n \to \mathbb{R}^{s_i}$, and $\mathcal{Z}_i : \mathbb{R}^n \to \mathbb{R}^{s_i \times m}$, such that $v_{\mathrm{s}i}(\cdot)$ is continuous, $v_{\mathrm{s}i}(0) = 0$, $i = 1, \ldots, q$, $w \in \mathcal{W}_{\mathrm{d}}$, $w(0) = 0$,

$$v_{\mathrm{s}i}(\mathcal{F}(x) + G(x)u) = v_{\mathrm{s}i}(\mathcal{F}(x)) + P_{1i}(x)u + u^{\mathrm{T}} P_{2i}(x)u, \quad x \in \mathbb{R}^n, \quad u \in \mathbb{R}^m, \tag{8.56}$$

the zero solution $r(k) \equiv 0$ to (8.54) is Lyapunov (respectively, asymptotically) stable, and, for all $x \in \mathbb{R}^n$ and $i = 1, \ldots, q$,

$$0 = v_{\mathrm{s}i}(\mathcal{F}(x)) - h^{\mathrm{T}}(x)\hat{Q}_i h(x) - w_i(V_{\mathrm{s}}(x)) + \ell_i^{\mathrm{T}}(x)\ell_i(x), \tag{8.57}$$

$$0 = \tfrac{1}{2} P_{1i}(x) - h^{\mathrm{T}}(x)(\hat{S}_i + \hat{Q}_i J(x)) + \ell_i^{\mathrm{T}}(x)\mathcal{Z}_i(x), \tag{8.58}$$

$$0 = \hat{R}_i + J^{\mathrm{T}}(x)\hat{S}_i + \hat{S}_i^{\mathrm{T}} J(x) + J^{\mathrm{T}}(x)\hat{Q}_i J(x) - P_{2i}(x) - \mathcal{Z}_i^{\mathrm{T}}(x)\mathcal{Z}_i(x), \tag{8.59}$$

then $\mathcal{G}$ is vector dissipative (respectively, geometrically vector dissipative) with respect to the vector quadratic supply rate $S(u, y)$, where $s_i(u_i, y_i) = u_i^{\mathrm{T}} R_i u_i + 2y_i^{\mathrm{T}} S_i u_i + y_i^{\mathrm{T}} Q_i y_i$, $i = 1, \ldots, q$.

Proof. Suppose that there exist functions $v_{\mathrm{s}i} : \mathbb{R}^n \to \overline{\mathbb{R}}_+$, $\ell_i : \mathbb{R}^n \to \mathbb{R}^{s_i}$, $\mathcal{Z}_i : \mathbb{R}^n \to \mathbb{R}^{s_i \times m}$, $w : \overline{\mathbb{R}}_+^q \to \mathbb{R}^q$, $P_{1i} : \mathbb{R}^n \to \mathbb{R}^{1 \times m}$, $P_{2i} : \mathbb{R}^n \to \mathbb{N}^m$, such that $v_{\mathrm{s}i}(\cdot)$ is continuous and nonnegative-definite, $v_{\mathrm{s}i}(0) = 0$, $i =$

$1, \ldots, q$, $w(0) = 0$, $w \in \mathcal{W}_{\mathrm{d}}$, the zero solution $r(k) \equiv 0$ to (8.54) is Lyapunov (respectively, asymptotically) stable, and (8.56)–(8.59) are satisfied. Then for every admissible input $u(\cdot) \in \mathcal{U}$ and $x \in \mathbb{R}^n$, $i = 1, \ldots, q$, it follows from (8.56)–(8.59) that

$$
\begin{aligned}
s_i(u_i, y_i) &= u^{\mathrm{T}} \hat{R}_i u + 2 y^{\mathrm{T}} \hat{S}_i u + y^{\mathrm{T}} \hat{Q}_i y \\
&= h^{\mathrm{T}}(x) \hat{Q}_i h(x) + 2 h^{\mathrm{T}}(x)(\hat{S}_i + \hat{Q}_i J(x)) u \\
&\quad + u^{\mathrm{T}}(J^{\mathrm{T}}(x) \hat{Q}_i J(x) + J^{\mathrm{T}}(x) \hat{S}_i + \hat{S}_i^{\mathrm{T}} J(x) + \hat{R}_i) u \\
&= v_{\mathrm{s}i}(\mathcal{F}(x)) - w_i(V_{\mathrm{s}}(x)) + P_{1i}(x) u + \ell_i^{\mathrm{T}}(x) \ell_i(x) \\
&\quad + 2 \ell_i^{\mathrm{T}}(x) \mathcal{Z}_i(x) u + u^{\mathrm{T}} P_{2i}(x) u + u^{\mathrm{T}} \mathcal{Z}_i^{\mathrm{T}}(x) \mathcal{Z}_i(x) u \\
&= v_{\mathrm{s}i}(\mathcal{F}(x) + G(x) u) - w_i(V_{\mathrm{s}}(x)) \\
&\quad + [\ell_i(x) + \mathcal{Z}_i(x) u]^{\mathrm{T}} [\ell_i(x) + \mathcal{Z}_i(x) u] \\
&\geq v_{\mathrm{s}i}(\mathcal{F}(x) + G(x) u) - w_i(V_{\mathrm{s}}(x)),
\end{aligned}
\tag{8.60}
$$

where $x(k)$, $k \geq k_0$, satisfies (8.6). Now, the result follows from (8.60) with vector storage function $V_{\mathrm{s}}(x) = [v_{\mathrm{s}1}(x), \ldots, v_{\mathrm{s}q}(x)]^{\mathrm{T}}$, $x \in \mathbb{R}^n$. $\square$

Using (8.57)–(8.59) it follows that for $k \geq k_0$ and $i = 1, \ldots, q$,

$$
\begin{aligned}
s_i(u_i(k), y_i(k)) + [w_i(V_{\mathrm{s}}(x(k))) - v_{\mathrm{s}i}(x(k))] &= \Delta v_{\mathrm{s}i}(x(k)) \\
&\quad + [\ell_i(x(k)) + \mathcal{Z}_i(x(k)) u(k)]^{\mathrm{T}} [\ell_i(x(k)) + \mathcal{Z}_i(x(k)) u(k)],
\end{aligned}
\tag{8.61}
$$

where $V_{\mathrm{s}}(x) = [v_{\mathrm{s}1}(x), \ldots, v_{\mathrm{s}q}(x)]^{\mathrm{T}}$, $x \in \mathbb{R}^n$, which can be interpreted as a *generalized energy* balance equation for the ith subsystem of $\mathcal{G}$ where $\Delta v_{\mathrm{s}i}(x(k))$ is the change in energy between consecutive discrete times, the two discrete terms on the left are, respectively, the external supplied energy to the ith subsystem and the energy gained by the ith subsystem from the net energy flow between all subsystems due to subsystem coupling, and the second discrete term on the right corresponds to the dissipated energy from the ith subsystem.

Note that if $\mathcal{G}$ with $u(k) \equiv 0$ is vector dissipative (respectively, geometrically vector dissipative) with respect to the vector quadratic supply rate where $Q_i \leq 0$, $i = 1, \ldots, q$, then it follows from the vector dissipation inequality that

$$
V_{\mathrm{s}}(x(k+1)) \leq\leq w(V_{\mathrm{s}}(x(k))) + S(0, y(k)) \leq\leq w(V_{\mathrm{s}}(x(k))), \quad k \geq k_0,
\tag{8.62}
$$

where $S(0, y) = [s_1(0, y_1), \ldots, s_q(0, y_q)]^{\mathrm{T}}$, $s_i(0, y_i(k)) = y_i^{\mathrm{T}}(k) Q_i y_i(k) \leq 0$, $k \geq k_0$, $i = 1, \ldots, q$, and $x(k)$, $k \geq k_0$, is the solution to (8.6) with $u(k) \equiv 0$. If, in addition, there exists $p \in \mathbb{R}_+^q$ such that $p^{\mathrm{T}} V_{\mathrm{s}}(x)$, $x \in \mathbb{R}^n$, is positive definite, then it follows from Corollary 2.6 that the undisturbed ($u(k) \equiv$

0) large-scale nonlinear dynamical system (8.6) is Lyapunov (respectively, asymptotically) stable.

Next, we extend the notions of passivity and nonexpansivity to vector passivity and vector nonexpansivity.

Definition 8.6. The discrete-time large-scale nonlinear dynamical system $\mathcal{G}$ given by (8.6) and (8.7) with $m_i = l_i$, $i = 1, \ldots, q$, is *vector passive* (respectively, *geometrically vector passive*) if it is vector dissipative (respectively, geometrically vector dissipative) with respect to the vector supply rate $S(u, y)$, where $s_i(u_i, y_i) = 2y_i^{\mathrm{T}} u_i$, $i = 1, \ldots, q$.

Definition 8.7. The discrete-time large-scale nonlinear dynamical system $\mathcal{G}$ given by (8.6) and (8.7) is *vector nonexpansive* (respectively, *geometrically vector nonexpansive*) if it is vector dissipative (respectively, geometrically vector dissipative) with respect to the vector supply rate $S(u, y)$, where $s_i(u_i, y_i) = \gamma_i^2 u_i^{\mathrm{T}} u_i - y_i^{\mathrm{T}} y_i$, $i = 1, \ldots, q$, and $\gamma_i > 0$, $i = 1, \ldots, q$, are given.

Note that a mixed vector passive-nonexpansive formulation of $\mathcal{G}$ can also be considered. Specifically, one can consider discrete-time large-scale nonlinear dynamical systems $\mathcal{G}$ that are vector dissipative with respect to vector supply rate $S(u, y)$, where $s_i(u_i, y_i) = 2y_i^{\mathrm{T}} u_i$, $i \in \mathbb{Z}_{\mathrm{p}}$, $s_j(u_j, y_j) = \gamma_j^2 u_j^{\mathrm{T}} u_j - y_j^{\mathrm{T}} y_j$, $\gamma_j > 0$, $j \in \mathbb{Z}_{\mathrm{ne}}$, and $\mathbb{Z}_{\mathrm{p}} \cup \mathbb{Z}_{\mathrm{ne}} = \{1, \ldots, q\}$. Furthermore, vector supply rates for vector input strict passivity, vector output strict passivity, and vector input-output strict passivity generalizing the passivity notions given in [89] can also be considered.

The next result presents constructive sufficient conditions guaranteeing vector dissipativity of $\mathcal{G}$ with respect to a vector quadratic supply rate for the case where the vector storage function $V_{\mathrm{s}}(x)$, $x \in \mathbb{R}^n$, is component decoupled, that is, $V_{\mathrm{s}}(x) = [v_{\mathrm{s}1}(x_1), \ldots, v_{\mathrm{s}q}(x_q)]^{\mathrm{T}}$, $x \in \mathbb{R}^n$.

Theorem 8.7. Consider the discrete-time large-scale nonlinear dynamical system $\mathcal{G}$ given by (8.6) and (8.7). Assume that there exist functions $V_{\mathrm{s}} = [v_{\mathrm{s}1}, \ldots, v_{\mathrm{s}q}]^{\mathrm{T}} : \mathbb{R}^n \to \overline{\mathbb{R}}_+^q$, $P_{1i} : \mathbb{R}^n \to \mathbb{R}^{1 \times m_i}$, $P_{2i} : \mathbb{R}^n \to \mathbb{N}^{m_i}$, $w = [w_1, \ldots, w_q]^{\mathrm{T}} : \overline{\mathbb{R}}_+^q \to \mathbb{R}^q$, $\ell_i : \mathbb{R}^n \to \mathbb{R}^{s_i}$, $\mathcal{Z}_i : \mathbb{R}^n \to \mathbb{R}^{s_i \times m_i}$ such that $v_{\mathrm{s}i}(\cdot)$ is continuous, $v_{\mathrm{s}i}(0) = 0$, $i = 1, \ldots, q$, $w \in \mathcal{W}_{\mathrm{d}}$, $w(0) = 0$, the zero solution $r(k) \equiv 0$ to (8.54) is Lyapunov (respectively, asymptotically) stable, and, for all $x \in \mathbb{R}^n$ and $i = 1, \ldots, q$,

$$0 \leq v_{\mathrm{s}i}(\mathcal{F}_i(x)) - v_{\mathrm{s}i}(\mathcal{F}_i(x) + G_i(x_i)u_i) + P_{1i}(x)u_i + u_i^{\mathrm{T}} P_{2i}(x)u_i, \qquad (8.63)$$

$$0 \geq v_{\mathrm{s}i}(\mathcal{F}_i(x)) - h_i^{\mathrm{T}}(x_i)Q_i h_i(x_i) - w_i(V_{\mathrm{s}}(x)) + \ell_i^{\mathrm{T}}(x_i)\ell_i(x_i), \qquad (8.64)$$

$$0 = \tfrac{1}{2}P_{1i}(x) - h_i^{\mathrm{T}}(x_i)(S_i + Q_i J_i(x_i)) + \ell_i^{\mathrm{T}}(x_i)\mathcal{Z}_i(x_i), \qquad (8.65)$$

$$0 \leq R_i + J_i^{\mathrm{T}}(x_i)S_i + S_i^{\mathrm{T}} J_i(x_i) + J_i^{\mathrm{T}}(x_i)Q_i J_i(x_i) - P_{2i}(x)$$
$$- \mathcal{Z}_i^{\mathrm{T}}(x_i)\mathcal{Z}_i(x_i). \qquad (8.66)$$

Then $\mathcal{G}$ is vector dissipative (respectively, geometrically vector dissipative) with respect to the vector supply rate $S(u, y)$, where $s_i(u_i, y_i) = u_i^{\mathrm{T}} R_i u_i + 2y_i^{\mathrm{T}} S_i u_i + y_i^{\mathrm{T}} Q_i y_i$, $i = 1, \ldots, q$.

Proof. For every admissible input $u = [u_1^{\mathrm{T}}, \ldots, u_q^{\mathrm{T}}]^{\mathrm{T}}$ such that $u_i \in \mathbb{R}^{m_i}$, $k \in \overline{\mathbb{Z}}_+$, and $i = 1, \ldots, q$, it follows from (8.63)–(8.66) that

$$
\begin{aligned}
s_i(u_i(k), y_i(k)) &= u_i^{\mathrm{T}}(k) R_i u_i(k) + 2y_i^{\mathrm{T}}(k) S_i u_i(k) + y_i^{\mathrm{T}}(k) Q_i y_i(k) \\
&= h_i^{\mathrm{T}}(x_i(k)) Q_i h_i(x_i(k)) \\
&\quad + 2h_i^{\mathrm{T}}(x_i(k))(S_i + Q_i J_i(x_i(k))) u_i(k) \\
&\quad + u_i^{\mathrm{T}}(k)(J_i^{\mathrm{T}}(x_i(k)) Q_i J_i(x_i(k)) + J_i^{\mathrm{T}}(x_i(k)) S_i \\
&\quad + S_i^{\mathrm{T}} J_i(x_i(k)) + R_i) u_i(k) \\
&\geq v_{\mathrm{s}i}(\mathcal{F}_i(x(k))) + P_{1i}(x(k)) u_i(k) + \ell_i^{\mathrm{T}}(x_i(k)) \ell_i(x_i(k)) \\
&\quad + 2\ell_i^{\mathrm{T}}(x_i(k)) \mathcal{Z}_i(x_i(k)) u_i(k) + u_i^{\mathrm{T}}(k) P_{2i}(x(k)) u_i(k) \\
&\quad + u_i^{\mathrm{T}}(k) \mathcal{Z}_i^{\mathrm{T}}(x_i(k)) \mathcal{Z}_i(x_i(k)) u_i(k) - w_i(V_{\mathrm{s}}(x(k))) \\
&\geq v_{\mathrm{s}i}(x_i(k+1)) - w_i(V_{\mathrm{s}}(x(k))) \\
&\quad + [\ell_i(x_i(k)) + \mathcal{Z}_i(x_i(k)) u_i(k)]^{\mathrm{T}} [\ell_i(x_i(k)) \\
&\quad + \mathcal{Z}_i(x_i(k)) u_i(k)] \\
&\geq v_{\mathrm{s}i}(x_i(k+1)) - w_i(V_{\mathrm{s}}(x(k))), \qquad\qquad (8.67)
\end{aligned}
$$

where $x(k)$, $k \geq k_0$, satisfies (8.6). Now, the result follows from (8.67) with vector storage function $V_{\mathrm{s}}(x) = [v_{\mathrm{s}1}(x_1), \ldots, v_{\mathrm{s}q}(x_q)]^{\mathrm{T}}$, $x \in \mathbb{R}^n$. $\qquad\square$

Finally, we provide necessary and sufficient conditions for the case where the discrete-time large-scale nonlinear dynamical system $\mathcal{G}$ is vector lossless with respect to a vector quadratic supply rate.

Theorem 8.8. Consider the discrete-time large-scale nonlinear dynamical system $\mathcal{G}$ given by (8.6) and (8.7). Let $R_i \in \mathbb{S}^{m_i}$, $S_i \in \mathbb{R}^{l_i \times m_i}$, and $Q_i \in \mathbb{S}^{l_i}$, $i = 1, \ldots, q$. Then $\mathcal{G}$ is vector lossless with respect to the vector quadratic supply rate $S(u, y)$, where $s_i(u_i, y_i) = u_i^{\mathrm{T}} R_i u_i + 2y_i^{\mathrm{T}} S_i u_i + y_i^{\mathrm{T}} Q_i y_i$, $i = 1, \ldots, q$, if and only if there exist functions $V_{\mathrm{s}} = [v_{\mathrm{s}1}, \ldots, v_{\mathrm{s}q}]^{\mathrm{T}} : \mathbb{R}^n \to \overline{\mathbb{R}}_+^q$, $P_{1i} : \mathbb{R}^n \to \mathbb{R}^{1 \times m}$, $P_{2i} : \mathbb{R}^n \to \mathbb{N}^m$, $w = [w_1, \ldots, w_q]^{\mathrm{T}} : \overline{\mathbb{R}}_+^q \to \mathbb{R}^q$ such that $v_{\mathrm{s}i}(\cdot)$ is continuous, $v_{\mathrm{s}i}(0) = 0$, $i = 1, \ldots, q$, $w \in \mathcal{W}_{\mathrm{d}}$, $w(0) = 0$, the zero solution $r(k) \equiv 0$ to (8.54) is Lyapunov stable, and, for all $x \in \mathbb{R}^n$, $i = 1, \ldots, q$, (8.56) holds and

$$
0 = v_{\mathrm{s}i}(\mathcal{F}(x)) - h^{\mathrm{T}}(x) \hat{Q}_i h(x) - w_i(V_{\mathrm{s}}(x)), \qquad\qquad (8.68)
$$

$$
0 = \tfrac{1}{2} P_{1i}(x) - h^{\mathrm{T}}(x)(\hat{S}_i + \hat{Q}_i J(x)), \qquad\qquad (8.69)
$$

$$
0 = \hat{R}_i + J^{\mathrm{T}}(x) \hat{S}_i + \hat{S}_i^{\mathrm{T}} J(x) + J^{\mathrm{T}}(x) \hat{Q}_i J(x) - P_{2i}(x). \qquad (8.70)
$$

Proof. Sufficiency follows as in the proof of Theorem 8.6. To show necessity, suppose that $\mathcal{G}$ is lossless with respect to the vector quadratic supply rate $S(u, y)$. Then, there exist continuous functions $V_{\mathrm{s}} = [v_{\mathrm{s}1}, \ldots, v_{\mathrm{s}q}]^{\mathrm{T}} :

$\mathbb{R}^n \to \overline{\mathbb{R}}_+^q$ and $w = [w_1, \ldots, w_q]^{\mathrm{T}} : \overline{\mathbb{R}}_+^q \to \mathbb{R}^q$ such that $V_{\mathrm{s}}(0) = 0$, the zero solution $r(k) \equiv 0$ to (8.54) is Lyapunov stable and

$$
\begin{aligned}
v_{\mathrm{s}i}(\mathcal{F}(x) + G(x)u) &= w_i(V_{\mathrm{s}}(x)) + s_i(u_i, y_i) \\
&= w_i(V_{\mathrm{s}}(x)) + u^{\mathrm{T}}\hat{R}_i u + 2y^{\mathrm{T}}\hat{S}_i u + y^{\mathrm{T}}\hat{Q}_i y \\
&= w_i(V_{\mathrm{s}}(x)) + h^{\mathrm{T}}(x)\hat{Q}_i h(x) \\
&\quad + 2h^{\mathrm{T}}(x)(\hat{Q}_i J(x) + \hat{S}_i)u \\
&\quad + u^{\mathrm{T}}(\hat{R}_i + \hat{S}_i^{\mathrm{T}} J(x) + J^{\mathrm{T}}(x)\hat{S}_i + J^{\mathrm{T}}(x)\hat{Q}_i J(x))u, \\
&\qquad\qquad\qquad\qquad x \in \mathbb{R}^n, \quad u \in \mathbb{R}^m. \qquad (8.71)
\end{aligned}
$$

Since the right-hand side of (8.71) is quadratic in u it follows that $v_{\mathrm{s}i}(\mathcal{F}(x) + G(x)u)$ is quadratic in u, and hence, there exist $P_{1i} : \mathbb{R}^n \to \mathbb{R}^{1 \times m}$ and $P_{2i} : \mathbb{R}^n \to \mathbb{N}^m$ such that

$$
\begin{aligned}
v_{\mathrm{s}i}(\mathcal{F}(x) + G(x)u) = v_{\mathrm{s}i}(\mathcal{F}(x)) + P_{1i}(x)u + u^{\mathrm{T}}P_{2i}(x)u, \\
x \in \mathbb{R}^n, \quad u \in \mathbb{R}^m. \qquad (8.72)
\end{aligned}
$$

Now, using (8.72) and equating coefficients of equal powers in (8.71) yields (8.68)–(8.70). $\qquad\square$

8.4 Specialization to Discrete-Time Large-Scale Linear Dynamical Systems

In this section, we specialize the results of Section 8.3 to the case of discrete-time large-scale linear dynamical systems. Specifically, we assume that $w \in \mathcal{W}_{\mathrm{d}}$ is linear so that $w(r) = Wr$, where $W \in \mathbb{R}^{q \times q}$ is nonnegative, and consider the discrete-time large-scale linear dynamical system $\mathcal{G}$ given by

$$
\begin{aligned}
x(k+1) &= Ax(k) + Bu(k), \quad x(k_0) = x_0, \quad k \geq k_0, \qquad (8.73) \\
y(k) &= Cx(k) + Du(k), \qquad\qquad\qquad\qquad\qquad\qquad (8.74)
\end{aligned}
$$

where $A \in \mathbb{R}^{n \times n}$ and A is partitioned as $A \triangleq [A_{ij}], i, j = 1, \ldots, q, A_{ij} \in \mathbb{R}^{n_i \times n_j}$, $\sum_{i=1}^{q} n_i = n$, $B = \text{block–diag}[B_1, \ldots, B_q]$, $C = \text{block–diag}[C_1, \ldots, C_q]$, $D = \text{block–diag}[D_1, \ldots, D_q]$, $B_i \in \mathbb{R}^{n_i \times m_i}$, $C_i \in \mathbb{R}^{l_i \times n_i}$, $D_i \in \mathbb{R}^{l_i \times m_i}$, and $i = 1, \ldots, q$.

Theorem 8.9. Consider the discrete-time large-scale linear dynamical system $\mathcal{G}$ given by (8.73) and (8.74). Let $R_i \in \mathbb{S}^{m_i}$, $S_i \in \mathbb{R}^{l_i \times m_i}$, $Q_i \in \mathbb{S}^{l_i}$, $i = 1, \ldots, q$. Then $\mathcal{G}$ is vector dissipative (respectively, geometrically vector dissipative) with respect to the vector supply rate $S(u, y)$, where $s_i(u_i, y_i) = u_i^{\mathrm{T}}R_i u_i + 2y_i^{\mathrm{T}}S_i u_i + y_i^{\mathrm{T}}Q_i y_i$, $i = 1, \ldots, q$, and with a three-times continuously differentiable vector storage function if and only if there exist $W \in \mathbb{R}^{q \times q}$, $P_i \in \mathbb{N}^n$, $L_i \in \mathbb{R}^{s_i \times n}$, and $Z_i \in \mathbb{R}^{s_i \times m}$, $i = 1, \ldots, q$, such that W

is nonnegative and semistable (respectively, asymptotically stable), and, for all $i = 1, \ldots, q$,

$$0 = A^{\mathrm{T}} P_i A - C^{\mathrm{T}} \hat{Q}_i C - \sum_{j=1}^{q} W_{(i,j)} P_j + L_i^{\mathrm{T}} L_i, \tag{8.75}$$

$$0 = A^{\mathrm{T}} P_i B - C^{\mathrm{T}} (\hat{S}_i + \hat{Q}_i D) + L_i^{\mathrm{T}} Z_i, \tag{8.76}$$

$$0 = \hat{R}_i + D^{\mathrm{T}} \hat{S}_i + \hat{S}_i^{\mathrm{T}} D + D^{\mathrm{T}} \hat{Q}_i D - B^{\mathrm{T}} P_i B - Z_i^{\mathrm{T}} Z_i. \tag{8.77}$$

Proof. Sufficiency follows from Theorem 8.6 with $\mathcal{F}(x) = Ax$, $G(x) = B$, $h(x) = Cx$, $J(x) = D$, $P_{1i}(x) = 2x^{\mathrm{T}} A^{\mathrm{T}} P_i B$, $P_{2i}(x) = B^{\mathrm{T}} P_i B$, $w(r) = Wr$, $\ell_i(x) = L_i x$, $\mathcal{Z}_i(x) = Z_i$, and $v_{\mathrm{s}i}(x) = x^{\mathrm{T}} P_i x$, $i = 1, \ldots, q$.

To show necessity, suppose $\mathcal{G}$ is vector dissipative with respect to the vector supply rate $S(u, y)$, where $s_i(u_i, y_i) = u_i^{\mathrm{T}} R_i u_i + 2y_i^{\mathrm{T}} S_i u_i + y_i^{\mathrm{T}} Q_i y_i$, $i = 1, \ldots, q$. Then, with $w(r) = Wr$, there exists $V_{\mathrm{s}} : \mathbb{R}^n \to \overline{\mathbb{R}}_+^q$ such that W is nonnegative and semistable (respectively, asymptotically stable), $V_{\mathrm{s}}(x) \triangleq [v_{\mathrm{s}1}(x), \ldots, v_{\mathrm{s}q}(x)]^{\mathrm{T}}$, $x \in \mathbb{R}^n$, $V_{\mathrm{s}}(0) = 0$, and for all $x \in \mathbb{R}^n$, $u \in \mathbb{R}^n$,

$$V_{\mathrm{s}}(Ax + Bu) - W V_{\mathrm{s}}(x) \leq\leq S(u, y). \tag{8.78}$$

Next, it follows from (8.78) that there exists a three-times continuously differentiable vector function $d = [d_1, \ldots, d_q]^{\mathrm{T}} : \mathbb{R}^n \times \mathbb{R}^m \to \mathbb{R}^q$ such that $d(x, u) \geq\geq 0$, $d(0, 0) = 0$, and

$$0 = V_{\mathrm{s}}(Ax + Bu) - W V_{\mathrm{s}}(x) - S(u, Cx + Du) + d(x, u). \tag{8.79}$$

Now, expanding $v_{\mathrm{s}i}(\cdot)$ and $d_i(\cdot, \cdot)$ via a Taylor series expansion about $x = 0$, $u = 0$, and using the fact that $v_{\mathrm{s}i}(\cdot)$ and $d_i(\cdot, \cdot)$ are nonnegative and $v_{\mathrm{s}i}(0) = 0$, $d_i(0, 0) = 0$, $i = 1, \ldots, q$, it follows that there exist $P_i \in \mathbb{N}^n$, $L_i \in \mathbb{R}^{s_i \times n}$, $Z_i \in \mathbb{R}^{s_i \times m}$, $i = 1, \ldots, q$, such that

$$v_{\mathrm{s}i}(x) = x^{\mathrm{T}} P_i x + v_{\mathrm{s}ri}(x), \tag{8.80}$$

$$d_i(x, u) = (L_i x + Z_i u)^{\mathrm{T}} (L_i x + Z_i u) + d_{ri}(x, u),$$

$$x \in \mathbb{R}^n, \quad u \in \mathbb{R}^m, \quad i = 1, \ldots, q, \tag{8.81}$$

where $v_{\mathrm{s}ri} : \mathbb{R}^n \to \mathbb{R}$ and $d_{ri} : \mathbb{R}^n \times \mathbb{R}^m \to \mathbb{R}$ contain the higher-order terms of $v_{\mathrm{s}i}(\cdot)$ and $d_i(\cdot, \cdot)$, respectively.

Using the above expressions, (8.79) can be written componentwise as

$$0 = (Ax + Bu)^{\mathrm{T}} P_i (Ax + Bu) - \sum_{j=1}^{q} W_{(i,j)} x^{\mathrm{T}} P_j x$$

$$- (x^{\mathrm{T}} C^{\mathrm{T}} \hat{Q}_i C x + 2x^{\mathrm{T}} C^{\mathrm{T}} \hat{Q}_i D u + u^{\mathrm{T}} D^{\mathrm{T}} \hat{Q}_i D u$$

$$+ 2x^{\mathrm{T}} C^{\mathrm{T}} \hat{S}_i u + 2u^{\mathrm{T}} D^{\mathrm{T}} \hat{S}_i u + u^{\mathrm{T}} \hat{R}_i u)$$

$$+ (L_i x + Z_i u)^{\mathrm{T}} (L_i x + Z_i u) + \delta(x, u), \tag{8.82}$$

where $\delta(x, u)$ is such that

$$\lim_{\|x\|^2 + \|u\|^2 \to 0} \frac{|\delta(x, u)|}{\|x\|^2 + \|u\|^2} = 0. \tag{8.83}$$

Now, viewing (8.82) as the componentwise Taylor series expansion of (8.79) about $x = 0$ and $u = 0$ it follows that for all $x \in \mathbb{R}^n$ and $u \in \mathbb{R}^m$,

$$0 = x^{\mathrm{T}}(A^{\mathrm{T}} P_i A - \sum_{j=1}^{q} W_{(i,j)} P_j - C^{\mathrm{T}} \hat{Q}_i C + L_i^{\mathrm{T}} L_i)x$$

$$+ 2x^{\mathrm{T}}(A^{\mathrm{T}} P_i B - C^{\mathrm{T}} \hat{S}_i - C^{\mathrm{T}} \hat{Q}_i D + L_i^{\mathrm{T}} Z_i)u$$
$$+ u^{\mathrm{T}}(Z_i^{\mathrm{T}} Z_i - D^{\mathrm{T}} \hat{Q}_i D - D^{\mathrm{T}} \hat{S}_i - \hat{S}_i^{\mathrm{T}} D - \hat{R}_i + B^{\mathrm{T}} P_i B)u,$$
$$i = 1, \ldots, q. \tag{8.84}$$

Now, equating coefficients of equal powers in (8.84) yields (8.75)–(8.77). $\square$

Note that (8.75)–(8.77) are equivalent to

$$\begin{bmatrix} \mathcal{A}_i & \mathcal{B}_i \\ \mathcal{B}_i^{\mathrm{T}} & \mathcal{C}_i \end{bmatrix} = -\begin{bmatrix} L_i^{\mathrm{T}} \\ Z_i^{\mathrm{T}} \end{bmatrix} \begin{bmatrix} L_i & Z_i \end{bmatrix} \leq 0, \quad i = 1, \ldots, q, \tag{8.85}$$

where, for all $i = 1, \ldots, q$,

$$\mathcal{A}_i = A^{\mathrm{T}} P_i A - C^{\mathrm{T}} \hat{Q}_i C - \sum_{j=1}^{q} W_{(i,j)} P_j, \tag{8.86}$$

$$\mathcal{B}_i = A^{\mathrm{T}} P_i B - C^{\mathrm{T}}(\hat{S}_i + \hat{Q}_i D), \tag{8.87}$$

$$\mathcal{C}_i = -(\hat{R}_i + D^{\mathrm{T}} \hat{S}_i + \hat{S}_i^{\mathrm{T}} D + D^{\mathrm{T}} \hat{Q}_i D - B^{\mathrm{T}} P_i B). \tag{8.88}$$

Hence, vector dissipativity of discrete-time large-scale linear dynamical systems with respect to vector quadratic supply rates can be characterized via (cascade) linear matrix inequalities (LMIs) [26]. A similar remark holds for Theorem 8.10 below.

The next result presents sufficient conditions guaranteeing vector dissipativity of $\mathcal{G}$ with respect to a vector quadratic supply rate in the case where the vector storage function is component decoupled.

Theorem 8.10. Consider the discrete-time large-scale linear dynamical system $\mathcal{G}$ given by (8.73) and (8.74). Let $R_i \in \mathbb{S}^{m_i}$, $S_i \in \mathbb{R}^{l_i \times m_i}$, $Q_i \in \mathbb{S}^{l_i}$, $i = 1, \ldots, q$, be given. Assume there exist matrices $W \in \mathbb{R}^{q \times q}$, $P_i \in \mathbb{N}^{n_i}$, $L_{ii} \in \mathbb{R}^{s_{ii} \times n_i}$, $Z_{ii} \in \mathbb{R}^{s_{ii} \times m_i}$, $i = 1, \ldots, q$, $L_{ij} \in \mathbb{R}^{s_{ij} \times n_i}$, and $Z_{ij} \in \mathbb{R}^{s_{ij} \times n_j}$, $i, j = 1, \ldots, q$, $i \neq j$, such that W is nonnegative and semistable (respectively, asymptotically stable), and, for all $i = 1, \ldots, q$,

$$0 \geq A_{ii}^{\mathrm{T}} P_i A_{ii} - C_i^{\mathrm{T}} Q_i C_i - W_{(i,i)} P_i + L_{ii}^{\mathrm{T}} L_{ii} + \sum_{j=1, j \neq i}^{q} L_{ij}^{\mathrm{T}} L_{ij}, \tag{8.89}$$

$$0 = A_{ii}^{\mathrm{T}} P_i B_i - C_i^{\mathrm{T}} S_i - C_i^{\mathrm{T}} Q_i D_i + L_{ii}^{\mathrm{T}} Z_{ii}, \tag{8.90}$$

$$0 \le R_i + D_i^{\mathrm{T}} S_i + S_i^{\mathrm{T}} D_i + D_i^{\mathrm{T}} Q_i D_i - B_i^{\mathrm{T}} P_i B_i - Z_{ii}^{\mathrm{T}} Z_{ii}, \tag{8.91}$$

and, for $j = 1, \ldots, q, l = 1, \ldots, q, j \ne i, l \ne i, l \ne j,$

$$0 = A_{ij}^{\mathrm{T}} P_i B_i, \tag{8.92}$$

$$0 = A_{ij}^{\mathrm{T}} P_i A_{il}, \tag{8.93}$$

$$0 = A_{ii}^{\mathrm{T}} P_i A_{ij} + L_{ij}^{\mathrm{T}} Z_{ij}, \tag{8.94}$$

$$0 \le W_{(i,j)} P_j - Z_{ij}^{\mathrm{T}} Z_{ij} - A_{ij}^{\mathrm{T}} P_i A_{ij}. \tag{8.95}$$

Then $\mathcal{G}$ is vector dissipative (respectively, geometrically vector dissipative) with respect to the vector supply rate $S(u, y) \triangleq [s_1(u_1, y_1), \ldots, s_q(u_q, y_q)]^{\mathrm{T}}$, where $s_i(u_i, y_i) = u_i^{\mathrm{T}} R_i u_i + 2 y_i^{\mathrm{T}} S_i u_i + y_i^{\mathrm{T}} Q_i y_i$, $i = 1, \ldots, q$.

Proof. Since $P_i \in \mathbb{N}^{n_i}$, the function $v_{si}(x_i) \triangleq x_i^{\mathrm{T}} P_i x_i$, $x_i \in \mathbb{R}^{n_i}$, is nonnegative definite and $v_{si}(0) = 0$. Moreover, since $v_{si}(\cdot)$ is continuous it follows from (8.89)–(8.95) that for all $u_i \in \mathbb{R}^{m_i}$, $i = 1, \ldots, q$, and $k \ge k_0$,

$$v_{si}(x_i(k+1)) = \left[\sum_{j=1}^{q} A_{ij} x_j(k) + B_i u_i(k) \right]^{\mathrm{T}} P_i \left[\sum_{j=1}^{q} A_{ij} x_j(k) + B_i u_i(k) \right]$$

$$\le x_i^{\mathrm{T}}(k) \left[W_{(i,i)} P_i + C_i^{\mathrm{T}} Q_i C_i - L_{ii}^{\mathrm{T}} L_{ii} - \sum_{j=1, j \ne i}^{q} L_{ij}^{\mathrm{T}} L_{ij} \right] x_i(k)$$

$$- \sum_{j=1, j \ne i}^{q} 2 x_i^{\mathrm{T}}(k) L_{ij}^{\mathrm{T}} Z_{ij} x_j(k) + 2 x_i^{\mathrm{T}}(k) C_i^{\mathrm{T}} S_i u_i(k)$$

$$+ 2 x_i^{\mathrm{T}}(k) C_i^{\mathrm{T}} Q_i D_i u_i(k) - 2 x_i^{\mathrm{T}}(k) L_{ii}^{\mathrm{T}} Z_{ii} u_i(k)$$

$$+ \sum_{j=1, j \ne i}^{q} x_j^{\mathrm{T}}(k) [W_{(i,j)} P_j - Z_{ij}^{\mathrm{T}} Z_{ij}] x_j(k) + u_i^{\mathrm{T}}(k) R_i u_i(k)$$

$$+ 2 u_i^{\mathrm{T}}(k) D_i^{\mathrm{T}} S_i u_i(k) + u_i^{\mathrm{T}}(k) D_i^{\mathrm{T}} Q_i D_i u_i(k)$$

$$- u_i^{\mathrm{T}}(k) Z_{ii}^{\mathrm{T}} Z_{ii} u_i(k)$$

$$= \sum_{j=1}^{q} W_{(i,j)} v_{sj}(x_j(k)) + u_i^{\mathrm{T}}(k) R_i u_i(k)$$

$$+ 2 y_i^{\mathrm{T}}(k) S_i u_i(k) + y_i^{\mathrm{T}}(k) Q_i y_i(k)$$

$$- [L_{ii} x_i(k) + Z_{ii} u_i(k)]^{\mathrm{T}} [L_{ii} x_i(k) + Z_{ii} u_i(k)]$$

$$- \sum_{j=1, j \ne i}^{q} (L_{ij} x_i(k) + Z_{ij} x_j(k))^{\mathrm{T}} (L_{ij} x_i(k) + Z_{ij} x_j(k))$$

$$\leq s_i(u_i(k), y_i(k)) + \sum_{j=1}^{q} W_{(i,j)} v_{sj}(x_j(k)), \tag{8.96}$$

or, equivalently, in vector form

$$V_{\mathrm{s}}(x(k+1)) \leq\leq W V_{\mathrm{s}}(x(k)) + S(u, y), \quad u \in \mathcal{U}, \quad k \geq k_0, \tag{8.97}$$

where $V_{\mathrm{s}}(x) \triangleq [v_{s1}(x_1), \ldots, v_{sq}(x_q)]^{\mathrm{T}}$, $x \in \mathbb{R}^n$. Now, it follows from Proposition 8.2 that $\mathcal{G}$ is vector dissipative (respectively, geometrically vector dissipative) with respect to the vector supply rate $S(u, y)$ and with vector storage function $V_{\mathrm{s}}(x)$, $x \in \mathbb{R}^n$. $\qquad\square$

8.5 Stability of Feedback Interconnections of Discrete-Time Large-Scale Nonlinear Dynamical Systems

In this section, we consider stability of feedback interconnections of discrete-time large-scale nonlinear dynamical systems. Specifically, for the discrete-time large-scale dynamical system $\mathcal{G}$ given by (8.6) and (8.7) we consider either a dynamic or static discrete-time large-scale feedback system $\mathcal{G}_{\mathrm{c}}$. Then by appropriately combining vector storage functions for each system we show stability of the feedback interconnection. We begin by considering the discrete-time large-scale nonlinear dynamical system (8.6) and (8.7) with the large-scale feedback system $\mathcal{G}_{\mathrm{c}}$ given by

$$x_{\mathrm{c}}(k+1) = F_{\mathrm{c}}(x_{\mathrm{c}}(k), u_{\mathrm{c}}(k)), \quad x_{\mathrm{c}}(k_0) = x_{\mathrm{c}0}, \quad k \geq k_0, \tag{8.98}$$

$$y_{\mathrm{c}}(k) = H_{\mathrm{c}}(x_{\mathrm{c}}(k), u_{\mathrm{c}}(k)), \tag{8.99}$$

where $F_{\mathrm{c}} : \mathbb{R}^{n_c} \times U_{\mathrm{c}} \to \mathbb{R}^{n_c}$, $H_{\mathrm{c}} : \mathbb{R}^{n_c} \times U_{\mathrm{c}} \to Y_{\mathrm{c}}$, $F_{\mathrm{c}} \triangleq [F_{\mathrm{c}1}^{\mathrm{T}}, \ldots, F_{\mathrm{c}q}^{\mathrm{T}}]^{\mathrm{T}}$, $H_{\mathrm{c}} \triangleq [H_{\mathrm{c}1}^{\mathrm{T}}, \ldots, H_{\mathrm{c}q}^{\mathrm{T}}]^{\mathrm{T}}$, $U_{\mathrm{c}} \subseteq \mathbb{R}^l$, $Y_{\mathrm{c}} \subseteq \mathbb{R}^m$. Moreover, for all $i = 1, \ldots, q$, we assume that

$$F_{\mathrm{c}i}(x_{\mathrm{c}}, u_{\mathrm{c}i}) = f_{\mathrm{c}i}(x_{\mathrm{c}i}) + \mathcal{I}_{\mathrm{c}i}(x_{\mathrm{c}}) + G_{\mathrm{c}i}(x_{\mathrm{c}i})u_{\mathrm{c}i}, \tag{8.100}$$

$$H_{\mathrm{c}i}(x_{\mathrm{c}i}, u_{\mathrm{c}i}) = h_{\mathrm{c}i}(x_{\mathrm{c}i}) + J_{\mathrm{c}i}(x_{\mathrm{c}i})u_{\mathrm{c}i}, \tag{8.101}$$

where $u_{\mathrm{c}i} \in U_{\mathrm{c}i} \subseteq \mathbb{R}^{l_i}$, $y_{\mathrm{c}i} \triangleq H_{\mathrm{c}i}(x_{\mathrm{c}i}, u_{\mathrm{c}i}) \in Y_i \subseteq \mathbb{R}^{m_i}$, $(u_{\mathrm{c}i}, y_{\mathrm{c}i})$ is the input-output pair for the ith subsystem of $\mathcal{G}_{\mathrm{c}}$, $f_{\mathrm{c}i} : \mathbb{R}^{n_{ci}} \to \mathbb{R}^{n_{ci}}$ and $\mathcal{I}_{\mathrm{c}i} : \mathbb{R}^{n_c} \to \mathbb{R}^{n_{ci}}$ satisfy $f_{\mathrm{c}i}(0) = 0$ and $\mathcal{I}_{\mathrm{c}i}(0) = 0$, $G_{\mathrm{c}i} : \mathbb{R}^{n_{ci}} \to \mathbb{R}^{n_{ci} \times l_i}$, $h_{\mathrm{c}i} : \mathbb{R}^{n_{ci}} \to \mathbb{R}^{m_i}$ and satisfies $h_{\mathrm{c}i}(0) = 0$, $J_{\mathrm{c}i} : \mathbb{R}^{n_{ci}} \to \mathbb{R}^{m_i \times l_i}$, and $\sum_{i=1}^{q} n_{ci} = n_c$.

Furthermore, we define the composite input and composite output for the system $\mathcal{G}_{\mathrm{c}}$ as $u_{\mathrm{c}} \triangleq [u_{\mathrm{c}1}^{\mathrm{T}}, \ldots, u_{\mathrm{c}q}^{\mathrm{T}}]^{\mathrm{T}}$ and $y_{\mathrm{c}} \triangleq [y_{\mathrm{c}1}^{\mathrm{T}}, \ldots, y_{\mathrm{c}q}^{\mathrm{T}}]^{\mathrm{T}}$, respectively. In this case, $U_{\mathrm{c}} = U_{\mathrm{c}1} \times \cdots \times U_{\mathrm{c}q}$ and $Y_{\mathrm{c}} = Y_{\mathrm{c}1} \times \cdots \times Y_{\mathrm{c}q}$. Note that with the feedback interconnection given by Figure 8.1, $u_{\mathrm{c}} = y$ and $y_{\mathrm{c}} = -u$. We assume that the negative feedback interconnection of $\mathcal{G}$ and $\mathcal{G}_{\mathrm{c}}$ is well posed; that is, $\det(I_{m_i} + J_{\mathrm{c}i}(x_{\mathrm{c}i})J_i(x_i)) \neq 0$ for all $x_i \in \mathbb{R}^{n_i}$, $x_{\mathrm{c}i} \in \mathbb{R}^{n_{ci}}$, and $i = 1, \ldots, q$. Furthermore, we assume that for the discrete-time large-scale

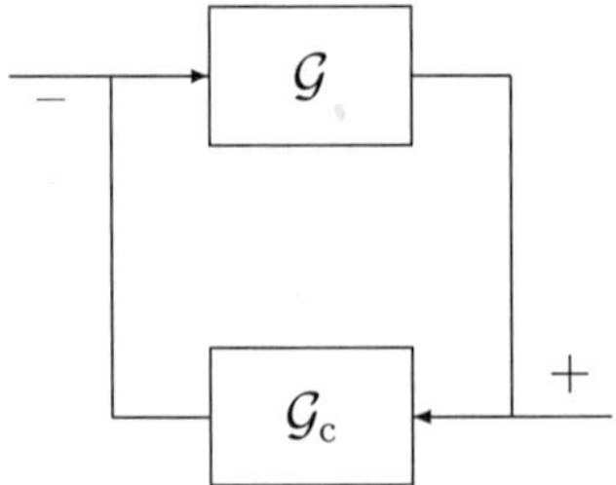

Figure 8.1 Feedback interconnection of large-scale systems $\mathcal{G}$ and $\mathcal{G}_{\mathrm{c}}$.

systems $\mathcal{G}$ and $\mathcal{G}_{\mathrm{c}}$, the conditions of Theorem 8.3 are satisfied; that is, if $V_{\mathrm{s}}(x)$, $x \in \mathbb{R}^n$, and $V_{\mathrm{cs}}(x_{\mathrm{c}})$, $x_{\mathrm{c}} \in \mathbb{R}^{n_c}$, are vector storage functions for $\mathcal{G}$ and $\mathcal{G}_{\mathrm{c}}$, respectively, then there exist $p \in \mathbb{R}_+^q$ and $p_{\mathrm{c}} \in \mathbb{R}_+^q$ such that the functions $v_{\mathrm{s}}(x) = p^{\mathrm{T}} V_{\mathrm{s}}(x)$, $x \in \mathbb{R}^n$, and $v_{\mathrm{cs}}(x_{\mathrm{c}}) = p_{\mathrm{c}}^{\mathrm{T}} V_{\mathrm{cs}}(x_{\mathrm{c}})$, $x_{\mathrm{c}} \in \mathbb{R}^{n_c}$, are positive definite. The following result gives sufficient conditions for Lyapunov and asymptotic stability of the feedback interconnection given by Figure 8.1.

Theorem 8.11. Consider the discrete-time large-scale nonlinear dynamical systems $\mathcal{G}$ and $\mathcal{G}_{\mathrm{c}}$ given by (8.6) and (8.7), and (8.98) and (8.99), respectively. Assume that $\mathcal{G}$ and $\mathcal{G}_{\mathrm{c}}$ are vector dissipative with respect to the vector supply rates $S(u, y)$ and $S_{\mathrm{c}}(u_{\mathrm{c}}, y_{\mathrm{c}})$, and with continuous vector storage functions $V_{\mathrm{s}}(\cdot)$ and $V_{\mathrm{cs}}(\cdot)$ and dissipation matrices $W \in \mathbb{R}^{q \times q}$ and $W_{\mathrm{c}} \in \mathbb{R}^{q \times q}$, respectively.

i) If there exists $\Sigma \triangleq \mathrm{diag}[\sigma_1, \ldots, \sigma_q] > 0$ such that $S(u, y) + \Sigma S_{\mathrm{c}}(u_{\mathrm{c}}, y_{\mathrm{c}}) \leq\leq 0$ and $\tilde{W} \in \mathbb{R}^{q \times q}$ is semistable (respectively, asymptotically stable), where

$$\tilde{W}_{(i,j)} \triangleq \max\{W_{(i,j)}, (\Sigma W_{\mathrm{c}} \Sigma^{-1})_{(i,j)}\}$$
$$= \max\{W_{(i,j)}, \frac{\sigma_i}{\sigma_j} W_{\mathrm{c}(i,j)}\}, \quad i, j = 1, \ldots, q, \quad (8.102)$$

then the negative feedback interconnection of $\mathcal{G}$ and $\mathcal{G}_{\mathrm{c}}$ is Lyapunov (respectively, asymptotically) stable.

ii) Let $Q_i \in \mathbb{S}^{l_i}$, $S_i \in \mathbb{R}^{l_i \times m_i}$, $R_i \in \mathbb{S}^{m_i}$, $Q_{ci} \in \mathbb{S}^{m_i}$, $S_{ci} \in \mathbb{R}^{m_i \times l_i}$, and $R_{ci} \in \mathbb{S}^{l_i}$, and suppose $S(u, y) = [s_1(u_1, y_1), \ldots, s_q(u_q, y_q)]^{\mathrm{T}}$ and $S_{\mathrm{c}}(u_{\mathrm{c}}, y_{\mathrm{c}}) = [s_{c1}(u_{c1}, y_{c1}), \ldots, s_q(u_{cq}, y_{cq})]^{\mathrm{T}}$, where $s_i(u_i, y_i) = u_i^{\mathrm{T}} R_i u_i + 2y_i^{\mathrm{T}} S_i u_i + y_i^{\mathrm{T}} Q_i y_i$ and $s_{ci}(u_{ci}, y_{ci}) = u_{ci}^{\mathrm{T}} R_{ci} u_{ci} + 2y_{ci}^{\mathrm{T}} S_{ci} u_{ci} + y_{ci}^{\mathrm{T}} Q_{ci} y_{ci}$, $i = 1, \ldots, q$. If there exists $\Sigma \triangleq \mathrm{diag}[\sigma_1, \ldots, \sigma_q] > 0$ such that for all

$i = 1, \ldots, q,$

$$\tilde{Q}_i \triangleq \begin{bmatrix} Q_i + \sigma_i R_{ci} & -S_i + \sigma_i S_{ci}^{\mathrm{T}} \\ -S_i^{\mathrm{T}} + \sigma_i S_{ci} & R_i + \sigma_i Q_{ci} \end{bmatrix} \leq 0 \qquad (8.103)$$

and $\tilde{W} \in \mathbb{R}^{q \times q}$ is semistable (respectively, asymptotically stable), where

$$\begin{aligned} \tilde{W}_{(i,j)} &\triangleq \max\{W_{(i,j)}, \ (\Sigma W_c \Sigma^{-1})_{(i,j)}\} \\ &= \max\{W_{(i,j)}, \ \frac{\sigma_i}{\sigma_j} W_{c(i,j)}\}, \quad i,j = 1, \ldots, q, \qquad (8.104) \end{aligned}$$

then the negative feedback interconnection of $\mathcal{G}$ and $\mathcal{G}_c$ is Lyapunov (respectively, asymptotically) stable.

Proof. i) Consider the vector Lyapunov function candidate $V(x, x_c) = V_s(x) + \Sigma V_{cs}(x_c)$, $(x, x_c) \in \mathbb{R}^n \times \mathbb{R}^{n_c}$, and note that

$$\begin{aligned} V(x(k+1), x_c(k+1)) &= V_s(x(k+1)) + \Sigma V_{cs}(x_c(k+1)) \\ &\leq\leq S(u(k), y(k)) + \Sigma S_c(u_c(k), y_c(k)) \\ &\quad + WV_s(x(k)) + \Sigma W_c V_{cs}(x_c(k)) \\ &\leq\leq WV_s(x(k)) + \Sigma W_c \Sigma^{-1} \Sigma V_{cs}(x_c(k)) \\ &\leq\leq \tilde{W}(V_s(x(k)) + \Sigma V_{cs}(x_c(k))) \\ &= \tilde{W} V(x(k), x_c(k)), \\ &\qquad\qquad (x(k), x_c(k)) \in \mathbb{R}^n \times \mathbb{R}^{n_c}, \quad k \geq k_0. \quad (8.105) \end{aligned}$$

Now, since for $V_s(x)$, $x \in \mathbb{R}^n$, and $V_{cs}(x_c)$, $x_c \in \mathbb{R}^{n_c}$, there exist, by assumption, $p \in \mathbb{R}_+^q$ and $p_c \in \mathbb{R}_+^q$ such that the functions $v_s(x) = p^{\mathrm{T}} V_s(x)$, $x \in \mathbb{R}^n$, and $v_{cs}(x_c) = p_c^{\mathrm{T}} V_{cs}(x_c)$, $x_c \in \mathbb{R}^{n_c}$, are positive definite and noting that $v_{cs}(x_c) \leq \max_{i=1,\ldots,q}\{p_{ci}\} \mathbf{e}^{\mathrm{T}} V_{cs}(x_c)$, where p_{ci} is the ith component of p_c and $\mathbf{e} = [1, \ldots, 1]^{\mathrm{T}}$, it follows that $\mathbf{e}^{\mathrm{T}} V_{cs}(x_c)$, $x_c \in \mathbb{R}^{n_c}$, is positive definite. Next, since $\min_{i=1,\ldots,q}\{p_i \sigma_i\} \mathbf{e}^{\mathrm{T}} V_{cs}(x_c) \leq p^{\mathrm{T}} \Sigma V_{cs}(x_c)$, it follows that $p^{\mathrm{T}} \Sigma V_{cs}(x_c)$, $x_c \in \mathbb{R}^{n_c}$, is positive definite. Hence, the function $v(x, x_c) = p^{\mathrm{T}} V(x, x_c)$, $(x, x_c) \in \mathbb{R}^n \times \mathbb{R}^{n_c}$, is positive definite. Now, the result is a direct consequence of Corollary 2.6.

ii) The proof follows from i) by noting that, for all $i = 1, \ldots, q$,

$$s_i(u_i, y_i) + \sigma_i s_{ci}(u_{ci}, y_{ci}) = \begin{bmatrix} y \\ y_c \end{bmatrix}^{\mathrm{T}} \tilde{Q}_i \begin{bmatrix} y \\ y_c \end{bmatrix}, \qquad (8.106)$$

and hence $S(u, y) + \Sigma S_c(u_c, y_c) \leq\leq 0$. $\qquad\square$

For the next result note that if the discrete-time large-scale nonlinear dynamical system $\mathcal{G}$ is vector dissipative with respect to the vector supply rate $S(u, y)$, where $s_i(u_i, y_i) = 2y_i^{\mathrm{T}} u_i$, $i = 1, \ldots, q$, then with $\kappa_i(y_i) = -\kappa_i y_i$, where $\kappa_i > 0$, $i = 1, \ldots, q$, it follows that $s_i(\kappa_i(y_i), y_i) = -\kappa_i y_i^{\mathrm{T}} y_i < 0$,

$y_i \neq 0$, $i = 1, \ldots, q$. Alternatively, if $\mathcal{G}$ is vector dissipative with respect to the vector supply rate $S(u, y)$, where $s_i(u_i, y_i) = \gamma_i^2 u_i^{\mathrm{T}} u_i - y_i^{\mathrm{T}} y_i$, where $\gamma_i > 0$, $i = 1, \ldots, q$, then with $\kappa_i(y_i) = 0$, it follows that $s_i(\kappa_i(y_i), y_i) = -y_i^{\mathrm{T}} y_i < 0$, $y_i \neq 0$, $i = 1, \ldots, q$. Hence, if $\mathcal{G}$ is zero-state observable and the dissipation matrix W is such that there exist $\alpha \geq 1$ and $p \in \mathbb{R}_+^q$ such that (2.97) holds, then it follows from Theorem 8.3 that (scalar) storage functions of the form $v_{\mathrm{s}}(x) = p^{\mathrm{T}} V_{\mathrm{s}}(x)$, $x \in \mathbb{R}^n$, where $V_{\mathrm{s}}(\cdot)$ is a vector storage function for $\mathcal{G}$, are positive definite. If $\mathcal{G}$ is geometrically vector dissipative, then p is positive.

Corollary 8.2. Consider the discrete-time large-scale nonlinear dynamical systems $\mathcal{G}$ and $\mathcal{G}_{\mathrm{c}}$ given by (8.6) and (8.7), and (8.98) and (8.99), respectively. Assume that $\mathcal{G}$ and $\mathcal{G}_{\mathrm{c}}$ are zero-state observable and the dissipation matrices $W \in \mathbb{R}^{q \times q}$ and $W_{\mathrm{c}} \in \mathbb{R}^{q \times q}$ are such that there exist, respectively, $\alpha \geq 1$, $p \in \mathbb{R}_+^q$, $\alpha_{\mathrm{c}} \geq 1$, and $p_{\mathrm{c}} \in \mathbb{R}_+^q$ such that (2.97) is satisfied. Then the following statements hold:

i) If $\mathcal{G}$ and $\mathcal{G}_{\mathrm{c}}$ are vector passive and $\tilde{W} \in \mathbb{R}^{q \times q}$ is asymptotically stable, where $\tilde{W}_{(i,j)} \triangleq \max\{W_{(i,j)}, W_{\mathrm{c}(i,j)}\}$, $i, j = 1, \ldots, q$, then the negative feedback interconnection of $\mathcal{G}$ and $\mathcal{G}_{\mathrm{c}}$ is asymptotically stable.

ii) If $\mathcal{G}$ and $\mathcal{G}_{\mathrm{c}}$ are vector nonexpansive and $\tilde{W} \in \mathbb{R}^{q \times q}$ is asymptotically stable, where $\tilde{W}_{(i,j)} \triangleq \max\{W_{(i,j)}, W_{\mathrm{c}(i,j)}\}$, $i, j = 1, \ldots, q$, then the negative feedback interconnection of $\mathcal{G}$ and $\mathcal{G}_{\mathrm{c}}$ is asymptotically stable.

Proof. The proof is a direct consequence of Theorem 8.11. Specifically, i) follows from Theorem 8.11 with $R_i = 0$, $S_i = I_{m_i}$, $Q_i = 0$, $R_{\mathrm{c}i} = 0$, $S_{\mathrm{c}i} = I_{m_i}$, $Q_{\mathrm{c}i} = 0$, $i = 1, \ldots, q$, and $\Sigma = I_q$, while ii) follows from Theorem 8.11 with $R_i = \gamma_i^2 I_{m_i}$, $S_i = 0$, $Q_i = -I_{l_i}$, $R_{\mathrm{c}i} = \gamma_{\mathrm{c}i}^2 I_{l_i}$, $S_{\mathrm{c}i} = 0$, $Q_{\mathrm{c}i} = -I_{m_i}$, $i = 1, \ldots, q$, and $\Sigma = I_q$. $\square$

Thermodynamic Modeling for Discrete-Time Large-Scale Dynamical Systems

9.1 Introduction

Thermodynamic principles have been repeatedly used in continuous-time dynamical system theory as well as information theory for developing models that capture the exchange of nonnegative quantities (e.g., mass and energy) between coupled subsystems [21, 27, 30, 69, 147, 170, 179]. In particular, conservation laws (e.g., mass and energy) are used to capture the exchange of material between coupled macroscopic subsystems known as compartments. Each compartment is assumed to be kinetically homogeneous, that is, any material entering the compartment is instantaneously mixed with the material in the compartment. These models are known as *compartmental* models and are widespread in engineering systems as well as biological and ecological sciences [4, 29, 62, 72, 100, 101, 156]. Even though the compartmental models developed in the literature are based on the first law of thermodynamics involving conservation of energy principles, they do not tell us whether any particular process can actually occur, that is, they do not address the second law of thermodynamics involving entropy notions in the energy flow between subsystems.

The goal of the present chapter is directed toward developing nonlinear discrete-time compartmental models that are consistent with thermodynamic principles. Specifically, since thermodynamic models are concerned with energy flow among subsystems, we develop a nonlinear compartmental dynamical system model that is characterized by energy conservation laws capturing the exchange of energy between coupled macroscopic subsystems. Furthermore, using graph-theoretic notions we state three thermodynamic assumptions consistent with the zeroth and second laws of thermodynamics that ensure that our large-scale dynamical system model gives rise to a thermodynamically consistent energy flow model. Specifically, using a large-scale dynamical systems theory perspective, we show that our compartmental dynamical system model leads to a precise formulation of the equivalence between work energy and heat in a large-scale dynamical system.

Next, we give a deterministic definition of entropy for a large-scale dynamical system that is consistent with the classical thermodynamic definition of entropy and show that it satisfies a Clausius-type inequality leading to the law of entropy nonconservation. Furthermore, we introduce a *new* and dual notion to entropy, namely, *ectropy*, as a measure of the tendency of a large-scale dynamical system to do useful work and grow more organized, and show that conservation of energy in an isolated thermodynamically consistent system necessarily leads to nonconservation of ectropy and entropy. Then, using the system ectropy as a Lyapunov function candidate, we show that our thermodynamically consistent large-scale nonlinear dynamical system model possesses a continuum of equilibria and is *semistable*; that is, it has convergent subsystem energies to Lyapunov stable energy equilibria determined by the large-scale system initial subsystem energies. In addition, we show that the steady-state distribution of the large-scale system energies is uniform, leading to system energy equipartitioning corresponding to a minimum ectropy and a maximum entropy equilibrium state.

9.2 Conservation of Energy and the First Law of Thermodynamics

To develop discrete-time compartmental models that are consistent with thermodynamic principles, consider the discrete-time large-scale dynamical system $\mathcal{G}$ shown in Figure 9.1 involving q interconnected subsystems. Let $E_i : \overline{\mathbb{Z}}_+ \to \overline{\mathbb{R}}_+$ denote the energy (and hence a nonnegative quantity) of the ith subsystem, let $S_i : \overline{\mathbb{Z}}_+ \to \mathbb{R}$ denote the external energy supplied to (or extracted from) the ith subsystem, let $\sigma_{ij} : \overline{\mathbb{R}}_+^q \to \overline{\mathbb{R}}_+$, $i \neq j$, $i, j = 1, \ldots, q$, denote the exchange of energy from the jth subsystem to the ith subsystem, and let $\sigma_{ii} : \overline{\mathbb{R}}_+^q \to \overline{\mathbb{R}}_+$, $i = 1, \ldots, q$, denote the energy loss from the ith subsystem.

An *energy balance* equation for the ith subsystem yields

$$\Delta E_i(k) = \sum_{j=1, j \neq i}^{q} [\sigma_{ij}(E(k)) - \sigma_{ji}(E(k))] - \sigma_{ii}(E(k)) + S_i(k), \quad k \geq k_0,$$

$$(9.1)$$

or, equivalently, in vector form,

$$E(k+1) = w(E(k)) - d(E(k)) + S(k), \quad k \geq k_0, \qquad (9.2)$$

where $E(k) = [E_1(k), \ldots, E_q(k)]^{\mathrm{T}}$, $S(k) = [S_1(k), \ldots, S_q(k)]^{\mathrm{T}}$, $d(E(k)) = [\sigma_{11}(E(k)), \ldots, \sigma_{qq}(E(k))]^{\mathrm{T}}$, $k \geq k_0$, and $w = [w_1, \ldots, w_q]^{\mathrm{T}} : \overline{\mathbb{R}}_+^q \to \mathbb{R}^q$ is such that

$$w_i(E) = E_i + \sum_{j=1, j \neq i}^{q} [\sigma_{ij}(E) - \sigma_{ji}(E)], \quad E \in \overline{\mathbb{R}}_+^q. \qquad (9.3)$$

Equation (9.1) yields a conservation of energy equation and implies that the change of energy stored in the ith subsystem is equal to the external

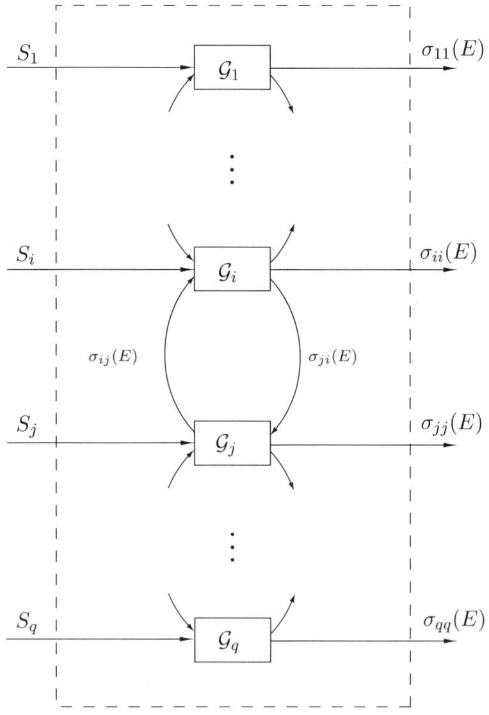

Figure 9.1 Large-scale dynamical system $\mathcal{G}$.

energy supplied to (or extracted from) the ith subsystem plus the energy gained by the ith subsystem from all other subsystems due to subsystem coupling minus the energy dissipated from the ith subsystem. Note that (9.2) or, equivalently, (9.1) is a statement reminiscent of the *first law of thermodynamics* for each of the subsystems, with $E_i(\cdot)$, $S_i(\cdot)$, $\sigma_{ij}(\cdot)$, $i \neq j$, and $\sigma_{ii}(\cdot)$, $i = 1,\ldots,q$, playing the role of the ith subsystem internal energy, energy supplied to (or extracted from) the ith subsystem, the energy exchange between subsystems due to coupling, and the energy dissipated to the environment, respectively.

To further elucidate that (9.2) is essentially the statement of the principle of the conservation of energy, let the total energy in the discrete-time large-scale dynamical system $\mathcal{G}$ be given by $U \triangleq \mathbf{e}^{\mathrm{T}}E$, $E \in \overline{\mathbb{R}}_+^q$, where $\mathbf{e}^{\mathrm{T}} \triangleq [1,\ldots,1]$, and let the energy received by the discrete-time large-scale dynamical system $\mathcal{G}$ (in forms other than work) over the discrete-time interval $\{k_1,\ldots,k_2\}$ be given by $Q \triangleq \sum_{k=k_1}^{k_2} \mathbf{e}^{\mathrm{T}}[S(k) - d(E(k))]$, where $E(k)$, $k \geq k_0$, is the solution to (9.2). Then, premultiplying (9.2) by $\mathbf{e}^{\mathrm{T}}$ and using the fact that $\mathbf{e}^{\mathrm{T}}w(E) = \mathbf{e}^{\mathrm{T}}E$, it follows that

$$\Delta U = Q, \tag{9.4}$$

where $\Delta U \triangleq U(k_2) - U(k_1)$ denotes the variation in the total energy of the discrete-time large-scale dynamical system $\mathcal{G}$ over the discrete-time interval $\{k_1, \ldots, k_2\}$. This is a statement of the first law of thermodynamics for the discrete-time large-scale dynamical system $\mathcal{G}$ and gives a precise formulation of the equivalence between variation in system internal energy and heat.

It is important to note that our discrete-time large-scale dynamical system model does not consider work done by the system on the environment nor work done by the environment on the system. Hence, Q can be interpreted physically as the amount of energy that is received by the system in forms other than work. The extension of addressing work performed by and on the system can be easily handled by including an additional state equation, coupled to the energy balance equation (9.2), involving volume states for each subsystem [81]. Since this slight extension does not alter any of the results in this chapter, it is not considered here for simplicity of exposition.

For our large-scale dynamical system model $\mathcal{G}$, we assume that $\sigma_{ij}(E) = 0$, $E \in \overline{\mathbb{R}}_+^q$, whenever $E_j = 0$, $i, j = 1, \ldots, q$. This constraint implies that if the energy of the jth subsystem of $\mathcal{G}$ is zero, then this subsystem cannot supply any energy to its surroundings nor dissipate energy to the environment. Furthermore, for the remainder of this chapter we assume that $E_i \geq \sigma_{ii}(E) - S_i - \sum_{j=1, j \neq i}^{q} [\sigma_{ij}(E) - \sigma_{ji}(E)] = -\Delta E_i$, $E \in \overline{\mathbb{R}}_+^q$, $S \in \mathbb{R}^q$, $i = 1, \ldots, q$. This constraint implies that the energy that can be dissipated, extracted, or exchanged by the ith subsystem cannot exceed the current energy in the subsystem. Note that this assumption implies that $E(k) \geq\geq 0$ for all $k \geq k_0$.

Next, premultiplying (9.2) by $\mathbf{e}^{\mathrm{T}}$ and using the fact that $\mathbf{e}^{\mathrm{T}}w(E) = \mathbf{e}^{\mathrm{T}}E$, it follows that

$$\mathbf{e}^{\mathrm{T}}E(k_1) = \mathbf{e}^{\mathrm{T}}E(k_0) + \sum_{k=k_0}^{k_1-1} \mathbf{e}^{\mathrm{T}}S(k) - \sum_{k=k_0}^{k_1-1} \mathbf{e}^{\mathrm{T}}d(E(k)), \quad k_1 \geq k_0. \tag{9.5}$$

Now, for the discrete-time large-scale dynamical system $\mathcal{G}$ define the input $u(k) \triangleq S(k)$ and the output $y(k) \triangleq d(E(k))$. Hence, it follows from (9.5) that the discrete-time large-scale dynamical system $\mathcal{G}$ is *lossless* [170] with respect to the *energy supply rate* $r(u, y) = \mathbf{e}^{\mathrm{T}}u - \mathbf{e}^{\mathrm{T}}y$ and with the *energy storage function* $U(E) \triangleq \mathbf{e}^{\mathrm{T}}E$, $E \in \overline{\mathbb{R}}_+^q$. This implies that (see [170] for details)

$$0 \leq U_{\mathrm{a}}(E_0) = U(E_0) = U_{\mathrm{r}}(E_0) < \infty, \quad E_0 \in \overline{\mathbb{R}}_+^q, \tag{9.6}$$

where

$$U_{\mathrm{a}}(E_0) \triangleq - \inf_{u(\cdot),\, K \geq k_0} \sum_{k=k_0}^{K-1} (\mathbf{e}^{\mathrm{T}} u(k) - \mathbf{e}^{\mathrm{T}} y(k)), \tag{9.7}$$

$$U_{\mathrm{r}}(E_0) \triangleq \inf_{u(\cdot),\, K \geq -k_0+1} \sum_{k=-K}^{k_0-1} (\mathbf{e}^{\mathrm{T}} u(k) - \mathbf{e}^{\mathrm{T}} y(k)), \tag{9.8}$$

and $E_0 = E(k_0) \in \overline{\mathbb{R}}_+^q$.

Since $U_{\mathrm{a}}(E_0)$ is the maximum amount of stored energy that can be extracted from the discrete-time large-scale dynamical system $\mathcal{G}$ at any discrete-time instant K, and $U_{\mathrm{r}}(E_0)$ is the minimum amount of energy that can be delivered to the discrete-time large-scale dynamical system $\mathcal{G}$ to transfer it from a state of minimum potential $E(-K) = 0$ to a given state $E(k_0) = E_0$, it follows from (9.6) that the discrete-time large-scale dynamical system $\mathcal{G}$ can deliver to its surroundings all of its stored subsystem energies and can store all of the work done to all of its subsystems. In the case where $S(k) \equiv 0$, it follows from (9.5) and the fact that $\sigma_{ii}(E) \geq 0$, $E \in \overline{\mathbb{R}}_+^q$, $i = 1, \ldots, q$, that the zero solution $E(k) \equiv 0$ of the discrete-time large-scale dynamical system $\mathcal{G}$ with the energy balance equation (9.2) is Lyapunov stable with Lyapunov function $U(E)$ corresponding to the total energy in the system.

The next result shows that the large-scale dynamical system $\mathcal{G}$ is locally controllable.

Proposition 9.1. Consider the discrete-time large-scale dynamical system $\mathcal{G}$ with energy balance equation (9.2). Then for every equilibrium state $E_{\mathrm{e}} \in \overline{\mathbb{R}}_+^q$ and every $\varepsilon > 0$ and $T \in \mathbb{Z}_+$, there exist $S_{\mathrm{e}} \in \mathbb{R}^q$, $\alpha > 0$, and $\hat{T} \in \{0, \cdots, T\}$ such that for every $\hat{E} \in \overline{\mathbb{R}}_+^q$ with $\|\hat{E} - E_{\mathrm{e}}\| \leq \alpha T$, there exists $S : \{0, \cdots, \hat{T}\} \to \mathbb{R}^q$ such that $\|S(k) - S_{\mathrm{e}}\| \leq \varepsilon$, $k \in \{0, \cdots, \hat{T}\}$, and $E(k) = E_{\mathrm{e}} + \frac{(\hat{E}-E_{\mathrm{e}})}{\hat{T}} k$, $k \in \{0, \cdots, \hat{T}\}$.

Proof. Note that with $S_{\mathrm{e}} = d(E_{\mathrm{e}}) - w(E_{\mathrm{e}}) + E_{\mathrm{e}}$, the state $E_{\mathrm{e}} \in \overline{\mathbb{R}}_+^q$ is an equilibrium state of (9.2). Let $\theta > 0$ and $T \in \mathbb{Z}_+$, and define

$$M(\theta, T) \triangleq \sup_{E \in \overline{\mathcal{B}}_1(0),\, k \in \{0, \cdots, T\}} \|w(E_{\mathrm{e}} + k\theta E) - w(E_{\mathrm{e}})$$
$$-d(E_{\mathrm{e}} + k\theta E) + d(E_{\mathrm{e}}) - k\theta E\|. \tag{9.9}$$

Note that for every $T \in \mathbb{Z}_+$, $\lim_{\theta \to 0^+} M(\theta, T) = 0$. Next, let $\varepsilon > 0$ and $T \in \mathbb{Z}_+$ be given, and let $\alpha > 0$ be such that $M(\alpha, T) + \alpha \leq \varepsilon$. (The existence of such an α is guaranteed since $M(\alpha, T) \to 0$ as $\alpha \to 0^+$). Now, let $\hat{E} \in \overline{\mathbb{R}}_+^q$ be such that $\|\hat{E} - E_{\mathrm{e}}\| \leq \alpha T$. With $\hat{T} \triangleq \lceil \frac{\|\hat{E}-E_{\mathrm{e}}\|}{\alpha} \rceil \leq T$, where $\lceil x \rceil$ denotes the *ceiling function* returning the smallest integer greater than or equal to x, and

$$S(k) = -w(E(k)) + d(E(k)) + E(k) + \frac{\hat{E} - E_{\mathrm{e}}}{\left\lceil \frac{\|\hat{E} - E_{\mathrm{e}}\|}{\alpha} \right\rceil}, \quad k \in \{0, \cdots, \hat{T}\},$$

(9.10)

it follows that

$$E(k) = E_{\mathrm{e}} + \frac{(\hat{E} - E_{\mathrm{e}})}{\left\lceil \frac{\|\hat{E} - E_{\mathrm{e}}\|}{\alpha} \right\rceil} k, \quad k \in \{0, \cdots, \hat{T}\}, \qquad (9.11)$$

is a solution to (9.2).

The result is now immediate by noting that $E(\hat{T}) = \hat{E}$ and

$$\|S(k) - S_{\mathrm{e}}\| \leq \left\| w\left(E_{\mathrm{e}} + \frac{(\hat{E} - E_{\mathrm{e}})}{\left\lceil \frac{\|\hat{E} - E_{\mathrm{e}}\|}{\alpha} \right\rceil} k\right) - w(E_{\mathrm{e}}) - d\left(E_{\mathrm{e}} + \frac{(\hat{E} - E_{\mathrm{e}})}{\left\lceil \frac{\|\hat{E} - E_{\mathrm{e}}\|}{\alpha} \right\rceil} k\right) \right.$$
$$\left. + d(E_{\mathrm{e}}) - \frac{(\hat{E} - E_{\mathrm{e}})}{\left\lceil \frac{\|\hat{E} - E_{\mathrm{e}}\|}{\alpha} \right\rceil} k \right\| + \alpha$$
$$\leq M(\alpha, T) + \alpha$$
$$\leq \varepsilon, \quad k \in \{0, \cdots, \hat{T}\}. \qquad (9.12)$$

This completes the proof. $\qquad\square$

It follows from Proposition 9.1 that the discrete-time large-scale dynamical system $\mathcal{G}$ with the energy balance equation (9.2) is *reachable* from and *controllable* to the origin in $\overline{\mathbb{R}}_+^q$. Recall that the discrete-time large-scale dynamical system $\mathcal{G}$ with the energy balance equation (9.2) is reachable from the origin in $\overline{\mathbb{R}}_+^q$ if, for all $E_0 = E(k_0) \in \overline{\mathbb{R}}_+^q$, there exists a finite time $k_{\mathrm{i}} \leq k_0$ and an input $S(k)$ defined on $\{k_{\mathrm{i}}, \ldots, k_0\}$ such that the state $E(k)$, $k \geq k_{\mathrm{i}}$, can be driven from $E(k_{\mathrm{i}}) = 0$ to $E(k_0) = E_0$. Alternatively, $\mathcal{G}$ is controllable to the origin in $\overline{\mathbb{R}}_+^q$ if, for all $E_0 = E(k_0) \in \overline{\mathbb{R}}_+^q$, there exists a finite time $k_{\mathrm{f}} \geq k_0$ and an input $S(k)$ defined on $\{k_0, \ldots, k_{\mathrm{f}}\}$ such that the state $E(k)$, $k \geq k_0$, can be driven from $E(k_0) = E_0$ to $E(k_{\mathrm{f}}) = 0$.

We let $\mathcal{U}_{\mathrm{r}}$ denote the set of all admissible bounded energy inputs to the discrete-time large-scale dynamical system $\mathcal{G}$ such that for every $K \geq -k_0$, the system energy state can be driven from $E(-K) = 0$ to $E(k_0) = E_0 \in \overline{\mathbb{R}}_+^q$ by $S(\cdot) \in \mathcal{U}_{\mathrm{r}}$, and we let $\mathcal{U}_{\mathrm{c}}$ denote the set of all admissible bounded energy inputs to the discrete-time large-scale dynamical system $\mathcal{G}$ such that for every $K \geq k_0$, the system energy state can be driven from $E(k_0) = E_0 \in \overline{\mathbb{R}}_+^q$ to $E(K) = 0$ by $S(\cdot) \in \mathcal{U}_{\mathrm{c}}$. Furthermore, let $\mathcal{U}$ be an input space that is a subset of bounded continuous $\mathbb{R}^q$-valued functions on $\mathbb{Z}$. The spaces $\mathcal{U}_{\mathrm{r}}$, $\mathcal{U}_{\mathrm{c}}$, and $\mathcal{U}$ are assumed to be closed under the shift operator, that is, if $S(\cdot) \in \mathcal{U}$ (respectively, $\mathcal{U}_{\mathrm{c}}$ or $\mathcal{U}_{\mathrm{r}}$), then the function S_K defined by $S_K(k) = S(k + K)$ is contained in $\mathcal{U}$ (respectively, $\mathcal{U}_{\mathrm{c}}$ or $\mathcal{U}_{\mathrm{r}}$) for all $K \geq 0$.

9.3 Nonconservation of Entropy and the Second Law of Thermodynamics

The nonlinear energy balance equation (9.2) can exhibit a full range of nonlinear behavior including bifurcations, limit cycles, and even chaos. However, a thermodynamically consistent energy flow model should ensure that the evolution of the system energy is diffusive (parabolic) in character with convergent subsystem energies. Hence, to ensure a thermodynamically consistent energy flow model we require the following assumptions. For the statement of these assumptions recall Definition 4.1 and let $\phi_{ij}(E) \triangleq \sigma_{ij}(E) - \sigma_{ji}(E)$, $E \in \overline{\mathbb{R}}_+^q$, denote the net energy exchange between subsystems $\mathcal{G}_i$ and $\mathcal{G}_j$ of the discrete-time large-scale dynamical system $\mathcal{G}$.

Assumption 9.1. The connectivity matrix $\mathcal{C} \in \mathbb{R}^{q \times q}$ associated with the large-scale dynamical system $\mathcal{G}$ is defined by

$$\mathcal{C}_{(i,j)} \triangleq \begin{cases} 0, & \text{if } \phi_{ij}(E) \equiv 0, \\ 1, & \text{otherwise,} \end{cases} \quad i \neq j, \quad i, j = 1, \ldots, q, \tag{9.13}$$

and

$$\mathcal{C}_{(i,i)} \triangleq - \sum_{k=1,\, k \neq i}^{q} \mathcal{C}_{(k,i)}, \quad i = j, \quad i = 1, \ldots, q, \tag{9.14}$$

and satisfies rank $\mathcal{C} = q - 1$. Moreover, for every $i \neq j$ such that $\mathcal{C}_{(i,j)} = 1$, $\phi_{ij}(E) = 0$ if and only if $E_i = E_j$.

Assumption 9.2. For $i, j = 1, \ldots, q$, $(E_i - E_j)\phi_{ij}(E) \leq 0$, $E \in \overline{\mathbb{R}}_+^q$.

Assumption 9.3. For $i, j = 1, \ldots, q$, $\frac{\Delta E_i - \Delta E_j}{E_i - E_j} \geq -1$, $E_i \neq E_j$.

The fact that $\phi_{ij}(E) = 0$ if and only if $E_i = E_j$, $i \neq j$, implies that subsystems $\mathcal{G}_i$ and $\mathcal{G}_j$ of $\mathcal{G}$ are *connected*; alternatively, $\phi_{ij}(E) \equiv 0$ implies that $\mathcal{G}_i$ and $\mathcal{G}_j$ are *disconnected*. Assumption 9.1 implies that if the energies in the connected subsystems $\mathcal{G}_i$ and $\mathcal{G}_j$ are equal, then energy exchange between these subsystems is not possible. This is a statement consistent with the *zeroth law of thermodynamics*, which postulates that temperature equality is a necessary and sufficient condition for thermal equilibrium. Furthermore, it follows from the fact that $\mathcal{C} = \mathcal{C}^{\mathrm{T}}$ and rank $\mathcal{C} = q - 1$ that the connectivity matrix $\mathcal{C}$ is irreducible, which implies that for any pair of subsystems $\mathcal{G}_i$ and $\mathcal{G}_j$, $i \neq j$, of $\mathcal{G}$ there exists a sequence of connected subsystems of $\mathcal{G}$ that connect $\mathcal{G}_i$ and $\mathcal{G}_j$.

Assumption 9.2 implies that energy is exchanged from more energetic subsystems to less energetic subsystems and is consistent with the *second law of thermodynamics*, which states that heat (energy) must flow in the direction of lower temperatures. Furthermore, note that $\phi_{ij}(E) = -\phi_{ji}(E)$, $E \in \overline{\mathbb{R}}_+^q$, $i \neq j$, $i,j = 1,\ldots,q$, which implies conservation of energy between lossless subsystems. With $S(k) \equiv 0$, Assumptions 9.1 and 9.2 along with the fact that $\phi_{ij}(E) = -\phi_{ji}(E)$, $E \in \overline{\mathbb{R}}_+^q$, $i \neq j$, $i,j = 1,\ldots,q$, imply that at a given instant of time energy can only be transported, stored, or dissipated but not created and the maximum amount of energy that can be transported and/or dissipated from a subsystem cannot exceed the energy in the subsystem.

Finally, Assumption 9.3 implies that for every pair of connected subsystems $\mathcal{G}_i$ and $\mathcal{G}_j$, $i \neq j$, the energy difference between consecutive time instants is monotonic, that is, $[E_i(k+1) - E_j(k+1)][E_i(k) - E_j(k)] \geq 0$ for all $E_i \neq E_j$, $k \geq k_0$, $i,j = 1,\ldots,q$.

Next, we give a deterministic definition of entropy for the discrete-time large-scale dynamical system $\mathcal{G}$ that is consistent with the classical thermodynamic definition of entropy.

Definition 9.1. For the discrete-time large-scale dynamical system $\mathcal{G}$ with energy balance equation (9.2), a function $\mathcal{S} : \overline{\mathbb{R}}_+^q \to \mathbb{R}$ satisfying

$$\mathcal{S}(E(k_2)) \geq \mathcal{S}(E(k_1)) + \sum_{k=k_1}^{k_2-1} \sum_{i=1}^{q} \frac{S_i(k) - \sigma_{ii}(E(k))}{c + E_i(k+1)}, \qquad (9.15)$$

for every $k_2 \geq k_1 \geq k_0$ and $S(\cdot) \in \mathcal{U}$, is called the *entropy* of $\mathcal{G}$.

The next proposition gives a closed-form expression for the entropy of $\mathcal{G}$.

Proposition 9.2. Consider the discrete-time large-scale dynamical system $\mathcal{G}$ with energy balance equation (9.2) and assume that Assumptions 9.2 and 9.3 hold. Then the function $\mathcal{S} : \overline{\mathbb{R}}_+^q \to \mathbb{R}$ given by

$$\mathcal{S}(E) = \mathbf{e}^{\mathrm{T}}\mathbf{log}_e(c\mathbf{e} + E) - q\log_e c, \qquad E \in \overline{\mathbb{R}}_+^q, \qquad (9.16)$$

where $c > 0$, is an entropy function of $\mathcal{G}$.

Proof. Since $E(k) \geq\geq 0$, $k \geq k_0$, and $\phi_{ij}(E) = -\phi_{ji}(E)$, $E \in \overline{\mathbb{R}}_+^q$, $i \neq j$, $i,j = 1,\ldots,q$, it follows that

$$\begin{aligned}
\Delta\mathcal{S}(E(k)) &= \sum_{i=1}^{q} \log_e\left[1 + \frac{\Delta E_i(k)}{c + E_i(k)}\right] \\
&\geq \sum_{i=1}^{q} \left[\frac{\Delta E_i(k)}{c + E_i(k)}\right]\left[1 + \frac{\Delta E_i(k)}{c + E_i(k)}\right]^{-1} \\
&= \sum_{i=1}^{q} \frac{\Delta E_i(k)}{c + E_i(k) + \Delta E_i(k)}
\end{aligned}$$

$$
= \sum_{i=1}^{q} \frac{\Delta E_i(k)}{c + E_i(k+1)}
$$

$$
= \sum_{i=1}^{q} \left[\frac{S_i(k) - \sigma_{ii}(E(k))}{c + E_i(k+1)} + \sum_{j=1, j \neq i}^{q} \frac{\phi_{ij}(E(k))}{c + E_i(k+1)} \right]
$$

$$
= \sum_{i=1}^{q} \frac{S_i(k) - \sigma_{ii}(E(k))}{c + E_i(k+1)}
$$

$$
+ \sum_{i=1}^{q-1} \sum_{j=i+1}^{q} \left(\frac{\phi_{ij}(E(k))}{c + E_i(k+1)} - \frac{\phi_{ij}(E(k))}{c + E_j(k+1)} \right)
$$

$$
= \sum_{i=1}^{q} \frac{S_i(k) - \sigma_{ii}(E(k))}{c + E_i(k+1)}
$$

$$
+ \sum_{i=1}^{q-1} \sum_{j=i+1}^{q} \frac{\phi_{ij}(E(k))(E_j(k+1) - E_i(k+1))}{(c + E_i(k+1))(c + E_j(k+1))}
$$

$$
\geq \sum_{i=1}^{q} \frac{S_i(k) - \sigma_{ii}(E(k))}{c + E_i(k+1)}, \quad k \geq k_0, \tag{9.17}
$$

where in (9.17) we used the fact that $\log_e(1 + x) \geq \frac{x}{x+1}, x > -1$. Now, summing (9.17) over $\{k_1, \ldots, k_2 - 1\}$ yields (9.15). $\qquad \square$

Note that it follows from the first equality in (9.17) that the entropy function given by (9.16) satisfies (9.15) as an equality for an equilibrium process and as a strict inequality for a nonequilibrium process. The entropy expression given by (9.16) is identical in form to the Boltzmann entropy for statistical thermodynamics. Due to the fact that the entropy is indeterminate to the extent of an additive constant, we can place the constant $q \log_e c$ to zero by taking $c = 1$. Since $\mathcal{S}(E)$ given by (9.16) achieves a maximum when all the subsystem energies E_i, $i = 1, \ldots, q$, are equal, entropy can be thought of as a measure of the tendency of a system to lose the ability to do useful work, lose order, and settle to a more homogeneous state.

9.4 Nonconservation of Ectropy

In this section, we introduce a *new* and dual notion to entropy, namely ectropy, describing the status quo of the discrete-time large-scale dynamical system $\mathcal{G}$. First, we present the definition of ectropy for the discrete-time large-scale dynamical system $\mathcal{G}$.

Definition 9.2. For the discrete-time large-scale dynamical system $\mathcal{G}$

with energy balance equation (9.2), a function $\mathcal{E} : \overline{\mathbb{R}}_+^q \to \mathbb{R}$ satisfying

$$\mathcal{E}(E(k_2)) \leq \mathcal{E}(E(k_1)) + \sum_{k=k_1}^{k_2-1} \sum_{i=1}^{q} E_i(k+1)[S_i(k) - \sigma_{ii}(E(k))], \quad (9.18)$$

for every $k_2 \geq k_1 \geq k_0$ and $S(\cdot) \in \mathcal{U}$, is called the *ectropy* of $\mathcal{G}$.

The next proposition gives a closed-form expression for the ectropy of $\mathcal{G}$.

Proposition 9.3. Consider the discrete-time large-scale dynamical system $\mathcal{G}$ with energy balance equation (9.2) and assume that Assumptions 9.2 and 9.3 hold. Then the function $\mathcal{E} : \overline{\mathbb{R}}_+^q \to \mathbb{R}$ given by

$$\mathcal{E}(E) = \tfrac{1}{2} E^{\mathrm{T}} E, \quad E \in \overline{\mathbb{R}}_+^q, \quad (9.19)$$

is an ectropy function of $\mathcal{G}$.

Proof. Since $E(k) \geq\geq 0$, $k \geq k_0$, and $\phi_{ij}(E) = -\phi_{ji}(E)$, $E \in \overline{\mathbb{R}}_+^q$, $i \neq j$, $i, j = 1, \ldots, q$, it follows that

$$\Delta\mathcal{E}(E(k)) = \tfrac{1}{2} E^{\mathrm{T}}(k+1)E(k+1) - \tfrac{1}{2} E^{\mathrm{T}}(k)E(k)$$

$$= \sum_{i=1}^{q} E_i(k+1)[S_i(k) - \sigma_{ii}(E(k))]$$

$$- \tfrac{1}{2} \sum_{i=1}^{q} \left[\sum_{j=1, j\neq i}^{q} \phi_{ij}(E(k)) + S_i(k) - \sigma_{ii}(E(k)) \right]^2$$

$$+ \sum_{i=1}^{q} \sum_{j=1, j\neq i}^{q} E_i(k+1)\phi_{ij}(E(k))$$

$$= \sum_{i=1}^{q} E_i(k+1)[S_i(k) - \sigma_{ii}(E(k))]$$

$$- \tfrac{1}{2} \sum_{i=1}^{q} \left[\sum_{j=1, j\neq i}^{q} \phi_{ij}(E(k)) + S_i(k) - \sigma_{ii}(E(k)) \right]^2$$

$$+ \sum_{i=1}^{q-1} \sum_{j=i+1}^{q} (E_i(k+1) - E_j(k+1))\phi_{ij}(E(k))$$

$$\leq \sum_{i=1}^{q} E_i(k+1)[S_i(k) - \sigma_{ii}(E(k))], \quad k \geq k_0. \quad (9.20)$$

Now, summing (9.20) over $\{k_1, \ldots, k_2 - 1\}$ yields (9.18). $\qquad\square$

Note that it follows from the last equality in (9.20) that the ectropy function given by (9.19) satisfies (9.18) as an equality for an equilibrium process and as a strict inequality for a nonequilibrium process. It follows from (9.19) that ectropy is a measure of the extent to which the system energy deviates from a homogeneous state. Thus, ectropy is the dual of entropy and is a measure of the tendency of the discrete-time large-scale dynamical system $\mathcal{G}$ to do useful work and grow more organized.

9.5 Semistability of Discrete-Time Thermodynamic Models

Inequality (9.15) is analogous to Clausius' inequality for equilibrium and nonequilibrium thermodynamics as applied to discrete-time large-scale dynamical systems, whereas inequality (9.18) is an anti-Clausius inequality. Moreover, for the ectropy function defined by (9.19), inequality (9.20) shows that a thermodynamically consistent discrete-time large-scale dynamical system is *dissipative* [170] with respect to the supply rate $E^{\mathrm{T}}S$ and with storage function corresponding to the system ectropy $\mathcal{E}(E)$. For the entropy function given by (9.16) note that $\mathcal{S}(0) = 0$, or, equivalently, $\lim_{E \to 0} \mathcal{S}(E) = 0$, which is consistent with the *third law of thermodynamics* (Nernst's theorem), which states that the entropy of every system at absolute zero can always be taken to be equal to zero.

For the isolated (i.e., $S(k) \equiv 0$ and $d(E(k)) \equiv 0$) discrete-time large-scale dynamical system $\mathcal{G}$, (9.15) yields the fundamental inequality

$$\mathcal{S}(E(k_2)) \geq \mathcal{S}(E(k_1)), \quad k_2 \geq k_1. \tag{9.21}$$

Inequality (9.21) implies that, for any dynamical change in an isolated discrete-time large-scale system, the entropy of the final state can never be less than the entropy of the initial state. It is important to stress that this result holds for an isolated dynamical system. It is, however, possible with energy supplied from an external dynamical system (e.g., a controller) to reduce the entropy of the discrete-time large-scale dynamical system. The entropy of both systems taken together, however, cannot decrease. The above observations imply that when an isolated discrete-time large-scale dynamical system with thermodynamically consistent energy flow characteristics (i.e., Assumptions 9.1, 9.2, and 9.3 hold) is at a state of maximum entropy consistent with its energy, it cannot be subject to any further dynamical change since any such change would result in a decrease of entropy. This of course implies that the state of *maximum entropy* is the stable state of an isolated system and this state has to be semistable.

Analogously, it follows from (9.18) that for an isolated discrete-time large-scale dynamical system $\mathcal{G}$ the fundamental inequality

$$\mathcal{E}(E(k_2)) \leq \mathcal{E}(E(k_1)), \quad k_2 \geq k_1, \tag{9.22}$$

is satisfied, which implies that the ectropy of the final state of $\mathcal{G}$ is always less than or equal to the ectropy of the initial state of $\mathcal{G}$. Hence, for the isolated large-scale dynamical system $\mathcal{G}$ the entropy increases if and only if the ectropy decreases. Thus, the state of *minimum ectropy* is the stable state of an isolated system and this equilibrium state has to be semistable. The next theorem concretizes the above observations.

Theorem 9.1. Consider the discrete-time large-scale dynamical system $\mathcal{G}$ with energy balance equation (9.2) with $S(k) \equiv 0$ and $d(E) \equiv 0$, and assume that Assumptions 9.1, 9.2, and 9.3 hold. Then for every $\alpha \geq 0$, $\alpha\mathbf{e}$ is a Lyapunov equilibrium state of (9.2). Furthermore, $E(k) \to \frac{1}{q}\mathbf{e}\mathbf{e}^{\mathrm{T}} E(k_0)$ as $k \to \infty$ and $\frac{1}{q}\mathbf{e}\mathbf{e}^{\mathrm{T}} E(k_0)$ is a semistable equilibrium state. Finally, if for some $m \in \{1, \dots, q\}$, $\sigma_{mm}(E) \geq 0$, $E \in \overline{\mathbb{R}}_+^q$, and $\sigma_{mm}(E) = 0$ if and only if $E_m = 0,$[1] then the zero solution $E(k) \equiv 0$ to (9.2) is a globally asymptotically stable equilibrium state of (9.2).

Proof. It follows from Assumption 9.1 that $\alpha\mathbf{e} \in \overline{\mathbb{R}}_+^q$, $\alpha \geq 0$, is an equilibrium state for (9.2). To show Lyapunov stability of the equilibrium state $\alpha\mathbf{e}$ consider the system shifted ectropy $\mathcal{E}_{\mathrm{s}}(E) = \frac{1}{2}(E - \alpha\mathbf{e})^{\mathrm{T}}(E - \alpha\mathbf{e})$ as a Lyapunov function candidate. Now, since $\phi_{ij}(E) = -\phi_{ji}(E)$, $E \in \overline{\mathbb{R}}_+^q$, $i \neq j$, $i, j = 1, \dots, q$, and $\mathbf{e}^{\mathrm{T}} E(k+1) = \mathbf{e}^{\mathrm{T}} E(k)$, $k \geq k_0$, it follows from Assumptions 9.2 and 9.3 that

$$\Delta\mathcal{E}_{\mathrm{s}}(E(k)) = \frac{1}{2}(E(k+1) - \alpha\mathbf{e})^{\mathrm{T}}(E(k+1) - \alpha\mathbf{e})$$
$$-\frac{1}{2}(E(k) - \alpha\mathbf{e})^{\mathrm{T}}(E(k) - \alpha\mathbf{e})$$
$$= \sum_{i=1}^{q} \sum_{j=1, j\neq i}^{q} E_i(k+1)\phi_{ij}(E(k))$$
$$-\frac{1}{2}\sum_{i=1}^{q}\left[\sum_{j=1, j\neq i}^{q} \phi_{ij}(E(k))\right]^2$$
$$= \sum_{i=1}^{q-1} \sum_{j=i+1}^{q} (E_i(k+1) - E_j(k+1))\phi_{ij}(E(k))$$
$$-\frac{1}{2}\sum_{i=1}^{q}\left[\sum_{j=1, j\neq i}^{q} \phi_{ij}(E(k))\right]^2$$
$$\leq 0, \quad E(k) \in \overline{\mathbb{R}}_+^q, \quad k \geq k_0, \tag{9.23}$$

which establishes Lyapunov stability of the equilibrium state $\alpha\mathbf{e}$.

---

[1] The assumption $\sigma_{mm}(E) \geq 0$, $E \in \overline{\mathbb{R}}_+^q$, and $\sigma_{mm}(E) = 0$ if and only if $E_m = 0$ for some $m \in \{1, \dots, q\}$ implies that if the mth subsystem possesses no energy, then this subsystem cannot dissipate energy to the environment. Conversely, if the mth subsystem does not dissipate energy to the environment, then this subsystem has no energy.

To show that $\alpha \mathbf{e}$ is semistable, note that

$$
\Delta \mathcal{E}_{\mathrm{s}}(E(k)) = \sum_{i=1}^{q} \sum_{j=1, j \neq i}^{q} E_i(k) \phi_{ij}(E(k)) + \tfrac{1}{2} \sum_{i=1}^{q} \left[\sum_{j=1, j \neq i}^{q} \phi_{ij}(E(k)) \right]^2
$$

$$
\geq \sum_{i=1}^{q-1} \sum_{j=i+1}^{q} (E_i(k) - E_j(k)) \phi_{ij}(E(k))
$$

$$
= \sum_{i=1}^{q-1} \sum_{j \in \mathcal{K}_i} (E_i(k) - E_j(k)) \phi_{ij}(E(k)), \quad E(k) \in \overline{\mathbb{R}}_+^q, \quad k \geq k_0,
$$

$$
(9.24)
$$

where $\mathcal{K}_i \triangleq \mathcal{N}_i \setminus \cup_{l=1}^{i-1} \{l\}$ and

$$
\mathcal{N}_i \triangleq \{j \in \{1, \ldots, q\} : \phi_{ij}(E) = 0 \text{ if and only if } E_i = E_j\}, \quad i = 1, \ldots, q.
$$

$$
(9.25)
$$

Next, we show that $\Delta \mathcal{E}_{\mathrm{s}}(E) = 0$ if and only if $(E_i - E_j) \phi_{ij}(E) = 0$, $i = 1, \ldots, q$, $j \in \mathcal{K}_i$. First, assume that $(E_i - E_j) \phi_{ij}(E) = 0$, $i = 1, \ldots, q$, $j \in \mathcal{K}_i$. Then it follows from (9.24) that $\Delta \mathcal{E}_{\mathrm{s}}(E) \geq 0$. However, it follows from (9.23) that $\Delta \mathcal{E}_{\mathrm{s}}(E) \leq 0$. Hence, $\Delta \mathcal{E}_{\mathrm{s}}(E) = 0$. Conversely, assume $\Delta \mathcal{E}_{\mathrm{s}}(E) = 0$. In this case, it follows from (9.23) that $(E_i(k+1) - E_j(k+1)) \phi_{ij}(E(k)) = 0$ and $\sum_{j=1, j \neq i}^{q} \phi_{ij}(E(k)) = 0$, $k \geq k_0$, $i, j = 1, \ldots, q$, $i \neq j$. Since

$$
[E_i(k+1) - E_j(k+1)] \phi_{ij}(E(k)) = [E_i(k) - E_j(k)] \phi_{ij}(E(k))
$$

$$
+ \left[\sum_{h=1, h \neq i}^{q} \phi_{ih}(E(k)) \right.
$$

$$
\left. - \sum_{l=1, l \neq j}^{q} \phi_{jl}(E(k)) \right] \phi_{ij}(E(k))
$$

$$
= [E_i(k) - E_j(k)] \phi_{ij}(E(k)),
$$

$$
k \geq k_0, \quad i, j = 1, \ldots, q, \quad i \neq j, \quad (9.26)
$$

it follows that $(E_i - E_j) \phi_{ij}(E) = 0$, $i = 1, \ldots, q$, $j \in \mathcal{K}_i$.

Let $\mathcal{R} \triangleq \{E \in \overline{\mathbb{R}}_+^q : \Delta \mathcal{E}_{\mathrm{s}}(E) = 0\} = \{E \in \overline{\mathbb{R}}_+^q : (E_i - E_j) \phi_{ij}(E) = 0, i = 1, \ldots, q, j \in \mathcal{K}_i\}$. Now, by Assumption 9.1 the directed graph associated with the connectivity matrix $\mathcal{C}$ for the discrete-time large-scale dynamical system $\mathcal{G}$ is strongly connected, which implies that $\mathcal{R} = \{E \in \overline{\mathbb{R}}_+^q : E_1 = \cdots = E_q\}$. Since the set $\mathcal{R}$ consists of the equilibrium states of (9.2), it follows that the largest invariant set $\mathcal{M}$ contained in $\mathcal{R}$ is given by $\mathcal{M} = \mathcal{R}$. Hence, it follows from the Krasovskii-LaSalle invariant set theorem that for every initial condition $E(k_0) \in \overline{\mathbb{R}}_+^q$, $E(k) \to \mathcal{M}$

as $k \to \infty$, and hence, $\alpha \mathbf{e}$ is a semistable equilibrium state of (9.2). Next, note that since $\mathbf{e}^{\mathrm{T}} E(k) = \mathbf{e}^{\mathrm{T}} E(k_0)$ and $E(k) \to \mathcal{M}$ as $k \to \infty$, it follows that $E(k) \to \frac{1}{q}\mathbf{e}\mathbf{e}^{\mathrm{T}} E(k_0)$ as $k \to \infty$. Hence, with $\alpha = \frac{1}{q}\mathbf{e}^{\mathrm{T}} E(k_0)$, $\alpha \mathbf{e} = \frac{1}{q}\mathbf{e}\mathbf{e}^{\mathrm{T}} E(k_0)$ is a semistable equilibrium state of (9.2).

Finally, to show that in the case where for some $m \in \{1, \ldots, q\}$, $\sigma_{mm}(E) \geq 0$, $E \in \overline{\mathbb{R}}_+^q$, and $\sigma_{mm}(E) = 0$ if and only if $E_m = 0$, the zero solution $E(k) \equiv 0$ to (9.2) is globally asymptotically stable, consider the system ectropy $\mathcal{E}(E) = \frac{1}{2}E^{\mathrm{T}}E$ as a candidate Lyapunov function. Note that $\mathcal{E}(0) = 0$, $\mathcal{E}(E) > 0$, $E \in \overline{\mathbb{R}}_+^q$, $E \neq 0$, and $\mathcal{E}(E)$ is radially unbounded. Now, the Lyapunov difference is given by

$$\Delta\mathcal{E}(E(k)) = \tfrac{1}{2}E^{\mathrm{T}}(k+1)E(k+1) - \tfrac{1}{2}E^{\mathrm{T}}(k)E(k)$$

$$= -E_m(k+1)\sigma_{mm}(E(k)) - \frac{1}{2}\left[\sum_{j=1, j\neq m}^{q} \phi_{mj}(E(k)) - \sigma_{mm}(E(k))\right]^2 - \frac{1}{2}\sum_{i=1, i\neq m}^{q}\left[\sum_{j=1, j\neq i}^{q} \phi_{ij}(E(k))\right]^2$$

$$+ \sum_{i=1}^{q}\sum_{j=1, j\neq i}^{q} E_i(k+1)\phi_{ij}(E(k))$$

$$= -E_m(k+1)\sigma_{mm}(E(k)) - \frac{1}{2}\left[\sum_{j=1, j\neq m}^{q} \phi_{mj}(E(k)) - \sigma_{mm}(E(k))\right]^2 - \frac{1}{2}\sum_{i=1, i\neq m}^{q}\left[\sum_{j=1, j\neq i}^{q} \phi_{ij}(E(k))\right]^2$$

$$+ \sum_{i=1}^{q-1}\sum_{j=i+1}^{q} (E_i(k+1) - E_j(k+1))\phi_{ij}(E(k))$$

$$\leq 0, \quad E(k) \in \overline{\mathbb{R}}_+^q, \quad k \geq k_0, \tag{9.27}$$

which shows that the zero solution $E(k) \equiv 0$ to (9.2) is Lyapunov stable.

To show global asymptotic stability of the zero equilibrium state, note that

$$\Delta\mathcal{E}(E(k)) = \sum_{i=1}^{q-1}\sum_{j=i+1}^{q} (E_i(k) - E_j(k))\phi_{ij}(E(k))$$

$$+ \frac{1}{2}\sum_{i=1, i\neq m}^{q}\left[\sum_{j=1, j\neq i}^{q} \phi_{ij}(E(k))\right]^2 - E_m(k)\sigma_{mm}(E(k))$$

$$+\frac{1}{2}\left[\sum_{j=1,j\neq m}^{q}\phi_{mj}(E(k))-\sigma_{mm}(E(k))\right]^2$$

$$\geq\sum_{i=1}^{q-1}\sum_{j\in\mathcal{K}_i}(E_i(k)-E_j(k))\phi_{ij}(E(k))-E_m(k)\sigma_{mm}(E(k)),$$

$$E(k)\in\overline{\mathbb{R}}_+^q,\quad k\geq k_0. \tag{9.28}$$

Next, we show that $\Delta\mathcal{E}(E)=0$ if and only if $(E_i-E_j)\phi_{ij}(E)=0$ and $\sigma_{mm}(E)=0$, $i=1,\ldots,q$, $j\in\mathcal{K}_i$, $m\in\{1,\ldots,q\}$. First, assume that $(E_i-E_j)\phi_{ij}(E)=0$ and $\sigma_{mm}(E)=0$, $i=1,\ldots,q$, $j\in\mathcal{K}_i$, $m\in\{1,\ldots,q\}$. Then it follows from (9.28) that $\Delta\mathcal{E}(E)\geq 0$. However, it follows from (9.27) that $\Delta\mathcal{E}(E)\leq 0$. Thus, $\Delta\mathcal{E}(E)=0$. Conversely, assume $\Delta\mathcal{E}(E)=0$. Then it follows from (9.27) that $(E_i(k+1)-E_j(k+1))\phi_{ij}(E(k))=0$, $i,j=1,\ldots,q$, $i\neq j$, $\sum_{j=1,j\neq i}^{q}\phi_{ij}(E(k))=0$, $i=1,\ldots,q$, $i\neq m$, $k\geq k_0$, and $\sigma_{mm}(E)=0$, $m\in\{1,\ldots,q\}$. Note that in this case it follows that $\sigma_{mm}(E)=\sum_{j=1,j\neq m}^{q}\phi_{mj}(E)=0$, and hence,

$$[E_i(k+1)-E_j(k+1)]\phi_{ij}(E(k))=[E_i(k)-E_j(k)]\phi_{ij}(E(k)),$$

$$k\geq k_0,\quad i,j=1,\ldots,q,\quad i\neq j, \tag{9.29}$$

which implies that $(E_i-E_j)\phi_{ij}(E)=0$, $i=1,\ldots,q$, $j\in\mathcal{K}_i$. Hence, $(E_i-E_j)\phi_{ij}(E)=0$ and $\sigma_{mm}(E)=0$, $i=1,\ldots,q$, $j\in\mathcal{K}_i$, $m\in\{1,\ldots,q\}$ if and only if $\Delta\mathcal{E}(E)=0$.

Let $\mathcal{R}\triangleq\{E\in\overline{\mathbb{R}}_+^q:\Delta\mathcal{E}(E)=0\}=\{E\in\overline{\mathbb{R}}_+^q:\sigma_{mm}(E)=0,\ m\in\{1,\ldots,q\}\}\cap\{E\in\overline{\mathbb{R}}_+^q:(E_i-E_j)\phi_{ij}(E)=0,\ i=1,\ldots,q,\ j\in\mathcal{K}_i\}$. Now, since Assumption 9.1 holds and $\sigma_{mm}(E)=0$ if and only if $E_m=0$, it follows that $\mathcal{R}=\{E\in\overline{\mathbb{R}}_+^q:E_m=0,\ m\in\{1,\ldots,q\}\}\cap\{E\in\overline{\mathbb{R}}_+^q:E_1=E_2=\cdots=E_q\}=\{0\}$ and the largest invariant set $\mathcal{M}$ contained in $\mathcal{R}$ is given by $\mathcal{M}=\{0\}$. Hence, it follows from the Krasovskii-LaSalle invariant set theorem that for every initial condition $E(k_0)\in\overline{\mathbb{R}}_+^q$, $E(k)\to\mathcal{M}=\{0\}$ as $k\to\infty$, which proves global asymptotic stability of the zero equilibrium state of (9.2). $\square$

It is important to note that Assumption 9.3 involving monotonicity of solutions is explicitly used to prove semistability for discrete-time compartmental dynamical systems. However, Assumption 9.3 is a sufficient condition and not necessary for guaranteeing semistability. Replacing the monotonicity condition with $\sum_{i=1,j=1,i\neq j}^{q}\alpha_{ij}(E)f_{ij}(E)\geq 0$, where

$$\alpha_{ij}(E)\triangleq\begin{cases}\frac{\phi_{ij}(E)}{E_j-E_i}, & E_i\neq E_j,\\ 0, & E_i=E_j,\end{cases} \tag{9.30}$$

$$f_{ij}(E)\triangleq[E_i(k)-E_j(k)][E_i(k+1)-E_j(k+1)], \tag{9.31}$$

provides a weaker sufficient condition for guaranteeing semistability. However, in this case, to ensure that the entropy of $\mathcal{G}$ is monotonically increasing, we additionally require that $\sum_{i=1,j=1,i\neq j}^{q} \beta_{ij}(E)f_{ij}(E) \geq 0$, where

$$\beta_{ij}(E) \triangleq \begin{cases} \frac{1}{(c+E_i(k+1))(c+E_j(k+1))} \cdot \frac{\phi_{ij}(E(k))}{E_j(k)-E_i(k)}, & E_i \neq E_j, \\ 0, & E_i = E_j, \end{cases} \tag{9.32}$$

Thus, a weaker condition for Assumption 9.3 that combines

$$\sum_{i=1,j=1,i\neq j}^{q} \alpha_{ij}(E)f_{ij}(E) \geq 0 \tag{9.33}$$

and

$$\sum_{i=1,j=1,i\neq j}^{q} \beta_{ij}(E)f_{ij}(E) \geq 0, \tag{9.34}$$

is

$$\sum_{i=1,j=1,i\neq j}^{q} \gamma_{ij}(E)f_{ij}(E) \geq 0, \tag{9.35}$$

where

$$\gamma_{ij}(E) \triangleq \alpha_{ij}(E) + \beta_{ij}(E) - \operatorname{sgn}(f_{ij}(E))|\alpha_{ij}(E) - \beta_{ij}(E)| \tag{9.36}$$

and

$$\operatorname{sgn}(f_{ij}(E)) \triangleq |f_{ij}(E)|/f_{ij}(E). \tag{9.37}$$

In Theorem 9.1 we used the shifted ectropy function to show that for the isolated (i.e., $S(k) \equiv 0$ and $d(E) \equiv 0$) discrete-time large-scale dynamical system $\mathcal{G}$ with Assumptions 9.1, 9.2, and 9.3, $E(k) \to \frac{1}{q}\mathbf{ee}^{\mathrm{T}}E(k_0)$ as $k \to \infty$ and $\frac{1}{q}\mathbf{ee}^{\mathrm{T}}E(k_0)$ is a semistable equilibrium state. This result can also be arrived at using the system entropy for the isolated discrete-time large-scale dynamical system $\mathcal{G}$ with Assumptions 9.1, 9.2, and 9.3. To see this, note that since $\mathbf{e}^{\mathrm{T}}w(E) = \mathbf{e}^{\mathrm{T}}E$, $E \in \overline{\mathbb{R}}_+^q$, it follows that $\mathbf{e}^{\mathrm{T}}\Delta E(k) = 0$, $k \geq k_0$. Hence, $\mathbf{e}^{\mathrm{T}}E(k) = \mathbf{e}^{\mathrm{T}}E(k_0)$, $k \geq k_0$. Furthermore, since $E(k) \geq\geq 0$, $k \geq k_0$, it follows that $0 \leq\leq E(k) \leq\leq \mathbf{ee}^{\mathrm{T}}E(k_0)$, $k \geq k_0$, which implies that all solutions to (9.2) are bounded. Next, since by (9.21) the entropy $\mathcal{S}(E(k))$, $k \geq k_0$, of $\mathcal{G}$ is monotonically increasing and $E(k)$, $k \geq k_0$, is bounded, the result follows by using similar arguments as in Theorem 9.1 and using the fact that $\frac{x}{1+x} \leq \log_e(1+x) \leq x$ for all $x > -1$.

Theorem 9.1 implies that the steady-state value of the energy in each subsystem $\mathcal{G}_i$ of the isolated large-scale dynamical system $\mathcal{G}$ is equal; that is, the steady-state energy of the isolated discrete-time large-scale dynamical

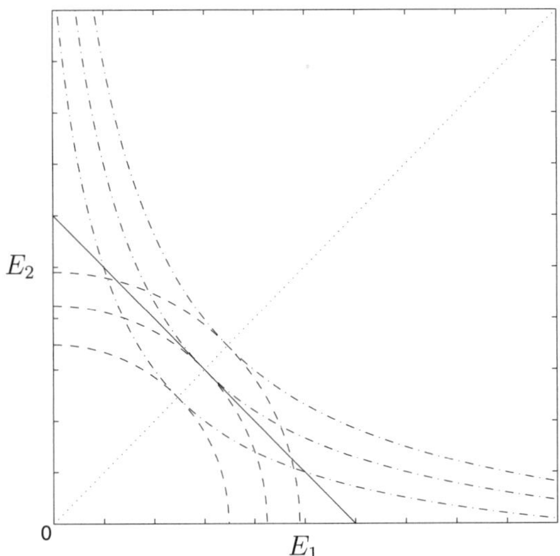

Figure 9.2 Thermodynamic equilibria ($\cdots$), constant energy surfaces (——),
constant ectropy surfaces ($-\,-\,-$), and constant entropy surfaces
($-\cdot-\cdot-$).

system $\mathcal{G}$ given by $E_\infty = \frac{1}{q}\mathbf{e}\mathbf{e}^{\mathrm{T}}E(k_0) = \left[\frac{1}{q}\sum_{i=1}^{q}E_i(k_0)\right]\mathbf{e}$ is uniformly distributed over all subsystems of $\mathcal{G}$. As noted in Chapter 4, this phenomenon is known as *equipartition of energy* [20, 21, 81, 88, 129, 148] and is an emergent behavior in thermodynamic systems. The next proposition shows that among all possible energy distributions in the discrete-time large-scale dynamical system $\mathcal{G}$, energy equipartition corresponds to the minimum value of the system's ectropy and the maximum value of the system's entropy (see Figure 9.2).

Proposition 9.4. Consider the discrete-time large-scale dynamical system $\mathcal{G}$ with energy balance equation (9.2), let $\mathcal{E} : \overline{\mathbb{R}}_+^q \to \mathbb{R}$ and $\mathcal{S} : \overline{\mathbb{R}}_+^q \to \mathbb{R}$ denote the ectropy and entropy of $\mathcal{G}$ given by (9.19) and (9.16), respectively, and define $\mathcal{D}_{\mathrm{c}} \triangleq \{E \in \overline{\mathbb{R}}_+^q : \mathbf{e}^{\mathrm{T}}E = \beta\}$, where $\beta \geq 0$. Then,

$$\arg\min_{E\in\mathcal{D}_{\mathrm{c}}}(\mathcal{E}(E)) = \arg\max_{E\in\mathcal{D}_{\mathrm{c}}}(\mathcal{S}(E)) = E^* = \frac{\beta}{q}\mathbf{e}. \qquad (9.38)$$

Furthermore, $\mathcal{E}_{\min} \triangleq \mathcal{E}(E^*) = \frac{1}{2}\frac{\beta^2}{q}$ and $\mathcal{S}_{\max} \triangleq \mathcal{S}(E^*) = q\log_e(c + \frac{\beta}{q}) - q\log_e c$.

Proof. This is a restatement of Proposition 4.1 for discrete-time large-scale dynamical systems. $\qquad\square$

It follows from (9.21), (9.22), and Proposition 9.4 that conservation of energy necessarily implies nonconservation of ectropy and entropy. Hence, in an isolated discrete-time large-scale dynamical system $\mathcal{G}$ all the energy, though always conserved, will eventually be degraded (diluted) to the point where it cannot produce any useful work. Hence, all motion would cease and the dynamical system would be fated to a state of eternal rest (semistability) wherein all subsystems would possess identical energies (energy equipartition). Ectropy would be a minimum and entropy would be a maximum giving rise to a state of absolute disorder.

9.6 Entropy Increase and the Second Law of Thermodynamics

In the preceding discussion it was assumed that our discrete-time large-scale nonlinear dynamical system model is such that energy is exchanged from more energetic subsystems to less energetic subsystems, that is, heat (energy) flows in the direction of lower temperatures. Although this universal phenomenon can be predicted with virtual certainty, it follows as a manifestation of entropy and ectropy nonconservation for the case of two subsystems. To see this, consider the isolated (i.e., $S(k) \equiv 0$ and $d(E) \equiv 0$) discrete-time large-scale dynamical system $\mathcal{G}$ with energy balance equation (9.2) and assume that the system entropy is monotonically increasing and hence $\Delta \mathcal{S}(E(k)) \geq 0$, $k \geq k_0$. Now, since

$$
\begin{aligned}
0 \leq \Delta \mathcal{S}(E(k)) \\
= \sum_{i=1}^{q} \log_e \left[1 + \frac{\Delta E_i(k)}{c + E_i(k)} \right] \\
\leq \sum_{i=1}^{q} \frac{\Delta E_i(k)}{c + E_i(k)} \\
= \sum_{i=1}^{q} \sum_{j=1, j \neq i}^{q} \frac{\phi_{ij}(E(k))}{c + E_i(k)} \\
= \sum_{i=1}^{q-1} \sum_{j=i+1}^{q} \left[\frac{\phi_{ij}(E(k))}{c + E_i(k)} - \frac{\phi_{ij}(E(k))}{c + E_j(k)} \right] \\
= \sum_{i=1}^{q-1} \sum_{j=i+1}^{q} \frac{\phi_{ij}(E(k))(E_j(k) - E_i(k))}{(c + E_i(k))(c + E_j(k))}, \quad k \geq k_0,
\end{aligned}
\tag{9.39}
$$

it follows that for $q = 2$, $(E_1 - E_2)\phi_{12}(E) \leq 0$, $E \in \overline{\mathbb{R}}_+^2$, which implies that energy (heat) flows naturally from a more energetic subsystem (hot object) to a less energetic subsystem (cooler object). The universality of this emergent behavior thus follows from the fact that entropy (respectively, ectropy) transfer, accompanying energy transfer, always increases (respectively, decreases).

In the case where we have multiple subsystems, it is clear from (9.39) that entropy and ectropy nonconservation does not necessarily imply Assumption 9.2. However, if we invoke the additional condition (Assumption 9.4) that if for any pair of connected subsystems $\mathcal{G}_k$ and $\mathcal{G}_l$, $k \neq l$, with energies $E_k \geq E_l$ (respectively, $E_k \leq E_l$), and for any other pair of connected subsystems $\mathcal{G}_m$ and $\mathcal{G}_n$, $m \neq n$, with energies $E_m \geq E_n$ (respectively, $E_m \leq E_n$) the inequality $\phi_{kl}(E)\phi_{mn}(E) \geq 0$, $E \in \overline{\mathbb{R}}_+^q$, holds, then nonconservation of entropy and ectropy in the isolated discrete-time large-scale dynamical system $\mathcal{G}$ implies Assumption 9.2. The above inequality postulates that the direction of energy exchange for any pair of *energy similar* subsystems is consistent; that is, if for a given pair of connected subsystems at given different energy levels the energy flows in a certain direction, then for any other pair of connected subsystems with the same energy level, the energy flow direction is consistent with the original pair of subsystems. Note that this assumption does *not* specify the direction of energy flow between subsystems.

To see that $\Delta\mathcal{S}(E(k)) \geq 0$, $k \geq k_0$, along with Assumption 9.4 implies Assumption 9.2, note that since (9.39) holds for all $k \geq k_0$ and $E(k_0) \in \overline{\mathbb{R}}_+^q$ is arbitrary, (9.39) implies

$$\sum_{i=1}^{q} \sum_{j \in \mathcal{K}_i} \frac{\phi_{ij}(E)(E_j - E_i)}{(c + E_i)(c + E_j)} \geq 0, \quad E \in \overline{\mathbb{R}}_+^q. \tag{9.40}$$

Now, it follows from (9.40) that for any fixed system energy level $E \in \overline{\mathbb{R}}_+^q$ there exists at least one pair of connected subsystems $\mathcal{G}_k$ and $\mathcal{G}_l$, $k \neq l$, such that $\phi_{kl}(E)(E_l - E_k) \geq 0$. Thus, if $E_k \geq E_l$ (respectively, $E_k \leq E_l$), then $\phi_{kl}(E) \leq 0$ (respectively, $\phi_{kl}(E) \geq 0$). Furthermore, it follows from Assumption 9.2 that for any other pair of connected subsystems $\mathcal{G}_m$ and $\mathcal{G}_n$, $m \neq n$, with $E_m \geq E_n$ (respectively, $E_m \leq E_n$) the inequality $\phi_{mn}(E) \leq 0$ (respectively, $\phi_{mn}(E) \geq 0$) holds, which implies that

$$\phi_{mn}(E)(E_n - E_m) \geq 0, \quad m \neq n. \tag{9.41}$$

Thus, it follows from (9.41) that energy (heat) flows naturally from more energetic subsystems (hot objects) to less energetic subsystems (cooler objects). Of course, since in the isolated discrete-time large-scale dynamical system $\mathcal{G}$ ectropy decreases if and only if entropy increases, the same result can be arrived at by considering the ectropy of $\mathcal{G}$. Since Assumption 9.2 holds, it follows from the conservation of energy and the fact that the discrete-time large-scale dynamical system $\mathcal{G}$ is strongly connected that nonconservation of entropy and ectropy necessarily implies energy equipartition.

Finally, we close this section by showing that our definition of entropy given by (9.16) satisfies the eight criteria established in [67] for the acceptance of an analytic expression for representing a system entropy function.

In particular, note that for a dynamical system $\mathcal{G}$: i) $\mathcal{S}(E)$ is well defined for every state $E \in \overline{\mathbb{R}}_+^q$ as long as $c > 0$. ii) If $\mathcal{G}$ is isolated, then $\mathcal{S}(E(k))$ is a nondecreasing function of time. iii) If $\mathcal{S}_i(E_i) = \log_e(c + E_i) - \log_e c$ is the entropy of the ith subsystem of the system $\mathcal{G}$, then $\mathcal{S}(E) = \sum_{i=1}^{q} \mathcal{S}_i(E_i) = \mathbf{e}^{\mathrm{T}} \log_e(c\mathbf{e} + E) - q \log_e c$, and hence, the system entropy $\mathcal{S}(E)$ is an additive quantity over all subsystems. iv) For the system $\mathcal{G}$, $\mathcal{S}(E) \geq 0$ for all $E \in \overline{\mathbb{R}}_+^q$. v) It follows from Proposition 9.4 that for a given value $\beta \geq 0$ of the total energy of the system $\mathcal{G}$, one and only one state, namely, $E^* = \frac{\beta}{q}\mathbf{e}$, corresponds to the largest value of $\mathcal{S}(E)$. vi) It follows from (9.16) that for the system $\mathcal{G}$, the graph of entropy versus energy is concave and smooth. vii) For a composite discrete-time large-scale dynamical system $\mathcal{G}_{\mathrm{C}}$ of two dynamical systems $\mathcal{G}_{\mathrm{A}}$ and $\mathcal{G}_{\mathrm{B}}$, the expression for the composite entropy $\mathcal{S}_{\mathrm{C}} = \mathcal{S}_{\mathrm{A}} + \mathcal{S}_{\mathrm{B}}$, where $\mathcal{S}_{\mathrm{A}}$ and $\mathcal{S}_{\mathrm{B}}$ are entropies of $\mathcal{G}_{\mathrm{A}}$ and $\mathcal{G}_{\mathrm{B}}$, respectively, is such that the expression for the equilibrium state where the composite maximum entropy is achieved is identical to those obtained for $\mathcal{G}_{\mathrm{A}}$ and $\mathcal{G}_{\mathrm{B}}$ individually. Specifically, if q_{A} and q_{B} denote the number of subsystems in $\mathcal{G}_{\mathrm{A}}$ and $\mathcal{G}_{\mathrm{B}}$, respectively, and β_{A} and β_{B} denote the total energies of $\mathcal{G}_{\mathrm{A}}$ and $\mathcal{G}_{\mathrm{B}}$, respectively, then the maximum entropy of $\mathcal{G}_{\mathrm{A}}$ and $\mathcal{G}_{\mathrm{B}}$ individually is achieved at $E_{\mathrm{A}}^* = \frac{\beta_{\mathrm{A}}}{q_{\mathrm{A}}}\mathbf{e}$ and $E_{\mathrm{B}}^* = \frac{\beta_{\mathrm{B}}}{q_{\mathrm{B}}}\mathbf{e}$, respectively, while the maximum entropy of the composite system $\mathcal{G}_{\mathrm{C}}$ is achieved at $E_{\mathrm{C}}^* = \frac{\beta_{\mathrm{A}} + \beta_{\mathrm{B}}}{q_{\mathrm{A}} + q_{\mathrm{B}}}\mathbf{e}$. $viii$) It follows from Theorem 9.1 that for a stable equilibrium state $E = \frac{\beta}{q}\mathbf{e}$, where $\beta \geq 0$ is the total energy of the system $\mathcal{G}$ and q is the number of subsystems of $\mathcal{G}$, the entropy is totally defined by β and q; that is, $\mathcal{S}(E) = q \log_e(c + \frac{\beta}{q}) - q \log_e c$. Dual criteria to the eight criteria outlined above can also be established for an analytic expression representing system ectropy.

9.7 Thermodynamic Models with Linear Energy Exchange

In this section, we specialize the results of Section 9.2 to the case of large-scale dynamical systems with linear energy exchange between subsystems, that is, $w(E) = WE$ and $d(E) = DE$, where $W \in \mathbb{R}^{q \times q}$ and $D \in \mathbb{R}^{q \times q}$. In this case, the vector form of the energy balance equation (9.2), with $k_0 = 0$, is given by

$$E(k + 1) = WE(k) - DE(k) + S(k), \quad E(0) = E_0, \quad k \geq 0. \tag{9.42}$$

Next, let the net energy exchange from the jth subsystem $\mathcal{G}_j$ to the ith subsystem $\mathcal{G}_i$ be parameterized as $\phi_{ij}(E) = \Phi_{ij}^{\mathrm{T}}E$, where $\Phi_{ij} \in \mathbb{R}^q$ and $E \in \overline{\mathbb{R}}_+^q$. In this case, since $w_i(E) = E_i + \sum_{i=1,j \neq i}^{q} \phi_{ij}(E)$, it follows that

$$W = I_q + \left[\sum_{j=2}^{q} \Phi_{1j}, \ldots, \sum_{j=1,j \neq i}^{q} \Phi_{ij}, \ldots, \sum_{j=1}^{q-1} \Phi_{qj} \right]^{\mathrm{T}}. \tag{9.43}$$

Since $\phi_{ij}(E) = -\phi_{ji}(E)$, $i, j = 1, \ldots, q$, $i \neq j$, $E \in \overline{\mathbb{R}}_+^q$, it follows that $\Phi_{ij} = -\Phi_{ji}$, $i \neq j$, $i, j = 1, \ldots, q$. The following proposition considers the special case where W is symmetric.

Proposition 9.5. Consider the large-scale dynamical system $\mathcal{G}$ with energy balance equation given by (9.42) and with $D = 0$. Then Assumptions 9.1 and 9.2 hold if and only if $W = W^{\mathrm{T}}$, $(W - I_q)\mathbf{e} = 0$, rank $(W - I_q) = q - 1$, and W is nonnegative. In addition, if $S = 0$ and Assumption 9.3 holds, then rank $(W + I_q) = q$ and rank $(W^2 - I_q) = q - 1$.

Proof. Assume Assumptions 9.1 and 9.2 hold. Since, by Assumption 9.2, $(E_i - E_j)\phi_{ij}(E) \leq 0$, $E \in \overline{\mathbb{R}}_+^q$, it follows that $E^{\mathrm{T}}\Phi_{ij}\mathbf{e}_{ij}^{\mathrm{T}}E \leq 0$, $i, j = 1, \ldots, q$, $i \neq j$, where $E \in \overline{\mathbb{R}}_+^q$ and $\mathbf{e}_{ij} \in \mathbb{R}^q$ is a vector whose ith entry is 1, jth entry is -1, and remaining entries are zero. Next, it can be shown that $E^{\mathrm{T}}\Phi_{ij}\mathbf{e}_{ij}^{\mathrm{T}}E \leq 0$, $E \in \overline{\mathbb{R}}_+^q$, $i \neq j$, $i, j = 1, \ldots, q$, if and only if $\Phi_{ij} \in \mathbb{R}^q$ is such that its ith entry is $-\sigma_{ij}$, its jth entry is σ_{ij}, and its remaining entries are zero, where $\sigma_{ij} \geq 0$. Furthermore, since $\Phi_{ij} = -\Phi_{ji}$, $i \neq j$, $i, j = 1, \ldots, q$, it follows that $\sigma_{ij} = \sigma_{ji}$, $i \neq j$, $i, j = 1, \ldots, q$. Hence, W is given by

$$W_{(i,j)} = \begin{cases} 1 - \sum_{k=1,k\neq j}^{q} \sigma_{kj}, & i = j, \\ \sigma_{ij}, & i \neq j, \end{cases} \tag{9.44}$$

which implies that W is symmetric (since $\sigma_{ij} = \sigma_{ji}$) and $(W - I_q)\mathbf{e} = 0$.

Note that since at any given instant of time energy can only be transported or stored but not created and the maximum amount of energy that can be transported cannot exceed the energy in a compartment, it follows that $1 \geq \sum_{k=1,k\neq j}^{q} \sigma_{kj}$. Thus, W is a nonnegative matrix. Now, since by Assumption 9.1, $\phi_{ij}(E) = 0$ if and only if $E_i = E_j$ for all $i, j = 1, \ldots, q$, $i \neq j$, such that $\mathcal{C}_{(i,j)} = 1$, it follows that $\sigma_{ij} > 0$ for all $i, j = 1, \ldots, q$, $i \neq j$, such that $\mathcal{C}_{(i,j)} = 1$. Hence, rank $(W - I_q) = $ rank $\mathcal{C} = q - 1$. The converse is immediate and, hence, is omitted.

Next, assume Assumption 9.3 holds. Since, by Assumption 9.3, $(E_i(k+1) - E_j(k+1))(E_i(k) - E_j(k)) \geq 0$, $i, j = 1, \ldots, q$, $i \neq j$, $k \geq k_0$, it follows that $E^{\mathrm{T}}(k+1)\mathbf{e}_{ij}\mathbf{e}_{ij}^{\mathrm{T}}E(k) \geq 0$ or, equivalently, $E^{\mathrm{T}}(k)W^{\mathrm{T}}\mathbf{e}_{ij}\mathbf{e}_{ij}^{\mathrm{T}}E(k) \geq 0$, $i, j = 1, \ldots, q$, $i \neq j$, $k \geq k_0$, where $E \in \overline{\mathbb{R}}_+^q$. Next, we show that $I_q + W$ is strictly diagonally dominant. Suppose, *ad absurdum*, that $1 + W_{(i,i)} \leq \sum_{l=1,l\neq i}^{q} W_{(i,l)}$ for some i, $1 \leq i \leq q$. Let $E(k_0) = \mathbf{e}_i$, $i = 1, \ldots, q$, where $\mathbf{e}_i \in \overline{\mathbb{R}}_+^q$ is a vector whose ith entry is 1 and remaining entries are zero. Then,

$$\begin{aligned} E^{\mathrm{T}}(k_0)W^{\mathrm{T}}\mathbf{e}_{ij}\mathbf{e}_{ij}^{\mathrm{T}}E(k_0) &= \mathbf{e}_i^{\mathrm{T}}W^{\mathrm{T}}\mathbf{e}_{ij}\mathbf{e}_{ij}^{\mathrm{T}}\mathbf{e}_i \\ &= W_{(i,i)} - W_{(i,j)} \\ &= 1 - \sum_{k=1,k\neq j}^{q} \sigma_{kj} - \sigma_{ij} \end{aligned}$$

$$\geq 0, \quad i, j = 1, \ldots, q, \quad i \neq j. \tag{9.45}$$

Now, it follows from (9.45) that

$$1 + W_{(i,j)} \leq 1 + W_{(i,i)} \leq \sum_{l=1, l \neq i}^{q} W_{(i,l)}, \quad j = 1, \ldots, q, \quad j \neq i, \quad 1 \leq i \leq q, \tag{9.46}$$

or, equivalently,

$$1 \leq \sum_{l=1, l \neq i, l \neq j}^{q} W_{(i,l)}, \quad j = 1, \ldots, q, \quad j \neq i, \quad 1 \leq i \leq q. \tag{9.47}$$

However, since W is compartmental and symmetric, it follows that

$$\sum_{l=1, l \neq i}^{q} W_{(i,l)} = \sum_{l=1, l \neq i}^{q} W_{(l,i)} = \sum_{l=1, l \neq i}^{q} \sigma_{l,i} \leq 1, \quad i = 1, \ldots, q. \tag{9.48}$$

Now, since $W_{(i,j)} = \sigma_{ij} > 0$ for all $i, j = 1, \ldots, q$, $i \neq j$, it follows that

$$\sum_{l=1, l \neq i, l \neq j}^{q} W_{(i,l)} < \sum_{l=1, l \neq i}^{q} W_{(i,l)} \leq 1, \quad i = 1, \ldots, q, \tag{9.49}$$

which contradicts (9.47).

Next, since $I_q + W$ is strictly diagonally dominant it follows from Theorem 6.1.10 of [92] that rank $(I_q + W) = q$. Furthermore, since rank $(W^2 - I_q) = $ rank $(W + I_q)(W - I_q)$, it follows from Sylvester's inequality that

$$\text{rank}\,(W + I_q) + \text{rank}\,(W - I_q) - q \leq \text{rank}\,(W^2 - I_q)$$
$$\leq \min\{\text{rank}\,(W + I_q), \text{rank}\,(W - I_q)\}. \tag{9.50}$$

Now, rank $(W^2 - I_q) = q - 1$ follows from (9.50) by noting that rank $(W - I_q) = q - 1$ and rank $(W + I_q) = q$. $\qquad\square$

Next, we specialize the energy balance equation (9.42) to the case where $D = \text{diag}[\sigma_{11}, \sigma_{22}, \ldots, \sigma_{qq}]$. In this case, the vector form of the energy balance equation (9.2), with $k_0 = 0$, is given by

$$E(k + 1) = AE(k) + S(k), \quad E(0) = E_0, \quad k \in \overline{\mathbb{Z}}_+, \tag{9.51}$$

where $A \triangleq W - D$ is such that

$$A_{(i,j)} = \left\{ \begin{array}{ll} 1 - \sum_{k=1}^{q} \sigma_{kj}, & i = j, \\ \sigma_{ij}, & i \neq j. \end{array} \right. \tag{9.52}$$

Note that (9.52) implies $\sum_{i=1}^{q} A_{(i,j)} = 1 - \sigma_{ii} \leq 1$, $j = 1, \ldots, q$, and hence A is a Lyapunov stable compartmental matrix. If $\sigma_{ii} > 0$, $i = 1, \ldots, q$, then A is an asymptotically stable compartmental matrix.

An important special case of (9.51) is the case where A is symmetric or, equivalently, $\sigma_{ij} = \sigma_{ji}$, $i \neq j$, $i, j = 1, \ldots, q$. In this case, it follows from (9.51) that for each subsystem the energy balance equation satisfies

$$\Delta E_i(k) + \sigma_{ii} E_i(k) + \sum_{j=1, \, j \neq i}^{q} \sigma_{ij}[E_i(k) - E_j(k)] = S_i(k), \quad k \in \overline{\mathbb{Z}}_+. \quad (9.53)$$

Note that $\phi_i(E) \triangleq \sum_{j=1, \, j \neq i}^{q} \sigma_{ij}(E_i - E_j)$, $i = 1, \ldots, q$, represents the energy exchange from the ith subsystem to all other subsystems and is given by the sum of the individual energy exchanges from the ith subsystem to the jth subsystem. Furthermore, these energy exchanges are proportional to the energy differences of the subsystems, that is, $E_i - E_j$. Hence, (9.53) is an energy balance equation that governs the energy exchange among coupled subsystems and is completely analogous to the equations of thermal transfer with subsystem energies playing the role of temperatures. Furthermore, note that since $\sigma_{ij} \geq 0$, $i, j = 1, \ldots, q$, energy is exchanged from more energetic subsystems to less energetic subsystems; this phenomenon is consistent with the second law of thermodynamics, which requires that heat (energy) *must* flow in the direction of lower temperatures.

The next lemma and proposition are needed for developing expressions for steady-state energy distributions of the discrete-time large-scale dynamical system $\mathcal{G}$ with linear energy balance equation (9.51).

Lemma 9.1. Let $A \in \mathbb{R}^{q \times q}$ be (discrete-time) compartmental and let $S \in \mathbb{R}^q$. Then the following properties hold:

i) $I_q - A$ is an M-matrix.

ii) $|\lambda| \leq 1$, $\lambda \in \mathrm{spec}\,(A)$.

iii) If A is semistable and $\lambda \in \mathrm{spec}\,(A)$, then either $|\lambda| < 1$ or $\lambda = 1$ and $\lambda = 1$ is semisimple.

iv) $\mathrm{ind}(I_q - A) \leq 1$ and $\mathrm{ind}(A) \leq 1$.

v) If A is semistable, then $\lim_{k \to \infty} A^k = I_q - (A - I_q)(A - I_q)^{\#} \geq\geq 0$.

vi) $\mathcal{R}(A - I_q) = \mathcal{N}(I_q - (A - I_q)(A - I_q)^{\#})$ and $\mathcal{N}(A - I_q) = \mathcal{R}(I_q - (A - I_q)(A - I_q)^{\#})$.

vii) $\sum_{i=0}^{k} A^i = (A - I_q)^{\#}(A^{k+1} - I_q) + (k+1)[I_q - (A - I_q)(A - I_q)^{\#}]$, $k \in \overline{\mathbb{Z}}_+$.

viii) If A is semistable, then $\sum_{i=0}^{\infty} A^i S$ exists if and only if $S \in \mathcal{R}(A - I_q)$, where $S \in \mathbb{R}^q$.

ix) If A is semistable and $S \in \mathcal{R}(A - I_q)$, then $\sum_{i=0}^{\infty} A^i S = -(A - I_q)^{\#} S$.

x) If A is semistable, $S \in \mathcal{R}(A - I_q)$, and $S \geq\geq 0$, then $-(A - I_q)^{\#} S \geq\geq 0$.

xi) $A - I_q$ is nonsingular if and only if $I_q - A$ is a nonsingular M-matrix.

xii) If A is semistable and $A - I_q$ is nonsingular, then A is asymptotically stable and $(I_q - A)^{-1} \geq\geq 0$.

Proof. i) Note that $A^{\mathrm{T}}\mathbf{e} = [-(1 - \sum_{i=1}^{q} A_{(i,\,1)}), -(1 - \sum_{i=1}^{q} A_{(i,\,2)}),$ $\ldots, -(1 - \sum_{i=1}^{q} A_{(i,\,q)})]^{\mathrm{T}} + \mathbf{e}$. Then $(I_q - A)^{\mathrm{T}}\mathbf{e} \geq\geq 0$ and $I_q - A$ is a Z-matrix. It follows from Theorem 1 of [16] that $(I_q - A)^{\mathrm{T}}$, and hence, $I_q - A$ is an M-matrix.

ii) The result follows from i) and Lemma 1 of [71].

iii) The result follows from Theorem 2 of [71].

iv) Since $(I_q - A)^{\mathrm{T}}\mathbf{e} \geq\geq 0$ it follows that $I_q - A$ is an M-matrix and has "property c" (see [15]). Hence, it follows from Lemma 4.11 of [15] that $I_q - A$ has "property c" if and only if $\mathrm{ind}(I_q - A) \leq 1$. Next, since $\mathrm{ind}(I_q - A) \leq 1$, it follows from the real Jordan decomposition that there exist invertible matrices $J \in \mathbb{R}^{r \times r}$, where $r = \mathrm{rank}(I_q - A)$, and $U \in \mathbb{R}^{q \times q}$ such that J is diagonal and

$$I_q - A = U \begin{bmatrix} J & 0 \\ 0 & 0 \end{bmatrix} U^{-1}, \tag{9.54}$$

which implies

$$A = U \begin{bmatrix} I_r - J & 0 \\ 0 & I_{q-r} \end{bmatrix} U^{-1}. \tag{9.55}$$

Hence, $\mathrm{ind}(A) \leq 1$.

v) The result follows from Theorem 2 of [71].

vi) Let $x \in \mathcal{R}(A - I_q)$, that is, there exists $y \in \mathbb{R}^q$ such that $x = (A - I_q)y$. Now, $(I_q - (A - I_q)(A - I_q)^{\#})x = x - (A - I_q)(A - I_q)^{\#}(A - I_q)y = x - (A - I_q)y = 0$, which implies that $\mathcal{R}(A - I_q) \subseteq \mathcal{N}(I_q - (A - I_q)(A - I_q)^{\#})$. Conversely, let $x \in \mathcal{N}(I_q - (A - I_q)(A - I_q)^{\#})$. Hence, $(I_q - (A - I_q)(A - I_q)^{\#})x = 0$, or, equivalently, $x = (A - I_q)(A - I_q)^{\#}x$, which implies that $x \in \mathcal{R}(A - I_q)$, and hence, proves $\mathcal{R}(A - I_q) = \mathcal{N}(I_q - (A - I_q)(A - I_q)^{\#})$. The equality $\mathcal{N}(A - I_q) = \mathcal{R}(I_q - (A - I_q)(A - I_q)^{\#})$ can be proved in an analogous manner.

vii) Note since $A = U \begin{bmatrix} I_r - J & 0 \\ 0 & I_{q-r} \end{bmatrix} U^{-1}$ and J is invertible it follows that

$$\sum_{i=0}^{k} A^i = \sum_{i=0}^{k} U \begin{bmatrix} (I_r - J)^i & 0 \\ 0 & I_{q-r} \end{bmatrix} U^{-1}$$

$$= U \begin{bmatrix} \sum_{i=0}^{k}(I_r - J)^i & 0 \\ 0 & (k+1)I_{q-r} \end{bmatrix} U^{-1}$$

$$= U \begin{bmatrix} -J^{-1}[(I_r - J)^{k+1} - I_r] & 0 \\ 0 & (k+1)I_{q-r} \end{bmatrix} U^{-1}$$

$$= U \begin{bmatrix} -J^{-1} & 0 \\ 0 & 0 \end{bmatrix} U^{-1} U \begin{bmatrix} (I_r - J)^{k+1} - I_r & 0 \\ 0 & 0 \end{bmatrix} U^{-1}$$

$$+ U \begin{bmatrix} 0 & 0 \\ 0 & (k+1)I_{q-r} \end{bmatrix} U^{-1}$$

$$= (A - I_q)^{\#}(A^{k+1} - I_q) + (k+1)$$

$$\cdot \left(I_q - U \begin{bmatrix} J - I_r & 0 \\ 0 & 0 \end{bmatrix} U^{-1} U \begin{bmatrix} (J - I_r)^{-1} & 0 \\ 0 & 0 \end{bmatrix} U^{-1} \right)$$

$$= (A - I_q)^{\#}(A^{k+1} - I_q) + (k+1)[I_q - (A - I_q)(A - I_q)^{\#}],$$
$$k \in \overline{\mathbb{Z}}_+. \qquad (9.56)$$

$viii$) The result is a direct consequence of v)–vii).

ix) The result follows from v) and vii).

x) The result follows from ix).

xi) The result follows from i).

xii) Asymptotic stability of A is a direct consequence of iii). $(I_q - A)^{-1} \geq\geq 0$ follows from Lemma 1 of [71]. $\qquad \square$

Proposition 9.6 ([71]). Consider the discrete-time large-scale dynamical system $\mathcal{G}$ with energy balance equation given by (9.51). Suppose $E_0 \geq\geq 0$, and $S(k) \geq\geq 0$, $k \in \overline{\mathbb{Z}}_+$. Then the solution $E(k)$, $k \in \overline{\mathbb{Z}}_+$, to (9.51) is nonnegative for all $k \in \overline{\mathbb{Z}}_+$ if and only if A is nonnegative.

Next, we develop expressions for the steady-state energy distribution for a discrete-time large-scale linear dynamical system $\mathcal{G}$ for the cases where A is semistable, and the supplied system energy $S(k)$ is a periodic function with period $\tau \in \overline{\mathbb{Z}}_+$, $\tau > 0$, that is, $S(k + \tau) = S(k)$, $k \in \overline{\mathbb{Z}}_+$, and $S(k)$ is constant, that is, $S(k) \equiv S$. Define $e(k) \triangleq E(k) - E(k + \tau)$, $k \in \overline{\mathbb{Z}}_+$, and note that

$$e(k + 1) = Ae(k), \quad e(0) = E(0) - E(\tau), \quad k \in \overline{\mathbb{Z}}_+. \qquad (9.57)$$

Hence, since

$$e(k) = A^k[E(0) - E(\tau)], \quad k \in \overline{\mathbb{Z}}_+, \qquad (9.58)$$

and A is semistable, it follows from v) of Lemma 9.1 that

$$\lim_{k \to \infty} e(k) = \lim_{k \to \infty} [E(k) - E(k + \tau)]$$
$$= [I_q - (A - I_q)(A - I_q)^{\#}][E(0) - E(\tau)], \qquad (9.59)$$

which represents a constant offset to the steady-state error energy distribution in the discrete-time large-scale nonlinear dynamical system $\mathcal{G}$. For the case where $S(k) \equiv S$, $\tau \to \infty$, and hence, the following result is immediate.

Proposition 9.7. Consider the discrete-time large-scale dynamical system $\mathcal{G}$ with energy balance equation given by (9.51). Suppose that A is semistable, $E_0 \geq\geq 0$, and $S(k) \equiv S \geq\geq 0$. Then $E_\infty \triangleq \lim_{k\to\infty} E(k)$ exists if and only if $S \in \mathcal{R}(A - I_q)$. In this case,

$$E_\infty = [I_q - (A - I_q)(A - I_q)^{\#}]E_0 - (A - I_q)^{\#}S \tag{9.60}$$

and $E_\infty \geq\geq 0$. If, in addition, $A - I_q$ is nonsingular, then E_∞ exists for all $S \geq\geq 0$ and is given by

$$E_\infty = (I_q - A)^{-1}S. \tag{9.61}$$

Proof. Note that the solution $E(k)$, $k \in \overline{\mathbb{Z}}_+$, to (9.51) is given by

$$E(k) = A^k E_0 + \sum_{i=0}^{k-1} A^{(k-1-i)}S(i), \quad k \in \overline{\mathbb{Z}}_+. \tag{9.62}$$

Now, the result is a direct consequence of Proposition 9.6 and $v)$, $viii)$, $ix)$, and $x)$ of Lemma 9.1. $\qquad\square$

Next, we specialize the result of Proposition 9.7 to the case where there is no energy dissipation from each subsystem $\mathcal{G}_i$ of $\mathcal{G}$, that is, $\sigma_{ii} = 0$, $i = 1, \ldots, q$. Note that in this case $\mathbf{e}^{\mathrm{T}}(A - I_q) = 0$ and hence $\operatorname{rank}(A - I_q) \leq q - 1$. Furthermore, if $S = 0$ it follows from (9.51) that $\mathbf{e}^{\mathrm{T}}\Delta E(k) = \mathbf{e}^{\mathrm{T}}(A - I_q)E(k) = 0$, $k \in \overline{\mathbb{Z}}_+$, and hence, the total energy of the isolated discrete-time large-scale nonlinear dynamical system $\mathcal{G}$ is conserved.

Proposition 9.8. Consider the discrete-time large-scale dynamical system $\mathcal{G}$ with energy balance equation given by (9.51). Assume $\operatorname{rank}(A - I_q) = \operatorname{rank}(A^2 - I_q) = q - 1$, $\sigma_{ii} = 0$, $i = 1, \ldots, q$, and $A = A^{\mathrm{T}}$. If $E_0 \geq\geq 0$, and $S = 0$, then the equilibrium state $\alpha\mathbf{e}$, $\alpha \geq 0$, of the isolated system $\mathcal{G}$ is semistable and the steady-state energy distribution E_∞ of the isolated discrete-time large-scale dynamical system $\mathcal{G}$ is given by

$$E_\infty = \left[\frac{1}{q}\sum_{i=1}^{q} E_{i0}\right]\mathbf{e}. \tag{9.63}$$

If, in addition, for some $m \in \{1, \ldots, q\}$, $\sigma_{mm} > 0$, then the zero solution $E(k) \equiv 0$ to (9.51) is globally asymptotically stable.

Proof. Note that since $\mathbf{e}^{\mathrm{T}}(A - I_q) = 0$ it follows from (9.51) with $S(k) \equiv 0$ that $\mathbf{e}^{\mathrm{T}}\Delta E(k) = 0$, $k \geq 0$, and hence, $\mathbf{e}^{\mathrm{T}}E(k) = \mathbf{e}^{\mathrm{T}}E_0$, $k \geq 0$.

Furthermore, since by Proposition 9.6 the solution $E(k)$, $k \geq k_0$, to (9.51) is nonnegative, it follows that $0 \leq E_i(k) \leq \mathbf{e}^{\mathrm{T}} E(k) = \mathbf{e}^{\mathrm{T}} E_0$, $k \geq 0$, $i = 1, \ldots, q$. Hence, the solution $E(k)$, $k \geq 0$, to (9.51) is bounded for all $E_0 \in \overline{\mathbb{R}}_+^q$. Next, note that $\phi_{ij}(E) = \sigma_{ij}(E_j - E_i)$ and $(E_i - E_j)\phi_{ij}(E) = -\sigma_{ij}(E_i - E_j)^2 \leq 0$, $E \in \overline{\mathbb{R}}_+^q$, $i \neq j$, $i, j = 1, \ldots, q$, which implies that Assumptions 9.1 and 9.2 are satisfied. Thus, $E = \alpha \mathbf{e}$, $\alpha \geq 0$, is the equilibrium state of the isolated large-scale dynamical system $\mathcal{G}$. Furthermore, define the Lyapunov function candidate $\mathcal{E}_{\mathrm{s}}(E) = \frac{1}{2}(E - \alpha \mathbf{e})^{\mathrm{T}}(E - \alpha \mathbf{e})$, $E \in \overline{\mathbb{R}}_+^q$. Since A is compartmental and symmetric, it follows from $ii)$ of Lemma 9.1 that

$$\begin{aligned}
\Delta \mathcal{E}_{\mathrm{s}}(E) &= \tfrac{1}{2}(AE - \alpha \mathbf{e})^{\mathrm{T}}(AE - \alpha \mathbf{e}) - \tfrac{1}{2}(E - \alpha \mathbf{e})^{\mathrm{T}}(E - \alpha \mathbf{e}) \\
&= \tfrac{1}{2} E^{\mathrm{T}}(A^2 - I_q) E \\
&\leq 0,
\end{aligned} \tag{9.64}$$

which implies Lyapunov stability of the equilibrium state $\alpha \mathbf{e}$, $\alpha \geq 0$.

Next, consider the set $\mathcal{R} \triangleq \{E \in \overline{\mathbb{R}}_+^q : \Delta \mathcal{E}_{\mathrm{s}}(E) = 0\} = \{E \in \overline{\mathbb{R}}_+^q : E^{\mathrm{T}}(A^2 - I_q)E = 0\}$. Since A is compartmental and symmetric it follows from $ii)$ of Lemma 9.1 that $A^2 - I_q$ is a negative semi-definite matrix and hence $E^{\mathrm{T}}(A^2 - I_q)E = 0$ if and only if $(A^2 - I_q)E = 0$. Furthermore, since, by assumption, $\mathrm{rank}\,(A - I_q) = \mathrm{rank}\,(A^2 - I_q) = q - 1$, it follows that there exists one and only one linearly independent solution to $(A^2 - I_q)E = 0$ given by $E = \mathbf{e}$. Hence, $\mathcal{R} = \{E \in \overline{\mathbb{R}}_+^q : E = \alpha \mathbf{e}, \alpha \geq 0\}$. Since $\mathcal{R}$ consists of only equilibrium states of (9.51) it follows that $\mathcal{M} = \mathcal{R}$, where $\mathcal{M}$ is the largest invariant set contained in $\mathcal{R}$. Hence, for every $E_0 \in \overline{\mathbb{R}}_+^q$, it follows from the Krasovskii-LaSalle invariant set theorem that $E(k) \to \alpha \mathbf{e}$ as $k \to \infty$ for some $\alpha \geq 0$ and, hence, $\alpha \mathbf{e}$, $\alpha \geq 0$, is a semistable equilibrium state of (9.51). Furthermore, since the energy is conserved in the isolated large-scale dynamical system $\mathcal{G}$ it follows that $q\alpha = \mathbf{e}^{\mathrm{T}} E_0$. Thus, $\alpha = \frac{1}{q} \sum_{i=1}^{q} E_{i0}$, which implies (9.63).

Finally, to show that in the case where $\sigma_{mm} > 0$ for some $m \in \{1, \ldots, q\}$, the zero solution $E(k) \equiv 0$ to (9.51) is globally asymptotically stable, consider the system ectropy $\mathcal{E}(E) = \frac{1}{2} E^{\mathrm{T}} E$, $E \in \overline{\mathbb{R}}_+^q$, as a candidate Lyapunov function. Note that Lyapunov stability of the zero equilibrium state follows from the previous analysis with $\alpha = 0$. Next, note that

$$\begin{aligned}
\Delta \mathcal{E}(E) &= \tfrac{1}{2} E^{\mathrm{T}}(A^2 - I_q) E \\
&= \tfrac{1}{2} E^{\mathrm{T}}[(W - D)^2 - I_q] E \\
&= \tfrac{1}{2} E^{\mathrm{T}}(W^2 - I_q) E - \tfrac{1}{2} E^{\mathrm{T}}(WD + DW - D^2) E \\
&= \tfrac{1}{2} E^{\mathrm{T}}(W^2 - I_q) E - \sum_{i=1, i \neq m}^{q} \sigma_{mm} \sigma_{mi} E_m E_i \\
&\quad - \sigma_{mm}(W_{(m,m)} - \sigma_{mm}) E_m^2 - \tfrac{1}{2} \sigma_{mm}^2 E_m^2, \quad E \in \overline{\mathbb{R}}_+^q. \tag{9.65}
\end{aligned}$$

Consider the set $\mathcal{R} \triangleq \{E \in \overline{\mathbb{R}}_+^q : \Delta \mathcal{E}(E) = 0\} = \{E \in \overline{\mathbb{R}}_+^q : E_1 = \cdots = E_q\} \cap \{E \in \overline{\mathbb{R}}_+^q : E_m = 0, m \in \{1, \ldots, q\}\} = \{0\}$. Hence, the largest invariant set contained in $\mathcal{R}$ is given by $\mathcal{M} = \mathcal{R} = \{0\}$, and thus, it follows from the Krasovskii-LaSalle invariant set theorem that the zero solution $E(k) \equiv 0$ to (9.51) is globally asymptotically stable. $\qquad\square$

Finally, we examine the steady-state energy distribution for large-scale nonlinear dynamical systems $\mathcal{G}$ in case of strong coupling between subsystems, that is, $\sigma_{ij} \to \infty$, $i \neq j$. For this analysis we assume that A given by (9.51) is symmetric, that is, $\sigma_{ij} = \sigma_{ji}$, $i \neq j$, $i,j = 1,\ldots,q$, and $\sigma_{ii} > 0$, $i = 1,\ldots,q$. Thus, $I_q - A$ is a nonsingular M-matrix for all values of σ_{ij}, $i \neq j$, $i,j = 1,\ldots,q$. Moreover, in this case it follows that if $\frac{\sigma_{ij}}{\sigma_{kl}} \to 1$ as $\sigma_{ij} \to \infty$, $i \neq j$, and $\sigma_{kl} \to \infty$, $k \neq l$, then

$$\lim_{\sigma_{ij} \to \infty, i \neq j} (I_q - A)^{-1} = \lim_{\sigma \to \infty} [D - \sigma(-qI_q + \mathbf{ee}^{\mathrm{T}})]^{-1}, \qquad (9.66)$$

where $D = \mathrm{diag}[\sigma_{11}, \ldots, \sigma_{qq}] > 0$. The following lemmas are needed for the next result.

Lemma 9.2. Let $Y \in \mathbb{R}^{q \times q}$ be such that $\mathrm{ind}\,(Y) \leq 1$. Then $\lim_{\sigma \to \infty} (I_q - \sigma Y)^{-1} = I_q - Y^{\#}Y$.

Proof. Note that

$$(I_q - \sigma Y)^{-1} = I_q + \sigma(I_q - \sigma Y)^{-1}Y$$

$$= I_q + \left(\frac{1}{\sigma}I_q - Y\right)^{-1} Y$$

$$= I_q - \left(Y - \frac{1}{\sigma}I_q\right)^{-1} Y. \qquad (9.67)$$

Now, using the fact that if $A \in \mathbb{R}^{q \times q}$ and $\mathrm{ind}\,A \leq 1$, then

$$\lim_{\alpha \to 0} (A + \alpha I)^{-1}A = AA^{\#} = A^{\#}A, \qquad (9.68)$$

it follows that

$$\lim_{\sigma \to \infty} (I_q - \sigma Y)^{-1} = I_q - \lim_{\frac{1}{\sigma} \to 0} \left(Y - \frac{1}{\sigma}I_q\right)^{-1} Y = I_q - Y^{\#}Y, \qquad (9.69)$$

which proves the result. $\qquad\square$

Lemma 9.3. Let $D \in \mathbb{R}^{q \times q}$ and $X \in \mathbb{R}^{q \times q}$ be such that $D > 0$ and $X = -qI_q + \mathbf{ee}^{\mathrm{T}}$. Then

$$I_q - Y^{\#}Y = \frac{D^{\frac{1}{2}}\mathbf{ee}^{\mathrm{T}}D^{\frac{1}{2}}}{\mathbf{e}^{\mathrm{T}}D\mathbf{e}}, \qquad (9.70)$$

where $Y \triangleq D^{-\frac{1}{2}}XD^{-\frac{1}{2}}$.

Proof. Note that

$$Y = D^{-\frac{1}{2}}(-qI_q + \mathbf{e}\mathbf{e}^{\mathrm{T}})D^{-\frac{1}{2}} = -qD^{-1} + D^{-\frac{1}{2}}\mathbf{e}\mathbf{e}^{\mathrm{T}}D^{-\frac{1}{2}}. \tag{9.71}$$

Now, using the fact that if $N \in \mathbb{R}^{q \times q}$ is nonsingular and symmetric and $b \in \mathbb{R}^q$ is a nonzero vector, then

$$(N + bb^{\mathrm{T}})^{+} = \left(I - \frac{1}{b^{\mathrm{T}}N^{-2}b}N^{-1}bb^{\mathrm{T}}N^{-1}\right)N^{-1}$$
$$\cdot \left(I - \frac{1}{b^{\mathrm{T}}N^{-2}b}N^{-1}bb^{\mathrm{T}}N^{-1}\right), \tag{9.72}$$

it follows that

$$-Y^{\#} = \frac{1}{q}\left(I_q - \frac{D^{\frac{1}{2}}\mathbf{e}\mathbf{e}^{\mathrm{T}}D^{\frac{1}{2}}}{\mathbf{e}^{\mathrm{T}}D\mathbf{e}}\right)D\left(I_q - \frac{D^{\frac{1}{2}}\mathbf{e}\mathbf{e}^{\mathrm{T}}D^{\frac{1}{2}}}{\mathbf{e}^{\mathrm{T}}D\mathbf{e}}\right). \tag{9.73}$$

Hence,

$$-Y^{\#}Y = -\left(I_q - \frac{D^{\frac{1}{2}}\mathbf{e}\mathbf{e}^{\mathrm{T}}D^{\frac{1}{2}}}{\mathbf{e}^{\mathrm{T}}D\mathbf{e}}\right)D\left(I_q - \frac{D^{\frac{1}{2}}\mathbf{e}\mathbf{e}^{\mathrm{T}}D^{\frac{1}{2}}}{\mathbf{e}^{\mathrm{T}}D\mathbf{e}}\right)$$
$$\cdot \left(D^{-1} - \frac{1}{q}D^{-\frac{1}{2}}\mathbf{e}\mathbf{e}^{\mathrm{T}}D^{-\frac{1}{2}}\right)$$
$$= -\left(I_q - \frac{D^{\frac{1}{2}}\mathbf{e}\mathbf{e}^{\mathrm{T}}D^{\frac{1}{2}}}{\mathbf{e}^{\mathrm{T}}D\mathbf{e}}\right). \tag{9.74}$$

Thus, $I_q - Y^{\#}Y = \frac{D^{\frac{1}{2}}\mathbf{e}\mathbf{e}^{\mathrm{T}}D^{\frac{1}{2}}}{\mathbf{e}^{\mathrm{T}}D\mathbf{e}}$. $\qquad\square$

Proposition 9.9. Consider the discrete-time large-scale dynamical system $\mathcal{G}$ with energy balance equation given by (9.51). Let $S(k) \equiv S$, $S \in \mathbb{R}^{q \times q}$, $A \in \mathbb{R}^{q \times q}$ be compartmental and assume A is symmetric, $\sigma_{ii} > 0$, $i = 1, \ldots, q$, and $\frac{\sigma_{ij}}{\sigma_{kl}} \to 1$ as $\sigma_{ij} \to \infty$, $i \neq j$, and $\sigma_{kl} \to \infty$, $k \neq l$. Then the steady-state energy distribution E_∞ of the discrete-time large-scale dynamical system $\mathcal{G}$ is given by

$$E_\infty = \left[\frac{\mathbf{e}^{\mathrm{T}}S}{\sum_{i=1}^{q}\sigma_{ii}}\right]\mathbf{e}. \tag{9.75}$$

Proof. Note that in the case where $\frac{\sigma_{ij}}{\sigma_{kl}} \to 1$ as $\sigma_{ij} \to \infty$, $i \neq j$, and $\sigma_{kl} \to \infty$, $k \neq l$, it follows that the corresponding limit of $(I_q - A)^{-1}$ can be equivalently taken as in (9.66). Next, with $D = \mathrm{diag}[\sigma_{11}, \ldots, \sigma_{qq}]$ and $X = -qI_q + \mathbf{e}\mathbf{e}^{\mathrm{T}}$, it follows that $I_q - A = D - \sigma X = D^{\frac{1}{2}}(I_q - \sigma D^{-\frac{1}{2}}XD^{-\frac{1}{2}})D^{\frac{1}{2}}$. Now, it follows from Lemmas 9.2 and 9.3 that

$$E_\infty = \lim_{\sigma_{ij} \to \infty, i \neq j}(I_q - A)^{-1}S = \frac{\mathbf{e}\mathbf{e}^{\mathrm{T}}}{\mathbf{e}^{\mathrm{T}}D\mathbf{e}}S = \left[\frac{\mathbf{e}^{\mathrm{T}}S}{\sum_{i=1}^{q}\sigma_{ii}}\right]\mathbf{e}, \tag{9.76}$$

which proves the result. $\qquad\square$

Proposition 9.9 shows that in the limit of strong coupling the steady-state energy distribution E_∞ given by (9.61) becomes

$$E_\infty = \lim_{\sigma_{ij}\to\infty,\, i\neq j}(I_q - A)^{-1}S = \left[\frac{\mathbf{e}^{\mathrm{T}}S}{\sum_{i=1}^{q}\sigma_{ii}}\right]\mathbf{e}, \qquad (9.77)$$

which implies energy equipartition.

Large-Scale Impulsive Dynamical Systems

10.1 Introduction

The complexity of modern controlled large-scale dynamical systems is further exacerbated by the use of hierarchical embedded control subsystems within the feedback control system, that is, abstract decision-making units performing logical checks that identify system mode operation and specify the continuous-variable subcontroller to be activated. As discussed in Chapter 1, such systems typically possess a multiechelon hierarchical hybrid decentralized control architecture characterized by continuous-time dynamics at the lower levels of the hierarchy and discrete-time dynamics at the higher levels of the hierarchy. The lower-level units directly interact with the dynamical system to be controlled, while the higher-level units receive information from the lower-level units as inputs and provide (possibly discrete) output commands that serve to coordinate and reconcile the (sometimes competing) actions of the lower-level units. The hierarchical controller organization reduces processor cost and controller complexity by breaking up the processing task into relatively small pieces and decomposing the fast and slow control functions. Typically, the higher-level units perform logical checks that determine system mode operation, and the lower-level units execute continuous-variable commands for a given system mode of operation.

In light of the fact that energy flow modeling arises naturally in large-scale dynamical systems, and vector Lyapunov functions provide a powerful stability analysis framework for these systems, it seems natural that hybrid dissipativity theory [68,74,75,82], on the subsystem level, should play a key role in analyzing large-scale impulsive dynamical systems. Specifically, hybrid dissipativity theory provides a fundamental framework for the analysis and design of impulsive dynamical systems using an input, state, and output description based on system energy-related considerations [74,75,82]. The hybrid dissipation hypothesis on impulsive dynamical systems results in a fundamental constraint on their dynamic behavior, wherein a dissipative impulsive dynamical system can deliver only a fraction of its energy to its surroundings and can store only a fraction of the work done to it. Such conservation laws are prevalent in large-scale impulsive dynamical systems such as aerospace, power, network, telecommunications, and transportation systems. Since these systems have numerous input-output properties related to conservation, dissipation, and transport of energy, extending hybrid dissipativity theory to capture conservation and dissipation notions on the subsystem level would provide a natural energy flow model for large-scale impulsive dynamical systems.

Aggregating the dissipativity properties of each of the impulsive subsystems by appropriate storage functions and hybrid supply rates would allow us to study the dissipativity properties of the composite large-scale impulsive system using *vector storage functions* and *vector hybrid supply rates*. Furthermore, since vector Lyapunov functions can be viewed as generalizations of composite energy functions for all of the impulsive subsystems, a generalized notion of hybrid dissipativity, namely, *vector hybrid dissipativity*, with appropriate vector storage functions and vector hybrid supply rates, can be used to construct vector Lyapunov functions for nonlinear feedback large-scale impulsive systems by appropriately combining vector storage functions for the forward and feedback large-scale impulsive systems. Finally, as in classical dynamical system theory, vector dissipativity theory can play a fundamental role in addressing robustness, disturbance rejection, stability of feedback interconnections, and optimality for large-scale impulsive dynamical systems.

In this chapter, we develop vector dissipativity notions for large-scale nonlinear impulsive dynamical systems. In particular, we introduce a generalized definition of dissipativity for large-scale nonlinear impulsive dynamical systems in terms of a *hybrid vector dissipation inequality* involving a vector hybrid supply rate, a vector storage function, and an essentially nonnegative, semistable dissipation matrix. Generalized notions of a vector available storage and a vector required supply are also defined and shown to be element-by-element ordered, nonnegative, and finite. On the impulsive subsystem level, the proposed approach provides an energy flow balance over the continuous-time dynamics and the resetting events in terms of the stored subsystem energy, the supplied subsystem energy, the subsystem energy gained from all other subsystems independent of the subsystem coupling strengths, and the subsystem energy dissipated. Furthermore, for large-scale impulsive dynamical systems decomposed into interconnected impulsive subsystems, dissipativity of the composite impulsive system is shown to be determined from the dissipativity properties of the individual impulsive subsystems and the nature of the interconnections.

In addition, we develop extended Kalman-Yakubovich-Popov conditions, in terms of the local impulsive subsystem dynamics and the interconnection constraints, for characterizing vector dissipativeness via vector storage functions for large-scale impulsive dynamical systems. Finally, using the concepts of vector dissipativity and vector storage functions as candidate vector Lyapunov functions, we develop feedback interconnection stability results of large-scale impulsive nonlinear dynamical systems. General stability criteria are given for Lyapunov and asymptotic stability of feedback large-scale impulsive dynamical systems. In the case of vector quadratic supply rates involving net subsystem powers and input-output subsystem energies, these results provide a positivity and small-gain theorem for large-scale impulsive systems predicated on vector Lyapunov functions.

10.2 Stability of Impulsive Systems via Vector Lyapunov Functions

In this section, we consider *state-dependent* impulsive dynamical systems [82] given by

$$\dot{x}(t) = f_{\mathrm{c}}(x(t)), \quad x(t_0) = x_0, \quad x(t) \notin \mathcal{Z}, \quad t \in \mathcal{I}_{x_0}, \qquad (10.1)$$

$$\Delta x(t) = f_{\mathrm{d}}(x(t)), \quad x(t) \in \mathcal{Z}, \qquad (10.2)$$

where $x(t) \in \mathcal{D} \subseteq \mathbb{R}^n$, $t \in \mathcal{I}_{x_0}$, is the system state vector, $\mathcal{I}_{x_0}$ is the maximal interval of existence of a solution $x(t)$ to (10.1) and (10.2), $\mathcal{D}$ is an open set, $0 \in \mathcal{D}$, $f_{\mathrm{c}} : \mathcal{D} \to \mathbb{R}^n$ is Lipschitz continuous and satisfies $f_{\mathrm{c}}(0) = 0$, $f_{\mathrm{d}} : \mathcal{D} \to \mathbb{R}^n$ is continuous, $\Delta x(t) \triangleq x(t^+) - x(t)$, where $x(t^+) \triangleq x(t) + f_{\mathrm{d}}(x(t)) = \lim_{\varepsilon \to 0} x(t + \varepsilon)$, $x(t) \in \mathcal{Z}$, and $\mathcal{Z} \subset \mathcal{D} \subseteq \mathbb{R}^n$ is the *resetting set*. For a particular trajectory $x(t)$, $t \geq 0$, we let $t_k = \tau_k(x_0)$, $x_0 \in \mathcal{D}$, denote the kth instant of time at which $x(t)$ intersects $\mathcal{Z}$. Note that $x_{\mathrm{e}} \in \mathcal{D}$ is an *equilibrium point* of (10.1) and (10.2) if and only if $f_{\mathrm{c}}(x_{\mathrm{e}}) = 0$ and $f_{\mathrm{d}}(x_{\mathrm{e}}) = 0$. To ensure the well-posedness of the resetting times we make the following assumptions [82]:

Assumption 10.1. If $x \in \overline{\mathcal{Z}} \backslash \mathcal{Z}$, then there exists $\varepsilon > 0$ such that, for all $0 < \delta < \varepsilon$, $\psi(\delta, x) \notin \mathcal{Z}$, where $\psi(\cdot, \cdot)$ denotes the solution to the continuous-time dynamics (10.1).

Assumption 10.2. If $x \in \mathcal{Z}$, then $x + f_{\mathrm{d}}(x) \notin \mathcal{Z}$.

Assumption 10.1 ensures that if a trajectory reaches the closure of $\mathcal{Z}$ at a point that does not belong to $\mathcal{Z}$, then the trajectory must be directed away from $\mathcal{Z}$, that is, a trajectory cannot enter $\mathcal{Z}$ through a point that belongs to the closure of $\mathcal{Z}$ but not to $\mathcal{Z}$. Furthermore, Assumption 10.2 ensures that when a trajectory intersects the resetting set $\mathcal{Z}$, it instantaneously exits $\mathcal{Z}$. Furthermore, note that if $x_0 \in \mathcal{Z}$, then the system initially resets to $x_0^+ = x_0 + f_{\mathrm{d}}(x_0) \notin \mathcal{Z}$, which serves as the initial condition for continuous-time dynamics (10.1). Note that if $x^* \in \mathcal{D}$ satisfies $f_{\mathrm{d}}(x^*) = 0$, then $x^* \notin \mathcal{Z}$. To see this, suppose $x^* \in \mathcal{Z}$. Then $x^* + f_{\mathrm{d}}(x^*) = x^* \in \mathcal{Z}$, contradicting Assumption 10.2. Thus, if $x = x_{\mathrm{e}}$ is an equilibrium point of (10.1) and (10.2), then $x_{\mathrm{e}} \notin \mathcal{Z}$, and hence, $x_{\mathrm{e}} \in \mathcal{D}$ is an equilibrium point of (10.1) and (10.2) if and only if $f_{\mathrm{c}}(x_{\mathrm{e}}) = 0$. In addition, note that it follows from the definition of $\tau_k(\cdot)$ that $\tau_1(x) > 0$, $x \notin \mathcal{Z}$, and $\tau_1(x) = 0$, $x \in \mathcal{Z}$. Finally, since $x + f_{\mathrm{d}}(x) \notin \mathcal{Z}$ for every $x \in \mathcal{Z}$, it follows that $\tau_2(x) = \tau_1(x) + \tau_1(x + f_{\mathrm{d}}(x)) > 0$. For further details, see [82].

A function $x : \mathcal{I}_{x_0} \to \mathcal{D}$ is a *solution* to the impulsive dynamical system (10.1) and (10.2) on the interval $\mathcal{I}_{x_0} \subseteq \mathbb{R}$ with initial condition $x(0) = x_0$, where $\mathcal{I}_{x_0}$ denotes the maximal interval of existence of a solution to (10.1) and (10.2), if $x(\cdot)$ is left-continuous and $x(t)$ satisfies (10.1) and (10.2) for all $t \in \mathcal{I}_{x_0}$. For further discussion on solutions to impulsive differential equations, see [11, 13, 28, 74, 82, 117, 137, 155, 175]. For convenience, we use the notation $s(t, x_0)$ to denote the solution $x(t)$ of (10.1) and (10.2) at time $t \geq 0$ with initial condition $x(0) = x_0$.

For a particular trajectory $x(t)$, we let $t_k \triangleq \tau_k(x_0)$ denote the kth instant of time at which $x(t)$ intersects $\mathcal{Z}$, and we call the times t_k the *resetting times*. Thus, the trajectory of the system (10.1) and (10.2) from the initial condition $x(0) = x_0$ is given by $\psi(t, x_0)$ for $0 < t \leq t_1$, where $\psi(\cdot, x_0)$ is the solution to (10.1) starting at x_0. If and when the trajectory reaches a state $x_1 \triangleq x(t_1)$ satisfying $x_1 \in \mathcal{Z}$, then the state is instantaneously transferred to $x_1^+ \triangleq x_1 + f_\mathrm{d}(x_1)$ according to the resetting law (10.2). The trajectory $x(t)$, $t_1 < t \leq t_2$, is then given by $\psi(t - t_1, x_1^+)$, and so on. Our convention here is that the solution $x(t)$ of (10.1) and (10.2) is left continuous, that is, it is continuous everywhere except at the resetting times t_k, and $x_k \triangleq x(t_k) = \lim_{\varepsilon \to 0^+} x(t_k - \varepsilon)$ and $x_k^+ \triangleq x(t_k) + f_\mathrm{d}(x(t_k)) = \lim_{\varepsilon \to 0^+} x(t_k + \varepsilon)$ for $k = 1, 2, \ldots$.

Since the resetting times are well defined and distinct, and since the solution to (10.1) exists and is unique, it follows that the solution of the impulsive dynamical system (10.1) and (10.2) also exists and is unique over a forward time interval. However, it is important to note that the analysis of impulsive dynamical systems can be quite involved. In particular, such systems can exhibit Zenoness and beating, as well as confluence, wherein solutions exhibit infinitely many resettings in a finite time, encounter the same resetting surface a finite or infinite number of times in zero time, and coincide after a certain point in time. In this chapter we allow for the possibility of confluence and Zeno solutions; however, Assumption 10.2 precludes the possibility of beating. Furthermore, since *not* every bounded solution of an impulsive dynamical system over a forward time interval can be extended to infinity due to Zeno solutions, we assume that existence and uniqueness of solutions are satisfied in forward time. For details, see [11, 13, 117].

The next result presents a generalization of the comparison principle given in Corollary 2.2 to impulsive dynamical systems. For this result and the remainder of the monograph we use the notation $\mathcal{W}$ and $\mathcal{W}_\mathrm{c}$ interchangeably.

Theorem 10.1. Consider the impulsive dynamical system (10.1) and (10.2). Assume there exists a continuously differentiable vector function $V : \mathcal{D} \to \mathcal{Q} \subseteq \mathbb{R}^q$ such that

$$V'(x)f_\mathrm{c}(x) \leq\leq w_\mathrm{c}(V(x), x), \quad x \notin \mathcal{Z}, \tag{10.3}$$

$$V(x + f_\mathrm{d}(x)) \leq\leq V(x) + w_\mathrm{d}(V(x), x), \quad x \in \mathcal{Z}, \tag{10.4}$$

where $w_c : \mathcal{Q} \times \mathcal{D} \to \mathbb{R}^q$ and $w_d : \mathcal{Q} \times \mathcal{Z} \to \mathbb{R}^q$ are continuous functions, $w_c(\cdot, x) \in \mathcal{W}_c$, $w_d(\cdot, x) \in \mathcal{W}_d$, and the comparison impulsive dynamical system

$$\dot{z}(t) = w_c(z(t), x(t)), \quad z(t_0) = z_0, \quad x(t) \notin \mathcal{Z}, \quad t \in \mathcal{I}_{z_0, x_0}, \qquad (10.5)$$

$$\Delta z(t) = w_d(z(t), x(t)), \quad x(t) \in \mathcal{Z}, \qquad (10.6)$$

has a unique solution $z(t)$, $t \in \mathcal{I}_{z_0, x_0}$, where $x(t)$, $t \in \mathcal{I}_{x_0}$, is a solution to (10.1) and (10.2). If $[t_0, t_0 + \tau] \subseteq \mathcal{I}_{x_0} \cap \mathcal{I}_{z_0, x_0}$, then

$$V(x_0) \leq\leq z_0, \quad z_0 \in \mathcal{Q}, \quad x_0 \in \mathcal{D}, \qquad (10.7)$$

implies

$$V(x(t)) \leq\leq z(t), \quad t \in [t_0, t_0 + \tau]. \qquad (10.8)$$

Proof. Without loss of generality, let $x_0 \notin \mathcal{Z}$, $x_0 \in \mathcal{D}$. If $x_0 \in \mathcal{Z}$, then by Assumption 10.2, $x_0 + f_d(x_0) \notin \mathcal{Z}$ serves as the initial condition for the continuous-time dynamics. If for $x_0 \notin \mathcal{Z}$ the solution $x(t) \notin \mathcal{Z}$ for all $t \in [t_0, t_0 + \tau]$, then the result follows from Corollary 2.2. Next, suppose the interval $[t_0, t_0 + \tau]$ contains the resetting times $\tau_k(x_0) < \tau_{k+1}(x_0)$, $k \in \{1, 2, \ldots, m\}$. Consider the compact interval $[t_0, \tau_1(x_0)]$ and let $V(x_0) \leq\leq z_0$. Then it follows from (10.3) and Corollary 2.2 that

$$V(x(t)) \leq\leq z(t), \quad t \in [t_0, \tau_1(x_0)], \qquad (10.9)$$

where $z(t)$, $t \in \mathcal{I}_{z_0}$, is the solution to (10.5). Now, since $w_d(\cdot, x) \in \mathcal{W}_d$ it follows from (10.4) and (10.9) that

$$\begin{aligned}
V(x(\tau_1^+(x_0))) &\leq\leq V(x(\tau_1(x_0))) + w_d(V(x(\tau_1(x_0))), x(\tau_1(x_0))) \\
&\leq\leq z(\tau_1(x_0)) + w_d(z(\tau_1(x_0)), x(\tau_1(x_0))) \\
&= z(\tau_1^+(x_0)).
\end{aligned} \qquad (10.10)$$

Consider the compact interval $[\tau_1^+(x_0), \tau_2(x_0)]$. Since $V(x(\tau_1^+(x_0))) \leq\leq z(\tau_1^+(x_0))$, it follows from (10.3) that

$$V(x(t)) \leq\leq z(t), \quad t \in [\tau_1^+(x_0), \tau_2(x_0)]. \qquad (10.11)$$

Repeating the above arguments for $t \in [\tau_k^+(x_0), \tau_{k+1}(x_0)]$, $k = 3, \ldots, m$, yields (10.8). Finally, in the case of infinitely many resettings over the time interval $[t_0, t_0 + \tau]$, let $\lim_{k \to \infty} \tau_k(x_0) = \tau_\infty(x_0) \in (t_0, t_0 + \tau]$. In this case, $[t_0, \tau_\infty(x_0)] = [t_0, \tau_1(x_0)] \cup \left[\bigcup_{k=1}^\infty [\tau_k(x_0), \tau_{k+1}(x_0)] \right]$. Repeating the above arguments, the result can be shown for the interval $[t_0, \tau_\infty(x_0)]$. $\qquad \square$

Note that if the solutions to (10.1), (10.2), (10.5), and (10.6) are globally defined for all $x_0 \in \mathcal{D}$ and $z_0 \in \mathcal{Q}$, then the result of Theorem 10.1 holds for every arbitrarily large but compact interval $[t_0, t_0 + \tau] \subset \overline{\mathbb{R}}_+$. For the remainder of this chapter we assume that the solutions to the systems (10.1), (10.2), (10.5), and (10.6) are defined for all $t \geq t_0$. Next, consider the cascade nonlinear impulsive dynamical system given by

$$\dot{z}(t) = w_{\mathrm{c}}(z(t), x(t)), \quad z(t_0) = z_0, \quad x(t) \notin \mathcal{Z}, \quad t \geq t_0, \tag{10.12}$$

$$\dot{x}(t) = f_{\mathrm{c}}(x(t)), \quad x(t_0) = x_0, \quad x(t) \notin \mathcal{Z}, \tag{10.13}$$

$$\Delta z(t) = w_{\mathrm{d}}(z(t), x(t)), \quad x(t) \in \mathcal{Z}, \tag{10.14}$$

$$\Delta x(t) = f_{\mathrm{d}}(x(t)), \quad x(t) \in \mathcal{Z}, \tag{10.15}$$

where $z_0 \in \mathcal{Q} \subseteq \mathbb{R}^q$, $x_0 \in \mathcal{D} \subseteq \mathbb{R}^n$, $[z^{\mathrm{T}}(t), x^{\mathrm{T}}(t)]^{\mathrm{T}}$, $t \geq t_0$, is the solution to (10.12)–(10.15), $w_{\mathrm{c}} : \mathcal{Q} \times \mathcal{D} \to \mathbb{R}^q$ and $w_{\mathrm{d}} : \mathcal{Q} \times \mathcal{Z} \to \mathbb{R}^q$ are continuous, $w_{\mathrm{c}}(\cdot, x) \in \mathcal{W}_{\mathrm{c}}$, $w_{\mathrm{d}}(\cdot, x) \in \mathcal{W}_{\mathrm{d}}$, $w_{\mathrm{c}}(0, 0) = 0$, $f_{\mathrm{c}} : \mathcal{D} \to \mathbb{R}^n$ is Lipschitz continuous on $\mathcal{D}$, $f_{\mathrm{c}}(0) = 0$, and $f_{\mathrm{d}} : \mathcal{D} \to \mathbb{R}^n$ is continuous.

The following definition introduces several types of partial stability of the nonlinear state-dependent impulsive dynamical system (10.12)–(10.15).

Definition 10.1 ([82]). *i*) The nonlinear impulsive dynamical system (10.12)–(10.15) is *Lyapunov stable with respect to z* if, for every $\varepsilon > 0$ and $x_0 \in \mathbb{R}^n$, there exists $\delta = \delta(\varepsilon, x_0) > 0$ such that $\|z_0\| < \delta$ implies that $\|z(t)\| < \varepsilon$ for all $t \geq 0$.

ii) The nonlinear impulsive dynamical system (10.12)–(10.15) is *Lyapunov stable with respect to z uniformly in x_0* if, for every $\varepsilon > 0$, there exists $\delta = \delta(\varepsilon) > 0$ such that $\|z_0\| < \delta$ implies that $\|z(t)\| < \varepsilon$ for all $t \geq 0$ and for all $x_0 \in \mathbb{R}^n$.

iii) The nonlinear impulsive dynamical system (10.12)–(10.15) is *asymptotically stable with respect to z* if it is Lyapunov stable with respect to z and, for every $x_0 \in \mathbb{R}^n$, there exists $\delta = \delta(x_0) > 0$ such that $\|z_0\| < \delta$ implies that $\lim_{t \to \infty} z(t) = 0$.

iv) The nonlinear impulsive dynamical system (10.12)–(10.15) is *asymptotically stable with respect to z uniformly in x_0* if it is Lyapunov stable with respect to z uniformly in x_0 and there exists $\delta > 0$ such that $\|z_0\| < \delta$ implies that $\lim_{t \to \infty} z(t) = 0$ uniformly in z_0 and x_0 for all $x_0 \in \mathbb{R}^n$.

v) The nonlinear impulsive dynamical system (10.12)–(10.15) is *globally asymptotically stable with respect to z* if it is Lyapunov stable with respect to z and $\lim_{t \to \infty} z(t) = 0$ for all $z_0 \in \mathbb{R}^q$ and $x_0 \in \mathbb{R}^n$.

vi) The nonlinear impulsive dynamical system (10.12)–(10.15) is *globally asymptotically stable with respect to z uniformly in x_0* if it is Lyapunov stable with respect to z uniformly in x_0 and $\lim_{t \to \infty} z(t) = 0$ uniformly in z_0 and x_0 for all $z_0 \in \mathbb{R}^q$ and $x_0 \in \mathbb{R}^n$.

vii) The nonlinear impulsive dynamical system (10.12)–(10.15) is *exponentially stable with respect to z uniformly in x_0* if there exist scalars $\alpha, \beta, \delta > 0$ such that $\|z_0\| < \delta$ implies that $\|z(t)\| \leq \alpha \|z_0\| e^{-\beta t}$, $t \geq 0$, for all $x_0 \in \mathbb{R}^n$.

viii) The nonlinear impulsive dynamical system (10.12)–(10.15) is *globally exponentially stable with respect to z uniformly in x_0 if there exist scalars $\alpha, \beta > 0$ such that $\|z(t)\| \leq \alpha \|z_0\| e^{-\beta t}$, $t \geq 0$, for all $z_0 \in \mathbb{R}^q$ and $x_0 \in \mathbb{R}^n$.*

Theorem 10.2. Consider the impulsive dynamical system (10.1) and (10.2). Assume that there exist a continuously differentiable vector function $V : \mathcal{D} \to \mathcal{Q} \cap \overline{\mathbb{R}}_+^q$ and a positive vector $p \in \mathbb{R}_+^q$ such that $V(0) = 0$, the scalar function $v : \mathcal{D} \to \overline{\mathbb{R}}_+$ defined by $v(x) \triangleq p^{\mathrm{T}} V(x)$, $x \in \mathcal{D}$, is such that $v(x) > 0$, $x \neq 0$, and

$$V'(x) f_{\mathrm{c}}(x) \leq\leq w_{\mathrm{c}}(V(x), x), \quad x \notin \mathcal{Z}, \tag{10.16}$$
$$V(x + f_{\mathrm{d}}(x)) \leq\leq V(x) + w_{\mathrm{d}}(V(x), x), \quad x \in \mathcal{Z}, \tag{10.17}$$

where $w_{\mathrm{c}} : \mathcal{Q} \times \mathcal{D} \to \mathbb{R}^q$ and $w_{\mathrm{d}} : \mathcal{Q} \times \mathcal{Z} \to \mathbb{R}^q$ are continuous, $w_{\mathrm{c}}(\cdot, x) \in \mathcal{W}_{\mathrm{c}}$, $w_{\mathrm{d}}(\cdot, x) \in \mathcal{W}_{\mathrm{d}}$, and $w_{\mathrm{c}}(0, 0) = 0$. Then the following statements hold:

i) If the nonlinear impulsive dynamical system (10.12)–(10.15) is Lyapunov stable with respect to z uniformly in x_0, then the zero solution $x(t) \equiv 0$ to (10.1) and (10.2) is Lyapunov stable.

ii) If the nonlinear impulsive dynamical system (10.12)–(10.15) is asymptotically stable with respect to z uniformly in x_0, then the zero solution $x(t) \equiv 0$ to (10.1) and (10.2) is asymptotically stable.

iii) If $\mathcal{D} = \mathbb{R}^n$, $\mathcal{Q} = \mathbb{R}^q$, $v : \mathbb{R}^n \to \overline{\mathbb{R}}_+$ is radially unbounded, and the nonlinear impulsive dynamical system (10.12)–(10.15) is globally asymptotically stable with respect to z uniformly in x_0, then the zero solution $x(t) \equiv 0$ to (10.1) and (10.2) is globally asymptotically stable.

iv) If there exist constants $\nu \geq 1$, $\alpha > 0$, and $\beta > 0$ such that $v : \mathcal{D} \to \overline{\mathbb{R}}_+$ satisfies

$$\alpha \|x\|^\nu \leq v(x) \leq \beta \|x\|^\nu, \quad x \in \mathcal{D}, \tag{10.18}$$

and the nonlinear impulsive dynamical system (10.12)–(10.15) is exponentially stable with respect to z uniformly in x_0, then the zero solution $x(t) \equiv 0$ to (10.1) and (10.2) is exponentially stable.

v) If $\mathcal{D} = \mathbb{R}^n$, $\mathcal{Q} = \mathbb{R}^q$, there exist constants $\nu \geq 1$, $\alpha > 0$, and $\beta > 0$ such that $v : \mathbb{R}^n \to \overline{\mathbb{R}}_+$ satisfies (10.18), and the nonlinear impulsive dynamical system (10.12)–(10.15) is globally exponentially stable with respect to z uniformly in x_0, then the zero solution $x(t) \equiv 0$ to (10.1) and (10.2) is globally exponentially stable.

Proof. Assume there exist a continuously differentiable vector function $V : \mathcal{D} \to \mathcal{Q} \cap \overline{\mathbb{R}}_+^q$ and a positive vector $p \in \mathbb{R}_+^q$ such that $v(x) = p^{\mathrm{T}} V(x)$, $x \in \mathcal{D}$, is positive definite, that is, $v(0) = 0$ and $v(x) > 0$, $x \neq$

0. Since $v(x) = p^{\mathrm{T}}V(x) \leq \max_{i=1,\ldots,q}\{p_i\}\mathbf{e}^{\mathrm{T}}V(x)$, $x \in \mathcal{D}$, where $\mathbf{e} \triangleq [1,\ldots,1]^{\mathrm{T}}$, the function $\mathbf{e}^{\mathrm{T}}V(x)$, $x \in \mathcal{D}$, is also positive definite. Thus, there exist $r > 0$ and class $\mathcal{K}$ functions $\alpha, \beta : [0, r] \to \overline{\mathbb{R}}_+$ such that $\mathcal{B}_r(0) \subset \mathcal{D}$ and

$$\alpha(\|x\|) \leq \mathbf{e}^{\mathrm{T}}V(x) \leq \beta(\|x\|), \quad x \in \mathcal{B}_r(0). \tag{10.19}$$

$i)$ Let $\varepsilon > 0$ and choose $0 < \hat{\varepsilon} < \min\{\varepsilon, r\}$. It follows from Lyapunov stability of the nonlinear impulsive dynamical system (10.12)–(10.15) with respect to z uniformly in x_0 that there exists $\mu = \mu(\hat{\varepsilon}) = \mu(\varepsilon) > 0$ such that if $\|z_0\|_1 < \mu$, where $\|z\|_1 \triangleq \sum_{i=1}^q |z_i|$ and z_i is the ith component of z, then $\|z(t)\|_1 < \alpha(\hat{\varepsilon})$, $t \geq t_0$, for every $x_0 \in \mathcal{D}$. Now, choose $z_0 = V(x_0) \geq\geq 0$, $x_0 \in \mathcal{D}$. Since $V(x)$, $x \in \mathcal{D}$, is continuous, the function $\mathbf{e}^{\mathrm{T}}V(x)$, $x \in \mathcal{D}$, is also continuous. Hence, for $\mu = \mu(\hat{\varepsilon}) > 0$ there exists $\delta = \delta(\mu(\hat{\varepsilon})) = \delta(\varepsilon) > 0$ such that $\delta < \hat{\varepsilon}$ and if $\|x_0\| < \delta$, then $\mathbf{e}^{\mathrm{T}}V(x_0) = \mathbf{e}^{\mathrm{T}}z_0 = \|z_0\|_1 < \mu$, which implies that $\|z(t)\|_1 < \alpha(\hat{\varepsilon})$, $t \geq t_0$. Now, with $z_0 = V(x_0) \geq\geq 0$, $x_0 \in \mathcal{D}$, and the assumption that $w_{\mathrm{c}}(\cdot, x) \in \mathcal{W}_{\mathrm{c}}$ and $w_{\mathrm{d}}(\cdot, x) \in \mathcal{W}_{\mathrm{d}}$, it follows from (10.16), (10.17), and Theorem 10.1 that $0 \leq\leq V(x(t)) \leq\leq z(t)$ on every compact interval $[t_0, t_0 + \tau]$, and hence, $\mathbf{e}^{\mathrm{T}}z(t) = \|z(t)\|_1$, $[t_0, t_0 + \tau]$. Let $\tau > t_0$ be such that $x(t) \in \mathcal{B}_r(0)$, $t \in [t_0, t_0 + \tau]$ for all $x_0 \in \mathcal{B}_\delta(0)$. Thus, using (10.19), it follows that for $\|x_0\| < \delta$,

$$\alpha(\|x(t)\|) \leq \mathbf{e}^{\mathrm{T}}V(x(t)) \leq \mathbf{e}^{\mathrm{T}}z(t) < \alpha(\hat{\varepsilon}), \quad t \in [t_0, t_0 + \tau], \tag{10.20}$$

which implies $\|x(t)\| < \hat{\varepsilon} < \varepsilon$, $t \in [t_0, t_0 + \tau]$.

Next, suppose, *ad absurdum*, that for some $x_0 \in \mathcal{B}_\delta(0)$ there exists $\hat{t} > t_0 + \tau$ such that $\|x(\hat{t})\| \geq \hat{\varepsilon}$. Then, for $z_0 = V(x_0)$ and the compact interval $[t_0, \hat{t}]$ it follows from (10.16), (10.17), and Theorem 10.1 that $V(x(\hat{t})) \leq\leq z(\hat{t})$, which implies that $\alpha(\hat{\varepsilon}) \leq \alpha(\|x(\hat{t})\|) \leq \mathbf{e}^{\mathrm{T}}V(x(\hat{t})) \leq \mathbf{e}^{\mathrm{T}}z(\hat{t}) < \alpha(\hat{\varepsilon})$. This is a contradiction, and hence, for a given $\varepsilon > 0$ there exists $\delta = \delta(\varepsilon) > 0$ such that for all $x_0 \in \mathcal{B}_\delta(0)$, $\|x(t)\| < \varepsilon$, $t \geq t_0$, which implies Lyapunov stability of the zero solution $x(t) \equiv 0$ to (10.1) and (10.2).

$ii)$ It follows from $i)$ and the asymptotic stability of the nonlinear impulsive dynamical system (10.12)-(10.15) with respect to z uniformly in x_0 that the zero solution $x(t) \equiv 0$ to (10.1) and (10.2) is Lyapunov stable, and there exists $\mu > 0$ such that if $\|z_0\|_1 < \mu$, then $\lim_{t\to\infty} z(t) = 0$ for any $x_0 \in \mathcal{D}$. As in $i)$, choose $z_0 = V(x_0) \geq\geq 0$, $x_0 \in \mathcal{D}$. It follows from Lyapunov stability of the zero solution $x(t) \equiv 0$ to (10.1) and (10.2), and the continuity of $V : \mathcal{D} \to \mathcal{Q} \cap \overline{\mathbb{R}}_+^q$ that there exists $\delta = \delta(\mu) > 0$ such that if $\|x_0\| < \delta$, then $\|x(t)\| < r$, $t \geq t_0$, and $\mathbf{e}^{\mathrm{T}}V(x_0) = \mathbf{e}^{\mathrm{T}}z_0 = \|z_0\|_1 < \mu$. Thus, by asymptotic stability of (10.12)-(10.15) with respect to z uniformly in x_0, for every arbitrary $\varepsilon > 0$ there exists $T = T(\varepsilon) > t_0$ such that $\|z(t)\|_1 < \alpha(\varepsilon)$, $t \geq T$. Thus, it follows from (10.16), (10.17), and Theorem 10.1 that $0 \leq\leq V(x(t)) \leq\leq z(t)$ on any compact interval $[T, T + \tau]$, and hence, $\mathbf{e}^{\mathrm{T}}z(t) = \|z(t)\|_1$, $t \in [T, T + \tau]$, and, by (10.19),

$$\alpha(\|x(t)\|) \leq \mathbf{e}^{\mathrm{T}}V(x(t)) \leq \mathbf{e}^{\mathrm{T}}z(t) < \alpha(\varepsilon), \quad t \in [T, T + \tau]. \tag{10.21}$$

Now, suppose, *ad absurdum*, that for some $x_0 \in \mathcal{B}_\delta(0)$, $\lim_{t\to\infty} x(t) \neq 0$, that is, there exists a sequence $\{t_n\}_{n=1}^\infty$, with $t_n \to \infty$ as $n \to \infty$, such that $\|x(t_n)\| \geq \hat{\varepsilon}$, $n \in \overline{\mathbb{Z}}_+$, for some $0 < \hat{\varepsilon} < r$. Choose $\varepsilon = \hat{\varepsilon}$ and the interval $[T, T+\tau]$ such that at least one $t_n \in [T, T+\tau]$. Then it follows from (10.21) that $\alpha(\varepsilon) \leq \alpha(\|x(t_n)\|) < \alpha(\varepsilon)$, which is a contradiction. Hence, there exists $\delta > 0$ such that for all $x_0 \in \mathcal{B}_\delta(0)$, $\lim_{t\to\infty} x(t) = 0$, which along with Lyapunov stability implies asymptotic stability of the zero solution $x(t) \equiv 0$ to (10.1) and (10.2).

iii) Suppose $\mathcal{D} = \mathbb{R}^n$, $\mathcal{Q} = \mathbb{R}^q$, $v : \mathbb{R}^n \to \overline{\mathbb{R}}_+$ is radially unbounded, and the nonlinear impulsive dynamical system (10.12)–(10.15) is globally asymptotically stable with respect to z uniformly in x_0. In this case, $V : \mathbb{R}^n \to \overline{\mathbb{R}}_+^q$ satisfies (10.19) for all $x \in \mathbb{R}^n$, where the functions $\alpha, \beta : \overline{\mathbb{R}}_+ \to \overline{\mathbb{R}}_+$ are of class $\mathcal{K}_\infty$ [110]. Furthermore, Lyapunov stability of the zero solution $x(t) \equiv 0$ to (10.1) and (10.2) follows from *i*). Next, for every $x_0 \in \mathbb{R}^n$ and $z_0 = V(x_0) \in \overline{\mathbb{R}}_+^q$, identical arguments to those in *ii*) can be used to show that $\lim_{t\to\infty} x(t) = 0$, which proves global asymptotic stability of the zero solution $x(t) \equiv 0$ to (10.1) and (10.2).

iv) Suppose (10.18) holds. Since $p \in \mathbb{R}_+^q$, then

$$\hat{\alpha}\|x\|^\nu \leq \mathbf{e}^{\mathrm{T}}V(x) \leq \hat{\beta}\|x\|^\nu, \quad x \in \mathcal{D}, \tag{10.22}$$

where $\hat{\alpha} \triangleq \alpha/\max_{i=1,\ldots,q}\{p_i\}$ and $\hat{\beta} \triangleq \beta/\min_{i=1,\ldots,q}\{p_i\}$. It follows from the exponential stability of the nonlinear impulsive dynamical system (10.12)–(10.15) with respect to z uniformly in x_0 that there exist positive constants γ, μ, and η such that if $\|z_0\|_1 < \mu$, then

$$\|z(t)\|_1 \leq \gamma\|z_0\|_1 e^{-\eta(t-t_0)}, \quad t \geq t_0, \tag{10.23}$$

for all $x_0 \in \mathcal{D}$. Choose $z_0 = V(x_0) \geq\geq 0$, $x_0 \in \mathcal{D}$. By continuity of $V : \mathcal{D} \to \mathcal{Q} \cap \overline{\mathbb{R}}_+^q$, there exists $\delta = \delta(\mu) > 0$ such that for all $x_0 \in \mathcal{B}_\delta(0)$, $\mathbf{e}^{\mathrm{T}}V(x_0) = \mathbf{e}^{\mathrm{T}}z_0 = \|z_0\|_1 < \mu$. Furthermore, it follows from (10.16), (10.17), (10.22), (10.23), and Theorem 10.1 that for all $x_0 \in \mathcal{B}_\delta(0)$ the inequality

$$\hat{\alpha}\|x(t)\|^\nu \leq \mathbf{e}^{\mathrm{T}}V(x(t)) \leq \mathbf{e}^{\mathrm{T}}z(t) \leq \gamma\|z_0\|_1 e^{-\eta(t-t_0)} \leq \gamma\hat{\beta}\|x_0\|^\nu e^{-\eta(t-t_0)}$$

holds on every compact interval $[t_0, t_0 + \tau]$. This in turn implies that for every $x_0 \in \mathcal{B}_\delta(0)$,

$$\|x(t)\| \leq \left(\frac{\gamma\hat{\beta}}{\hat{\alpha}}\right)^{\frac{1}{\nu}} \|x_0\| e^{-\frac{\eta}{\nu}(t-t_0)}, \quad t \in [t_0, t_0 + \tau]. \tag{10.24}$$

Now, suppose, *ad absurdum*, that for some $x_0 \in \mathcal{B}_\delta(0)$ there exists $\hat{t} > t_0 + \tau$ such that

$$\|x(\hat{t})\| > \left(\frac{\gamma\hat{\beta}}{\hat{\alpha}}\right)^{\frac{1}{\nu}} \|x_0\| e^{-\frac{\eta}{\nu}(\hat{t}-t_0)}. \tag{10.25}$$

Then for the compact interval $[t_0, \hat{t}]$, it follows from (10.24) that $\|x(\hat{t})\| \leq \left(\frac{\gamma\hat{\beta}}{\hat{\alpha}}\right)^{\frac{1}{\nu}} \|x_0\| e^{-\frac{\eta}{\nu}(\hat{t}-t_0)}$, which is a contradiction. Thus, inequality (10.24) holds for all $t \geq t_0$, establishing exponential stability of the zero solution $x(t) \equiv 0$ to (10.1) and (10.2).

$v)$ The proof is identical to the proof of $iv)$. $\qquad\square$

If $V : \mathcal{D} \to \mathcal{Q} \cap \overline{\mathbb{R}}_+^q$ satisfies the conditions of Theorem 10.2 we say that $V(x)$, $x \in \mathcal{D}$, is a *vector Lyapunov function* [159]. Note that for stability analysis each component of a vector Lyapunov function need not be positive definite, nor does it need to have a negative definite time derivative along the trajectories of (10.1) and (10.2) between resettings and negative semi-definite difference across the resettings. This provides more flexibility in searching for a vector Lyapunov function as compared to a scalar Lyapunov function for addressing the stability of impulsive dynamical systems. The next corollary is immediate from Theorem 10.2.

Corollary 10.1. Consider the impulsive dynamical system (10.1) and (10.2). Assume that there exist a continuously differentiable vector function $V : \mathcal{D} \to \mathcal{Q} \cap \overline{\mathbb{R}}_+^q$ and a positive vector $p \in \mathbb{R}_+^q$ such that $V(0) = 0$, the scalar function $v : \mathcal{D} \to \overline{\mathbb{R}}_+$ defined by $v(x) \triangleq p^{\mathrm{T}} V(x)$, $x \in \mathcal{D}$, is such that $v(x) > 0$, $x \neq 0$, and

$$V'(x)f_{\mathrm{c}}(x) \leq\leq w(V(x)), \quad x \notin \mathcal{Z}, \tag{10.26}$$
$$V(x + f_{\mathrm{d}}(x)) \leq\leq V(x), \quad x \in \mathcal{Z}, \tag{10.27}$$

where $w : \mathcal{Q} \to \mathbb{R}^q$ is continuous, $w(\cdot) \in \mathcal{W}$, and $w(0) = 0$. Then the following statements hold:

$i)$ If the zero solution $z(t) \equiv 0$ to

$$\dot{z}(t) = w(z(t)), \quad z(t_0) = z_0, \quad t \geq t_0, \tag{10.28}$$

is Lyapunov stable, then the zero solution $x(t) \equiv 0$ to (10.1) and (10.2) is Lyapunov stable.

$ii)$ If the zero solution $z(t) \equiv 0$ to (10.28) is asymptotically stable, then the zero solution $x(t) \equiv 0$ to (10.1) and (10.2) is asymptotically stable.

$iii)$ If $\mathcal{D} = \mathbb{R}^n$, $\mathcal{Q} = \mathbb{R}^q$, $v : \mathbb{R}^n \to \overline{\mathbb{R}}_+$ is radially unbounded, and the zero solution $z(t) \equiv 0$ to (10.28) is globally asymptotically stable, then the zero solution $x(t) \equiv 0$ to (10.1) and (10.2) is globally asymptotically stable.

$iv)$ If there exist constants $\nu \geq 1$, $\alpha > 0$, and $\beta > 0$ such that $v : \mathcal{D} \to \overline{\mathbb{R}}_+$ satisfies

$$\alpha\|x\|^\nu \leq v(x) \leq \beta\|x\|^\nu, \quad x \in \mathcal{D}, \tag{10.29}$$

and the zero solution $z(t) \equiv 0$ to (10.28) is exponentially stable, then the zero solution $x(t) \equiv 0$ to (10.1) and (10.2) is exponentially stable.

$v)$ If $\mathcal{D} = \mathbb{R}^n$, $\mathcal{Q} = \mathbb{R}^q$, there exist constants $\nu \geq 1$, $\alpha > 0$, and $\beta > 0$ such that $v : \mathbb{R}^n \to \overline{\mathbb{R}}_+$ satisfies (10.29), and the zero solution $z(t) \equiv 0$ to (10.28) is globally exponentially stable, then the zero solution $x(t) \equiv 0$ to (10.1) and (10.2) is globally exponentially stable.

Proof. The proof is a direct consequence of Theorem 10.2 with $w_{\mathrm{c}}(z, x) = w(z)$ and $w_{\mathrm{d}}(z, x) \equiv 0$. $\qquad\square$

Next, we use the vector Lyapunov stability results of Theorem 10.2 to develop partial stability analysis results for nonlinear impulsive dynamical systems [41, 78]. Specifically, consider the nonlinear impulsive dynamical system (10.1) and (10.2) with partitioned dynamics[1] given by

$$\dot{x}_{\mathrm{I}}(t) = f_{\mathrm{Ic}}(x_{\mathrm{I}}(t), x_{\mathrm{II}}(t)), \quad x_{\mathrm{I}}(t_0) = x_{\mathrm{I}0}, \quad x(t) \notin \mathcal{Z}, \quad t \geq t_0, \quad (10.30)$$

$$\dot{x}_{\mathrm{II}}(t) = f_{\mathrm{IIc}}(x_{\mathrm{I}}(t), x_{\mathrm{II}}(t)), \quad x_{\mathrm{II}}(t_0) = x_{\mathrm{II}0}, \quad x(t) \notin \mathcal{Z}, \quad (10.31)$$

$$\Delta x_{\mathrm{I}}(t) = f_{\mathrm{Id}}(x_{\mathrm{I}}(t), x_{\mathrm{II}}(t)), \quad x(t) \in \mathcal{Z}, \quad (10.32)$$

$$\Delta x_{\mathrm{II}}(t) = f_{\mathrm{IId}}(x_{\mathrm{I}}(t), x_{\mathrm{II}}(t)), \quad x(t) \in \mathcal{Z}, \quad (10.33)$$

where $x_{\mathrm{I}}(t) \in \mathcal{D}_{\mathrm{I}}$, $t \geq t_0$, $\mathcal{D}_{\mathrm{I}} \subseteq \mathbb{R}^{n_{\mathrm{I}}}$ is an open set such that $0 \in \mathcal{D}_{\mathrm{I}}$, $x_{\mathrm{II}}(t) \in \mathbb{R}^{n_{\mathrm{II}}}$, $t \geq t_0$, $\Delta x_{\mathrm{I}}(t) = x_{\mathrm{I}}(t^+) - x_{\mathrm{I}}(t)$, $\Delta x_{\mathrm{II}}(t) = x_{\mathrm{II}}(t^+) - x_{\mathrm{II}}(t)$, $f_{\mathrm{Ic}} : \mathcal{D}_{\mathrm{I}} \times \mathbb{R}^{n_{\mathrm{II}}} \to \mathbb{R}^{n_{\mathrm{I}}}$ is such that, for all $x_{\mathrm{II}} \in \mathbb{R}^{n_{\mathrm{II}}}$, $f_{\mathrm{Ic}}(0, x_{\mathrm{II}}) = 0$ and $f_{\mathrm{Ic}}(\cdot, x_{\mathrm{II}})$ is locally Lipschitz in x_{I}, $f_{\mathrm{IIc}} : \mathcal{D}_{\mathrm{I}} \times \mathbb{R}^{n_{\mathrm{II}}} \to \mathbb{R}^{n_{\mathrm{II}}}$ is such that, for every $x_{\mathrm{I}} \in \mathcal{D}_{\mathrm{I}}$, $f_{\mathrm{IIc}}(x_{\mathrm{I}}, \cdot)$ is locally Lipschitz in x_{II}, $f_{\mathrm{Id}} : \mathcal{D}_{\mathrm{I}} \times \mathbb{R}^{n_{\mathrm{II}}} \to \mathbb{R}^{n_{\mathrm{I}}}$ is continuous and $f_{\mathrm{Id}}(0, x_{\mathrm{II}}) = 0$ for all $x_{\mathrm{II}} \in \mathbb{R}^{n_{\mathrm{II}}}$, $f_{\mathrm{IId}} : \mathcal{D}_{\mathrm{I}} \times \mathbb{R}^{n_{\mathrm{II}}} \to \mathbb{R}^{n_{\mathrm{II}}}$ is continuous, $\mathcal{Z} \in \mathcal{D}_{\mathrm{I}} \times \mathbb{R}^{n_{\mathrm{II}}}$, $x(t) \triangleq [x_{\mathrm{I}}^{\mathrm{T}}(t), x_{\mathrm{II}}^{\mathrm{T}}(t)]^{\mathrm{T}} \in \mathcal{D} = \mathcal{D}_{\mathrm{I}} \times \mathbb{R}^{n_{\mathrm{II}}} \subseteq \mathbb{R}^n$, $t \geq t_0$, $x_0 \triangleq [x_{\mathrm{I}0}^{\mathrm{T}}, x_{\mathrm{II}0}^{\mathrm{T}}]^{\mathrm{T}}$, and $n_{\mathrm{I}} + n_{\mathrm{II}} = n$. For the nonlinear impulsive dynamical system (10.30)–(10.33) the definitions of partial stability given in [78] hold. Furthermore, for a particular trajectory $x(t) = (x_{\mathrm{I}}(t), x_{\mathrm{II}}(t))$, $t \geq 0$, we let $t_k (= \tau_k(x_{\mathrm{I}0}, x_{\mathrm{II}0}))$ denote the kth instant of time at which $x(t)$ intersects $\mathcal{Z}$, and we assume that Assumptions 10.1 and 10.2 hold for $x(t) = (x_{\mathrm{I}}(t), x_{\mathrm{II}}(t))$, $t \geq 0$. Note that for the impulsive dynamical system (10.1) and (10.2), $f_{\mathrm{c}}(x_{\mathrm{I}}, x_{\mathrm{II}}) = [f_{\mathrm{Ic}}^{\mathrm{T}}(x_{\mathrm{I}}, x_{\mathrm{II}}), f_{\mathrm{IIc}}^{\mathrm{T}}(x_{\mathrm{I}}, x_{\mathrm{II}})]^{\mathrm{T}}$, $(x_{\mathrm{I}}, x_{\mathrm{II}}) \in \mathcal{D}_{\mathrm{I}} \times \mathbb{R}^{n_{\mathrm{II}}}$, and $f_{\mathrm{d}}(x_{\mathrm{I}}, x_{\mathrm{II}}) = [f_{\mathrm{Id}}^{\mathrm{T}}(x_{\mathrm{I}}, x_{\mathrm{II}}), f_{\mathrm{IId}}^{\mathrm{T}}(x_{\mathrm{I}}, x_{\mathrm{II}})]^{\mathrm{T}}$, $(x_{\mathrm{I}}, x_{\mathrm{II}}) \in \mathcal{D}_{\mathrm{I}} \times \mathbb{R}^{n_{\mathrm{II}}}$.

For the following result define

$$\Delta V(x_{\mathrm{I}}, x_{\mathrm{II}}) \triangleq V(x_{\mathrm{I}} + f_{\mathrm{Id}}(x_{\mathrm{I}}, x_{\mathrm{II}}), x_{\mathrm{II}} + f_{\mathrm{IId}}(x_{\mathrm{I}}, x_{\mathrm{II}})) - V(x_{\mathrm{I}}, x_{\mathrm{II}}),$$

for a given vector function $V : \mathcal{D}_{\mathrm{I}} \times \mathbb{R}^{n_{\mathrm{II}}} \to \mathbb{R}^q$.

---

[1] Here we use the Roman subscripts I and II as opposed to Arabic subscripts 1 and 2 for denoting the partial states of x not to confuse the partial states with the component states of the vector Lyapunov function.

Theorem 10.3. Consider the nonlinear impulsive dynamical system (10.30)–(10.33). Assume that there exist a continuously differentiable vector function $V : \mathcal{D}_\mathrm{I} \times \mathbb{R}^{n_\mathrm{II}} \to \mathcal{Q} \cap \overline{\mathbb{R}}_+^q$, a positive vector $p \in \mathbb{R}_+^q$, and class $\mathcal{K}$ functions $\alpha(\cdot)$ and $\beta(\cdot)$ such that the scalar function $v : \mathcal{D}_\mathrm{I} \times \mathbb{R}^{n_\mathrm{II}} \to \overline{\mathbb{R}}_+$ defined by $v(x_\mathrm{I}, x_\mathrm{II}) \triangleq p^\mathrm{T} V(x_\mathrm{I}, x_\mathrm{II})$ satisfies

$$\alpha(\|x_\mathrm{I}\|) \leq v(x_\mathrm{I}, x_\mathrm{II}) \leq \beta(\|x_\mathrm{I}\|), \quad (x_\mathrm{I}, x_\mathrm{II}) \in \mathcal{D}_\mathrm{I} \times \mathbb{R}^{n_\mathrm{II}}, \qquad (10.34)$$

and

$$V'(x_\mathrm{I}, x_\mathrm{II})f(x_\mathrm{I}, x_\mathrm{II}) \leq\leq w_\mathrm{c}(V(x_\mathrm{I}, x_\mathrm{II}), x_\mathrm{I}, x_\mathrm{II}), \quad (x_\mathrm{I}, x_\mathrm{II}) \notin \mathcal{Z}, \qquad (10.35)$$
$$\Delta V(x_\mathrm{I}, x_\mathrm{II}) \leq\leq w_\mathrm{d}(V(x_\mathrm{I}, x_\mathrm{II}), x_\mathrm{I}, x_\mathrm{II}), \quad (x_\mathrm{I}, x_\mathrm{II}) \in \mathcal{Z}, \qquad (10.36)$$

where $w_\mathrm{c} : \mathcal{Q} \times \mathcal{D}_\mathrm{I} \times \mathbb{R}^{n_\mathrm{II}} \to \mathbb{R}^q$ and $w_\mathrm{d} : \mathcal{Q} \times \mathcal{Z} \to \mathbb{R}^q$, where $\mathcal{Z} \subset \mathcal{D}_\mathrm{I} \times \mathbb{R}^{n_\mathrm{II}}$, are continuous, $w_\mathrm{c}(\cdot, x_\mathrm{I}, x_\mathrm{II}) \in \mathcal{W}_\mathrm{c}$, $w_\mathrm{d}(\cdot, x_\mathrm{I}, x_\mathrm{II}) \in \mathcal{W}_\mathrm{d}$, and $w_\mathrm{c}(0,0,0) = 0$. Then the following statements hold:

   i) If the nonlinear impulsive dynamical system (10.12), (10.14), (10.30)–(10.33) is Lyapunov (respectively, asymptotically) stable with respect to z uniformly in $(x_\mathrm{I0}, x_\mathrm{II0})$, then the nonlinear impulsive dynamical system (10.30)–(10.33) is Lyapunov (respectively, asymptotically) stable with respect to x_I uniformly in x_II0.

   ii) If $\mathcal{D}_\mathrm{I} = \mathbb{R}^{n_\mathrm{I}}$, $\mathcal{Q} = \mathbb{R}^q$, the functions $\alpha(\cdot)$ and $\beta(\cdot)$ are class $\mathcal{K}_\infty$, and the nonlinear impulsive dynamical system (10.12), (10.14), (10.30)–(10.33) is globally asymptotically stable with respect to z uniformly in $(x_\mathrm{I0}, x_\mathrm{II0})$, then the nonlinear impulsive dynamical system (10.30)–(10.33) is globally asymptotically stable with respect to x_I uniformly in x_II0.

   iii) If there exist constants $\nu \geq 1$, $\alpha > 0$, and $\beta > 0$ such that $v : \mathcal{D}_\mathrm{I} \times \mathbb{R}^{n_\mathrm{II}} \to \overline{\mathbb{R}}_+$ satisfies

$$\alpha\|x_\mathrm{I}\|^\nu \leq v(x_\mathrm{I}, x_\mathrm{II}) \leq \beta\|x_\mathrm{I}\|^\nu, \quad (x_\mathrm{I}, x_\mathrm{II}) \in \mathcal{D}_\mathrm{I} \times \mathbb{R}^{n_\mathrm{II}}, \qquad (10.37)$$

and the nonlinear impulsive dynamical system (10.12), (10.14), (10.30)–(10.33) is exponentially stable with respect to z uniformly in $(x_\mathrm{I0}, x_\mathrm{II0})$, then the nonlinear impulsive dynamical system (10.30)–(10.33) is exponentially stable with respect to x_I uniformly in x_II0.

   iv) If $\mathcal{D}_\mathrm{I} = \mathbb{R}^{n_\mathrm{I}}$, $\mathcal{Q} = \mathbb{R}^q$, there exist constants $\nu \geq 1$, $\alpha > 0$, and $\beta > 0$ such that $v : \mathbb{R}^{n_\mathrm{I}} \times \mathbb{R}^{n_\mathrm{II}} \to \overline{\mathbb{R}}_+$ satisfies (10.37), and the nonlinear impulsive dynamical system (10.12), (10.14), (10.30)–(10.33) is globally exponentially stable with respect to z uniformly in $(x_\mathrm{I0}, x_\mathrm{II0})$, then the nonlinear impulsive dynamical system (10.30)–(10.33) is globally exponentially stable with respect to x_I uniformly in x_II0.

Proof. Since $p \in \mathbb{R}_+^q$ is a positive vector it follows from (10.34) that

$$\alpha(\|x_\mathrm{I}\|)/\max_{i=1,\ldots,q}\{p_i\} \leq \mathbf{e}^\mathrm{T} V(x_\mathrm{I}, x_\mathrm{II}) \leq \beta(\|x_\mathrm{I}\|)/\min_{i=1,\ldots,q}\{p_i\},$$
$$(x_\mathrm{I}, x_\mathrm{II}) \in \mathcal{D}_\mathrm{I} \times \mathbb{R}^{n_\mathrm{II}}. \qquad (10.38)$$

Next, let $\varepsilon > 0$ and note that it follows from Lyapunov stability of the nonlinear impulsive dynamical system (10.12), (10.14), (10.30)–(10.33) with respect to z uniformly in $(x_\mathrm{I0}, x_\mathrm{II0})$ that there exists $\mu = \mu(\varepsilon) > 0$ such that if $\|z_0\|_1 < \mu$, where $\|z\|_1 \triangleq \sum_{i=1}^q |z_i|$ and z_i is the ith component of z, then $\|z(t)\|_1 < \alpha(\varepsilon)/\max_{i=1,\ldots,q}\{p_i\}$, $t \geq t_0$, for every $(x_\mathrm{I0}, x_\mathrm{II0}) \in \mathcal{D}_\mathrm{I} \times \mathbb{R}^{n_\mathrm{II}}$. Now, choose $z_0 = V(x_\mathrm{I0}, x_\mathrm{II0}) \geq\geq 0$, $(x_\mathrm{I0}, x_\mathrm{II0}) \in \mathcal{D}_\mathrm{I} \times \mathbb{R}^{n_\mathrm{II}}$. Since $V(\cdot, \cdot)$ is continuous, the function $\mathbf{e}^\mathrm{T} V(\cdot, \cdot)$ is also continuous. Moreover, it follows from the continuity of $\beta(\cdot)$ that for $\mu = \mu(\varepsilon)$ there exists $\delta = \delta(\mu(\varepsilon)) = \delta(\varepsilon) > 0$ such that $\delta < \varepsilon$ and if $\|x_\mathrm{I0}\| < \delta$, then $\beta(\|x_\mathrm{I0}\|)/\min_{i=1,\ldots,q}\{p_i\} < \mu$, which, by (10.38), implies that $\mathbf{e}^\mathrm{T} V(x_\mathrm{I0}, x_\mathrm{II0}) = \mathbf{e}^\mathrm{T} z_0 = \|z_0\|_1 < \mu$ for all $x_\mathrm{II0} \in \mathbb{R}^{n_\mathrm{II}}$, and hence, $\|z(t)\|_1 < \alpha(\varepsilon)/\max_{i=1,\ldots,q}\{p_i\}$, $t \geq t_0$. In addition, it follows from (10.35), (10.36), and Theorem 10.1 that $0 \leq\leq V(x_\mathrm{I}(t), x_\mathrm{II}(t)) \leq\leq z(t)$ on every compact interval $[t_0, t_0 + \tau]$, and hence, $\mathbf{e}^\mathrm{T} z(t) = \|z\|_1$, $[t_0, t_0 + \tau]$. Thus, it follows from (10.38) that for all $\|x_\mathrm{I0}\| < \delta$, $x_\mathrm{II0} \in \mathbb{R}^{n_\mathrm{II}}$, and $t \in [t_0, t_0 + \tau]$,

$$\alpha(\|x_\mathrm{I}(t)\|)/\max_{i=1,\ldots,q}\{p_i\} \leq \mathbf{e}^\mathrm{T} V(x_\mathrm{I}(t), x_\mathrm{II}(t))$$
$$\leq \mathbf{e}^\mathrm{T} z(t)$$
$$< \alpha(\varepsilon)/\max_{i=1,\ldots,q}\{p_i\}, \qquad (10.39)$$

which implies that $\|x_\mathrm{I}(t)\| < \varepsilon$, $t \in [t_0, t_0 + \tau]$.

Next, suppose, *ad absurdum*, that for some $x_\mathrm{I0} \in \mathcal{D}_\mathrm{I}$ with $\|x_\mathrm{I0}\| < \delta$ and for some $x_\mathrm{II0} \in \mathbb{R}^{n_\mathrm{II}}$ there exists $\hat{t} > t_0 + \tau$ such that $\|x_\mathrm{I}(\hat{t})\| \geq \varepsilon$. Then, for $z_0 = V(x_\mathrm{I0}, x_\mathrm{II0})$ and the compact interval $[t_0, \hat{t}]$ it follows from Theorem 10.1 that $V(x_\mathrm{I}(\hat{t}), x_\mathrm{II}(\hat{t})) \leq\leq z(\hat{t})$, which implies that

$$\frac{\alpha(\varepsilon)}{\max_{i=1,\ldots,q}\{p_i\}} \leq \frac{\alpha(\|x_\mathrm{I}(\hat{t})\|)}{\max_{i=1,\ldots,q}\{p_i\}}$$
$$\leq \mathbf{e}^\mathrm{T} V(x_\mathrm{I}(\hat{t}), x_\mathrm{II}(\hat{t}))$$
$$\leq \mathbf{e}^\mathrm{T} z(\hat{t})$$
$$< \frac{\alpha(\varepsilon)}{\max_{i=1,\ldots,q}\{p_i\}}. \qquad (10.40)$$

This is a contradiction and hence, for a given $\varepsilon > 0$ there exists $\delta = \delta(\varepsilon) > 0$ such that for all $x_\mathrm{I0} \in \mathcal{D}_\mathrm{I}$ with $\|x_\mathrm{I0}\| < \delta$ and for all $x_\mathrm{II0} \in \mathbb{R}^{n_\mathrm{II}}$, $\|x_\mathrm{I}(t)\| < \varepsilon$, $t \geq t_0$, which implies Lyapunov stability of the nonlinear impulsive dynamical system (10.30)–(10.33) with respect to x_I uniformly in x_II0.

The remainder of the proof involves similar arguments as those above and as those in the proof of parts $ii) - v)$ of Theorem 10.2 and, hence, is omitted. $\square$

Note that Theorem 10.3 allows us to address stability of *time-dependent* nonlinear impulsive dynamical systems via vector Lyapunov functions. In particular, with $x_{\mathrm{I}}(t) \equiv x(t)$, $x_{\mathrm{II}}(t) \equiv t$, $n_{\mathrm{I}} = n$, $n_{\mathrm{II}} = 1$, $f_{\mathrm{Ic}}(x_{\mathrm{I}}, x_{\mathrm{II}}) = f_{\mathrm{c}}(x(t), t)$, $f_{\mathrm{IIc}}(x_{\mathrm{I}}(t), x_{\mathrm{II}}(t)) = 1$, $f_{\mathrm{Id}}(x_{\mathrm{I}}, x_{\mathrm{II}}) = f_{\mathrm{d}}(x(t), t)$, $f_{\mathrm{IId}}(x_{\mathrm{I}}(t), x_{\mathrm{II}}(t)) = 0$, and $\mathcal{Z} = \mathcal{D} \times \mathcal{T}$, with $\mathcal{T} \triangleq \{t_1, t_2, \ldots\}$, Theorem 10.3 can be used to establish stability results for the nonlinear time-dependent impulsive dynamical system given by

$$\dot{x}(t) = f_{\mathrm{c}}(x(t), t), \quad x(t_0) = x_0, \quad t \neq t_k, \quad t \geq t_0, \tag{10.41}$$
$$\Delta x(t) = f_{\mathrm{d}}(x(t), t), \quad t = t_k, \tag{10.42}$$

where $x_0 \in \mathcal{D} \subseteq \mathbb{R}^n$. For details on the unification between partial stability of state-dependent impulsive systems and stability theory for time-dependent impulsive systems see [78, 82].

10.3 Vector Dissipativity Theory for Large-Scale Impulsive Dynamical Systems

In this section, we develop vector dissipativity theory for impulsive large-scale dynamical systems. For that, first recall the standard notions of dissipativity and exponential dissipativity [68, 74] for *input/state-dependent* impulsive dynamical systems $\mathcal{G}$ of the form [82]

$$\dot{x}(t) = f_{\mathrm{c}}(x(t)) + G_{\mathrm{c}}(x(t))u_{\mathrm{c}}(t), \quad x(t_0) = x_0, \quad (x(t), u_{\mathrm{c}}(t)) \notin \mathcal{Z}, \tag{10.43}$$
$$\Delta x(t) = f_{\mathrm{d}}(x(t)) + G_{\mathrm{d}}(x(t))u_{\mathrm{d}}(t), \quad (x(t), u_{\mathrm{c}}(t)) \in \mathcal{Z}, \tag{10.44}$$
$$y_{\mathrm{c}}(t) = h_{\mathrm{c}}(x(t)) + J_{\mathrm{c}}(x(t))u_{\mathrm{c}}(t), \quad (x(t), u_{\mathrm{c}}(t)) \notin \mathcal{Z}, \tag{10.45}$$
$$y_{\mathrm{d}}(t) = h_{\mathrm{d}}(x(t)) + J_{\mathrm{d}}(x(t))u_{\mathrm{d}}(t), \quad (x(t), u_{\mathrm{c}}(t)) \in \mathcal{Z}, \tag{10.46}$$

where $t \geq t_0$, $x(t) \in \mathcal{D} \subseteq \mathbb{R}^n$, $u_{\mathrm{c}}(t) \in U_{\mathrm{c}} \subseteq \mathbb{R}^{m_{\mathrm{c}}}$, $u_{\mathrm{d}}(t_k) \in U_{\mathrm{d}} \subseteq \mathbb{R}^{m_{\mathrm{d}}}$, t_k denotes the kth instant of time at which $(x(t), u_{\mathrm{c}}(t))$ intersects $\mathcal{Z} \subset \mathcal{D} \times U_{\mathrm{c}}$ for a particular trajectory $x(t)$ and input $u_{\mathrm{c}}(t)$, $y_{\mathrm{c}}(t) \in Y_{\mathrm{c}} \subseteq \mathbb{R}^{l_{\mathrm{c}}}$, $y_{\mathrm{d}}(t_k) \in Y_{\mathrm{d}} \subseteq \mathbb{R}^{l_{\mathrm{d}}}$, $f_{\mathrm{c}} : \mathcal{D} \to \mathbb{R}^n$ is Lipschitz continuous and satisfies $f_{\mathrm{c}}(0) = 0$, $G_{\mathrm{c}} : \mathcal{D} \to \mathbb{R}^{n \times m_{\mathrm{c}}}$, $f_{\mathrm{d}} : \mathcal{D} \to \mathbb{R}^n$ is continuous, $G_{\mathrm{d}} : \mathcal{D} \to \mathbb{R}^{n \times m_{\mathrm{d}}}$, $h_{\mathrm{c}} : \mathcal{D} \to \mathbb{R}^{l_{\mathrm{c}}}$ satisfies $h_{\mathrm{c}}(0) = 0$, $J_{\mathrm{c}} : \mathcal{D} \to \mathbb{R}^{l_{\mathrm{c}} \times m_{\mathrm{c}}}$, $h_{\mathrm{d}} : \mathcal{D} \to \mathbb{R}^{l_{\mathrm{d}}}$, and $J_{\mathrm{d}} : \mathcal{D} \to \mathbb{R}^{l_{\mathrm{d}} \times m_{\mathrm{d}}}$. For the impulsive dynamical system $\mathcal{G}$ we assume that the required properties for the existence and uniqueness of solutions are satisfied, that is, $u_{\mathrm{c}}(\cdot)$ satisfies sufficient regularity conditions such that (10.43) has a unique solution forward in time.

For the impulsive dynamical system $\mathcal{G}$ given by (10.43)–(10.46) defined on the state space $\mathcal{D} \subseteq \mathbb{R}^n$, $\mathcal{U} \triangleq \mathcal{U}_{\mathrm{c}} \times \mathcal{U}_{\mathrm{d}}$ and $\mathcal{Y} \triangleq \mathcal{Y}_{\mathrm{c}} \times \mathcal{Y}_{\mathrm{d}}$ define an input and output space, respectively, consisting of left-continuous bounded

U-valued and Y-valued functions on the semi-infinite interval $[0, \infty)$. The set $U \triangleq U_{\mathrm{c}} \times U_{\mathrm{d}}$, where $U_{\mathrm{c}} \subseteq \mathbb{R}^{m_{\mathrm{c}}}$ and $U_{\mathrm{d}} \subseteq \mathbb{R}^{m_{\mathrm{d}}}$, contains the set of input values, that is, for every $u = (u_{\mathrm{c}}, u_{\mathrm{d}}) \in \mathcal{U}$ and $t \in [0, \infty)$, $u(t) \in U$, $u_{\mathrm{c}}(t) \in U_{\mathrm{c}}$, and $u_{\mathrm{d}}(t_k) \in U_{\mathrm{d}}$. The set $Y \triangleq Y_{\mathrm{c}} \times Y_{\mathrm{d}}$, where $Y_{\mathrm{c}} \subseteq \mathbb{R}^{l_{\mathrm{c}}}$ and $Y_{\mathrm{d}} \subseteq \mathbb{R}^{l_{\mathrm{d}}}$, contains the set of output values, that is, for every $y = (y_{\mathrm{c}}, y_{\mathrm{d}}) \in \mathcal{Y}$ and $t \in [0, \infty)$, $y(t) \in Y$, $y_{\mathrm{c}}(t) \in Y_{\mathrm{c}}$, and $y_{\mathrm{d}}(t_k) \in Y_{\mathrm{d}}$. The spaces $\mathcal{U}$ and $\mathcal{Y}$ are assumed to be closed under the shift operator, that is, if $u(\cdot) \in \mathcal{U}$ (respectively, $y(\cdot) \in \mathcal{Y}$), then the function u_T (respectively, y_T) defined by $u_T \triangleq u(t + T)$ (respectively, $y_T \triangleq y(t + T)$) is contained in $\mathcal{U}$ (respectively, $\mathcal{Y}$) for all $T \geq 0$.

For convenience, we use the notation $s(t, \tau, x_0, u)$ to denote the solution $x(t)$ of (10.43) and (10.44) at time $t \geq \tau$ with initial condition $x(\tau) = x_0$, where $u = (u_{\mathrm{c}}, u_{\mathrm{d}}) : \mathbb{R} \times \mathcal{T} \to U_{\mathrm{c}} \times U_{\mathrm{d}}$ and $\mathcal{T} \triangleq \{t_1, t_2, \ldots\}$. Thus, the trajectory of the system (10.43) and (10.44) from the initial condition $x(0) = x_0$ is given by $\psi(t, 0, x_0, u_{\mathrm{c}})$ for $0 < t \leq t_1$, where $\psi(\cdot, 0, x_0, u_{\mathrm{c}})$ is the solution to (10.43) with the input $u_{\mathrm{c}}(\cdot) \in \mathcal{U}_{\mathrm{c}}$. If and when the trajectory reaches a state $x_1 \triangleq x(t_1)$ satisfying $(x_1, u_1) \in \mathcal{Z}$, where $u_1 \triangleq u_{\mathrm{c}}(t_1)$, then the state is instantaneously transferred to $x_1^+ \triangleq x_1 + f_{\mathrm{d}}(x_1) + G_{\mathrm{d}}(x_1)u_{\mathrm{d}}(t_1)$, where $u_{\mathrm{d}}(\cdot) \in \mathcal{U}_{\mathrm{d}}$ is a given input, according to the resetting law (10.44). The trajectory $x(t)$, $t_1 < t \leq t_2$, is then given by $\psi(t, t_1, x_1^+, u_{\mathrm{c}})$, and so on. As in the uncontrolled case, the solution $x(t)$ of (10.43) and (10.44) is left-continuous, that is, it is continuous everywhere except at the resetting times t_k, and

$$x_k \triangleq x(t_k) = \lim_{\varepsilon \to 0^+} x(t_k - \varepsilon), \tag{10.47}$$

$$\begin{aligned} x_k^+ &\triangleq x(t_k) + f_{\mathrm{d}}(x(t_k)) + G_{\mathrm{d}}(x(t_k))u_{\mathrm{d}}(t_k) \\ &= \lim_{\varepsilon \to 0^+} x(t_k + \varepsilon), \quad u_{\mathrm{d}}(t_k) \in U_{\mathrm{d}}, \end{aligned} \tag{10.48}$$

for $k = 1, 2, \ldots$. Furthermore, the analogs to Assumptions 10.1 and 10.2 become:

Assumption 10.3. If $(x(t), u_{\mathrm{c}}(t)) \in \overline{\mathcal{Z}} \backslash \mathcal{Z}$, then there exists $\varepsilon > 0$ such that, for all $0 < \delta < \varepsilon$,

$$\psi(t + \delta, t, x(t), u_{\mathrm{c}}(t + \delta)) \notin \mathcal{Z}. \tag{10.49}$$

Assumption 10.4. If $(x(t_k), u_{\mathrm{c}}(t_k)) \in \partial\mathcal{Z} \cap \mathcal{Z}$, then there exists $\varepsilon > 0$ such that, for all $0 \leq \delta < \varepsilon$ and $u_{\mathrm{d}}(t_k) \in U_{\mathrm{d}}$,

$$\psi(t_k + \delta, t_k, x(t_k) + f_{\mathrm{d}}(x(t_k)) + G_{\mathrm{d}}(x(t_k))u_{\mathrm{d}}(t_k), u_{\mathrm{c}}(t_k + \delta)) \notin \mathcal{Z}. \tag{10.50}$$

Thus, it follows from Assumptions 10.3 and 10.4 that if $(x, u_{\mathrm{c}}) \in \mathcal{Z}$, then $(x + f_{\mathrm{d}}(x) + G_{\mathrm{d}}(x)u_{\mathrm{d}}, u_{\mathrm{c}}) \notin \mathcal{Z}$, $u_{\mathrm{d}} \in U_{\mathrm{d}}$. In addition, if at time t the trajectory $(x(t), u_{\mathrm{c}}(t)) \in \overline{\mathcal{Z}} \backslash \mathcal{Z}$, then there exists $\varepsilon > 0$ such that for

$0 < \delta < \varepsilon$, $(x(t+\delta), u_{\mathrm{c}}(t+\delta)) \notin \mathcal{Z}$. Finally, in the case where $\mathcal{Z} \triangleq \mathcal{Z}_x \times U_{\mathrm{c}}$ and $\mathcal{Z}_x \subset \mathcal{D}$, we refer to (10.43)–(10.46) as a *state-dependent impulsive dynamical system*. Alternatively, for $\mathcal{Z} \triangleq \mathcal{D} \times \mathcal{Z}_{u_{\mathrm{c}}}$, where $\mathcal{Z}_{u_{\mathrm{c}}} \subset U_{\mathrm{c}}$, we refer to (10.43)–(10.46) as an *input-dependent impulsive dynamical system*, while in the case where $\mathcal{Z} \triangleq (\mathcal{Z}_x \times U_{\mathrm{c}}) \cup (\mathcal{D} \times \mathcal{Z}_{u_{\mathrm{c}}})$ we refer to (10.43)–(10.46) as an *input/state-dependent impulsive dynamical system*.

For the impulsive dynamical system $\mathcal{G}$ given by (10.43)–(10.46) a function $(s_{\mathrm{c}}(u_{\mathrm{c}}, y_{\mathrm{c}}), s_{\mathrm{d}}(u_{\mathrm{d}}, y_{\mathrm{d}}))$, where $s_{\mathrm{c}} : U_{\mathrm{c}} \times Y_{\mathrm{c}} \to \mathbb{R}$ and $s_{\mathrm{d}} : U_{\mathrm{d}} \times Y_{\mathrm{d}} \to \mathbb{R}$ are such that $s_{\mathrm{c}}(0, 0) = 0$ and $s_{\mathrm{d}}(0, 0) = 0$, is called a *hybrid supply rate* [68, 74] if it is locally integrable for all input-output pairs satisfying (10.43) and (10.45), that is, for all input-output pairs $u_{\mathrm{c}}(\cdot) \in \mathcal{U}_{\mathrm{c}}$ and $y_{\mathrm{c}}(\cdot) \in \mathcal{Y}_{\mathrm{c}}$ satisfying (10.43) and (10.45), $s_{\mathrm{c}}(\cdot, \cdot)$ satisfies $\int_t^{\hat{t}} |s_{\mathrm{c}}(u_{\mathrm{c}}(\sigma), y_{\mathrm{c}}(\sigma))| \mathrm{d}\sigma < \infty$, $t, \hat{t} \geq 0$. Note that since all input-output pairs $u_{\mathrm{d}}(\cdot) \in \mathcal{U}_{\mathrm{d}}$ and $y_{\mathrm{d}}(\cdot) \in \mathcal{Y}_{\mathrm{d}}$ satisfying (10.44) and (10.46) are defined for discrete instants, $s_{\mathrm{d}}(\cdot, \cdot)$ satisfies $\sum_{k \in \mathbb{Z}_{[t,\hat{t})}} |s_{\mathrm{d}}(u_{\mathrm{d}}(t_k), y_{\mathrm{d}}(t_k))| < \infty$, where $\mathbb{Z}_{[t,\hat{t})} \triangleq \{k : t \leq t_k < \hat{t}\}$.

Definition 10.2 ([82]). The impulsive dynamical system $\mathcal{G}$ given by (10.43)–(10.46) is *exponentially dissipative* (respectively, *dissipative*) *with respect to the hybrid supply rate* $(s_{\mathrm{c}}, s_{\mathrm{d}})$ if there exist a continuous, nonnegative-definite function $v_{\mathrm{s}} : \mathcal{D} \to \mathbb{R}$ and a scalar $\varepsilon > 0$ (respectively, $\varepsilon = 0$) such that $v_{\mathrm{s}}(0) = 0$, called a *storage function*, and the *hybrid dissipation inequality*

$$
e^{\varepsilon T} v_{\mathrm{s}}(x(T)) \leq e^{\varepsilon t_0} v_{\mathrm{s}}(x(t_0)) + \int_{t_0}^{T} e^{\varepsilon t} s_{\mathrm{c}}(u_{\mathrm{c}}(t), y_{\mathrm{c}}(t)) \mathrm{d}t
$$

$$
+ \sum_{k \in \mathbb{Z}_{[t_0, T)}} e^{\varepsilon t_k} s_{\mathrm{d}}(u_{\mathrm{d}}(t_k), y_{\mathrm{d}}(t_k)), \quad T \geq t_0, \qquad (10.51)
$$

is satisfied for all $T \geq t_0$, where $x(t)$, $t \geq t_0$, is the solution of (10.43)–(10.46) with $(u_{\mathrm{c}}(\cdot), u_{\mathrm{d}}(\cdot)) \in \mathcal{U}_{\mathrm{c}} \times \mathcal{U}_{\mathrm{d}}$. The impulsive dynamical system $\mathcal{G}$ given by (10.43)–(10.46) is *lossless with respect to the hybrid supply rate* $(s_{\mathrm{c}}, s_{\mathrm{d}})$ if the hybrid dissipation inequality is satisfied as an equality with $\varepsilon = 0$ for all $T \geq t_0$ and $(u_{\mathrm{c}}(\cdot), u_{\mathrm{d}}(\cdot)) \in \mathcal{U}_{\mathrm{c}} \times \mathcal{U}_{\mathrm{d}}$.

The following result gives necessary and sufficient conditions for dissipativity over an interval $t \in (t_k, t_{k+1}]$ involving the consecutive resetting times t_k and t_{k+1}. First, however, the following definition is required.

Definition 10.3 ([82]). A large-scale impulsive dynamical system $\mathcal{G}$ given by (10.43)–(10.46) is *completely reachable* if for all $(t_0, x_{\mathrm{i}}) \in \mathbb{R} \times \mathcal{D}$, there exist a finite time $t_{\mathrm{i}} < t_0$, a square integrable input $u_{\mathrm{c}}(t)$ defined on $[t_{\mathrm{i}}, t_0]$, and inputs $u_{\mathrm{d}}(t_k)$ defined on $k \in \mathbb{Z}_{[t_{\mathrm{i}}, t_0)}$, such that the state $x(t)$, $t \geq t_{\mathrm{i}}$, can be driven from $x(t_{\mathrm{i}}) = 0$ to $x(t_0) = x_{\mathrm{i}}$. A large-scale impulsive dynamical system $\mathcal{G}$ given by (10.43)–(10.46) is *zero-state observable* if $(u_{\mathrm{c}}(t), u_{\mathrm{d}}(t_k)) \equiv 0$ and $(y_{\mathrm{c}}(t), y_{\mathrm{d}}(t_k)) \equiv 0$ implies $x(t) \equiv 0$.

Theorem 10.4 ([82]). Assume $\mathcal{G}$ is completely reachable. Then $\mathcal{G}$ is exponentially dissipative (respectively, dissipative) with respect to the hybrid supply rate $(s_\mathrm{c}, s_\mathrm{d})$ if and only if there exist a continuous nonnegative-definite function $v_\mathrm{s} : \mathcal{D} \to \mathbb{R}$ and a scalar $\varepsilon > 0$ (respectively, $\varepsilon = 0$) such that $v_\mathrm{s}(0) = 0$ and for all $k \in \overline{\mathbb{Z}}_+$,

$$e^{\varepsilon \hat{t}} v_\mathrm{s}(x(\hat{t})) \le e^{\varepsilon t} v_\mathrm{s}(x(t)) + \int_t^{\hat{t}} e^{\varepsilon s} s_\mathrm{c}(u_\mathrm{c}(s), y_\mathrm{c}(s)) \mathrm{d}s, \quad t_k < t \le \hat{t} \le t_{k+1},$$
$$(10.52)$$
$$v_\mathrm{s}(x(t_k) + f_\mathrm{d}(x(t_k)) + G_\mathrm{d}(x(t_k)) u_\mathrm{d}(t_k)) \le v_\mathrm{s}(x(t_k)) + s_\mathrm{d}(u_\mathrm{d}(t_k), y_\mathrm{d}(t_k)).$$
$$(10.53)$$

Finally, $\mathcal{G}$ given by (10.43)–(10.46) is lossless with respect to the hybrid supply rate $(s_\mathrm{c}, s_\mathrm{d})$ if and only if (10.52) and (10.53) are satisfied as equalities with $\varepsilon = 0$ for all $k \in \overline{\mathbb{Z}}_+$.

To develop vector dissipativity notions for large-scale impulsive dynamical systems we consider input/state-dependent impulsive dynamical systems $\mathcal{G}$ of the form

$$\dot{x}(t) = F_\mathrm{c}(x(t), u_\mathrm{c}(t)), \quad x(t_0) = x_0, \quad (x(t), u_\mathrm{c}(t)) \notin \mathcal{Z}, \quad t \ge t_0, \quad (10.54)$$
$$\Delta x(t) = F_\mathrm{d}(x(t), u_\mathrm{d}(t)), \quad (x(t), u_\mathrm{c}(t)) \in \mathcal{Z}, \quad (10.55)$$
$$y_\mathrm{c}(t) = H_\mathrm{c}(x(t), u_\mathrm{c}(t)), \quad (x(t), u_\mathrm{c}(t)) \notin \mathcal{Z}, \quad (10.56)$$
$$y_\mathrm{d}(t) = H_\mathrm{d}(x(t), u_\mathrm{d}(t)), \quad (x(t), u_\mathrm{c}(t)) \in \mathcal{Z}, \quad (10.57)$$

where $x(t) \in \mathcal{D} \subseteq \mathbb{R}^n$, $t \ge t_0$, $u_\mathrm{c}(t) \in U_\mathrm{c} \subseteq \mathbb{R}^{m_\mathrm{c}}$, $u_\mathrm{d}(t_k) \in U_\mathrm{d} \subseteq \mathbb{R}^{m_\mathrm{d}}$, $y_\mathrm{c}(t) \in Y_\mathrm{c} \subseteq \mathbb{R}^{l_\mathrm{c}}$, $y_\mathrm{d}(t_k) \in Y_\mathrm{d} \subseteq \mathbb{R}^{l_\mathrm{d}}$, $F_\mathrm{c} : \mathcal{D} \times U_\mathrm{c} \to \mathbb{R}^n$, $F_\mathrm{d} : \mathcal{D} \times U_\mathrm{d} \to \mathbb{R}^n$, $H_\mathrm{c} : \mathcal{D} \times U_\mathrm{c} \to Y_\mathrm{c}$, $H_\mathrm{d} : \mathcal{D} \times U_\mathrm{d} \to Y_\mathrm{d}$, $\mathcal{D}$ is an open set with $0 \in \mathcal{D}$, $\mathcal{Z} \subset \mathcal{D} \times U_\mathrm{c}$, and $F_\mathrm{c}(0, 0) = 0$. Here, we assume that $\mathcal{G}$ represents a large-scale impulsive dynamical system composed of q interconnected controlled impulsive subsystems $\mathcal{G}_i$ such that, for all $i = 1, \ldots, q$,

$$F_{ci}(x, u_{ci}) = f_{ci}(x_i) + \mathcal{I}_{ci}(x) + G_{ci}(x_i) u_{ci}, \quad (10.58)$$
$$F_{di}(x, u_{di}) = f_{di}(x_i) + \mathcal{I}_{di}(x) + G_{di}(x_i) u_{di}, \quad (10.59)$$
$$H_{ci}(x_i, u_{ci}) = h_{ci}(x_i) + J_{ci}(x_i) u_{ci}, \quad (10.60)$$
$$H_{di}(x_i, u_{di}) = h_{di}(x_i) + J_{di}(x_i) u_{di}, \quad (10.61)$$

where $x_i \in \mathcal{D}_i \subseteq \mathbb{R}^{n_i}$, $u_{ci} \in U_{ci} \subseteq \mathbb{R}^{m_{ci}}$, $u_{di} \in U_{di} \subseteq \mathbb{R}^{m_{di}}$, $y_{ci} \triangleq H_{ci}(x_i, u_{ci}) \in Y_{ci} \subseteq \mathbb{R}^{l_{ci}}$, $y_{di} \triangleq H_{di}(x_i, u_{di}) \in Y_{di} \subseteq \mathbb{R}^{l_{di}}$, $((u_{ci}, u_{di}), (y_{ci}, y_{di}))$ is the hybrid input-output pair for the ith subsystem such that $u_{ci}(\cdot) \in \mathcal{U}_{ci}$, $u_{di}(\cdot) \in \mathcal{U}_{di}$, $y_{ci}(\cdot) \in \mathcal{Y}_{ci}$, $y_{di}(\cdot) \in \mathcal{Y}_{di}$, where $(\mathcal{U}_{ci}, \mathcal{U}_{di})$ and $(\mathcal{Y}_{ci}, \mathcal{Y}_{di})$ denote the ith subsystem input and output spaces, $f_{ci} : \mathbb{R}^{n_i} \to \mathbb{R}^{n_i}$ and $\mathcal{I}_{ci} : \mathcal{D} \to \mathbb{R}^{n_i}$ are Lipschitz continuous and satisfy $f_{ci}(0) = 0$ and $\mathcal{I}_{ci}(0) = 0$, $f_{di} : \mathbb{R}^{n_i} \to \mathbb{R}^{n_i}$ and $\mathcal{I}_{di} : \mathcal{D} \to \mathbb{R}^{n_i}$ are continuous, $G_{ci} : \mathbb{R}^{n_i} \to \mathbb{R}^{n_i \times m_{ci}}$ and $G_{di} : \mathbb{R}^{n_i} \to \mathbb{R}^{n_i \times m_{di}}$ are continuous, $h_{ci} : \mathbb{R}^{n_i} \to \mathbb{R}^{l_{ci}}$ and satisfies $h_{ci}(0) = 0$, $h_{di} : \mathbb{R}^{n_i} \to \mathbb{R}^{l_{di}}$, $J_{ci} : \mathbb{R}^{n_i} \to \mathbb{R}^{l_{ci} \times m_{ci}}$, $J_{di} : \mathbb{R}^{n_i} \to \mathbb{R}^{l_{di} \times m_{di}}$, $\sum_{i=1}^{q} n_i = n$, $\sum_{i=1}^{q} m_{ci} = m_{\mathrm{c}}$, $\sum_{i=1}^{q} m_{di} = m_{\mathrm{d}}$, $\sum_{i=1}^{q} l_{ci} = l_{\mathrm{c}}$, and $\sum_{i=1}^{q} l_{di} = l_{\mathrm{d}}$.

Here, $f_{ci} : \mathcal{D}_i \subseteq \mathbb{R}^{n_i} \to \mathbb{R}^{n_i}$ and $f_{di} : \mathcal{D}_i \subseteq \mathbb{R}^{n_i} \to \mathbb{R}^{n_i}$ define vector fields of each isolated subsystem of (10.54) and (10.55), and $\mathcal{I}_{ci} : \mathcal{D} \to \mathbb{R}^{n_i}$ and $\mathcal{I}_{di} : \mathcal{D} \to \mathbb{R}^{n_i}$ define the structure of the interconnection dynamics of the ith impulsive subsystem with all other impulsive subsystems. Furthermore, for the large-scale dynamical system $\mathcal{G}$ we assume that the required properties for the existence and uniqueness of solutions are satisfied, that is, for each $i \in \{1, \ldots, q\}$, $u_{ci}(\cdot)$ and $u_{di}(\cdot)$ satisfy sufficient regularity conditions such that the system (10.54) and (10.55) has a unique solution forward in time. We define the composite input and composite output for the large-scale impulsive dynamical system $\mathcal{G}$ as $u_{\mathrm{c}} \triangleq [u_{c1}^{\mathrm{T}}, \ldots, u_{cq}^{\mathrm{T}}]^{\mathrm{T}}$, $u_{\mathrm{d}} \triangleq [u_{d1}^{\mathrm{T}}, \ldots, u_{dq}^{\mathrm{T}}]^{\mathrm{T}}$, $y_{\mathrm{c}} \triangleq [y_{c1}^{\mathrm{T}}, \ldots, y_{cq}^{\mathrm{T}}]^{\mathrm{T}}$, and $y_{\mathrm{d}} \triangleq [y_{d1}^{\mathrm{T}}, \ldots, y_{dq}^{\mathrm{T}}]^{\mathrm{T}}$, respectively. In addition, we define $\mathcal{U} \triangleq \mathcal{U}_{\mathrm{c}} \times \mathcal{U}_{\mathrm{d}}$ and $\mathcal{Y} \triangleq \mathcal{Y}_{\mathrm{c}} \times \mathcal{Y}_{\mathrm{d}}$, where $\mathcal{U}_{\mathrm{c}} \triangleq \mathcal{U}_{c1} \times \cdots \times \mathcal{U}_{cq}$, $\mathcal{U}_{\mathrm{d}} \triangleq \mathcal{U}_{d1} \times \cdots \times \mathcal{U}_{dq}$, $\mathcal{Y}_{\mathrm{c}} \triangleq \mathcal{Y}_{c1} \times \cdots \times \mathcal{Y}_{cq}$, $\mathcal{Y}_{\mathrm{d}} \triangleq \mathcal{Y}_{d1} \times \cdots \times \mathcal{Y}_{dq}$, to be input and output spaces, respectively, for the system (10.54)–(10.57) consisting of left-continuous bounded U-valued and Y-valued functions on the semi-infinite interval $[0, \infty)$.

Definition 10.4. For the large-scale impulsive dynamical system $\mathcal{G}$ given by (10.54)–(10.57) a vector function $(S_{\mathrm{c}}(u_{\mathrm{c}}, y_{\mathrm{c}}), S_{\mathrm{d}}(u_{\mathrm{d}}, y_{\mathrm{d}}))$, where

$$S_{\mathrm{c}}(u_{\mathrm{c}}, y_{\mathrm{c}}) \triangleq [s_{c1}(u_{c1}, y_{c1}), \ldots, s_{cq}(u_{cq}, y_{cq})]^{\mathrm{T}}, \tag{10.62}$$

$$S_{\mathrm{d}}(u_{\mathrm{d}}, y_{\mathrm{d}}) \triangleq [s_{d1}(u_{d1}, y_{d1}), \ldots, s_{dq}(u_{dq}, y_{dq})]^{\mathrm{T}}, \tag{10.63}$$

$s_{ci} : U_{ci} \times Y_{ci} \to \mathbb{R}$, and $s_{di} : U_{di} \times Y_{di} \to \mathbb{R}$, $i = 1, \ldots, q$, such that $S_{\mathrm{c}}(0, 0) = 0$ and $S_{\mathrm{d}}(0, 0) = 0$, is called a *vector hybrid supply rate* if it is locally componentwise integrable for all input-output pairs satisfying (10.54)–(10.57), that is, for every $i \in \{1, \ldots, q\}$ and for all input-output pairs $u_{ci}(\cdot) \in \mathcal{U}_{ci}$ and $y_{ci}(\cdot) \in \mathcal{Y}_{ci}$ satisfying (10.54)–(10.57), $s_{ci}(\cdot, \cdot)$ satisfies $\int_t^{\hat{t}} |s_{ci}(u_{ci}(s), y_{ci}(s))| \mathrm{d}s < \infty$, $t, \hat{t} \geq t_0$.

Note that since all input-output pairs $u_{di}(\cdot) \in \mathcal{U}_{di}$ and $y_{di}(\cdot) \in \mathcal{Y}_{di}$ are defined for discrete instants, $s_{di}(\cdot, \cdot)$ in Definition 10.4 satisfies

$$\sum_{k \in \mathbb{Z}_{[t,\hat{t})}} |s_{di}(u_{di}(t_k), y_{di}(t_k))| < \infty, \tag{10.64}$$

where $\mathbb{Z}_{[t,\hat{t})} \triangleq \{k : t \leq t_k < \hat{t}\}$. For the statement of the next definition, recall that a matrix $W \in \mathbb{R}^{q \times q}$ is *semistable* if and only if $\lim_{t \to \infty} e^{Wt}$ exists [21,69], while W is *asymptotically stable* if and only if $\lim_{t \to \infty} e^{Wt} = 0$.

Definition 10.5. The large-scale impulsive dynamical system $\mathcal{G}$ given by (10.54)–(10.57) is *vector dissipative* (respectively, *exponentially vector* *dissipative) with respect to the vector hybrid supply rate* $(S_{\mathrm{c}}, S_{\mathrm{d}})$ if there exist a continuous, nonnegative definite vector function $V_{\mathrm{s}} = [v_{\mathrm{s}1}, \ldots, v_{\mathrm{s}q}]^{\mathrm{T}} :$ $\mathcal{D} \to \overline{\mathbb{R}}_{+}^{q}$, called a *vector storage function*, and an essentially nonnegative *dissipation matrix* $W \in \mathbb{R}^{q \times q}$ such that $V_{\mathrm{s}}(0) = 0$, W is semistable (respectively, asymptotically stable), and the *vector hybrid dissipation inequality*

$$
V_{\mathrm{s}}(x(T)) \leq\leq e^{W(T-t_0)} V_{\mathrm{s}}(x(t_0)) + \int_{t_0}^{T} e^{W(T-t)} S_{\mathrm{c}}(u_{\mathrm{c}}(t), y_{\mathrm{c}}(t)) \mathrm{d}t
$$

$$
+ \sum_{k \in \mathbb{Z}_{[t_0,T)}} e^{W(T-t_k)} S_{\mathrm{d}}(u_{\mathrm{d}}(t_k), y_{\mathrm{d}}(t_k)), \quad T \geq t_0, \quad (10.65)
$$

is satisfied, where $x(t)$, $t \geq t_0$, is the solution to (10.54)–(10.57) with $(u_{\mathrm{c}}(\cdot),$ $u_{\mathrm{d}}(\cdot)) \in \mathcal{U}_{\mathrm{c}} \times \mathcal{U}_{\mathrm{d}}$ and $x(t_0) = x_0$. The large-scale impulsive dynamical system $\mathcal{G}$ given by (10.54)–(10.57) is *vector lossless with respect to the vector hybrid supply rate* $(S_{\mathrm{c}}, S_{\mathrm{d}})$ if the vector hybrid dissipation inequality is satisfied as an equality with W semistable.

Note that if the subsystems $\mathcal{G}_i$ of $\mathcal{G}$ are *disconnected*, that is, $\mathcal{I}_{ci}(x) \equiv 0$ and $\mathcal{I}_{di}(x) \equiv 0$ for all $i = 1, \ldots, q$, and $-W \in \mathbb{R}^{q \times q}$ is diagonal and nonnegative definite, then it follows from Definition 10.5 that each of the disconnected subsystems $\mathcal{G}_i$ is dissipative or exponentially dissipative in the sense of Definition 10.2. A similar remark holds in the case where $q = 1$.

Next, define the *vector available storage* of the large-scale impulsive dynamical system $\mathcal{G}$ by

$$
V_{\mathrm{a}}(x_0) \triangleq - \inf_{(u_{\mathrm{c}}(\cdot),\, u_{\mathrm{d}}(\cdot)),\, T \geq t_0} \left[\int_{t_0}^{T} e^{-W(t-t_0)} S_{\mathrm{c}}(u_{\mathrm{c}}(t), y_{\mathrm{c}}(t)) \mathrm{d}t \right.
$$

$$
\left. + \sum_{k \in \mathbb{Z}_{[t_0,T)}} e^{-W(t_k - t_0)} S_{\mathrm{d}}(u_{\mathrm{d}}(t_k), y_{\mathrm{d}}(t_k)) \right],
$$

$$
(10.66)
$$

where $x(t)$, $t \geq t_0$, is the solution to (10.54)–(10.57) with $x(t_0) = x_0$ and admissible inputs $(u_{\mathrm{c}}(\cdot), u_{\mathrm{d}}(\cdot)) \in \mathcal{U}_{\mathrm{c}} \times \mathcal{U}_{\mathrm{d}}$. The infimum in (10.66) is taken componentwise, which implies that for each element of $V_{\mathrm{a}}(\cdot)$ the infimum is calculated separately. Note that $V_{\mathrm{a}}(x_0) \geq\geq 0$, $x_0 \in \mathcal{D}$, since $V_{\mathrm{a}}(x_0)$ is the infimum over a set of vectors containing the zero vector $(T = t_0)$.

Theorem 10.5. Consider the large-scale impulsive dynamical system $\mathcal{G}$ given by (10.54)–(10.57) and assume that $\mathcal{G}$ is completely reachable. Then $\mathcal{G}$ is vector dissipative (respectively, exponentially vector dissipative) with respect to the vector hybrid supply rate $(S_{\mathrm{c}}, S_{\mathrm{d}})$ if and only if there exist a continuous, nonnegative-definite vector function $V_{\mathrm{s}} : \mathcal{D} \to \overline{\mathbb{R}}_{+}^{q}$ and an essentially nonnegative dissipation matrix $W \in \mathbb{R}^{q \times q}$ such that $V_{\mathrm{s}}(0) = 0$, W is semistable (respectively, asymptotically stable), and for all $k \in \overline{\mathbb{Z}}_{+}$,

$$V_{\mathrm{s}}(x(\hat{t})) \leq\leq e^{W(\hat{t}-t)} V_{\mathrm{s}}(x(t)) + \int_{t}^{\hat{t}} e^{W(\hat{t}-s)} S_{\mathrm{c}}(u_{\mathrm{c}}(s), y_{\mathrm{c}}(s)) \mathrm{d}s,$$

$$t_k < t \leq \hat{t} \leq t_{k+1}, \tag{10.67}$$

$$V_{\mathrm{s}}(x(t_k) + F_{\mathrm{d}}(x(t_k), u_{\mathrm{d}}(t_k))) \leq\leq V_{\mathrm{s}}(x(t_k)) + S_{\mathrm{d}}(u_{\mathrm{d}}(t_k), y_{\mathrm{d}}(t_k)). \tag{10.68}$$

Alternatively, $\mathcal{G}$ is vector lossless with respect to the vector hybrid supply rate $(S_{\mathrm{c}}, S_{\mathrm{d}})$ if and only if there exists a continuous, nonnegative-definite vector function $V_{\mathrm{s}} : \mathcal{D} \to \overline{\mathbb{R}}_{+}^{q}$ such that (10.67) and (10.68) are satisfied as equalities with W semistable.

Proof. Let $k \in \overline{\mathbb{Z}}_{+}$ and suppose $\mathcal{G}$ is vector dissipative (respectively, exponentially vector dissipative) with respect to the vector hybrid supply rate $(S_{\mathrm{c}}, S_{\mathrm{d}})$. Then, there exist a continuous nonnegative-definite vector function $V_{\mathrm{s}} : \mathcal{D} \to \overline{\mathbb{R}}_{+}^{q}$ and an essentially nonnegative matrix $W \in \mathbb{R}^{q \times q}$ such that (10.65) holds. Now, since for $t_k < t \leq \hat{t} \leq t_{k+1}$, $\mathbb{Z}_{[t,\hat{t})} = \emptyset$, (10.67) is immediate. Next, it follows from (10.65) that

$$V_{\mathrm{s}}(x(t_k^{+})) \leq\leq e^{W(t_k^{+}-t_k)} V_{\mathrm{s}}(x(t_k)) + \int_{t_k}^{t_k^{+}} e^{W(t_k^{+}-s)} S_{\mathrm{c}}(u_{\mathrm{c}}(s), y_{\mathrm{c}}(s)) \mathrm{d}s$$

$$+ \sum_{k \in \mathbb{Z}_{[t_k, t_k^{+})}} e^{W(t_k^{+}-t_k)} S_{\mathrm{d}}(u_{\mathrm{d}}(t_k), y_{\mathrm{d}}(t_k)) \tag{10.69}$$

which, since $\mathbb{Z}_{[t_k, t_k^{+})} = k$, implies (10.68).

Conversely, suppose (10.67) and (10.68) hold and let $\hat{t} \geq t \geq t_0$ and $\mathbb{Z}_{[t,\hat{t})} = \{i, i+1, \ldots, j\}$. (Note that if $\mathbb{Z}_{[t,\hat{t})} = \emptyset$, then the converse result is a direct consequence of (10.67).) If $\mathbb{Z}_{[t,\hat{t})} \neq \emptyset$, then it follows from (10.67) and (10.68) that

$$\begin{aligned}
V_{\mathrm{s}}(x(\hat{t})) &- e^{W(\hat{t}-t)} V_{\mathrm{s}}(x(t)) \\
&= V_{\mathrm{s}}(x(\hat{t})) - e^{W(\hat{t}-t_j^{+})} V_{\mathrm{s}}(x(t_j^{+})) \\
&\quad + e^{W(\hat{t}-t_j^{+})} V_{\mathrm{s}}(x(t_j^{+})) - e^{W(\hat{t}-t_{j-1}^{+})} V_{\mathrm{s}}(x(t_{j-1}^{+})) \\
&\quad + e^{W(\hat{t}-t_{j-1}^{+})} V_{\mathrm{s}}(x(t_{j-1}^{+})) - \cdots - e^{W(\hat{t}-t_i^{+})} V_{\mathrm{s}}(x(t_i^{+})) \\
&\quad + e^{W(\hat{t}-t_i^{+})} V_{\mathrm{s}}(x(t_i^{+})) - e^{W(\hat{t}-t)} V_{\mathrm{s}}(x(t)) \\
&= V_{\mathrm{s}}(x(\hat{t})) - e^{W(\hat{t}-t_j)} V_{\mathrm{s}}(x(t_j^{+})) \\
&\quad + e^{W(\hat{t}-t_j)} V_{\mathrm{s}}(x(t_j) + F_{\mathrm{d}}(x(t_j), u_{\mathrm{d}}(t_j))) - e^{W(\hat{t}-t_j)} V_{\mathrm{s}}(x(t_j)) \\
&\quad + e^{W(\hat{t}-t_j)} V_{\mathrm{s}}(x(t_j)) - e^{W(\hat{t}-t_{j-1}^{+})} V_{\mathrm{s}}(x(t_{j-1}^{+})) + \cdots
\end{aligned}$$

$$+e^{W(\hat{t}-t_i)}V_{\mathrm{s}}(x(t_i)+F_{\mathrm{d}}(x(t_i),u_{\mathrm{d}}(t_i)))-e^{W(\hat{t}-t_i)}V_{\mathrm{s}}(x(t_i))$$

$$+e^{W(\hat{t}-t_i)}V_{\mathrm{s}}(x(t_i))-e^{W(\hat{t}-t)}V_{\mathrm{s}}(x(t))$$

$$=\ V_{\mathrm{s}}(x(\hat{t}))-e^{W(\hat{t}-t_j)}V_{\mathrm{s}}(x(t_j^+))$$

$$+e^{W(\hat{t}-t_j)}[V_{\mathrm{s}}(x(t_j)+F_{\mathrm{d}}(x(t_j),u_{\mathrm{d}}(t_j)))-V_{\mathrm{s}}(x(t_j))]$$

$$+e^{W(\hat{t}-t_j)}[V_{\mathrm{s}}(x(t_j))-e^{W(t_j-t_{j-1})}V_{\mathrm{s}}(x(t_{j-1}^+))]+\cdots$$

$$+e^{W(\hat{t}-t_i)}[V_{\mathrm{s}}(x(t_i)+F_{\mathrm{d}}(x(t_i),u_{\mathrm{d}}(t_i)))-V_{\mathrm{s}}(x(t_i))]$$

$$+e^{W(\hat{t}-t_i)}[V_{\mathrm{s}}(x(t_i))-e^{W(t_i-t)}V_{\mathrm{s}}(x(t))]$$

$$\leq\leq\ \int_{t_j}^{\hat{t}}e^{W(\hat{t}-s)}S_{\mathrm{c}}(u_{\mathrm{c}}(s),y_{\mathrm{c}}(s))\mathrm{d}s+e^{W(\hat{t}-t_j)}S_{\mathrm{d}}(u_{\mathrm{d}}(t_j),y_{\mathrm{d}}(t_j))$$

$$+e^{W(\hat{t}-t_j)}\int_{t_{j-1}}^{t_j}e^{W(t_j-s)}S_{\mathrm{c}}(u_{\mathrm{c}}(s),y_{\mathrm{c}}(s))\mathrm{d}s+\cdots$$

$$+e^{W(\hat{t}-t_i)}S_{\mathrm{d}}(u_{\mathrm{d}}(t_i),y_{\mathrm{d}}(t_i))$$

$$+e^{W(\hat{t}-t_i)}\int_{t}^{t_i}e^{W(t_i-s)}S_{\mathrm{c}}(u_{\mathrm{c}}(s),y_{\mathrm{c}}(s))\mathrm{d}s$$

$$=\ \int_{t}^{\hat{t}}e^{W(\hat{t}-s)}S_{\mathrm{c}}(u_{\mathrm{c}}(s),y_{\mathrm{c}}(s))\mathrm{d}s$$

$$+\sum_{k\in\mathbb{Z}_{[t,\hat{t})}}e^{W(\hat{t}-t_k)}S_{\mathrm{d}}(u_{\mathrm{d}}(t_k),y_{\mathrm{d}}(t_k)),\tag{10.70}$$

which implies that $\mathcal{G}$ is vector dissipative (respectively, exponentially vector dissipative) with respect to the vector hybrid supply rate $(S_{\mathrm{c}},S_{\mathrm{d}})$.

Finally, similar constructions show that $\mathcal{G}$ is vector lossless with respect to the vector hybrid supply rate $(S_{\mathrm{c}},S_{\mathrm{d}})$ if and only if (10.67) and (10.68) are satisfied as equalities with W semistable. $\qquad\square$

Theorem 10.6. Consider the large-scale impulsive dynamical system $\mathcal{G}$ given by (10.54)–(10.57) and assume that $\mathcal{G}$ is completely reachable. Let $W\in\mathbb{R}^{q\times q}$ be essentially nonnegative and semistable (respectively, asymptotically stable). Then

$$\int_{t_0}^{T}e^{-W(t-t_0)}S_{\mathrm{c}}(u_{\mathrm{c}}(t),y_{\mathrm{c}}(t))\mathrm{d}t+\sum_{k\in\mathbb{Z}_{[t_0,T)}}e^{-W(t_k-t_0)}S_{\mathrm{d}}(u_{\mathrm{d}}(t_k),y_{\mathrm{d}}(t_k))\geq\geq 0,$$

$$T\geq t_0,\tag{10.71}$$

for $x(t_0)=0$ and $(u_{\mathrm{c}}(\cdot),u_{\mathrm{d}}(\cdot))\in\mathcal{U}_{\mathrm{c}}\times\mathcal{U}_{\mathrm{d}}$ if and only if $V_{\mathrm{a}}(0)=0$ and $V_{\mathrm{a}}(x)$ is finite for all $x\in\mathcal{D}$. Moreover, if (10.71) holds, then $V_{\mathrm{a}}(x)$, $x\in\mathcal{D}$, is a vector storage function for $\mathcal{G}$, and hence, $\mathcal{G}$ is vector dissipative (respectively, exponentially vector dissipative) with respect to the vector hybrid supply rate $(S_c(u_c, y_c), S_d(u_d, y_d))$.

Proof. Suppose $V_a(0) = 0$ and $V_a(x)$, $x \in \mathcal{D}$, is finite. Then

$$
0 = V_a(0) = - \inf_{(u_c(\cdot),\, u_d(\cdot)),\, T \geq t_0} \left[\int_{t_0}^{T} e^{-W(t-t_0)} S_c(u_c(t), y_c(t)) dt \right.
$$
$$
\left. + \sum_{k \in \mathbb{Z}_{[t_0, T)}} e^{-W(t_k - t_0)} S_d(u_d(t_k), y_d(t_k)) \right], \tag{10.72}
$$

which implies (10.71).

Next, suppose (10.71) holds. Then for $x(t_0) = 0$,

$$
- \inf_{(u_c(\cdot),\, u_d(\cdot)),\, T \geq t_0} \left[\int_{t_0}^{T} e^{-W(t-t_0)} S_c(u_c(t), y_c(t)) dt \right.
$$
$$
\left. + \sum_{k \in \mathbb{Z}_{[t_0, T)}} e^{-W(t_k - t_0)} S_d(u_d(t_k), y_d(t_k)) \right] \leq\leq 0,
$$
$$
\tag{10.73}
$$

which implies that $V_a(0) \leq\leq 0$. However, since $V_a(x_0) \geq\geq 0$, $x_0 \in \mathcal{D}$, it follows that $V_a(0) = 0$. Moreover, since $\mathcal{G}$ is completely reachable it follows that for every $x_0 \in \mathcal{D}$ there exists $\hat{t} > t_0$ and an admissible input $u(\cdot)$ defined on $[t_0, \hat{t}]$ such that $x(\hat{t}) = x_0$.

Now, since (10.71) holds for $x(t_0) = 0$ it follows that for all admissible $(u_c(\cdot), y_c(\cdot)) \in \mathcal{U}_c \times \mathcal{Y}_c$ and $(u_d(\cdot), y_d(\cdot)) \in \mathcal{U}_d \times \mathcal{Y}_d$,

$$
\int_{t_0}^{T} e^{-W(t-t_0)} S_c(u_c(t), y_c(t)) dt + \sum_{k \in \mathbb{Z}_{[t_0, T)}} e^{-W(t_k - t_0)} S_d(u_d(t_k), y_d(t_k)) \geq\geq 0,
$$
$$
T \geq \hat{t}, \tag{10.74}
$$

or, equivalently, multiplying (10.74) by the nonnegative matrix $e^{W(\hat{t}-t_0)}$, $\hat{t} \geq t_0$, (see Corollary 2.1) yields

$$
- \int_{\hat{t}}^{T} e^{-W(t-\hat{t})} S_c(u_c(t), y_c(t)) dt - \sum_{k \in \mathbb{Z}_{[\hat{t}, T)}} e^{-W(t_k - \hat{t})} S_d(u_d(t_k), u_d(t_k))
$$
$$
\leq\leq \int_{t_0}^{\hat{t}} e^{-W(t-\hat{t})} S_c(u_c(t), y_c(t)) dt
$$
$$
+ \sum_{k \in \mathbb{Z}_{[t_0, \hat{t})}} e^{-W(t_k - \hat{t})} S_d(u_d(t_k), u_d(t_k))
$$

$$\leq\leq Q(x_0)$$
$$<<\infty, \quad T \geq \hat{t}, \quad (u_{\rm c}(\cdot), u_{\rm d}(\cdot)) \in \mathcal{U}_{\rm c} \times \mathcal{U}_{\rm d}, \tag{10.75}$$

where $Q : \mathcal{D} \to \mathbb{R}^q$. Hence,

$$
V_{\rm a}(x_0) = - \inf_{(u_{\rm c}(\cdot), u_{\rm d}(\cdot)),\, T \geq \hat{t}} \left[\int_{\hat{t}}^{T} e^{-W(t-\hat{t})} S_{\rm c}(u_{\rm c}(t), y_{\rm c}(t)) {\rm d}t \right.
$$
$$
\left. + \sum_{k \in \mathbb{Z}_{[\hat{t}, T)}} e^{-W(t_k - \hat{t})} S_{\rm d}(u_{\rm d}(t_k), u_{\rm d}(t_k)) \right]
$$
$$\leq\leq Q(x_0)$$
$$<<\infty, \quad x_0 \in \mathcal{D}, \tag{10.76}$$

which implies that $V_{\rm a}(x_0)$, $x_0 \in \mathcal{D}$, is finite.

Finally, since (10.71) implies that $V_{\rm a}(0) = 0$ and $V_{\rm a}(x)$, $x \in \mathcal{D}$, is finite it follows from the definition of the vector available storage that

$$
-V_{\rm a}(x_0) \leq\leq \int_{t_0}^{T} e^{-W(t-t_0)} S_{\rm c}(u_{\rm c}(t), y_{\rm c}(t)) {\rm d}t
$$
$$
+ \sum_{k \in \mathbb{Z}_{[t_0, T)}} e^{-W(t_k - t_0)} S_{\rm d}(u_{\rm d}(t_k), u_{\rm d}(t_k))
$$
$$
= \int_{t_0}^{t_{\rm f}} e^{-W(t-t_0)} S_{\rm c}(u_{\rm c}(t), y_{\rm c}(t)) {\rm d}t
$$
$$
+ \sum_{k \in \mathbb{Z}_{[t_0, t_{\rm f})}} e^{-W(t_k - t_0)} S_{\rm d}(u_{\rm d}(t_k), u_{\rm d}(t_k))
$$
$$
+ \int_{t_{\rm f}}^{T} e^{-W(t-t_0)} S_{\rm c}(u_{\rm c}(t), y_{\rm c}(t)) {\rm d}t
$$
$$
+ \sum_{k \in \mathbb{Z}_{[t_{\rm f}, T)}} e^{-W(t_k - t_0)} S_{\rm d}(u_{\rm d}(t_k), u_{\rm d}(t_k)), \quad T \geq t_0. \tag{10.77}
$$

Now, multiplying (10.77) by the nonnegative matrix $e^{W(t_{\rm f} - t_0)}$, $t_{\rm f} \geq t_0$, (see Corollary 2.1) it follows that

$$
e^{W(t_{\rm f} - t_0)} V_{\rm a}(x_0) + \int_{t_0}^{t_{\rm f}} e^{W(t_{\rm f} - t)} S_{\rm c}(u_{\rm c}(t), y_{\rm c}(t)) {\rm d}t
$$
$$
+ \sum_{k \in \mathbb{Z}_{[t_0, t_{\rm f})}} e^{W(t_{\rm f} - t_k)} S_{\rm d}(u_{\rm d}(t_k), u_{\rm d}(t_k))
$$
$$
\geq\geq - \inf_{(u_{\rm c}(\cdot), u_{\rm d}(\cdot)),\, T \geq t_{\rm f}} \left[\int_{t_{\rm f}}^{T} e^{-W(t-t_{\rm f})} S_{\rm c}(u_{\rm c}(t), y_{\rm d}(t)) {\rm d}t \right.
$$

$$+ \sum_{k \in \mathbb{Z}_{[t_{\mathrm{f}}, T)}} e^{-W(t_k - t_{\mathrm{f}})} S_{\mathrm{d}}(u_{\mathrm{d}}(t_k), u_{\mathrm{d}}(t_k)) \Bigg]$$

$$= V_{\mathrm{a}}(x(t_{\mathrm{f}})), \tag{10.78}$$

which implies that $V_{\mathrm{a}}(x)$, $x \in \mathcal{D}$, is a vector storage function, and hence, $\mathcal{G}$ is vector dissipative (respectively, exponentially vector dissipative) with respect to the vector hybrid supply rate $(S_{\mathrm{c}}(u_{\mathrm{c}}, y_{\mathrm{c}}), S_{\mathrm{d}}(u_{\mathrm{d}}, y_{\mathrm{d}}))$. $\qquad\square$

Recall that it follows from Lemma 2.1 that if $W \in \mathbb{R}^{q \times q}$ is essentially nonnegative and semistable (respectively, asymptotically stable), then there exist a scalar $\alpha \geq 0$ (respectively, $\alpha > 0$) and a nonnegative vector $p \in \overline{\mathbb{R}}_+^q$, $p \neq 0$, (respectively, positive vector $p \in \mathbb{R}_+^q$) such that (2.4) holds. In this case,

$$\begin{aligned} p^{\mathrm{T}} e^{Wt} &= p^{\mathrm{T}}[I_q + Wt + \tfrac{1}{2} W^2 t^2 + \cdots] \\ &= p^{\mathrm{T}}[I_q - \alpha t I_q + \tfrac{1}{2} \alpha^2 t^2 I_q + \cdots] \\ &= e^{-\alpha t} p^{\mathrm{T}}, \ t \in \mathbb{R}. \end{aligned} \tag{10.79}$$

Using (10.79), we define the (scalar) *available storage* for the large-scale impulsive dynamical system $\mathcal{G}$ by

$$\begin{aligned} v_{\mathrm{a}}(x_0) &\triangleq - \inf_{(u_{\mathrm{c}}(\cdot), u_{\mathrm{d}}(\cdot)), \, T \geq t_0} \Bigg[\int_{t_0}^{T} p^{\mathrm{T}} e^{-W(t - t_0)} S_{\mathrm{c}}(u_{\mathrm{c}}(t), y_{\mathrm{c}}(t)) \mathrm{d}t \\ &\qquad + \sum_{k \in \mathbb{Z}_{[t_0, T)}} p^{\mathrm{T}} e^{-W(t_k - t_0)} S_{\mathrm{d}}(u_{\mathrm{d}}(t_k), y_{\mathrm{d}}(t_k)) \Bigg] \\ &= - \inf_{(u_{\mathrm{c}}(\cdot), u_{\mathrm{d}}(\cdot)), \, T \geq t_0} \Bigg[\int_{t_0}^{T} e^{\alpha(t - t_0)} s_{\mathrm{c}}(u_{\mathrm{c}}(t), y_{\mathrm{c}}(t)) \mathrm{d}t \\ &\qquad + \sum_{k \in \mathbb{Z}_{[t_0, T)}} e^{\alpha(t_k - t_0)} s_{\mathrm{d}}(u_{\mathrm{d}}(t_k), y_{\mathrm{d}}(t_k)) \Bigg], \end{aligned} \tag{10.80}$$

where $s_{\mathrm{c}} : U_{\mathrm{c}} \times Y_{\mathrm{c}} \to \mathbb{R}$ and $s_{\mathrm{d}} : U_{\mathrm{d}} \times Y_{\mathrm{d}} \to \mathbb{R}$ defined as $s_{\mathrm{c}}(u_{\mathrm{c}}, y_{\mathrm{c}}) \triangleq p^{\mathrm{T}} S_{\mathrm{c}}(u_{\mathrm{c}}, y_{\mathrm{c}})$ and $s_{\mathrm{d}}(u_{\mathrm{d}}, y_{\mathrm{d}}) \triangleq p^{\mathrm{T}} S_{\mathrm{d}}(u_{\mathrm{d}}, y_{\mathrm{d}})$ form the (scalar) hybrid supply rate $(s_{\mathrm{c}}, s_{\mathrm{d}})$ for the large-scale impulsive dynamical system $\mathcal{G}$. Clearly, $v_{\mathrm{a}}(x) \geq 0$ for all $x \in \mathcal{D}$. As in standard hybrid dissipativity theory [74, 82], the available storage $v_{\mathrm{a}}(x)$, $x \in \mathcal{D}$, denotes the maximum amount of (scaled) energy that can be extracted from the large-scale impulsive dynamical system $\mathcal{G}$ at any time T.

The following theorem relates vector storage functions and vector hybrid supply rates to scalar storage functions and scalar hybrid supply rates of large-scale impulsive dynamical systems.

Theorem 10.7. Consider the large-scale impulsive dynamical system $\mathcal{G}$ given by (10.54)–(10.57). Suppose $\mathcal{G}$ is vector dissipative (respectively, exponentially vector dissipative) with respect to the vector hybrid supply rate $(S_{\mathrm{c}}(u_{\mathrm{c}}, y_{\mathrm{c}}), S_{\mathrm{d}}(u_{\mathrm{d}}, y_{\mathrm{d}})) : (\mathcal{U}_{\mathrm{c}} \times Y_{\mathrm{c}}, \mathcal{U}_{\mathrm{d}} \times Y_{\mathrm{d}}) \to \mathbb{R}^q \times \mathbb{R}^q$ and with vector storage function $V_{\mathrm{s}} : \mathcal{D} \to \overline{\mathbb{R}}_+^q$. Then there exists $p \in \overline{\mathbb{R}}_+^q$, $p \neq 0$, (respectively, $p \in \mathbb{R}_+^q$) such that $\mathcal{G}$ is dissipative (respectively, exponentially dissipative) with respect to the scalar hybrid supply rate $(s_{\mathrm{c}}(u_{\mathrm{c}}, y_{\mathrm{c}}), s_{\mathrm{d}}(u_{\mathrm{d}}, y_{\mathrm{d}})) = (p^{\mathrm{T}} S_{\mathrm{c}}(u_{\mathrm{c}}, y_{\mathrm{c}}), p^{\mathrm{T}} S_{\mathrm{d}}(u_{\mathrm{d}}, y_{\mathrm{d}}))$ and with storage function $v_{\mathrm{s}}(x) = p^{\mathrm{T}} V_{\mathrm{s}}(x)$, $x \in \mathcal{D}$. Moreover, in this case $v_{\mathrm{a}}(x)$, $x \in \mathcal{D}$, is a storage function for $\mathcal{G}$ and

$$0 \leq v_{\mathrm{a}}(x) \leq v_{\mathrm{s}}(x), \quad x \in \mathcal{D}. \tag{10.81}$$

Proof. Suppose $\mathcal{G}$ is vector dissipative (respectively, exponentially vector dissipative) with respect to the vector hybrid supply rate $(S_{\mathrm{c}}(u_{\mathrm{c}}, y_{\mathrm{c}}), S_{\mathrm{d}}(u_{\mathrm{d}}, y_{\mathrm{d}}))$. Then there exist an essentially nonnegative, semistable (respectively, asymptotically stable) dissipation matrix W and a vector storage function $V_{\mathrm{s}} : \mathcal{D} \to \overline{\mathbb{R}}_+^q$ such that the dissipation inequality (10.65) holds. Furthermore, it follows from Lemma 2.1 that there exist $\alpha \geq 0$ (respectively, $\alpha > 0$) and a nonzero vector $p \in \overline{\mathbb{R}}_+^q$ (respectively, $p \in \mathbb{R}_+^q$) satisfying (10.79). Hence, premultiplying (10.65) by p^{T} and using (10.79) it follows that

$$e^{\alpha T} v_{\mathrm{s}}(x(T)) \leq e^{\alpha t_0} v_{\mathrm{s}}(x(t_0)) + \int_{t_0}^{T} e^{\alpha t} s_{\mathrm{c}}(u_{\mathrm{c}}(t), y_{\mathrm{c}}(t)) \mathrm{d}t$$

$$+ \sum_{k \in \mathbb{Z}_{[t_0, T)}} e^{\alpha t_k} s_{\mathrm{d}}(u_{\mathrm{d}}(t_k), y_{\mathrm{d}}(t_k)),$$

$$T \geq t_0, \quad (u_{\mathrm{c}}(\cdot), u_{\mathrm{d}}(\cdot)) \in \mathcal{U}_{\mathrm{c}} \times \mathcal{U}_{\mathrm{d}}, \tag{10.82}$$

where $v_{\mathrm{s}}(x) = p^{\mathrm{T}} V_{\mathrm{s}}(x)$, $x \in \mathcal{D}$, which implies dissipativity (respectively, exponential dissipativity) of $\mathcal{G}$ with respect to the scalar hybrid supply rate $(s_{\mathrm{c}}(u_{\mathrm{c}}, y_{\mathrm{c}}), s_{\mathrm{d}}(u_{\mathrm{d}}, y_{\mathrm{d}}))$ and with storage function $v_{\mathrm{s}}(x)$, $x \in \mathcal{D}$. Moreover, since $v_{\mathrm{s}}(0) = 0$, it follows from (10.82) that for $x(t_0) = 0$,

$$\int_{t_0}^{T} e^{\alpha(t - t_0)} s_{\mathrm{c}}(u_{\mathrm{c}}(t), y_{\mathrm{c}}(t)) \mathrm{d}t + \sum_{k \in \mathbb{Z}_{[t_0, T)}} e^{\alpha(t_k - t_0)} s_{\mathrm{d}}(u_{\mathrm{d}}(t_k), y_{\mathrm{d}}(t_k)) \geq 0,$$

$$T \geq t_0, \quad (u_{\mathrm{c}}(\cdot), u_{\mathrm{d}}(\cdot)) \in \mathcal{U}_{\mathrm{c}} \times \mathcal{U}_{\mathrm{d}}, \tag{10.83}$$

which, using (10.80), implies that $v_{\mathrm{a}}(0) = 0$. Now, it can be easily shown that $v_{\mathrm{a}}(x)$, $x \in \mathcal{D}$, satisfies (10.82), and hence, the available storage defined by (10.80) is a storage function for $\mathcal{G}$.

Finally, it follows from (10.82) that

$$v_{\mathrm{s}}(x(t_0)) \geq e^{\alpha(T - t_0)} v_{\mathrm{s}}(x(T)) - \int_{t_0}^{T} e^{\alpha(t - t_0)} s_{\mathrm{c}}(u_{\mathrm{c}}(t), y_{\mathrm{c}}(t)) \mathrm{d}t$$

$$- \sum_{k \in \mathbb{Z}_{[t_0,T)}} e^{\alpha(t_k - t_0)} s_\mathrm{d}(u_\mathrm{d}(t_k), y_\mathrm{d}(t_k))$$

$$\geq - \int_{t_0}^{T} e^{\alpha(t - t_0)} s_\mathrm{c}(u(t), y(t)) \mathrm{d}t$$

$$- \sum_{k \in \mathbb{Z}_{[t_0,T)}} e^{\alpha(t_k - t_0)} s_\mathrm{d}(u_\mathrm{d}(t_k), y_\mathrm{d}(t_k)),$$

$$T \geq t_0, \quad (u_\mathrm{c}(\cdot), u_\mathrm{d}(\cdot)) \in \mathcal{U}_\mathrm{c} \times \mathcal{U}_\mathrm{d}, \quad (10.84)$$

which implies

$$v_\mathrm{s}(x(t_0)) \geq - \inf_{(u_\mathrm{c}(\cdot), u_\mathrm{d}(\cdot)), \, T \geq t_0} \left[\int_{t_0}^{T} e^{\alpha(t - t_0)} s_\mathrm{c}(u_\mathrm{c}(t), y_\mathrm{c}(t)) \mathrm{d}t \right.$$

$$\left. + \sum_{k \in \mathbb{Z}_{[t_0,T)}} e^{\alpha(t_k - t_0)} s_\mathrm{d}(u_\mathrm{d}(t_k), y_\mathrm{d}(t_k)) \right]$$

$$= v_\mathrm{a}(x(t_0)), \quad (10.85)$$

and hence, (10.81) holds. $\qquad\qquad\qquad\qquad\qquad\qquad\qquad\qquad\qquad\square$

It follows from Theorem 10.6 that if (10.71) holds for $x(t_0) = 0$, then the vector available storage $V_\mathrm{a}(x)$, $x \in \mathcal{D}$, is a vector storage function for $\mathcal{G}$. In this case, it follows from Theorem 10.7 that there exists $p \in \overline{\mathbb{R}}_+^q$, $p \neq 0$, such that $v_\mathrm{s}(x) \triangleq p^\mathrm{T} V_\mathrm{a}(x)$ is a storage function for $\mathcal{G}$ that satisfies (10.82), and hence, by (10.81), $v_\mathrm{a}(x) \leq p^\mathrm{T} V_\mathrm{a}(x)$, $x \in \mathcal{D}$. Furthermore, it is important to note that it follows from Theorem 10.7 that if $\mathcal{G}$ is vector dissipative, then $\mathcal{G}$ can either be (scalar) dissipative or (scalar) exponentially dissipative.

The following theorem provides sufficient conditions guaranteeing that all scalar storage functions defined in terms of vector storage functions, that is, $v_\mathrm{s}(x) = p^\mathrm{T} V_\mathrm{s}(x)$, of a given vector dissipative large-scale impulsive nonlinear dynamical system are positive definite.

Theorem 10.8. Consider the large-scale impulsive dynamical system $\mathcal{G}$ given by (10.54)–(10.57) and assume that $\mathcal{G}$ is zero-state observable. Furthermore, assume that $\mathcal{G}$ is vector dissipative (respectively, exponentially vector dissipative) with respect to the vector hybrid supply rate $(S_\mathrm{c}(u_\mathrm{c}, y_\mathrm{c}),$ $S_\mathrm{d}(u_\mathrm{d}, y_\mathrm{d}))$ and there exist $\alpha \geq 0$ and $p \in \mathbb{R}_+^q$ such that (2.4) holds. In addition, assume that there exist functions $\kappa_{ci} : Y_{ci} \to U_{ci}$ and $\kappa_{di} : Y_{di} \to U_{di}$ such that $\kappa_{ci}(0) = 0$, $\kappa_{di}(0) = 0$, $s_{ci}(\kappa_{ci}(y_{ci}), y_{ci}) < 0$, $y_{ci} \neq 0$, and $s_{di}(\kappa_{di}(y_{di}), y_{di}) < 0$, $y_{di} \neq 0$, for all $i = 1, \ldots, q$. Then for all vector storage functions $V_\mathrm{s} : \mathcal{D} \to \overline{\mathbb{R}}_+^q$ the storage function $v_\mathrm{s}(x) \triangleq p^\mathrm{T} V_\mathrm{s}(x)$, $x \in \mathcal{D}$, is positive definite, that is, $v_\mathrm{s}(0) = 0$ and $v_\mathrm{s}(x) > 0$, $x \in \mathcal{D}$, $x \neq 0$.

Proof. It follows from Theorem 10.7 that $v_a(x)$, $x \in \mathcal{D}$, is a storage function for $\mathcal{G}$ that satisfies (10.82). Next, suppose, *ad absurdum*, that there exists $x \in \mathcal{D}$ such that $v_a(x) = 0$, $x \neq 0$. Then it follows from the definition of $v_a(x)$, $x \in \mathcal{D}$, that for $x(t_0) = x$,

$$\int_{t_0}^{T} e^{\alpha(t-t_0)} s_c(u_c(t), y_c(t))dt + \sum_{k \in \mathbb{Z}_{[t_0,T)}} e^{\alpha(t_k - t_0)} s_d(u_d(t_k), y_d(t_k)) \geq 0,$$

$$T \geq t_0, \quad (u_c(\cdot), u_d(\cdot)) \in \mathcal{U}_c \times \mathcal{U}_d. \tag{10.86}$$

However, for $u_{ci} = \kappa_{ci}(y_{ci})$ and $u_{di} = \kappa_{di}(y_{di})$ we have $s_{ci}(\kappa_{ci}(y_{ci}), y_{ci}) < 0$, $s_{di}(\kappa_{di}(y_{di}), y_{di}) < 0$, $y_{ci} \neq 0$, $y_{di} \neq 0$ for all $i = 1, \ldots, q$, and since $p >> 0$, it follows that $y_{ci}(t) = 0, t_k < t \leq t_{k+1}, y_{di}(t_k) = 0, k \in \overline{\mathbb{Z}}_+, i = 1, \ldots, q$, which further implies that $u_{ci}(t) = 0, t_k < t \leq t_{k+1}$, and $u_{di}(t_k) = 0, k \in \overline{\mathbb{Z}}_+, i = 1, \ldots, q$. Since $\mathcal{G}$ is zero-state observable it follows that $x = 0$, and hence, $v_a(x) = 0$ if and only if $x = 0$. The result now follows from (10.81). Finally, for the exponentially vector dissipative case it follows from Lemma 2.1 that $p >> 0$, with the rest of the proof being identical to that above. $\square$

Next, we introduce the concept of *vector required supply* of a large-scale impulsive dynamical system. Specifically, define the vector required supply of the large-scale impulsive dynamical system $\mathcal{G}$ by

$$V_r(x_0) \triangleq \inf_{(u_c(\cdot), u_d(\cdot)), \, T \leq t_0} \left[\int_T^{t_0} e^{-W(t-t_0)} S_c(u_c(t), y_c(t))dt \right.$$

$$\left. + \sum_{k \in \mathbb{Z}_{[T,t_0)}} e^{-W(t_k - t_0)} S_d(u_d(t_k), y_d(t_k)) \right],$$

$$\tag{10.87}$$

where $x(t)$, $t \geq T$, is the solution to (10.54)–(10.57) with $x(T) = 0$ and $x(t_0) = x_0$. Note that since, with $x(t_0) = 0$, the infimum in (10.87) is the zero vector, it follows that $V_r(0) = 0$. Moreover, since $\mathcal{G}$ is completely reachable it follows that $V_r(x) << \infty$, $x \in \mathcal{D}$. Using the notion of the vector required supply we present necessary and sufficient conditions for vector dissipativity of a large-scale impulsive dynamical system with respect to a vector hybrid supply rate.

Theorem 10.9. Consider the large-scale impulsive dynamical system $\mathcal{G}$ given by (10.54)–(10.57) and assume that $\mathcal{G}$ is completely reachable. Then $\mathcal{G}$ is vector dissipative (respectively, exponentially vector dissipative) with respect to the vector hybrid supply rate $(S_c(u_c, y_c), S_d(u_d, y_d))$ if and only if

$$0 \leq\leq V_r(x) << \infty, \quad x \in \mathcal{D}. \tag{10.88}$$

Moreover, if (10.88) holds, then $V_{\mathrm{r}}(x)$, $x \in \mathcal{D}$, is a vector storage function for $\mathcal{G}$. Finally, if the vector available storage $V_{\mathrm{a}}(x)$, $x \in \mathcal{D}$, is a vector storage function for $\mathcal{G}$, then

$$0 \leq\leq V_{\mathrm{a}}(x) \leq\leq V_{\mathrm{r}}(x) << \infty, \quad x \in \mathcal{D}. \tag{10.89}$$

Proof. Suppose (10.88) holds and let $x(t)$, $t \in \mathbb{R}$, satisfy (10.54)–(10.57) with admissible inputs $(u_{\mathrm{c}}(\cdot), u_{\mathrm{d}}(\cdot)) \in \mathcal{U}_{\mathrm{c}} \times \mathcal{U}_{\mathrm{d}}$ and $x(t_0) = x_0$. Then, it follows from the definition of $V_{\mathrm{r}}(\cdot)$ that for $T \leq t_{\mathrm{f}} \leq t_0$, $u_{\mathrm{c}}(\cdot) \in \mathcal{U}_{\mathrm{c}}$, and $u_{\mathrm{d}}(\cdot) \in \mathcal{U}_{\mathrm{d}}$,

$$
\begin{aligned}
V_{\mathrm{r}}(x_0) \leq\leq &\int_T^{t_0} e^{-W(t-t_0)} S_{\mathrm{c}}(u_{\mathrm{c}}(t), y_{\mathrm{c}}(t)) \mathrm{d}t \\
&+ \sum_{k \in \mathbb{Z}_{[T,t_0)}} e^{-W(t_k-t_0)} S_{\mathrm{d}}(u_{\mathrm{d}}(t_k), y_{\mathrm{d}}(t_k)) \\
= &\int_T^{t_{\mathrm{f}}} e^{-W(t-t_0)} S_{\mathrm{c}}(u_{\mathrm{c}}(t), y_{\mathrm{c}}(t)) \mathrm{d}t \\
&+ \sum_{k \in \mathbb{Z}_{[T,t_{\mathrm{f}})}} e^{-W(t_k-t_0)} S_{\mathrm{d}}(u_{\mathrm{d}}(t_k), y_{\mathrm{d}}(t_k)) \\
&+ \int_{t_{\mathrm{f}}}^{t_0} e^{-W(t-t_0)} S_{\mathrm{c}}(u_{\mathrm{c}}(t), y_{\mathrm{c}}(t)) \mathrm{d}t \\
&+ \sum_{k \in \mathbb{Z}_{[t_{\mathrm{f}},t_0)}} e^{-W(t_k-t_0)} S_{\mathrm{d}}(u_{\mathrm{d}}(t_k), y_{\mathrm{d}}(t_k)), \tag{10.90}
\end{aligned}
$$

and hence,

$$
\begin{aligned}
V_{\mathrm{r}}(x_0) \leq\leq\; & e^{W(t_0-t_{\mathrm{f}})} \inf_{(u_{\mathrm{c}}(\cdot), u_{\mathrm{d}}(\cdot)),\, T \leq t_{\mathrm{f}}} \left[\int_T^{t_{\mathrm{f}}} e^{-W(t-t_{\mathrm{f}})} S_{\mathrm{c}}(u_{\mathrm{c}}(t), y_{\mathrm{c}}(t)) \mathrm{d}t \right. \\
& \left. + \sum_{k \in \mathbb{Z}_{[T,t_{\mathrm{f}})}} e^{-W(t_k-t_{\mathrm{f}})} S_{\mathrm{d}}(u_{\mathrm{d}}(t_k), y_{\mathrm{d}}(t_k)) \right] \\
& + \int_{t_{\mathrm{f}}}^{t_0} e^{-W(t-t_0)} S_{\mathrm{c}}(u_{\mathrm{c}}(t), y_{\mathrm{c}}(t)) \mathrm{d}t \\
& + \sum_{k \in \mathbb{Z}_{[t_{\mathrm{f}},t_0)}} e^{-W(t_k-t_0)} S_{\mathrm{d}}(u_{\mathrm{d}}(t_k), y_{\mathrm{d}}(t_k)) \\
=\; & e^{W(t_0-t_{\mathrm{f}})} V_{\mathrm{r}}(x(t_{\mathrm{f}})) + \int_{t_{\mathrm{f}}}^{t_0} e^{-W(t-t_0)} S_{\mathrm{c}}(u_{\mathrm{c}}(t), y_{\mathrm{c}}(t)) \mathrm{d}t \\
& + \sum_{k \in \mathbb{Z}_{[t_{\mathrm{f}},t_0)}} e^{-W(t_k-t_0)} S_{\mathrm{d}}(u_{\mathrm{d}}(t_k), y_{\mathrm{d}}(t_k)), \tag{10.91}
\end{aligned}
$$

which shows that $V_{\rm r}(x)$, $x \in \mathcal{D}$, is a vector storage function for $\mathcal{G}$, and hence, $\mathcal{G}$ is vector dissipative with respect to the vector hybrid supply rate $(S_{\rm c}(u_{\rm c}, y_{\rm c}), S_{\rm d}(u_{\rm d}, y_{\rm d}))$.

Conversely, suppose that $\mathcal{G}$ is vector dissipative with respect to the vector hybrid supply rate $(S_{\rm c}(u_{\rm c}, y_{\rm c}), S_{\rm d}(u_{\rm d}, y_{\rm d}))$. Then there exists a nonnegative vector storage function $V_{\rm s}(x)$, $x \in \mathcal{D}$, such that $V_{\rm s}(0) = 0$. Since $\mathcal{G}$ is completely reachable it follows that for $x(t_0) = x_0$ there exist $T < t_0$ and $u_{\rm c}(t)$, $t \in [T, t_0]$, and $u_{\rm d}(t_k)$, $k \in \mathbb{Z}_{[T,t_0]}$, such that $x(T) = 0$. Hence, it follows from the vector hybrid dissipation inequality (10.65) that

$$0 \leq\leq V_{\rm s}(x(t_0)) \leq\leq e^{W(t_0-T)} V_{\rm s}(x(T)) + \int_T^{t_0} e^{W(t_0-t)} S_{\rm c}(u_{\rm c}(t), y_{\rm c}(t))\mathrm{d}t$$
$$+ \sum_{k \in \mathbb{Z}_{[T,t_0)}} e^{W(t_0-t_k)} S_{\rm d}(u_{\rm d}(t_k), y_{\rm d}(t_k)), \qquad (10.92)$$

which implies that for all $T \leq t_0$, $u_{\rm c}(t) \in U_{\rm c}$, and $u_{\rm d}(t_k) \in U_{\rm d}$,

$$0 \leq\leq \int_T^{t_0} e^{W(t_0-t)} S_{\rm c}(u_{\rm c}(t), y_{\rm c}(t))\mathrm{d}t$$
$$+ \sum_{k \in \mathbb{Z}_{[T,t_0)}} e^{W(t_0-t_k)} S_{\rm d}(u_{\rm d}(t_k), y_{\rm d}(t_k)) \qquad (10.93)$$

or, equivalently,

$$0 \leq\leq \inf_{(u_{\rm c}(\cdot), u_{\rm d}(\cdot)),\, T \leq t_0} \left[\int_T^{t_0} e^{W(t_0-t)} S_{\rm c}(u_{\rm c}(t), y_{\rm c}(t))\mathrm{d}t \right.$$
$$\left. + \sum_{k \in \mathbb{Z}_{[T,t_0)}} e^{W(t_0-t_k)} S_{\rm d}(u_{\rm d}(t_k), y_{\rm d}(t_k)) \right]$$
$$= V_{\rm r}(x_0). \qquad (10.94)$$

Since, by complete reachability, $V_{\rm r}(x) << \infty$, $x \in \mathcal{D}$, it follows that (10.88) holds.

Finally, suppose that $V_{\rm a}(x)$, $x \in \mathcal{D}$, is a vector storage function. Then for $x(T) = 0$, $x(t_0) = x_0$, $u_{\rm c}(\cdot) \in \mathcal{U}_{\rm c}$, and $u_{\rm d}(\cdot) \in \mathcal{U}_{\rm d}$, it follows that

$$V_{\rm a}(x(t_0)) \leq\leq e^{W(t_0-T)} V_{\rm a}(x(T)) + \int_T^{t_0} e^{W(t_0-t)} S_{\rm c}(u_{\rm c}(t), y_{\rm c}(t))\mathrm{d}t$$
$$+ \sum_{k \in \mathbb{Z}_{[T,t_0)}} e^{W(t_0-t_k)} S_{\rm d}(u_{\rm d}(t_k), y_{\rm d}(t_k)), \qquad (10.95)$$

which implies that

$$0 \leq\leq V_{\rm a}(x(t_0))$$

$$\leq\leq \inf_{(u_{\mathrm c}(\cdot),u_{\mathrm d}(\cdot)),\,T\leq t_0}\left[\int_T^{t_0} e^{W(t_0-t)}S_{\mathrm c}(u_{\mathrm c}(t),y_{\mathrm c}(t))\mathrm{d}t\right.$$

$$\left. +\sum_{k\in\mathbb{Z}_{[T,t_0)}} e^{W(t_0-t_k)}S_{\mathrm d}(u_{\mathrm d}(t_k),y_{\mathrm d}(t_k))\right]$$

$$= V_{\mathrm r}(x(t_0)),\quad x\in\mathcal{D}. \tag{10.96}$$

Since $x(t_0)=x_0\in\mathcal{D}$ is arbitrary and, by complete reachability, $V_{\mathrm r}(x)<<\infty$, $x\in\mathcal{D}$, (10.96) implies (10.89). $\qquad\square$

The next result is a direct consequence of Theorems 10.6 and 10.9.

Proposition 10.1. Consider the large-scale impulsive dynamical system $\mathcal{G}$ given by (10.54)–(10.57) and assume $\mathcal{G}$ is completely reachable. Let $M=\operatorname{diag}[\mu_1,\ldots,\mu_q]$ be such that $0\leq\mu_i\leq 1$, $i=1,\ldots,q$. If $V_{\mathrm a}(x)$, $x\in\mathcal{D}$, and $V_{\mathrm r}(x)$, $x\in\mathcal{D}$, are vector storage functions for $\mathcal{G}$, then

$$V_{\mathrm s}(x)=MV_{\mathrm a}(x)+(I_q-M)V_{\mathrm r}(x),\quad x\in\mathcal{D}, \tag{10.97}$$

is a vector storage function for $\mathcal{G}$.

Proof. First note that $M\geq\geq 0$ and $I_q-M\geq\geq 0$ if and only if $M=\operatorname{diag}[\mu_1,\ldots,\mu_q]$ and $\mu_i\in[0,1]$, $i=1,\ldots,q$. Now, the result is a direct consequence of the complete reachability of $\mathcal{G}$ along with vector hybrid dissipation inequality (10.65) by noting that if $V_{\mathrm a}(x)$ and $V_{\mathrm r}(x)$ satisfy (10.65), then $V_{\mathrm s}(x)$ satisfies (10.65). $\qquad\square$

Next, recall that if $\mathcal{G}$ is vector dissipative (respectively, exponentially vector dissipative), then there exist $p\in\overline{\mathbb{R}}_+^q$, $p\neq 0$, and $\alpha\geq 0$ (respectively, $p\in\mathbb{R}_+^q$ and $\alpha>0$) such that (2.4) and (10.79) hold. Now, define the (scalar) *required supply* for the large-scale impulsive dynamical system $\mathcal{G}$ by

$$v_{\mathrm r}(x_0)\triangleq \inf_{(u_{\mathrm c}(\cdot),u_{\mathrm d}(\cdot)),\,T\leq t_0}\left[\int_T^{t_0} p^{\mathrm T}e^{-W(t-t_0)}S_{\mathrm c}(u_{\mathrm c}(t),y_{\mathrm c}(t))\mathrm{d}t\right.$$

$$\left. +\sum_{k\in\mathbb{Z}_{[T,t_0)}} e^{-W(t_k-t_0)}S_{\mathrm d}(u_{\mathrm d}(t_k),y_{\mathrm d}(t_k))\right]$$

$$= \inf_{(u_{\mathrm c}(\cdot),u_{\mathrm d}(\cdot)),\,T\leq t_0}\left[\int_T^{t_0} e^{\alpha(t-t_0)}s_{\mathrm c}(u_{\mathrm c}(t),y_{\mathrm c}(t))\mathrm{d}t\right.$$

$$\left. +\sum_{k\in\mathbb{Z}_{[T,t_0)}} e^{\alpha(t_k-t_0)}s_{\mathrm d}(u_{\mathrm d}(t_k),y_{\mathrm d}(t_k))\right],\quad x_0\in\mathcal{D}, \tag{10.98}$$

where $s_c(u_c, y_c) = p^{\mathrm{T}} S_c(u_c, y_c)$, $s_d(u_d, y_d) = p^{\mathrm{T}} S_d(u_d, y_d)$, and $x(t)$, $t \geq T$, is the solution to (10.54)–(10.57) with $x(T) = 0$ and $x(t_0) = x_0$. It follows from (10.98) that the required supply of a large-scale impulsive dynamical system is the minimum amount of generalized energy that can be delivered to the large-scale system to transfer it from an initial state $x(T) = 0$ to a given state $x(t_0) = x_0$. Using the same arguments as in the case of the vector required supply, it follows that $v_r(0) = 0$ and $v_r(x) < \infty$, $x \in \mathcal{D}$.

Next, using the notion of the required supply, we show that all storage functions of the form $v_s(x) = p^{\mathrm{T}} V_s(x)$, where $p \in \overline{\mathbb{R}}_+^q$, $p \neq 0$, are bounded from above by the required supply and bounded from below by the available storage. Hence, a dissipative large-scale impulsive dynamical system can deliver to its surroundings only a fraction of all of its stored subsystem energies and can store only a fraction of the work done to all of its subsystems.

Corollary 10.2. Consider the large-scale impulsive dynamical system $\mathcal{G}$ given by (10.54)–(10.57). Assume that $\mathcal{G}$ is vector dissipative with respect to the vector hybrid supply rate $(S_c(u_c, y_c), S_d(u_d, y_d))$ and with vector storage function $V_s : \mathcal{D} \to \overline{\mathbb{R}}_+^q$. Then $v_r(x)$, $x \in \mathcal{D}$, is a storage function for $\mathcal{G}$. Moreover, if $v_s(x) \triangleq p^{\mathrm{T}} V_s(x)$, $x \in \mathcal{D}$, where $p \in \overline{\mathbb{R}}_+^q$, $p \neq 0$, then

$$0 \leq v_a(x) \leq v_s(x) \leq v_r(x) < \infty, \quad x \in \mathcal{D}. \tag{10.99}$$

Proof. It follows from Theorem 10.7 that if $\mathcal{G}$ is vector dissipative with respect to the vector hybrid supply rate $(S_c(u_c, y_c), S_d(u_d, y_d))$ and with a vector storage function $V_s : \mathcal{D} \to \overline{\mathbb{R}}_+^q$, then there exists $p \in \overline{\mathbb{R}}_+^q$, $p \neq 0$, such that $\mathcal{G}$ is dissipative with respect to the hybrid supply rate $(s_c(u_c, y_c), s_d(u_d, y_d)) = (p^{\mathrm{T}} S_c(u_c, y_c), p^{\mathrm{T}} S_d(u_d, y_d))$ and with storage function $v_s(x) = p^{\mathrm{T}} V_s(x)$, $x \in \mathcal{D}$. Hence, it follows from (10.82), with $x(T) = 0$ and $x(t_0) = x_0$, that

$$\int_T^{t_0} e^{\alpha(t-t_0)} s_c(u_c(t), y_c(t)) \mathrm{d}t + \sum_{k \in \mathbb{Z}_{[T,t_0)}} e^{\alpha(t_k - t_0)} s_d(u_d(t_k), y_d(t_k)) \geq 0,$$

$$T \leq t_0, \quad (u_c(\cdot), u_d(\cdot)) \in \mathcal{U}_c \times \mathcal{U}_d, \tag{10.100}$$

which implies that $v_r(x_0) \geq 0$, $x_0 \in \mathcal{D}$.

Furthermore, it is easy to see from the definition of the required supply that $v_r(x)$, $x \in \mathcal{D}$, satisfies the dissipation inequality (10.82). Hence, $v_r(x)$, $x \in \mathcal{D}$, is a storage function for $\mathcal{G}$. Moreover, it follows from the dissipation inequality (10.82), with $x(T) = 0$, $x(t_0) = x_0$, $u_c(\cdot) \in \mathcal{U}_c$, and $u_d(\cdot) \in \mathcal{U}_d$, that

$$e^{\alpha t_0} v_s(x(t_0)) \leq e^{\alpha T} v_s(x(T)) + \int_T^{t_0} e^{\alpha t} s_c(u_c(t), y_c(t)) \mathrm{d}t$$

$$+ \sum_{k \in \mathbb{Z}_{[T,t_0)}} e^{\alpha t_k} s_d(u_d(t_k), y_d(t_k))$$

$$= \int_T^{t_0} e^{\alpha t} s_{\mathrm{c}}(u_{\mathrm{c}}(t), y_{\mathrm{c}}(t)) \mathrm{d}t$$

$$+ \sum_{k \in \mathbb{Z}_{[T,t_0)}} e^{\alpha t_k} s_{\mathrm{d}}(u_{\mathrm{d}}(t_k), y_{\mathrm{d}}(t_k)), \tag{10.101}$$

which implies that

$$v_{\mathrm{s}}(x(t_0)) \leq \inf_{(u_{\mathrm{c}}(\cdot), u_{\mathrm{d}}(\cdot)), \, T \leq t_0} \left[\int_T^{t_0} e^{\alpha(t-t_0)} s_{\mathrm{c}}(u_{\mathrm{c}}(t), y_{\mathrm{c}}(t)) \mathrm{d}t \right.$$

$$\left. + \sum_{k \in \mathbb{Z}_{[T,t_0)}} e^{\alpha(t_k - t_0)} s_{\mathrm{d}}(u_{\mathrm{d}}(t_k), y_{\mathrm{d}}(t_k)) \right]$$

$$= v_{\mathrm{r}}(x(t_0)). \tag{10.102}$$

Finally, it follows from Theorem 10.7 that $v_{\mathrm{a}}(x)$, $x \in \mathcal{D}$, is a storage function for $\mathcal{G}$, and hence, using (10.81) and (10.102), (10.99) holds. $\qquad\square$

It follows from Theorem 10.9 that if $\mathcal{G}$ is vector dissipative with respect to the vector hybrid supply rate $(S_{\mathrm{c}}(u_{\mathrm{c}}, y_{\mathrm{c}}), S_{\mathrm{d}}(u_{\mathrm{d}}, y_{\mathrm{d}}))$, then $V_{\mathrm{r}}(x)$, $x \in \mathcal{D}$, is a vector storage function for $\mathcal{G}$ and, by Theorem 10.7, there exists $p \in \overline{\mathbb{R}}_+^q$, $p \neq 0$, such that $v_{\mathrm{s}}(x) \triangleq p^{\mathrm{T}} V_{\mathrm{r}}(x)$, $x \in \mathcal{D}$, is a storage function for $\mathcal{G}$ satisfying (10.82). Hence, it follows from Corollary 10.2 that $p^{\mathrm{T}} V_{\mathrm{r}}(x) \leq v_{\mathrm{r}}(x)$, $x \in \mathcal{D}$. The next result relates the vector (respectively, scalar) available storage and the vector (respectively, scalar) required supply for vector lossless large-scale impulsive dynamical systems.

Theorem 10.10. Consider the large-scale impulsive dynamical system $\mathcal{G}$ given by (10.54)–(10.57). Assume that $\mathcal{G}$ is completely reachable to and from the origin. If $\mathcal{G}$ is vector lossless with respect to the vector hybrid supply rate $(S_{\mathrm{c}}(u_{\mathrm{c}}, y_{\mathrm{c}}), S_{\mathrm{d}}(u_{\mathrm{d}}, y_{\mathrm{d}}))$ and $V_{\mathrm{a}}(x)$, $x \in \mathcal{D}$, is a vector storage function, then $V_{\mathrm{a}}(x) = V_{\mathrm{r}}(x)$, $x \in \mathcal{D}$. Moreover, if $V_{\mathrm{s}}(x)$, $x \in \mathcal{D}$, is a vector storage function, then all (scalar) storage functions of the form $v_{\mathrm{s}}(x) = p^{\mathrm{T}} V_{\mathrm{s}}(x)$, $x \in \mathcal{D}$, where $p \in \overline{\mathbb{R}}_+^q$, $p \neq 0$, are given by

$$v_{\mathrm{s}}(x_0) = v_{\mathrm{a}}(x_0) = v_{\mathrm{r}}(x_0) = - \int_{t_0}^{T_+} e^{\alpha(t-t_0)} s_{\mathrm{c}}(u_{\mathrm{c}}(t), y_{\mathrm{c}}(t)) \mathrm{d}t$$

$$- \sum_{k \in \mathbb{Z}_{[t_0, T_+)}} e^{\alpha(t_k - t_0)} s_{\mathrm{d}}(u_{\mathrm{d}}(t_k), y_{\mathrm{d}}(t_k))$$

$$= \int_{T_-}^{t_0} e^{\alpha(t-t_0)} s_{\mathrm{c}}(u_{\mathrm{c}}(t), y_{\mathrm{c}}(t)) \mathrm{d}t$$

$$+ \sum_{k \in \mathbb{Z}_{[T_-, t_0)}} e^{\alpha(t_k - t_0)} s_{\mathrm{d}}(u_{\mathrm{d}}(t_k), y_{\mathrm{d}}(t_k)), \tag{10.103}$$

where $x(t)$, $t \geq t_0$, is the solution to (10.54)–(10.57) with $u_c(\cdot) \in \mathcal{U}_c$, $u_d(\cdot) \in \mathcal{U}_d$, $x(t_0) = x_0 \in \mathcal{D}$, $s_c(u_c, y_c) = p^T S_c(u_c, y_c)$, and $s_d(u_d, y_d) = p^T S_d(u_d, y_d)$, for every $T_+ > t_0$ and $T_- < t_0$ such that $x(T_+) = 0$ and $x(T_-) = 0$.

Proof. Suppose $\mathcal{G}$ is vector lossless with respect to the vector hybrid supply rate $(S_c(u_c, y_c), S_d(u_d, y_d))$. Since $\mathcal{G}$ is completely reachable to and from the origin it follows that for every $x_0 = x(t_0) \in \mathcal{D}$ there exist $T_+ > t_0$, $T_- < t_0$, $u_c(t)$, $t \in [T_-, T_+]$, and $u_d(t_k)$, $k \in \mathbb{Z}_{[T_-, T_+]}$, such that $x(T_-) = 0$, $x(T_+) = 0$, and $x(t_0) = x_0$. Now, it follows from the dissipation inequality (10.65), which is satisfied as an equality, that

$$
0 = \int_{T_-}^{T_+} e^{W(T_+ - t)} S_c(u_c(t), y_c(t)) \mathrm{d}t
$$
$$
+ \sum_{k \in \mathbb{Z}_{[T_-, T_+)}} e^{W(T_+ - t_k)} S_d(u_d(t_k), y_d(t_k)), \tag{10.104}
$$

or, equivalently,

$$
0 = \int_{T_-}^{T_+} e^{-W(t - t_0)} S_c(u_c(t), y_c(t)) \mathrm{d}t
$$
$$
+ \sum_{k \in \mathbb{Z}_{[T_-, T_+)}} e^{-W(t_k - t_0)} S_d(u_d(t_k), y_d(t_k))
$$
$$
= \int_{T_-}^{t_0} e^{-W(t - t_0)} S_c(u_c(t), y_c(t)) \mathrm{d}t
$$
$$
+ \sum_{k \in \mathbb{Z}_{[T_-, t_0)}} e^{-W(t_k - t_0)} S_d(u_d(t_k), y_d(t_k))
$$
$$
+ \int_{t_0}^{T_+} e^{-W(t - t_0)} S_c(u_c(t), y_c(t)) \mathrm{d}t
$$
$$
+ \sum_{k \in \mathbb{Z}_{[t_0, T_+)}} e^{-W(t_k - t_0)} S_d(u_d(t_k), y_d(t_k))
$$
$$
\geq\geq \inf_{(u_c(\cdot), u_d(\cdot)),\, T_- \leq t_0} \left[\int_{T_-}^{t_0} e^{-W(t - t_0)} S_c(u_c(t), y_c(t)) \mathrm{d}t \right.
$$
$$
\left. + \sum_{k \in \mathbb{Z}_{[T_-, t_0)}} e^{-W(t_k - t_0)} S_d(u_d(t_k), y_d(t_k)) \right]
$$
$$
+ \inf_{(u_c(\cdot), u_d(\cdot)),\, T_+ \geq t_0} \left[\int_{t_0}^{T_+} e^{-W(t - t_0)} S_c(u_c(t), y_c(t)) \mathrm{d}t \right.
$$

$$+ \sum_{k\in\mathbb{Z}_{[t_0,T_+)}} e^{-W(t_k-t_0)} S_{\mathrm{d}}(u_{\mathrm{d}}(t_k), y_{\mathrm{d}}(t_k)) \Bigg]$$

$$= V_{\mathrm{r}}(x_0) - V_{\mathrm{a}}(x_0), \tag{10.105}$$

which implies that $V_{\mathrm{r}}(x_0) \leq\leq V_{\mathrm{a}}(x_0)$, $x_0 \in \mathcal{D}$. However, it follows from Theorem 10.9 that if $\mathcal{G}$ is vector dissipative and $V_{\mathrm{a}}(x)$, $x \in \mathcal{D}$, is a vector storage function, then $V_{\mathrm{a}}(x) \leq\leq V_{\mathrm{r}}(x)$, $x \in \mathcal{D}$, which along with (10.105) implies that $V_{\mathrm{a}}(x) = V_{\mathrm{r}}(x)$, $x \in \mathcal{D}$.

Next, since $\mathcal{G}$ is vector lossless there exist a nonzero vector $p \in \overline{\mathbb{R}}_+^q$ and a scalar $\alpha \geq 0$ satisfying (2.4). Now, it follows from (10.104) that

$$\begin{aligned}
0 =& \int_{T_-}^{T_+} p^{\mathrm{T}} e^{-W(t-t_0)} S_{\mathrm{c}}(u_{\mathrm{c}}(t), y_{\mathrm{c}}(t))\mathrm{d}t \\
&+ \sum_{k\in\mathbb{Z}_{[T_-,T_+)}} p^{\mathrm{T}} e^{-W(t_k-t_0)} S_{\mathrm{d}}(u_{\mathrm{d}}(t_k), y_{\mathrm{d}}(t_k)) \\
=& \int_{T_-}^{T_+} e^{\alpha(t-t_0)} s_{\mathrm{c}}(u_{\mathrm{c}}(t), y_{\mathrm{c}}(t))\mathrm{d}t \\
&+ \sum_{k\in\mathbb{Z}_{[T_-,T_+)}} e^{\alpha(t_k-t_0)} s_{\mathrm{d}}(u_{\mathrm{d}}(t_k), y_{\mathrm{d}}(t_k)) \\
=& \int_{T_-}^{t_0} e^{\alpha(t-t_0)} s_{\mathrm{c}}(u_{\mathrm{c}}(t), y_{\mathrm{c}}(t))\mathrm{d}t \\
&+ \sum_{k\in\mathbb{Z}_{[T_-,t_0)}} e^{\alpha(t_k-t_0)} s_{\mathrm{d}}(u_{\mathrm{d}}(t_k), y_{\mathrm{d}}(t_k)) \\
&+ \int_{t_0}^{T_+} e^{\alpha(t-t_0)} s_{\mathrm{c}}(u_{\mathrm{c}}(t), y_{\mathrm{c}}(t))\mathrm{d}t \\
&+ \sum_{k\in\mathbb{Z}_{[t_0,T_+)}} e^{\alpha(t_k-t_0)} s_{\mathrm{d}}(u_{\mathrm{d}}(t_k), y_{\mathrm{d}}(t_k)) \\
\geq& \inf_{(u_{\mathrm{c}}(\cdot),u_{\mathrm{d}}(\cdot)),\, T_-\leq t_0} \Bigg[\int_{T_-}^{t_0} e^{\alpha(t-t_0)} s_{\mathrm{c}}(u_{\mathrm{c}}(t), y_{\mathrm{c}}(t))\mathrm{d}t \\
&+ \sum_{k\in\mathbb{Z}_{[T_-,t_0)}} e^{\alpha(t_k-t_0)} s_{\mathrm{d}}(u_{\mathrm{d}}(t_k), y_{\mathrm{d}}(t_k)) \Bigg] \\
&+ \inf_{(u_{\mathrm{c}}(\cdot),u_{\mathrm{d}}(\cdot)),\, T_+\geq t_0} \Bigg[\int_{t_0}^{T_+} e^{\alpha(t-t_0)} s_{\mathrm{c}}(u_{\mathrm{c}}(t), y_{\mathrm{c}}(t))\mathrm{d}t
\end{aligned}$$

$$+ \sum_{k \in \mathbb{Z}_{[t_0, T_+)}} e^{\alpha(t_k - t_0)} s_\mathrm{d}(u_\mathrm{d}(t_k), y_\mathrm{d}(t_k)) \Bigg]$$

$$= v_\mathrm{r}(x_0) - v_\mathrm{a}(x_0), \quad x_0 \in \mathcal{D}, \tag{10.106}$$

which along with (10.99) implies that for any (scalar) storage function of the form $v_\mathrm{s}(x) = p^\mathrm{T} V_\mathrm{s}(x)$, $x \in \mathcal{D}$, the equality $v_\mathrm{a}(x) = v_\mathrm{s}(x) = v_\mathrm{r}(x)$, $x \in \mathcal{D}$, holds. Moreover, since $\mathcal{G}$ is vector lossless the inequalities (10.82) and (10.101) are satisfied as equalities and

$$v_\mathrm{s}(x_0) = - \int_{t_0}^{T_+} e^{\alpha(t - t_0)} s_\mathrm{c}(u_\mathrm{c}(t), y_\mathrm{c}(t)) \mathrm{d}t$$

$$- \sum_{k \in \mathbb{Z}_{[t_0, T_+)}} e^{\alpha(t_k - t_0)} s_\mathrm{d}(u_\mathrm{d}(t_k), y_\mathrm{d}(t_k))$$

$$= \int_{T_-}^{t_0} e^{\alpha(t - t_0)} s_\mathrm{c}(u_\mathrm{c}(t), y_\mathrm{c}(t)) \mathrm{d}t$$

$$+ \sum_{k \in \mathbb{Z}_{[T_-, t_0)}} e^{\alpha(t_k - t_0)} s_\mathrm{d}(u_\mathrm{d}(t_k), y_\mathrm{d}(t_k)), \tag{10.107}$$

where $x(t)$, $t \geq t_0$, is the solution to (10.54)–(10.57) with $u_\mathrm{c}(\cdot) \in \mathcal{U}_\mathrm{c}$, $u_\mathrm{d}(\cdot) \in \mathcal{U}_\mathrm{d}$, $x(T_-) = 0$, $x(T_+) = 0$, and $x(t_0) = x_0 \in \mathcal{D}$. $\square$

The next proposition presents a characterization for vector dissipativity of large-scale impulsive dynamical systems in the case where $V_\mathrm{s}(\cdot)$ is continuously differentiable.

Proposition 10.2. Consider the large-scale impulsive dynamical system $\mathcal{G}$ given by (10.54)–(10.57), assume $V_\mathrm{s} = [v_{\mathrm{s}1}, \ldots, v_{\mathrm{s}q}]^\mathrm{T} : \mathcal{D} \to \overline{\mathbb{R}}_+^q$ is a continuously differentiable vector storage function for $\mathcal{G}$, and assume $\mathcal{G}$ is completely reachable. Then $\mathcal{G}$ is vector dissipative with respect to the vector hybrid supply rate $(S_\mathrm{c}(u_\mathrm{c}, y_\mathrm{c}), S_\mathrm{d}(u_\mathrm{d}, y_\mathrm{d}))$ if and only if

$$\dot{V}_\mathrm{s}(x(t)) \leq\leq W V_\mathrm{s}(x(t)) + S_\mathrm{c}(u_\mathrm{c}(t), y_\mathrm{c}(t)), \quad t_k < t \leq t_{k+1}, \tag{10.108}$$

$$V_\mathrm{s}(x(t_k) + F_\mathrm{d}(x(t_k), u_\mathrm{d}(t_k))) \leq\leq V_\mathrm{s}(x(t_k)) + S_\mathrm{d}(u_\mathrm{d}(t_k), y_\mathrm{d}(t_k)), \quad k \in \overline{\mathbb{Z}}_+, \tag{10.109}$$

where $\dot{V}_\mathrm{s}(x(t))$ denotes the total time derivative of each component of $V_\mathrm{s}(\cdot)$ along the state trajectories $x(t)$, $t_k < t \leq t_{k+1}$, of $\mathcal{G}$.

Proof. Suppose $\mathcal{G}$ is vector dissipative with respect to the vector hybrid supply rate $(S_\mathrm{c}(u_\mathrm{c}, y_\mathrm{c}), S_\mathrm{d}(u_\mathrm{d}, y_\mathrm{d}))$ and with a continuously differentiable vector storage function $V_\mathrm{s} : \mathcal{D} \to \overline{\mathbb{R}}_+^q$. Then, with $T = \hat{t}$ and $t_0 = t$, it follows from (10.67) that there exists a nonnegative vector function $l(t, \hat{t}, x_0, u_{\mathrm{c}}(\cdot)) \geq\geq 0, t_{k+1} \geq \hat{t} \geq t > t_k, x_0 \in \mathcal{D}, u_{\mathrm{c}}(\cdot) \in \mathcal{U}_{\mathrm{c}}$, such that

$$V_{\mathrm{s}}(x(\hat{t})) = e^{W(\hat{t}-t)} V_{\mathrm{s}}(x(t)) + \int_t^{\hat{t}} e^{W(\hat{t}-\sigma)} S_{\mathrm{c}}(u_{\mathrm{c}}(\sigma), y_{\mathrm{c}}(\sigma)) \mathrm{d}\sigma$$
$$-l(t, \hat{t}, x_0, u_{\mathrm{c}}(\cdot)), \tag{10.110}$$

or, equivalently,

$$e^{-W\hat{t}} V_{\mathrm{s}}(x(\hat{t})) - e^{-Wt} V_{\mathrm{s}}(x(t)) = \int_t^{\hat{t}} e^{-W\sigma} S_{\mathrm{c}}(u_{\mathrm{c}}(\sigma), y_{\mathrm{c}}(\sigma)) \mathrm{d}\sigma$$
$$-e^{-W\hat{t}} l(t, \hat{t}, x_0, u_{\mathrm{c}}(\cdot)). \tag{10.111}$$

Now, dividing (10.111) by $\hat{t} - t$ and letting $\hat{t} \to t^+$, (10.111) is equivalent to

$$\frac{\mathrm{d}}{\mathrm{d}\sigma} \left[e^{-W\sigma} V_{\mathrm{s}}(x(\sigma)) \right]\big|_{\sigma=t} = e^{-Wt} S_{\mathrm{c}}(u_{\mathrm{c}}(t), y_{\mathrm{c}}(t))$$
$$-e^{-Wt} \lim_{\hat{t}\to t^+} \frac{l(t, \hat{t}, x_0, u_{\mathrm{c}}(\cdot))}{\hat{t} - t}, \tag{10.112}$$

where the limit in (10.112) exists since $V_{\mathrm{s}}(\cdot)$ is assumed to be continuously differentiable. Next, premultiplying (10.112) by e^{Wt}, $t \geq 0$, yields

$$\dot{V}_{\mathrm{s}}(x(t)) - W V_{\mathrm{s}}(x(t)) = S_{\mathrm{c}}(u_{\mathrm{c}}(t), y_{\mathrm{c}}(t)) - \lim_{\hat{t}\to t^+} \frac{l(t, \hat{t}, x_0, u_{\mathrm{c}}(\cdot))}{\hat{t} - t}, \tag{10.113}$$

which, since $\lim_{\hat{t}\to t^+} \frac{l(t,\hat{t},x_0,u_{\mathrm{c}}(\cdot))}{\hat{t}-t} \geq\geq 0$ and t is arbitrary, gives (10.108). Inequality (10.109) is a restatement of (10.68).

The converse is immediate from Theorem 10.5. $\qquad\square$

Recall that if a disconnected subsystem $\mathcal{G}_i$ (i.e., $\mathcal{I}_{ci}(x) \equiv 0$ and $\mathcal{I}_{di}(x) \equiv 0, i \in \{1, \ldots, q\}$) of a large-scale impulsive dynamical system $\mathcal{G}$ is exponentially dissipative (respectively, dissipative) with respect to a hybrid supply rate $(s_{ci}(u_{ci}, y_{ci}), s_{di}(u_{di}, y_{di}))$, then there exist a storage function $v_{si} : \mathbb{R}^{n_i} \to \overline{\mathbb{R}}_+$ and a constant $\varepsilon_i > 0$ (respectively, $\varepsilon_i = 0$), $i = 1, \ldots, q$, such that the dissipation inequality

$$e^{\varepsilon_i T} v_{si}(x(T)) \leq e^{\varepsilon_i t_0} v_{si}(x(t_0)) + \int_{t_0}^T e^{\varepsilon_i t} s_{ci}(u_{ci}(t), y_{ci}(t)) \mathrm{d}t$$
$$+ \sum_{k \in \mathbb{Z}_{[t_0, T)}} e^{\varepsilon_i t_k} s_{di}(u_{di}(t_k), y_{di}(t_k)), \quad T \geq t_0, \tag{10.114}$$

holds. In the case where $v_{si} : \mathbb{R}^{n_i} \to \overline{\mathbb{R}}_+$ is continuously differentiable and $\mathcal{G}$ is completely reachable, (10.114) yields

$$v_{si}'(x_i)(f_{ci}(x_i) + G_{ci}(x_i)u_{ci}) \leq -\varepsilon_i v_{si}(x_i) + s_{ci}(u_{ci}, y_{ci}),$$

$$x \notin \mathcal{Z}_i, \quad u_{ci} \in U_{ci}, \quad (10.115)$$

$$v_{si}(x_i + f_{di}(x_i) + G_{di}(x_i)u_{di}) \le v_{si}(x_i) + s_{di}(u_{di}, y_{di}), \quad x \in \mathcal{Z}_i, \quad u_{di} \in U_{di},$$
$$(10.116)$$

where $\mathcal{Z}_i \triangleq \mathbb{R}^{n_1} \times \cdots \times \mathbb{R}^{n_{i-1}} \times \mathcal{Z}_{x_i} \times \mathbb{R}^{n_{i+1}} \times \cdots \times \mathbb{R}^q \subset \mathbb{R}^n$ and $\mathcal{Z}_{x_i} \subset \mathbb{R}^{n_i}$, $i = 1, \ldots, q$. The next result relates exponential dissipativity with respect to a scalar hybrid supply rate of each disconnected subsystem $\mathcal{G}_i$ of $\mathcal{G}$ with vector dissipativity (or, possibly, exponential vector dissipativity) of $\mathcal{G}$ with respect to a vector hybrid supply rate.

Proposition 10.3. Consider the large-scale impulsive dynamical system $\mathcal{G}$ given by (10.54)–(10.57) with $\mathcal{Z}_x = \cup_{i=1}^q \mathcal{Z}_i$. Assume that $\mathcal{G}$ is completely reachable and each disconnected subsystem $\mathcal{G}_i$ of $\mathcal{G}$ is exponentially dissipative with respect to the hybrid supply rate $(s_{ci}(u_{ci}, y_{ci}), s_{di}(u_{di}, y_{di}))$ and with a continuously differentiable storage function $v_{si} : \mathbb{R}^{n_i} \to \overline{\mathbb{R}}_+$, $i = 1, \ldots, q$. Furthermore, assume that the interconnection functions $\mathcal{I}_{ci} : \mathcal{D} \to \mathbb{R}^{n_i}$ and $\mathcal{I}_{di} : \mathcal{D} \to \mathbb{R}^{n_i}$, $i = 1, \ldots, q$, of $\mathcal{G}$ are such that

$$v_{si}'(x_i)\mathcal{I}_{ci}(x) \le \sum_{j=1}^q \xi_{ij}(x)v_{sj}(x_j), \quad x \notin \mathcal{Z}_x, \quad (10.117)$$
$$v_{si}(x_i + f_{di}(x_i) + \mathcal{I}_{di}(x) + G_{di}(x_i)u_{di}) \le v_{si}(x_i + f_{di}(x_i) + G_{di}(x_i)u_{di}),$$
$$x \in \mathcal{Z}_x, \quad u_{di} \in U_{di}, \quad i = 1, \ldots, q, \quad (10.118)$$

where $\xi_{ij} : \mathcal{D} \to \mathbb{R}$, $i, j = 1, \ldots, q$, are given bounded functions. If $W \in \mathbb{R}^{q \times q}$ is semistable (respectively, asymptotically stable), with

$$W_{(i,j)} = \begin{cases} -\varepsilon_i + \alpha_{ii}, & i = j, \\ \alpha_{ij}, & i \ne j, \end{cases} \quad (10.119)$$

where $\varepsilon_i > 0$ and $\alpha_{ij} \triangleq \max\{0, \sup_{x \in \mathcal{D}} \xi_{ij}(x)\}$, for all $i, j = 1, \ldots, q$, then $\mathcal{G}$ is vector dissipative (respectively, exponentially vector dissipative) with respect to the vector hybrid supply rate

$$(S_c(u_c, y_c), S_d(u_d, y_d)) \triangleq \left(\begin{bmatrix} s_{c1}(u_{c1}, y_{c1}) \\ \vdots \\ s_{cq}(u_{cq}, y_{cq}) \end{bmatrix}, \begin{bmatrix} s_{d1}(u_{d1}, y_{d1}) \\ \vdots \\ s_{dq}(u_{dq}, y_{dq}) \end{bmatrix} \right) \quad (10.120)$$

and with vector storage function $V_s(x) \triangleq [v_{s1}(x_1), \ldots, v_{sq}(x_q)]^{\mathrm{T}}$, $x \in \mathcal{D}$.

Proof. Since each disconnected subsystem $\mathcal{G}_i$ of $\mathcal{G}$ is exponentially dissipative with respect to the hybrid supply rate $s_{ci}(u_{ci}, y_{ci})$, $i = 1, \ldots, q$, it follows from (10.115)–(10.118) that, for all $u_{ci} \in U_{ci}$ and $i = 1, \ldots, q$,

$$\dot{v}_{si}(x_i(t)) = v_{si}'(x_i(t))[f_{ci}(x_i(t)) + \mathcal{I}_{ci}(x(t)) + G_{ci}(x_i(t))u_{ci}(t)]$$

$$\le -\varepsilon_i v_{si}(x_i(t)) + s_{ci}(u_{ci}(t), y_{ci}(t)) + \sum_{j=1}^q \xi_{ij}(x(t))v_{sj}(x_j(t))$$

$$\leq -\varepsilon_i v_{si}(x_i(t)) + s_{ci}(u_{ci}(t), y_{ci}(t)) + \sum_{j=1}^{q} \alpha_{ij} v_{sj}(x_j(t)),$$
$$t_k < t \leq t_{k+1}, \quad (10.121)$$

and

$$v_{si}(x_i(t_k) + f_{di}(x_i(t_k)) + \mathcal{I}_{di}(x(t_k)) + G_{di}(x_i(t_k))u_{di}(t_k))$$
$$\leq v_{si}(x_i(t_k) + f_{di}(x_i(t_k)) + G_{di}(x_i(t_k))u_{di}(t_k))$$
$$\leq v_{si}(x_i(t_k)) + s_{di}(u_{di}(t_k), y_{di}(t_k)), \quad k \in \overline{\mathbb{Z}}_+. \quad (10.122)$$

Now, the result follows from Proposition 10.2 by noting that for all subsystems $\mathcal{G}_i$ of $\mathcal{G}$,

$$\dot{V}_s(x(t)) \leq\leq WV_s(x(t)) + S_c(u_c(t), y_c(t)), \quad t_k < t \leq t_{k+1}, \quad u_c(\cdot) \in \mathcal{U}_c,$$
$$(10.123)$$
$$V_s(x(t_k) + F_d(x(t_k), u_d(t_k))) \leq\leq V_s(x(t_k)) + S_d(u_d(t_k), y_d(t_k)),$$
$$k \in \overline{\mathbb{Z}}_+, \quad u_d(\cdot) \in \mathcal{U}_d, \quad (10.124)$$

where W is essentially nonnegative and, by assumption, semistable (respectively, asymptotically stable), and the vector function $V_s(x) \triangleq [v_{s1}(x_1), \ldots, v_{sq}(x_q)]^T$, for all $x \in \mathcal{D}$, is a vector storage function for $\mathcal{G}$. $\qquad \square$

As a special case of vector dissipativity theory we can analyze the stability of large-scale impulsive dynamical systems. Specifically, assume that the large-scale impulsive dynamical system $\mathcal{G}$ is vector dissipative (respectively, exponentially vector dissipative) with respect to the vector hybrid supply rate $(S_c(u_c, y_c), S_d(u_d, y_d))$ and with a continuously differentiable vector storage function $V_s : \mathcal{D} \to \overline{\mathbb{R}}_+^q$. Moreover, assume that the conditions of Theorem 10.8 are satisfied. Then it follows from Proposition 10.2, with $u_c(t) \equiv 0$, $u_d(t_k) \equiv 0$, $y_c(t) \equiv 0$, and $y_d(t_k) \equiv 0$, that

$$\dot{V}_s(x(t)) \leq\leq WV_s(x(t)), \quad t_k < t \leq t_{k+1} \qquad (10.125)$$
$$V_s(x(t_k) + f_d(x(t_k)) + \mathcal{I}_d(x(t_k))) \leq\leq V_s(x(t_k)), \quad k \in \overline{\mathbb{Z}}_+, \qquad (10.126)$$

where $x(t)$, $t \geq t_0$, is a solution to (10.54)–(10.57) with $x(t_0) = x_0$, $u_c(t) \equiv 0$, and $u_d(t_k) \equiv 0$. Now, it follows from Theorem 10.2, with $w_c(z) = Wz$ and $w_d(z) = 0$, that the zero solution $x(t) \equiv 0$ to (10.54)–(10.57), with $u_c(t) \equiv 0$ and $u_d(t_k) \equiv 0$, is Lyapunov (respectively, asymptotically) stable.

More generally, the problem of control system design for large-scale impulsive dynamical systems can be addressed within the framework of vector dissipativity theory. In particular, suppose that there exists a continuously differentiable vector function $V_s : \mathcal{D} \to \overline{\mathbb{R}}_+^q$ such that $V_s(0) = 0$ and

$$\dot{V}_s(x(t)) \leq\leq \mathcal{F}_c(V_s(x(t)), u_c(t)), \quad t_k < t \leq t_{k+1}, \quad u_c(\cdot) \in \mathcal{U}_c, \quad (10.127)$$

$$V_{\mathrm{s}}(x(t_k)) + F_{\mathrm{d}}(x(t_k), u_{\mathrm{d}}(t_k))) \leq\leq V_{\mathrm{s}}(x(t_k)), \quad k \in \overline{\mathbb{Z}}_+, \quad u_{\mathrm{d}}(\cdot) \in \mathcal{U}_{\mathrm{d}},$$
$$(10.128)$$

where $\mathcal{F}_{\mathrm{c}} : \overline{\mathbb{R}}_+^q \times \mathbb{R}^{m_{\mathrm{c}}} \to \mathbb{R}^q$ and $\mathcal{F}_{\mathrm{c}}(0,0) = 0$. Then the control system design problem for a large-scale impulsive dynamical system reduces to constructing a hybrid *energy* feedback control law $(\phi_{\mathrm{c}}, \phi_{\mathrm{d}}) : \overline{\mathbb{R}}_+^q \times \overline{\mathbb{R}}_+^q \to U_{\mathrm{c}} \times U_{\mathrm{d}}$ of the form

$$u_{\mathrm{c}} = \phi_{\mathrm{c}}(V_{\mathrm{s}}(x)) \triangleq [\phi_{\mathrm{c}1}^{\mathrm{T}}(V_{\mathrm{s}}(x)), \ldots, \phi_{\mathrm{c}q}^{\mathrm{T}}(V_{\mathrm{s}}(x))]^{\mathrm{T}}, \quad x \notin \mathcal{Z}_x, \qquad (10.129)$$

$$u_{\mathrm{d}} = \phi_{\mathrm{d}}(V_{\mathrm{s}}(x)) \triangleq [\phi_{\mathrm{d}1}^{\mathrm{T}}(V_{\mathrm{s}}(x)), \ldots, \phi_{\mathrm{d}q}^{\mathrm{T}}(V_{\mathrm{s}}(x))]^{\mathrm{T}}, \quad x \in \mathcal{Z}_x, \qquad (10.130)$$

where $\phi_{\mathrm{c}i} : \overline{\mathbb{R}}_+^q \to U_{\mathrm{c}i}$, $\phi_{\mathrm{c}i}(0) = 0$, $\phi_{\mathrm{d}i} : \overline{\mathbb{R}}_+^q \to U_{\mathrm{d}i}$, $i = 1, \ldots, q$, such that the zero solution $z(t) \equiv 0$ to the comparison system

$$\dot{z}(t) = w_{\mathrm{c}}(z(t)), \quad z(t_0) = V_{\mathrm{s}}(x(t_0)), \quad t \geq t_0, \qquad (10.131)$$

is rendered asymptotically stable, where $w_{\mathrm{c}}(z) \triangleq \mathcal{F}_{\mathrm{c}}(z, \phi_{\mathrm{c}}(z))$ is of class $\mathcal{W}$, and Assumptions 10.1 and 10.2 hold. In this case, if there exists $p \in \mathbb{R}_+^q$ such that $v_{\mathrm{s}}(x) \triangleq p^{\mathrm{T}} V_{\mathrm{s}}(x)$, $x \in \mathcal{D}$, is positive definite, then it follows from Theorem 10.2 that the zero solution $x(t) \equiv 0$ to (10.54)–(10.57), with u_{c} and u_{d} given by (10.129) and (10.130), respectively, is asymptotically stable.

As can be seen from the above discussion, using an energy feedback control architecture and exploiting the comparison system within the control design for large-scale impulsive dynamical systems can significantly reduce the dimensionality of a control synthesis problem in terms of the number of states that need to be stabilized. It should be noted, however, that for stability analysis of large-scale impulsive dynamical systems the comparison system need not be linear as implied by (10.125). A nonlinear comparison system would still guarantee stability of a large-scale impulsive dynamical system provided that the conditions of Theorem 10.2 are satisfied.

10.4 Extended Kalman-Yakubovich-Popov Conditions for Large-Scale Impulsive Dynamical Systems

In this section, we show that vector dissipativeness (respectively, exponential vector dissipativeness) of a large-scale impulsive dynamical system $\mathcal{G}$ of the form (10.54)–(10.57) can be characterized in terms of the local subsystem functions $f_{\mathrm{c}i}(\cdot)$, $G_{\mathrm{c}i}(\cdot)$, $h_{\mathrm{c}i}(\cdot)$, $J_{\mathrm{c}i}(\cdot)$, $f_{\mathrm{d}i}(\cdot)$, $G_{\mathrm{d}i}(\cdot)$, $h_{\mathrm{d}i}(\cdot)$, and $J_{\mathrm{d}i}(\cdot)$, along with the interconnection structures $\mathcal{I}_{\mathrm{c}i}(\cdot)$ and $\mathcal{I}_{\mathrm{d}i}(\cdot)$ for $i = 1, \ldots, q$. For the results in this section we consider the special case of dissipative systems with quadratic vector hybrid supply rates and set $\mathcal{D} = \mathbb{R}^n$, $U_{\mathrm{c}i} = \mathbb{R}^{m_{\mathrm{c}i}}$, $U_{\mathrm{d}i} = \mathbb{R}^{m_{\mathrm{d}i}}$, $Y_{\mathrm{c}i} = \mathbb{R}^{l_{\mathrm{c}i}}$, and $Y_{\mathrm{d}i} = \mathbb{R}^{l_{\mathrm{d}i}}$. Furthermore, we assume that $\mathcal{Z} = \mathcal{Z}_x \times \mathbb{R}^{m_{\mathrm{c}}}$, where $\mathcal{Z}_x \subset \mathcal{D}$, so that resetting occurs only when $x(t)$ intersects $\mathcal{Z}_x$. Specifically, let $R_{\mathrm{c}i} \in \mathbb{S}^{m_{\mathrm{c}i}}$, $S_{\mathrm{c}i} \in \mathbb{R}^{l_{\mathrm{c}i} \times m_{\mathrm{c}i}}$, $Q_{\mathrm{c}i} \in \mathbb{S}^{l_{\mathrm{c}i}}$, $R_{\mathrm{d}i} \in \mathbb{S}^{m_{\mathrm{d}i}}$, $S_{\mathrm{d}i} \in \mathbb{R}^{l_{\mathrm{d}i} \times m_{\mathrm{d}i}}$, and $Q_{\mathrm{d}i} \in \mathbb{S}^{l_{\mathrm{d}i}}$ be given, and assume $S_{\mathrm{c}}(u_{\mathrm{c}}, y_{\mathrm{c}})$ is such that $s_{ci}(u_{ci}, y_{ci}) = y_{ci}^T Q_{ci} y_{ci} + 2 y_{ci}^T S_{ci} u_{ci} + u_{ci}^T R_{ci} u_{ci}$ and $S_d(u_d, y_d)$ is such that $s_{di}(u_{di}, y_{di}) = y_{di}^T Q_{di} y_{di} + 2 y_{di}^T S_{di} u_{di} + u_{di}^T R_{di} u_{di}$, $i = 1, \ldots, q$. Furthermore, for the remainder of this chapter we assume that there exists a continuously differentiable vector storage function $V_s(x)$, $x \in \mathbb{R}^n$, for the large-scale impulsive dynamical system $\mathcal{G}$.

For the statement of the next result recall that $x = [x_1^T, \ldots, x_q^T]^T$, $u_c = [u_{c1}^T, \ldots, u_{cq}^T]^T$, $y_c = [y_{c1}^T, \ldots, y_{cq}^T]^T$, $u_d = [u_{d1}^T, \ldots, u_{dq}^T]^T$, $y_d = [y_{d1}^T, \ldots, y_{dq}^T]^T$, $x_i \in \mathbb{R}^{n_i}$, $u_{ci} \in \mathbb{R}^{m_{ci}}$, $y_{ci} \in \mathbb{R}^{l_{ci}}$, $u_{di} \in \mathbb{R}^{m_{di}}$, $y_{di} \in \mathbb{R}^{l_{di}}$, $i = 1, \ldots, q$, $\sum_{i=1}^q n_i = n$, $\sum_{i=1}^q m_{ci} = m_c$, $\sum_{i=1}^q m_{di} = m_d$, $\sum_{i=1}^q l_{ci} = l_c$, and $\sum_{i=1}^q l_{di} = l_d$. Furthermore, for (10.54)–(10.57) define $\mathcal{F}_c : \mathbb{R}^n \to \mathbb{R}^n$, $G_c : \mathbb{R}^n \to \mathbb{R}^{n \times m_c}$, $h_c : \mathbb{R}^n \to \mathbb{R}^{l_c}$, $J_c : \mathbb{R}^n \to \mathbb{R}^{l_c \times m_c}$, $\mathcal{F}_d : \mathbb{R}^n \to \mathbb{R}^n$, $G_d : \mathbb{R}^n \to \mathbb{R}^{n \times m_d}$, $h_d : \mathbb{R}^n \to \mathbb{R}^{l_d}$, and $J_d : \mathbb{R}^n \to \mathbb{R}^{l_d \times m_d}$ by $\mathcal{F}_c(x) \triangleq [\mathcal{F}_{c1}^T(x), \ldots, \mathcal{F}_{cq}^T(x)]^T$, $\mathcal{F}_d(x) \triangleq [\mathcal{F}_{d1}^T(x), \ldots, \mathcal{F}_{dq}^T(x)]^T$, where $\mathcal{F}_{ci}(x) \triangleq f_{ci}(x_i) + \mathcal{I}_{ci}(x)$, $\mathcal{F}_{di}(x) \triangleq f_{di}(x_i) + \mathcal{I}_{di}(x)$, $i = 1, \ldots, q$, $G_c(x) \triangleq \text{block–diag}[G_{c1}(x_1), \ldots, G_{cq}(x_q)]$, $G_d(x) \triangleq \text{block–diag}[G_{d1}(x_1), \ldots, G_{dq}(x_q)]$, $h_c(x) \triangleq [h_{c1}^T(x_1), \ldots, h_{cq}^T(x_q)]^T$, $h_d(x) \triangleq [h_{d1}^T(x_1), \ldots, h_{dq}^T(x_q)]^T$, $J_c(x) \triangleq \text{block–diag}[J_{c1}(x_1), \ldots, J_{cq}(x_q)]$, and $J_d(x) \triangleq \text{block–diag}[J_{d1}(x_1), \ldots, J_{dq}(x_q)]$. Moreover, for all $i = 1, \ldots, q$, define $\hat{R}_{ci} \in \mathbb{S}^{m_c}$, $\hat{S}_{ci} \in \mathbb{R}^{l_c \times m_c}$, $\hat{Q}_{ci} \in \mathbb{S}^{l_c}$, $\hat{R}_{di} \in \mathbb{S}^{m_d}$, $\hat{S}_{di} \in \mathbb{R}^{l_d \times m_d}$, and $\hat{Q}_{di} \in \mathbb{S}^{l_d}$ such that each of these block matrices consists of zero blocks except, respectively, for the matrix blocks $R_{ci} \in \mathbb{S}^{m_{ci}}$, $S_{ci} \in \mathbb{R}^{l_{ci} \times m_{ci}}$, $Q_{ci} \in \mathbb{S}^{l_{ci}}$, $R_{di} \in \mathbb{S}^{m_{di}}$, $S_{di} \in \mathbb{R}^{l_{di} \times m_{di}}$, and $Q_{di} \in \mathbb{S}^{l_{di}}$ on (i, i) position.

The next result introduces a more general definition of vector dissipativity involving an underlying nonlinear comparison system.

Definition 10.6. The large-scale impulsive dynamical system $\mathcal{G}$ given by (10.54)–(10.57) is *vector dissipative* (respectively, *exponentially vector dissipative*) *with respect to the vector hybrid supply rate* $(S_c(u_c, y_c), S_d(u_d, y_d))$ if there exist a continuous, nonnegative definite vector function $V_s = [v_{s1}, \ldots, v_{sq}]^T : \mathcal{D} \to \overline{\mathbb{R}}_+^q$, called a *vector storage function*, and a class $\mathcal{W}$ function $w_c : \overline{\mathbb{R}}_+^q \to \mathbb{R}^q$ such that $V_s(0) = 0$, $w_c(0) = 0$, the zero solution $z(t) \equiv 0$ to the comparison system

$$\dot{z}(t) = w_c(z(t)), \quad z(t_0) = z_0, \quad t \geq t_0, \tag{10.132}$$

is Lyapunov (respectively, asymptotically) stable, and the *vector hybrid dissipation inequality*

$$V_s(x(T)) \leq\leq V_s(x(t_0)) + \int_{t_0}^T w_c(V_s(x(t)))\mathrm{d}t + \int_{t_0}^T S_c(u_c(t), y_c(t))\mathrm{d}t$$

$$+ \sum_{k \in \mathbb{Z}_{[t_0, T)}} S_d(u_d(t_k), y_d(t_k)), \quad T \geq t_0, \tag{10.133}$$

is satisfied, where $x(t)$, $t \geq t_0$, is the solution to (10.54)–(10.57) with $u_c(\cdot) \in$

$\mathcal{U}_c$ and $u_d(\cdot) \in \mathcal{U}_d$. The large-scale impulsive dynamical system $\mathcal{G}$ given by (10.54)–(10.57) is *vector lossless with respect to the vector hybrid supply rate* $(S_c(u_c, y_c), S_d(u_d, y_d))$ if the vector hybrid dissipation inequality is satisfied as an equality with the zero solution $z(t) \equiv 0$ to (10.132) being Lyapunov stable.

If $\mathcal{G}$ is completely reachable and $V_s(\cdot)$ is continuously differentiable, then (10.133) can be equivalently written as

$$\dot{V}_s(x(t)) \leq\leq w_c(V_s(x(t))) + S_c(u_c(t), y_c(t)), \quad t_k < t \leq t_{k+1}, \quad (10.134)$$

$$V_s(x(t_k) + F_d(x(t_k), u_d(t_k))) \leq\leq V_s(x(t_k)) + S_d(u_d(t_k), y_d(t_k)), \quad k \in \overline{\mathbb{Z}}_+, \quad (10.135)$$

with $u_c(\cdot) \in \mathcal{U}_c$ and $u_d(\cdot) \in \mathcal{U}_d$. If in Definition 10.6 the function $w_c : \overline{\mathbb{R}}_+^q \to \mathbb{R}^q$ is such that $w_c(z) = Wz$, where $W \in \mathbb{R}^{q \times q}$, then W is essentially nonnegative and Definition 10.6 collapses to Definition 10.5.

Theorem 10.11. Consider the large-scale impulsive dynamical system $\mathcal{G}$ given by (10.54)–(10.57). Let $R_{ci} \in \mathbb{S}^{m_{ci}}$, $S_{ci} \in \mathbb{R}^{l_{ci} \times m_{ci}}$, $Q_{ci} \in \mathbb{S}^{l_{ci}}$, $R_{di} \in \mathbb{S}^{m_{di}}$, $S_{di} \in \mathbb{R}^{l_{di} \times m_{di}}$, and $Q_{di} \in \mathbb{S}^{l_{di}}, i = 1, \ldots, q$. Then $\mathcal{G}$ is vector dissipative (respectively, exponentially vector dissipative) with respect to the quadratic hybrid supply rate $(S_c(u_c, y_c), S_d(u_d, y_d))$, where $s_{ci}(u_{ci}, y_{ci}) = y_{ci}^T Q_{ci} y_{ci} + 2 y_{ci}^T S_{ci} u_{ci} + u_{ci}^T R_{ci} u_{ci}$ and $s_{di}(u_{di}, y_{di}) = y_{di}^T Q_{di} y_{di} + 2 y_{di}^T S_{di} u_{di} + u_{di}^T R_{di} u_{di}$, $i = 1, \ldots, q$, if there exist functions $V_s = [v_{s1}, \ldots, v_{sq}]^T : \mathbb{R}^n \to \overline{\mathbb{R}}_+^q$, $w_c = [w_{c1}, \ldots, w_{cq}]^T : \overline{\mathbb{R}}_+^q \to \mathbb{R}^q$, $\ell_{ci} : \mathbb{R}^n \to \mathbb{R}^{s_{ci}}$, $\mathcal{Z}_{ci} : \mathbb{R}^n \to \mathbb{R}^{s_{ci} \times m_c}$, $\ell_{di} : \mathbb{R}^n \to \mathbb{R}^{s_{di}}$, $\mathcal{Z}_{di} : \mathbb{R}^n \to \mathbb{R}^{s_{di} \times m_d}$, $P_{1i} : \mathbb{R}^n \to \mathbb{R}^{1 \times m_d}$, and $P_{2i} : \mathbb{R}^n \to \mathbb{N}^{m_d}$ such that $v_{si}(\cdot)$ is continuously differentiable, $v_{si}(0) = 0$, $i = 1, \ldots, q$, $w_c(\cdot) \in \mathcal{W}$, $w_c(0) = 0$, the zero solution $z(t) \equiv 0$ to (10.132) is Lyapunov (respectively, asymptotically) stable,

$$v_{si}(x + \mathcal{F}_d(x) + G_d(x)u_d) = v_{si}(x + \mathcal{F}_d(x)) + P_{1i}(x)u_d + u_d^T P_{2i}(x)u_d,$$
$$x \in \mathcal{Z}_x, \quad u_d \in \mathbb{R}^{m_d}, \quad (10.136)$$

and, for all $i = 1, \ldots, q$,

$$0 = v_{si}'(x)\mathcal{F}_c(x) - h_c^T(x)\hat{Q}_{ci}h_c(x) - w_{ci}(V_s(x)) + \ell_{ci}^T(x)\ell_{ci}(x), \quad x \notin \mathcal{Z}_x, \quad (10.137)$$

$$0 = \tfrac{1}{2}v_{si}'(x)G_c(x) - h_c^T(x)(\hat{S}_{ci} + \hat{Q}_{ci}J_c(x)) + \ell_{ci}^T(x)\mathcal{Z}_{ci}(x), \quad x \notin \mathcal{Z}_x, \quad (10.138)$$

$$0 = \hat{R}_{ci} + J_c^T(x)\hat{S}_{ci} + \hat{S}_{ci}^T J_c(x) + J_c^T(x)\hat{Q}_{ci}J_c(x) - \mathcal{Z}_{ci}^T(x)\mathcal{Z}_{ci}(x), \quad x \notin \mathcal{Z}_x, \quad (10.139)$$

$$0 = v_{si}(x + \mathcal{F}_d(x)) - h_d^T(x)\hat{Q}_{di}h_d(x) - v_{si}(x) + \ell_{di}^T(x)\ell_{di}(x), \quad x \in \mathcal{Z}_x, \quad (10.140)$$

$$0 = \tfrac{1}{2}P_{1i}(x) - h_d^T(x)(\hat{S}_{di} + \hat{Q}_{di}J_d(x)) + \ell_{di}^T(x)\mathcal{Z}_{di}(x), \quad x \in \mathcal{Z}_x, \quad (10.141)$$

$$0 = \hat{R}_{\mathrm{d}i} + J_{\mathrm{d}}^{\mathrm{T}}(x)\hat{S}_{\mathrm{d}i} + \hat{S}_{\mathrm{d}i}^{\mathrm{T}}J_{\mathrm{d}}(x) + J_{\mathrm{d}}^{\mathrm{T}}(x)\hat{Q}_{\mathrm{d}i}J_{\mathrm{d}}(x) - P_{2i}(x)$$
$$- \mathcal{Z}_{\mathrm{d}i}^{\mathrm{T}}(x)\mathcal{Z}_{\mathrm{d}i}(x), \quad x \in \mathcal{Z}_x. \tag{10.142}$$

Proof. Suppose that there exist functions $v_{\mathrm{s}i} : \mathbb{R}^n \to \overline{\mathbb{R}}_+$, $\ell_{\mathrm{c}i} : \mathbb{R}^n \to \mathbb{R}^{s_{\mathrm{c}i}}$, $\mathcal{Z}_{\mathrm{c}i} : \mathbb{R}^n \to \mathbb{R}^{s_{\mathrm{c}i} \times m_{\mathrm{c}}}$, $\ell_{\mathrm{d}i} : \mathbb{R}^n \to \mathbb{R}^{s_{\mathrm{d}i}}$, $\mathcal{Z}_{\mathrm{d}i} : \mathbb{R}^n \to \mathbb{R}^{s_{\mathrm{d}i} \times m_{\mathrm{d}}}$, $w_{\mathrm{c}} : \overline{\mathbb{R}}_+^q \to \mathbb{R}^q$, $P_{1i} : \mathbb{R}^n \to \mathbb{R}^{1 \times m_{\mathrm{d}}}$, and $P_{2i} : \mathbb{R}^n \to \mathbb{N}^{m_{\mathrm{d}}}$ such that $v_{\mathrm{s}i}(\cdot)$ is continuously differentiable and nonnegative definite, $v_{\mathrm{s}i}(0) = 0$, $i = 1,\ldots,q$, $w_{\mathrm{c}}(0) = 0$, $w_{\mathrm{c}}(\cdot) \in \mathcal{W}$, the zero solution $z(t) \equiv 0$ to (10.132) is Lyapunov (respectively, asymptotically) stable, and (10.136)–(10.142) are satisfied. Then for every $u_{\mathrm{c}}(t) \in \mathbb{R}^{m_{\mathrm{c}}}$, $t, \hat{t} \in \mathbb{R}$, $t_k < t \leq \hat{t} \leq t_{k+1}$, $k \in \overline{\mathbb{Z}}_+$, and $i = 1,\ldots,q$, it follows from (10.137)–(10.139) that

$$\begin{aligned}
\int_t^{\hat{t}} s_{\mathrm{c}i}(u_{\mathrm{c}i}(\sigma), y_{\mathrm{c}i}(\sigma))\mathrm{d}\sigma &= \int_t^{\hat{t}} [u_{\mathrm{c}}^{\mathrm{T}}(\sigma)\hat{R}_{\mathrm{c}i}u_{\mathrm{c}}(\sigma) + 2y_{\mathrm{c}}^{\mathrm{T}}(\sigma)\hat{S}_{\mathrm{c}i}u_{\mathrm{c}}(\sigma) \\
&\quad + y_{\mathrm{c}}^{\mathrm{T}}(\sigma)\hat{Q}_{\mathrm{c}i}y_{\mathrm{c}}(\sigma)]\mathrm{d}\sigma \\
&= \int_t^{\hat{t}} [h_{\mathrm{c}}^{\mathrm{T}}(x(\sigma))\hat{Q}_{\mathrm{c}i}h_{\mathrm{c}}(x(\sigma)) \\
&\quad + 2h_{\mathrm{c}}^{\mathrm{T}}(x(\sigma))(\hat{S}_{\mathrm{c}i} + \hat{Q}_{\mathrm{c}i}J_{\mathrm{c}}(x(\sigma)))u_{\mathrm{c}}(\sigma) \\
&\quad + u_{\mathrm{c}}^{\mathrm{T}}(\sigma)(J_{\mathrm{c}}^{\mathrm{T}}(x(\sigma))\hat{Q}_{\mathrm{c}i}J_{\mathrm{c}}(x(\sigma)) + J_{\mathrm{c}}^{\mathrm{T}}(x(\sigma))\hat{S}_{\mathrm{c}i} \\
&\quad + \hat{S}_{\mathrm{c}i}^{\mathrm{T}}J_{\mathrm{c}}(x(\sigma)) + \hat{R}_{\mathrm{c}i})u_{\mathrm{c}}(\sigma)]\mathrm{d}\sigma \\
&= \int_t^{\hat{t}} [v_{\mathrm{s}i}'(x(\sigma))(\mathcal{F}_{\mathrm{c}}(x(\sigma)) + G_{\mathrm{c}}(x(\sigma))u_{\mathrm{c}}(\sigma)) \\
&\quad + \ell_{\mathrm{c}i}^{\mathrm{T}}(x(\sigma))\ell_{\mathrm{c}i}(x(\sigma)) \\
&\quad + 2\ell_{\mathrm{c}i}^{\mathrm{T}}(x(\sigma))\mathcal{Z}_{\mathrm{c}i}(x(\sigma))u_{\mathrm{c}}(\sigma) \\
&\quad + u_{\mathrm{c}}^{\mathrm{T}}(\sigma)\mathcal{Z}_{\mathrm{c}i}^{\mathrm{T}}(x(\sigma))\mathcal{Z}_{\mathrm{c}i}(x(\sigma))u_{\mathrm{c}}(\sigma) \\
&\quad - w_{\mathrm{c}i}(V_{\mathrm{s}}(x(\sigma)))]\mathrm{d}\sigma \\
&= \int_t^{\hat{t}} [\dot{v}_{\mathrm{s}i}(x(\sigma)) \\
&\quad + [\ell_{\mathrm{c}i}(x(\sigma)) + \mathcal{Z}_{\mathrm{c}i}(x(\sigma))u_{\mathrm{c}}(\sigma)]^{\mathrm{T}}[\ell_{\mathrm{c}i}(x(\sigma)) \\
&\quad + \mathcal{Z}_{\mathrm{c}i}(x(\sigma))u_{\mathrm{c}}(\sigma)] - w_{\mathrm{c}i}(V_{\mathrm{s}}(x(\sigma)))]\mathrm{d}\sigma \\
&\geq v_{\mathrm{s}i}(x(\hat{t})) - v_{\mathrm{s}i}(x(t)) - \int_t^{\hat{t}} w_{\mathrm{c}i}(V_{\mathrm{s}}(x(\sigma)))\mathrm{d}\sigma,
\end{aligned} \tag{10.143}$$

where $x(\sigma)$, $\sigma \in (t_k, t_{k+1}]$, satisfies (10.54).

Next, for every $u_{\mathrm{d}}(t_k) \in \mathbb{R}^{m_{\mathrm{d}}}$, $t_k \in \mathbb{R}$, and $k \in \overline{\mathbb{Z}}_+$, it follows from (10.136) and (10.140)–(10.142) that

$$\begin{aligned}
v_{\mathrm{s}i}(x + \mathcal{F}_{\mathrm{d}}(x) &+ \hat{G}_{\mathrm{d}}(x)u_{\mathrm{d}}) - v_{\mathrm{s}i}(x) \\
&= v_{\mathrm{s}i}(x + \mathcal{F}_{\mathrm{d}}(x)) - v_{\mathrm{s}i}(x) + P_{1i}(x)u_{\mathrm{d}} + u_{\mathrm{d}}^{\mathrm{T}}P_{2i}(x)u_{\mathrm{d}}
\end{aligned}$$

$$\begin{aligned}
&= h_{\mathrm{d}}^{\mathrm{T}}(x)\hat{Q}_{\mathrm{d}i}h_{\mathrm{d}}(x) - \ell_{\mathrm{d}i}^{\mathrm{T}}(x)\ell_{\mathrm{d}i}(x) + 2[h_{\mathrm{d}}^{\mathrm{T}}(x)(\hat{Q}_{\mathrm{d}i}J_{\mathrm{d}}(x) \\
&\quad + \hat{S}_{\mathrm{d}i}) - \ell_{\mathrm{d}i}^{\mathrm{T}}(x)\mathcal{Z}_{\mathrm{d}i}(x)]u_{\mathrm{d}} + u_{\mathrm{d}}^{\mathrm{T}}[\hat{R}_{\mathrm{d}i} + \hat{S}_{\mathrm{d}i}^{\mathrm{T}}J_{\mathrm{d}}(x) \\
&\quad + J_{\mathrm{d}}^{\mathrm{T}}(x)\hat{S}_{\mathrm{d}i} + J_{\mathrm{d}}^{\mathrm{T}}(x)\hat{Q}_{\mathrm{d}i}J_{\mathrm{d}}(x) - \mathcal{Z}_{\mathrm{d}i}^{\mathrm{T}}(x)\mathcal{Z}_{\mathrm{d}i}(x)]u_{\mathrm{d}} \\
&= s_{\mathrm{d}i}(u_{\mathrm{d}i}, y_{\mathrm{d}i}) - [\ell_{\mathrm{d}i}(x) + \mathcal{Z}_{\mathrm{d}i}(x)u_{\mathrm{d}}]^{\mathrm{T}}[\ell_{\mathrm{d}i}(x) + \mathcal{Z}_{\mathrm{d}i}(x)u_{\mathrm{d}}] \\
&\leq s_{\mathrm{d}i}(u_{\mathrm{d}i}, y_{\mathrm{d}i}).
\end{aligned} \tag{10.144}$$

Now, using (10.143) and (10.144) the result is immediate with vector storage function $V_{\mathrm{s}}(x) = [v_{\mathrm{s}1}(x), \ldots, v_{\mathrm{s}q}(x)]^{\mathrm{T}}$, $x \in \mathbb{R}^n$. $\qquad\square$

Using (10.137)–(10.142) it follows that for $T \geq t_0 \geq 0$, $k \in \mathbb{Z}_{[t_0, T)}$, and $i = 1, \ldots, q$,

$$\int_{t_0}^{T} s_{\mathrm{c}i}(u_{\mathrm{c}i}(t), y_{\mathrm{c}i}(t))\mathrm{d}t + \int_{t_0}^{T} w_{\mathrm{c}i}(V_{\mathrm{s}}(x(t)))\mathrm{d}t + \sum_{k \in \mathbb{Z}_{[t_0, T)}} s_{\mathrm{d}i}(u_{\mathrm{d}}(t_k), y_{\mathrm{d}}(t_k))$$

$$= v_{\mathrm{s}i}(x(T)) - v_{\mathrm{s}i}(x(t_0)) + \int_{t_0}^{T} [\ell_{\mathrm{c}i}(x(t)) + \mathcal{Z}_{\mathrm{c}i}(x(t))u_{\mathrm{c}}(t)]^{\mathrm{T}}$$

$$\cdot [\ell_{\mathrm{c}i}(x(t)) + \mathcal{Z}_{\mathrm{c}i}(x(t))u_{\mathrm{c}}(t)]\mathrm{d}t$$

$$+ \sum_{k \in \mathbb{Z}_{[t_0, T)}} [\ell_{\mathrm{d}i}(x(t_k)) + \mathcal{Z}_{\mathrm{d}i}(x(t_k))u_{\mathrm{d}}(t_k)]^{\mathrm{T}}[\ell_{\mathrm{d}i}(x(t_k)) + \mathcal{Z}_{\mathrm{d}i}(x(t_k))u_{\mathrm{d}}(t_k)],$$

$$\tag{10.145}$$

where $V_{\mathrm{s}}(x) = [v_{\mathrm{s}1}(x), \ldots, v_{\mathrm{s}q}(x)]^{\mathrm{T}}$, $x \in \mathbb{R}^n$, which can be interpreted as a *generalized energy* balance equation for the ith impulsive subsystem of $\mathcal{G}$, where $v_{\mathrm{s}i}(x(T)) - v_{\mathrm{s}i}(x(t_0))$ is the stored or accumulated generalized energy of the ith impulsive subsystem; the two path-dependent terms on the left are, respectively, the external supplied energy to the ith subsystem over the continuous-time dynamics and the energy gained over the continuous-time dynamics by the ith subsystem from the net energy flow between all subsystems due to subsystem coupling; the last discrete term on the left corresponds to the external supplied energy to the ith subsystem at the resetting instants; the second path-dependent term on the right corresponds to the dissipated energy from the ith impulsive subsystem over the continuous-time dynamics; and the last discrete term on the right corresponds to the dissipated energy from the ith impulsive subsystem at the resetting instants.

Equivalently, (10.145) can be rewritten as

$$\dot{v}_{\mathrm{s}i}(x(t)) = s_{\mathrm{c}i}(u_{\mathrm{c}i}(t), y_{\mathrm{c}i}(t)) + w_{\mathrm{c}i}(V_{\mathrm{s}}(x(t)))$$

$$- [\ell_{\mathrm{c}i}(x(t)) + \mathcal{Z}_{\mathrm{c}i}(x(t))u_{\mathrm{c}}(t)]^{\mathrm{T}}[\ell_{\mathrm{c}i}(x(t)) + \mathcal{Z}_{\mathrm{c}i}(x(t))u_{\mathrm{c}}(t)],$$

$$t_k < t \leq t_{k+1}, \quad i = 1, \ldots, q, \tag{10.146}$$

$$v_{\mathrm{s}i}(x(t_k) + \mathcal{F}_{\mathrm{d}}(x(t_k)) + G_{\mathrm{d}}(x(t_k))u_{\mathrm{d}}(t_k)) - v_{\mathrm{s}i}(x(t_k))$$

$$= s_{\mathrm{d}i}(u_{\mathrm{d}}(t_k), y_{\mathrm{d}}(t_k)) - [\ell_{\mathrm{d}i}(x(t_k)) + \mathcal{Z}_{\mathrm{d}i}(x(t_k))u_{\mathrm{d}}(t_k)]^{\mathrm{T}}$$

$$\cdot [\ell_{\mathrm{d}i}(x(t_k)) + \mathcal{Z}_{\mathrm{d}i}(x(t_k))u_{\mathrm{d}}(t_k)], \quad k \in \overline{\mathbb{Z}}_{+}, \tag{10.147}$$

which yields a set of q generalized energy conservation equations for the large-scale impulsive dynamical system $\mathcal{G}$. Specifically, (10.146) shows that the rate of change in generalized energy, or generalized power, over the time interval $t \in (t_k, t_{k+1}]$ for the ith subsystem of $\mathcal{G}$ is equal to the generalized system power input to the ith subsystem plus the instantaneous rate of energy supplied to the ith subsystem from the net energy flow between all subsystems minus the internal generalized system power dissipated from the ith subsystem; (10.147) shows that the change of energy at the resetting times t_k, $k \in \overline{\mathbb{Z}}_+$, is equal to the external generalized system supplied energy at the resetting times minus the generalized dissipated energy at the resetting times.

Note that if $\mathcal{G}$, with $(u_{\mathrm c}(t), u_{\mathrm d}(t_k)) \equiv (0,0)$, is vector dissipative (respectively, exponentially vector dissipative) with respect to the quadratic hybrid supply rate, and $Q_{ci} \leq 0$ and $Q_{di} \leq 0$, $i = 1, \ldots, q$, then it follows from the vector hybrid dissipation inequality that for all $k \in \overline{\mathbb{Z}}_+$,

$$\dot{V}_{\mathrm s}(x(t)) \leq\leq w_{\mathrm c}(V_{\mathrm s}(x(t))) + S_{\mathrm c}(0, y_{\mathrm c}(t)) \leq\leq w_{\mathrm c}(V_{\mathrm s}(x(t))), \quad t_k < t \leq t_{k+1}, \tag{10.148}$$

$$V_{\mathrm s}(x(t_k) + \mathcal{F}_{\mathrm d}(x(t_k))) - V_{\mathrm s}(x(t_k)) \leq\leq S_{\mathrm d}(0, y_{\mathrm d}(t_k)) \leq\leq 0, \tag{10.149}$$

where $S_{\mathrm c}(0, y_{\mathrm c}) = [s_{c1}(0, y_{c1}), \ldots, s_{cq}(0, y_{cq})]^{\mathrm T}$, $S_{\mathrm d}(0, y_{\mathrm d}) = [s_{d1}(0, y_{d1}), \ldots, s_{dq}(0, y_{dq})]^{\mathrm T}$,

$$s_{ci}(0, y_{ci}(t)) = y_{ci}^{\mathrm T}(t) Q_{ci} y_{ci}(t) \leq 0, \quad i = 1, \ldots, q, \tag{10.150}$$

$$s_{di}(0, y_{di}(t_k)) = y_{di}^{\mathrm T}(t_k) Q_{di} y_{di}(t_k) \leq 0, \quad t_k < t \leq t_{k+1}, \quad k \in \overline{\mathbb{Z}}_+,$$
$$i = 1, \ldots, q, \tag{10.151}$$

and $x(t)$, $t \geq t_0$, is the solution to (10.54)–(10.57) with $(u_{\mathrm c}(t), u_{\mathrm d}(t_k)) \equiv (0,0)$. If, in addition, there exists $p \in \mathbb{R}_+^q$ such that $p^{\mathrm T} V_{\mathrm s}(x)$, $x \in \mathbb{R}^n$, is positive definite, then it follows from Theorem 10.2 that the undisturbed $((u_{\mathrm c}(t), u_{\mathrm d}(t_k)) \equiv (0,0))$ large-scale impulsive dynamical system (10.54)–(10.57) is Lyapunov (respectively, asymptotically) stable.

Next, we extend the notions of passivity and nonexpansivity to vector passivity and vector nonexpansivity.

Definition 10.7. The large-scale impulsive dynamical system $\mathcal{G}$ given by (10.54)–(10.57) with $m_{ci} = l_{ci}$, $m_{di} = l_{di}$, $i = 1, \ldots, q$, is *vector passive* (respectively, *vector exponentially passive*) if it is vector dissipative (respectively, exponentially vector dissipative) with respect to the vector hybrid supply rate $(S_{\mathrm c}(u_{\mathrm c}, y_{\mathrm c}), S_{\mathrm d}(u_{\mathrm d}, y_{\mathrm d}))$, where $s_{ci}(u_{ci}, y_{ci}) = 2y_{ci}^{\mathrm T} u_{ci}$ and $s_{di}(u_{di}, y_{di}) = 2y_{di}^{\mathrm T} u_{di}$, $i = 1, \ldots, q$.

Definition 10.8. The large-scale impulsive dynamical system $\mathcal{G}$ given by (10.54)–(10.57) is *vector nonexpansive* (respectively, *vector exponentially nonexpansive*) if it is vector dissipative (respectively, exponentially vector dissipative) with respect to the vector hybrid supply rate $(S_c(u_c, y_c), S_d(u_d, y_d))$, where $s_{ci}(u_{ci}, y_{ci}) = \gamma_{ci}^2 u_{ci}^T u_{ci} - y_{ci}^T y_{ci}$ and $s_{di}(u_{di}, y_{di}) = \gamma_{di}^2 u_{di}^T u_{di} - y_{di}^T y_{di}$, $i = 1, \ldots, q$, and $\gamma_{ci} > 0$, $\gamma_{di} > 0$, $i = 1, \ldots, q$, are given.

Note that a mixed vector passive-nonexpansive formulation of $\mathcal{G}$ can also be considered. Specifically, one can consider large-scale impulsive dynamical systems $\mathcal{G}$ that are vector dissipative with respect to vector hybrid supply rates $(S_c(u_c, y_c), S_d(u_d, y_d))$, where $s_{ci}(u_{ci}, y_{ci}) = 2y_{ci}^T u_{ci}$, $s_{di}(u_{di}, y_{di}) = 2y_{di}^T u_{di}$, $i \in \mathbb{Z}_p$, $s_{cj}(u_{cj}, y_{cj}) = \gamma_{cj}^2 u_{cj}^T u_{cj} - y_{cj}^T y_{cj}$, $\gamma_{cj} > 0$, $s_{dj}(u_{dj}, y_{dj}) = \gamma_{dj}^2 u_{dj}^T u_{dj} - y_{dj}^T y_{dj}$, $\gamma_{dj} > 0$, $j \in \mathbb{Z}_{ne}$, $\mathbb{Z}_p \cap \mathbb{Z}_{ne} = \varnothing$, and $\mathbb{Z}_p \cup \mathbb{Z}_{ne} = \{1, \ldots, q\}$. Furthermore, hybrid supply rates for vector input strict passivity, vector output strict passivity, and vector input-output strict passivity generalizing the dissipativity notions given in [89] can also be considered.

The next result presents constructive sufficient conditions guaranteeing vector dissipativity of $\mathcal{G}$ with respect to a quadratic hybrid supply rate for the case where the vector storage function $V_s(x)$, $x \in \mathbb{R}^n$, is component decoupled, that is, $V_s(x) = [v_{s1}(x_1), \ldots, v_{sq}(x_q)]^T$, $x \in \mathbb{R}^n$.

Theorem 10.12. Consider the large-scale impulsive dynamical system $\mathcal{G}$ given by (10.54)–(10.57). Assume that there exist functions $V_s = [v_{s1}, \ldots, v_{sq}]^T : \mathbb{R}^n \to \overline{\mathbb{R}}_+^q$, $w_c = [w_{c1}, \ldots, w_{cq}]^T : \overline{\mathbb{R}}_+^q \to \mathbb{R}^q$, $\ell_{ci} : \mathbb{R}^n \to \mathbb{R}^{s_{ci}}$, $\mathcal{Z}_{ci} : \mathbb{R}^n \to \mathbb{R}^{s_{ci} \times m_{ci}}$, $\ell_{di} : \mathbb{R}^n \to \mathbb{R}^{s_{di}}$, $\mathcal{Z}_{di} : \mathbb{R}^n \to \mathbb{R}^{s_{di} \times m_{di}}$, $P_{1i} : \mathbb{R}^n \to \mathbb{R}^{1 \times m_{di}}$, and $P_{2i} : \mathbb{R}^n \to \mathbb{N}^{m_{di}}$ such that $V_s(x) = [v_{s1}(x_1), \ldots, v_{sq}(x_q)]^T$, $v_{si}(\cdot)$ is continuously differentiable, $v_{si}(0) = 0$, $i = 1, \ldots, q$, $w_c(\cdot) \in \mathcal{W}$, $w_c(0) = 0$, the zero solution $z(t) \equiv 0$ to (10.132) is Lyapunov (respectively, asymptotically) stable, and, for all $x \in \mathbb{R}^n$ and $i = 1, \ldots, q$,

$$0 \leq v_{si}(x_i + \mathcal{F}_{di}(x)) - v_{si}(x_i + \mathcal{F}_{di}(x) + G_{di}(x_i)u_{di}) + P_{1i}(x)u_{di}$$
$$+ u_{di}^T P_{2i}(x)u_{di}, \quad x \in \mathcal{Z}_x, \quad u_{di} \in \mathbb{R}^{m_{di}}, \tag{10.152}$$

$$0 \geq v_{si}'(x_i)\mathcal{F}_{ci}(x) - h_{ci}^T(x_i)Q_{ci}h_{ci}(x_i) - w_{ci}(V_s(x)) + \ell_{ci}^T(x_i)\ell_{ci}(x_i),$$
$$x \notin \mathcal{Z}_x, \tag{10.153}$$

$$0 = \tfrac{1}{2}v_{si}'(x_i)G_{ci}(x_i) - h_{ci}^T(x_i)(S_{ci} + Q_{ci}J_{ci}(x_i)) + \ell_{ci}^T(x_i)\mathcal{Z}_{ci}(x_i),$$
$$x \notin \mathcal{Z}_x, \tag{10.154}$$

$$0 \leq R_{ci} + J_{ci}^T(x_i)S_{ci} + S_{ci}^T J_{ci}(x_i) + J_{ci}^T(x_i)Q_{ci}J_{ci}(x_i) - \mathcal{Z}_{ci}^T(x_i)\mathcal{Z}_{ci}(x_i),$$
$$x \notin \mathcal{Z}_x, \tag{10.155}$$

$$0 \geq v_{si}(x_i + \mathcal{F}_{di}(x)) - h_{di}^T(x_i)Q_{di}h_{di}(x_i) - v_{si}(x_i) + \ell_{di}^T(x_i)\ell_{di}(x_i),$$
$$x \in \mathcal{Z}_x, \tag{10.156}$$

$$0 = \tfrac{1}{2}P_{1i}(x) - h_{di}^T(x_i)(S_{di} + Q_{di}J_{di}(x_i)) + \ell_{di}^T(x_i)\mathcal{Z}_{di}(x_i), \quad x \in \mathcal{Z}_x, \tag{10.157}$$

$$0 \leq R_{di} + J_{di}^T(x_i)S_{di} + S_{di}^T J_{di}(x_i) + J_{di}^T(x_i)Q_{di}J_{di}(x_i) - P_{2i}(x)$$
$$- \mathcal{Z}_{di}^T(x_i)\mathcal{Z}_{di}(x_i), \quad x \in \mathcal{Z}_x. \tag{10.158}$$

Then $\mathcal{G}$ is vector dissipative (respectively, exponentially vector dissipative) with respect to the vector hybrid supply rate $(S_c(u_c, y_c), S_d(u_d, y_d))$, where $s_{ci}(u_{ci}, y_{ci}) = u_{ci}^{\mathrm{T}} R_{ci} u_{ci} + 2y_{ci}^{\mathrm{T}} S_{ci} u_{ci} + y_{ci}^{\mathrm{T}} Q_{ci} y_{ci}$ and $s_{di}(u_{di}, y_{di}) = u_{di}^{\mathrm{T}} R_{di} u_{di} + 2y_{di}^{\mathrm{T}} S_{di} u_{di} + y_{di}^{\mathrm{T}} Q_{di} y_{di}$, $i = 1, \ldots, q$.

Proof. For every admissible input $u_c(t) = [u_{c1}^{\mathrm{T}}(t), \ldots, u_{cq}^{\mathrm{T}}(t)]^{\mathrm{T}}$ such that $u_{ci}(t) \in \mathbb{R}^{m_{ci}}$, $t, \hat{t} \in \mathbb{R}$, $t_k < t \le \hat{t} \le t_{k+1}$, $k \in \overline{\mathbb{Z}}_+$, and $i = 1, \ldots, q$, it follows from (10.153)–(10.155) that

$$
\begin{aligned}
\int_t^{\hat{t}} s_{ci}(u_{ci}(\sigma), y_{ci}(\sigma))\mathrm{d}\sigma = & \int_t^{\hat{t}} [u_{ci}^{\mathrm{T}}(\sigma) R_{ci} u_{ci}(\sigma) + 2y_{ci}^{\mathrm{T}}(\sigma) S_{ci} u_{ci}(\sigma) \\
& \quad + y_{ci}^{\mathrm{T}}(\sigma) Q_{ci} y_{ci}(\sigma)]\mathrm{d}\sigma \\
= & \int_t^{\hat{t}} [h_{ci}^{\mathrm{T}}(x_i(\sigma)) Q_{ci} h_{ci}(x_i(\sigma)) \\
& \quad + 2h_{ci}^{\mathrm{T}}(x_i(\sigma))(S_{ci} + Q_{ci} J_{ci}(x_i(\sigma))) \\
& \quad u_{ci}(\sigma) + u_{ci}^{\mathrm{T}}(\sigma)(J_{ci}^{\mathrm{T}}(x_i(\sigma)) Q_{ci} J_{ci}(x_i(\sigma)) \\
& \quad + J_{ci}^{\mathrm{T}}(x_i(\sigma)) S_{ci} + S_{ci}^{\mathrm{T}} J_{ci}(x_i(\sigma)) \\
& \quad + R_{ci}) u_{ci}(\sigma)]\mathrm{d}\sigma \\
\ge & \int_t^{\hat{t}} [v_{si}'(x_i(\sigma))[\mathcal{F}_{ci}(x(\sigma)) + G_{ci}(x_i(\sigma)) u_{ci}(\sigma)] \\
& \quad + \ell_{ci}^{\mathrm{T}}(x_i(\sigma)) \ell_{ci}(x_i(\sigma)) \\
& \quad + 2\ell_{ci}^{\mathrm{T}}(x_i(\sigma)) \mathcal{Z}_{ci}(x_i(\sigma)) u_{ci}(\sigma) \\
& \quad + u_{ci}^{\mathrm{T}}(\sigma) \mathcal{Z}_{ci}^{\mathrm{T}}(x_i(\sigma)) \mathcal{Z}_{ci}(x_i(\sigma)) u_{ci}(\sigma) \\
& \quad - w_{ci}(V_s(x(\sigma)))]\mathrm{d}\sigma \\
= & \int_t^{\hat{t}} [\dot{v}_{si}(x_i(\sigma)) + [\ell_{ci}(x_i(\sigma)) \\
& \quad + \mathcal{Z}_{ci}(x_i(\sigma)) u_{ci}(\sigma)]^{\mathrm{T}} [\ell_{ci}(x_i(\sigma)) \\
& \quad + \mathcal{Z}_{ci}(x_i(\sigma)) u_{ci}(\sigma)] - w_{ci}(V_s(x(\sigma)))]\mathrm{d}\sigma \\
\ge & \; v_{si}(x_i(\hat{t})) - v_{si}(x_i(t)) - \int_t^{\hat{t}} w_{ci}(V_s(x(\sigma)))\mathrm{d}\sigma,
\end{aligned}
$$

$$(10.159)$$

where $x(\sigma)$, $t_k < \sigma \le t_{k+1}$, satisfies (10.54).

Next, for every admissible input $u_d(t_k) = [u_{d1}^{\mathrm{T}}(t_k), \ldots, u_{dq}^{\mathrm{T}}(t_k)]^{\mathrm{T}}$ such that $u_{di}(t_k) \in \mathbb{R}^{m_{di}}$, $t_k \in \mathbb{R}$, $k \in \overline{\mathbb{Z}}_+$, and $i = 1, \ldots, q$, it follows from (10.152) and (10.156)–(10.158) that

$$
\begin{aligned}
s_{di}(u_{di}(t_k), y_{di}(t_k)) = & \; u_{di}^{\mathrm{T}}(t_k) R_{di} u_{di}(t_k) + 2y_{di}^{\mathrm{T}}(t_k) S_{di} u_{di}(t_k) \\
& \quad + y_{di}^{\mathrm{T}}(t_k) Q_{di} y_{di}(t_k) \\
= & \; h_{di}^{\mathrm{T}}(x_i(t_k)) Q_{di} h_{di}(x_i(t_k)) \\
& \quad + 2h_{di}^{\mathrm{T}}(x_i(t_k))(S_{di} + Q_{di} J_{di}(x_i(t_k))) u_{di}(t_k)
\end{aligned}
$$

$$+u_{\mathrm{d}i}^{\mathrm{T}}(t_k)(J_{\mathrm{d}i}^{\mathrm{T}}(x_i(t_k))Q_{\mathrm{d}i}J_{\mathrm{d}i}(x_i(t_k))$$
$$+J_{\mathrm{d}i}^{\mathrm{T}}(x_i(t_k))S_{\mathrm{d}i} + S_{\mathrm{d}i}^{\mathrm{T}}J_{\mathrm{d}i}(x_i(t_k)) + R_{\mathrm{d}i})u_{\mathrm{d}i}(t_k)$$
$$\geq v_{\mathrm{s}i}(x_i(t_k) + \mathcal{F}_{\mathrm{d}i}(x(t_k))) + P_{1i}(x(t_k))u_{\mathrm{d}i}(t_k)$$
$$+\ell_{\mathrm{d}i}^{\mathrm{T}}(x_i(t_k))\ell_{\mathrm{d}i}(x_i(t_k)) + 2\ell_{\mathrm{d}i}^{\mathrm{T}}(x_i(t_k))\mathcal{Z}_{\mathrm{d}i}(x_i(t_k))u_{\mathrm{d}i}(t_k)$$
$$+u_{\mathrm{d}i}^{\mathrm{T}}(t_k)P_{2i}(x(t_k))u_{\mathrm{d}i}(t_k)$$
$$+u_{\mathrm{d}i}^{\mathrm{T}}(t_k)\mathcal{Z}_{\mathrm{d}i}^{\mathrm{T}}(x_i(t_k))\mathcal{Z}_{\mathrm{d}i}(x_i(t_k))u_{\mathrm{d}i}(t_k) - v_{\mathrm{s}i}(x_i(t_k))$$
$$\geq v_{\mathrm{s}i}(x_i(t_k) + \mathcal{F}_{\mathrm{d}i}(x(t_k)) + G_{\mathrm{d}i}(x_i(t_k))u_{\mathrm{d}i}(t_k))$$
$$+[\ell_{\mathrm{d}i}(x_i(t_k)) + \mathcal{Z}_{\mathrm{d}i}(x_i(t_k))u_{\mathrm{d}i}(t_k)]^{\mathrm{T}}[\ell_{\mathrm{d}i}(x_i(t_k))$$
$$+\mathcal{Z}_{\mathrm{d}i}(x_i(t_k))u_{\mathrm{d}i}(t_k)] - v_{\mathrm{s}i}(x_i(t_k))$$
$$\geq v_{\mathrm{s}i}(x_i(t_k) + \mathcal{F}_{\mathrm{d}i}(x(t_k)) + G_{\mathrm{d}i}(x_i(t_k))u_{\mathrm{d}i}(t_k))$$
$$-v_{\mathrm{s}i}(x_i(t_k)), \tag{10.160}$$

where $x(t_k)$, $k \in \overline{\mathbb{Z}}_+$, satisfies (10.55). Now, the result follows from (10.159) and (10.160) with vector storage function $V_{\mathrm{s}}(x) = [v_{\mathrm{s}1}(x_1), \ldots, v_{\mathrm{s}q}(x_q)]^{\mathrm{T}}$, $x \in \mathbb{R}^n$. $\qquad\square$

Finally, we provide necessary and sufficient conditions for the case where the large-scale impulsive dynamical system $\mathcal{G}$ is vector lossless with respect to a quadratic hybrid supply rate.

Theorem 10.13. Consider the large-scale impulsive dynamical system $\mathcal{G}$ given by (10.54)–(10.57). Let $R_{\mathrm{c}i} \in \mathbb{S}^{m_{\mathrm{c}i}}$, $S_{\mathrm{c}i} \in \mathbb{R}^{l_{\mathrm{c}i} \times m_{\mathrm{c}i}}$, $Q_{\mathrm{c}i} \in \mathbb{S}^{l_{\mathrm{c}i}}$, $R_{\mathrm{d}i} \in \mathbb{S}^{m_{\mathrm{d}i}}$, $S_{\mathrm{d}i} \in \mathbb{R}^{l_{\mathrm{d}i} \times m_{\mathrm{d}i}}$, and $Q_{\mathrm{d}i} \in \mathbb{S}^{l_{\mathrm{d}i}}$, $i = 1, \ldots, q$. Then $\mathcal{G}$ is vector loss-less with respect to the quadratic hybrid supply rate $(S_{\mathrm{c}}(u_{\mathrm{c}}, y_{\mathrm{c}}), S_{\mathrm{d}}(u_{\mathrm{d}}, y_{\mathrm{d}}))$, where $s_{\mathrm{c}i}(u_{\mathrm{c}i}, y_{\mathrm{c}i}) = u_{\mathrm{c}i}^{\mathrm{T}}R_{\mathrm{c}i}u_{\mathrm{c}i} + 2y_{\mathrm{c}i}^{\mathrm{T}}S_{\mathrm{c}i}u_{\mathrm{c}i} + y_{\mathrm{c}i}^{\mathrm{T}}Q_{\mathrm{c}i}y_{\mathrm{c}i}$ and $s_{\mathrm{d}i}(u_{\mathrm{d}i}, y_{\mathrm{d}i}) = u_{\mathrm{d}i}^{\mathrm{T}}R_{\mathrm{d}i}u_{\mathrm{d}i} + 2y_{\mathrm{d}i}^{\mathrm{T}}S_{\mathrm{d}i}u_{\mathrm{d}i} + y_{\mathrm{d}i}^{\mathrm{T}}Q_{\mathrm{d}i}y_{\mathrm{d}i}$, $i = 1, \ldots, q$, if and only if there exist functions $V_{\mathrm{s}} = [v_{\mathrm{s}1}, \ldots, v_{\mathrm{s}q}]^{\mathrm{T}} : \mathbb{R}^n \to \overline{\mathbb{R}}_+^q$, $P_{1i} : \mathbb{R}^n \to \mathbb{R}^{1 \times m_{\mathrm{d}}}$, $P_{2i} : \mathbb{R}^n \to \mathbb{N}^{m_{\mathrm{d}}}$, and $w_{\mathrm{c}} = [w_{\mathrm{c}1}, \ldots, w_{\mathrm{c}q}]^{\mathrm{T}} : \overline{\mathbb{R}}_+^q \to \mathbb{R}^q$ such that $v_{\mathrm{s}i}(\cdot)$ is continuously differentiable, $v_{\mathrm{s}i}(0) = 0$, $i = 1, \ldots, q$, $w_{\mathrm{c}} \in \mathcal{W}$, $w_{\mathrm{c}}(0) = 0$, the zero solution $z(t) \equiv 0$ to (10.132) is Lyapunov stable, and, for all $x \in \mathbb{R}^n$, $i = 1, \ldots, q$, (10.136) holds and

$$0 = v_{\mathrm{s}i}'(x)\mathcal{F}_{\mathrm{c}}(x) - h_{\mathrm{c}}^{\mathrm{T}}(x)\hat{Q}_{\mathrm{c}i}h_{\mathrm{c}}(x) - w_{\mathrm{c}i}(V_{\mathrm{s}}(x)), \quad x \notin \mathcal{Z}_x, \tag{10.161}$$

$$0 = \tfrac{1}{2}v_{\mathrm{s}i}'(x)G_{\mathrm{c}}(x) - h_{\mathrm{c}}^{\mathrm{T}}(x)(\hat{S}_{\mathrm{c}i} + \hat{Q}_{\mathrm{c}i}J_{\mathrm{c}}(x)), \quad x \notin \mathcal{Z}_x, \tag{10.162}$$

$$0 = \hat{R}_{\mathrm{c}i} + J_{\mathrm{c}}^{\mathrm{T}}(x)\hat{S}_{\mathrm{c}i} + \hat{S}_{\mathrm{c}i}^{\mathrm{T}}J_{\mathrm{c}}(x) + J_{\mathrm{c}}^{\mathrm{T}}(x)\hat{Q}_{\mathrm{c}i}J_{\mathrm{c}}(x), \quad x \notin \mathcal{Z}_x, \tag{10.163}$$

$$0 = v_{\mathrm{s}i}(x + \mathcal{F}_{\mathrm{d}}(x)) - h_{\mathrm{d}}^{\mathrm{T}}(x)\hat{Q}_{\mathrm{d}i}h_{\mathrm{d}}(x) - v_{\mathrm{s}i}(x), \quad x \in \mathcal{Z}_x, \tag{10.164}$$

$$0 = \tfrac{1}{2}P_{1i}(x) - h_{\mathrm{d}}^{\mathrm{T}}(x)(\hat{S}_{\mathrm{d}i} + \hat{Q}_{\mathrm{d}i}J_{\mathrm{d}}(x)), \quad x \in \mathcal{Z}_x, \tag{10.165}$$

$$0 = \hat{R}_{\mathrm{d}i} + J_{\mathrm{d}}^{\mathrm{T}}(x)\hat{S}_{\mathrm{d}i} + \hat{S}_{\mathrm{d}i}^{\mathrm{T}}J_{\mathrm{d}}(x) + J_{\mathrm{d}}^{\mathrm{T}}(x)\hat{Q}_{\mathrm{d}i}J_{\mathrm{d}}(x) - P_{2i}(x), \quad x \in \mathcal{Z}_x. \tag{10.166}$$

Proof. Sufficiency follows as in the proof of Theorem 10.11. To show necessity, suppose that $\mathcal{G}$ is lossless with respect to the quadratic hybrid supply rate $(S_c(u_c, y_c), S_d(u_d, y_d))$. Then, there exist continuous functions $V_s = [v_{s1}, \ldots, v_{sq}]^T : \mathbb{R}^n \to \overline{\mathbb{R}}_+^q$ and $w_c = [w_{c1}, \ldots, w_{cq}]^T : \overline{\mathbb{R}}_+^q \to \mathbb{R}^q$ such that $V_s(0) = 0$, the zero solution $z(t) \equiv 0$ to (10.132) is Lyapunov stable, and for all $k \in \overline{\mathbb{Z}}_+$, $i = 1, \ldots, q$,

$$v_{si}(x(\hat{t})) - v_{si}(x(t)) = \int_t^{\hat{t}} s_{ci}(u_{ci}(\sigma), y_{ci}(\sigma)) \mathrm{d}\sigma + \int_t^{\hat{t}} w_{ci}(V_s(x(\sigma))) \mathrm{d}\sigma,$$

$$t_k < t \le \hat{t} \le t_{k+1}, \qquad (10.167)$$

and

$$v_{si}(x(t_k) + \mathcal{F}_d(x(t_k)) + G_d(x(t_k))u_d(t_k)) = v_{si}(x(t_k)) + s_{di}(u_{di}(t_k), y_{di}(t_k)).$$

$$(10.168)$$

Now, dividing (10.167) by $\hat{t} - t^+$ and letting $\hat{t} \to t^+$, (10.167) is equivalent to

$$\dot{v}_{si}(x(t)) = v'_{si}(x(t))[\mathcal{F}_c(x(t)) + G_c(x(t))u_c(t)]$$
$$= s_{ci}(u_{ci}(t), y_{ci}(t)) + w_{ci}(V_s(x(t))), \quad t_k < t \le t_{k+1}. \qquad (10.169)$$

Next, with $t = t_0$, it follows from (10.169) that

$$v'_{si}(x_0)[\mathcal{F}_c(x_0) + G_c(x_0)u_c(t_0)] = s_{ci}(u_{ci}(t_0), y_{ci}(t_0)) + w_{ci}(V_s(x_0)),$$

$$x_0 \notin \mathcal{Z}_x, \quad u_c(t_0) \in \mathbb{R}^{m_c}. \qquad (10.170)$$

Since $x_0 \notin \mathcal{Z}_x$ is arbitrary, it follows that

$$v'_{si}(x)[\mathcal{F}_c(x) + G_c(x)u_c] = w_{ci}(V_s(x)) + u_c^T \hat{R}_{ci} u_c + 2 y_c^T \hat{S}_{ci} u_c + y_c^T \hat{Q}_{ci} y_c$$
$$= w_{ci}(V_s(x)) + h_c^T(x) \hat{Q}_{ci} h_c(x)$$
$$+ 2 h_c^T(x)(\hat{Q}_{ci} J_c(x) + \hat{S}_{ci}) u_c$$
$$+ u_c^T(\hat{R}_{ci} + \hat{S}_{ci}^T J_c(x) + J_c^T(x) \hat{S}_{ci}$$
$$+ J_c^T(x) \hat{Q}_{ci} J_c(x)) u_c, \quad x \in \mathbb{R}^n, \quad u_c \in \mathbb{R}^{m_c}.$$

$$(10.171)$$

Now, equating coefficients of equal powers yields (10.161)–(10.163).

Next, it follows from (10.168) with $k = 1$ that

$$v_{si}(x(t_1) + \mathcal{F}_d(x(t_1)) + G_d(x(t_1))u_d(t_1)) = v_{si}(x(t_1)) + s_{di}(u_{di}(t_1), y_{di}(t_1)).$$

$$(10.172)$$

Now, since the continuous-time dynamics (10.54) are Lipschitz, it follows that for arbitrary $x \in \mathcal{Z}_x$ there exists $x_0 \notin \mathcal{Z}_x$ such that $x(t_1) = x$. Hence, it follows from (10.172) that

$$
\begin{aligned}
v_{si}(x + \mathcal{F}_d(x) + G_d(x)u_d) &= v_{si}(x) + u_d^{\mathrm{T}}\hat{R}_{di}u_d + 2y_d^{\mathrm{T}}\hat{S}_{di}u_d + y_d^{\mathrm{T}}\hat{Q}_{di}y_d \\
&= v_{si}(x) + h_d^{\mathrm{T}}(x)\hat{Q}_{di}h_d(x) \\
&\quad + 2h_d^{\mathrm{T}}(x)(\hat{Q}_{di}J_d(x) + \hat{S}_{di})u_d \\
&\quad + u_d^{\mathrm{T}}(\hat{R}_{di} + \hat{S}_{di}^{\mathrm{T}}J_d(x) + J_d^{\mathrm{T}}(x)\hat{S}_{di} \\
&\quad + J_d^{\mathrm{T}}(x)\hat{Q}_{di}J_d(x))u_d, \quad x \in \mathbb{R}^n, \quad u_d \in \mathbb{R}^{m_d}.
\end{aligned}
$$

$$(10.173)$$

Since the right-hand side of (10.173) is quadratic in u_d it follows that $v_{si}(x + \mathcal{F}_d(x) + G_d(x)u_d)$ is quadratic in u_d, and hence, there exist $P_{1i} : \mathbb{R}^n \to \mathbb{R}^{1 \times m_d}$ and $P_{2i} : \mathbb{R}^n \to \mathbb{N}^{m_d}$, $i = 1, \ldots, q$, such that

$$
v_{si}(x + \mathcal{F}_d(x) + G_d(x)u_d) = v_{si}(x + \mathcal{F}_d(x)) + P_{1i}(x)u_d + u_d^{\mathrm{T}}P_{2i}(x)u_d,
$$

$$x \in \mathbb{R}^n, \quad u_d \in \mathbb{R}^{m_d}. \qquad (10.174)$$

Now, using (10.174) and equating coefficients of equal powers in (10.173) yields (10.164)–(10.166). $\qquad\square$

10.5 Specialization to Large-Scale Linear Impulsive Dynamical Systems

In this section, we specialize the results of Section 10.4 to the case of large-scale linear[2] impulsive dynamical systems. Specifically, we assume that $w_c(\cdot) \in \mathcal{W}$ is linear so that $w_c(z) = Wz$, where $W \in \mathbb{R}^{q \times q}$ is essentially nonnegative, and consider the large-scale linear impulsive dynamical system $\mathcal{G}$ given by

$$
\begin{aligned}
\dot{x}(t) &= A_c x(t) + B_c u_c(t), & x(t) \notin \mathcal{Z}_x, & \qquad (10.175) \\
\Delta x(t) &= (A_d - I_n)x(t) + B_d u_d(t), & x(t) \in \mathcal{Z}_x, & \qquad (10.176) \\
y_c(t) &= C_c x(t) + D_c u_c(t), & x(t) \notin \mathcal{Z}_x, & \qquad (10.177) \\
y_d(t) &= C_d x(t) + D_d u_d(t), & x(t) \in \mathcal{Z}_x, & \qquad (10.178)
\end{aligned}
$$

where $A_c \in \mathbb{R}^{n \times n}$ is partitioned as $A_c \triangleq [A_{cij}]$, $i,j = 1, \ldots, q$, $A_{cij} \in \mathbb{R}^{n_i \times n_j}$, $\sum_{i=1}^{q} n_i = n$, $B_c = \mathrm{block\text{–}diag}[B_{c1}, \ldots, B_{cq}]$, $C_c = \mathrm{block\text{–}diag}[C_{c1}, \ldots, C_{cq}]$, $D_c = \mathrm{block\text{–}diag}[D_{c1}, \ldots, D_{cq}]$, $B_{ci} \in \mathbb{R}^{n_i \times m_{ci}}$, $C_{ci} \in \mathbb{R}^{l_{ci} \times n_i}$, $D_{ci} \in \mathbb{R}^{l_{ci} \times m_{ci}}$, $A_d \in \mathbb{R}^{n \times n}$ is partitioned as $A_d \triangleq [A_{dij}]$, $i,j = 1, \ldots, q$, $A_{dij} \in \mathbb{R}^{n_i \times n_j}$, $B_d = \mathrm{block\text{–}diag}[B_{d1}, \ldots, B_{dq}]$, $C_d = \mathrm{block\text{–}diag}[C_{d1}, \ldots,$

---

[2]Impulsive dynamical systems with $f_c(x) = A_c x$, $G_c(x) = B_c$, $f_d(x) = (A_d - I)x$, $G_d(x) = B_d$, $h_c(x) = C_c x$, $J_c(x) = D_c$, $h_d(x) = C_d x$, and $J_d(x) = D_d$ are *not* linear. However, this minor abuse in terminology provides a natural way of differentiating between impulsive dynamical systems with nonlinear vector fields and nonlinear input functions versus impulsive dynamical systems with linear vector fields and linear input functions.

$C_{\mathrm{d}q}]$, $D_{\mathrm{d}} = \mathrm{block\text{-}diag}[D_{\mathrm{d}1}, \ldots, D_{\mathrm{d}q}]$, $B_{\mathrm{d}i} \in \mathbb{R}^{n_i \times m_{\mathrm{d}i}}$, $C_{\mathrm{d}i} \in \mathbb{R}^{l_{\mathrm{d}i} \times n_i}$, $D_{\mathrm{d}i} \in \mathbb{R}^{l_{\mathrm{d}i} \times m_{\mathrm{d}i}}$, and $i = 1, \ldots, q$.

Theorem 10.14. Consider the large-scale linear impulsive dynamical system $\mathcal{G}$ given by (10.175)–(10.178). Let $R_{ci} \in \mathbb{S}^{m_{ci}}$, $S_{ci} \in \mathbb{R}^{l_{ci} \times m_{ci}}$, $Q_{ci} \in \mathbb{S}^{l_{ci}}$, $R_{\mathrm{d}i} \in \mathbb{S}^{m_{\mathrm{d}i}}$, $S_{\mathrm{d}i} \in \mathbb{R}^{l_{\mathrm{d}i} \times m_{\mathrm{d}i}}$, $Q_{\mathrm{d}i} \in \mathbb{S}^{l_{\mathrm{d}i}}$, $i = 1, \ldots, q$. Then $\mathcal{G}$ is vector dissipative (respectively, exponentially vector dissipative) with respect to the vector hybrid supply rate $(S_{\mathrm{c}}(u_{\mathrm{c}}, y_{\mathrm{c}}), S_{\mathrm{d}}(u_{\mathrm{d}}, y_{\mathrm{d}}))$, where $s_{ci}(u_{ci}, y_{ci}) = u_{ci}^{\mathrm{T}} R_{ci} u_{ci} + 2 y_{ci}^{\mathrm{T}} S_{ci} u_{ci} + y_{ci}^{\mathrm{T}} Q_{ci} y_{ci}$ and $s_{\mathrm{d}i}(u_{\mathrm{d}i}, y_{\mathrm{d}i}) = u_{\mathrm{d}i}^{\mathrm{T}} R_{\mathrm{d}i} u_{\mathrm{d}i} + 2 y_{\mathrm{d}i}^{\mathrm{T}} S_{\mathrm{d}i} u_{\mathrm{d}i} + y_{\mathrm{d}i}^{\mathrm{T}} Q_{\mathrm{d}i} y_{\mathrm{d}i}$, $i = 1, \ldots, q$, if there exist $W \in \mathbb{R}^{q \times q}$, $P_i \in \mathbb{N}^n$, $L_{ci} \in \mathbb{R}^{s_{ci} \times n}$, $Z_{ci} \in \mathbb{R}^{s_{ci} \times m_{\mathrm{c}}}$, $L_{\mathrm{d}i} \in \mathbb{R}^{s_{\mathrm{d}i} \times n}$, and $Z_{\mathrm{d}i} \in \mathbb{R}^{s_{\mathrm{d}i} \times m_{\mathrm{d}}}$, $i = 1, \ldots, q$, such that W is essentially nonnegative and semistable (respectively, asymptotically stable), and, for all $i = 1, \ldots, q$,

$$0 = x^{\mathrm{T}}\Big(A_{\mathrm{c}}^{\mathrm{T}} P_i + P_i A_{\mathrm{c}} - C_{\mathrm{c}}^{\mathrm{T}} \hat{Q}_{ci} C_{\mathrm{c}} - \sum_{j=1}^{q} W_{(i,j)} P_j + L_{ci}^{\mathrm{T}} L_{ci}\Big) x, \quad x \notin \mathcal{Z}_x,$$

$$(10.179)$$

$$0 = x^{\mathrm{T}}\big(P_i B_{\mathrm{c}} - C_{\mathrm{c}}^{\mathrm{T}}(\hat{S}_{ci} + \hat{Q}_{ci} D_{\mathrm{c}}) + L_{ci}^{\mathrm{T}} Z_{ci}\big), \quad x \notin \mathcal{Z}_x, \tag{10.180}$$

$$0 = \hat{R}_{ci} + D_{\mathrm{c}}^{\mathrm{T}} \hat{S}_{ci} + \hat{S}_{ci}^{\mathrm{T}} D_{\mathrm{c}} + D_{\mathrm{c}}^{\mathrm{T}} \hat{Q}_{ci} D_{\mathrm{c}} - Z_{ci}^{\mathrm{T}} Z_{ci}, \tag{10.181}$$

$$0 = x^{\mathrm{T}}\big(A_{\mathrm{d}}^{\mathrm{T}} P_i A_{\mathrm{d}} - C_{\mathrm{d}}^{\mathrm{T}} \hat{Q}_{\mathrm{d}i} C_{\mathrm{d}} - P_i + L_{\mathrm{d}i}^{\mathrm{T}} L_{\mathrm{d}i}\big) x, \quad x \in \mathcal{Z}_x, \tag{10.182}$$

$$0 = x^{\mathrm{T}}\big(A_{\mathrm{d}}^{\mathrm{T}} P_i B_{\mathrm{d}} - C_{\mathrm{d}}^{\mathrm{T}}(\hat{S}_{\mathrm{d}i} + \hat{Q}_{\mathrm{d}i} D_{\mathrm{d}}) + L_{\mathrm{d}i}^{\mathrm{T}} Z_{\mathrm{d}i}\big), \quad x \in \mathcal{Z}_x, \tag{10.183}$$

$$0 = \hat{R}_{\mathrm{d}i} + D_{\mathrm{d}}^{\mathrm{T}} \hat{S}_{\mathrm{d}i} + \hat{S}_{\mathrm{d}i}^{\mathrm{T}} D_{\mathrm{d}} + D_{\mathrm{d}}^{\mathrm{T}} \hat{Q}_{\mathrm{d}i} D_{\mathrm{d}} - B_{\mathrm{d}}^{\mathrm{T}} P_i B_{\mathrm{d}} - Z_{\mathrm{d}i}^{\mathrm{T}} Z_{\mathrm{d}i}. \tag{10.184}$$

Proof. The proof follows from Theorem 10.11 with $\mathcal{F}_{\mathrm{c}}(x) = A_{\mathrm{c}} x$, $G_{\mathrm{c}}(x) = B_{\mathrm{c}}$, $h_{\mathrm{c}}(x) = C_{\mathrm{c}} x$, $J_{\mathrm{c}}(x) = D_{\mathrm{c}}$, $w_{\mathrm{c}}(r) = W r$, $\ell_{ci}(x) = L_{ci} x$, $\mathcal{Z}_{ci}(x) = Z_{ci}$, $\mathcal{F}_{\mathrm{d}}(x) = A_{\mathrm{d}} x$, $G_{\mathrm{d}}(x) = B_{\mathrm{d}}$, $h_{\mathrm{d}}(x) = C_{\mathrm{d}} x$, $J_{\mathrm{d}}(x) = D_{\mathrm{d}}$, $\ell_{\mathrm{d}i}(x) = L_{\mathrm{d}i} x$, $\mathcal{Z}_{\mathrm{d}i}(x) = Z_{\mathrm{d}i}$, $P_{1i}(x) = 2 x^{\mathrm{T}} A_{\mathrm{d}}^{\mathrm{T}} P_i B_{\mathrm{d}}$, $P_{2i}(x) = B_{\mathrm{d}}^{\mathrm{T}} P_i B_{\mathrm{d}}$, and $v_{si}(x) = x^{\mathrm{T}} P_i x$, $i = 1, \ldots, q$. $\qquad\square$

Note that (10.179)–(10.184) are implied by

$$\begin{bmatrix} \mathcal{A}_{ci} & \mathcal{B}_{ci} \\ \mathcal{B}_{ci}^{\mathrm{T}} & \mathcal{C}_{ci} \end{bmatrix} = -\begin{bmatrix} L_{ci}^{\mathrm{T}} \\ Z_{ci}^{\mathrm{T}} \end{bmatrix} \begin{bmatrix} L_{ci} & Z_{ci} \end{bmatrix} \le 0, \tag{10.185}$$

$$\begin{bmatrix} \mathcal{A}_{\mathrm{d}i} & \mathcal{B}_{\mathrm{d}i} \\ \mathcal{B}_{\mathrm{d}i}^{\mathrm{T}} & \mathcal{C}_{\mathrm{d}i} \end{bmatrix} = -\begin{bmatrix} L_{\mathrm{d}i}^{\mathrm{T}} \\ Z_{\mathrm{d}i}^{\mathrm{T}} \end{bmatrix} \begin{bmatrix} L_{\mathrm{d}i} & Z_{\mathrm{d}i} \end{bmatrix} \le 0, \quad i = 1, \ldots, q, \tag{10.186}$$

where, for all $i = 1, \ldots, q$,

$$\mathcal{A}_{ci} = A_{\mathrm{c}}^{\mathrm{T}} P_i + P_i A_{\mathrm{c}} - C_{\mathrm{c}}^{\mathrm{T}} \hat{Q}_{ci} C_{\mathrm{c}} - \sum_{j=1}^{q} W_{(i,j)} P_j, \tag{10.187}$$

$$\mathcal{B}_{ci} = P_i B_{\mathrm{c}} - C_{\mathrm{c}}^{\mathrm{T}}(\hat{S}_{ci} + \hat{Q}_{ci} D_{\mathrm{c}}), \tag{10.188}$$

$$\mathcal{C}_{ci} = -(\hat{R}_{ci} + D_{\mathrm{c}}^{\mathrm{T}} \hat{S}_{ci} + \hat{S}_{ci}^{\mathrm{T}} D_{\mathrm{c}} + D_{\mathrm{c}}^{\mathrm{T}} \hat{Q}_{ci} D_{\mathrm{c}}), \tag{10.189}$$

$$\mathcal{A}_{\mathrm{d}i} = A_{\mathrm{d}}^{\mathrm{T}} P_i A_{\mathrm{d}} - C_{\mathrm{d}}^{\mathrm{T}} \hat{Q}_{\mathrm{d}i} C_{\mathrm{d}} - P_i, \tag{10.190}$$

$$\mathcal{B}_{\mathrm{d}i} = A_{\mathrm{d}}^{\mathrm{T}} P_i B_{\mathrm{d}} - C_{\mathrm{d}}^{\mathrm{T}} (\hat{S}_{\mathrm{d}i} + \hat{Q}_{\mathrm{d}i} D_{\mathrm{d}}), \tag{10.191}$$

$$\mathcal{C}_{\mathrm{d}i} = -(\hat{R}_{\mathrm{d}i} + D_{\mathrm{d}}^{\mathrm{T}} \hat{S}_{\mathrm{d}i} + \hat{S}_{\mathrm{d}i}^{\mathrm{T}} D_{\mathrm{d}} + D_{\mathrm{d}}^{\mathrm{T}} \hat{Q}_{\mathrm{d}i} D_{\mathrm{d}} - B_{\mathrm{d}}^{\mathrm{T}} P_i B_{\mathrm{d}}). \tag{10.192}$$

Hence, vector dissipativity of large-scale linear impulsive dynamical systems with respect to quadratic hybrid supply rates can be characterized via (cascade) linear matrix inequalities (LMIs) [26]. A similar remark holds for Theorem 10.15 below.

The next result presents sufficient conditions guaranteeing vector dissipativity of $\mathcal{G}$ with respect to a quadratic hybrid supply rate in the case where the vector storage function is component decoupled.

Theorem 10.15. Consider the large-scale linear impulsive dynamical system $\mathcal{G}$ given by (10.175)–(10.178). Let $R_{\mathrm{c}i} \in \mathbb{S}^{m_{\mathrm{c}i}}$, $S_{\mathrm{c}i} \in \mathbb{R}^{l_{\mathrm{c}i} \times m_{\mathrm{c}i}}$, $Q_{\mathrm{c}i} \in \mathbb{S}^{l_{\mathrm{c}i}}$, $R_{\mathrm{d}i} \in \mathbb{S}^{m_{\mathrm{d}i}}$, $S_{\mathrm{d}i} \in \mathbb{R}^{l_{\mathrm{d}i} \times m_{\mathrm{d}i}}$, and $Q_{\mathrm{d}i} \in \mathbb{S}^{l_{\mathrm{d}i}}$, $i = 1, \ldots, q$, be given. Assume there exist matrices $W \in \mathbb{R}^{q \times q}$, $P_i \in \mathbb{N}^{n_i}$, $L_{\mathrm{c}ii} \in \mathbb{R}^{s_{\mathrm{c}ii} \times n_i}$, $Z_{\mathrm{c}ii} \in \mathbb{R}^{s_{\mathrm{c}ii} \times m_{\mathrm{c}i}}$, $L_{\mathrm{d}ii} \in \mathbb{R}^{s_{\mathrm{d}ii} \times n_i}$, $Z_{\mathrm{d}ii} \in \mathbb{R}^{s_{\mathrm{d}ii} \times m_{\mathrm{c}i}}$, $i = 1, \ldots, q$, $L_{\mathrm{c}ij} \in \mathbb{R}^{s_{\mathrm{c}ij} \times n_i}$, $Z_{\mathrm{c}ij} \in \mathbb{R}^{s_{\mathrm{c}ij} \times n_j}$, $L_{\mathrm{d}ij} \in \mathbb{R}^{s_{\mathrm{d}ij} \times n_i}$, and $Z_{\mathrm{d}ij} \in \mathbb{R}^{s_{\mathrm{d}ij} \times n_j}$, $i, j = 1, \ldots, q$, $i \neq j$, such that W is essentially nonnegative and semistable (respectively, asymptotically stable), and, for all $i = 1, \ldots, q$,

$$0 \geq x_i^{\mathrm{T}} \left(A_{\mathrm{c}ii}^{\mathrm{T}} P_i + P_i A_{\mathrm{c}ii} - C_{\mathrm{c}i}^{\mathrm{T}} Q_{\mathrm{c}i} C_{\mathrm{c}i} - W_{(i,i)} P_i + L_{\mathrm{c}ii}^{\mathrm{T}} L_{\mathrm{c}ii} \right.$$

$$\left. - \sum_{j=1, j \neq i}^{q} L_{\mathrm{c}ij}^{\mathrm{T}} L_{\mathrm{c}ij} \right) x_i, \quad x \notin \mathcal{Z}_x, \tag{10.193}$$

$$0 = x_i^{\mathrm{T}} (P_i B_{\mathrm{c}i} - C_{\mathrm{c}i}^{\mathrm{T}} S_{\mathrm{c}i} - C_{\mathrm{c}i}^{\mathrm{T}} Q_{\mathrm{c}i} D_{\mathrm{c}i} + L_{\mathrm{c}ii}^{\mathrm{T}} Z_{\mathrm{c}ii}), \quad x \notin \mathcal{Z}_x \tag{10.194}$$

$$0 \leq R_{\mathrm{c}i} + D_{\mathrm{c}i}^{\mathrm{T}} S_{\mathrm{c}i} + S_{\mathrm{c}i}^{\mathrm{T}} D_{\mathrm{c}i} + D_{\mathrm{c}i}^{\mathrm{T}} Q_{\mathrm{c}i} D_{\mathrm{c}i} - Z_{\mathrm{c}ii}^{\mathrm{T}} Z_{\mathrm{c}ii}, \tag{10.195}$$

$$0 \geq x_i^{\mathrm{T}} \left(A_{\mathrm{d}ii}^{\mathrm{T}} P_i A_{\mathrm{d}ii} - C_{\mathrm{d}i}^{\mathrm{T}} Q_{\mathrm{d}i} C_{\mathrm{d}i} - P_i + L_{\mathrm{d}ii}^{\mathrm{T}} L_{\mathrm{d}ii} \right.$$

$$\left. + \sum_{j=1, j \neq i}^{q} L_{\mathrm{d}ij}^{\mathrm{T}} L_{\mathrm{d}ij} \right) x_i, \quad x \in \mathcal{Z}_x, \tag{10.196}$$

$$0 = x_i^{\mathrm{T}} (A_{\mathrm{d}ii}^{\mathrm{T}} P_i B_{\mathrm{d}i} - C_{\mathrm{d}i}^{\mathrm{T}} S_{\mathrm{d}i} - C_{\mathrm{d}i}^{\mathrm{T}} Q_{\mathrm{d}i} D_{\mathrm{d}i} + L_{\mathrm{d}ii}^{\mathrm{T}} Z_{\mathrm{d}ii}), \quad x \in \mathcal{Z}_x, \tag{10.197}$$

$$0 \leq R_{\mathrm{d}i} + D_{\mathrm{d}i}^{\mathrm{T}} S_{\mathrm{d}i} + S_{\mathrm{d}i}^{\mathrm{T}} D_{\mathrm{d}i} + D_{\mathrm{d}i}^{\mathrm{T}} Q_{\mathrm{d}i} D_{\mathrm{d}i} - B_{\mathrm{d}i}^{\mathrm{T}} P_i B_{\mathrm{d}i} - Z_{\mathrm{d}ii}^{\mathrm{T}} Z_{\mathrm{d}ii}, \tag{10.198}$$

and for $j = 1, \ldots, q$, $l = 1, \ldots, q$, $j \neq i$, $l \neq i$,

$$0 = x_i^{\mathrm{T}} (P_i A_{\mathrm{c}ij} + L_{\mathrm{c}ij}^{\mathrm{T}} Z_{\mathrm{c}ij}), \quad x \notin \mathcal{Z}_x, \tag{10.199}$$

$$0 \leq x_j^{\mathrm{T}}(W_{(i,j)}P_j - Z_{\mathrm{c}ij}^{\mathrm{T}}Z_{\mathrm{c}ij})x_j, \quad x \notin \mathcal{Z}_x, \tag{10.200}$$

$$0 = x_j^{\mathrm{T}}(A_{\mathrm{d}ij}^{\mathrm{T}}P_iB_{\mathrm{d}i}), \quad x \in \mathcal{Z}_x, \tag{10.201}$$

$$0 \geq x_j^{\mathrm{T}}(A_{\mathrm{d}ij}^{\mathrm{T}}P_iA_{\mathrm{d}il})x_l, \quad x \in \mathcal{Z}_x, \tag{10.202}$$

$$0 \geq x_i^{\mathrm{T}}(A_{\mathrm{d}ii}^{\mathrm{T}}P_iA_{\mathrm{d}ij} + L_{\mathrm{d}ij}^{\mathrm{T}}Z_{\mathrm{d}ij})x_j, \quad x \in \mathcal{Z}_x, \tag{10.203}$$

$$0 = x_j^{\mathrm{T}}(Z_{\mathrm{d}ij}^{\mathrm{T}}Z_{\mathrm{d}ij})x_j, \quad x \in \mathcal{Z}_x. \tag{10.204}$$

Then $\mathcal{G}$ is vector dissipative (respectively, exponentially vector dissipative) with respect to the vector hybrid supply rate $(S_{\mathrm{c}}(u_{\mathrm{c}}, y_{\mathrm{c}}), S_{\mathrm{d}}(u_{\mathrm{d}}, y_{\mathrm{d}}))$, where $s_{\mathrm{c}i}(u_{\mathrm{c}i}, y_{\mathrm{c}i}) = u_{\mathrm{c}i}^{\mathrm{T}}R_{\mathrm{c}i}u_{\mathrm{c}i} + 2y_{\mathrm{c}i}^{\mathrm{T}}S_{\mathrm{c}i}u_{\mathrm{c}i} + y_{\mathrm{c}i}^{\mathrm{T}}Q_{\mathrm{c}i}y_{\mathrm{c}i}$ and $s_{\mathrm{d}i}(u_{\mathrm{d}i}, y_{\mathrm{d}i}) = u_{\mathrm{d}i}^{\mathrm{T}}R_{\mathrm{d}i}u_{\mathrm{d}i} + 2y_{\mathrm{d}i}^{\mathrm{T}}S_{\mathrm{d}i}u_{\mathrm{d}i} + y_{\mathrm{d}i}^{\mathrm{T}}Q_{\mathrm{d}i}y_{\mathrm{d}i}$, $i = 1, \ldots, q$.

Proof. Since P_i is nonnegative definite, the function $v_{\mathrm{s}i}(x_i) \triangleq x_i^{\mathrm{T}}P_ix_i$, $x_i \in \mathbb{R}^{n_i}$, is nonnegative definite and $v_{\mathrm{s}i}(0) = 0$. Moreover, since $v_{\mathrm{s}i}(\cdot)$ is continuously differentiable it follows from (10.193)–(10.204) that for all $u_{\mathrm{c}i}(t) \in \mathbb{R}^{m_{\mathrm{c}i}}$, $u_{\mathrm{d}i}(t_k) \in \mathbb{R}^{m_{\mathrm{d}i}}$, $i = 1, \ldots, q$, and $t_k < t \leq t_{k+1}$, $k \in \overline{\mathbb{Z}}_+$,

$$\dot{v}_{\mathrm{s}i}(x_i(t))$$

$$= 2x_i^{\mathrm{T}}(t)P_i\left[\sum_{j=1}^{q}A_{\mathrm{c}ij}x_j(t) + B_{\mathrm{c}i}u_{\mathrm{c}i}(t)\right]$$

$$\leq x_i^{\mathrm{T}}(t)\left[W_{(i,i)}P_i + C_{\mathrm{c}i}^{\mathrm{T}}Q_{\mathrm{c}i}C_{\mathrm{c}i} - L_{\mathrm{c}ii}^{\mathrm{T}}L_{\mathrm{c}ii} - \sum_{j=1, j\neq i}^{q}L_{\mathrm{c}ij}^{\mathrm{T}}L_{\mathrm{c}ij}\right]x_i(t)$$

$$- \sum_{j=1, j\neq i}^{q}2x_i^{\mathrm{T}}(t)L_{\mathrm{c}ij}^{\mathrm{T}}Z_{\mathrm{c}ij}x_j(t)$$

$$+2x_i^{\mathrm{T}}(t)C_{\mathrm{c}i}^{\mathrm{T}}S_{\mathrm{c}i}u_{\mathrm{c}i}(t) + 2x_i^{\mathrm{T}}(t)C_{\mathrm{c}i}^{\mathrm{T}}Q_{\mathrm{c}i}D_{\mathrm{c}i}u_{\mathrm{c}i}(t)$$

$$-2x_i^{\mathrm{T}}(t)L_{\mathrm{c}ii}^{\mathrm{T}}Z_{\mathrm{c}ii}u_{\mathrm{c}i}(t) + \sum_{j=1, j\neq i}^{q}x_j^{\mathrm{T}}(t)[W_{(i,j)}P_j - Z_{\mathrm{c}ij}^{\mathrm{T}}Z_{\mathrm{c}ij}]x_j(t)$$

$$+u_{\mathrm{c}i}^{\mathrm{T}}(t)R_{\mathrm{c}i}u_{\mathrm{c}i}(t) + 2u_{\mathrm{c}i}^{\mathrm{T}}(t)D_{\mathrm{c}i}^{\mathrm{T}}S_{\mathrm{c}i}u_{\mathrm{c}i}(t) + u_{\mathrm{c}i}^{\mathrm{T}}(t)D_{\mathrm{c}i}^{\mathrm{T}}Q_{\mathrm{c}i}D_{\mathrm{c}i}u_{\mathrm{c}i}(t)$$

$$-u_{\mathrm{c}i}^{\mathrm{T}}(t)Z_{\mathrm{c}ii}^{\mathrm{T}}Z_{\mathrm{c}ii}u_{\mathrm{c}i}(t)$$

$$= \sum_{j=1}^{q}W_{(i,j)}v_{\mathrm{s}j}(x_j(t)) + u_{\mathrm{c}i}^{\mathrm{T}}(t)R_{\mathrm{c}i}u_{\mathrm{c}i}(t) + 2y_{\mathrm{c}i}^{\mathrm{T}}(t)S_{\mathrm{c}i}u_{\mathrm{c}i}(t)$$

$$+y_{\mathrm{c}i}^{\mathrm{T}}(t)Q_{\mathrm{c}i}y_{\mathrm{c}i}(t)$$

$$-[L_{\mathrm{c}ii}x_i(t) + Z_{\mathrm{c}ii}u_{\mathrm{c}i}(t)]^{\mathrm{T}}[L_{\mathrm{c}ii}x_i(t) + Z_{\mathrm{c}ii}u_{\mathrm{c}i}(t)]$$

$$- \sum_{j=1, j\neq i}^{q}(L_{\mathrm{c}ij}x_i(t) + Z_{\mathrm{c}ij}x_j(t))^{\mathrm{T}}(L_{\mathrm{c}ij}x_i(t) + Z_{\mathrm{c}ij}x_j(t))$$

$$\leq s_{\mathrm{c}i}(u_{\mathrm{c}i}(t), y_{\mathrm{c}i}(t)) + \sum_{j=1}^{q} W_{(i,j)} v_{\mathrm{s}j}(x_j(t)). \tag{10.205}$$

Furthermore,

$$v_{\mathrm{s}i}\left(\sum_{j=1}^{q} A_{\mathrm{d}ij}x_j(t_k) + B_{\mathrm{d}i}u_{\mathrm{d}i}(t_k)\right) = \left[\sum_{j=1}^{q} A_{\mathrm{d}ij}x_j(t_k) + B_{\mathrm{d}i}u_{\mathrm{d}i}(t_k)\right]^{\mathrm{T}}$$

$$\cdot P_i\left[\sum_{j=1}^{q} A_{\mathrm{d}ij}x_j(t_k) + B_{\mathrm{d}i}u_{\mathrm{d}i}(t_k)\right]$$

$$\leq x_i^{\mathrm{T}}(t_k)\left[P_i + C_{\mathrm{d}i}^{\mathrm{T}}Q_{\mathrm{d}i}C_{\mathrm{d}i} - L_{\mathrm{d}ii}^{\mathrm{T}}L_{\mathrm{d}ii} - \sum_{j=1,\,j\neq i}^{q} L_{\mathrm{d}ij}^{\mathrm{T}}L_{\mathrm{d}ij}\right]x_i(t_k)$$

$$- \sum_{j=1,\,j\neq i}^{q} 2x_i^{\mathrm{T}}(t_k)L_{\mathrm{d}ij}^{\mathrm{T}}Z_{\mathrm{d}ij}x_j(t_k) + 2x_i^{\mathrm{T}}(t_k)C_{\mathrm{d}i}^{\mathrm{T}}S_{\mathrm{d}i}u_{\mathrm{d}i}(t_k)$$

$$+ 2x_i^{\mathrm{T}}(t_k)C_{\mathrm{d}i}^{\mathrm{T}}Q_{\mathrm{d}i}D_{\mathrm{d}i}u_{\mathrm{d}i}(t_k) - 2x_i^{\mathrm{T}}(t_k)L_{\mathrm{d}ii}^{\mathrm{T}}Z_{\mathrm{d}ii}u_{\mathrm{d}i}(t_k)$$

$$- \sum_{j=1,\,j\neq i}^{q} x_j^{\mathrm{T}}(t_k)Z_{\mathrm{d}ij}^{\mathrm{T}}Z_{\mathrm{d}ij}x_j(t_k) + u_{\mathrm{d}i}^{\mathrm{T}}(t_k)R_{\mathrm{d}i}u_{\mathrm{d}i}(t_k)$$

$$+ 2u_{\mathrm{d}i}^{\mathrm{T}}(t_k)D_{\mathrm{d}i}^{\mathrm{T}}S_{\mathrm{d}i}u_{\mathrm{d}i}(t_k) + u_{\mathrm{d}i}^{\mathrm{T}}(t_k)D_{\mathrm{d}i}^{\mathrm{T}}Q_{\mathrm{d}i}D_{\mathrm{d}i}u_{\mathrm{d}i}(t_k)$$

$$- u_{\mathrm{d}i}^{\mathrm{T}}(t_k)Z_{\mathrm{d}ii}^{\mathrm{T}}Z_{\mathrm{d}ii}u_{\mathrm{d}i}(t_k)$$

$$= v_{\mathrm{s}i}(x_i(t_k)) + u_{\mathrm{d}i}^{\mathrm{T}}(t_k)R_{\mathrm{d}i}u_{\mathrm{d}i}(t_k) + 2y_{\mathrm{d}i}^{\mathrm{T}}(t_k)S_{\mathrm{d}i}u_{\mathrm{d}i}(t_k) + y_{\mathrm{d}i}^{\mathrm{T}}(t_k)Q_{\mathrm{d}i}y_{\mathrm{d}i}(t_k)$$

$$- [L_{\mathrm{d}ii}x_i(t_k) + Z_{\mathrm{d}ii}u_{\mathrm{d}i}(t_k)]^{\mathrm{T}}[L_{\mathrm{d}ii}x_i(t_k) + Z_{\mathrm{d}ii}u_{\mathrm{d}i}(t_k)]$$

$$- \sum_{j=1,\,j\neq i}^{q} [L_{\mathrm{d}ij}x_i(t_k) + Z_{\mathrm{d}ij}x_j(t_k)]^{\mathrm{T}}[L_{\mathrm{d}ij}x_i(t_k) + Z_{\mathrm{d}ij}x_j(t_k)]$$

$$\leq s_{\mathrm{d}i}(u_{\mathrm{d}i}(t_k), y_{\mathrm{d}i}(t_k)) + v_{\mathrm{s}i}(x_i(t_k)). \tag{10.206}$$

Writing (10.205) and (10.206) in vector form yields

$$\dot{V}_{\mathrm{s}}(x) \leq\leq W V_{\mathrm{s}}(x) + S_{\mathrm{c}}(u_{\mathrm{c}}, y_{\mathrm{c}}), \quad u_{\mathrm{c}} \in \mathbb{R}^{m_{\mathrm{c}}}, \quad x \notin \mathcal{Z}_x, \tag{10.207}$$

$$V_{\mathrm{s}}(A_{\mathrm{d}}x + B_{\mathrm{d}}u_{\mathrm{d}}) \leq\leq V_{\mathrm{s}}(x) + S_{\mathrm{d}}(u_{\mathrm{d}}, y_{\mathrm{d}}), \quad u_{\mathrm{d}} \in \mathbb{R}^{m_{\mathrm{d}}}, \quad x \in \mathcal{Z}_x, \tag{10.208}$$

where $V_{\mathrm{s}}(x) \triangleq [v_{\mathrm{s}1}(x_1), \ldots, v_{\mathrm{s}q}(x_q)]^{\mathrm{T}}$, $x \in \mathbb{R}^n$. Now, it follows from Definition 10.6 that $\mathcal{G}$ is vector dissipative (respectively, exponentially vector dissipative) with respect to the vector hybrid supply rate $(S_{\mathrm{c}}(u_{\mathrm{c}}, y_{\mathrm{c}}), S_{\mathrm{d}}(u_{\mathrm{d}}, y_{\mathrm{d}}))$ and with vector storage function $V_{\mathrm{s}}(x)$, $x \in \mathbb{R}^n$. $\qquad\square$

10.6 Stability of Feedback Interconnections of Large-Scale Impulsive Dynamical Systems

In this section, we use the concepts of vector dissipativity and vector storage functions as candidate vector Lyapunov functions to develop feedback interconnection stability results of large-scale impulsive dynamical systems. General stability criteria are given for Lyapunov and asymptotic stability of feedback large-scale impulsive dynamical systems. Specifically, we consider input/state-dependent impulsive large-scale dynamical systems $\mathcal{G}$ of the form

$$\dot{x}(t) = F_{\mathrm{c}}(x(t), u_{\mathrm{c}}(t)), \quad x(t_0) = x_0, \quad (x(t), u_{\mathrm{c}}(t)) \notin \mathcal{Z}, \quad t \geq t_0, \quad (10.209)$$

$$\Delta x(t) = F_{\mathrm{d}}(x(t), u_{\mathrm{d}}(t)), \quad (x(t), u_{\mathrm{c}}(t)) \in \mathcal{Z}, \quad (10.210)$$

$$y_{\mathrm{c}}(t) = H_{\mathrm{c}}(x(t), u_{\mathrm{c}}(t)), \quad (x(t), u_{\mathrm{c}}(t)) \notin \mathcal{Z}, \quad (10.211)$$

$$y_{\mathrm{d}}(t) = H_{\mathrm{d}}(x(t), u_{\mathrm{d}}(t)), \quad (x(t), u_{\mathrm{c}}(t)) \in \mathcal{Z}, \quad (10.212)$$

where $x(t) \in \mathcal{D} \subseteq \mathbb{R}^n$, $t \geq t_0$, $u_{\mathrm{c}}(t) \in U_{\mathrm{c}} \subseteq \mathbb{R}^{m_{\mathrm{c}}}$, $u_{\mathrm{d}}(t_k) \in U_{\mathrm{d}} \subseteq \mathbb{R}^{m_{\mathrm{d}}}$, $y_{\mathrm{c}}(t) \in Y_{\mathrm{c}} \subseteq \mathbb{R}^{l_c}$, $y_{\mathrm{d}}(t_k) \in Y_{\mathrm{d}} \subseteq \mathbb{R}^{l_d}$, $F_{\mathrm{c}} : \mathcal{D} \times U_{\mathrm{c}} \to \mathbb{R}^n$, $F_{\mathrm{d}} : \mathcal{D} \times U_{\mathrm{d}} \to \mathbb{R}^n$, $H_{\mathrm{c}} : \mathcal{D} \times U_{\mathrm{c}} \to Y_{\mathrm{c}}$, $H_{\mathrm{d}} : \mathcal{D} \times U_{\mathrm{d}} \to Y_{\mathrm{d}}$, $\mathcal{D}$ is an open set with $0 \in \mathcal{D}$, $\mathcal{Z} \subset \mathcal{D} \times U_{\mathrm{c}}$, and $F_{\mathrm{c}}(0,0) = 0$.

Here, we assume that $\mathcal{G}$ represents a large-scale impulsive dynamical system composed of q interconnected controlled impulsive subsystems $\mathcal{G}_i$ such that, for all $i = 1, \ldots, q$,

$$F_{\mathrm{c}i}(x, u_{\mathrm{c}i}) = f_{\mathrm{c}i}(x_i) + \mathcal{I}_{\mathrm{c}i}(x) + G_{\mathrm{c}i}(x_i)u_{\mathrm{c}i}, \quad (10.213)$$

$$F_{\mathrm{d}i}(x, u_{\mathrm{d}i}) = f_{\mathrm{d}i}(x_i) + \mathcal{I}_{\mathrm{d}i}(x) + G_{\mathrm{d}i}(x_i)u_{\mathrm{d}i}, \quad (10.214)$$

$$H_{\mathrm{c}i}(x_i, u_{\mathrm{c}i}) = h_{\mathrm{c}i}(x_i) + J_{\mathrm{c}i}(x_i)u_{\mathrm{c}i}, \quad (10.215)$$

$$H_{\mathrm{d}i}(x_i, u_{\mathrm{d}i}) = h_{\mathrm{d}i}(x_i) + J_{\mathrm{d}i}(x_i)u_{\mathrm{d}i}, \quad (10.216)$$

where $x_i \in \mathcal{D}_i \subseteq \mathbb{R}^{n_i}$, $u_{\mathrm{c}i} \in U_{\mathrm{c}i} \subseteq \mathbb{R}^{m_{\mathrm{c}i}}$, $u_{\mathrm{d}i} \in U_{\mathrm{d}i} \subseteq \mathbb{R}^{m_{\mathrm{d}i}}$, $y_{\mathrm{c}i} \triangleq H_{\mathrm{c}i}(x_i, u_{\mathrm{c}i}) \in Y_{\mathrm{c}i} \subseteq \mathbb{R}^{l_{\mathrm{c}i}}$, $y_{\mathrm{d}i} \triangleq H_{\mathrm{d}i}(x_i, u_{\mathrm{d}i}) \in Y_{\mathrm{d}i} \subseteq \mathbb{R}^{l_{\mathrm{d}i}}$, $((u_{\mathrm{c}i}, u_{\mathrm{d}i}), (y_{\mathrm{c}i}, y_{\mathrm{d}i}))$ is the hybrid input-output pair for the ith subsystem, $f_{\mathrm{c}i} : \mathbb{R}^{n_i} \to \mathbb{R}^{n_i}$ and $\mathcal{I}_{\mathrm{c}i} : \mathcal{D} \to \mathbb{R}^{n_i}$ are Lipschitz continuous and satisfy $f_{\mathrm{c}i}(0) = 0$ and $\mathcal{I}_{\mathrm{c}i}(0) = 0$, $f_{\mathrm{d}i} : \mathbb{R}^{n_i} \to \mathbb{R}^{n_i}$ and $\mathcal{I}_{\mathrm{d}i} : \mathcal{D} \to \mathbb{R}^{n_i}$ are continuous, $G_{\mathrm{c}i} : \mathbb{R}^{n_i} \to \mathbb{R}^{n_i \times m_{\mathrm{c}i}}$ and $G_{\mathrm{d}i} : \mathbb{R}^{n_i} \to \mathbb{R}^{n_i \times m_{\mathrm{d}i}}$ are continuous, $h_{\mathrm{c}i} : \mathbb{R}^{n_i} \to \mathbb{R}^{l_{\mathrm{c}i}}$ and satisfies $h_{\mathrm{c}i}(0) = 0$, $h_{\mathrm{d}i} : \mathbb{R}^{n_i} \to \mathbb{R}^{l_{\mathrm{d}i}}$, $J_{\mathrm{c}i} : \mathbb{R}^{n_i} \to \mathbb{R}^{l_{\mathrm{c}i} \times m_{\mathrm{c}i}}$, $J_{\mathrm{d}i} : \mathbb{R}^{n_i} \to \mathbb{R}^{l_{\mathrm{d}i} \times m_{\mathrm{d}i}}$, $\sum_{i=1}^{q} n_i = n$, $\sum_{i=1}^{q} m_{\mathrm{c}i} = m_{\mathrm{c}}$, $\sum_{i=1}^{q} m_{\mathrm{d}i} = m_{\mathrm{d}}$, $\sum_{i=1}^{q} l_{\mathrm{c}i} = l_{\mathrm{c}}$, and $\sum_{i=1}^{q} l_{\mathrm{d}i} = l_{\mathrm{d}}$. Furthermore, for the large-scale dynamical system $\mathcal{G}$ we assume that the required properties for the existence and uniqueness of solutions are satisfied, that is, for each $i \in \{1, \ldots, q\}$, $u_{\mathrm{c}i}(\cdot)$ and $u_{\mathrm{d}i}(\cdot)$ satisfy sufficient regularity conditions such that the system (10.209) and (10.210) has a unique solution forward in time. We define the composite input and composite output for the large-scale impulsive dynamical system $\mathcal{G}$ as $u_{\mathrm{c}} \triangleq [u_{\mathrm{c}1}^{\mathrm{T}}, \ldots, u_{\mathrm{c}q}^{\mathrm{T}}]^{\mathrm{T}}$, $u_{\mathrm{d}} \triangleq [u_{\mathrm{d}1}^{\mathrm{T}}, \ldots, u_{\mathrm{d}q}^{\mathrm{T}}]^{\mathrm{T}}$, $y_{\mathrm{c}} \triangleq [y_{\mathrm{c}1}^{\mathrm{T}}, \ldots, y_{\mathrm{c}q}^{\mathrm{T}}]^{\mathrm{T}}$, and $y_{\mathrm{d}} \triangleq [y_{\mathrm{d}1}^{\mathrm{T}}, \ldots, y_{\mathrm{d}q}^{\mathrm{T}}]^{\mathrm{T}}$, respectively.

Next, we consider a dynamical large-scale impulsive feedback system $\mathcal{G}_{\mathrm{c}}$ given by

$$\dot{x}_{\mathrm{c}}(t) = F_{\mathrm{cc}}(x_{\mathrm{c}}(t), u_{\mathrm{cc}}(t)), \quad x_{\mathrm{c}}(t_0) = x_{\mathrm{c}0}, \quad (x_{\mathrm{c}}(t), u_{\mathrm{cc}}(t)) \notin \mathcal{Z}_{\mathrm{c}}, \quad (10.217)$$

$$\Delta x_{\mathrm{c}}(t) = F_{\mathrm{dc}}(x_{\mathrm{c}}(t), u_{\mathrm{dc}}(t)), \quad (x_{\mathrm{c}}(t), u_{\mathrm{cc}}(t)) \in \mathcal{Z}_{\mathrm{c}}, \quad (10.218)$$

$$y_{\mathrm{cc}}(t) = H_{\mathrm{cc}}(x_{\mathrm{c}}(t), u_{\mathrm{cc}}(t)), \quad (x_{\mathrm{c}}(t), u_{\mathrm{cc}}(t)) \notin \mathcal{Z}_{\mathrm{c}}, \quad (10.219)$$

$$y_{\mathrm{dc}}(t) = H_{\mathrm{dc}}(x_{\mathrm{c}}(t), u_{\mathrm{dc}}(t)), \quad (x_{\mathrm{c}}(t), u_{\mathrm{cc}}(t)) \in \mathcal{Z}_{\mathrm{c}}, \quad (10.220)$$

where $F_{\mathrm{cc}} : \mathbb{R}^{n_c} \times U_{\mathrm{cc}} \to \mathbb{R}^{n_c}$, $F_{\mathrm{dc}} : \mathbb{R}^{n_c} \times U_{\mathrm{dc}} \to \mathbb{R}^{n_c}$, $H_{\mathrm{cc}} : \mathbb{R}^{n_c} \times U_{\mathrm{cc}} \to Y_{\mathrm{cc}}$, $H_{\mathrm{dc}} : \mathbb{R}^{n_c} \times U_{\mathrm{dc}} \to Y_{\mathrm{dc}}$, $F_{\mathrm{cc}} \triangleq [F_{\mathrm{cc}1}^{\mathrm{T}}, \ldots, F_{\mathrm{cc}q}^{\mathrm{T}}]^{\mathrm{T}}$, $F_{\mathrm{dc}} \triangleq [F_{\mathrm{dc}1}^{\mathrm{T}}, \ldots, F_{\mathrm{dc}q}^{\mathrm{T}}]^{\mathrm{T}}$, $H_{\mathrm{cc}} \triangleq [H_{\mathrm{cc}1}^{\mathrm{T}}, \ldots, H_{\mathrm{cc}q}^{\mathrm{T}}]^{\mathrm{T}}$, $H_{\mathrm{dc}} \triangleq [H_{\mathrm{dc}1}^{\mathrm{T}}, \ldots, H_{\mathrm{dc}q}^{\mathrm{T}}]^{\mathrm{T}}$, $U_{\mathrm{cc}} \subseteq \mathbb{R}^{l_c}$, $U_{\mathrm{dc}} \subseteq \mathbb{R}^{l_d}$, $Y_{\mathrm{cc}} \subseteq \mathbb{R}^{m_c}$, $Y_{\mathrm{dc}} \subseteq \mathbb{R}^{m_d}$. Moreover, for all $i = 1, \ldots, q$, we assume that

$$F_{\mathrm{cc}i}(x_{\mathrm{c}}, u_{\mathrm{cc}i}) = f_{\mathrm{cc}i}(x_{\mathrm{c}i}) + \mathcal{I}_{\mathrm{cc}i}(x_{\mathrm{c}}) + G_{\mathrm{cc}i}(x_{\mathrm{c}i})u_{\mathrm{cc}i}, \quad (10.221)$$

$$F_{\mathrm{dc}i}(x_{\mathrm{c}}, u_{\mathrm{dc}i}) = f_{\mathrm{dc}i}(x_{\mathrm{c}i}) + \mathcal{I}_{\mathrm{dc}i}(x_{\mathrm{c}}) + G_{\mathrm{dc}i}(x_{\mathrm{c}i})u_{\mathrm{dc}i}, \quad (10.222)$$

$$H_{\mathrm{cc}i}(x_{\mathrm{c}i}, u_{\mathrm{cc}i}) = h_{\mathrm{cc}i}(x_{\mathrm{c}i}) + J_{\mathrm{cc}i}(x_{\mathrm{c}i})u_{\mathrm{cc}i}, \quad (10.223)$$

$$H_{\mathrm{dc}i}(x_{\mathrm{c}i}, u_{\mathrm{dc}i}) = h_{\mathrm{dc}i}(x_{\mathrm{c}i}) + J_{\mathrm{dc}i}(x_{\mathrm{c}i})u_{\mathrm{dc}i}, \quad (10.224)$$

where $u_{\mathrm{cc}i} \in U_{\mathrm{cc}i} \subseteq \mathbb{R}^{l_{ci}}$, $u_{\mathrm{dc}i} \in U_{\mathrm{dc}i} \subseteq \mathbb{R}^{l_{di}}$, $y_{\mathrm{cc}i} \triangleq H_{\mathrm{cc}i}(x_{\mathrm{c}i}, u_{\mathrm{cc}i}) \in Y_{\mathrm{cc}i} \subseteq \mathbb{R}^{m_{ci}}$, $y_{\mathrm{dc}i} \triangleq H_{\mathrm{dc}i}(x_{\mathrm{c}i}, u_{\mathrm{dc}i}) \in Y_{\mathrm{dc}i} \subseteq \mathbb{R}^{m_{di}}$, $f_{\mathrm{cc}i} : \mathbb{R}^{n_{ci}} \to \mathbb{R}^{n_{ci}}$ and $\mathcal{I}_{\mathrm{cc}i} : \mathbb{R}^{n_c} \to \mathbb{R}^{n_{ci}}$ satisfy $f_{\mathrm{cc}i}(0) = 0$ and $\mathcal{I}_{\mathrm{cc}i}(0) = 0$, $f_{\mathrm{dc}i} : \mathbb{R}^{n_{ci}} \to \mathbb{R}^{n_{ci}}$, $\mathcal{I}_{\mathrm{dc}i} : \mathbb{R}^{n_c} \to \mathbb{R}^{n_{ci}}$, $G_{\mathrm{cc}i} : \mathbb{R}^{n_{ci}} \to \mathbb{R}^{n_{ci} \times l_{ci}}$, $G_{\mathrm{dc}i} : \mathbb{R}^{n_{ci}} \to \mathbb{R}^{n_{ci} \times l_{di}}$, $h_{\mathrm{cc}i} : \mathbb{R}^{n_{ci}} \to \mathbb{R}^{m_{ci}}$ and satisfies $h_{\mathrm{cc}i}(0) = 0$, $h_{\mathrm{dc}i} : \mathbb{R}^{n_{ci}} \to \mathbb{R}^{m_{di}}$, $J_{\mathrm{cc}i} : \mathbb{R}^{n_{ci}} \to \mathbb{R}^{m_{ci} \times l_{ci}}$, $J_{\mathrm{dc}i} : \mathbb{R}^{n_{ci}} \to \mathbb{R}^{m_{di} \times l_{di}}$, and $\sum_{i=1}^{q} n_{ci} = n_{\mathrm{c}}$. Furthermore, we define the composite input and composite output for the system $\mathcal{G}_{\mathrm{c}}$ as $u_{\mathrm{cc}} \triangleq [u_{\mathrm{cc}1}^{\mathrm{T}}, \ldots, u_{\mathrm{cc}q}^{\mathrm{T}}]^{\mathrm{T}}$, $u_{\mathrm{dc}} \triangleq [u_{\mathrm{dc}1}^{\mathrm{T}}, \ldots, u_{\mathrm{dc}q}^{\mathrm{T}}]^{\mathrm{T}}$, $y_{\mathrm{cc}} \triangleq [y_{\mathrm{cc}1}^{\mathrm{T}}, \ldots, y_{\mathrm{cc}q}^{\mathrm{T}}]^{\mathrm{T}}$, and $y_{\mathrm{dc}} \triangleq [y_{\mathrm{dc}1}^{\mathrm{T}}, \ldots, y_{\mathrm{dc}q}^{\mathrm{T}}]^{\mathrm{T}}$, respectively. In this case, $U_{\mathrm{cc}} = U_{\mathrm{cc}1} \times \cdots \times U_{\mathrm{cc}q}$, $U_{\mathrm{dc}} = U_{\mathrm{dc}1} \times \cdots \times U_{\mathrm{dc}q}$, $Y_{\mathrm{cc}} = Y_{\mathrm{cc}1} \times \cdots \times Y_{\mathrm{cc}q}$, and $Y_{\mathrm{dc}} = Y_{\mathrm{dc}1} \times \cdots \times Y_{\mathrm{dc}q}$. Note that with the negative feedback interconnection given by Figure 10.1, $(u_{\mathrm{cc}}, u_{\mathrm{dc}}) = (y_{\mathrm{c}}, y_{\mathrm{d}})$ and $(y_{\mathrm{cc}}, y_{\mathrm{dc}}) = (-u_{\mathrm{c}}, -u_{\mathrm{d}})$.

We assume that the negative feedback interconnection of $\mathcal{G}$ and $\mathcal{G}_{\mathrm{c}}$ is well posed, that is, $\det[I_{m_{ci}} + J_{\mathrm{cc}i}(x_{\mathrm{c}i})J_{\mathrm{c}i}(x_i)] \neq 0$, $\det[I_{m_{di}} + J_{\mathrm{dc}i}(x_{\mathrm{c}i})J_{\mathrm{d}i}(x_i)] \neq 0$ for all $x_i \in \mathbb{R}^{n_i}$, $x_{\mathrm{c}i} \in \mathbb{R}^{n_{ci}}$, and $i = 1, \ldots, q$. Next, we assume that $\mathcal{Z}_{\mathrm{c}} \triangleq \mathcal{Z}_{\mathrm{c}x_{\mathrm{c}}} \times \mathcal{Z}_{\mathrm{c}u_{\mathrm{cc}}} = \{(x_{\mathrm{c}}, u_{\mathrm{cc}}) : \mathcal{X}_{\mathrm{c}}(x_{\mathrm{c}}, u_{\mathrm{cc}}) = 0\}$, where $\mathcal{X}_{\mathrm{c}} : \mathbb{R}^{n_c} \times U_{\mathrm{cc}} \to \mathbb{R}$, and define the closed-loop resetting set

$$\tilde{\mathcal{Z}}_{\tilde{x}} \triangleq \mathcal{Z}_{x} \times \mathcal{Z}_{\mathrm{c}x_{\mathrm{c}}} \cup \{(x, x_{\mathrm{c}}) : (\mathcal{L}_{\mathrm{cc}}(x, x_{\mathrm{c}}), \mathcal{L}_{\mathrm{c}}(x, x_{\mathrm{c}})) \in \mathcal{Z}_{\mathrm{c}u_{\mathrm{cc}}} \times \mathcal{Z}_{u_{\mathrm{c}}}\}, \quad (10.225)$$

where $\mathcal{L}_{\mathrm{cc}}(\cdot, \cdot)$ and $\mathcal{L}_{\mathrm{c}}(\cdot, \cdot)$ are functions of x and x_{c} arising from the algebraic loops due to u_{cc} and u_{c}, respectively. Note that since the feedback interconnection of $\mathcal{G}$ and $\mathcal{G}_{\mathrm{c}}$ is well posed, it follows that $\tilde{\mathcal{Z}}_{\tilde{x}}$ is well defined and depends on the closed-loop states $\tilde{x} \triangleq [x^{\mathrm{T}} x_{\mathrm{c}}^{\mathrm{T}}]^{\mathrm{T}}$. Furthermore, we

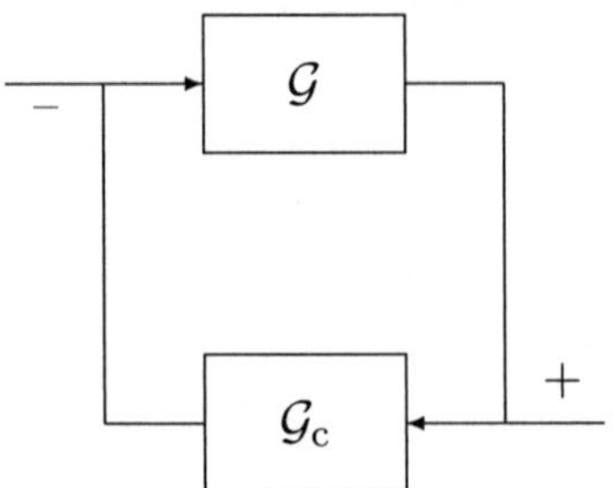

Figure 10.1 Feedback interconnection of large-scale systems $\mathcal{G}$ and $\mathcal{G}_c$.

assume that for the large-scale systems $\mathcal{G}$ and $\mathcal{G}_c$, the conditions of Theorem 10.8 are satisfied, that is, if $V_s(x)$, $x \in \mathbb{R}^n$, and $V_{cs}(x_c)$, $x_c \in \mathbb{R}^{n_c}$, are vector storage functions for $\mathcal{G}$ and $\mathcal{G}_c$, respectively, then there exist $p \in \mathbb{R}_+^q$ and $p_c \in \mathbb{R}_+^q$ such that the functions $v_s(x) = p^T V_s(x)$, $x \in \mathbb{R}^n$, and $v_{cs}(x_c) = p_c^T V_{cs}(x_c)$, $x_c \in \mathbb{R}^{n_c}$, are positive definite. The following result gives sufficient conditions for Lyapunov and asymptotic stability of the feedback interconnection given by Figure 10.1.

Theorem 10.16. Consider the large-scale impulsive dynamical systems $\mathcal{G}$ and $\mathcal{G}_c$ given by (10.209)–(10.212) and (10.217)–(10.220), respectively. Assume that $\mathcal{G}$ and $\mathcal{G}_c$ are vector dissipative with respect to the vector hybrid supply rates $(S_c(u_c, y_c), S_d(u_d, y_d))$ and $(S_{cc}(u_{cc}, y_{cc}), S_{dc}(u_{dc}, y_{dc}))$, and with continuously differentiable vector storage functions $V_s(\cdot)$ and $V_{cs}(\cdot)$, and dissipation matrices $W \in \mathbb{R}^{q \times q}$ and $W_c \in \mathbb{R}^{q \times q}$, respectively.

i) If there exists $\Sigma \triangleq \mathrm{diag}[\sigma_1, \ldots, \sigma_q] > 0$ such that

$$S_c(u_c, y_c) + \Sigma S_{cc}(u_{cc}, y_{cc}) \leqq 0, \qquad (10.226)$$
$$S_d(u_d, y_d) + \Sigma S_{dc}(u_{dc}, y_{dc}) \leqq 0, \qquad (10.227)$$

and $\tilde{W} \in \mathbb{R}^{q \times q}$ is semistable (respectively, asymptotically stable), where

$$\tilde{W}_{(i,j)} \triangleq \max\left\{ W_{(i,j)}, (\Sigma W_c \Sigma^{-1})_{(i,j)} \right\}$$
$$= \max\left\{ W_{(i,j)}, \frac{\sigma_i}{\sigma_j} W_{c(i,j)} \right\}, \quad i, j = 1, \ldots, q, \quad (10.228)$$

then the negative feedback interconnection of $\mathcal{G}$ and $\mathcal{G}_c$ is Lyapunov (respectively, asymptotically) stable.

ii) Let $Q_{ci} \in \mathbb{S}^{l_{ci}}$, $S_{ci} \in \mathbb{R}^{l_{ci} \times m_{ci}}$, $R_{ci} \in \mathbb{S}^{m_{ci}}$, $Q_{di} \in \mathbb{S}^{l_{di}}$, $S_{di} \in \mathbb{R}^{l_{di} \times m_{di}}$, $R_{di} \in \mathbb{S}^{m_{di}}$, $Q_{cci} \in \mathbb{S}^{m_{ci}}$, $S_{cci} \in \mathbb{R}^{m_{ci} \times l_{ci}}$, $R_{cci} \in \mathbb{S}^{l_{ci}}$, $Q_{dci} \in \mathbb{S}^{m_{di}}$,

$S_{\mathrm{dc}i} \in \mathbb{R}^{m_{\mathrm{d}i} \times l_{\mathrm{d}i}}$, and $R_{\mathrm{dc}i} \in \mathbb{S}^{l_{\mathrm{d}i}}$, and suppose

$$S_{\mathrm{c}}(u_{\mathrm{c}}, y_{\mathrm{c}}) = [s_{\mathrm{c}1}(u_{\mathrm{c}1}, y_{\mathrm{c}1}), \ldots, s_{\mathrm{c}q}(u_{\mathrm{c}q}, y_{\mathrm{c}q})]^{\mathrm{T}}, \qquad (10.229)$$

$$S_{\mathrm{d}}(u_{\mathrm{d}}, y_{\mathrm{d}}) = [s_{\mathrm{d}1}(u_{\mathrm{d}1}, y_{\mathrm{d}1}), \ldots, s_{\mathrm{d}q}(u_{\mathrm{d}q}, y_{\mathrm{d}q})]^{\mathrm{T}}, \qquad (10.230)$$

$$S_{\mathrm{cc}}(u_{\mathrm{cc}}, y_{\mathrm{cc}}) = [s_{\mathrm{cc}1}(u_{\mathrm{cc}1}, y_{\mathrm{cc}1}), \ldots, s_{\mathrm{cc}q}(u_{\mathrm{cc}q}, y_{\mathrm{cc}q})]^{\mathrm{T}}, \qquad (10.231)$$

$$S_{\mathrm{dc}}(u_{\mathrm{dc}}, y_{\mathrm{dc}}) = [s_{\mathrm{dc}1}(u_{\mathrm{dc}1}, y_{\mathrm{dc}1}), \ldots, s_{\mathrm{dc}q}(u_{\mathrm{dc}q}, y_{\mathrm{dc}q})]^{\mathrm{T}}, \qquad (10.232)$$

where

$$s_{\mathrm{c}i}(u_{\mathrm{c}i}, y_{\mathrm{c}i}) = u_{\mathrm{c}i}^{\mathrm{T}} R_{\mathrm{c}i} u_{\mathrm{c}i} + 2 y_{\mathrm{c}i}^{\mathrm{T}} S_{\mathrm{c}i} u_{\mathrm{c}i} + y_{\mathrm{c}i}^{\mathrm{T}} Q_{\mathrm{c}i} y_{\mathrm{c}i}, \qquad (10.233)$$

$$s_{\mathrm{d}i}(u_{\mathrm{d}i}, y_{\mathrm{d}i}) = u_{\mathrm{d}i}^{\mathrm{T}} R_{\mathrm{d}i} u_{\mathrm{d}i} + 2 y_{\mathrm{d}i}^{\mathrm{T}} S_{\mathrm{d}i} u_{\mathrm{d}i} + y_{\mathrm{d}i}^{\mathrm{T}} Q_{\mathrm{d}i} y_{\mathrm{d}i}, \qquad (10.234)$$

$$s_{\mathrm{cc}i}(u_{\mathrm{cc}i}, y_{\mathrm{cc}i}) = u_{\mathrm{cc}i}^{\mathrm{T}} R_{\mathrm{cc}i} u_{\mathrm{cc}i} + 2 y_{\mathrm{cc}i}^{\mathrm{T}} S_{\mathrm{cc}i} u_{\mathrm{cc}i} + y_{\mathrm{cc}i}^{\mathrm{T}} Q_{\mathrm{cc}i} y_{\mathrm{cc}i}, \qquad (10.235)$$

$$s_{\mathrm{dc}i}(u_{\mathrm{dc}i}, y_{\mathrm{dc}i}) = u_{\mathrm{dc}i}^{\mathrm{T}} R_{\mathrm{dc}i} u_{\mathrm{dc}i} + 2 y_{\mathrm{dc}i}^{\mathrm{T}} S_{\mathrm{dc}i} u_{\mathrm{dc}i} + y_{\mathrm{dc}i}^{\mathrm{T}} Q_{\mathrm{dc}i} y_{\mathrm{dc}i}, \qquad (10.236)$$

for all $i = 1, \ldots, q$. If there exists $\Sigma \triangleq \mathrm{diag}[\sigma_1, \ldots, \sigma_q] > 0$ such that for all $i = 1, \ldots, q$,

$$\tilde{Q}_{\mathrm{c}i} \triangleq \begin{bmatrix} Q_{\mathrm{c}i} + \sigma_i R_{\mathrm{cc}i} & -S_{\mathrm{c}i} + \sigma_i S_{\mathrm{cc}i}^{\mathrm{T}} \\ -S_{\mathrm{c}i}^{\mathrm{T}} + \sigma_i S_{\mathrm{cc}i} & R_{\mathrm{c}i} + \sigma_i Q_{\mathrm{cc}i} \end{bmatrix} \leq 0, \qquad (10.237)$$

$$\tilde{Q}_{\mathrm{d}i} \triangleq \begin{bmatrix} Q_{\mathrm{d}i} + \sigma_i R_{\mathrm{dc}i} & -S_{\mathrm{d}i} + \sigma_i S_{\mathrm{dc}i}^{\mathrm{T}} \\ -S_{\mathrm{d}i}^{\mathrm{T}} + \sigma_i S_{\mathrm{dc}i} & R_{\mathrm{d}i} + \sigma_i Q_{\mathrm{dc}i} \end{bmatrix} \leq 0, \qquad (10.238)$$

and $\tilde{W} \in \mathbb{R}^{q \times q}$ is semistable (respectively, asymptotically stable), where

$$\tilde{W}_{(i,j)} \triangleq \max \left\{ W_{(i,j)}, (\Sigma W_{\mathrm{c}} \Sigma^{-1})_{(i,j)} \right\}$$

$$= \max \left\{ W_{(i,j)}, \frac{\sigma_i}{\sigma_j} \cdot W_{\mathrm{c}(i,j)} \right\}, \quad i, j = 1, \ldots, q, \qquad (10.239)$$

then the negative feedback interconnection of $\mathcal{G}$ and $\mathcal{G}_{\mathrm{c}}$ is Lyapunov (respectively, asymptotically) stable.

Proof. Let $\tilde{\mathcal{T}}^{\mathrm{c}} \triangleq \mathcal{T}_{x_0, u_{\mathrm{c}}}^{\mathrm{c}} \cup \mathcal{T}_{x_{\mathrm{c}0}, u_{\mathrm{cc}}}^{\mathrm{c}}$ and $t_k \in \tilde{\mathcal{T}}^{\mathrm{c}}$, $k \in \overline{\mathbb{Z}}_+$. First, note that it follows from Assumptions 10.1 and 10.2 that the resetting times $t_k (= \tau_k(\tilde{x}_0))$ for the feedback system are well defined and distinct for every closed-loop trajectory.

$i)$ Consider the vector Lyapunov function candidate $V(x, x_{\mathrm{c}}) = V_{\mathrm{s}}(x) + \Sigma V_{\mathrm{cs}}(x_{\mathrm{c}})$, $(x, x_{\mathrm{c}}) \in \mathbb{R}^n \times \mathbb{R}^{n_{\mathrm{c}}}$, and note that the corresponding vector Lyapunov derivative of $V(x, x_{\mathrm{c}})$ along the state trajectories $(x(t), x_{\mathrm{c}}(t))$, $t \in (t_k, t_{k+1})$, is given by

$$\dot{V}(x(t), x_{\mathrm{c}}(t)) = \dot{V}_{\mathrm{s}}(x(t)) + \Sigma \dot{V}_{\mathrm{cs}}(x_{\mathrm{c}}(t))$$

$$\leq\leq S_{\mathrm{c}}(u_{\mathrm{c}}(t), y_{\mathrm{c}}(t)) + \Sigma S_{\mathrm{cc}}(u_{\mathrm{cc}}(t), y_{\mathrm{cc}}(t)) + W V_{\mathrm{s}}(x(t))$$

$$+ \Sigma W_{\mathrm{c}} V_{\mathrm{cs}}(x_{\mathrm{c}}(t))$$

$$
\begin{aligned}
&\overset{<}{\scriptstyle<} W V_{\mathrm{s}}(x(t)) + \Sigma W_{\mathrm{c}} \Sigma^{-1} \Sigma V_{\mathrm{cs}}(x_{\mathrm{c}}(t)) \\
&\overset{<}{\scriptstyle<} \tilde{W}(V_{\mathrm{s}}(x(t)) + \Sigma V_{\mathrm{cs}}(x_{\mathrm{c}}(t))) \\
&= \tilde{W} V(x(t), x_{\mathrm{c}}(t)), \quad (x(t), x_{\mathrm{c}}(t)) \notin \tilde{\mathcal{Z}}_{\tilde{x}},
\end{aligned}
\tag{10.240}
$$

and the Lyapunov difference of $V(x, x_{\mathrm{c}})$ at the resetting times t_k, $k \in \overline{\mathbb{Z}}_+$, is given by

$$
\begin{aligned}
\Delta V(x(t_k), x_{\mathrm{c}}(t_k)) &= \Delta V_{\mathrm{s}}(x(t_k)) + \Sigma \Delta V_{\mathrm{cs}}(x_{\mathrm{c}}(t_k)) \\
&\overset{<}{\scriptstyle<} S_{\mathrm{d}}(u_{\mathrm{d}}(t_k), y_{\mathrm{d}}(t_k)) + \Sigma S_{\mathrm{dc}}(u_{\mathrm{dc}}(t_k), y_{\mathrm{dc}}(t_k)) \\
&\overset{<}{\scriptstyle<} 0, \quad (x(t), x_{\mathrm{c}}(t)) \in \tilde{\mathcal{Z}}_{\tilde{x}}.
\end{aligned}
\tag{10.241}
$$

Next, since for $V_{\mathrm{s}}(x)$, $x \in \mathbb{R}^n$, and $V_{\mathrm{cs}}(x_{\mathrm{c}})$, $x_{\mathrm{c}} \in \mathbb{R}^{n_c}$, there exist, by assumption, $p \in \mathbb{R}_+^q$ and $p_{\mathrm{c}} \in \mathbb{R}_+^q$ such that the functions $v_{\mathrm{s}}(x) = p^{\mathrm{T}} V_{\mathrm{s}}(x)$, $x \in \mathbb{R}^n$, and $v_{\mathrm{cs}}(x_{\mathrm{c}}) = p_{\mathrm{c}}^{\mathrm{T}} V_{\mathrm{cs}}(x_{\mathrm{c}})$, $x_{\mathrm{c}} \in \mathbb{R}^{n_c}$, are positive definite, and noting that $v_{\mathrm{cs}}(x_{\mathrm{c}}) \leq \max_{i=1,\dots,q}\{p_{ci}\} \mathbf{e}^{\mathrm{T}} V_{\mathrm{cs}}(x_{\mathrm{c}})$, where p_{ci} is the ith element of p_{c} and $\mathbf{e} \triangleq [1, \dots, 1]^{\mathrm{T}}$, it follows that $\mathbf{e}^{\mathrm{T}} V_{\mathrm{cs}}(x_{\mathrm{c}})$, $x_{\mathrm{c}} \in \mathbb{R}^{n_c}$, is positive definite. Now, since

$$
\min_{i=1,\dots,q} \{p_i \sigma_i\} \mathbf{e}^{\mathrm{T}} V_{\mathrm{cs}}(x_{\mathrm{c}}) \leq p^{\mathrm{T}} \Sigma V_{\mathrm{cs}}(x_{\mathrm{c}}),
\tag{10.242}
$$

it follows that $p^{\mathrm{T}} \Sigma V_{\mathrm{cs}}(x_{\mathrm{c}})$, $x_{\mathrm{c}} \in \mathbb{R}^{n_c}$, is positive definite. Hence, the function $v(x, x_{\mathrm{c}}) \triangleq p^{\mathrm{T}} V(x, x_{\mathrm{c}})$, $(x, x_{\mathrm{c}}) \in \mathbb{R}^n \times \mathbb{R}^{n_c}$, is positive definite. Now, the result is a direct consequence of Theorem 10.2.

$ii)$ The proof follows from $i)$ by noting that, for all $i = 1, \dots, q$,

$$
s_{ci}(u_{ci}, y_{ci}) + \sigma_i s_{cci}(u_{cci}, y_{cci}) = \begin{bmatrix} y_{\mathrm{c}} \\ y_{\mathrm{cc}} \end{bmatrix}^{\mathrm{T}} \tilde{Q}_{ci} \begin{bmatrix} y_{\mathrm{c}} \\ y_{\mathrm{cc}} \end{bmatrix},
\tag{10.243}
$$

$$
s_{di}(u_{di}, y_{di}) + \sigma_i s_{dci}(u_{dci}, y_{dci}) = \begin{bmatrix} y_{\mathrm{d}} \\ y_{\mathrm{dc}} \end{bmatrix}^{\mathrm{T}} \tilde{Q}_{di} \begin{bmatrix} y_{\mathrm{d}} \\ y_{\mathrm{dc}} \end{bmatrix},
\tag{10.244}
$$

and hence, $S_{\mathrm{c}}(u_{\mathrm{c}}, y_{\mathrm{c}}) + \Sigma S_{\mathrm{cc}}(u_{\mathrm{cc}}, y_{\mathrm{cc}}) \overset{<}{\scriptstyle<} 0$ and $S_{\mathrm{d}}(u_{\mathrm{d}}, y_{\mathrm{d}}) + \Sigma S_{\mathrm{dc}}(u_{\mathrm{dc}}, y_{\mathrm{dc}}) \overset{<}{\scriptstyle<} 0$. $\qquad\square$

For the next result note that if the large-scale impulsive dynamical system $\mathcal{G}$ is vector dissipative with respect to the vector hybrid supply rate $(S_{\mathrm{c}}(u_{\mathrm{c}}, y_{\mathrm{c}}), S_{\mathrm{d}}(u_{\mathrm{d}}, y_{\mathrm{d}}))$, where $s_{ci}(u_{ci}, y_{ci}) = 2y_{ci}^{\mathrm{T}} u_{ci}$ and $s_{di}(u_{di}, y_{di}) = 2y_{di}^{\mathrm{T}} u_{di}$, $i = 1, \dots, q$, then, with $\kappa_{ci}(y_{ci}) = -\kappa_{ci} y_{ci}$ and $\kappa_{di}(y_{di}) = -\kappa_{di} y_{di}$, where $\kappa_{ci} > 0$, $\kappa_{di} > 0$, $i = 1, \dots, q$, it follows that $s_{ci}(\kappa_{ci}(y_{ci}), y_{ci}) = -2\kappa_{ci} y_{ci}^{\mathrm{T}} y_{ci} < 0$ and $s_{di}(\kappa_{di}(y_{di}), y_{di}) = -2\kappa_{di} y_{di}^{\mathrm{T}} y_{di} < 0$, $y_{ci} \neq 0$, $y_{di} \neq 0$, $i = 1, \dots, q$. Alternatively, if $\mathcal{G}$ is vector dissipative with respect to the vector hybrid supply rate $(S_{\mathrm{c}}(u_{\mathrm{c}}, y_{\mathrm{c}}), S_{\mathrm{d}}(u_{\mathrm{d}}, y_{\mathrm{d}}))$, where $s_{ci}(u_{ci}, y_{ci}) = \gamma_{ci}^2 u_{ci}^{\mathrm{T}} u_{ci} - y_{ci}^{\mathrm{T}} y_{ci}$ and $s_{di}(u_{di}, y_{di}) = \gamma_{di}^2 u_{di}^{\mathrm{T}} u_{di} - y_{di}^{\mathrm{T}} y_{di}$, where $\gamma_{ci} > 0$, $\gamma_{di} > 0$, $i = 1, \dots, q$, then, with $\kappa_{ci}(y_{ci}) = 0$ and $\kappa_{di}(y_{di}) = 0$, it follows that $s_{ci}(\kappa_{ci}(y_{ci}), y_{ci}) = -y_{ci}^{\mathrm{T}} y_{ci} < 0$ and $s_{di}(\kappa_{di}(y_{di}), y_{di}) = -y_{di}^{\mathrm{T}} y_{di} < 0$, $y_{ci} \neq 0$,

$y_{\mathrm{d}i} \neq 0$, $i = 1, \ldots, q$. Hence, if $\mathcal{G}$ is zero-state observable and the dissipation matrix W is such that there exist $\alpha \geq 0$ and $p \in \mathbb{R}_+^q$ such that (2.4) holds, then it follows from Theorem 10.8 that (scalar) storage functions of the form $v_{\mathrm{s}}(x) = p^{\mathrm{T}} V_{\mathrm{s}}(x)$, $x \in \mathbb{R}^n$, where $V_{\mathrm{s}}(\cdot)$ is a vector storage function for $\mathcal{G}$, are positive definite. If $\mathcal{G}$ is exponentially vector dissipative, then p is positive.

Corollary 10.3. Consider the large-scale impulsive dynamical systems $\mathcal{G}$ and $\mathcal{G}_{\mathrm{c}}$ given by (10.209)–(10.212) and (10.217)–(10.220), respectively. Assume that $\mathcal{G}$ and $\mathcal{G}_{\mathrm{c}}$ are zero-state observable and the dissipation matrices $W \in \mathbb{R}^{q \times q}$ and $W_{\mathrm{c}} \in \mathbb{R}^{q \times q}$ are such that there exist, respectively, $\alpha \geq 0$, $p \in \mathbb{R}_+^q$, $\alpha_{\mathrm{c}} \geq 0$, and $p_{\mathrm{c}} \in \mathbb{R}_+^q$ such that (2.4) is satisfied. Then the following statements hold:

i) If $\mathcal{G}$ and $\mathcal{G}_{\mathrm{c}}$ are vector passive and $\tilde{W} \in \mathbb{R}^{q \times q}$ is asymptotically stable, where $\tilde{W}_{(i,j)} \triangleq \max\{W_{(i,j)}, W_{\mathrm{c}(i,j)}\}$, $i,j = 1, \ldots, q$, then the negative feedback interconnection of $\mathcal{G}$ and $\mathcal{G}_{\mathrm{c}}$ is asymptotically stable.

ii) If $\mathcal{G}$ and $\mathcal{G}_{\mathrm{c}}$ are vector nonexpansive and $\tilde{W} \in \mathbb{R}^{q \times q}$ is asymptotically stable, where $\tilde{W}_{(i,j)} \triangleq \max\{W_{(i,j)}, W_{\mathrm{c}(i,j)}\}$, $i,j = 1, \ldots, q$, then the negative feedback interconnection of $\mathcal{G}$ and $\mathcal{G}_{\mathrm{c}}$ is asymptotically stable.

Proof. The proof is a direct consequence of Theorem 10.16. Specifically, statement i) follows from Theorem 10.16 with $R_{\mathrm{c}i} = 0$, $S_{\mathrm{c}i} = I_{m_{\mathrm{c}i}}$, $Q_{\mathrm{c}i} = 0$, $R_{\mathrm{d}i} = 0$, $S_{\mathrm{d}i} = I_{m_{\mathrm{d}i}}$, $Q_{\mathrm{d}i} = 0$, $R_{\mathrm{cc}i} = 0$, $S_{\mathrm{cc}i} = I_{m_{\mathrm{c}i}}$, $Q_{\mathrm{cc}i} = 0$, $R_{\mathrm{dc}i} = 0$, $S_{\mathrm{dc}i} = I_{m_{\mathrm{d}i}}$, $Q_{\mathrm{dc}i} = 0$, $i = 1, \ldots, q$, and $\Sigma = I_q$. Statement ii) follows from Theorem 10.16 with $R_{\mathrm{c}i} = \gamma_{\mathrm{c}i}^2 I_{m_{\mathrm{c}i}}$, $S_{\mathrm{c}i} = 0$, $Q_{\mathrm{c}i} = -I_{l_{\mathrm{c}i}}$, $R_{\mathrm{d}i} = \gamma_{\mathrm{d}i}^2 I_{m_{\mathrm{d}i}}$, $S_{\mathrm{d}i} = 0$, $Q_{\mathrm{d}i} = -I_{l_{\mathrm{d}i}}$, $R_{\mathrm{cc}i} = \gamma_{\mathrm{cc}i}^2 I_{l_{\mathrm{c}i}}$, $S_{\mathrm{cc}i} = 0$, $Q_{\mathrm{cc}i} = -I_{m_{\mathrm{c}i}}$, $R_{\mathrm{dc}i} = \gamma_{\mathrm{dc}i}^2 I_{l_{\mathrm{d}i}}$, $S_{\mathrm{dc}i} = 0$, $Q_{\mathrm{dc}i} = -I_{m_{\mathrm{d}i}}$, $i = 1, \ldots, q$, and $\Sigma = I_q$. $\square$

Control Vector Lyapunov Functions for Large-Scale Impulsive Systems

11.1 Introduction

The mathematical descriptions of many hybrid dynamical systems can be characterized by impulsive differential equations [11, 13, 82, 94, 117, 155]. As shown in Chapter 10, impulsive dynamical systems can be viewed as a subclass of hybrid systems and consist of three elements—namely, a continuous-time differential equation, which governs the motion of the dynamical system between impulsive or resetting events; a difference equation, which governs the way the system states are instantaneously changed when a resetting event occurs; and a criterion for determining when the states of the system are to be reset. Since impulsive systems can involve impulses at variable times, they are in general time-varying systems, wherein the resetting events are both a function of time and the system's state. In the case where the resetting events are defined by a prescribed sequence of times that are independent of the system state, the equations are known as *time-dependent differential equations* [11, 13, 32, 74, 75, 117]. Alternatively, in the case where the resetting events are defined by a manifold in the state space that is independent of time, the equations are autonomous and are known as *state-dependent differential equations* [11, 13, 32, 74, 75, 117].

Even though impulsive dynamical systems were first formulated by Mil'man and Myshkis [138, 139], the fundamental theory of impulsive differential equations is developed in the monographs by Bainov, Lakshmikantham, Perestyuk, Samoilenko, and Simeonov [11–13, 117, 155]. These monographs develop qualitative solution properties, existence of solutions, asymptotic properties of solutions, and stability theory of impulsive dynamical systems. In a recent series of papers [37, 74, 75, 78], stability, dissipativity, and optimality results have been developed for impulsive dynamical systems that include invariant set stability theorems, partial stability, dissipativity theory, and optimal control design.

In this chapter, we provide generalizations to vector Lyapunov theory for continuous-time systems presented in Chapter 5 to address stability and control design of impulsive dynamical systems via vector Lyapunov functions. Vector Lyapunov theory has been developed to weaken the hypothesis of standard Lyapunov theory to enlarge the class of Lyapunov functions that can be used for analyzing system stability. In particular, as shown in

Chapters 5 and 6 the use of vector Lyapunov functions in dynamical system theory offers a very flexible framework since each component of the vector Lyapunov function can satisfy less rigid requirements as compared to a single scalar Lyapunov function. Weakening the hypothesis on the Lyapunov function enlarges the class of Lyapunov functions that can be used for analyzing system stability. In particular, each component of a vector Lyapunov function need not be positive definite with a negative or even negative-semidefinite derivative. Alternatively, the time derivative of the vector Lyapunov function need only satisfy an element-by-element inequality involving a vector field of a certain comparison system. Since in this case the stability properties of the comparison system imply the stability properties of the dynamical system, the use of vector Lyapunov theory can significantly reduce the complexity (i.e., dimensionality) of the dynamical system being analyzed. Extensions of vector Lyapunov function theory that include relaxed conditions on standard vector Lyapunov functions as well as matrix Lyapunov functions appear in [52, 131, 132].

In this chapter, we extend the notion of *control vector Lyapunov functions* presented in Chapter 5 to impulsive dynamical systems and show that in the case of a scalar comparison system the definition of a control vector Lyapunov function collapses into a combination of the classical definition of a control Lyapunov function for continuous-time dynamical systems given in [6] and the definition of a control Lyapunov function for discrete-time dynamical systems given in [3, 38]. In addition, using control vector Lyapunov functions, we present a universal hybrid decentralized feedback stabilizer for a decentralized affine in the control nonlinear impulsive dynamical system with guaranteed gain and sector margins. These results are then used to develop hybrid decentralized controllers for large-scale impulsive dynamical systems with robustness guarantees against full modeling and input uncertainty.

11.2 Control Vector Lyapunov Functions for Impulsive Systems

In this section, we consider a feedback control problem and generalize the notion of a control vector Lyapunov function introduced in Chapter 5 to nonlinear impulsive dynamical systems. Specifically, consider the nonlinear controlled impulsive dynamical system given by

$$\dot{x}(t) = F_{\mathrm{c}}(x(t), u_{\mathrm{c}}(t)), \quad x(t_0) = x_0, \quad x(t) \notin \mathcal{Z}, \quad t \geq t_0, \tag{11.1}$$

$$\Delta x(t) = F_{\mathrm{d}}(x(t), u_{\mathrm{d}}(t)), \quad x(t) \in \mathcal{Z}, \tag{11.2}$$

where $x_0 \in \mathcal{D}$, $\mathcal{D} \subseteq \mathbb{R}^n$ is an open set with $0 \in \mathcal{D}$, $u_{\mathrm{c}}(t) \in U_{\mathrm{c}} \subseteq \mathbb{R}^{m_{\mathrm{c}}}$, $t \geq t_0$, $u_{\mathrm{d}}(t_k) \in U_{\mathrm{d}} \subseteq \mathbb{R}^{m_{\mathrm{d}}}$, t_k denotes the kth instant of time at which $x(t)$ intersects $\mathcal{Z}$ for a particular trajectory $x(t)$ and input $(u_{\mathrm{c}}(\cdot), u_{\mathrm{d}}(\cdot))$, $F_{\mathrm{c}} : \mathcal{D} \times U_{\mathrm{c}} \to \mathbb{R}^n$ is Lipschitz continuous for all $(x, u_{\mathrm{c}}) \in \mathcal{D} \times U_{\mathrm{c}}$ and satisfies $F_{\mathrm{c}}(0, 0) = 0$, and $F_{\mathrm{d}} : \mathcal{D} \times U_{\mathrm{d}} \to \mathbb{R}^n$ is continuous. Here, we assume that $u_{\mathrm{c}}(\cdot)$

and $u_{\mathrm{d}}(\cdot)$ are restricted to the class of *admissible control inputs* consisting of measurable functions such that $(u_{\mathrm{c}}(t), u_{\mathrm{d}}(t_k)) \in U_{\mathrm{c}} \times U_{\mathrm{d}}$ for all $t \geq t_0$ and $k \in \mathbb{Z}_{[t_0,t)} \triangleq \{k : t_0 \leq t_k < t\}$, where the constraint set $U_{\mathrm{c}} \times U_{\mathrm{d}}$ is given with $(0,0) \in U_{\mathrm{c}} \times U_{\mathrm{d}}$. Furthermore, we assume that $u_{\mathrm{c}}(\cdot)$ and $u_{\mathrm{d}}(\cdot)$ satisfy sufficient regularity conditions such that the nonlinear impulsive dynamical system (11.1) and (11.2) has a unique solution forward in time. Let $\phi_{\mathrm{c}} : \mathcal{D} \to U_{\mathrm{c}}$ be such that $\phi_{\mathrm{c}}(0) = 0$ and let $\phi_{\mathrm{d}} : \mathcal{Z} \to U_{\mathrm{d}}$. If $(u_{\mathrm{c}}(t), u_{\mathrm{d}}(t_k)) = (\phi_{\mathrm{c}}(x(t)), \phi_{\mathrm{d}}(x(t_k)))$, where $x(t)$, $t \geq t_0$, satisfies (11.1) and (11.2), then $(u_{\mathrm{c}}(\cdot), u_{\mathrm{d}}(\cdot))$ is called a *hybrid feedback control*.

Definition 11.1. If there exist a continuously differentiable vector function $V = [v_1, \ldots, v_q]^{\mathrm{T}} : \mathcal{D} \to \mathcal{Q} \cap \overline{\mathbb{R}}_+^q$, continuous functions $w_{\mathrm{c}} = [w_{\mathrm{c}1}, \ldots, w_{\mathrm{c}q}]^{\mathrm{T}} : \mathcal{Q} \times \mathcal{D} \to \mathbb{R}^q$ and $w_{\mathrm{d}} = [w_{\mathrm{d}1}, \ldots, w_{\mathrm{d}q}]^{\mathrm{T}} : \mathcal{Q} \times \mathcal{Z} \to \mathbb{R}^q$, and a positive vector $p \in \mathbb{R}_+^q$ such that $V(0) = 0$, $v(x) \triangleq p^{\mathrm{T}} V(x)$, $x \in \mathcal{D}$, is positive definite, $w_{\mathrm{c}}(\cdot, x) \in \mathcal{W}_{\mathrm{c}}$, $w_{\mathrm{d}}(\cdot, x) \in \mathcal{W}_{\mathrm{d}}$, $w_{\mathrm{c}}(0,0) = 0$, $\mathcal{F}_{\mathrm{c}}(x) \triangleq \cap_{i=1}^q \mathcal{F}_{\mathrm{c}i}(x) \neq \varnothing$, $x \in \mathcal{D}$, $x \notin \mathcal{Z}$, $x \neq 0$, $\mathcal{F}_{\mathrm{d}}(x) \triangleq \cap_{i=1}^q \mathcal{F}_{\mathrm{d}i}(x) \neq \varnothing$, $x \in \mathcal{Z}$, where $\mathcal{F}_{\mathrm{c}i}(x) \triangleq \{u_{\mathrm{c}} \in U_{\mathrm{c}} : v_i'(x) F_{\mathrm{c}}(x, u_{\mathrm{c}}) < w_{\mathrm{c}i}(V(x), x)\}$, $x \in \mathcal{D}$, $x \notin \mathcal{Z}$, $x \neq 0$, $i = 1, \ldots, q$, and $\mathcal{F}_{\mathrm{d}i}(x) \triangleq \{u_{\mathrm{d}} \in U_{\mathrm{d}} : v_i(x + F_{\mathrm{d}}(x, u_{\mathrm{d}})) - v_i(x) \leq w_{\mathrm{d}i}(V(x), x)\}$, $x \in \mathcal{Z}$, $i = 1, \ldots, q$, then the vector function $V : \mathcal{D} \to \mathcal{Q} \cap \overline{\mathbb{R}}_+^q$ is called a *control vector Lyapunov function candidate*.

It follows from Definition 11.1 that if there exists a control vector Lyapunov function candidate, then there exists a hybrid feedback control law $\phi_{\mathrm{c}} : \mathcal{D} \to U_{\mathrm{c}}$ and $\phi_{\mathrm{d}} : \mathcal{Z} \to U_{\mathrm{d}}$ such that

$$V'(x) F_{\mathrm{c}}(x, \phi_{\mathrm{c}}(x)) << w_{\mathrm{c}}(V(x), x), \quad x \in \mathcal{D}, \quad x \notin \mathcal{Z}, \quad x \neq 0,$$
$$V(x + F_{\mathrm{d}}(x, \phi_{\mathrm{d}}(x))) \leq\leq V(x) + w_{\mathrm{d}}(V(x), x), \quad x \in \mathcal{Z}.$$

Moreover, if the nonlinear impulsive dynamical system

$$\dot{z}(t) = w_{\mathrm{c}}(z(t), x(t)), \quad z(t_0) = z_0, \quad x(t) \notin \mathcal{Z}, \quad t \geq t_0, \tag{11.3}$$
$$\dot{x}(t) = F_{\mathrm{c}}(x(t), \phi_{\mathrm{c}}(x(t))), \quad x(t_0) = x_0, \quad x(t) \notin \mathcal{Z}, \tag{11.4}$$
$$\Delta z(t) = w_{\mathrm{d}}(z(t), x(t)), \quad x(t) \in \mathcal{Z}, \tag{11.5}$$
$$\Delta x(t) = F_{\mathrm{d}}(x(t), \phi_{\mathrm{d}}(x(t))), \quad x(t) \in \mathcal{Z}, \tag{11.6}$$

where $z_0 \in \mathcal{Q}$ and $x_0 \in \mathcal{D}$, is asymptotically stable with respect to z uniformly in x_0, then it follows from Theorem 10.2 that the zero solution $x(t) \equiv 0$ to (11.4) and (11.6) is asymptotically stable. In this case, the vector function $V : \mathcal{D} \to \overline{\mathbb{R}}_+^q$ given in Definition 11.1 is called a *control vector Lyapunov function* for impulsive dynamical system (11.1) and (11.2). This is a generalization of the notion of control vector Lyapunov functions introduced in Chapter 5 for continuous-time systems. Furthermore, if $\mathcal{D} = \mathbb{R}^n$, $\mathcal{Q} = \mathbb{R}^q$, $U_{\mathrm{c}} = \mathbb{R}^{m_{\mathrm{c}}}$, $U_{\mathrm{d}} = \mathbb{R}^{m_{\mathrm{d}}}$, $v : \mathbb{R}^n \to \overline{\mathbb{R}}_+$ is radially unbounded, and the system (11.3)–(11.6) is globally asymptotically stable with respect to

z uniformly in x_0, then the zero solution $x(t) \equiv 0$ to (11.1) and (11.2) is globally asymptotically stabilizable.

If in Definition 11.1 $w_{\mathrm{c}}(z,x) = w_{\mathrm{c}}(z)$, $w_{\mathrm{d}}(z,x) = w_{\mathrm{d}}(z)$, and the zero solution $z(t) \equiv 0$ to

$$\dot{z}(t) = w_{\mathrm{c}}(z(t)), \quad z(t_0) = z_0, \quad x(t) \notin \mathcal{Z}, \quad t \geq t_0, \tag{11.7}$$

$$\Delta z(t) = w_{\mathrm{d}}(z(t)), \quad x(t) \in \mathcal{Z}, \tag{11.8}$$

where $z_0 \in \mathcal{Q}$, is asymptotically stable, then it follows from Theorem 10.2, with $w_{\mathrm{c}}(z,x) = w_{\mathrm{c}}(z)$, $w_{\mathrm{d}}(z,x) = w_{\mathrm{d}}(z)$, that $V : \mathcal{D} \to \mathcal{Q} \cap \overline{\mathbb{R}}_+^q$ is a control vector Lyapunov function. In this case, the nonlinear impulsive comparison system (11.7) and (11.8) is a time-dependent impulsive dynamical system [82]. To see this, note that the resetting times $\tau_k(x_0)$, $k \in \overline{\mathbb{Z}}_+$, where $x(\tau_k(x_0)) \in \mathcal{Z}$ and $x(t)$ is the solution to (11.1) and (11.2), are determined by the state of (11.1) and (11.2), and hence, provide a prescribed sequence of the resetting times for (11.7) and (11.8) since the dynamics of the impulsive system (11.1) and (11.2) and the comparison system (11.7) and (11.8) are decoupled.

In the case where $q = 1$, $w_{\mathrm{c}}(z,x) \equiv w_{\mathrm{c}}(z)$, and $w_{\mathrm{d}}(z,x) \equiv w_{\mathrm{d}}(z)$, Definition 11.1 implies the existence of a positive-definite continuously differentiable function $v : \mathcal{D} \to \mathcal{Q} \cap \overline{\mathbb{R}}_+$ and continuous functions $w_{\mathrm{c}} : \mathcal{Q} \to \mathbb{R}$ and $w_{\mathrm{d}} : \mathcal{Q} \to \mathbb{R}$, where $\mathcal{Q} \subseteq \mathbb{R}$, such that $w_{\mathrm{c}}(0) = 0$, $\mathcal{F}_{\mathrm{c}}(x) = \{u_{\mathrm{c}} \in U_{\mathrm{c}} : v'(x)F_{\mathrm{c}}(x,u_{\mathrm{c}}) < w_{\mathrm{c}}(v(x))\} \neq \varnothing$, $x \in \mathcal{D}$, $x \notin \mathcal{Z}$, $x \neq 0$, and $\mathcal{F}_{\mathrm{d}}(x) = \{u_{\mathrm{d}} \in U_{\mathrm{d}} : v(x + F_{\mathrm{d}}(x,u_{\mathrm{d}})) - v(x) \leq w_{\mathrm{d}}(v(x))\} \neq \varnothing$, $x \in \mathcal{Z}$, which implies

$$\inf_{u_{\mathrm{c}} \in U_{\mathrm{c}}} v'(x)F_{\mathrm{c}}(x,u_{\mathrm{c}}) < w_{\mathrm{c}}(v(x)), \quad x \notin \mathcal{Z}, \quad x \neq 0, \tag{11.9}$$

$$\inf_{u_{\mathrm{d}} \in U_{\mathrm{d}}} [v(x + F_{\mathrm{d}}(x,u_{\mathrm{d}})) - v(x)] \leq w_{\mathrm{d}}(v(x)), \quad x \in \mathcal{Z}. \tag{11.10}$$

Now, the fact that $\mathcal{F}_{\mathrm{c}}(x) \neq \varnothing$, $x \in \mathcal{D}$, $x \notin \mathcal{Z}$, $x \neq 0$, and $\mathcal{F}_{\mathrm{d}}(x) \neq \varnothing$, $x \in \mathcal{Z}$, implies the existence of a hybrid feedback control law $\phi_{\mathrm{c}} : \mathcal{D} \to U_{\mathrm{c}}$ and $\phi_{\mathrm{d}} : \mathcal{Z} \to U_{\mathrm{d}}$ such that $v'(x)F_{\mathrm{c}}(x,\phi_{\mathrm{c}}(x)) < w_{\mathrm{c}}(v(x))$, $x \in \mathcal{D}$, $x \notin \mathcal{Z}$, $x \neq 0$, and $v(x + F_{\mathrm{d}}(x,\phi_{\mathrm{d}}(x))) - v(x) \leq w_{\mathrm{d}}(v(x))$, $x \in \mathcal{Z}$. Moreover, if $v : \mathcal{D} \to \overline{\mathbb{R}}_+$ is a control vector Lyapunov function (with $q = 1$), then it follows from the discussion above that the zero solution $z(t) \equiv 0$ to the system (11.7) and (11.8) is asymptotically stable and, since $q = 1$, this implies that $w_{\mathrm{c}}(z) < 0$, $z \in \mathcal{Q} \cap \overline{\mathbb{R}}_+$, $z \neq 0$. Thus, since $v(\cdot)$ is positive definite, (11.9) and (11.10) can be rewritten as

$$\inf_{u_{\mathrm{c}} \in U_{\mathrm{c}}} v'(x)F_{\mathrm{c}}(x,u_{\mathrm{c}}) < 0, \quad x \notin \mathcal{Z}, \quad x \neq 0, \tag{11.11}$$

$$\inf_{u_{\mathrm{d}} \in U_{\mathrm{d}}} [v(x + F_{\mathrm{d}}(x,u_{\mathrm{d}})) - v(x)] \leq w_{\mathrm{d}}(v(x)), \quad x \in \mathcal{Z}, \tag{11.12}$$

where (11.11) is similar to the definition of a scalar control Lyapunov function for continuous-time dynamical systems.

Since the system (11.7) and (11.8) is time-dependent it is impossible to determine the sign of $w_{\mathrm{d}}(\cdot)$, which, in this case, can vary for the different values of the argument. The sign definiteness of $w_{\mathrm{c}}(\cdot)$ follows from the fact that the system (11.7) and (11.8) is asymptotically stable and might exhibit no resettings for a particular trajectory. However, asymptotic stability of the zero solution to scalar state-dependent impulsive dynamical systems necessarily implies the negative definiteness of both continuous-time and discrete-time dynamics in a sufficiently small neighborhood of the origin.

Next, consider the case where the control input to (11.1) and (11.2) possesses a decentralized control architecture so that the dynamics of (11.1) and (11.2) are given by

$$\dot{x}_i(t) = F_{ci}(x(t), u_{ci}(t)), \quad x(t) \notin \mathcal{Z}, \quad t \geq t_0, \quad i = 1, \dots, q, \quad (11.13)$$
$$\Delta x_i(t) = F_{di}(x(t), u_{di}(t)), \quad x(t) \in \mathcal{Z}, \quad i = 1, \dots, q, \quad (11.14)$$

where $x_i(t) \in \mathbb{R}^{n_i}$, $x(t) = [x_1^{\mathrm{T}}(t), \dots, x_q^{\mathrm{T}}(t)]^{\mathrm{T}}$, $u_{ci}(t) \in U_{ci} \subseteq \mathbb{R}^{m_{ci}}$, $t \geq t_0$, $u_{di}(t_k) \in U_{di} \subseteq \mathbb{R}^{m_{di}}$, $k \in \overline{\mathbb{Z}}_+$, t_k denotes the kth resetting time for a particular trajectory of (11.13) and (11.14), $\sum_{i=1}^{q} n_i = n$, $\sum_{i=1}^{q} m_{ci} = m_{\mathrm{c}}$, and $\sum_{i=1}^{q} m_{di} = m_{\mathrm{d}}$. Note that $x_i(t) \in \mathbb{R}^{n_i}$, $t \geq t_0$, $i = 1, \dots, q$, as long as $x(t) \in \mathcal{D}$, $t \geq t_0$, and the sets of control inputs are given by $U_{\mathrm{c}} = U_{c1} \times \cdots \times U_{cq} \subseteq \mathbb{R}^{m_{\mathrm{c}}}$ and $U_{\mathrm{d}} = U_{d1} \times \cdots \times U_{dq} \subseteq \mathbb{R}^{m_{\mathrm{d}}}$.

In the case of a component decoupled control vector Lyapunov function candidate, that is, $V(x) = [v_1(x_1), \dots, v_q(x_q)]^{\mathrm{T}}$, $x \in \mathcal{D}$, it suffices to require in Definition 11.1 that, for all $i = 1, \dots, q$,

$$\mathcal{F}_{ci}(x) = \{u_{\mathrm{c}} \in U_{\mathrm{c}} : v_i'(x_i) F_{ci}(x, u_{ci}) < w_{ci}(V(x), x)\} \neq \varnothing, \quad x \in \mathcal{D},$$
$$x \notin \mathcal{Z}, \quad x \neq 0, \quad (11.15)$$
$$\mathcal{F}_{di}(x) = \{u_{\mathrm{d}} \in U_{\mathrm{d}} : v_i(x_i + F_{di}(x, u_{di})) - v_i(x) \leq w_{di}(V(x), x)\} \neq \varnothing,$$
$$x \in \mathcal{Z}, \quad (11.16)$$

to ensure that $\mathcal{F}_{\mathrm{c}}(x) = \cap_{i=1}^{q} \mathcal{F}_{ci}(x) \neq \varnothing$, $x \in \mathcal{D}$, $x \notin \mathcal{Z}$, $x \neq 0$, and $\mathcal{F}_{\mathrm{d}}(x) = \cap_{i=1}^{q} \mathcal{F}_{di}(x) \neq \varnothing$, $x \in \mathcal{Z}$. Note that for a component decoupled control vector Lyapunov function $V : \mathcal{D} \to \mathcal{Q} \cap \overline{\mathbb{R}}_+^q$, (11.15) is equivalent to

$$\inf_{u_{\mathrm{c}} \in U_{\mathrm{c}}} V'(x) F_{\mathrm{c}}(x, u_{\mathrm{c}}) << w_{\mathrm{c}}(V(x), x), \quad x \in \mathcal{D}, \quad x \notin \mathcal{Z}, \quad x \neq 0, \quad (11.17)$$

and (11.16) implies

$$\inf_{u_{\mathrm{d}} \in U_{\mathrm{d}}} [V(x + F_{\mathrm{d}}(x, u_{\mathrm{d}})) - V(x)] \leq\leq w_{\mathrm{d}}(V(x), x), \quad x \in \mathcal{Z}, \quad (11.18)$$

where the infimum in (11.17) and (11.18) is taken componentwise, that is, for each component of (11.17) and (11.18) the infimum is calculated separately. It follows from the fact that $\mathcal{F}_{\mathrm{c}}(x) \neq \varnothing$, $x \in \mathcal{D}$, $x \notin \mathcal{Z}$, $x \neq 0$, and $\mathcal{F}_{\mathrm{d}}(x) \neq \varnothing$, $x \in \mathcal{Z}$, that there exists a hybrid feedback control law $\phi_{\mathrm{c}} : \mathcal{D} \to U_{\mathrm{c}}$ and $\phi_{\mathrm{d}} : \mathcal{Z} \to U_{\mathrm{d}}$ such that $\phi_{\mathrm{c}}(x) = [\phi_{c1}^{\mathrm{T}}(x), \dots, \phi_{cq}^{\mathrm{T}}(x)]^{\mathrm{T}}$, $x \in \mathcal{D}$,

$\phi_{\mathrm{d}}(x) = [\phi_{\mathrm{d}1}^{\mathrm{T}}(x), \ldots, \phi_{\mathrm{d}q}^{\mathrm{T}}(x)]^{\mathrm{T}}$, $x \in \mathcal{Z}$, where $\phi_{\mathrm{c}i} : \mathcal{D} \to U_{\mathrm{c}i}$, $\phi_{\mathrm{d}i} : \mathcal{Z} \to U_{\mathrm{d}i}$, $v_i'(x_i)F_{\mathrm{c}i}(x, \phi_{\mathrm{c}i}(x)) < w_{\mathrm{c}i}(V(x), x)$, $x \in \mathcal{D}$, $x \notin \mathcal{Z}$, $x \neq 0$, and $v_i(x_i + F_{\mathrm{d}i}(x, \phi_{\mathrm{d}i}(x))) - v_i(x_i) \leq w_{\mathrm{d}i}(V(x), x)$, $x \in \mathcal{Z}$, $i = 1, \ldots, q$. Finally, note that if $w_{\mathrm{c}i}(V(x), x) = 0$ for $x \in \mathcal{D}$ with $x_i = 0$, then condition (11.15) holds for all $x \in \mathcal{D}$ such that $x_i \neq 0$.

Next, we consider the special case of a nonlinear impulsive dynamical system of the form (11.13) and (11.14) with affine control inputs given by

$$\dot{x}_i(t) = f_{\mathrm{c}i}(x(t)) + G_{\mathrm{c}i}(x(t))u_{\mathrm{c}i}(t), \quad x(t) \notin \mathcal{Z}, \quad t \geq t_0, \quad i = 1, \ldots, q,$$
$$\tag{11.19}$$
$$\Delta x_i(t) = f_{\mathrm{d}i}(x(t)) + G_{\mathrm{d}i}(x(t))u_{\mathrm{d}i}(t), \quad x(t) \in \mathcal{Z}, \quad i = 1, \ldots, q, \tag{11.20}$$

where $f_{\mathrm{c}i} : \mathbb{R}^n \to \mathbb{R}^{n_i}$ satisfying $f_{\mathrm{c}i}(0) = 0$ and $G_{\mathrm{c}i} : \mathbb{R}^n \to \mathbb{R}^{n_i \times m_{\mathrm{c}i}}$ are smooth functions (at least continuously differentiable mappings) for all $i = 1, \ldots, q$, $f_{\mathrm{d}i} : \mathbb{R}^n \to \mathbb{R}^{n_i}$ and $G_{\mathrm{d}i} : \mathbb{R}^n \to \mathbb{R}^{n_i \times m_{\mathrm{d}i}}$ are continuous for all $i = 1, \ldots, q$, $u_{\mathrm{c}i}(t) \in \mathbb{R}^{m_{\mathrm{c}i}}$, $t \geq t_0$, and $u_{\mathrm{d}i}(t_k) \in \mathbb{R}^{m_{\mathrm{d}i}}$, $k \in \overline{\mathbb{Z}}_+$, for all $i = 1, \ldots, q$.

Theorem 11.1. Consider the controlled nonlinear impulsive dynamical system given by (11.19) and (11.20). If there exist a continuously differentiable, component decoupled vector function $V : \mathbb{R}^n \to \overline{\mathbb{R}}_+^q$, continuous functions $P_{1ui} : \mathbb{R}^n \to \mathbb{R}^{1 \times m_{\mathrm{d}i}}$, $P_{2ui} : \mathbb{R}^n \to \mathbb{R}^{m_{\mathrm{d}i} \times m_{\mathrm{d}i}}$, $i = 1, \ldots, q$, $w_{\mathrm{c}} = [w_{\mathrm{c}1}, \ldots, w_{\mathrm{c}q}]^{\mathrm{T}} : \overline{\mathbb{R}}_+^q \times \mathbb{R}^n \to \mathbb{R}^q$, $w_{\mathrm{d}} = [w_{\mathrm{d}1}, \ldots, w_{\mathrm{d}q}]^{\mathrm{T}} : \overline{\mathbb{R}}_+^q \times \mathcal{Z} \to \mathbb{R}^q$, and a positive vector $p \in \mathbb{R}_+^q$ such that $V(0) = 0$, the scalar function $v : \mathbb{R}^n \to \overline{\mathbb{R}}_+$ defined by $v(x) \triangleq p^{\mathrm{T}}V(x)$, $x \in \mathbb{R}^n$, is positive definite and radially unbounded, $w_{\mathrm{c}}(\cdot, x) \in \mathcal{W}_{\mathrm{c}}$, $w_{\mathrm{d}}(\cdot, x) \in \mathcal{W}_{\mathrm{d}}$, $w_{\mathrm{c}}(0, 0) = 0$, and, for all $i = 1, \ldots, q$,

$$v_i(x_i + f_{\mathrm{d}i}(x) + G_{\mathrm{d}i}(x)u_{\mathrm{d}i}) = v_i(x_i + f_{\mathrm{d}i}(x)) + P_{1ui}(x)u_{\mathrm{d}i} + u_{\mathrm{d}i}^{\mathrm{T}}P_{2ui}(x)u_{\mathrm{d}i},$$
$$x \in \mathbb{R}^n, \quad u_{\mathrm{d}i} \in \mathbb{R}^{m_{\mathrm{d}i}}, \quad (11.21)$$
$$v_i'(x_i)f_{\mathrm{c}i}(x) < w_{\mathrm{c}i}(V(x), x), \quad x \in \mathcal{R}_i, \tag{11.22}$$
$$v_i(x_i + f_{\mathrm{d}i}(x)) - v_i(x_i) - \tfrac{1}{4}P_{1ui}(x)P_{2ui}^+(x)P_{1ui}^{\mathrm{T}}(x) \leq w_{\mathrm{d}i}(V(x), x), \quad x \in \mathcal{Z},$$
$$\tag{11.23}$$

where $\mathcal{R}_i \triangleq \{x \in \mathbb{R}^n, x \neq 0 : v_i'(x_i)G_{\mathrm{c}i}(x) = 0\}$, $i = 1, \ldots, q$, and $P_{2ui}^+(\cdot)$ is the Moore-Penrose generalized inverse of $P_{2ui}(\cdot)$, then $V : \mathbb{R}^n \to \overline{\mathbb{R}}_+^q$ is a control vector Lyapunov function candidate. If, in addition, there exist $\phi_{\mathrm{c}} : \mathbb{R}^n \to \mathbb{R}^{m_{\mathrm{c}}}$ and $\phi_{\mathrm{d}} : \mathcal{Z} \to \mathbb{R}^{m_{\mathrm{d}}}$ such that $\phi_{\mathrm{c}}(x) = [\phi_{\mathrm{c}1}^{\mathrm{T}}(x), \ldots, \phi_{\mathrm{c}q}^{\mathrm{T}}(x)]^{\mathrm{T}}$, $x \in \mathbb{R}^n$, $\phi_{\mathrm{d}}(x) = [\phi_{\mathrm{d}1}^{\mathrm{T}}(x), \ldots, \phi_{\mathrm{d}q}^{\mathrm{T}}(x)]^{\mathrm{T}}$, $x \in \mathcal{Z}$, and the system (11.3)–(11.6) is globally asymptotically stable with respect to z uniformly in x_0, then the zero solution $x(t) \equiv 0$ to (11.4) and (11.6) is globally asymptotically stable and $V : \mathbb{R}^n \to \overline{\mathbb{R}}_+^q$ is a control vector Lyapunov function.

Proof. Note that for all $i = 1, \ldots, q$,

$$\inf_{u_{ci} \in \mathbb{R}^{m_{ci}}} v_i'(x_i)(f_{ci}(x) + G_{ci}(x)u_{ci}) = \begin{cases} -\infty, & x \notin \mathcal{R}_i, \\ v_i'(x_i)f_{ci}(x), & x \in \mathcal{R}_i, \end{cases}$$
$$< w_{ci}(V(x), x), \quad x \in \mathbb{R}^n, \quad x \neq 0,$$

which implies that $\mathcal{F}_{ci}(x) \neq \emptyset$, $x \in \mathbb{R}^n$, $x \neq 0$, $i = 1, \ldots, q$. Next, note that it follows from a Taylor series expansion about $x_i^* = x_i + f_{di}(x)$ that $P_{1ui}(x) = v_i'(x_i^*)G_{di}(x)$ and

$$P_{2ui}(x) = \frac{1}{2} G_{di}^{\mathrm{T}}(x) \left. \frac{\partial^2 v_i}{\partial x_i^2} \right|_{x_i = x_i^*} G_{di}(x). \tag{11.24}$$

Since $V(\cdot)$ is continuously differentiable it follows that $\left. \frac{\partial^2 v_i}{\partial x_i^2} \right|_{x_i = x_i^*}$ is symmetric, and hence, $P_{2ui}(\cdot)$ is symmetric for all $i = 1, \ldots, q$.

Next, with $u_{di} = -\frac{1}{2} P_{2ui}^+(x) P_{1ui}^{\mathrm{T}}(x)$, $x \in \mathcal{Z}$, $i = 1, \ldots, q$, it follows from (11.21) and (11.23) that

$$v_i(x_i + f_{di}(x) + G_{di}(x)u_{di}) - v_i(x_i)$$
$$= v_i(x_i + f_{di}(x)) - v_i(x_i) + P_{1ui}(x)u_{di} + u_{di}^{\mathrm{T}} P_{2ui}(x)u_{di}$$
$$= v_i(x_i + f_{di}(x)) - v_i(x_i) - \frac{1}{4} P_{1ui}(x) P_{2ui}^+(x) P_{1ui}^{\mathrm{T}}(x)$$
$$\leq w_{di}(V(x), x), \quad x \in \mathcal{Z}, \quad i = 1, \ldots, q, \tag{11.25}$$

which implies that $\mathcal{F}_{di}(x) \neq \emptyset$, $x \in \mathcal{Z}$. Now, the result is a direct consequence of the definition of a control vector Lyapunov function by noting that, for component decoupled vector Lyapunov functions, (11.15) and (11.16) are equivalent to $\mathcal{F}_c(x) \neq \emptyset$, $x \in \mathbb{R}^n$, $x \neq 0$, and $\mathcal{F}_d(x) \neq \emptyset$, $x \in \mathcal{Z}$, respectively. $\square$

Using Theorem 11.1 we can construct an explicit feedback control law that is a function of the control vector Lyapunov function $V(\cdot)$. Specifically, consider the hybrid feedback control law $\phi_c(x) = [\phi_{c1}^{\mathrm{T}}(x), \ldots, \phi_{cq}^{\mathrm{T}}(x)]^{\mathrm{T}}$, $x \in \mathbb{R}^n$, and $\phi_d(x) = [\phi_{d1}^{\mathrm{T}}(x), \ldots, \phi_{dq}^{\mathrm{T}}(x)]^{\mathrm{T}}$, $x \in \mathcal{Z}$, given by

$$\phi_{ci}(x) = \begin{cases} -\left(c_{0i} + \frac{(\alpha_i(x) - w_{ci}(V(x),x)) + \mu_i(x)}{\beta_i^{\mathrm{T}}(x)\beta_i(x)} \right) \beta_i(x), & \beta_i(x) \neq 0, \\ 0, & \beta_i(x) = 0, \end{cases} \tag{11.26}$$

and

$$\phi_{di}(x) = -\frac{1}{2} P_{2ui}^+(x) P_{1ui}^{\mathrm{T}}(x), \quad x \in \mathcal{Z}, \tag{11.27}$$

where

$$\mu_i(x) \triangleq \sqrt{(\alpha_i(x) - w_{ci}(V(x), x))^2 + (\beta_i^{\mathrm{T}}(x)\beta_i(x))^2}, \tag{11.28}$$

$$\alpha_i(x) \triangleq v_i'(x_i) f_{ci}(x), \tag{11.29}$$

$$\beta_i(x) \triangleq G_{ci}^{\mathrm{T}}(x) v_i'^{\mathrm{T}}(x_i), \quad x \in \mathbb{R}^n, \tag{11.30}$$

and $c_{0i} > 0$, $i = 1, \ldots, q$. The derivative $\dot{V}(\cdot)$ along the trajectories of the dynamical system (11.19), with $u_c = \phi_c(x)$, $x \in \mathbb{R}^n$, given by (11.26), is given by

$$\begin{aligned}
\dot{v}_i(x_i) &= v_i'(x_i)(f_{ci}(x) + G_{ci}(x)\phi_{ci}(x)) \\
&= \alpha_i(x) + \beta_i^{\mathrm{T}}(x)\phi_{ci}(x) \\
&= \begin{cases} -c_{0i}\beta_i^{\mathrm{T}}(x)\beta_i(x) - \mu_i(x) + w_{ci}(V(x),x), & \beta_i(x) \neq 0, \\ \alpha_i(x), & \beta_i(x) = 0, \end{cases} \\
&< w_{ci}(V(x),x), \quad x \in \mathbb{R}^n.
\end{aligned} \tag{11.31}$$

In addition, using (11.21), the difference of $V(x)$ at the resetting instants with $u_d = \phi_d(x)$, $x \in \mathcal{Z}$, given by (11.27), is given by

$$\begin{aligned}
\Delta v_i(x_i) &= v_i(x_i + f_{di}(x) + G_{di}(x)\phi_{di}(x)) - v_i(x_i) \\
&= v_i(x_i + f_{di}(x)) - v_i(x_i) - \frac{1}{4} P_{1ui}(x) P_{2ui}^+(x) P_{1ui}^{\mathrm{T}}(x) \\
&\leq w_{di}(V(x),x), \quad x \in \mathcal{Z}.
\end{aligned} \tag{11.32}$$

Thus, if the zero solution $z(t) \equiv 0$ to (11.3)–(11.6) is globally asymptotically stable with respect to z uniformly in x_0, then it follows from Theorem 10.2 that the zero solution $x(t) \equiv 0$ to (11.19) and (11.20) with $u_c = \phi_c(x) = [\phi_{c1}^{\mathrm{T}}(x), \ldots, \phi_{cq}^{\mathrm{T}}(x)]^{\mathrm{T}}$, $x \in \mathbb{R}^n$, given by (11.26) and $u_d = \phi_d(x) = [\phi_{d1}^{\mathrm{T}}(x), \ldots, \phi_{dq}^{\mathrm{T}}(x)]^{\mathrm{T}}$, $x \in \mathcal{Z}$, given by (11.27) is globally asymptotically stable.

If in Theorem 11.1 $w_c(z,x) = w_c(z)$, $w_d(z,x) = w_d(z)$, and the zero solution $z(t) \equiv 0$ to (11.7) and (11.8) is globally asymptotically stable, then it follows from Theorem 10.2 that the hybrid feedback control law given by (11.26) and (11.27) is a globally asymptotically stabilizing controller for the nonlinear impulsive dynamical system (11.19) and (11.20).

In the case where $q = 1$, the functions $w(\cdot,\cdot)$ and $w_d(\cdot,\cdot)$ in Theorem 11.1 can be set to be identically zero, that is, $w_c(z,x) \equiv 0$ and $w_d(z,x) \equiv 0$. In this case, the feedback control law (11.26) specializes to Sontag's universal formula [165], the feedback control law (11.27) specializes to the discrete-time control law given in [38], and the hybrid feedback control law (11.26) and (11.27) is a global stabilizer for the nonlinear impulsive dynamical system (11.19) and (11.20).

Since $f_{ci}(\cdot)$ and $G_{ci}(\cdot)$ are smooth and $v_i(\cdot)$ is continuously differentiable for all $i = 1, \ldots, q$, it follows that $\alpha_i(x)$ and $\beta_i(x)$, $x \in \mathbb{R}^n$, $i = 1, \ldots, q$, are continuous functions, and hence, $\phi_{ci}(x)$ given by (11.26) is continuous for all $x \in \mathbb{R}^n$ if either $\beta_i(x) \neq 0$ or $\alpha_i(x) - w_{ci}(V(x),x) < 0$ for all $i = 1, \ldots, q$. Hence, the feedback control law given by (11.26) is continuous everywhere except for the origin. The following result provides necessary and sufficient conditions under which the feedback control law given by (11.26) is guaranteed to be continuous at the origin in addition to being continuous everywhere else.

Proposition 11.1. The feedback control law $\phi_c(x)$ given by (11.26) is continuous on $\mathbb{R}^n$ if and only if for every $\varepsilon > 0$, there exists $\delta > 0$ such that for all $0 < \|x\| < \delta$ there exists $u_{ci} \in \mathbb{R}^{m_{ci}}$ such that $\|u_{ci}\| < \varepsilon$ and $\alpha_i(x) + \beta_i^{\mathrm{T}}(x)u_{ci} < w_{ci}(V(x), x)$, $i = 1, \ldots, q$.

Proof. This is a restatement of Proposition 5.1. $\qquad\square$

If the conditions of Proposition 11.1 are satisfied, then the feedback control law $\phi_c(x)$ given by (11.26) is continuous on $\mathbb{R}^n$. However, it is important to note that for a particular trajectory $x(t)$, $t \geq 0$, of (11.19) and (11.20), $\phi_c(x(t))$ is a piecewise continuous function of time due to state resettings.

11.3 Stability Margins and Inverse Optimality

In this section, we construct a hybrid feedback control law that is robust to sector bounded input nonlinearities. Specifically, we consider the nonlinear impulsive dynamical system (11.19) and (11.20) with nonlinear uncertainties in the input so that the impulsive dynamics of the system are given by

$$\dot{x}_i(t) = f_{ci}(x(t)) + G_{ci}(x(t))\sigma_{ci}(u_{ci}(t)), \quad x(t) \notin \mathcal{Z}, \quad t \geq t_0, \quad i = 1, \ldots, q,$$
$$(11.33)$$

$$\Delta x_i(t) = f_{di}(x(t)) + G_{di}(x(t))\sigma_{di}(u_{di}(t)), \quad x(t) \in \mathcal{Z}, \quad i = 1, \ldots, q, \quad (11.34)$$

where $\sigma_{ci}(\cdot) \in \Phi_{ci} \triangleq \{\sigma_{ci} : \mathbb{R}^{m_{ci}} \to \mathbb{R}^{m_{ci}} : \sigma_{ci}(0) = 0 \text{ and } \frac{1}{2}u_{ci}^{\mathrm{T}}u_{ci} \leq \sigma_{ci}^{\mathrm{T}}(u_{ci})u_{ci} < \infty, u_{ci} \in \mathbb{R}^{m_{ci}}\}$, $\sigma_{di}(\cdot) \in \Phi_{di} \triangleq \{\sigma_{di} : \mathbb{R}^{m_{di}} \to \mathbb{R}^{m_{di}} : \sigma_{di}(0) = 0 \text{ and } \alpha_d u_{dij}^2 \leq \sigma_{dij}(u_{di})u_{dij} < \beta_d u_{dij}^2, u_{di} \in \mathbb{R}^{m_{di}}\}$, $i = 1, \ldots, q$, $\sigma_{dij}(\cdot) \in \mathbb{R}$ and $u_{dij} \in \mathbb{R}$, $j = 1, \ldots, m_{di}$, are the jth components of $\sigma_{di}(\cdot)$ and u_{di}, respectively, and $0 \leq \alpha_d < 1 < \beta_d < \infty$. In addition, we show that for the dynamical system (11.19) and (11.20) the hybrid feedback control law to be defined in Theorem 11.2 is inverse optimal in the sense that it minimizes a derived hybrid performance functional over the set of stabilizing controllers $\mathcal{S}(x_0) \triangleq \{(u_c(\cdot), u_d(\cdot)) : u_c(\cdot) \text{ and } u_d(\cdot) \text{ are admissible and } x(t) \to 0 \text{ as } t \to \infty\}$.

Theorem 11.2. Consider the nonlinear impulsive dynamical system (11.19) and (11.20) and assume that the conditions of Theorem 11.1 hold with (11.23) replaced by

$$v_i(x_i + f_{di}(x)) - v_i(x_i) - \frac{1}{4}P_{1ui}(x)(R_{2di}(x) + P_{2ui}(x))^{-1}P_{1ui}^{\mathrm{T}}(x)$$

$$\leq w_{\mathrm{d}i}(V(x)), \quad x \in \mathcal{Z}, \qquad (11.35)$$

where $R_{2\mathrm{d}i} : \mathcal{Z} \to \mathbb{R}^{m_{\mathrm{d}i} \times m_{\mathrm{d}i}}$ is positive definite, $w_{\mathrm{c}}(z,x) \equiv w_{\mathrm{c}}(z)$, $w_{\mathrm{d}}(z,x) \equiv w_{\mathrm{d}}(z)$, and with the zero solution $z(t) \equiv 0$ to (11.7) and (11.8) being globally asymptotically stable. Then the hybrid feedback control law $(\phi_{\mathrm{c}i}(\cdot), \phi_{\mathrm{d}i}(\cdot))$ given by (11.26) and

$$\phi_{\mathrm{d}i}(x) = -\frac{1}{2}(R_{2\mathrm{d}i}(x) + P_{2ui}(x))^{-1} P_{1ui}^{\mathrm{T}}(x), \quad x \in \mathcal{Z}, \quad i = 1, \ldots, q, \quad (11.36)$$

minimizes the performance functional given by

$$J(x_0, u_{\mathrm{c}}(\cdot), u_{\mathrm{d}}(\cdot)) = \int_{t_0}^{\infty} \sum_{i=1}^{q} [L_{1\mathrm{c}i}(x(t)) + u_{\mathrm{c}i}^{\mathrm{T}}(t) R_{2\mathrm{c}i}(x(t)) u_{\mathrm{c}i}(t)] \mathrm{d}t$$

$$+ \sum_{k \in \mathbb{Z}_{[t_0,\infty)}} \sum_{i=1}^{q} [L_{1\mathrm{d}i}(x(t_k)) + u_{\mathrm{d}i}^{\mathrm{T}}(t_k) R_{2\mathrm{d}i}(x(t_k)) u_{\mathrm{d}i}(t_k)]$$

$$(11.37)$$

in the sense that

$$J(x_0, \phi_{\mathrm{c}}(x(\cdot)), \phi_{\mathrm{d}}(x(\cdot))) = \min_{(u_{\mathrm{c}}(\cdot), u_{\mathrm{d}}(\cdot)) \in \mathcal{S}(x_0)} J(x_0, u_{\mathrm{c}}(\cdot), u_{\mathrm{d}}(\cdot)), \quad x_0 \in \mathbb{R}^n,$$

$$(11.38)$$

where $\mathbb{Z}_{[t_0,\infty)} = \{k : t_0 \leq t_k < \infty\}$,

$$L_{1\mathrm{c}i}(x) \triangleq -\alpha_i(x) + \frac{\gamma_i(x)}{2} \beta_i^{\mathrm{T}}(x)\beta_i(x), \quad x \in \mathbb{R}^n, \qquad (11.39)$$

$$R_{2\mathrm{c}i}(x) \triangleq \begin{cases} \frac{1}{2\gamma_i(x)} I_{m_{\mathrm{c}i}}, & \beta_i(x) \neq 0, \\ 0, & \beta_i(x) = 0, \end{cases} \qquad (11.40)$$

$$\gamma_i(x) \triangleq \begin{cases} c_{0i} + \frac{(\alpha_i(x) - w_{\mathrm{c}i}(V(x))) + \mu_i(x)}{\beta_i^{\mathrm{T}}(x)\beta_i(x)} > 0, & \beta_i(x) \neq 0, \\ 0, & \beta_i(x) = 0, \end{cases} \qquad (11.41)$$

$L_{1\mathrm{d}i}(x) \triangleq \phi_{\mathrm{d}i}^{\mathrm{T}}(x)(R_{2\mathrm{d}i}(x) + P_{2ui}(x))\phi_{\mathrm{d}i}(x) - v_i(x_i + f_{\mathrm{d}i}(x)) + v_i(x_i)$, $x \in \mathcal{Z}$, $\mu_i(x) \triangleq \sqrt{(\alpha_i(x) - w_{\mathrm{c}i}(V(x)))^2 + (\beta_i^{\mathrm{T}}(x)\beta_i(x))^2}$, $x \in \mathbb{R}^n$, and $P_{2ui}(\cdot)$ and $v_i(\cdot)$ are given in Theorem 11.1. Furthermore, $J(x_0, \phi_{\mathrm{c}}(x(\cdot)), \phi_{\mathrm{d}}(x(\cdot))) = \mathbf{e}^{\mathrm{T}}V(x_0)$, $x_0 \in \mathbb{R}^n$, where $V : \mathbb{R}^n \to \overline{\mathbb{R}}_+^q$ is a control vector Lyapunov function for the impulsive dynamical system (11.19) and (11.20). In addition, the nonlinear impulsive dynamical system (11.33) and (11.34) with the hybrid feedback control law given by (11.26) and (11.36) is globally asymptotically stable for all $\sigma_{\mathrm{c}i}(\cdot) \in \Phi_{\mathrm{c}i}$ and $\sigma_{\mathrm{d}i}(\cdot) \in \Phi_{\mathrm{d}i}$, $i = 1, \ldots, q$, where $\alpha_{\mathrm{d}} = \max_{i=1,\ldots,q}\left(\frac{1}{1+\theta_{\mathrm{d}i}}\right)$, $\beta_{\mathrm{d}} = \min_{i=1,\ldots,q}\left(\frac{1}{1-\theta_{\mathrm{d}i}}\right)$, $\theta_{\mathrm{d}i} \triangleq \sqrt{\frac{\overline{\gamma}_{\mathrm{d}i}}{\underline{\gamma}_{\mathrm{d}i}}}$, $\underline{\gamma}_{\mathrm{d}i} \triangleq \inf_{x \in \mathcal{Z}} \sigma_{\min}(R_{2\mathrm{d}i}(x))$, $\overline{\gamma}_{\mathrm{d}i} \triangleq \sup_{x \in \mathcal{Z}} \sigma_{\max}(R_{2\mathrm{d}i}(x) + P_{2ui}(x))$, $i = 1, \ldots, q$.

Proof. To show that the hybrid feedback control law (11.26) and (11.36) minimizes (11.37) in the sense of (11.38), define the Hamiltonian

$$\mathcal{H}(x, u_{\mathrm{c}}, u_{\mathrm{d}}) = \mathcal{H}_{\mathrm{c}}(x, u_{\mathrm{c}}) + \mathcal{H}_{\mathrm{d}}(x, u_{\mathrm{d}}), \tag{11.42}$$

where

$$\mathcal{H}_{\mathrm{c}}(x, u_{\mathrm{c}}) \triangleq \sum_{i=1}^{q} [L_{1\mathrm{c}i}(x) + u_{\mathrm{c}i}^{\mathrm{T}} R_{2\mathrm{c}i}(x) u_{\mathrm{c}i} + v_i'(x_i)(f_{\mathrm{c}i}(x) + G_{\mathrm{c}i}(x) u_{\mathrm{c}i})],$$

$$\mathcal{H}_{\mathrm{d}}(x, u_{\mathrm{d}}) \triangleq \sum_{i=1}^{q} [L_{1\mathrm{d}i}(x) + u_{\mathrm{d}i}^{\mathrm{T}} R_{2\mathrm{d}i}(x) u_{\mathrm{d}i}$$
$$+ v_i(x_i + f_{\mathrm{d}i}(x) + G_{\mathrm{d}i}(x) u_{\mathrm{d}i}) - v_i(x_i)],$$

and note that $\mathcal{H}(x, \phi_{\mathrm{c}}(\cdot), \phi_{\mathrm{d}}(\cdot)) = 0$ and $\mathcal{H}(x, u_{\mathrm{c}}, u_{\mathrm{d}}) \geq 0$, $x \in \mathbb{R}^n$, $u_{\mathrm{c}} \in \mathbb{R}^{m_{\mathrm{c}}}$, $u_{\mathrm{d}} \in \mathbb{R}^{m_{\mathrm{d}}}$, since

$$\mathcal{H}(x, u_{\mathrm{c}}, u_{\mathrm{d}}) = \sum_{i=1}^{q} (u_{\mathrm{c}i} - \phi_{\mathrm{c}i}(x))^{\mathrm{T}} R_{2\mathrm{c}i}(x)(u_{\mathrm{c}i} - \phi_{\mathrm{c}i}(x))$$

$$+ \sum_{i=1}^{q} (u_{\mathrm{d}i} - \phi_{\mathrm{d}i}(x))^{\mathrm{T}} (R_{2\mathrm{d}i}(x) + P_{2ui}(x))(u_{\mathrm{d}i} - \phi_{\mathrm{d}i}(x)),$$

$$x \in \mathbb{R}^n, \quad u_{\mathrm{c}} \in \mathbb{R}^{m_{\mathrm{c}}}, \quad u_{\mathrm{d}} \in \mathbb{R}^{m_{\mathrm{d}}}. \tag{11.43}$$

Thus,

$$J(x_0, u_{\mathrm{c}}(\cdot), u_{\mathrm{d}}(\cdot)) = \int_{t_0}^{\infty} \left[\mathcal{H}_{\mathrm{c}}(x(t), u_{\mathrm{c}}(t)) \right.$$

$$\left. - \sum_{i=1}^{q} v_i'(x_i(t))(f_{\mathrm{c}i}(x(t)) + G_{\mathrm{c}i}(x(t)) u_{\mathrm{c}i}(t)) \right] \mathrm{d}t$$

$$+ \sum_{k \in \mathbb{Z}_{[t_0, \infty)}} [\mathcal{H}_{\mathrm{d}}(x(t_k), u_{\mathrm{d}}(t_k))$$

$$+ \sum_{i=1}^{q} (v_i(x_i(t_k)) - v_i(x_i(t_k) + f_{\mathrm{d}i}(x(t_k))$$

$$+ G_{\mathrm{d}i}(x(t_k)) u_{\mathrm{d}i}(t_k))]$$

$$= - \lim_{t \to \infty} \mathbf{e}^{\mathrm{T}} V(x(t)) + \mathbf{e}^{\mathrm{T}} V(x_0)$$

$$+ \int_{t_0}^{\infty} \mathcal{H}_{\mathrm{c}}(x(t), u_{\mathrm{c}}(t)) \mathrm{d}t$$

$$+ \sum_{k \in \mathbb{Z}_{[t_0, \infty)}} \mathcal{H}_{\mathrm{d}}(x(t_k), u_{\mathrm{d}}(t_k))$$

$$\geq \mathbf{e}^{\mathrm{T}} V(x_0)$$

$$= J(x_0, \phi_{\mathrm{c}}(x(\cdot)), \phi_{\mathrm{d}}(x(\cdot))), \tag{11.44}$$

which yields (11.38).

Next, we show that the uncertain nonlinear impulsive dynamical system (11.33) and (11.34) with the hybrid feedback control law (11.26) and (11.36) is globally asymptotically stable for all $\sigma_{ci}(\cdot) \in \Phi_{ci}(\cdot)$ and $\sigma_{di}(\cdot) \in \Phi_{di}(\cdot)$. It follows from Theorem 11.1 that the hybrid feedback control law (11.26) and (11.36) globally asymptotically stabilizes the impulsive dynamical system (11.19) and (11.20) and the vector function $V : \mathbb{R}^n \to \overline{\mathbb{R}}_+^q$ is a control vector Lyapunov function for the impulsive dynamical system (11.19) and (11.20). Note that with (11.41) the continuous-time feedback control law (11.26) can be rewritten as $\phi_{ci}(x) = -\gamma_i(x)\beta_i(x)$, $x \in \mathbb{R}^n$, $i = 1, \ldots, q$. Let the control vector Lyapunov function $V : \mathbb{R}^n \to \overline{\mathbb{R}}_+^q$ for (11.19) and (11.20) be a vector Lyapunov function candidate for (11.33) and (11.34). Then the vector Lyapunov derivative components along the trajectories of (11.33) are given by

$$\dot{v}_i(x_i) = v_i'(x_i)(f_{ci}(x) + G_{ci}(x)\sigma_{ci}(\phi_{ci}(x)))$$
$$= \alpha_i(x) + \beta_i^{\mathrm{T}}(x)\sigma_{ci}(\phi_{ci}(x)), \quad x \notin \mathcal{Z}, \quad i = 1, \ldots, q. \qquad (11.45)$$

Note that $\phi_{ci}(x) = 0$, and hence, $\sigma_{ci}(\phi_{ci}(x)) = 0$ whenever $\beta_i(x) = 0$ for all $i = 1, \ldots, q$. In this case, it follows from (11.22) that $\dot{v}_i(x_i) < w_{ci}(V(x))$, $x \notin \mathcal{Z}$, $\beta_i(x) = 0$, $x \neq 0$, $i = 1, \ldots, q$.

Next, consider the case where $\beta_i(x) \neq 0$, $i = 1, \ldots, q$. In this case, note that, for all $i = 1, \ldots, q$,

$$\alpha_i(x) - w_{ci}(V(x)) - \frac{\gamma_i(x)}{2}\beta_i^{\mathrm{T}}(x)\beta_i(x)$$
$$= \frac{-c_{0i}\beta_i^{\mathrm{T}}(x)\beta_i(x)}{2} + \frac{(\alpha_i(x) - w_{ci}(V(x))) - \mu_i(x)}{2}$$
$$< 0, \quad x \notin \mathcal{Z}, \quad \beta_i(x) \neq 0, \qquad (11.46)$$

where $\mu_i(x) \triangleq \sqrt{(\alpha_i(x) - w_{ci}(V(x)))^2 + (\beta_i^{\mathrm{T}}(x)\beta_i(x))^2}$. Thus, the vector Lyapunov derivative components given by (11.45) satisfy

$$\dot{v}_i(x_i) < w_{ci}(V(x)) + \frac{\gamma_i(x)}{2}\beta_i^{\mathrm{T}}(x)\beta_i(x) + \beta_i^{\mathrm{T}}(x)\sigma_{ci}(\phi_{ci}(x))$$
$$= w_{ci}(V(x)) + \frac{1}{2\gamma_i(x)}\phi_{ci}^{\mathrm{T}}(x)\phi_{ci}(x) - \frac{1}{\gamma_i(x)}\phi_{ci}^{\mathrm{T}}(x)\sigma_{ci}(\phi_{ci}(x))$$
$$= w_{ci}(V(x)) + \frac{1}{\gamma_i(x)}\left[\frac{\phi_{ci}^{\mathrm{T}}(x)\phi_{ci}(x)}{2} - \phi_{ci}^{\mathrm{T}}(x)\sigma_{ci}(\phi_{ci}(x))\right]$$
$$\leq w_{ci}(V(x)), \quad x \notin \mathcal{Z}, \quad \beta_i(x) \neq 0, \qquad (11.47)$$

for all $\sigma_{ci}(\cdot) \in \Phi_{ci}$ and $i = 1, \ldots, q$.

Next, consider the Lyapunov difference of each component of $V(\cdot)$ at the resetting instants for the resetting dynamics (11.34) with $u_{di} = \phi_{di}(x)$ given by (11.36). Note that it follows from (11.35) that $L_{1di}(x) +$

$w_{\mathrm{d}i}(V(x)) \geq 0$, $x \in \mathcal{Z}$, $i = 1, \ldots, q$, and hence, since $\sigma_{\mathrm{d}i}(\cdot) \in \Phi_{\mathrm{d}i}$ it follows that

$$
\begin{aligned}
\Delta v_i(x_i) &= v_i(x_i + f_{\mathrm{d}i}(x) + G_{\mathrm{d}i}(x)\sigma_{\mathrm{d}i}(\phi_{\mathrm{d}i}(x))) - v_i(x_i) \\
&\leq v_i(x_i + f_{\mathrm{d}i}(x)) - v_i(x_i) + P_{1ui}(x)\sigma_{\mathrm{d}i}(\phi_{\mathrm{d}i}(x)) \\
&\quad + \sigma_{\mathrm{d}i}^{\mathrm{T}}(\phi_{\mathrm{d}i}(x))P_{2ui}(x)\sigma_{\mathrm{d}i}(\phi_{\mathrm{d}i}(x)) + L_{1\mathrm{d}i}(x) + w_{\mathrm{d}i}(V(x)) \\
&= P_{1ui}(x)\sigma_{\mathrm{d}i}(\phi_{\mathrm{d}i}(x)) + \sigma_{\mathrm{d}i}^{\mathrm{T}}(\phi_{\mathrm{d}i}(x))P_{2ui}(x)\sigma_{\mathrm{d}i}(\phi_{\mathrm{d}i}(x)) \\
&\quad + \phi_{\mathrm{d}i}^{\mathrm{T}}(x)(R_{2\mathrm{d}i}(x) + P_{2ui}(x))\phi_{\mathrm{d}i}(x) + w_{\mathrm{d}i}(V(x)) \\
&= \phi_{\mathrm{d}i}^{\mathrm{T}}(x)(R_{2\mathrm{d}i}(x) + P_{2ui}(x))\phi_{\mathrm{d}i}(x) \\
&\quad + \sigma_{\mathrm{d}i}^{\mathrm{T}}(\phi_{\mathrm{d}i}(x))P_{2ui}(x)\sigma_{\mathrm{d}i}(\phi_{\mathrm{d}i}(x)) \\
&\quad - 2\phi_{\mathrm{d}i}^{\mathrm{T}}(x)(R_{2\mathrm{d}i}(x) + P_{2ui}(x))\sigma_{\mathrm{d}i}(\phi_{\mathrm{d}i}(x)) + w_{\mathrm{d}i}(V(x)) \\
&= [\sigma_{\mathrm{d}i}(\phi_{\mathrm{d}i}(x)) - \phi_{\mathrm{d}i}(x)]^{\mathrm{T}}(R_{2\mathrm{d}i}(x) + P_{2ui}(x))[\sigma_{\mathrm{d}i}(\phi_{\mathrm{d}i}(x)) \\
&\quad - \phi_{\mathrm{d}i}(x)] - \sigma_{\mathrm{d}i}^{\mathrm{T}}(\phi_{\mathrm{d}i}(x))R_{2\mathrm{d}i}(x)\sigma_{\mathrm{d}i}(\phi_{\mathrm{d}i}(x)) + w_{\mathrm{d}i}(V(x)) \\
&\leq \overline{\gamma}_{\mathrm{d}i}[\sigma_{\mathrm{d}i}(\phi_{\mathrm{d}i}(x)) - \phi_{\mathrm{d}i}(x)]^{\mathrm{T}}[\sigma_{\mathrm{d}i}(\phi_{\mathrm{d}i}(x)) - \phi_{\mathrm{d}i}(x)] \\
&\quad - \underline{\gamma}_{\mathrm{d}i}\sigma_{\mathrm{d}i}^{\mathrm{T}}(\phi_{\mathrm{d}i}(x))\sigma_{\mathrm{d}i}(\phi_{\mathrm{d}i}(x)) + w_{\mathrm{d}i}(V(x)) \\
&= \overline{\gamma}_{\mathrm{d}i}(1 - \theta_{\mathrm{d}i}^2)\left[\sigma_{\mathrm{d}i}(\phi_{\mathrm{d}i}(x)) - \frac{1}{1 + \theta_{\mathrm{d}i}}\phi_{\mathrm{d}i}(x)\right]^{\mathrm{T}} \\
&\quad \cdot \left[\sigma_{\mathrm{d}i}(\phi_{\mathrm{d}i}(x)) - \frac{1}{1 - \theta_{\mathrm{d}i}}\phi_{\mathrm{d}i}(x)\right] + w_{\mathrm{d}i}(V(x)) \\
&= \overline{\gamma}_{\mathrm{d}i}(1 - \theta_{\mathrm{d}i}^2)\sum_{j=1}^{m_{\mathrm{d}i}}\left(\sigma_{\mathrm{d}ij}(\phi_{\mathrm{d}i}(x)) - \frac{1}{1 + \theta_{\mathrm{d}i}}\phi_{\mathrm{d}ij}(x)\right) \\
&\quad \cdot \left(\sigma_{\mathrm{d}ij}(\phi_{\mathrm{d}i}(x)) - \frac{1}{1 - \theta_{\mathrm{d}i}}\phi_{\mathrm{d}ij}(x)\right) + w_{\mathrm{d}i}(V(x)) \\
&\leq w_{\mathrm{d}i}(V(x)), \quad x \in \mathcal{Z}, \quad i = 1, \ldots, q.
\end{aligned}
\tag{11.48}
$$

Since the impulsive dynamical system (11.7) and (11.8) is globally asymptotically stable it follows from Theorem 10.2 that the nonlinear impulsive dynamical system (11.33) and (11.34) is globally asymptotically stable for all $\sigma_{ci}(\cdot) \in \Phi_{ci}$ and $\sigma_{\mathrm{d}i}(\cdot) \in \Phi_{\mathrm{d}i}$, $i = 1, \ldots, q$. $\qquad\square$

It follows from Theorem 11.2 that with the hybrid feedback stabilizing control law (11.26) and (11.36) the nonlinear impulsive dynamical system (11.19) and (11.20) has a sector (and hence gain) margin $((\frac{1}{2}, \infty), (\frac{1}{1+\theta_{\mathrm{d}i}}, \frac{1}{1-\theta_{\mathrm{d}i}}))$, $i = 1, \ldots, q$, in each decentralized input channel. For further details on gain, sector, and disk margins for nonlinear impulsive dynamical systems controlled by optimal hybrid regulators along with connections to the equivalence between dissipativity and optimality of nonlinear hybrid controllers, see [76].

11.4 Decentralized Control for Large-Scale Impulsive Dynamical Systems

In this section, we apply the proposed hybrid control framework to decentralized control of large-scale nonlinear impulsive dynamical systems. Specifically, we consider the large-scale dynamical system $\mathcal{G}$ involving energy exchange between n interconnected subsystems. Let $x_i : [0, \infty) \to \overline{\mathbb{R}}_+$ denote the energy (and hence a nonnegative quantity) of the ith subsystem, let $u_{ci} : [0, \infty) \to \mathbb{R}$ denote the control input to the ith subsystem, let $\sigma_{cij} : \overline{\mathbb{R}}_+^n \to \overline{\mathbb{R}}_+$, $i \neq j$, $i, j = 1, \ldots, n$, denote the instantaneous rate of energy flow from the jth subsystem to the ith subsystem between resettings, let $\sigma_{dij} : \overline{\mathbb{R}}_+^n \to \overline{\mathbb{R}}_+$, $i \neq j$, $i, j = 1, \ldots, n$, denote the amount of energy transferred from the jth subsystem to the ith subsystem at the resetting instant, and let $\mathcal{Z} \subset \overline{\mathbb{R}}_+^n$ be a resetting set for the large-scale impulsive dynamical system $\mathcal{G}$.

An energy balance for each subsystem $\mathcal{G}_i$, $i = 1, \ldots q$, yields [81,82]

$$\dot{x}_i(t) = \sum_{j=1, j \neq i}^{n} [\sigma_{cij}(x(t)) - \sigma_{cji}(x(t))] + G_{ci}(x(t))u_{ci}(t), \quad x(t_0) = x_0,$$

$$x(t) \notin \mathcal{Z}, \quad t \geq t_0, \quad (11.49)$$

$$\Delta x_i(t) = \sum_{j=1, j \neq i}^{n} [\sigma_{dij}(x(t)) - \sigma_{dji}(x(t))] + G_{di}(x(t))u_{di}(t), \quad x(t) \in \mathcal{Z},$$

$$(11.50)$$

or, equivalently, in vector form for the large-scale impulsive dynamical system $\mathcal{G}$

$$\dot{x}(t) = f_c(x(t)) + G_c(x(t))u_c(t), \quad x(t_0) = x_0, \quad x(t) \notin \mathcal{Z}, \quad t \geq t_0, \quad (11.51)$$

$$\Delta x(t) = f_d(x(t)) + G_d(x(t))u_d(t), \quad x(t) \in \mathcal{Z}, \quad (11.52)$$

where $x(t) = [x_1(t), \ldots, x_n(t)]^{\mathrm{T}}$, $t \geq t_0$, $f_{ci}(x) = \sum_{j=1, j \neq i}^{n} \phi_{cij}(x)$, where $\phi_{cij}(x) \triangleq \sigma_{cij}(x) - \sigma_{cji}(x)$, $x \in \overline{\mathbb{R}}_+^n$, $i \neq j$, $i, j = 1, \ldots, q$, denotes the net energy flow from the jth subsystem to the ith subsystem between resettings, $G_c(x) = \mathrm{diag}[G_{c1}(x), \ldots, G_{cn}(x)] = \mathrm{diag}[x_1, \ldots, x_n]$, $x \in \overline{\mathbb{R}}_+^n$, $G_d(x) = \mathrm{diag}[G_{d1}(x), \ldots, G_{dn}(x)]$, $x \in \mathcal{Z}$, $G_{di} : \mathbb{R}^n \to \mathbb{R}$, $i = 1, \ldots, n$, $u_c(t) \in \mathbb{R}^n$, $t \geq t_0$, $u_d(t_k) \in \mathbb{R}^n$, $k \in \overline{\mathbb{Z}}_+$, $f_{di}(x) = \sum_{j=1, j \neq i}^{n} \phi_{dij}(x)$, where $\phi_{dij}(x) \triangleq \sigma_{dij}(x) - \sigma_{dji}(x)$, $x \in \mathcal{Z}$, $i \neq j$, $i, j = 1, \ldots, q$, denotes the net amount of energy transferred from the jth subsystem to the ith subsystem at the instant of resetting. Here, we assume that $\sigma_{cij} : \overline{\mathbb{R}}_+^n \to \overline{\mathbb{R}}_+$, $i \neq j$, $i, j = 1, \ldots, n$, are locally Lipschitz continuous on $\overline{\mathbb{R}}_+^n$, $\sigma_{cij}(0) = 0$, $i \neq j$, $i, j = 1, \ldots, n$, and $u_c = [u_{c1}, \ldots, u_{cn}]^{\mathrm{T}} : \mathbb{R} \to \mathbb{R}^n$ is such that $u_{ci} : \mathbb{R} \to \mathbb{R}$, $i = 1, \ldots, n$, are bounded piecewise continuous functions of time. Furthermore, we assume that $\sigma_{cij}(x) = 0$, $x \in \overline{\mathbb{R}}_+^n$, whenever $x_j = 0$, $i \neq j$, $i, j = 1, \ldots, n$, and $x_i + \sum_{j=1,j\neq i}^{n} \phi_{\mathrm{d}ij}(x) \geq 0$, $x \in \mathcal{Z}$. In this case, $f_{\mathrm{c}}(\cdot)$ is *essentially nonnegative* [69, 81] (that is, $f_{\mathrm{c}i}(x) \geq 0$ for all $x \in \overline{\mathbb{R}}_+^n$ such that $x_i = 0$, $i = 1, \ldots, n$) and $x + f_{\mathrm{d}}(x)$, $x \in \mathcal{Z} \subset \overline{\mathbb{R}}_+^n$, is *nonnegative* (that is, $x_i + f_{\mathrm{d}i}(x) \geq 0$ for all $x \in \mathcal{Z}$, $i = 1, \ldots, n$).

The above constraints imply that if the energy of the jth subsystem of $\mathcal{G}$ is zero, then this subsystem cannot supply any energy to its surroundings between resettings and the ith subsystem of $\mathcal{G}$ cannot transfer more energy to its surroundings than it possesses at the instant of resetting. Finally, to ensure that the trajectories of the closed-loop system remain in the nonnegative orthant of the state space for all nonnegative initial conditions, we seek a hybrid feedback control law $(u_{\mathrm{c}}(\cdot), u_{\mathrm{d}}(\cdot))$ that guarantees the continuous-time closed-loop system dynamics (11.51) are essentially nonnegative and the closed-loop system states after the resettings are nonnegative [77].

For the dynamical system $\mathcal{G}$, consider the control vector Lyapunov function candidate $V(x) = [v_1(x_1), \ldots, v_n(x_n)]^{\mathrm{T}}$, $x \in \overline{\mathbb{R}}_+^n$, given by

$$V(x) = [x_1, \ldots, x_n]^{\mathrm{T}}, \quad x \in \overline{\mathbb{R}}_+^n. \tag{11.53}$$

Note that $V(0) = 0$ and $v(x) \triangleq \mathbf{e}^{\mathrm{T}} V(x)$, $x \in \overline{\mathbb{R}}_+^n$, is such that $v(0) = 0$, $v(x) > 0$, $x \neq 0$, $x \in \overline{\mathbb{R}}_+^n$, and $v(x) \to \infty$ as $\|x\| \to \infty$ with $x \in \overline{\mathbb{R}}_+^n$. In addition, note that since $v_i(x_i) = x_i$, $x \in \mathbb{R}^n$, $i = 1, \ldots, n$, are linear functions of x, it follows that $P_{2ui}(x) \equiv 0$, $i = 1, \ldots, n$, and hence, by (11.27), $\phi_{\mathrm{d}i}(x) \equiv 0$, $i = 1, \ldots, n$. Furthermore, consider the functions

$$w_{\mathrm{c}}(V(x), x) = \begin{bmatrix} -\sigma_{\mathrm{c}11}(v_1(x_1)) + \sum_{j=1,j\neq1}^{n} \phi_{\mathrm{c}1j}(x) \\ \vdots \\ -\sigma_{\mathrm{c}nn}(v_n(x_n)) + \sum_{j=1,j\neq n}^{n} \phi_{\mathrm{c}nj}(x) \end{bmatrix}, \quad x \in \overline{\mathbb{R}}_+^n, \tag{11.54}$$

$$w_{\mathrm{d}}(V(x), x) = \left[\sum_{j=1,j\neq1}^{n} \phi_{\mathrm{d}1j}(x), \ldots, \sum_{j=1,j\neq n}^{n} \phi_{\mathrm{d}nj}(x) \right]^{\mathrm{T}}, \quad x \in \mathcal{Z}, \tag{11.55}$$

where $\sigma_{\mathrm{c}ii} : \overline{\mathbb{R}}_+ \to \overline{\mathbb{R}}_+$, $i = 1, \ldots, n$, are positive definite functions, and note that $w_{\mathrm{c}}(\cdot, x) \in \mathcal{W}_{\mathrm{c}}$, $x \in \overline{\mathbb{R}}_+^n$, $w_{\mathrm{c}}(0, 0) = 0$, $w_{\mathrm{d}}(\cdot, \cdot)$ does not depend on $V(\cdot)$, and hence, $w_{\mathrm{d}}(\cdot, x) \in \mathcal{W}_{\mathrm{d}}$, $x \in \mathcal{Z}$. Note that $\mathcal{R}_i \triangleq \{x \in \overline{\mathbb{R}}_+^n, x_i \neq 0 : V_i'(x_i)G_{\mathrm{c}i}(x) = 0\} = \{x \in \overline{\mathbb{R}}_+^n, x_i \neq 0 : x_i = 0\} = \varnothing$, and hence, condition (11.22) is satisfied for $V(\cdot)$ and $w_{\mathrm{c}}(\cdot, \cdot)$ given by (11.53) and (11.54), respectively, and condition (11.23) is satisfied as an equality for $w_{\mathrm{d}}(\cdot, \cdot)$ given by (11.55) and $P_{2ui}(x) \equiv 0$, $i = 1, \ldots, n$.

To show that the impulsive dynamical system

$$\dot{z}(t) = w_{\mathrm{c}}(z(t), x(t)), \quad z(t_0) = z_0, \quad x(t) \notin \mathcal{Z}, \quad t \geq t_0, \tag{11.56}$$

$$\Delta z(t) = w_{\mathrm{d}}(z(t), x(t)), \quad x(t) \in \mathcal{Z}, \tag{11.57}$$

where $z(t) \in \overline{\mathbb{R}}_+^n$, $t \geq t_0$, $x(t)$, $t \geq t_0$, is the solution to (11.51) and (11.52), the ith component of $w_{\mathrm{c}}(z, x)$ is given by $w_{\mathrm{c}i}(z, x) = -\sigma_{\mathrm{c}ii}(z_i) +$

$\sum_{j=1, j\neq i}^{n} \phi_{cij}(x)$, $z \in \overline{\mathbb{R}}_{+}^{n}$, $x \in \overline{\mathbb{R}}_{+}^{n}$, and the ith component of $w_{\mathrm{d}}(z, x)$ is given by $w_{\mathrm{d}i}(z, x) = \sum_{j=1, j\neq i}^{n} \phi_{\mathrm{d}ij}(x)$, $x \in \mathcal{Z}$, is globally asymptotically stable with respect to z uniformly in x_0, consider the partial Lyapunov function candidate $\tilde{v}(z) = \mathbf{e}^{\mathrm{T}} z$, $z \in \overline{\mathbb{R}}_{+}^{n}$. Note that $\tilde{v}(\cdot)$ is radially unbounded, $\tilde{v}(0) = 0$, $\tilde{v}(z) > 0$, $z \in \overline{\mathbb{R}}_{+}^{n}$, $z \neq 0$, $\dot{\tilde{v}}(z) = -\sum_{i=1}^{n} \sigma_{cii}(z_i) + \sum_{i=1}^{n} \sum_{j=1, j\neq i}^{n} \phi_{cij}(x) = -\sum_{i=1}^{n} \sigma_{cii}(z_i) < 0$, $z \in \overline{\mathbb{R}}_{+}^{n}$, $z \neq 0$, $x \in \overline{\mathbb{R}}_{+}^{n}$, and $\Delta\tilde{v}(z) = \sum_{i=1}^{n} \sum_{j=1, j\neq i}^{n} \phi_{\mathrm{d}ij}(x) = 0$, $z \in \overline{\mathbb{R}}_{+}^{n}$, $x \in \mathcal{Z}$. Thus, it follows from Theorem 2.1 of [78] that the impulsive dynamical system (11.56), (11.57), (11.51), and (11.52) is globally asymptotically stable with respect to z uniformly in x_0. Hence, it follows from Theorem 11.1 that $V(x)$, $x \in \overline{\mathbb{R}}_{+}^{n}$, given by (11.53) is a control vector Lyapunov function for the dynamical system (11.51) and (11.52).

Next, using (11.26) with $\alpha_i(x) = v_i'(x_i) f_{ci}(x) = \sum_{j=1, j\neq i}^{n} \phi_{cij}(x)$, $\beta_i(x) = x_i$, $x \in \overline{\mathbb{R}}_{+}^{n}$, and $c_{0i} > 0$, $i = 1, \ldots, n$, we construct a globally stabilizing hybrid decentralized feedback controller for (11.51) and (11.52) given by

$$
\phi_{\mathrm{c}i}(x) = \begin{cases} -\left(c_{0i} + \dfrac{\sigma_{cii}(x_i) + \sqrt{\sigma_{cii}^2(x_i) + x_i^2}}{x_i^2} \right) x_i, & x_i \neq 0, \\[2ex] 0, & x_i = 0, \end{cases} \tag{11.58}
$$

and

$$
\phi_{\mathrm{d}i}(x) \equiv 0, \tag{11.59}
$$

for all $i = 1, \ldots, n$. It can be seen from the structure of the feedback control law (11.58) and (11.59) that the continuous-time closed-loop system dynamics are essentially nonnegative and the closed-loop system states after the resettings are nonnegative. Furthermore, since $\alpha_i(x) - w_{ci}(V(x), x) = \sigma_{cii}(v_i(x_i))$, $x \in \overline{\mathbb{R}}_{+}^{n}$, $i = 1, \ldots, n$, the continuous-time feedback controller $\phi_{\mathrm{c}}(\cdot)$ is fully independent from $f_{\mathrm{c}}(x)$, which represents the internal interconnections of the large-scale system dynamics, and hence, is robust against full modeling uncertainty in $f_{\mathrm{c}}(x)$.

For the following simulation we consider (11.51) and (11.52) with $\sigma_{cij}(x) = \sigma_{cij} x_i x_j$, $\sigma_{cii}(x) = \sigma_{cii} x_i^2$, and $\sigma_{\mathrm{d}ij}(x) = \sigma_{\mathrm{d}ij} x_j$, where $\sigma_{cij} \geq 0$, $i \neq j$, $i, j = 1, \ldots, n$, $\sigma_{cii} > 0$, $i = 1, \ldots, n$, and $\sigma_{\mathrm{d}ij} \geq 0$, $i \neq j$, $i, j = 1, \ldots, n$, and $1 \geq \sum_{j=1, j\neq i}^{n} \sigma_{\mathrm{d}ji}$, $i = 1, \ldots, n$. Note that in this case the conditions of Proposition 11.1 are satisfied, and hence, the continuous-time feedback control law (11.58) is continuous on $\overline{\mathbb{R}}_{+}^{n}$. For our simulation we set $n = 3$, $\sigma_{c11} = 0.1$, $\sigma_{c22} = 0.2$, $\sigma_{c33} = 0.01$, $\sigma_{c12} = 2$, $\sigma_{c13} = 3$, $\sigma_{c21} = 1.5$, $\sigma_{c23} = 0.3$, $\sigma_{c31} = 4.4$, $\sigma_{c32} = 0.6$, $\sigma_{\mathrm{d}12} = 0.75$, $\sigma_{\mathrm{d}13} = 0.33$, $\sigma_{\mathrm{d}21} = 0.2$, $\sigma_{\mathrm{d}23} = 0.5$, $\sigma_{\mathrm{d}31} = 0.66$, $\sigma_{\mathrm{d}32} = 0.2$, $c_{01} = 1$, $c_{02} = 1$, $c_{03} = 0.25$, the resetting set $\mathcal{Z} = \{x \in \mathbb{R}^3 : x_1 + x_2 - 0.75 = 0\}$, with initial condition $x_0 = [3, 4, 1]^{\mathrm{T}}$. Figure 11.1 shows the states of the closed-loop system versus time and Figure 11.2 shows continuous-time control signal for each decentralized control channel as a function of time.

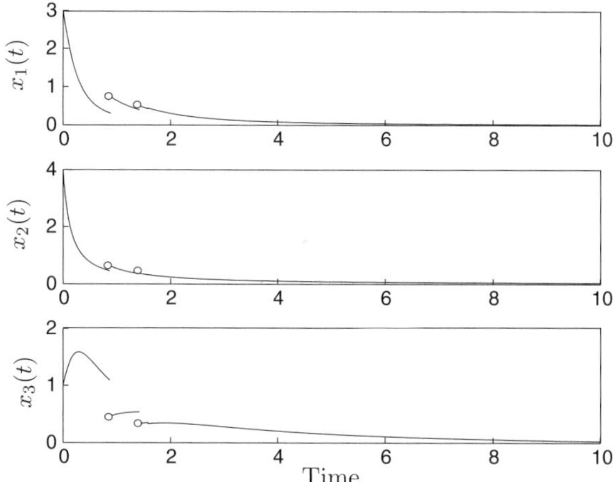

Figure 11.1 Controlled system states versus time.

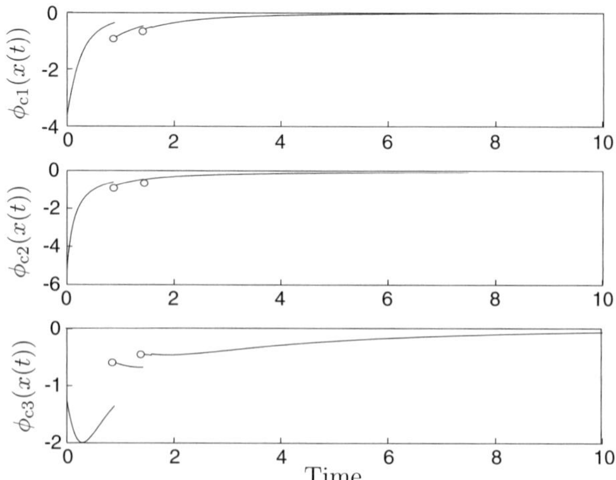

Figure 11.2 Control signals in each decentralized control channel versus time.

Finite-Time Stabilization of Large-Scale Impulsive Dynamical Systems

12.1 Introduction

As noted in Chapter 6, finite-time stability implies Lyapunov stability and convergence of system trajectories to an equilibrium state in finite time, and hence, is a stronger notion than asymptotic stability. For continuous-time dynamical systems, finite-time stability implies non-Lipschitzian dynamics [24,86] giving rise to non-uniqueness of solutions in reverse time. Uniqueness of solutions in forward time, however, can be preserved in the case of finite-time convergence. Sufficient conditions that ensure uniqueness of solutions in forward time in the absence of Lipschitz continuity are given in [1, 58, 108, 176]. Finite-time convergence to a Lyapunov stable equilibrium for continuous-time systems, that is, finite-time stability, was rigorously studied in [24, 25] using Hölder continuous Lyapunov functions.

Finite-time stability of impulsive dynamical systems has not been studied in the literature. For impulsive dynamical systems, it may be possible to reset the system states to an equilibrium state, in which case finite-time convergence of the system trajectories can be achieved without the requirement of non-Lipschitzian dynamics. In addition, due to system resettings, impulsive dynamical systems may exhibit non-uniqueness of solutions in reverse time even when the continuous-time dynamics are Lipschitz continuous.

In this chapter, we develop sufficient conditions for finite-time stability of nonlinear impulsive dynamical systems. Furthermore, we present stability results using vector Lyapunov functions wherein finite-time stability of the impulsive system is guaranteed via finite-time stability of a hybrid vector comparison system. We use these results to construct hybrid finite-time stabilizing controllers for impulsive dynamical systems. In addition, we construct decentralized finite-time stabilizers for large-scale impulsive dynamical systems. Finally, we present a numerical example to show the utility of the proposed framework.

12.2 Finite-Time Stability of Impulsive Dynamical Systems

Consider the nonlinear *state-dependent* impulsive dynamical system $\mathcal{G}$ [82] given by

$$\dot{x}(t) = f_{\mathrm{c}}(x(t)), \quad x(0) = x_0, \quad x(t) \notin \mathcal{Z}, \quad t \in \mathcal{I}_{x_0}, \qquad (12.1)$$

$$\Delta x(t) = f_{\mathrm{d}}(x(t)), \quad x(t) \in \mathcal{Z}. \tag{12.2}$$

where $x(t) \in \mathcal{D} \subseteq \mathbb{R}^n$, $t \in \mathcal{I}_{x_0}$, is the system state vector, $\mathcal{I}_{x_0}$ is the maximal interval of existence of a solution $x(t)$ to (12.1) and (12.2), $\mathcal{D}$ is an open set, $0 \in \mathcal{D}$, $f_{\mathrm{c}} : \mathcal{D} \to \mathbb{R}^n$ is continuous and satisfies $f_{\mathrm{c}}(0) = 0$, $f_{\mathrm{d}} : \mathcal{D} \to \mathbb{R}^n$ is continuous, $\Delta x(t) \triangleq x(t^+) - x(t)$, $x(t^+) \triangleq x(t) + f_{\mathrm{d}}(x(t)) = \lim_{\varepsilon \to 0} x(t + \varepsilon)$, $x(t) \in \mathcal{Z}$, and $\mathcal{Z} \subset \mathcal{D} \subseteq \mathbb{R}^n$ is the resetting set. A function $x : \mathcal{I}_{x_0} \to \mathcal{D}$ is a *solution* to the impulsive dynamical system (12.1) and (12.2) on the interval $\mathcal{I}_{x_0} \subseteq \mathbb{R}$ with initial condition $x(0) = x_0$ if $x(\cdot)$ is left-continuous and $x(t)$ satisfies (12.1) and (12.2) for all $t \in \mathcal{I}_{x_0}$. For a particular trajectory $x(t)$, $t \geq 0$, we let $t_k = \tau_k(x_0)$, $x_0 \in \mathcal{D}$, denote the kth instant of time at which $x(t)$ intersects $\mathcal{Z}$ and we let $x_k^+ \triangleq x(\tau_k^+(x_0)) \triangleq x(\tau_k(x_0)) + f_{\mathrm{d}}(x(\tau_k(x_0)))$ denote the state of (12.1) and (12.2) after the kth resetting. To ensure well-posedness of the resetting times we assume that Assumptions 10.1 and 10.2 hold.

Furthermore, we assume that (12.1) possesses unique solutions in forward time for all initial conditions in $\mathcal{D}$ except possibly the origin in the following sense. For every $x \in \mathcal{D} \backslash \{0\}$ there exists $\tau_x > 0$ such that, if $y_1 : [0, \tau_1) \to \mathcal{D}$ and $y_2 : [0, \tau_2) \to \mathcal{D}$ are two solutions of (12.1) with $y_1(0) = y_2(0) = x$, then $\tau_x \leq \min\{\tau_1, \tau_2\}$ and $y_1(t) = y_2(t)$ for all $t \in [0, \tau_x)$. Without loss of generality, we assume that for each $x \in \mathcal{D}$, τ_x is chosen to be the largest such number in $\overline{\mathbb{R}}_+$. Sufficient conditions for forward uniqueness of solutions to continuous-time dynamical systems in the absence of Lipschitz continuity of the system dynamics can be found in [1], [58, Sect. 10], [108], and [176, Sect. 1].

Since the resetting times are well defined and distinct, and since the solution to (12.1) exists and is unique, it follows that the solution of the impulsive dynamical system (12.1) and (12.2) also exists and is unique over a forward time interval. Furthermore, as in Chapter 10, we allow for the possibility of confluence and Zeno solutions; however, Assumption 10.2 precludes the possibility of beating. In addition, since *not* every bounded solution of an impulsive dynamical system over a forward time interval can be extended to infinity due to Zeno solutions, we assume that existence and uniqueness of solutions are satisfied in forward time. For details, see [11, 13, 117]. Finally, we denote the *trajectory* or *solution curve* of (12.1) and (12.2) satisfying $x(0) = x$ by $s(\cdot, x)$ or $s^x(\cdot)$.

The following definition introduces the notion of finite-time stability for impulsive dynamical systems.

Definition 12.1. Consider the nonlinear impulsive dynamical system $\mathcal{G}$ given by (12.1) and (12.2). The zero solution $x(t) \equiv 0$ to (12.1) and (12.2) is *finite-time stable* if there exist an open neighborhood $\mathcal{N} \subseteq \mathcal{D}$ of the origin and a function $T : \mathcal{N} \backslash \{0\} \to (0, \infty)$, called the *settling-time function*, such that the following statements hold:

i) Finite-time convergence. For every $x \in \mathcal{N}\backslash\{0\}$, $s^x(t)$ is defined on $[0, T(x))$, $s^x(t) \in \mathcal{N}\backslash\{0\}$ for all $t \in [0, T(x))$, and $\lim_{t \to T(x)} s(x, t) = 0$.

ii) Lyapunov stability. For every $\varepsilon > 0$ there exists $\delta > 0$ such that $\mathcal{B}_\delta(0) \subset \mathcal{N}$ and for every $x \in \mathcal{B}_\delta(0)\backslash\{0\}$, $s(t, x) \in \mathcal{B}_\varepsilon(0)$ for all $t \in [0, T(x))$.

The zero solution $x(t) \equiv 0$ to (12.1) and (12.2) is *globally finite-time stable* if it is finite-time stable with $\mathcal{N} = \mathcal{D} = \mathbb{R}^n$.

Note that if the zero solution $x(t) \equiv 0$ to (12.1) and (12.2) is finite-time stable, then it is asymptotically stable, and hence, finite-time stability is a stronger notion than asymptotic stability. In the case of impulsive dynamical systems it may be possible to reset the states to the origin, and hence, $s(t, x_0) = 0$, $t > \tau_k(x_0) = T(x_0)$. The following result provides sufficient conditions for finite-time stability of impulsive systems using a Lyapunov function involving a scalar differential inequality.

Theorem 12.1. Consider the nonlinear impulsive dynamical system $\mathcal{G}$ given by (12.1) and (12.2). Assume there exists a continuously differentiable function $V : \mathcal{D} \to \overline{\mathbb{R}}_+$ satisfying $V(0) = 0$, $V(x) > 0$, $x \in \mathcal{D}$, $x \neq 0$, and

$$V'(x)f_{\rm c}(x) \leq -c(V(x))^\alpha, \quad x \notin \mathcal{Z}, \tag{12.3}$$
$$V(x + f_{\rm d}(x)) \leq V(x), \quad x \in \mathcal{Z}, \tag{12.4}$$

where $c > 0$ and $\alpha \in (0, 1)$. Then the zero solution $x(t) \equiv 0$ to (12.1) and (12.2) is finite-time stable. If, in addition, $\mathcal{D} = \mathbb{R}^n$ and $V(\cdot)$ is radially unbounded, then the zero solution $x(t) \equiv 0$ to (12.1) and (12.2) is globally finite-time stable.

Proof. Note that it follows from Theorem 2.1 of [82] that the zero solution to (12.1) and (12.2) is asymptotically stable. Thus, it remains to be shown that for all initial conditions in some neighborhood $\mathcal{N} \subseteq \mathcal{D}$ of the origin the trajectories of (12.1) and (12.2) converge to the origin in finite time. Since the system (12.1) and (12.2) is asymptotically stable, it follows that there exists $\delta > 0$ such that for all $x_0 \in \mathcal{B}_\delta(0) \subset \mathcal{D}$ the trajectory $s(t, x_0) \to 0$ as $t \to \infty$. Next, we separately consider the cases when the trajectories of (12.1) and (12.2) have a finite and infinite number of resettings.

Assume that for some $x_0 \in \mathcal{B}_\delta(0)$ the trajectory $s(t, x_0)$, $t \geq 0$, exhibits a finite number of resettings with resetting times $\tau_k(x_0)$, $k = 1, \ldots, m$. If $s(\tau_m^+(x_0), x_0) = 0$, then since $f_{\rm c}(0) = 0$ it follows that $s(t, x_0) = 0$, $t \geq \tau_m(x_0)$, which implies that $s(t, x_0)$, $t \geq t_0$, converges to the origin in finite time with a settling-time function $T(x_0) = \tau_m(x_0)$. Alternatively, if $s(\tau_m^+(x_0), x_0) \neq 0$, then for all $t > \tau_m(x_0)$ the continuous time dynamics are active and it follows from (12.3) and Theorem 4.2 of [24] that the trajectory $s(t, s(\tau_m^+(x_0), x_0))$, $t \geq 0$, converges to the origin in finite time that is less than or equal to $\frac{1}{c(1-\alpha)}[V(s(\tau_m^+(x_0),x_0))]^{1-\alpha}$. In this case, the settling-time function for $s(t,x_0)$, $t \geq 0$, is $T(x_0) \leq \tau_m(x_0) + \frac{1}{c(1-\alpha)}[V(s(\tau_m^+(x_0),x_0))]^{1-\alpha}$.

Alternatively, assume that for some $x_0 \in \mathcal{B}_\delta(0)$ the trajectory $s(t,x_0)$, $t \geq 0$, exhibits an infinite number of resettings with the resetting times $\tau_k(x_0)$, $k = 0,1,\ldots$, where $\tau_0(x_0) \triangleq 0$. Let $x_k^+ \triangleq s(\tau_k^+(x_0),x_0)$, $k = 0,1,\ldots$, where $x_0^+ \triangleq x_0$, and note that since (12.1) and (12.2) is asymptotically stable it follows that $\tau_1(x_k^+) \to 0$ as $k \to \infty$. It was shown in [24] that with (12.3), the continuous-time dynamics are finite-time stable for the case when $\mathcal{Z} = \emptyset$. Furthermore, note that $\tau_1(x_k^+) < T_{\mathrm{c}}(x_k^+)$, $k = 0,1,\ldots$, since (12.1) and (12.2) exhibit an infinite number of resettings, where $T_{\mathrm{c}}(\cdot)$ is the settling-time function when $\mathcal{Z} = \emptyset$. Moreover, as shown in [24],

$$V(s(t,y)) \leq [(V(y))^{1-\alpha} - c(1-\alpha)t]^{\frac{1}{1-\alpha}}, \quad t \in [0, T_{\mathrm{c}}(y)), \quad y \in \mathcal{B}_\delta(0),$$
$$(12.5)$$

and hence, since $\tau_1(x_0) < T_{\mathrm{c}}(x_0)$, it follows that

$$V(x_1) \leq [(V(x_0))^{1-\alpha} - c(1-\alpha)\tau_1(x_0)]^{\frac{1}{1-\alpha}}. \tag{12.6}$$

Thus, since $V(x + f_{\mathrm{d}}(x)) \leq V(x)$, $x \in \mathcal{Z}$, it follows from (12.6) that

$$\begin{aligned}
\tau_1(x_1^+) &< T_{\mathrm{c}}(x_1^+) \\
&\leq \frac{1}{c(1-\alpha)}(V(x_1^+))^{1-\alpha} \\
&\leq \frac{1}{c(1-\alpha)}(V(x_1))^{1-\alpha} \\
&\leq \frac{1}{c(1-\alpha)}[(V(x_0))^{1-\alpha} - c(1-\alpha)\tau_1(x_0)]. \tag{12.7}
\end{aligned}$$

Similarly, it follows from (12.5) that, for $y = x_2^+$,

$$\begin{aligned}
\tau_1(x_2^+) &< T_{\mathrm{c}}(x_2^+) \\
&\leq \frac{1}{c(1-\alpha)}(V(x_2^+))^{1-\alpha} \\
&\leq \frac{1}{c(1-\alpha)}(V(x_2))^{1-\alpha} \\
&\leq \frac{1}{c(1-\alpha)}[(V(x_1^+))^{1-\alpha} - c(1-\alpha)\tau_1(x_1^+)] \\
&\leq \frac{1}{c(1-\alpha)}[(V(x_1))^{1-\alpha} - c(1-\alpha)\tau_1(x_1^+)] \\
&\leq \frac{1}{c(1-\alpha)}[(V(x_0))^{1-\alpha} - c(1-\alpha)\tau_1(x_0) - c(1-\alpha)\tau_1(x_1^+)]. \tag{12.8}
\end{aligned}$$

Recursively repeating this procedure for $k = 3, 4, \ldots$, it follows that, with $\tau_1(x_0^+) = \tau_1(x_0)$,

$$\tau_1(x_k^+) < \frac{1}{c(1-\alpha)}[(V(x_0))^{1-\alpha} - c(1-\alpha)\sum_{i=0}^{k-1}\tau_1(x_i^+)]. \qquad (12.9)$$

Next, let $k \to \infty$ and note that since $x(t) \equiv 0$ is asymptotically stable, it follows that $\lim_{k\to\infty}\tau_1(x_k^+) = 0$. Hence,

$$0 = \lim_{k\to\infty}\tau_1(x_k^+) < \frac{1}{c(1-\alpha)}[(V(x_0))^{1-\alpha} - c(1-\alpha)\sum_{i=0}^{\infty}\tau_1(x_i^+)].$$

Thus,

$$T(x_0) = \sum_{i=0}^{\infty}\tau_1(x_i^+) < \frac{1}{c(1-\alpha)}(V(x_0))^{1-\alpha} < \infty, \qquad (12.10)$$

which implies that the trajectory $s(\cdot, x_0)$ is Zeno [82] and converges to the origin in finite time with an infinite number of resettings, that is, $s(t, x_0) \to 0$ as $t \to T(x_0)$.

Finally, suppose, *ad absurdum*, that $s(t', x_0) \neq 0$ for some $t' > T(x_0)$, $x_0 \in \mathcal{B}_\delta(0)$. Then, since $V(\cdot)$ is positive definite, $V(s(t', x_0)) = \beta > 0$. Furthermore, since $s(t, x_0) \to 0$ as $t \to T(x_0)$, there exists $t'' < T(x_0)$ such that $V(s(t'', x_0)) < \beta$. Now, since $V(s(t, x_0))$ is a decreasing function of time, it follows that for $t'' < T(x_0) < t'$,

$$\beta = V(s(t', x_0)) < V(s(t'', x_0)) < \beta, \qquad (12.11)$$

which leads to a contradiction. Hence, $s(t, x_0) = 0$, $t \geq T(x_0)$, $x_0 \in \mathcal{B}_\delta(0)$, which implies convergence in finite time with $\mathcal{N} \triangleq \mathcal{B}_\delta(0)$. This completes the proof of finite-time stability.

Finally, if $\mathcal{D} = \mathbb{R}^n$ and $V(\cdot)$ is radially unbounded, then global finite-time stability follows using standard arguments. See, for instance, [82]. $\square$

Conditions (12.3) and (12.4) are only sufficient conditions for guaranteeing finite-time stability of impulsive dynamical systems. Alternatively, finite-time stability can also be achieved by imposing additional conditions on the discrete-time dynamics. For example, if for some $x_0 \in \mathcal{D}$, $x(t_k) \in \mathcal{Z} \cap \{x \in \mathcal{D} : x - f_d(x) = 0\}$, then the trajectory $x(\cdot)$ resets to the origin and, since $f_c(0) = 0$, finite-time convergence is achieved. A convergent Zeno solution is yet another example of finite-time convergence to the equilibrium point.

The next theorem generalizes Theorem 12.1 to the case of vector Lyapunov functions involving a vector differential inequality. For the remainder of this chapter, recall the notions of quasi-monotone increasing and nondecreasing functions given in Definitions 2.2 and 2.5, respectively.

Theorem 12.2. Consider the nonlinear impulsive dynamical system given by (12.1) and (12.2). Assume there exist a continuously differentiable vector function $V : \mathcal{D} \to \mathcal{Q} \cap \overline{\mathbb{R}}_+^q$, where $\mathcal{Q} \subset \mathbb{R}^q$ and $0 \in \mathcal{Q}$, continuous functions $w_{\mathrm{c}} : \mathcal{Q} \to \mathbb{R}^q$ and $w_{\mathrm{d}} : \mathcal{Q} \to \mathbb{R}^q$, and a positive vector $p \in \mathbb{R}_+^q$ such that $V(0) = 0$, $w_{\mathrm{c}}(\cdot) \in \mathcal{W}_{\mathrm{c}}$, $w_{\mathrm{d}}(\cdot) \in \mathcal{W}_{\mathrm{d}}$, $w_{\mathrm{c}}(0) = 0$, $w_{\mathrm{d}}(0) = 0$, the scalar function $p^{\mathrm{T}} V(x)$, $x \in \mathcal{D}$, is positive definite, and

$$V'(x) f_{\mathrm{c}}(x) \leq\leq w_{\mathrm{c}}(V(x)), \quad x \notin \mathcal{Z}, \tag{12.12}$$
$$V(x + f_{\mathrm{d}}(x)) \leq\leq V(x) + w_{\mathrm{d}}(V(x)), \quad x \in \mathcal{Z}. \tag{12.13}$$

In addition, assume that the hybrid vector comparison system

$$\dot{z}(t) = w_{\mathrm{c}}(z(t)), \quad z(0) = 0, \quad x(t) \notin \mathcal{Z}, \tag{12.14}$$
$$\Delta z(t) = w_{\mathrm{d}}(z(t)), \quad x(t) \in \mathcal{Z}, \tag{12.15}$$

has a unique solution $z(t)$, $t \geq 0$, in forward time, and there exist a continuously differentiable function $v : \mathcal{Q} \to \mathbb{R}$, real numbers $c > 0$ and $\alpha \in (0,1)$, and a neighborhood $\mathcal{M} \subseteq \mathcal{Q}$ of the origin such that $v(\cdot)$ is positive definite and

$$v'(z) w_{\mathrm{c}}(z) \leq -c(v(z))^\alpha, \quad z \in \mathcal{M}, \tag{12.16}$$
$$v(z + w_{\mathrm{d}}(z)) \leq v(z), \quad z \in \mathcal{M}. \tag{12.17}$$

Then the zero solution $x(t) \equiv 0$ to (12.1) and (12.2) is finite-time stable. If, in addition, $\mathcal{D} = \mathbb{R}^n$, $v(\cdot)$ is radially unbounded, and (12.16) and (12.17) hold on $\mathbb{R}^q$, then the zero solution $x(t) \equiv 0$ to (12.1) and (12.2) is globally finite-time stable.

Proof. First, note that since the dynamics of the impulsive dynamical system (12.1) and (12.2), and the impulsive comparison system (12.14) and (12.15) are decoupled, the comparison system (12.14) and (12.15) is a time-dependent impulsive dynamical system [82]; that is, the resetting times for the trajectories of (12.14) and (12.15) are predetermined by the trajectories of (12.1) and (12.2). Hence, the stability analysis for the comparison system (12.14) and (12.15) involves a *time-dependent* impulsive dynamical system [82].

Since $v(\cdot)$ is positive definite, there exist $r > 0$ and class $\mathcal{K}$ functions [85] $\alpha, \beta : [0, r] \to \overline{\mathbb{R}}_+$ such that $\mathcal{B}_r(0) \subset \mathcal{M}$ and

$$\alpha(\|z\|) \leq v(z) \leq \beta(\|z\|), \quad z \in \mathcal{B}_r(0), \tag{12.18}$$

and hence,

$$v'(z) w_{\mathrm{c}}(z) \leq -c(v(z))^\alpha \leq -c(\alpha(\|z\|))^\alpha, \quad z \in \mathcal{B}_r(0). \tag{12.19}$$

Thus, it follows from (12.17), (12.18), (12.19), and Theorem 2.6 of [82] that the zero solution $z(t) \equiv 0$ to the impulsive comparison system (12.14) and (12.15) is uniformly asymptotically stable. Furthermore, using identical arguments as in the proof of Theorem 12.1, it can be shown that the trajectories of (12.14) and (12.15) converge to the origin in finite time for all $z_0 \in \mathcal{B}_r(0)$, and hence, the zero solution $z(t) \equiv 0$ to (12.14) and (12.15) is finite-time stable. Moreover, it follows from (12.12) and (12.13), the asymptotic stability of the comparison system (12.14) and (12.15), and Theorem 2.11 of [82] that the nonlinear impulsive dynamical system $\mathcal{G}$ given by (12.1) and (12.2) is asymptotically stable. Hence, it remains to be shown that there exists a neighborhood $\mathcal{N} \subseteq \mathcal{D}$ of the origin such that the trajectories of (12.1) and (12.2) converge to the origin in finite time for all $x_0 \in \mathcal{N}$.

Since $V(\cdot)$ is continuous there exists $\delta > 0$ such that $V(x_0) \in \mathcal{B}_r(0)$ for all $x_0 \in \mathcal{B}_\delta(0)$. Let $x_0 \in \mathcal{B}_\delta(0)$ and $z_0 = V(x_0) \in \mathcal{B}_r(0)$. It follows from (12.12) and (12.13), and Theorem 10.1 that

$$0 \leq\leq V(x(t)) \leq\leq z(t), \quad t \in [0, \tau], \tag{12.20}$$

where $x(t)$, $t \geq t_0$, is the solution to (12.1) and (12.2) with the initial condition $x_0 \in \mathcal{B}_\delta(0)$, $z(t)$, $t \geq t_0$, is the solution to (12.14) and (12.15) with the initial condition $z_0 = V(x_0) \in \mathcal{B}_r(0)$, and $[0, \tau]$ is any arbitrarily large compact time interval. Now, forming p^{T}(12.20) yields

$$0 \leq p^{\mathrm{T}} V(x(t)) \leq p^{\mathrm{T}} z(t), \quad t \in [0, \tau]. \tag{12.21}$$

Since $z(\cdot)$ converges to the origin in finite time and $p^{\mathrm{T}} V(\cdot)$ is positive definite it follows that $x(t)$, $t \geq t_0$, converges to the origin in finite time. Hence, the nonlinear impulsive dynamical system $\mathcal{G}$ given by (12.1) and (12.2) is finite-time stable with $\mathcal{N} \triangleq \mathcal{B}_\delta(0)$.

Finally, if $\mathcal{D} = \mathbb{R}^n$, $v(\cdot)$ is radially unbounded, and (12.16) and (12.17) hold on $\mathbb{R}^q$, then global finite-time stability follows using standard arguments. $\qquad\square$

The next result is a specialization of Theorem 12.2 to the case where the structure of the comparison dynamics directly guarantees finite-time stability of the impulsive comparison system. That is, there is *no* need to require the existence of a scalar function $v(\cdot)$ such that (12.16) and (12.17) hold to guarantee finite-time stability of the nonlinear impulsive dynamical system (12.1) and (12.2).

Corollary 12.1. Consider the nonlinear impulsive dynamical system given by (12.1) and (12.2). Assume there exist a continuously differentiable vector function $V = [V_1, \ldots, V_q]^{\mathrm{T}} : \mathcal{D} \to \mathcal{Q} \cap \overline{\mathbb{R}}_+^q$, where $\mathcal{Q} \subset \mathbb{R}^q$ and $0 \in \mathcal{Q}$, and a positive vector $p \in \mathbb{R}_+^q$ such that $V(0) = 0$, the scalar function $p^{\mathrm{T}} V(x)$, $x \in \mathcal{D}$, is positive definite, and

$$V'(x) f(x) \leq\leq W(V(x))^{[\alpha]}, \quad x \notin \mathcal{Z}, \tag{12.22}$$

$$V(x + f_{\mathrm{d}}(x)) \leq\leq V(x), \quad x \in \mathcal{Z}, \tag{12.23}$$

where $\alpha \in (0,1)$, $W \in \mathbb{R}^{q \times q}$ is essentially nonnegative and Hurwitz, and $(V(x))^{[\alpha]} \triangleq [(V_1(x))^{\alpha}, \ldots, (V_q(x))^{\alpha}]^{\mathrm{T}}$. Then the zero solution $x(t) \equiv 0$ to (12.1) and (12.2) is finite-time stable. If, in addition, $\mathcal{D} = \mathbb{R}^n$, then the zero solution $x(t) \equiv 0$ to (12.1) and (12.2) is globally finite-time stable.

Proof. Consider the impulsive comparison system given by

$$\dot{z}(t) = W(z(t))^{[\alpha]}, \quad z(0) = z_0, \quad x(t) \notin \mathcal{Z}, \qquad (12.24)$$

$$\Delta z(t) = 0, \quad x(t) \in \mathcal{Z}, \qquad (12.25)$$

where $z_0 \in \overline{\mathbb{R}}_+^q$. Note that the right-hand side of (12.24) is of class $\mathcal{W}_c$ and is essentially nonnegative and, hence, the solutions to (12.24) and (12.25) are nonnegative for all nonnegative initial conditions [77]. Since $W \in \mathbb{R}^{q \times q}$ is essentially nonnegative and Hurwitz, it follows from Lemma 2.1 that there exist positive vectors $\hat{p} \in \mathbb{R}_+^q$ and $r \in \mathbb{R}_+^q$ such that

$$0 = W^{\mathrm{T}}\hat{p} + r. \qquad (12.26)$$

Now, consider the Lyapunov function candidate $v(z) = \hat{p}^{\mathrm{T}} z$, $z \in \mathbb{R}_+^q$. Note that $v(0) = 0$, $v(z) > 0$, $z \in \mathbb{R}_+^q$, $z \neq 0$, and $v(\cdot)$ is radially unbounded. Let $\beta \triangleq \min_{i=1,\ldots,q} r_i$ and $\gamma \triangleq \max_{i=1,\ldots,q} \hat{p}_i^{\alpha}$, where r_i and $\hat{p}_i$ are the ith components of $r \in \mathbb{R}_+^q$ and $\hat{p} \in \mathbb{R}_+^q$, respectively. Then

$$
\begin{aligned}
\dot{v}(z) &= \hat{p}^{\mathrm{T}} W z^{[\alpha]} \\
&= -r^{\mathrm{T}} z^{[\alpha]} \\
&\leq -\frac{\beta}{\gamma}\gamma \left(\sum_{i=1}^{q} z_i^{\alpha} \right) \\
&\leq -\frac{\beta}{\gamma} \left(\sum_{i=1}^{q} \hat{p}_i^{\alpha} z_i^{\alpha} \right) \\
&\leq -\frac{\beta}{\gamma} \left(\sum_{i=1}^{q} \hat{p}_i z_i \right)^{\alpha} \\
&\leq -\frac{\beta}{\gamma} (v(z))^{\alpha} \\
&= -c(v(z))^{\alpha}, \quad z \in \overline{\mathbb{R}}_+^q, \qquad (12.27)
\end{aligned}
$$

and

$$\Delta v(z) = 0, \quad z \in \overline{\mathbb{R}}_+^q, \qquad (12.28)$$

where $c \triangleq \frac{\beta}{\gamma}$. Thus, it follows from Theorem 12.1 that the impulsive comparison system (12.24) and (12.25) is finite-time stable. In addition, it follows from Corollary 10.1 that the nonlinear impulsive dynamical system (12.1) and (12.2) is asymptotically stable with the domain of attraction $\mathcal{N} \subset \mathcal{D}$. Now, the result is a direct consequence of Theorem 12.2. $\qquad \square$

12.3 Finite-Time Stabilization of Impulsive Dynamical Systems

In this section, we design hybrid finite-time stabilizing controllers for nonlinear affine in the control impulsive dynamical systems. In addition, for large-scale impulsive dynamical systems we design decentralized hybrid finite-time stabilizers predicated on a control vector Lyapunov function [141]. Consider the controlled nonlinear impulsive dynamical system given by

$$\dot{x}(t) = f_{\mathrm{c}}(x(t)) + G_{\mathrm{c}}(x(t))u_{\mathrm{c}}(t), \quad x(t) \notin \mathcal{Z}, \quad t \geq t_0, \tag{12.29}$$

$$\Delta x(t) = f_{\mathrm{d}}(x(t)) + G_{\mathrm{d}}(x(t))u_{\mathrm{d}}(t), \quad x(t) \in \mathcal{Z}, \tag{12.30}$$

where $f_{\mathrm{c}} : \mathbb{R}^n \to \mathbb{R}^n$ satisfying $f_{\mathrm{c}}(0) = 0$ and $G_{\mathrm{c}} : \mathbb{R}^n \to \mathbb{R}^{n \times m_{\mathrm{c}}}$ are continuous functions, $f_{\mathrm{d}} : \mathbb{R}^n \to \mathbb{R}^n$ and $G_{\mathrm{d}} : \mathbb{R}^n \to \mathbb{R}^{n \times m_{\mathrm{d}}}$ are continuous, $u_{\mathrm{c}}(t) \in \mathbb{R}^{m_{\mathrm{c}}}$, $t \geq t_0$, and $u_{\mathrm{d}}(t_k) \in \mathbb{R}^{m_{\mathrm{d}}}$, $k \in \overline{\mathbb{Z}}_+$.

Theorem 12.3. Consider the controlled nonlinear impulsive dynamical system given by (12.29) and (12.30). If there exist a continuously differentiable function $V : \mathbb{R}^n \to \overline{\mathbb{R}}_+$ and continuous functions $P_{1u} : \mathbb{R}^n \to \mathbb{R}^{1 \times m_{\mathrm{d}}}$ and $P_{2u} : \mathbb{R}^n \to \mathbb{R}^{m_{\mathrm{d}} \times m_{\mathrm{d}}}$ such that $V(\cdot)$ is positive definite and

$$V(x + f_{\mathrm{d}}(x) + G_{\mathrm{d}}(x)u_{\mathrm{d}}) = V(x + f_{\mathrm{d}}(x)) + P_{1u}(x)u_{\mathrm{d}} + u_{\mathrm{d}}^{\mathrm{T}} P_{2u}(x)u_{\mathrm{d}},$$
$$x \in \mathbb{R}^n, \quad u_{\mathrm{d}} \in \mathbb{R}^{m_{\mathrm{d}}}, \tag{12.31}$$

$$V'(x)f_{\mathrm{c}}(x) \leq -c(V(x))^{\alpha}, \quad x \in \mathcal{R}, \tag{12.32}$$

$$V(x + f_{\mathrm{d}}(x)) - V(x) - \tfrac{1}{4}P_{1u}(x)P_{2u}^{+}(x)P_{1u}^{\mathrm{T}}(x) \leq 0, \quad x \in \mathcal{Z}, \tag{12.33}$$

where $c > 0$, $\alpha \in (0,1)$, and $\mathcal{R} \triangleq \{x \in \mathbb{R}^n, x \notin \mathcal{Z} : V'(x)G_{\mathrm{c}}(x) = 0\}$, then the nonlinear impulsive dynamical system (12.29) and (12.30) with the hybrid feedback control law $(u_{\mathrm{c}}, u_{\mathrm{d}}) = (\phi_{\mathrm{c}}(\cdot), \phi_{\mathrm{d}}(\cdot))$ given by

$$\phi_{\mathrm{c}}(x) = \begin{cases} -\left(c_0 + \frac{(\alpha(x)-w_{\mathrm{c}}(V(x)))+\mu(x)}{\beta^{\mathrm{T}}(x)\beta(x)}\right)\beta(x), & \beta(x) \neq 0, \quad x \notin \mathcal{Z}, \\ 0, & \beta(x) = 0, \quad x \notin \mathcal{Z}, \end{cases}$$
$$\tag{12.34}$$

and

$$\phi_{\mathrm{d}}(x) = -\frac{1}{2}P_{2u}^{+}(x)P_{1u}^{\mathrm{T}}(x), \quad x \in \mathcal{Z}, \tag{12.35}$$

where $\alpha(x) \triangleq V'(x)f_{\mathrm{c}}(x)$, $x \in \mathbb{R}^n$, $\beta(x) \triangleq G_{\mathrm{c}}^{\mathrm{T}}(x)V'^{\mathrm{T}}(x)$, $x \in \mathbb{R}^n$, $\mu(x) \triangleq \sqrt{(\alpha(x) - w_{\mathrm{c}}(V(x)))^2 + (\beta^{\mathrm{T}}(x)\beta(x))^2}$, $x \in \mathbb{R}^n$, $w_{\mathrm{c}}(V(x)) \triangleq -c(V(x))^{\alpha}$, $x \in \mathbb{R}^n$, and $c_0 > 0$, is finite-time stable.

Proof. Note that between resettings the time derivative of $V(\cdot)$ along the trajectories of (12.29), with $u_{\mathrm{c}} = \phi_{\mathrm{c}}(x)$, $x \in \mathbb{R}^n$, given by (12.34), is given by

$$\dot{V}(x) = V'(x)(f_{\mathrm{c}}(x) + G_{\mathrm{c}}(x)\phi_{\mathrm{c}}(x))$$

$$= \alpha(x) + \beta^{\mathrm{T}}(x)\phi_{\mathrm{c}}(x)$$

$$= \begin{cases} -c_0\beta^{\mathrm{T}}(x)\beta(x) - \mu(x) + w_{\mathrm{c}}(V(x)), & \beta(x) \neq 0, \\ \alpha(x), & \beta(x) = 0, \end{cases}$$

$$\leq w_{\mathrm{c}}(V(x)), \quad x \notin \mathcal{Z}. \tag{12.36}$$

In addition, using (12.31) and (12.33), the difference of $V(\cdot)$ at the resetting instants, with $u_{\mathrm{d}} = \phi_{\mathrm{d}}(x)$, $x \in \mathcal{Z}$, given by (12.35), is given by

$$\begin{aligned} \Delta V(x) &= V(x + f_{\mathrm{d}}(x) + G_{\mathrm{d}}(x)\phi_{\mathrm{d}}(x)) - V(x) \\ &= V(x + f_{\mathrm{d}}(x)) - V(x) + P_{1u}(x)\phi_{\mathrm{d}}(x) + \phi_{\mathrm{d}}(x)P_{2u}^+(x)\phi_{\mathrm{d}}(x) \\ &= V(x + f_{\mathrm{d}}(x)) - V(x) - \frac{1}{4}P_{1u}(x)P_{2u}^+(x)P_{1u}^{\mathrm{T}}(x) \\ &\leq 0, \quad x \in \mathcal{Z}. \end{aligned} \tag{12.37}$$

Hence, it follows from Theorem 12.1 that the zero solution $x(t) \equiv 0$ to (12.29) and (12.30) with $u_{\mathrm{c}} = \phi_{\mathrm{c}}(x)$, $x \notin \mathcal{Z}$, given by (12.34) and $u_{\mathrm{d}} = \phi_{\mathrm{d}}(x)$, $x \in \mathcal{Z}$, given by (12.35), is finite-time stable, which proves the result. $\qquad\square$

In the next result we develop decentralized finite-time stabilizing hybrid controllers for a large-scale impulsive dynamical system composed of q interconnected subsystems given by

$$\dot{x}_i(t) = f_{\mathrm{c}i}(x(t)) + G_{\mathrm{c}i}(x(t))u_{\mathrm{c}i}(t), \quad x(t) \notin \mathcal{Z}, \quad i = 1,\ldots,q, \tag{12.38}$$

$$\Delta x_i(t) = f_{\mathrm{d}i}(x(t)) + G_{\mathrm{d}i}(x(t))u_{\mathrm{d}i}(t), \quad x(t) \in \mathcal{Z}, \quad i = 1,\ldots,q, \tag{12.39}$$

where $t \geq 0$, $f_{\mathrm{c}i} : \mathbb{R}^n \to \mathbb{R}^{n_i}$ satisfying $f_{\mathrm{c}i}(0) = 0$ and $G_{\mathrm{c}i} : \mathbb{R}^n \to \mathbb{R}^{n_i \times m_{\mathrm{c}i}}$ are continuous functions for all $i = 1,\ldots,q$, $f_{\mathrm{d}i} : \mathbb{R}^n \to \mathbb{R}^{n_i}$ and $G_{\mathrm{d}i} : \mathbb{R}^n \to \mathbb{R}^{n_i \times m_{\mathrm{d}i}}$ are continuous for all $i = 1,\ldots,q$, $u_{\mathrm{c}i}(t) \in \mathbb{R}^{m_{\mathrm{c}i}}$, $t \geq 0$, and $u_{\mathrm{d}i}(t_k) \in \mathbb{R}^{m_{\mathrm{d}i}}$, $k \in \overline{\mathbb{Z}}_+$, for all $i = 1,\ldots,q$.

Theorem 12.4. Consider the controlled nonlinear impulsive system given by (12.38) and (12.39). Assume there exist a continuously differentiable, component decoupled vector function $V = [V_1(x_1),\ldots,V_q(x_q)]^{\mathrm{T}} : \mathbb{R}^n \to \overline{\mathbb{R}}_+^q$, continuous functions $P_{1ui} : \mathbb{R}^n \to \mathbb{R}^{1 \times m_{\mathrm{d}i}}$, $P_{2ui} : \mathbb{R}^n \to \mathbb{R}^{m_{\mathrm{d}i} \times m_{\mathrm{d}i}}$, $i = 1,\ldots,q$, $w_{\mathrm{c}} = [w_{\mathrm{c}1},\ldots,w_{\mathrm{c}q}]^{\mathrm{T}} : \overline{\mathbb{R}}_+^q \to \mathbb{R}^q$, $w_{\mathrm{d}} = [w_{\mathrm{d}1},\ldots,w_{\mathrm{d}q}]^{\mathrm{T}} : \overline{\mathbb{R}}_+^q \to \mathbb{R}^q$, and a positive vector $p \in \mathbb{R}_+^q$ such that $V(0) = 0$, the scalar function $p^{\mathrm{T}}V(x)$, $x \in \mathbb{R}^n$, is positive definite, $w_{\mathrm{c}}(\cdot) \in \mathcal{W}_{\mathrm{c}}$, $w_{\mathrm{d}}(\cdot) \in \mathcal{W}_{\mathrm{d}}$, $w_{\mathrm{c}}(0) = 0$, $w_{\mathrm{d}}(0) = 0$, and, for all $i = 1,\ldots,q$,

$$V_i(x_i + f_{\mathrm{d}i}(x) + G_{\mathrm{d}i}(x)u_{\mathrm{d}i}) = V_i(x_i + f_{\mathrm{d}i}(x)) + P_{1ui}(x)u_{\mathrm{d}i} + u_{\mathrm{d}i}^{\mathrm{T}}P_{2ui}(x)u_{\mathrm{d}i},$$
$$x \in \mathbb{R}^n, \quad u_{\mathrm{d}i} \in \mathbb{R}^{m_{\mathrm{d}i}}, \tag{12.40}$$

$$V_i'(x_i)f_{\mathrm{c}i}(x) \leq w_{\mathrm{c}i}(V(x)), \quad x \in \mathcal{R}_i, \tag{12.41}$$

$$V_i(x_i + f_{\mathrm{d}i}(x)) - V_i(x_i) - \tfrac{1}{4}P_{1ui}(x)P_{2ui}^+(x)P_{1ui}^{\mathrm{T}}(x) \leq w_{\mathrm{d}i}(V(x)), \quad x \in \mathcal{Z}, \tag{12.42}$$

where $\mathcal{R}_i \triangleq \{x \in \mathbb{R}^n, x \notin \mathcal{Z} : V_i'(x_i)G_{ci}(x) = 0\}$, $i = 1, \ldots, q$. In addition, assume there exist a positive definite function $v : \overline{\mathbb{R}}_+^q \to \overline{\mathbb{R}}_+$, real numbers $c > 0$ and $\alpha \in (0, 1)$, and a neighborhood $\mathcal{M} \subseteq \mathbb{R}^q$ of the origin such that

$$v'(z)w_c(z) \leq -c(v(z))^\alpha, \quad z \in \mathcal{M} \cap \overline{\mathbb{R}}_+^q, \tag{12.43}$$

$$v(z + w_d(z)) \leq v(z), \quad z \in \mathcal{M} \cap \overline{\mathbb{R}}_+^q. \tag{12.44}$$

Then the nonlinear impulsive dynamical system (12.38) and (12.39) is finite-time stable with the hybrid feedback control law $u_c = \phi_c(x) = [\phi_{c1}^{\mathrm{T}}(x), \ldots, \phi_{cq}^{\mathrm{T}}(x)]^{\mathrm{T}}$, $x \notin \mathcal{Z}$, and $u_d = \phi_d(x) = [\phi_{d1}^{\mathrm{T}}(x), \ldots, \phi_{dq}^{\mathrm{T}}(x)]^{\mathrm{T}}$, $x \in \mathcal{Z}$, where, for $i = 1, \ldots, q$,

$$\phi_{ci}(x) = \begin{cases} -\left(c_{0i} + \dfrac{(\alpha_i(x) - w_{ci}(V(x))) + \mu_i(x)}{\beta_i^{\mathrm{T}}(x)\beta_i(x)}\right)\beta_i(x), & \beta_i(x) \neq 0, \\ 0, & \beta_i(x) = 0, \end{cases} \tag{12.45}$$

for $x \notin \mathcal{Z}$, and

$$\phi_{di}(x) = -\frac{1}{2}P_{2ui}^+(x)P_{1ui}^{\mathrm{T}}(x), \quad x \in \mathcal{Z}, \tag{12.46}$$

where $\alpha_i(x) \triangleq V_i'(x_i)f_{ci}(x)$, $x \in \mathbb{R}^n$, $\beta_i(x) \triangleq G_{ci}^{\mathrm{T}}(x)V_i'^{\mathrm{T}}(x_i)$, $x \in \mathbb{R}^n$, $\mu_i(x) \triangleq \sqrt{(\alpha_i(x) - w_{ci}(V(x)))^2 + (\beta_i^{\mathrm{T}}(x)\beta_i(x))^2}$, $x \in \mathbb{R}^n$, and $c_{0i} > 0$, $i = 1, \ldots, q$.

Proof. Using identical arguments as in the proof of Theorem 12.3 it can be shown that for the closed-loop system (12.38), (12.39), (12.45), and (12.46) the time derivative and the difference of the vector function $V(\cdot)$ between resettings and resetting instants satisfy, respectively,

$$\dot{V}(x) \leq\leq w_c(V(x)), \quad x \notin \mathcal{Z}, \tag{12.47}$$

$$V(x + f_d(x) + G_d(x)\phi_d(x)) \leq\leq V(x) + w_d(V(x)), \quad x \in \mathcal{Z}, \tag{12.48}$$

where $G_d(x) \triangleq \text{block-diag}\,[G_{d1}(x), \ldots, G_{dq}(x)]$, $x \in \mathcal{Z}$. The result now follows immediately from (12.43), (12.44), and Theorem 12.2. $\square$

If, in Theorem 12.4, $\mathcal{R}_i = \varnothing$, $i = 1, \ldots, q$, and (12.42) is satisfied with $w_d(z) \equiv 0$, then the function $w_c(\cdot)$ in (12.45) can be chosen to be

$$w_c(z) = Wz^{[\alpha]}, \quad z \in \overline{\mathbb{R}}_+^q, \tag{12.49}$$

where $W \in \mathbb{R}^{q \times q}$ is essentially nonnegative and asymptotically stable, $\alpha \in (0, 1)$, and $z^{[\alpha]} \triangleq [z_1^\alpha, \ldots, z_q^\alpha]^{\mathrm{T}}$. In this case, conditions (12.43) and (12.44) need *not* be verified and it follows from Corollary 12.1 that the close-loop system (12.38), (12.39), (12.45), and (12.46) with $w_c(\cdot)$ given by (12.49) is finite-time stable and, hence, the hybrid controller (12.45) and (12.46) is a finite-time stabilizing controller for (12.38) and (12.39).

Since $f_{ci}(\cdot)$ and $G_{ci}(\cdot)$ are continuous and $V_i(\cdot)$ is continuously differentiable for all $i = 1, \ldots, q$, it follows that $\alpha_i(x)$ and $\beta_i(x)$, $x \in \mathbb{R}^n$,

$i = 1, \ldots, q$, are continuous functions, and hence, $\phi_{ci}(x)$ given by (12.45) is continuous for all $x \in \mathbb{R}^n$ if either $\beta_i(x) \neq 0$ or $\alpha_i(x) - w_{ci}(V(x)) < 0$ for all $i = 1, \ldots, q$. Hence, the feedback control law given by (12.45) is continuous everywhere except for the origin. However, as shown in Proposition 11.1, the feedback control law $\phi_c(x)$ given by (12.45) is continuous on $\mathbb{R}^n$ if and only if for every $\varepsilon > 0$, there exists $\delta > 0$ such that for all $0 < \|x\| < \delta$ there exists $u_{ci} \in \mathbb{R}^{m_{ci}}$ such that $\|u_{ci}\| < \varepsilon$ and $\alpha_i(x) + \beta_i^{\mathrm{T}}(x)u_{ci} < w_{ci}(V(x))$, $i = 1, \ldots, q$.

Identical necessary and sufficient conditions apply in the case where $q = 1$ to ensure continuity of the feedback control law given by (12.34). It is important to note that even though the feedback control law $\phi_c(x)$ given by (12.45) is continuous on $\mathbb{R}^n$, for a particular trajectory $x(t)$, $t \geq 0$, of (12.38) and (12.39), $\phi_c(x(t))$ is left-continuous on $[0, \infty)$ and is continuous everywhere on $[0, \infty)$ except on an unbounded closed discrete set of times when the resettings occur for $x(t)$, $t \geq 0$.

12.4 Finite-Time Stabilizing Control for Large-Scale Impulsive Dynamical Systems

In this section, we apply the proposed hybrid control framework to decentralized control of large-scale nonlinear impulsive dynamical systems. Specifically, we consider the large-scale dynamical system $\mathcal{G}$ shown in Figure 12.1 involving energy exchange between n interconnected subsystems. Let $x_i : [0, \infty) \to \overline{\mathbb{R}}_+$ denote the energy (and hence a nonnegative quantity) of the ith subsystem, let $u_{ci} : [0, \infty) \to \mathbb{R}$ denote the control input to the ith subsystem, let $\sigma_{cij} : \overline{\mathbb{R}}_+^n \to \overline{\mathbb{R}}_+$, $i \neq j$, $i, j = 1, \ldots, n$, denote the instantaneous rate of energy flow from the jth subsystem to the ith subsystem between resettings, let $\sigma_{dij} : \overline{\mathbb{R}}_+^n \to \overline{\mathbb{R}}_+$, $i \neq j$, $i, j = 1, \ldots, n$, denote the amount of energy transferred from the jth subsystem to the ith subsystem at the resetting instant, and let $\mathcal{Z} \subset \overline{\mathbb{R}}_+^n$ be a resetting set for the large-scale impulsive dynamical system $\mathcal{G}$. In Figure 12.1, the solid arrows correspond to the energy exchange among subsystems of $\mathcal{G}$ between resettings and the dashed lines correspond to the energy exchange among subsystems of $\mathcal{G}$ at the instants of resettings.

An energy balance for each subsystem $\mathcal{G}_i$, $i = 1, \ldots q$, yields [81,82]

$$\dot{x}_i(t) = \sum_{j=1, j \neq i}^{n} [\sigma_{cij}(x(t)) - \sigma_{cji}(x(t))] + G_{ci}(x_i(t))u_{ci}(t), \quad x(t_0) = x_0,$$

$$x(t) \notin \mathcal{Z}, \quad t \geq t_0, \quad (12.50)$$

$$\Delta x_i(t) = \sum_{j=1, j \neq i}^{n} [\sigma_{dij}(x(t)) - \sigma_{dji}(x(t))] + G_{di}(x(t))u_{di}(t), \quad x(t) \in \mathcal{Z},$$

$$(12.51)$$

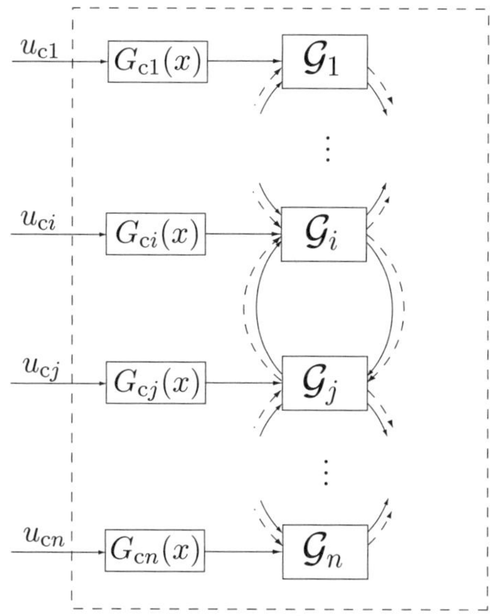

Figure 12.1 Large-scale dynamical system $\mathcal{G}$.

or, equivalently, in vector form

$$\dot{x}(t) = f_{\mathrm{c}}(x(t)) + G_{\mathrm{c}}(x(t))u_{\mathrm{c}}(t), \quad x(t_0) = x_0, \quad x(t) \notin \mathcal{Z}, \quad t \geq t_0, \quad (12.52)$$
$$\Delta x(t) = f_{\mathrm{d}}(x(t)) + G_{\mathrm{d}}(x(t))u_{\mathrm{d}}(t), \quad x(t) \in \mathcal{Z}, \quad (12.53)$$

where $x(t) = [x_1(t), \ldots, x_n(t)]^{\mathrm{T}}$, $t \geq t_0$, $f_{ci}(x) = \sum_{j=1, j \neq i}^{n} \phi_{cij}(x)$, where $\phi_{cij}(x) \triangleq \sigma_{cij}(x) - \sigma_{cji}(x)$, $x \in \overline{\mathbb{R}}_+^n$, $i \neq j$, $i, j = 1, \ldots, q$, denotes the net energy flow from the jth subsystem to the ith subsystem between resettings, $G_{\mathrm{c}}(x) = \mathrm{diag}[G_{c1}(x_1), \ldots, G_{cn}(x_n)]$, $x \in \overline{\mathbb{R}}_+^n$, $G_{ci} : \mathbb{R} \to \mathbb{R}$, $i = 1, \ldots, n$, is such that $G_{ci}(x_i) = 0$ if and only if $x_i = 0$ for all $i = 1, \ldots, n$, $G_{\mathrm{d}}(x) = \mathrm{diag}[G_{d1}(x), \ldots, G_{dn}(x)]$, $x \in \mathcal{Z}$, $G_{di} : \mathbb{R}^n \to \mathbb{R}$, $i = 1, \ldots, n$, $u_{\mathrm{c}}(t) \in \mathbb{R}^n$, $t \geq t_0$, $u_{\mathrm{d}}(t_k) \in \mathbb{R}^n$, $k \in \overline{\mathbb{Z}}_+$, and $f_{di}(x) = \sum_{j=1, j \neq i}^{n} \phi_{dij}(x)$, where $\phi_{dij}(x) \triangleq \sigma_{dij}(x) - \sigma_{dji}(x)$, $x \in \mathcal{Z}$, $i \neq j$, $i, j = 1, \ldots, q$, denotes the net amount of energy transferred from the jth subsystem to the ith subsystem at the instant of resetting. Here, we assume that $\sigma_{cij}(0) = 0$, $i \neq j$, $i, j = 1, \ldots, n$, and $u_{\mathrm{c}} = [u_{c1}, \ldots, u_{cn}]^{\mathrm{T}} : \mathbb{R} \to \mathbb{R}^n$ is such that $u_{ci} : \mathbb{R} \to \mathbb{R}$, $i = 1, \ldots, n$, are bounded piecewise continuous functions of time. Furthermore, we assume that $\sigma_{cij}(x) = 0$, $x \in \overline{\mathbb{R}}_+^n$, whenever $x_j = 0$, $i \neq j$, $i, j = 1, \ldots, n$, and $x_i + \sum_{j=1, j \neq i}^{n} \phi_{dij}(x) \geq 0$, $x \in \mathcal{Z}$. In this case, $f_{\mathrm{c}}(\cdot)$ is essentially nonnegative and $x + f_{\mathrm{d}}(x)$, $x \in \mathcal{Z} \subset \overline{\mathbb{R}}_+^n$, is nonnegative.

The above constraints imply that if the energy of the jth subsystem of $\mathcal{G}$ is zero, then this subsystem cannot supply any energy to its surroundings between resettings and the ith subsystem of $\mathcal{G}$ cannot transfer more energy to its surroundings than it already possesses at the instant of resetting. Finally, to ensure that the trajectories of the closed-loop system remain in the nonnegative orthant of the state space for all nonnegative initial conditions, we seek a hybrid feedback control law $(u_{\mathrm{c}}(\cdot), u_{\mathrm{d}}(\cdot))$ that guarantees the continuous-time closed-loop system dynamics (12.52) are essentially nonnegative and the closed-loop system states after resettings are nonnegative [77, 82].

For the dynamical system $\mathcal{G}$, consider the Lyapunov function candidate $V(x) = \mathbf{e}^{\mathrm{T}} x$, $x \in \overline{\mathbb{R}}_+^n$. Note that $V(0) = 0$ and $V(x) > 0$, $x \neq 0$, $x \in \overline{\mathbb{R}}_+^n$. Furthermore, note that since $V(x) = \mathbf{e}^{\mathrm{T}} x$, $x \in \overline{\mathbb{R}}_+^n$, is a linear function of x, it follows from (12.31) that $P_{1u}(x) = \mathbf{e}^{\mathrm{T}} G_{\mathrm{d}}(x)$, $x \in \overline{\mathbb{R}}_+^n$, and $P_{2u}(x) \equiv 0$, and hence, by (12.35), $\phi_{\mathrm{d}}(x) \equiv 0$. Define $w_{\mathrm{c}}(V(x)) \triangleq -(V(x))^{\frac{1}{2}}$, $x \in \overline{\mathbb{R}}_+^n$, and note that $\mathcal{R} \triangleq \{x \in \overline{\mathbb{R}}_+^n, x \notin \mathcal{Z} : V'(x)G_{\mathrm{c}}(x) = 0\} = \{x \in \overline{\mathbb{R}}_+^n, x \notin \mathcal{Z} : x = 0\} = \{0\}$, since $0 \notin \mathcal{Z}$, and hence, condition (12.32) is satisfied. In addition, note that since $V(\cdot)$ is linear in x, condition (12.31) is trivially satisfied and inequality (12.33) is satisfied as an equality.

Next, with $\alpha(x) = V'(x)f_{\mathrm{c}}(x) = \sum_{i=1}^{n} \sum_{j=1, j\neq i}^{n} \phi_{cij}(x) = 0$, $\beta(x) = [G_{c1}(x_1), \ldots, G_{cn}(x_n)]^{\mathrm{T}}$, $x \in \overline{\mathbb{R}}_+^n$, and $c_0 > 0$, we construct a finite-time stabilizing hybrid feedback controller for (12.52) and (12.53) given by

$$\phi_{\mathrm{c}}(x) = \begin{cases} -\left(c_0 + \dfrac{(\mathbf{e}^{\mathrm{T}} x)^{1/2} + \sqrt{(\mathbf{e}^{\mathrm{T}} x) + (\beta^{\mathrm{T}}(x)\beta(x))^2}}{\beta^{\mathrm{T}}(x)\beta(x)}\right)\beta(x), & x \neq 0, \\ 0, & x = 0, \end{cases} \tag{12.54}$$

and

$$\phi_{\mathrm{d}}(x) \equiv 0. \tag{12.55}$$

It can be seen from the structure of the feedback control law (12.54) and (12.55) that the continuous-time closed-loop system dynamics are essentially nonnegative and the closed-loop system states after resettings are nonnegative. Furthermore, since $\alpha(x) - w_{\mathrm{c}}(V(x)) = (V(x))^{1/2}$, $x \in \overline{\mathbb{R}}_+^n$, the continuous-time feedback controller $\phi_{\mathrm{c}}(\cdot)$ is fully independent from $f_{\mathrm{c}}(x)$, which represents the internal interconnections of the large-scale system dynamics, and hence, is robust against full modeling uncertainty in $f_{\mathrm{c}}(x)$. Finally, it follows from Theorem 12.3 that the closed-loop system (12.52)–(12.55) is finite-time stable.

For the following simulation we consider (12.52) and (12.53) with $\sigma_{cij}(x) = \sigma_{cij} x_i x_j$ and $\sigma_{dij}(x) = \sigma_{dij} x_j$, and $G_{ci}(x_i) = x_i^{1/4}$, where $\sigma_{cij} \geq 0$, $i \neq j$, $i, j = 1, \ldots, n$, $\sigma_{dij} \geq 0$, $i \neq j$, $i, j = 1, \ldots, n$, and $1 \geq \sum_{j=1, j\neq i}^{n} \sigma_{dji}$, $i = 1, \ldots, n$. To show that the conditions of Proposition 11.1 are satisfied for the case when $q = 1$, let $u_{\mathrm{c}} = [-x_1^{1/4}, \ldots, -x_n^{1/4}]^{\mathrm{T}}$ and note that

$$\alpha(x) + \beta(x)u_{\mathrm{c}} = -\sum_{i=1}^{n} x_i^{1/2}$$

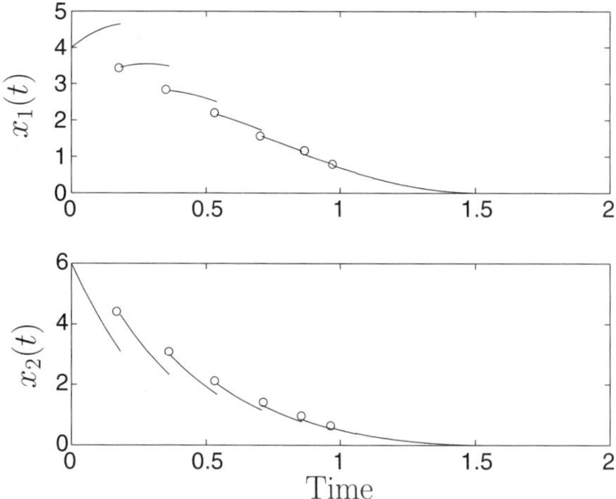

Figure 12.2 Controlled system states versus time.

$$< -\left(\sum_{i=1}^{n} x_i\right)^{1/2}$$

$$= -(V(x))^{1/2}$$

$$= w_{\mathrm{c}}(V(x)), \quad x \in \overline{\mathbb{R}}_{+}^{n}, \quad x \neq 0. \qquad (12.56)$$

Thus, for every $\varepsilon > 0$ there exists $\delta > 0$, such that for all $0 < \|x\| < \delta$ there exists $u_{\mathrm{c}} \in \mathbb{R}^{m_{\mathrm{c}}}$ such that $\|u_{\mathrm{c}}\| < \varepsilon$ and $\alpha(x) + \beta^{\mathrm{T}}(x)u_{\mathrm{c}} < w_{\mathrm{c}}(V(x))$, and hence, the continuous-time feedback control law (12.54) is continuous on $\overline{\mathbb{R}}_{+}^{n}$. For our simulation we set $n = 2$, $\sigma_{\mathrm{c}12} = 2$, $\sigma_{\mathrm{c}21} = 1.5$, $\sigma_{\mathrm{d}12} = 0.25$, $\sigma_{\mathrm{d}21} = 0.33$, $c_0 = 1$, and $\mathcal{Z} = \{x \in \mathbb{R}^2, x \neq 0 : x_2 - \frac{2}{3}x_1 = 0\}$, with initial condition $x_0 = [4, 6]^{\mathrm{T}}$. Figure 12.2 shows the states of the closed-loop system versus time, and Figure 12.3 shows continuous-time control signals as a function of time. Figure 12.4 shows the phase portrait of the closed-loop system; the dashed line denotes the resetting set $\mathcal{Z}$.

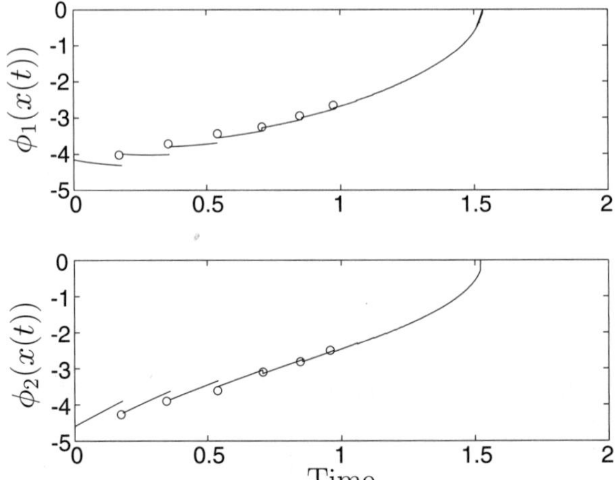

Figure 12.3 Control signals in each control channel versus time.

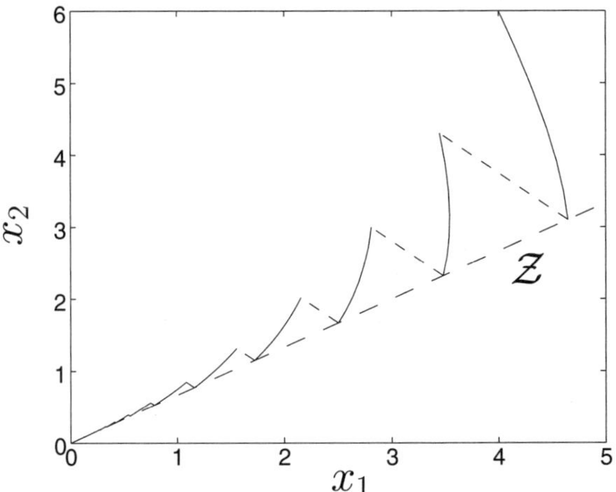

Figure 12.4 Phase portrait of the controlled system.

Hybrid Decentralized Maximum Entropy Control for Large-Scale Systems

13.1 Introduction

In this chapter, we develop a novel energy- and entropy-based hybrid decentralized control framework for vector lossless and vector dissipative large-scale dynamical systems based on subsystem decomposition. The notion of energy here refers to abstract energy notions for which a physical system energy interpretation is not necessary. These dynamical systems cover a broad spectrum of applications, including mechanical systems, fluid systems, electromechanical systems, electrical systems, combustion systems, structural vibration systems, biological systems, physiological systems, power systems, telecommunications systems, and economic systems, to cite but a few examples. The concept of an energy-based hybrid decentralized controller can be viewed as a feedback control technique that exploits the coupling between a physical large-scale dynamical system and an energy-based decentralized controller to efficiently remove energy from the physical large-scale system.

Specifically, if a vector dissipative or vector lossless large-scale system is at high energy level, and a lossless feedback decentralized controller at a low energy level is attached to it, then subsystem energy generally tends to flow from each subsystem into the corresponding subcontroller, decreasing the subsystem energy and increasing the subcontroller energy [113]. Of course, emulated energy, and not physical energy, is accumulated by each subcontroller. Conversely, if each attached subcontroller is at a high energy level and the corresponding subsystem is at a low energy level, then energy can flow from each subcontroller to each corresponding subsystem, since each subcontroller can generate real, physical energy to effect the required energy flow. Hence, if and when the subcontroller states coincide with a high emulated energy level, then we can *reset* these states to remove the emulated energy so that the emulated energy is not returned to the plant. In this case, the overall closed-loop system consisting of the plant and the controller possesses discontinuous flows since it combines logical switchings with continuous dynamics, leading to impulsive differential equations [11, 13, 74, 75, 82, 117, 155].

13.2 Hybrid Decentralized Control and Large-Scale Impulsive Dynamical Systems

In this chapter, we consider continuous-time nonlinear dynamical systems $\mathcal{G}$ of the form

$$\dot{x}(t) = F(x(t), u(t)), \quad x(0) = x_0, \quad t \geq 0, \tag{13.1}$$
$$y(t) = H(x(t)), \tag{13.2}$$

where $t \geq 0$, $x(t) \in \mathcal{D} \subseteq \mathbb{R}^n$, $u(t) \in \mathbb{R}^m$, $y(t) \in \mathbb{R}^l$, $F : \mathcal{D} \times \mathbb{R}^m \to \mathbb{R}^n$, $H : \mathcal{D} \to \mathbb{R}^l$, and $\mathcal{D}$ is an open set with $0 \in \mathcal{D}$. Here, we assume that $\mathcal{G}$ represents a large-scale dynamical system composed of q interconnected controlled subsystems $\mathcal{G}_i$ so that, for all $i = 1, \ldots, q$,

$$F_i(x, u) = f_i(x_i) + \mathcal{I}_i(x) + G_i(x_i)u_i, \tag{13.3}$$
$$H_i(x) = h_i(x_i), \tag{13.4}$$

where $x_i \in \mathcal{D}_i \subseteq \mathbb{R}^{n_i}$, $u_i \in \mathbb{R}^{m_i}$, $y_i \triangleq h_i(x_i) \in \mathbb{R}^{l_i}$, (u_i, y_i) is the input-output pair for the ith subsystem, $f_i : \mathbb{R}^{n_i} \to \mathbb{R}^{n_i}$ and $\mathcal{I}_i : \mathcal{D} \to \mathbb{R}^{n_i}$ are smooth (i.e., infinitely differentiable) and satisfy $f_i(0) = 0$ and $\mathcal{I}_i(0) = 0$, $G_i : \mathbb{R}^{n_i} \to \mathbb{R}^{n_i \times m_i}$ is smooth, $h_i : \mathbb{R}^{n_i} \to \mathbb{R}^{l_i}$ and satisfies $h_i(0) = 0$, $\sum_{i=1}^q n_i = n$, $\sum_{i=1}^q m_i = m$, and $\sum_{i=1}^q l_i = l$. Here, $f_i : \mathcal{D}_i \subseteq \mathbb{R}^{n_i} \to \mathbb{R}^{n_i}$ defines the vector field of each isolated subsystem of (13.1) and $\mathcal{I}_i : \mathcal{D} \to \mathbb{R}^{n_i}$ defines the structure of the interconnection dynamics of the ith subsystem with all other subsystems. Furthermore, for the large-scale dynamical system $\mathcal{G}$ we assume that the required properties for the existence and uniqueness of solutions are satisfied, that is, for every $i \in \{1, \ldots, q\}$, $u_i(\cdot)$ satisfies sufficient regularity conditions such that the system (13.1) has a unique solution forward in time. We define the composite input and composite output for the large-scale system $\mathcal{G}$ as $u \triangleq [u_1^{\mathrm{T}}, \ldots, u_q^{\mathrm{T}}]^{\mathrm{T}}$ and $y \triangleq [y_1^{\mathrm{T}}, \ldots, y_q^{\mathrm{T}}]^{\mathrm{T}}$, respectively.

Next, we consider state-dependent hybrid (resetting) decentralized dynamic controllers $\mathcal{G}_{ci}$, $i = 1, \ldots, q$, of the form

$$\dot{x}_{ci}(t) = f_{ci}(x_{ci}(t), y_i(t)), \quad x_{ci}(0) = x_{ci0}, \quad (x_{ci}(t), y_i(t)) \notin \mathcal{Z}_{ci}, \quad t \geq 0, \tag{13.5}$$
$$\Delta x_{ci}(t) = f_{di}(x_{ci}(t), y_i(t)), \quad (x_{ci}(t), y_i(t)) \in \mathcal{Z}_{ci}, \tag{13.6}$$
$$u_i(t) = h_{ci}(x_{ci}(t), y_i(t)), \tag{13.7}$$

where $x_{ci} \in \mathcal{D}_{ci} \subseteq \mathbb{R}^{n_{ci}}$, $\mathcal{D}_{ci}$ is an open set with $0 \in \mathcal{D}_{ci}$, $y_{ci} \triangleq h_{ci}(x_{ci}, y_i) \in \mathbb{R}^{m_i}$, $f_{ci} : \mathcal{D}_{ci} \times \mathbb{R}^{l_i} \to \mathbb{R}^{n_{ci}}$ is smooth and satisfies $f_{ci}(0, 0) = 0$, $f_{di} : \mathcal{D}_{ci} \times \mathbb{R}^{l_i} \to \mathbb{R}^{n_{ci}}$ is continuous, $h_{ci} : \mathcal{D}_{ci} \times \mathbb{R}^{l_i} \to \mathbb{R}^{m_i}$ is smooth and satisfies $h_{ci}(0, 0) = 0$, $\Delta x_{ci}(t) \triangleq x_{ci}(t^+) - x_{ci}(t)$, $\mathcal{Z}_{ci} \subset \mathcal{D}_{ci} \times \mathbb{R}^{l_i}$ is the resetting set, and $\sum_{i=1}^q n_{ci} = n_c$. Note that the hybrid decentralized controller (13.5)–(13.7) represents an impulsive dynamical system $\mathcal{G}_c$ composed of q impulsive subsystems $\mathcal{G}_{ci}$ involving multiple hybrid processors operating independently, with each processor receiving a subset of the available system measurements and updating a subset of the system actuators. Furthermore, for generality, we allow the hybrid decentralized dynamic controller to be of fixed dimension n_c, which may be less than the plant order n. In addition, we define the composite input and composite output for the impulsive decentralized dynamic compensator $\mathcal{G}_c$ as $u_c \triangleq y = [u_{c1}^T, \ldots, u_{cq}^T]^T$ and $y_c \triangleq u = [y_{c1}^T, \ldots, y_{cq}^T]^T$, respectively.

The equations of motion for each closed-loop dynamical subsystem $\tilde{\mathcal{G}}_i$, $i = 1, \ldots, q$, have the form

$$\dot{\tilde{x}}_i(t) = \tilde{f}_{ci}(\tilde{x}_i(t)) + \tilde{\mathcal{I}}_i(x), \quad \tilde{x}_i(0) = \tilde{x}_{i0}, \quad \tilde{x}_i(t) \notin \tilde{\mathcal{Z}}_i, \quad t \geq 0, \quad (13.8)$$
$$\Delta \tilde{x}_i(t) = \tilde{f}_{di}(\tilde{x}_i(t)), \quad \tilde{x}_i(t) \in \tilde{\mathcal{Z}}_i, \quad (13.9)$$

where

$$\tilde{x}_i \triangleq \begin{bmatrix} x_i \\ x_{ci} \end{bmatrix} \in \mathbb{R}^{\tilde{n}_i}, \quad \tilde{f}_{ci}(\tilde{x}_i) \triangleq \begin{bmatrix} f_i(x_i) + G_i(x_i)h_{ci}(x_{ci}, h_i(x_i)) \\ f_{ci}(x_{ci}, h_i(x_i)) \end{bmatrix},$$
$$(13.10)$$

$$\tilde{\mathcal{I}}_i(x) \triangleq \begin{bmatrix} \mathcal{I}_i(x) \\ 0 \end{bmatrix}, \quad \tilde{f}_{di}(\tilde{x}_i) \triangleq \begin{bmatrix} 0 \\ f_{di}(x_{ci}, h_i(x_i)) \end{bmatrix}, \quad (13.11)$$

and $\tilde{\mathcal{Z}}_i \triangleq \{\tilde{x}_i \in \tilde{\mathcal{D}}_i : (x_{ci}, h_i(x_i)) \in \mathcal{Z}_{ci}\}$, with $\tilde{n}_i \triangleq n_i + n_{ci}$ and $\tilde{\mathcal{D}}_i \triangleq \mathcal{D}_i \times \mathcal{D}_{ci}$, $i = 1, \ldots, q$. Hence, the equations of motion for the closed-loop dynamical system $\tilde{\mathcal{G}}$ have the form

$$\dot{\tilde{x}}(t) = \tilde{f}_c(\tilde{x}(t)), \quad \tilde{x}(0) = \tilde{x}_0, \quad \tilde{x}(t) \notin \tilde{\mathcal{Z}}, \quad t \geq 0, \quad (13.12)$$
$$\Delta \tilde{x}(t) = \tilde{f}_d(\tilde{x}(t)), \quad \tilde{x}(t) \in \tilde{\mathcal{Z}}, \quad (13.13)$$

where $\tilde{x}(t) = [\tilde{x}_1^T(t), \ldots, \tilde{x}_q^T(t)]^T$, $\tilde{f}_c(\tilde{x}) \triangleq [\tilde{f}_{c1}^T(\tilde{x}_1) + \tilde{\mathcal{I}}_1^T(x), \ldots, \tilde{f}_{cq}^T(\tilde{x}_q) + \tilde{\mathcal{I}}_q^T(x)]^T$, $\tilde{\mathcal{Z}} \triangleq \cup_{i=1}^q \{\tilde{x} \in \tilde{\mathcal{D}} : \tilde{x}_i \in \tilde{\mathcal{Z}}_i\}$, $\tilde{\mathcal{D}} \triangleq \cup_{i=1}^q \tilde{\mathcal{D}}_i$, and

$$\tilde{f}_d(\tilde{x}) \triangleq \begin{bmatrix} \tilde{f}_{d1}(\tilde{x}_1)\chi_{\tilde{\mathcal{Z}}_1}(\tilde{x}_1) \\ \vdots \\ \tilde{f}_{dq}(\tilde{x}_q)\chi_{\tilde{\mathcal{Z}}_q}(\tilde{x}_q) \end{bmatrix}, \quad \chi_{\tilde{\mathcal{Z}}_i}(\tilde{x}_i) = \begin{cases} 1, & \tilde{x}_i \in \tilde{\mathcal{Z}}_i \\ 0, & \tilde{x}_i \notin \tilde{\mathcal{Z}}_i \end{cases}, \quad i = 1, \ldots, q.$$
$$(13.14)$$

We refer to the differential equation (13.12) as the *continuous-time dynamics*, and we refer to the difference equation (13.13) as the *resetting law*. Note that although the closed-loop state vector consists of plant states and controller states, it is clear from (13.11) that only those states associated with the controller are reset. A function $\tilde{x} : \mathcal{I}_{\tilde{x}_0} \to \tilde{\mathcal{D}}$ is a *solution* to the impulsive dynamical system (13.12) and (13.13) on the interval $\mathcal{I}_{\tilde{x}_0} \subseteq \mathbb{R}$ with initial condition $\tilde{x}(0) = \tilde{x}_0$ if $\tilde{x}(\cdot)$ is left-continuous and $\tilde{x}(t)$ satisfies (13.12) and (13.13) for all $t \in \mathcal{I}_{\tilde{x}_0}$. For further discussion on solutions to impulsive differential equations, see [11, 13, 28, 74, 75, 82, 117, 137, 155, 175].

For convenience, we use the notation $\tilde{s}(t, \tilde{x}_0)$ to denote the solution $\tilde{x}(t)$ of (13.12) and (13.13) at time $t \geq 0$ with initial condition $\tilde{x}(0) = \tilde{x}_0$.

For a particular closed-loop trajectory $\tilde{x}(t)$, we let $t_k \triangleq \tau_k(\tilde{x}_0)$ denote the kth instant of time at which $\tilde{x}(t)$ intersects $\tilde{\mathcal{Z}}$, and we call the times t_k the *resetting times*. Thus, the trajectory of the closed-loop system $\tilde{\mathcal{G}}$ from the initial condition $\tilde{x}(0) = \tilde{x}_0$ is given by $\tilde{\psi}(t, \tilde{x}_0)$ for $0 < t \leq t_1$, where $\tilde{\psi}(t, \tilde{x}_0)$ denotes the solution to the continuous-time dynamics of the closed-loop system $\tilde{\mathcal{G}}$. If and when the trajectory reaches a state $\tilde{x}(t_1)$ satisfying $\tilde{x}(t_1) \in \tilde{\mathcal{Z}}$, then the state is instantaneously transferred to $\tilde{x}(t_1^+) \triangleq \tilde{x}(t_1) + \tilde{f}_{\mathrm{d}}(\tilde{x}(t_1))$ according to the resetting law (13.13). The trajectory $\tilde{x}(t)$, $t_1 < t \leq t_2$, is then given by $\tilde{\psi}(t - t_1, \tilde{x}(t_1^+))$, and so on. Our convention here is that the solution $\tilde{x}(t)$ of $\tilde{\mathcal{G}}$ is left-continuous, that is, it is continuous everywhere except at the resetting times t_k, and

$$\tilde{x}_k \triangleq \tilde{x}(t_k) = \lim_{\varepsilon \to 0^+} \tilde{x}(t_k - \varepsilon), \tag{13.15}$$

$$\tilde{x}_k^+ \triangleq \tilde{x}(t_k) + \tilde{f}_{\mathrm{d}}(\tilde{x}(t_k)) = \lim_{\varepsilon \to 0^+} \tilde{x}(t_k + \varepsilon), \tag{13.16}$$

for $k = 1, 2, \ldots$.

To ensure the well-posedness of the resetting times, we make the following additional assumptions:

Assumption 13.1. If $\tilde{x} \in \overline{\tilde{\mathcal{Z}}} \setminus \tilde{\mathcal{Z}}$, then there exists $\varepsilon > 0$ such that, for all $0 < \delta < \varepsilon$, $\tilde{s}(\delta, \tilde{x}) \notin \tilde{\mathcal{Z}}$.

Assumption 13.2. If $\tilde{x} \in \tilde{\mathcal{Z}}$, then $\tilde{x} + \tilde{f}_{\mathrm{d}}(\tilde{x}) \notin \tilde{\mathcal{Z}}$.

Assumptions 13.1 and 13.2 are a restatement of Assumptions 10.1 and 10.2 as applied to the closed-loop system (13.8) and (13.9). Hence, it follows from Assumptions 13.1 and 13.2 that for a particular initial condition, the resetting times $t_k = \tau_k(\tilde{x}_0)$ are distinct and well defined [74]. Since the resetting set $\tilde{\mathcal{Z}}$ is a subset of the state space and is independent of time, impulsive dynamical systems of the form (13.12) and (13.13) are time-invariant systems.

For the statement of the next result the following key assumption is needed.

Assumption 13.3. Consider the closed-loop impulsive dynamical system $\tilde{\mathcal{G}}$. Then for every $\tilde{x}_0 \notin \tilde{\mathcal{Z}}$ and every $\varepsilon > 0$ and $t \neq t_k$, there exists $\delta(\varepsilon, \tilde{x}_0, t) > 0$ such that if $\|\tilde{x}_0 - y\| < \delta(\varepsilon, \tilde{x}_0, t)$, $y \in \tilde{\mathcal{D}}$, then $\|\tilde{s}(t, \tilde{x}_0) - \tilde{s}(t, y)\| < \varepsilon$.

Assumption 13.3 is a weakened version of the quasi-continuous dependence assumption given in [74, 82], and is a generalization of the standard continuous dependence property for dynamical systems with continuous flows to dynamical systems with left-continuous flows. Specifically, by letting $t \in [0, \infty)$, Assumption 13.3 specializes to the classical continuous dependence of solutions of a given dynamical system with respect to the system's initial conditions $\tilde{x}_0 \in \tilde{\mathcal{D}}$. Since solutions of impulsive dynamical systems are not continuous in time and solutions are not continuous functions of the system initial conditions, Assumption 13.3 involving pointwise continuous dependence is needed to apply the hybrid invariance principle developed in [74, 82] to hybrid closed-loop systems. Sufficient conditions that guarantee that the impulsive dynamical system $\tilde{\mathcal{G}}$ satisfies a stronger version of Assumption 13.3 are given in [82] (see also [64]). The following result provides a generalization of the results given in [82] for establishing sufficient conditions for guaranteeing that the impulsive dynamical system $\tilde{\mathcal{G}}$ satisfies Assumption 13.3.

Proposition 13.1. Consider the large-scale impulsive dynamical system $\tilde{\mathcal{G}}$ given by the feedback interconnection of $\mathcal{G}$ and $\mathcal{G}_c$. Assume that Assumptions 13.1 and 13.2 hold, $\tau_1(\cdot)$ is continuous at every $\tilde{x} \notin \tilde{\mathcal{Z}}$ such that $0 < \tau_1(\tilde{x}) < \infty$, and if $\tilde{x} \in \tilde{\mathcal{Z}}$, then $\tilde{x} + \tilde{f}_{\mathrm{d}}(\tilde{x}) \in \overline{\tilde{\mathcal{Z}}} \backslash \tilde{\mathcal{Z}}$. Furthermore, for every $\tilde{x} \in \overline{\tilde{\mathcal{Z}}} \backslash \tilde{\mathcal{Z}}$ such that $0 < \tau_1(\tilde{x}) < \infty$, assume that the following statements hold:

  i) If a sequence $\{\tilde{x}_{(i)}\}_{i=1}^{\infty} \in \tilde{\mathcal{D}}$ is such that $\lim_{i \to \infty} \tilde{x}_{(i)} = \tilde{x}$ and $\lim_{i \to \infty} \tau_1(\tilde{x}_{(i)})$ exists, then either $\tilde{f}_{\mathrm{d}}(\tilde{x}) = 0$ and $\lim_{i \to \infty} \tau_1(\tilde{x}_{(i)}) = 0$, or $\lim_{i \to \infty} \tau_1(\tilde{x}_{(i)}) = \tau_1(\tilde{x})$.

  ii) If a sequence $\{\tilde{x}_{(i)}\}_{i=1}^{\infty} \in \overline{\tilde{\mathcal{Z}}} \backslash \tilde{\mathcal{Z}}$ is such that $\lim_{i \to \infty} \tilde{x}_{(i)} = \tilde{x}$ and $\lim_{i \to \infty} \tau_1(\tilde{x}_{(i)})$ exists, then $\lim_{i \to \infty} \tau_1(\tilde{x}_{(i)}) = \tau_1(\tilde{x})$.

Then $\tilde{\mathcal{G}}$ satisfies Assumption 13.3.

Proof. Let $\tilde{x}_0 \in \overline{\tilde{\mathcal{Z}}} \backslash \tilde{\mathcal{Z}}$ and let $\{\tilde{x}_{(i)}\}_{i=1}^{\infty} \in \tilde{\mathcal{D}}$ be such that $\tilde{f}_{\mathrm{d}}(\tilde{x}_0) = 0$ and $\lim_{i \to \infty} \tau_1(\tilde{x}_{(i)}) = 0$ hold. Define $\tilde{z}_{(i)} \triangleq \tilde{s}(\tau_1(\tilde{x}_{(i)}), \tilde{x}_{(i)}) + \tilde{f}_{\mathrm{d}}(\tilde{s}(\tau_1(\tilde{x}_{(i)}), \tilde{x}_{(i)})) = \tilde{\psi}(\tau_1(\tilde{x}_{(i)}), \tilde{x}_{(i)}) + \tilde{f}_{\mathrm{d}}(\tilde{\psi}(\tau_1(\tilde{x}_{(i)}), \tilde{x}_{(i)}))$, $i = 1, 2, \ldots$, where $\tilde{\psi}(t, \tilde{x}_0)$ denotes the solution to the continuous-time dynamics (13.8), and note that, since $\tilde{f}_{\mathrm{d}}(\tilde{x}_0) = 0$ and $\lim_{i \to \infty} \tau_1(\tilde{x}_{(i)}) = 0$, it follows that $\lim_{i \to \infty} \tilde{z}_{(i)} = \tilde{x}_0$. Hence, since by assumption $\tilde{z}_{(i)} \in \overline{\tilde{\mathcal{Z}}} \backslash \tilde{\mathcal{Z}}$, $i = 1, 2, \ldots$, it follows from $ii)$ that $\lim_{i \to \infty} \tau_1(\tilde{z}_{(i)}) = \tau_1(\tilde{x}_0)$, or, equivalently, $\lim_{i \to \infty} \tau_2(\tilde{x}_{(i)}) = \tau_1(\tilde{x}_0)$. Similarly, it can be shown that $\lim_{i \to \infty} \tau_{k+1}(\tilde{x}_{(i)}) = \tau_k(\tilde{x}_0)$, $k = 2, 3, \ldots$. Next, note that

$$\lim_{i \to \infty} \tilde{s}(\tau_2(\tilde{x}_{(i)}), \tilde{x}_{(i)}) = \lim_{i \to \infty} \tilde{\psi}(\tau_2(\tilde{x}_{(i)}) - \tau_1(\tilde{x}_{(i)}), \tilde{s}(\tau_1(\tilde{x}_{(i)}), \tilde{x}_{(i)})$$
$$+ \tilde{f}_{\mathrm{d}}(\tilde{s}(\tau_1(\tilde{x}_{(i)}), \tilde{x}_{(i)})))$$
$$= \tilde{\psi}(\tau_1(\tilde{x}_0), \tilde{x}_0)$$
$$= \tilde{s}(\tau_1(\tilde{x}_0), \tilde{x}_0).$$

Now, using mathematical induction it can be shown that $\lim_{i\to\infty} \tilde{s}(\tau_{k+1}(\tilde{x}_{(i)}), \tilde{x}_{(i)}) = \tilde{s}(\tau_k(\tilde{x}_0), \tilde{x}_0)$, $k = 2, 3, \ldots$.

Next, let $k \in \{1, 2, \ldots\}$ and let $t \in (\tau_k(\tilde{x}_0), \tau_{k+1}(\tilde{x}_0))$. Since $\lim_{i\to\infty} \tau_{k+1}(\tilde{x}_{(i)}) = \tau_k(\tilde{x}_0)$, it follows that there exists $I \in \{1, 2, \ldots\}$ such that $\tau_{k+1}(\tilde{x}_{(i)}) < t$ and $\tau_{k+2}(\tilde{x}_{(i)}) > t$ for all $i > I$. Hence, it follows that for every $t \in (\tau_k(\tilde{x}_0), \tau_{k+1}(\tilde{x}_0))$,

$$\lim_{i\to\infty} \tilde{s}(t, \tilde{x}_{(i)}) = \lim_{i\to\infty} \tilde{\psi}(t - \tau_{k+1}(\tilde{x}_{(i)}), \tilde{s}(\tau_{k+1}(\tilde{x}_{(i)}), \tilde{x}_{(i)})$$
$$+ \tilde{f}_{\mathrm{d}}(\tilde{s}(\tau_{k+1}(\tilde{x}_{(i)}), \tilde{x}_{(i)})))$$
$$= \tilde{\psi}(t - \tau_k(\tilde{x}_0), \tilde{s}(\tau_k(\tilde{x}_0), \tilde{x}_0) + \tilde{f}_{\mathrm{d}}(\tilde{s}(\tau_k(\tilde{x}_0), \tilde{x}_0)))$$
$$= \tilde{s}(t, \tilde{x}_0).$$

Alternatively, if $\tilde{x}_0 \in \overline{\tilde{\mathcal{Z}}} \backslash \tilde{\mathcal{Z}}$ is such that $\lim_{i\to\infty} \tau_1(\tilde{x}_{(i)}) = \tau_1(\tilde{x}_0)$ for $\{\tilde{x}_{(i)}\}_{i=1}^\infty \in \overline{\tilde{\mathcal{Z}}} \backslash \tilde{\mathcal{Z}}$, then using identical arguments as above, it can be shown that $\lim_{i\to\infty} \tilde{s}(t, \tilde{x}_{(i)}) = \tilde{s}(t, \tilde{x}_0)$ for every $t \in (\tau_k(\tilde{x}_0), \tau_{k+1}(\tilde{x}_0))$, $k = 1, 2, \ldots$.

Finally, let $\tilde{x}_0 \notin \overline{\tilde{\mathcal{Z}}}$, $0 < \tau_1(\tilde{x}_0) < \infty$, and assume $\tau_1(\cdot)$ is continuous. In this case, it follows from the definition of $\tau_1(\tilde{x}_0)$ that for every $\tilde{x}_0 \notin \overline{\tilde{\mathcal{Z}}}$ and $t \in (\tau_1(\tilde{x}_0), \tau_2(\tilde{x}_0)]$,

$$\tilde{s}(t, \tilde{x}_0) = \tilde{\psi}(t - \tau_1(\tilde{x}_0), \tilde{s}(\tau_1(\tilde{x}_0), \tilde{x}_0) + \tilde{f}_{\mathrm{d}}(\tilde{s}(\tau_1(\tilde{x}_0), \tilde{x}_0))). \tag{13.17}$$

Since $\tilde{\psi}(\cdot, \cdot)$ is continuous in both its arguments, $\tau_1(\cdot)$ is continuous at $\tilde{x}_0$, and $\tilde{f}_{\mathrm{d}}(\cdot)$ is continuous, it follows that $\tilde{s}(t, \cdot)$ is continuous at $\tilde{x}_0$ for every $t \in (\tau_1(\tilde{x}_0), \tau_2(\tilde{x}_0))$. Next, for every sequence $\{\tilde{x}_{(i)}\}_{i=1}^\infty \in \tilde{\mathcal{D}}$ such that $\lim_{i\to\infty} \tilde{x}_{(i)} = \tilde{x}_0$, it follows that $\lim_{i\to\infty} \tilde{s}(\tau_1(\tilde{x}_{(i)}), \tilde{x}_{(i)}) = \lim_{i\to\infty} \tilde{\psi}(\tau_1(\tilde{x}_{(i)}), \tilde{x}_{(i)}) = \tilde{\psi}(\tau_1(\tilde{x}_0), \tilde{x}_0) = \tilde{s}(\tau_1(\tilde{x}_0), \tilde{x}_0)$. Furthermore, note that by assumption $\tilde{z}_{(i)} \triangleq \tilde{s}(\tau_1(\tilde{x}_{(i)}), \tilde{x}_{(i)}) + \tilde{f}_{\mathrm{d}}(\tilde{s}(\tau_1(\tilde{x}_{(i)}), \tilde{x}_{(i)})) \in \overline{\tilde{\mathcal{Z}}} \backslash \tilde{\mathcal{Z}}$, $i = 0, 1, \ldots$. Hence, it follows that for all $t \in (\tau_k(\tilde{z}_{(0)}), \tau_{k+1}(\tilde{z}_{(0)}))$, $k = 1, 2, \ldots$, $\lim_{i\to\infty} \tilde{s}(t, \tilde{z}_{(i)}) = \tilde{s}(t, \tilde{z}_{(0)})$, or, equivalently, for all $t \in (\tau_k(\tilde{x}_0), \tau_{k+1}(\tilde{x}_0))$, $k = 2, 3, \ldots$, $\lim_{i\to\infty} \tilde{s}(t, \tilde{x}_{(i)}) = \tilde{s}(t, \tilde{x}_0)$, which proves the result. $\square$

Proposition 13.1 presents a generalization of Proposition 4.1 of [82] to the case where the resetting set $\tilde{\mathcal{Z}}$ is not necessarily closed. This generalization is key in developing energy and entropy-based hybrid controllers.

The following result provides sufficient conditions for establishing continuity of $\tau_1(\cdot)$ at $\tilde{x}_0 \notin \overline{\tilde{\mathcal{Z}}}$ and *sequential continuity* of $\tau_1(\cdot)$ at $\tilde{x}_0 \in \overline{\tilde{\mathcal{Z}}} \backslash \tilde{\mathcal{Z}}$, that is, $\lim_{i\to\infty} \tau_1(\tilde{x}_{(i)}) = \tau_1(\tilde{x}_0)$ for $\{\tilde{x}_{(i)}\}_{i=1}^\infty \notin \tilde{\mathcal{Z}}$ and $\lim_{i\to\infty} \tilde{x}_{(i)} = \tilde{x}_0$. For this result, the following definition is needed. First, however, recall that the *Lie derivative* of a smooth function $\mathcal{X} : \tilde{\mathcal{D}} \to \mathbb{R}$ along the vector field of the continuous-time dynamics $\tilde{f}_{\mathrm{c}}(\tilde{x})$ is given by

$$L_{\tilde{f}_{\mathrm{c}}} \mathcal{X}(\tilde{x}) \triangleq \frac{\mathrm{d}}{\mathrm{d}t} \mathcal{X}(\tilde{\psi}(t, \tilde{x}))|_{t=0} = \frac{\partial \mathcal{X}(\tilde{x})}{\partial \tilde{x}} \tilde{f}_{\mathrm{c}}(\tilde{x}), \tag{13.18}$$

and the *zeroth* and *higher-order Lie derivatives* are, respectively, defined by $L_{\tilde{f}_c}^0 \mathcal{X}(\tilde{x}) \triangleq \mathcal{X}(\tilde{x})$ and $L_{\tilde{f}_c}^k \mathcal{X}(\tilde{x}) \triangleq L_{\tilde{f}_c}(L_{\tilde{f}_c}^{k-1}\mathcal{X}(\tilde{x}))$, where $k \geq 1$.

Definition 13.1. Let $\mathcal{M} \triangleq \cup_{i=1}^q \{\tilde{x} \in \tilde{\mathcal{D}} : \mathcal{X}_i(\tilde{x}) = 0\}$, where $\mathcal{X}_i : \tilde{\mathcal{D}} \to \mathbb{R}$, $i = 1, \ldots, q$, are infinitely differentiable functions. A point $\tilde{x} \in \mathcal{M}$ such that $\tilde{f}_c(\tilde{x}) \neq 0$ is k_i-*transversal* to (13.12) if there exist $k_i \in \{1, 2, \ldots\}$, $i = 1, \ldots, q$, such that

$$L_{\tilde{f}_c}^r \mathcal{X}_i(\tilde{x}) = 0, \quad r = 0, \ldots, 2k_i - 2, \quad L_{\tilde{f}_c}^{2k_i - 1} \mathcal{X}_i(\tilde{x}) \neq 0, \quad i = 1, \ldots, q.$$

(13.19)

Proposition 13.2. Consider the large-scale impulsive dynamical system $\tilde{\mathcal{G}}$ given by the feedback interconnection of $\mathcal{G}$ and $\mathcal{G}_c$. Let $\mathcal{X}_i : \tilde{\mathcal{D}} \to \mathbb{R}$, $i = 1, \ldots, q$, be infinitely differentiable functions such that $\overline{\tilde{\mathcal{Z}}} = \cup_{i=1}^q \{\tilde{x} \in \tilde{\mathcal{D}} : \mathcal{X}_i(\tilde{x}) = 0\}$, and assume that every $\tilde{x} \in \overline{\tilde{\mathcal{Z}}}$ is k_i-transversal to (13.12). Then at every $\tilde{x}_0 \notin \overline{\tilde{\mathcal{Z}}}$ such that $0 < \tau_1(\tilde{x}_0) < \infty$, $\tau_1(\cdot)$ is continuous. Furthermore, if $\tilde{x}_0 \in \overline{\tilde{\mathcal{Z}}} \backslash \tilde{\mathcal{Z}}$ is such that $\tau_1(\tilde{x}_0) \in (0, \infty)$ and i) $\{\tilde{x}_{(i)}\}_{i=1}^\infty \in \overline{\tilde{\mathcal{Z}}} \backslash \tilde{\mathcal{Z}}$ or ii) $\lim_{i \to \infty} \tau_1(\tilde{x}_{(i)}) > 0$, where $\{\tilde{x}_{(i)}\}_{i=1}^\infty \notin \overline{\tilde{\mathcal{Z}}}$ is such that $\lim_{i \to \infty} \tilde{x}_{(i)} = \tilde{x}_0$ and $\lim_{i \to \infty} \tau_1(\tilde{x}_{(i)})$ exists, then $\lim_{i \to \infty} \tau_1(\tilde{x}_{(i)}) = \tau_1(\tilde{x}_0)$.

Proof. Let $\tilde{x}_0 \notin \overline{\tilde{\mathcal{Z}}}$ be such that $0 < \tau_1(\tilde{x}_0) < \infty$. It follows from the definition of $\tau_1(\cdot)$ that $\tilde{s}(t, \tilde{x}_0) = \tilde{\psi}(t, \tilde{x}_0)$, $t \in [0, \tau_1(\tilde{x}_0)]$, $\Pi_{i=1}^q \mathcal{X}_i(\tilde{s}(t, \tilde{x}_0)) \neq 0$, $t \in (0, \tau_1(\tilde{x}_0))$, and $\Pi_{i=1}^q \mathcal{X}_i(\tilde{s}(\tau_1(\tilde{x}_0), \tilde{x}_0)) = 0$. Without loss of generality, let $\Pi_{i=1}^q \mathcal{X}_i(\tilde{s}(t, \tilde{x}_0)) > 0$, $t \in (0, \tau_1(\tilde{x}_0))$. Since $\hat{x} \triangleq \tilde{\psi}(\tau_1(\tilde{x}_0), \tilde{x}_0) \in \overline{\tilde{\mathcal{Z}}}$ is k_i-transversal to (13.8), it follows that there exist $\theta > 0$ and $i \in \{1, \ldots, q\}$ such that $\mathcal{X}_i(\tilde{\psi}(t, \hat{x})) > 0$, $t \in [-\theta, 0)$, and $\mathcal{X}_i(\tilde{\psi}(t, \hat{x})) < 0$, $t \in (0, \theta]$. (This fact can be easily shown by expanding $\mathcal{X}_i(\tilde{\psi}(t, x))$ via a Taylor series expansion about $\hat{x}$ and using the fact that $\hat{x}$ is k_i-transversal to (13.8).) Hence, $\mathcal{X}_i(\tilde{\psi}(t, \tilde{x}_0)) > 0$, $t \in [\hat{t}_1, \tau_1(\tilde{x}_0))$, and $\mathcal{X}_i(\tilde{\psi}(t, \tilde{x}_0)) < 0$, $t \in (\tau_1(\tilde{x}_0), \hat{t}_2]$, where $\hat{t}_1 \triangleq \tau_1(\tilde{x}_0) - \theta$ and $\hat{t}_2 \triangleq \tau_1(\tilde{x}_0) + \theta$.

Next, let $\varepsilon \triangleq \min\{|\mathcal{X}_i(\tilde{\psi}(\hat{t}_1, \tilde{x}_0))|, |\mathcal{X}_i(\tilde{\psi}(\hat{t}_2, \tilde{x}_0))|\}$. Now, it follows from the continuity of $\mathcal{X}_i(\cdot)$ and the continuous dependence of $\tilde{\psi}(\cdot, \cdot)$ on the system initial conditions that there exists $\delta > 0$ such that

$$\sup_{0 \leq t \leq \hat{t}_2} |\mathcal{X}_i(\tilde{\psi}(t, x)) - \mathcal{X}_i(\tilde{\psi}(t, \tilde{x}_0))| < \varepsilon, \quad x \in \mathcal{B}_\delta(\tilde{x}_0), \qquad (13.20)$$

which implies that $\mathcal{X}_i(\tilde{\psi}(\hat{t}_1, x)) > 0$ and $\mathcal{X}_i(\tilde{\psi}(\hat{t}_2, x)) < 0$, $x \in \mathcal{B}_\delta(\tilde{x}_0)$. Hence, it follows that $\hat{t}_1 < \tau_1(x) < \hat{t}_2$, $x \in \mathcal{B}_\delta(\tilde{x}_0)$. The continuity of $\tau_1(\cdot)$ at $\tilde{x}_0$ now follows immediately by noting that θ can be chosen arbitrarily small.

Finally, let $\tilde{x}_0 \in \overline{\tilde{\mathcal{Z}}} \backslash \tilde{\mathcal{Z}}$ be such that $\lim_{i \to \infty} \tilde{x}_{(i)} = \tilde{x}_0$ for some sequence $\{\tilde{x}_{(i)}\}_{i=1}^\infty \in \overline{\tilde{\mathcal{Z}}} \backslash \tilde{\mathcal{Z}}$. Then using similar arguments as above it can be shown that $\lim_{i \to \infty} \tau_1(\tilde{x}_{(i)}) = \tau_1(\tilde{x}_0)$. Alternatively, if $\tilde{x}_0 \in \overline{\tilde{\mathcal{Z}}} \backslash \tilde{\mathcal{Z}}$ is such that

$\lim_{i\to\infty}\tilde{x}_{(i)} = \tilde{x}_0$ and $\lim_{i\to\infty}\tau_1(\tilde{x}_{(i)}) > 0$ for some sequence $\{\tilde{x}_{(i)}\}_{i=1}^{\infty} \notin \tilde{\mathcal{Z}}$, then it follows that there exists sufficiently small $\hat{t} > 0$ and $I \in \mathbb{Z}_+$ such that $\tilde{s}(\hat{t}, \tilde{x}_{(i)}) = \tilde{\psi}(\hat{t}, \tilde{x}_{(i)})$, $i = I, I+1, \ldots$, which implies that $\lim_{i\to\infty}\tilde{s}(\hat{t}, \tilde{x}_{(i)}) = \tilde{s}(\hat{t}, \tilde{x}_0)$. Next, define $\tilde{z}_{(i)} \triangleq \tilde{\psi}(\hat{t}, \tilde{x}_{(i)})$, $i = 0, 1, \ldots$, so that $\lim_{i\to\infty}\tilde{z}_{(i)} = \tilde{z}_{(0)}$, and note that it follows from the k_i-transversality assumption that $\tilde{z}_{(0)} \notin \overline{\tilde{\mathcal{Z}}}$, which implies that $\tau_1(\cdot)$ is continuous at $\tilde{z}_{(0)}$. Hence, $\lim_{i\to\infty}\tau_1(\tilde{z}_{(i)}) = \tau_1(\tilde{z}_{(0)})$. The result now follows by noting that $\tau_1(\tilde{x}_{(i)}) = \hat{t} + \tau_1(\tilde{z}_{(i)})$, $i = 1, 2, \ldots$. $\qquad\square$

Let $\tilde{x}_0 \notin \tilde{\mathcal{Z}}$ be such that $\lim_{i\to\infty}\tau_1(\tilde{x}_{(i)}) \neq \tau_1(\tilde{x}_0)$ for some sequence $\{\tilde{x}_{(i)}\}_{i=1}^{\infty} \notin \tilde{\mathcal{Z}}$. Then it follows from Proposition 13.2 that $\lim_{i\to\infty}\tau_1(\tilde{x}_{(i)}) = 0$. Proposition 13.2 is a nontrivial generalization of Proposition 4.2 of [82] and Lemma 3 of [64]. Specifically, Proposition 13.2 establishes the continuity of $\tau_1(\cdot)$ in the case where the resetting set $\tilde{\mathcal{Z}}$ is not a closed set. In addition, the k_i-transversality condition given in Definition 13.1 is also a generalization of the conditions given in [82] and [64] by considering higher-order derivatives of the function $\mathcal{X}_i(\cdot)$ rather than simply considering the first-order derivative as in [64, 82]. This condition guarantees that the solution of the closed-loop system (13.8) and (13.9) is not tangent to the closure of the resetting set $\tilde{\mathcal{Z}}$ at the intersection with $\overline{\tilde{\mathcal{Z}}}$.

The next result characterizes impulsive dynamical system limit sets in terms of continuously differentiable functions. In particular, we show that the system trajectories of a state-dependent impulsive dynamical system converge to an invariant set contained in a union of level surfaces characterized by the continuous-time system dynamics and the resetting system dynamics. For the next result assume that $\tilde{f}_c(\cdot)$, $\tilde{f}_d(\cdot)$, $\tilde{\mathcal{I}}(\cdot)$, and $\tilde{\mathcal{Z}}$ are such that the dynamical system $\tilde{\mathcal{G}}$ given by (13.12) and (13.13) satisfies Assumptions 13.1–13.3 and $\tilde{\mathcal{Z}} \cap \{x : \tilde{f}_d(x) = x\}$ is empty. Note that for addressing the stability of the zero solution of an impulsive dynamical system the usual stability definitions are valid. For details, see [11, 13, 74, 75, 82, 117, 155].

Theorem 13.1 ([73]). Consider the impulsive dynamical system (13.12) and (13.13) and assume Assumptions 13.1–13.3 hold. Assume $\tilde{\mathcal{D}}_{ci} \subset \tilde{\mathcal{D}}$ is a compact positively invariant set with respect to (13.12) and (13.13), assume that if $\tilde{x}_0 \in \tilde{\mathcal{Z}}$ then $\tilde{x}_0 + \tilde{f}_d(\tilde{x}_0) \in \overline{\tilde{\mathcal{Z}}} \backslash \tilde{\mathcal{Z}}$, and assume that there exist a continuously differentiable function $V : \tilde{\mathcal{D}}_{ci} \to \mathbb{R}$ such that

$$V'(\tilde{x})\tilde{f}_c(\tilde{x}) \leq 0, \quad \tilde{x} \in \tilde{\mathcal{D}}_{ci}, \quad \tilde{x} \notin \tilde{\mathcal{Z}}, \tag{13.21}$$

$$V(\tilde{x} + \tilde{f}_d(\tilde{x})) \leq V(\tilde{x}), \quad \tilde{x} \in \tilde{\mathcal{D}}_{ci}, \quad \tilde{x} \in \tilde{\mathcal{Z}}. \tag{13.22}$$

Let $\mathcal{R} \triangleq \{\tilde{x} \in \tilde{\mathcal{D}}_{ci} : \tilde{x} \notin \tilde{\mathcal{Z}}, V'(\tilde{x})\tilde{f}_c(\tilde{x}) = 0\} \cup \{\tilde{x} \in \tilde{\mathcal{D}}_{ci} : \tilde{x} \in \tilde{\mathcal{Z}}, V(\tilde{x} + \tilde{f}_d(\tilde{x})) - V(\tilde{x}) = 0\}$ and let $\mathcal{M}$ denote the largest invariant set contained in $\mathcal{R}$. If $\tilde{x}_0 \in \tilde{\mathcal{D}}_{ci}$, then $\tilde{x}(t) \to \mathcal{M}$ as $t \to \infty$. Furthermore, if $0 \in \overset{\circ}{\tilde{\mathcal{D}}}_{ci}$, $V(0) = 0$,

$V(\tilde{x}) > 0$, $\tilde{x} \neq 0$, and the set $\mathcal{R}$ contains no invariant set other than the set $\{0\}$, then the zero solution $\tilde{x}(t) \equiv 0$ to (13.12) and (13.13) is asymptotically stable and $\tilde{\mathcal{D}}_{ci}$ is a subset of the domain of attraction of (13.12) and (13.13).

Setting $\tilde{\mathcal{D}} = \mathbb{R}^n$ and requiring $V(\tilde{x}) \to \infty$ as $\|\tilde{x}\| \to \infty$ in Theorem 13.1, it follows that the zero solution $\tilde{x}(t) \equiv 0$ to (13.12) and (13.13) is globally asymptotically stable. A similar remark holds for Theorem 13.2 below.

Theorem 13.2. Consider the impulsive dynamical system $\tilde{\mathcal{G}}$ (13.12) and (13.13) and assume Assumptions 13.1–13.3 hold. Assume $\tilde{\mathcal{D}}_{ci} \subset \tilde{\mathcal{D}}$ is a compact positively invariant set with respect to (13.12) and (13.13) such that $0 \in \overset{\circ}{\tilde{\mathcal{D}}}_{ci}$, assume that if $\tilde{x}_0 \in \tilde{\mathcal{Z}}$ then $\tilde{x}_0 + \tilde{f}_d(\tilde{x}_0) \in \overline{\tilde{\mathcal{Z}}} \backslash \tilde{\mathcal{Z}}$, and assume that for all $\tilde{x}_0 \in \tilde{\mathcal{D}}_{ci}$, $\tilde{x}_0 \neq 0$, there exists $\tau \geq 0$ such that $\tilde{x}(\tau) \in \tilde{\mathcal{Z}}$, where $\tilde{x}(t)$, $t \geq 0$, denotes the solution to (13.12) and (13.13) with the initial condition $\tilde{x}_0$. Furthermore, assume that there exist a continuously differentiable vector function $V = [v_1, \ldots, v_q]^{\mathrm{T}} : \tilde{\mathcal{D}} \to \overline{\mathbb{R}}_+^q$ and a positive vector $p \in \mathbb{R}_+^q$ such that $V(0) = 0$, the scalar function $v : \tilde{\mathcal{D}} \to \overline{\mathbb{R}}_+$ defined by $v(\tilde{x}) \triangleq p^{\mathrm{T}} V(\tilde{x})$, $\tilde{x} \in \tilde{\mathcal{D}}$, is such that $v(\tilde{x}) > 0$, $\tilde{x} \in \tilde{\mathcal{D}}$, $\tilde{x} \neq 0$, and

$$v'(\tilde{x})\tilde{f}_c(\tilde{x}) \leq 0, \quad \tilde{x} \in \tilde{\mathcal{D}}_{ci}, \quad \tilde{x} \notin \tilde{\mathcal{Z}}, \tag{13.23}$$

$$v(\tilde{x} + \tilde{f}_d(\tilde{x})) < v(\tilde{x}), \quad \tilde{x} \in \tilde{\mathcal{D}}_{ci}, \quad \tilde{x} \in \tilde{\mathcal{Z}}. \tag{13.24}$$

Then the zero solution $\tilde{x}(t) \equiv 0$ to (13.12) and (13.13) is asymptotically stable and $\tilde{\mathcal{D}}_{ci}$ is a subset of the domain of attraction of (13.12) and (13.13).

Proof. It follows from (13.24) that $\mathcal{R} = \{\tilde{x} \in \tilde{\mathcal{D}}_{ci} : \tilde{x} \notin \tilde{\mathcal{Z}}, v'(\tilde{x})\tilde{f}_c(\tilde{x}) = 0\}$. Since for all $\tilde{x}_0 \in \tilde{\mathcal{D}}_{ci}$, $\tilde{x}_0 \neq 0$, there exists $\tau \geq 0$ such that $\tilde{x}(\tau) \in \tilde{\mathcal{Z}}$, it follows that the largest invariant set contained in $\mathcal{R}$ is $\{0\}$. Now, the result is a direct consequence of Theorem 13.1. $\qquad\square$

13.3 Hybrid Decentralized Control for Large-Scale Dynamical Systems

In this section, we present a hybrid decentralized controller design framework for large-scale dynamical systems. Specifically, we consider nonlinear large-scale dynamical systems $\mathcal{G}$ of the form given by (13.1) and (13.2) where $u(\cdot)$ satisfies sufficient regularity conditions such that (13.1) has a unique solution forward in time. Furthermore, we consider hybrid decentralized dynamic controllers $\mathcal{G}_{ci}$, $i = 1, \ldots, q$, of the form

$$\dot{x}_{ci}(t) = f_{ci}(x_{ci}(t), y_i(t)), \quad x_{ci}(0) = x_{c0i}, \quad (x_{ci}(t), y_i(t)) \notin \mathcal{Z}_{ci}, \tag{13.25}$$

$$\Delta x_{ci}(t) = \eta_i(y_i(t)) - x_{ci}(t), \quad (x_{ci}(t), y_i(t)) \in \mathcal{Z}_{ci}, \tag{13.26}$$

$$y_{ci}(t) = h_{ci}(x_{ci}(t), y_i(t)), \tag{13.27}$$

where $x_{ci}(t) \in \mathcal{D}_{ci} \subseteq \mathbb{R}^{n_{ci}}$, $\mathcal{D}_{ci}$ is an open set with $0 \in \mathcal{D}_{ci}$, $y_i(t) \in \mathbb{R}^{l_i}$, $y_{ci}(t) \in \mathbb{R}^{m_i}$, $f_{ci} : \mathcal{D}_{ci} \times \mathbb{R}^{l_i} \to \mathbb{R}^{n_{ci}}$ is smooth on $\mathcal{D}_{ci}$ and satisfies $f_{ci}(0,0) = 0$, $\eta_i : \mathbb{R}^{l_i} \to \mathcal{D}_{ci}$ is continuous and satisfies $\eta_i(0) = 0$, $h_{ci} : \mathcal{D}_{ci} \times \mathbb{R}^{l_i} \to \mathbb{R}^{m_i}$ is smooth and satisfies $h_{ci}(0,0) = 0$, $\sum_{i=1}^{q} l_i = l$, and $\sum_{i=1}^{q} m_i = m$.

Recall that for the dynamical system $\mathcal{G}$ given by (13.1) and (13.2), a vector function $S(u,y) \triangleq [s_1(u_1,y_1), \ldots, s_q(u_q,y_q)]^{\mathrm{T}}$, where $S : U \times Y \to \mathbb{R}^q$ is such that $S(0,0) = 0$, is called a *vector supply rate* if it is componentwise locally integrable for all input-output pairs satisfying (13.1) and (13.2), that is, for every $i \in \{1, \ldots, q\}$ and for all input-output pairs $(u_i(\cdot), y_i(\cdot)) \in \mathcal{U}_i \times \mathcal{Y}_i$ satisfying (13.1) and (13.2), $s_i(\cdot, \cdot)$ satisfies $\int_{t_1}^{t_2} |s_i(u_i(\sigma), y_i(\sigma))| \mathrm{d}\sigma < \infty$, $t_2 \geq t_1 \geq 0$. Here, $\mathcal{U} = \mathcal{U}_1 \times \cdots \times \mathcal{U}_q$ and $\mathcal{Y} = \mathcal{Y}_1 \times \cdots \times \mathcal{Y}_q$ are input and output spaces, respectively, that are assumed to be closed under the shift operator. Furthermore, we assume that $\mathcal{G}$ is *vector lossless with respect to the vector supply rate* $S(u,y)$, and hence, there exist a continuous, nonnegative definite *vector storage function* $V_{\mathrm{s}} = [v_{\mathrm{s}1}, \ldots, v_{\mathrm{s}q}]^{\mathrm{T}} : \mathcal{D} \to \overline{\mathbb{R}}_{+}^{q}$ and a *Kamke function* $w : \overline{\mathbb{R}}_{+}^{q} \to \mathbb{R}^q$ such that $V_{\mathrm{s}}(0) = 0$, $w(0) = 0$, the zero solution $z(t) \equiv 0$ to the comparison system

$$\dot{z}(t) = w(z(t)), \quad z(0) = z_0, \quad t \geq 0, \tag{13.28}$$

is Lyapunov stable, and the *vector dissipation equality*

$$V_{\mathrm{s}}(x(t)) = V_{\mathrm{s}}(x(t_0)) + \int_{t_0}^{t} w(V_{\mathrm{s}}(x(\sigma)))\mathrm{d}\sigma + \int_{t_0}^{t} S(u(\sigma), y(\sigma))\mathrm{d}\sigma,$$

$$\tag{13.29}$$

is satisfied for all $t \geq t_0 \geq 0$, where $x(t)$, $t \geq t_0$, is the solution to $\mathcal{G}$ with $u(\cdot) \in \mathcal{U}$.

In this case, it follows from Theorem 3.2 that there exists a nonnegative vector $p \in \overline{\mathbb{R}}_{+}^{q}$, $p \neq 0$, such that $\mathcal{G}$ is lossless with respect to the supply rate $p^{\mathrm{T}} S(u,y)$ and with the storage function $v_{\mathrm{s}}(x) = p^{\mathrm{T}} V_{\mathrm{s}}(x)$, $x \in \mathcal{D}$. In addition, we assume that the nonlinear large-scale dynamical system $\mathcal{G}$ is *completely reachable* and *zero-state observable*, and there exist functions $\kappa_i : Y_i \to U_i$ such that $\kappa_i(0) = 0$ and $s_i(\kappa_i(y_i), y_i) < 0$, $y_i \neq 0$, for all $i = 1, \ldots, q$, so that all storage functions $v_{\mathrm{s}}(x) = p^{\mathrm{T}} V_{\mathrm{s}}(x)$, $x \in \mathcal{D}$, are positive definite, that is, $p^{\mathrm{T}} V_{\mathrm{s}}(x) > 0$, $x \in \mathcal{D}$, $x \neq 0$ [80]. Finally, we assume that $V_{\mathrm{s}}(\cdot)$ is component decoupled, that is, $V_{\mathrm{s}}(x) = [v_{\mathrm{s}1}(x_1), \ldots, v_{\mathrm{s}q}(x_q)]^{\mathrm{T}}$, $x \in \mathcal{D}$, and continuously differentiable. Note that if each *disconnected* subsystem $\mathcal{G}_i$ (i.e., $\mathcal{I}_i(x) \equiv 0$, $i \in \{1, \ldots, q\}$) of $\mathcal{G}$ is lossless with respect to the supply rate $s_i(u_i, y_i)$, then $V_{\mathrm{s}}(\cdot)$ is component decoupled.

Consider the negative feedback interconnection of $\mathcal{G}$ and $\mathcal{G}_{\mathrm{c}}$ given by $y_i = u_{ci}$ and $u_i = -y_{ci}$, $i = 1, \ldots, q$. In this case, the closed-loop system $\tilde{\mathcal{G}}$ can be written in terms of the subsystems $\tilde{\mathcal{G}}_i$, $i = 1, \ldots, q$, given by

$$\dot{\tilde{x}}_i(t) = \tilde{f}_{ci}(\tilde{x}_i(t)) + \tilde{\mathcal{I}}_i(x), \quad \tilde{x}_i(0) = \tilde{x}_{i0}, \quad \tilde{x}_i(t) \notin \tilde{\mathcal{Z}}_i, \quad t \geq 0, \tag{13.30}$$

$$\Delta \tilde{x}_i(t) = \tilde{f}_{di}(\tilde{x}_i(t)), \quad \tilde{x}_i(t) \in \tilde{\mathcal{Z}}_i, \tag{13.31}$$

where $t \geq 0$, $\tilde{x}_i(t) \triangleq [x_i^{\mathrm{T}}(t), x_{\mathrm{c}i}^{\mathrm{T}}(t)]^{\mathrm{T}}$, $\tilde{\mathcal{Z}}_i \triangleq \{\tilde{x}_i \in \tilde{\mathcal{D}}_i : (x_{\mathrm{c}i}, h_i(x_i)) \in \mathcal{Z}_{\mathrm{c}i}\}$,

$$\tilde{f}_{\mathrm{c}i}(\tilde{x}_i) \triangleq \begin{bmatrix} f_i(x_i) - G_i(x_i)h_{\mathrm{c}i}(x_{\mathrm{c}i}, h_i(x_i)) \\ f_{\mathrm{c}i}(x_{\mathrm{c}i}, h_i(x_i)) \end{bmatrix}, \quad \tilde{\mathcal{I}}_i(x) \triangleq \begin{bmatrix} \mathcal{I}_i(x) \\ 0 \end{bmatrix}, \quad (13.32)$$

$$\tilde{f}_{\mathrm{d}i}(\tilde{x}_i) \triangleq \begin{bmatrix} 0 \\ \eta_i(h_i(x_i)) - x_{\mathrm{c}i} \end{bmatrix}. \tag{13.33}$$

Hence, the equations of the motion for the closed-loop system $\tilde{\mathcal{G}}$ have the form

$$\dot{\tilde{x}}(t) = \tilde{f}_{\mathrm{c}}(\tilde{x}(t)), \quad \tilde{x}(t_0) = \tilde{x}_0, \quad \tilde{x}(t) \notin \tilde{\mathcal{Z}}, \quad t \geq t_0, \tag{13.34}$$

$$\Delta \tilde{x}(t) = \tilde{f}_{\mathrm{d}}(\tilde{x}(t)), \quad \tilde{x}(t) \in \tilde{\mathcal{Z}}, \tag{13.35}$$

where $\tilde{x}(t) = [\tilde{x}_1^{\mathrm{T}}(t), \ldots, \tilde{x}_q^{\mathrm{T}}(t)]^{\mathrm{T}}$, $\tilde{f}_{\mathrm{c}}(\tilde{x}) \triangleq [\tilde{f}_{\mathrm{c}1}^{\mathrm{T}}(\tilde{x}_1) + \tilde{\mathcal{I}}_1^{\mathrm{T}}(x), \ldots, \tilde{f}_{\mathrm{c}q}^{\mathrm{T}}(\tilde{x}_q) + \tilde{\mathcal{I}}_q^{\mathrm{T}}(x)]^{\mathrm{T}}$, $\tilde{\mathcal{Z}} \triangleq \cup_{i=1}^{q}\{\tilde{x} \in \tilde{\mathcal{D}} : \tilde{x}_i \in \tilde{\mathcal{Z}}_i\}$, $\tilde{\mathcal{D}} \triangleq \cup_{i=1}^{q}\tilde{\mathcal{D}}_i$, and

$$\tilde{f}_{\mathrm{d}}(\tilde{x}) \triangleq \begin{bmatrix} \tilde{f}_{\mathrm{d}1}(\tilde{x}_1)\chi_{\tilde{\mathcal{Z}}_1}(\tilde{x}_1) \\ \vdots \\ \tilde{f}_{\mathrm{d}q}(\tilde{x}_q)\chi_{\tilde{\mathcal{Z}}_q}(\tilde{x}_q) \end{bmatrix}, \quad \chi_{\tilde{\mathcal{Z}}_i}(\tilde{x}_i) = \begin{cases} 1, & \tilde{x}_i \in \tilde{\mathcal{Z}}_i \\ 0, & \tilde{x}_i \notin \tilde{\mathcal{Z}}_i \end{cases}, \quad i = 1, \ldots, q.$$

$$\tag{13.36}$$

Assume that there exist infinitely differentiable functions $v_{\mathrm{c}i} : \mathcal{D}_{\mathrm{c}i} \times \mathbb{R}^{l_i} \to \overline{\mathbb{R}}_+$, $i = 1, \ldots, q$, such that $v_{\mathrm{c}i}(x_{\mathrm{c}i}, y_i) \geq 0$, $x_{\mathrm{c}i} \in \mathcal{D}_{\mathrm{c}i}$, $y_i \in \mathbb{R}^{l_i}$, and $v_{\mathrm{c}i}(x_{\mathrm{c}i}, y_i) = 0$ if and only if $x_{\mathrm{c}i} = \eta_i(y_i)$ and

$$\dot{v}_{\mathrm{c}i}(x_{\mathrm{c}i}(t), y_i(t)) = s_{\mathrm{c}i}(u_{\mathrm{c}i}(t), y_{\mathrm{c}i}(t)), \quad (x_{\mathrm{c}i}(t), y_i(t)) \notin \tilde{\mathcal{Z}}_i, \quad t \geq 0, \tag{13.37}$$

where $s_{\mathrm{c}i} : \mathbb{R}^{l_i} \times \mathbb{R}^{m_i} \to \mathbb{R}$ is such that $s_{\mathrm{c}i}(0, 0) = 0$, $i = 1, \ldots, q$.

We associate with the plant a positive-definite, continuously differentiable function $v_{\mathrm{p}}(x) \triangleq p^{\mathrm{T}}V_{\mathrm{s}}(x)$, which we will refer to as the *plant energy* composed of the *subsystem energies* $v_{\mathrm{s}i}(x_i)$, $i = 1, \ldots, q$. Furthermore, we associate with the controller a nonnegative-definite, infinitely differentiable function $v_{\mathrm{c}}(x_{\mathrm{c}}, y) \triangleq p^{\mathrm{T}}V_{\mathrm{c}}(x_{\mathrm{c}}, y)$, where

$$V_{\mathrm{c}}(x_{\mathrm{c}}, y) \triangleq [v_{\mathrm{c}1}(x_{\mathrm{c}1}, y_1), \ldots, v_{\mathrm{c}q}(x_{\mathrm{c}q}, y_q)]^{\mathrm{T}},$$

called the controller *emulated energy* composed of the *subcontroller emulated energies* $v_{\mathrm{c}i}(x_{\mathrm{c}i}, y_i)$, $i = 1, \ldots, q$. Finally, we associate with the closed-loop system the function

$$v(\tilde{x}) \triangleq v_{\mathrm{p}}(x) + v_{\mathrm{c}}(x_{\mathrm{c}}, H(x)), \tag{13.38}$$

called the *total energy* composed of the *total subsystem energies* $v_{\mathrm{s}i}(x_i) + v_{\mathrm{c}i}(x_{\mathrm{c}i}, y_i)$, $i = 1, \ldots, q$.

Next, we construct the resetting set for each subsystem $\tilde{\mathcal{G}}_i$, $i = 1 \ldots, q$, of the closed-loop system $\tilde{\mathcal{G}}$ in the following form

$$
\tilde{\mathcal{Z}}_i = \Big\{ (x_i, x_{ci}) \in \mathcal{D} \times \mathcal{D}_{ci} : L_{\tilde{f}_c} v_{ci}(x_{ci}, h_i(x_i)) = 0 \text{ and } v_{ci}(x_{ci}, h_i(x_i)) > 0 \Big\}
$$
$$
= \{ (x_i, x_{ci}) \in \mathcal{D} \times \mathcal{D}_{ci} : s_{ci}(h_i(x_i), h_{ci}(x_{ci}, h_i(x_i))) = 0
$$
$$
\text{and } v_{ci}(x_{ci}, h_i(x_i)) > 0\}, \tag{13.39}
$$

where $i = 1, \ldots, q$. The resetting sets $\tilde{\mathcal{Z}}_i$, $i = 1, \ldots, q$, are thus defined to be the sets of all points in the closed-loop state space that correspond to decreasing subcontroller emulated energy. By resetting the subcontroller states, the subsystem energy can never increase after the first resetting event. Furthermore, if the closed-loop subsystem total energy is conserved between resetting events, then a decrease in subsystem energy is accompanied by a corresponding increase in subsystem emulated energy. Hence, this approach allows the subsystem energy to flow to the subcontroller, where it increases the subcontroller emulated energy but does not allow the subcontroller emulated energy to flow back to the subsystem after the first resetting event. This energy dissipating hybrid decentralized controller effectively enforces a one-way energy transfer between each subsystem and corresponding subcontroller after the first resetting event. For practical implementation, knowledge of x_{ci} and y_i is sufficient to determine whether or not the closed-loop state vector is in the set $\tilde{\mathcal{Z}}_i$, $i = 1, \ldots, q$.

The next theorem gives sufficient conditions for asymptotic stability of the closed-loop system $\tilde{\mathcal{G}}$ using state-dependent hybrid decentralized controllers.

Theorem 13.3. Consider the closed-loop impulsive dynamical system $\tilde{\mathcal{G}}$ given by (13.34) and (13.35). Assume that $\tilde{\mathcal{D}}_{ci} \subset \tilde{\mathcal{D}}$ is a compact positively invariant set with respect to $\tilde{\mathcal{G}}$ such that $0 \in \overset{\circ}{\tilde{\mathcal{D}}}_{ci}$, assume that $\mathcal{G}$ is vector lossless with respect to the vector supply rate $S(u, y) \triangleq [s_1(u_1, y_1), \ldots, s_q(u_q, y_q)]^{\mathrm{T}}$ and with a positive, continuously differentiable vector storage function $V_{\mathrm{s}}(x) = [v_{\mathrm{s}1}(x_1), \ldots, v_{\mathrm{s}q}(x_q)]^{\mathrm{T}}$, $x \in \mathcal{D}$. In addition, assume there exist smooth functions $v_{ci} : \mathcal{D}_{ci} \times \mathbb{R}^{l_i} \to \overline{\mathbb{R}}_+$ such that $v_{ci}(x_{ci}, y_i) \geq 0$, $x_{ci} \in \mathcal{D}_{ci}$, $y_i \in \mathbb{R}^{l_i}$, $v_{ci}(x_{ci}, y_i) = 0$ if and only if $x_{ci} = \eta_i(y_i)$, and (13.37) holds. Finally, assume that every $\tilde{x}_0 \in \overline{\tilde{\mathcal{Z}}}$ is k_i-transversal to (13.30) and

$$
s_i(u_i, y_i) + s_{ci}(u_{ci}, y_{ci}) = 0, \quad \tilde{x}_i \notin \tilde{\mathcal{Z}}_i, \quad i = 1, \ldots, q, \tag{13.40}
$$

where $y_i = u_{ci} = h_i(x_i)$, $u_i = -y_{ci} = -h_{ci}(x_{ci}, h_i(x_i))$, and $\tilde{\mathcal{Z}}_i$, $i = 1, \ldots, q$, is given by (13.39). Then the zero solution $\tilde{x}(t) \equiv 0$ to the closed-loop system $\tilde{\mathcal{G}}$ is asymptotically stable. In addition, the total energy function $v(\tilde{x})$ of $\tilde{\mathcal{G}}$ given by (13.38) is strictly decreasing across resetting events. Finally, if $\mathcal{D} = \mathbb{R}^n$, $\mathcal{D}_{\mathrm{c}} = \mathbb{R}^{n_{\mathrm{c}}}$, and $v(\cdot)$ is radially unbounded, then the zero solution $\tilde{x}(t) \equiv 0$ to $\tilde{\mathcal{G}}$ is globally asymptotically stable.

Proof. First, note that since $v_{ci}(x_{ci}, y_i) \geq 0$, $x_{ci} \in \mathcal{D}_{ci}$, $y_i \in \mathbb{R}^{l_i}$, $i = 1, \ldots, q$, it follows that

$$\overline{\tilde{\mathcal{Z}}}_i = \left\{ (x_i, x_{ci}) \in \mathcal{D} \times \mathcal{D}_{ci} : L_{\tilde{f}_c} v_{ci}(x_{ci}, h_i(x_i)) = 0 \text{ and } v_{ci}(x_{ci}, h_i(x_i)) \geq 0 \right\}$$
$$= \left\{ (x_i, x_{ci}) \in \mathcal{D} \times \mathcal{D}_{ci} : \mathcal{X}_i(\tilde{x}_i) = 0 \right\}, \tag{13.41}$$

where $\mathcal{X}_i(\tilde{x}_i) = L_{\tilde{f}_c} v_{ci}(x_{ci}, h_i(x_i))$, $i = 1, \ldots, q$. Next, we show that if the k_i-transversality condition (13.19) holds, then Assumptions 13.1–13.3 hold and, for every $\tilde{x}_0 \in \tilde{\mathcal{D}}_{ci}$, there exists $\tau \geq 0$ such that $\tilde{x}(\tau) \in \tilde{\mathcal{Z}}$. Note that if $\tilde{x}_0 \in \overline{\tilde{\mathcal{Z}}} \setminus \tilde{\mathcal{Z}}$, that is, $v_{ci}(x_{ci}(0), h_i(x_i(0))) = 0$ and $L_{\tilde{f}_c} v_{ci}(x_{ci}(0), h_i(x_i(0))) = 0$, $i \in \{1, \ldots, q\}$, it follows from the k_i-transversality condition that there exists $\delta_i > 0$ such that for all $t \in (0, \delta_i]$, $L_{\tilde{f}_c} v_{ci}(x_{ci}(t), h_i(x_i(t))) \neq 0$. Hence, since

$$v_{ci}(x_{ci}(t), h_i(x_i(t))) = v_{ci}(x_{ci}(0), h_i(x_i(0))) + t L_{\tilde{f}_c} v_{ci}(x_{ci}(\tau), h_i(x_i(\tau)))$$

for some $\tau \in (0, t]$ and $v_{ci}(x_{ci}, y_i) \geq 0$, $x_{ci} \in \mathcal{D}_{ci}$, $y_i \in \mathbb{R}^{l_i}$, $i \in \{1, \ldots, q\}$, it follows that $v_{ci}(x_{ci}(t), h_i(x_i(t))) > 0$, $t \in (0, \delta]$, which implies that Assumption 13.1 is satisfied. Furthermore, if $\tilde{x} \in \tilde{\mathcal{Z}}$ then, since $v_{ci}(x_{ci}, y_i) = 0$ if and only if $x_{ci} = \eta(y_i)$, it follows from (13.37) that $\tilde{x}_i + \tilde{f}_{di}(\tilde{x}_i) \in \overline{\tilde{\mathcal{Z}}}_i \setminus \tilde{\mathcal{Z}}_i$, $i \in \{1, \ldots, q\}$. Hence, Assumption 13.2 holds.

Next, consider the set

$$\mathcal{M}_\gamma \triangleq \cup_{i=1}^q \left\{ \tilde{x} \in \tilde{\mathcal{D}}_{ci} : v_{ci}(x_{ci}, h_i(x_i)) = \gamma_i \right\}, \tag{13.42}$$

where $\gamma_i \geq 0$, $i = 1, \ldots, q$, and $\gamma \triangleq [\gamma_1, \ldots, \gamma_q]^{\mathrm{T}}$. It follows from the k_i-transversality condition that for every $\gamma_i \geq 0$, $\mathcal{M}_\gamma$ does not contain any nontrivial trajectory of $\tilde{\mathcal{G}}$, $i = 1, \ldots, q$. To see this, suppose, *ad absurdum*, that there exists a nontrivial trajectory $\tilde{x}(t) \in \mathcal{M}_\gamma$, $t \geq 0$, for some $\gamma_i \geq 0$ and for some $i \in \{1, \ldots, q\}$. In this case, it follows that

$$\frac{\mathrm{d}^k}{\mathrm{d}t^k} v_{ci}(x_{ci}(t), h_i(x_i(t))) = L_{\tilde{f}_c}^k v_{ci}(x_{ci}(t), h_i(x_i(t))) \equiv 0, \quad k = 1, 2, \ldots,$$
$$i \in \{1, \ldots, q\}, \tag{13.43}$$

which contradicts the k_i-transversality condition.

Next, we show that for every $\tilde{x}_0 \notin \tilde{\mathcal{Z}}$, $\tilde{x}_0 \neq 0$, there exists $\tau > 0$ such that $\tilde{x}(\tau) \in \tilde{\mathcal{Z}}$. To see this, suppose, *ad absurdum*, that $\tilde{x}_i(t) \notin \tilde{\mathcal{Z}}_i$ for all $i = 1, \ldots, q$, $t \geq 0$, which implies that

$$\frac{\mathrm{d}}{\mathrm{d}t} v_{ci}(x_{ci}(t), h_i(x_i(t))) \neq 0, \quad t \geq 0, \quad i = 1, \ldots, q, \tag{13.44}$$

or

$$v_{ci}(x_{ci}(t), h_i(x_i(t))) = 0, \quad t \geq 0, \quad i = 1, \ldots, q. \tag{13.45}$$

If (13.44) holds, then it follows that $v_{\mathrm{c}i}(x_{\mathrm{c}i}(t), h_i(x_i(t)))$ is a (decreasing or increasing) monotonic function of time. Hence, $v_{\mathrm{c}i}(x_{\mathrm{c}i}(t), h_i(x_i(t))) \to \gamma_i$ as $t \to \infty$, where $\gamma_i \geq 0$ is a constant for $i = 1, \ldots, q$, which implies that the positive limit set of the closed-loop system is contained in $\mathcal{M}_\gamma$ for some $\gamma_i \geq 0$, $i = 1, \ldots, q$, and hence, is a contradiction.

Similarly, if (13.45) holds, then $\mathcal{M}_0$ contains a nontrivial trajectory of $\tilde{\mathcal{G}}$ also leading to a contradiction. Hence, for every $\tilde{x}_0 \notin \tilde{\mathcal{Z}}$, there exists $\tau > 0$ such that $\tilde{x}(\tau) \in \tilde{\mathcal{Z}}$. Thus, it follows that for every $\tilde{x}_0 \notin \tilde{\mathcal{Z}}$, $0 < \tau_1(\tilde{x}_0) < \infty$. Now, it follows from Proposition 13.2 that $\tau_1(\cdot)$ is continuous at $\tilde{x}_0 \notin \overline{\tilde{\mathcal{Z}}}$. Furthermore, for all $\tilde{x}_0 \in \overline{\tilde{\mathcal{Z}}} \backslash \tilde{\mathcal{Z}}$ and for every sequence $\{\tilde{x}_{(i)}\}_{i=1}^\infty \in \overline{\tilde{\mathcal{Z}}} \backslash \tilde{\mathcal{Z}}$ converging to $\tilde{x}_0 \in \overline{\tilde{\mathcal{Z}}} \backslash \tilde{\mathcal{Z}}$, it follows from the k_i-transversality condition and Proposition 13.2 that $\lim_{i \to \infty} \tau_1(\tilde{x}_{(i)}) = \tau_1(\tilde{x}_0)$.

Next, let $\tilde{x}_0 \in \overline{\tilde{\mathcal{Z}}} \backslash \tilde{\mathcal{Z}}$ and let $\{\tilde{x}_{(i)}\}_{i=1}^\infty \in \tilde{\mathcal{D}}_{\mathrm{c}i}$ be such that $\lim_{i \to \infty} \tilde{x}_{(i)} = \tilde{x}_0$ and $\lim_{i \to \infty} \tau_1(\tilde{x}_{(i)})$ exists. In this case, it follows from Proposition 13.2 that either $\lim_{i \to \infty} \tau_1(\tilde{x}_{(i)}) = 0$ or $\lim_{i \to \infty} \tau_1(\tilde{x}_{(i)}) = \tau_1(\tilde{x}_0)$. Furthermore, since $\tilde{x}_0 \in \overline{\tilde{\mathcal{Z}}} \backslash \tilde{\mathcal{Z}}$ corresponds to the case where $v_{\mathrm{c}i}(x_{\mathrm{c}i0}, h_i(x_{i0})) = 0$, $i \in \{1, \ldots, q\}$, it follows that $x_{\mathrm{c}i0} = \eta_i(h_i(x_{i0}))$, and hence, $\tilde{f}_{\mathrm{d}i}(\tilde{x}_{i0}) = 0$, $i \in \{1, \ldots, q\}$. Now, it follows from Proposition 13.1 that Assumption 13.3 holds.

To show that the zero solution $\tilde{x}(t) \equiv 0$ to $\tilde{\mathcal{G}}$ is asymptotically stable, consider the Lyapunov function candidate corresponding to the total energy function $v(\tilde{x})$ given by (13.38). Since $\mathcal{G}$ is vector lossless with respect to the vector supply rate $S(u, y)$, and hence, lossless with respect to the supply rate $p^{\mathrm{T}} S(u, y)$, where $p \in \mathbb{R}_+^q$, and (13.37) and (13.40) hold, it follows that

$$\dot{v}(\tilde{x}(t)) = \sum_{i=1}^q p_i [s_i(u_i(t), y_i(t)) + s_{\mathrm{c}i}(u_{\mathrm{c}i}(t), y_{\mathrm{c}i}(t))] = 0, \quad \tilde{x}(t) \notin \tilde{\mathcal{Z}}, \quad (13.46)$$

where p_i, $i = 1, \ldots, q$, denotes the ith component of $p \in \mathbb{R}_+^q$. Furthermore, it follows from (13.33) and (13.41) that

$$\Delta v(\tilde{x}(t_k)) = v_{\mathrm{c}}(x_{\mathrm{c}}(t_k^+), H(x(t_k^+))) - v_{\mathrm{c}}(x_{\mathrm{c}}(t_k), H(x(t_k)))$$

$$= -\sum_{i=1}^q p_i v_{\mathrm{c}i}(x_{\mathrm{c}i}(t_k), h_i(x_i(t_k))) \chi_{\tilde{\mathcal{Z}}_i}(\tilde{x}_i(t_k))$$

$$< 0, \quad \tilde{x}(t_k) \in \tilde{\mathcal{Z}}, \quad k \in \overline{\mathbb{Z}}_+. \quad (13.47)$$

Thus, it follows from Theorem 13.2 that the zero solution $\tilde{x}(t) \equiv 0$ to $\tilde{\mathcal{G}}$ is asymptotically stable. Finally, if $\mathcal{D} = \mathbb{R}^n$, $\mathcal{D}_{\mathrm{c}} = \mathbb{R}^{n_{\mathrm{c}}}$, and $v(\cdot)$ is radially unbounded, then global asymptotic stability is immediate. $\square$

If $v_{\mathrm{c}i} = v_{\mathrm{c}i}(x_{\mathrm{c}i}, y_i)$ is only a function of $x_{\mathrm{c}i}$ and $v_{\mathrm{c}i}(x_{\mathrm{c}i})$ is a positive-definite function, $i \in \{1, \ldots, q\}$, then we can choose $\eta_i(y_i) \equiv 0$. In this case, $v_{ci}(x_{ci}) = 0$ if and only if $x_{ci} = 0$. In the proof of Theorem 13.3, we assume that $\tilde{x}_0 \notin \tilde{\mathcal{Z}}$ for $\tilde{x}_0 \neq 0$. This proviso is necessary since it may be possible to reset the states of the closed-loop system to the origin, in which case $\tilde{x}(s) = 0$ for a finite value of s. In this case, for $t > s$, we have $v(\tilde{x}(t)) = v(\tilde{x}(s)) = v(0) = 0$. This situation does not present a problem, however, since reaching the origin in finite time is a stronger condition than reaching the origin as $t \to \infty$.

Theorem 13.3 can be generalized to the case where $\mathcal{G}$ is *vector dissipative* with respect to the vector supply rate $S(u, y)$ with the component decoupled vector storage function $V_{\rm s}(x) = [v_{\rm s1}(x_1), \ldots, v_{\rm sq}(x_q)]^{\rm T}$, $x \in \mathcal{D}$. Specifically, in this case (13.46) becomes

$$\dot{v}(\tilde{x}(t)) = \sum_{i=1}^{q} p_i d_i(x_i(t)) \leq 0, \quad \tilde{x}(t) \in \tilde{\mathcal{Z}}, \tag{13.48}$$

where $d_i : \mathcal{D}_i \to \mathbb{R}$, $i = 1, \ldots, q$, is a continuous, nonnegative-definite dissipation rate function. Now, Theorem 13.3 holds with the additional assumption that the only invariant set contained in $\mathcal{R} \triangleq \cap_{i=1}^{q} \{\tilde{x} \in \tilde{\mathcal{D}}_{\rm ci} : d_i(x_i) = 0\}$ is $\mathcal{M} = \{0\}$.

13.4 Interconnected Euler-Lagrange Dynamical Systems

In this section, we specialize the control framework developed in Section 13.3 to interconnected Euler-Lagrange dynamical systems. For this, we consider the governing equations of motion of an n-degree-of-freedom dynamical system given by the *Euler-Lagrange* equation

$$\frac{\mathrm{d}}{\mathrm{d}t}\left[\frac{\partial \mathcal{L}}{\partial \dot{q}}(q(t), \dot{q}(t))\right]^{\rm T} - \left[\frac{\partial \mathcal{L}}{\partial q}(q(t), \dot{q}(t))\right]^{\rm T} = u(t), \quad q(0) = q_0, \quad \dot{q}(0) = \dot{q}_0,$$
$$\tag{13.49}$$

where $t \geq 0$, $q \in \mathbb{R}^n$ represents the generalized system positions, $\dot{q} \in \mathbb{R}^n$ represents the generalized system velocities, $\mathcal{L} : \mathbb{R}^n \times \mathbb{R}^n \to \mathbb{R}$ denotes the system Lagrangian given by $\mathcal{L}(q, \dot{q}) = T(q, \dot{q}) - U(q)$, where $T : \mathbb{R}^n \times \mathbb{R}^n \to \mathbb{R}$ is the system total kinetic energy and $U : \mathbb{R}^n \to \mathbb{R}$ is the system total potential energy, and $u \in \mathbb{R}^n$ is the vector of generalized control forces acting on the system. We assume that (13.49) represents an interconnected Euler-Lagrange dynamical system composed of s subsystems given by

$$\frac{\mathrm{d}}{\mathrm{d}t}\left[\frac{\partial \mathcal{L}}{\partial \dot{q}_i}(q(t), \dot{q}(t))\right]^{\rm T} - \left[\frac{\partial \mathcal{L}}{\partial q_i}(q(t), \dot{q}(t))\right]^{\rm T} = u_i(t), \quad i = 1, \ldots, s, \tag{13.50}$$

where $q_i \in \mathbb{R}^{n_i}$, $u_i \in \mathbb{R}^{n_i}$, $i = 1, \ldots, s$, $\sum_{i=1}^{s} n_i = n$, $q = [q_1^{\rm T}, \ldots, q_s^{\rm T}]^{\rm T}$, $u = [u_1^{\rm T}, \ldots u_s^{\rm T}]^{\rm T}$, q_i and $\dot{q}_i$ represent, respectively, generalized subsystem positions and velocities, and u_i denotes the vector of decentralized control input for the ith subsystem.

Furthermore, let $\mathcal{H} : \mathbb{R}^n \times \mathbb{R}^n \to \mathbb{R}$ denote the *Legendre transformation* of the Lagrangian function $\mathcal{L}(q, \dot{q})$ with respect to the generalized velocity $\dot{q}$ defined by $\mathcal{H}(q, p) \triangleq \dot{q}^{\mathrm{T}} p - \mathcal{L}(q, \dot{q})$, where p denotes the vector of generalized momenta given by

$$p(q, \dot{q}) = \left[\frac{\partial \mathcal{L}}{\partial \dot{q}}(q, \dot{q}) \right]^{\mathrm{T}}, \tag{13.51}$$

and where the map from the generalized velocities $\dot{q}$ to the generalized momenta p is assumed to be *bijective* (i.e., one-to-one and onto). Note that $p = [p_1^{\mathrm{T}}, \ldots, p_s^{\mathrm{T}}]^{\mathrm{T}}$, where

$$p_i(q, \dot{q}) \triangleq \left[\frac{\partial \mathcal{L}}{\partial \dot{q}_i}(q, \dot{q}) \right]^{\mathrm{T}}, \quad i = 1, \ldots, s, \tag{13.52}$$

denotes the vector of the ith subsystem generalized momenta. We assume that the system total kinetic energy is such that $T(q, \dot{q}) = \frac{1}{2} \dot{q}^{\mathrm{T}} [\frac{\partial T}{\partial \dot{q}}(q, \dot{q})]^{\mathrm{T}}$, $T(q, 0) = 0$, and $T(q, \dot{q}) > 0$, $\dot{q} \neq 0$, $\dot{q} \in \mathbb{R}^n$. We also assume that the system total potential energy $U(\cdot)$ is such that $U(0) = 0$ and $U(q) > 0$, $q \neq 0$, $q \in \mathcal{D}_{\mathrm{q}} \subseteq \mathbb{R}^n$, which implies that $\mathcal{H}(q, p) = T(q, \dot{q}) + U(q) > 0$, $(q, \dot{q}) \neq 0$, $(q, \dot{q}) \in \mathcal{D}_{\mathrm{q}} \times \mathbb{R}^n$.

Next, we present a decentralized hybrid feedback control framework for Euler-Lagrange dynamical systems. Specifically, consider the ith subsystem (13.50) with output

$$y_i = \begin{bmatrix} h_{1i}(q_i) \\ h_{2i}(\dot{q}_i) \end{bmatrix} = \begin{bmatrix} h_{1i}(q_i) \\ h_{2i}\left(\frac{\partial \mathcal{H}}{\partial p_i}(q, p) \right) \end{bmatrix}, \tag{13.53}$$

where $i = 1, \ldots, s$, $y_i \in \mathbb{R}^{l_i}$, $h_{1i} : \mathbb{R}^{n_i} \to \mathbb{R}^{l_{1i}}$ and $h_{2i} : \mathbb{R}^{n_i} \to \mathbb{R}^{l_i - l_{1i}}$ are continuously differentiable, $h_{1i}(0) = 0$, $h_{2i}(0) = 0$, and $h_{1i}(q_i) \neq 0$. Next, consider the decentralized energy-based hybrid controller for the ith subsystem

$$\frac{\mathrm{d}}{\mathrm{d}t} \left[\frac{\partial \mathcal{L}_{ci}}{\partial \dot{q}_{ci}}(q_{ci}(t), \dot{q}_{ci}(t), y_{q_i}(t)) \right]^{\mathrm{T}} - \left[\frac{\partial \mathcal{L}_{ci}}{\partial q_{ci}}(q_{ci}(t), \dot{q}_{ci}(t), y_{q_i}(t)) \right]^{\mathrm{T}} = 0,$$
$$q_{ci}(0) = q_{ci0}, \quad \dot{q}_{ci}(0) = \dot{q}_{ci0}, \quad (q_{ci}(t), \dot{q}_{ci}(t), y_i(t)) \notin \mathcal{Z}_{ci}, \tag{13.54}$$

$$\begin{bmatrix} \Delta q_{ci}(t) \\ \Delta \dot{q}_{ci}(t) \end{bmatrix} = \begin{bmatrix} \eta_i(y_{q_i}(t)) - q_{ci}(t) \\ -\dot{q}_{ci}(t) \end{bmatrix}, \quad (q_{ci}(t), \dot{q}_{ci}(t), y_i(t)) \in \mathcal{Z}_{ci},$$

$$\tag{13.55}$$

$$u_i(t) = \left[\frac{\partial \mathcal{L}_{ci}}{\partial q_i}(q_{ci}(t), \dot{q}_{ci}(t), y_{q_i}(t)) \right]^{\mathrm{T}}, \tag{13.56}$$

where $t \geq 0$, $i = 1, \ldots, s$, $q_{ci} \in \mathbb{R}^{n_{ci}}$ represents virtual subcontroller positions, $\dot{q}_{ci} \in \mathbb{R}^{n_{ci}}$ represents virtual subcontroller velocities, $n_{\mathrm{c}} \triangleq \sum_{i=1}^{s} n_{ci}$,

$y_{q_i} \triangleq h_{1i}(q_i)$, $\mathcal{L}_{ci} : \mathbb{R}^{n_{ci}} \times \mathbb{R}^{n_{ci}} \times \mathbb{R}^{l_{1i}} \to \mathbb{R}$ denotes the subcontroller Lagrangian given by $\mathcal{L}_{ci}(q_{ci}, \dot{q}_{ci}, y_{q_i}) \triangleq T_{ci}(q_{ci}, \dot{q}_{ci}) - U_{ci}(q_{ci}, y_{q_i})$, where $T_{ci} : \mathbb{R}^{n_{ci}} \times \mathbb{R}^{n_{ci}} \to \mathbb{R}$ is the subcontroller kinetic energy and $U_{ci} : \mathbb{R}^{n_{ci}} \times \mathbb{R}^{l_{1i}} \to \mathbb{R}$ is the subcontroller potential energy, $\eta_i(\cdot)$ is a continuously differentiable function such that $\eta_i(0) = 0$, $\mathcal{Z}_{ci} \subset \mathbb{R}^{n_{ci}} \times \mathbb{R}^{n_{ci}} \times \mathbb{R}^{l_i}$ is the ith subcontroller resetting set, $\Delta q_{ci}(t) \triangleq q_{ci}(t^+) - q_{ci}(t)$, $\Delta \dot{q}_{ci}(t) \triangleq \dot{q}_{ci}(t^+) - \dot{q}_{ci}(t)$, and t_k, $k \in \overline{\mathbb{Z}}_+$, denotes a resetting instant. We assume that the subcontroller kinetic energy $T_{ci}(q_{ci}, \dot{q}_{ci})$ is such that $T_{ci}(q_{ci}, \dot{q}_{ci}) = \frac{1}{2}\dot{q}_{ci}^{\mathrm{T}}[\frac{\partial T_{ci}}{\partial \dot{q}_{ci}}(q_{ci}, \dot{q}_{ci})]^{\mathrm{T}}$, with $T_{ci}(q_{ci}, 0) = 0$ and $T_{ci}(q_{ci}, \dot{q}_{ci}) > 0$, $\dot{q}_{ci} \neq 0$, $\dot{q}_{ci} \in \mathbb{R}^{n_{ci}}$. Furthermore, we assume that $U_{ci}(\eta_i(y_{q_i}), y_{q_i}) = 0$ and $U_{ci}(q_{ci}, y_{q_i}) > 0$ for $q_{ci} \neq \eta_i(y_{q_i})$, $q_{ci} \in \mathbb{R}^{n_{ci}}$.

As in Section 13.3, note that

$$V_{\mathrm{p}}(q, \dot{q}) \triangleq T(q, \dot{q}) + U(q) \tag{13.57}$$

is the total energy of the interconnected system (13.49) and

$$\begin{aligned} V_{\mathrm{c}}(q_{\mathrm{c}}, \dot{q}_{\mathrm{c}}, y_q) &\triangleq \sum_{i=1}^{s} T_{ci}(q_{ci}, \dot{q}_{ci}) + U_{ci}(q_{ci}, y_{q_i}) \\ &= \sum_{i=1}^{s} V_{ci}(q_{ci}, \dot{q}_{ci}, y_{q_i}) \end{aligned} \tag{13.58}$$

is the sum of subcontroller emulated energies, where $q_{\mathrm{c}} \triangleq [q_{c1}^{\mathrm{T}}, \ldots, q_{cs}^{\mathrm{T}}]^{\mathrm{T}}$, $\dot{q}_{\mathrm{c}} \triangleq [\dot{q}_{c1}^{\mathrm{T}}, \ldots, \dot{q}_{cs}^{\mathrm{T}}]^{\mathrm{T}}$, and $y_q \triangleq [y_{q_1}^{\mathrm{T}}, \ldots, y_{q_s}^{\mathrm{T}}]^{\mathrm{T}}$. Finally,

$$V(q, \dot{q}, q_{\mathrm{c}}, \dot{q}_{\mathrm{c}}) \triangleq V_{\mathrm{p}}(q, \dot{q}) + V_{\mathrm{c}}(q_{\mathrm{c}}, \dot{q}_{\mathrm{c}}, y_q) \tag{13.59}$$

is the total energy of the interconnected closed-loop system (13.50)–(13.56).

Next, we study the behavior of the total energy function $V(q, \dot{q}, q_{\mathrm{c}}, \dot{q}_{\mathrm{c}})$ along the trajectories of the closed-loop system dynamics. For the interconnected closed-loop system, we define our resetting set as

$$\mathcal{Z} \triangleq \cup_{i=1}^{s}\{(q, \dot{q}, q_{\mathrm{c}}, \dot{q}_{\mathrm{c}}) \in \mathcal{D}_{\mathrm{q}} \times \mathbb{R}^n \times \mathbb{R}^{n_c} \times \mathbb{R}^{n_c} : (q_{ci}, \dot{q}_{ci}, y_i) \in \mathcal{Z}_{ci}\}. \tag{13.60}$$

Note that $\frac{\mathrm{d}}{\mathrm{d}t}V_{\mathrm{p}}(q, \dot{q}) = \frac{\mathrm{d}}{\mathrm{d}t}\mathcal{H}(q, p) = u^{\mathrm{T}}\dot{q}$, $(q, \dot{q}, q_{\mathrm{c}}, \dot{q}_{\mathrm{c}}) \notin \mathcal{Z}$. Furthermore, we define the ith subcontroller Hamiltonian by

$$\mathcal{H}_{ci}(q_{ci}, \dot{q}_{ci}, p_{ci}, y_{q_i}) \triangleq \dot{q}_{ci}^{\mathrm{T}} p_{ci} - \mathcal{L}_{ci}(q_{ci}, \dot{q}_{ci}, y_{q_i}), \quad i = 1, \ldots, s, \tag{13.61}$$

where the subcontroller momentum p_{ci} is given by

$$p_{ci}(q_{ci}, \dot{q}_{ci}, y_{q_i}) = \left[\frac{\partial \mathcal{L}_{ci}}{\partial \dot{q}_{ci}}(q_{ci}, \dot{q}_{ci}, y_{q_i})\right]^{\mathrm{T}}, \tag{13.62}$$

and it follows from the structure of $T_{ci}(q_{ci}, \dot{q}_{ci})$ that $\mathcal{H}_{ci}(q_{ci}, \dot{q}_{ci}, p_{ci}, y_{q_i}) = V_{ci}(q_{ci}, \dot{q}_{ci}, y_{q_i}) = T_{ci}(q_{ci}, \dot{q}_{ci}) + U_{ci}(q_{ci}, y_{q_i})$.

Now, it follows from (13.54), (13.56), and (13.61) that, for $t \in (t_k, t_{k+1}]$,

$$\frac{\mathrm{d}}{\mathrm{d}t} V_{ci}(q_{ci}(t), \dot{q}_{ci}(t), y_{q_i}(t)) = \ddot{q}_{ci}^{\mathrm{T}}(t) p_{ci}(t) + \dot{q}_{ci}^{\mathrm{T}}(t) \dot{p}_{ci}(t)$$

$$- \frac{\mathrm{d}}{\mathrm{d}t} \mathcal{L}_{ci}(q_{ci}(t), \dot{q}_{ci}(t), y_{q_i}(t))$$

$$= \dot{q}_{ci}^{\mathrm{T}}(t) \frac{\mathrm{d}}{\mathrm{d}t} \left[\frac{\partial \mathcal{L}_{ci}}{\partial \dot{q}_{ci}} (q_{ci}(t), \dot{q}_{ci}(t), y_{q_i}(t)) \right]^{\mathrm{T}}$$

$$- \frac{\partial \mathcal{L}_{ci}}{\partial q_{ci}} (q_{ci}(t), \dot{q}_{ci}(t), y_{q_i}(t)) \dot{q}_{ci}(t)$$

$$- \frac{\partial \mathcal{L}_{ci}}{\partial q_i} (q_{ci}(t), \dot{q}_{ci}(t), y_{q_i}(t)) \dot{q}_i(t)$$

$$= -u_i^{\mathrm{T}}(t) \dot{q}_i(t), \quad (q(t), \dot{q}(t), q_c(t), \dot{q}_c(t)) \notin \mathcal{Z}$$

$$(13.63)$$

Hence,

$$\frac{\mathrm{d}}{\mathrm{d}t} V(q(t), \dot{q}(t), q_c(t), \dot{q}_c(t)) = u(t)^{\mathrm{T}} \dot{q}(t) - \sum_{i=1}^{s} u_i^{\mathrm{T}}(t) q_i(t)$$

$$= 0, \quad (q(t), \dot{q}(t), q_c(t), \dot{q}_c(t)) \notin \mathcal{Z},$$

$$t_k < t \le t_{k+1}, \qquad (13.64)$$

which implies that the total energy of the interconnected closed-loop system between resetting events is conserved.

The total energy difference across resetting events is given by

$$\Delta V(q(t_k), \dot{q}(t_k), q_c(t_k), \dot{q}_c(t_k)) = V_c(q_c(t_k^+), \dot{q}_c(t_k^+), y_q(t_k^+))$$

$$- V_c(q_c(t_k), \dot{q}_c(t_k), y_q(t_k))$$

$$= \sum_{i=1}^{s} [V_{ci}(q_{ci}(t_k^+), \dot{q}_{ci}(t_k^+), y_{q_i}(t_k^+))$$

$$- V_{ci}(q_{ci}(t_k), \dot{q}_{ci}(t_k), y_{q_i}(t_k))]$$

$$= - \sum_{i=1}^{s} V_{ci}(q_{ci}(t_k), \dot{q}_{ci}(t_k), y_{q_i}(t_k))$$

$$\cdot \chi_{\mathcal{Z}_{ci}}(q_{ci}(t_k), \dot{q}_{ci}(t_k), y_i(t_k))$$

$$< 0,$$

$$(q(t_k), \dot{q}(t_k), q_c(t_k), \dot{q}_c(t_k)) \in \mathcal{Z}, \quad k \in \overline{\mathbb{Z}}_+, \quad (13.65)$$

where

$$\chi_{\mathcal{Z}_{ci}}(q_{ci}, \dot{q}_{ci}, y_{q_i}) \triangleq \begin{cases} 1, & (q_{ci}, \dot{q}_{ci}, y_i) \in \mathcal{Z}_{ci}, \\ 0, & (q_{ci}, \dot{q}_{ci}, y_i) \notin \mathcal{Z}_{ci}, \end{cases} \qquad (13.66)$$

which implies that the resetting law (13.55) ensures the total energy decrease across resetting events by an amount equal to the accumulated emulated subcontroller energy.

Here, we consider decentralized energy-dissipating state-dependent resetting controllers that affect a one-way energy transfer between the corresponding subsystem and the subcontroller. Specifically, consider the closed-loop system (13.50)–(13.56), where $\mathscr{Z}_{ci}$, $i = 1, \ldots, s$, are defined by

$$\mathscr{Z}_{ci} \triangleq \left\{ (q_i, \dot{q}_i, q_{ci}, \dot{q}_{ci}) : \frac{\mathrm{d}}{\mathrm{d}t} V_{ci}(q_{ci}, \dot{q}_{ci}, y_{q_i}) = 0 \text{ and } V_{ci}(q_{ci}, \dot{q}_{ci}, y_{q_i}) > 0 \right\}.$$

(13.67)

Once again, for practical implementation, knowledge of q_{c}, $\dot{q}_{\mathrm{c}}$, and y_q is sufficient to determine whether or not the closed-loop state vector is in the set $\mathscr{Z}$ given by (13.60), where $\mathscr{Z}_{ci}$, $i = 1, \ldots, s$, are defined by (13.67).

The next theorem gives sufficient conditions for stabilization of interconnected Euler-Lagrange dynamical systems using decentralized energy-based hybrid controllers. For this result define the closed-loop system states $x \triangleq [q^{\mathrm{T}}, \dot{q}^{\mathrm{T}}, q_{\mathrm{c}}^{\mathrm{T}}, \dot{q}_{\mathrm{c}}^{\mathrm{T}}]^{\mathrm{T}}$.

Theorem 13.4. Consider the interconnected closed-loop dynamical system $\mathcal{G}$ given by (13.50)–(13.56), with the resetting set $\mathscr{Z}$ given by (13.60), where $\mathscr{Z}_{ci}$, $i = 1, \ldots, s$, are defined by (13.67). Assume that $\mathcal{D}_{ci} \subset \mathcal{D}_{\mathrm{q}} \times \mathbb{R}^n \times \mathbb{R}^{n_{\mathrm{c}}} \times \mathbb{R}^{n_{\mathrm{c}}}$ is a compact positively invariant set with respect to $\mathcal{G}$ such that $0 \in \overset{\circ}{\mathcal{D}}_{ci}$. Furthermore, assume that the k_i-transversality condition (13.19) holds with $\mathcal{X}_i(x) = \frac{\mathrm{d}}{\mathrm{d}t} V_{ci}(q_{ci}, \dot{q}_{ci}, y_{q_i})$, $i = 1, \ldots, s$. Then the zero solution $x(t) \equiv 0$ to $\mathcal{G}$ is asymptotically stable. Finally, if $\mathcal{D}_{\mathrm{q}} = \mathbb{R}^n$ and the total energy function $V(x)$ is radially unbounded, then the above asymptotic stability results are global.

Proof. Note that the interconnected Euler-Lagrange dynamical system (13.49) and (13.53) is lossless with respect to the supply rate

$$\mathbf{e}^{\mathrm{T}} S(u, y) = \mathbf{e}^{\mathrm{T}} [u_1^{\mathrm{T}} \dot{q}_1, \ldots, u_s^{\mathrm{T}} \dot{q}_s]^{\mathrm{T}} = u^{\mathrm{T}} \dot{q} = u^{\mathrm{T}} \rho(y), \qquad (13.68)$$

where $\rho(y(q, \dot{q})) = \dot{q}$, $\mathbf{e} \triangleq [1, \ldots, 1]^{\mathrm{T}}$, and $\mathbf{e} \in \mathbb{R}^s$. Furthermore, it follows from (13.63) and (13.64) that (13.40) is satisfied. Now, the result is a direct consequence of Theorem 13.3 with $v_{\mathrm{p}}(x) = V_{\mathrm{p}}(q, \dot{q})$, $v_{\mathrm{c}}(x_{\mathrm{c}}, y) = V_{\mathrm{c}}(q_{\mathrm{c}}, \dot{q}_{\mathrm{c}}, y_q)$, $u = \left[\frac{\partial \mathcal{L}_{c1}}{\partial q_1}, \ldots, \frac{\partial \mathcal{L}_{cs}}{\partial q_s} \right]^{\mathrm{T}}$, and $s_{ci}(u_{ci}, y_{ci}) = -u_i^{\mathrm{T}} \dot{q}_i = y_{ci}^{\mathrm{T}} \hat{\rho}_i(y_i)$, where $\hat{\rho}_i(y_i(q_i, \dot{q}_i)) = \dot{q}_i$, $i = 1, \ldots, s$. $\qquad \square$

13.5 Hybrid Decentralized Control Design

In this section, we apply the decentralized hybrid control framework developed in Section 13.4 to the multi-rotational/translational proof-mass actua-

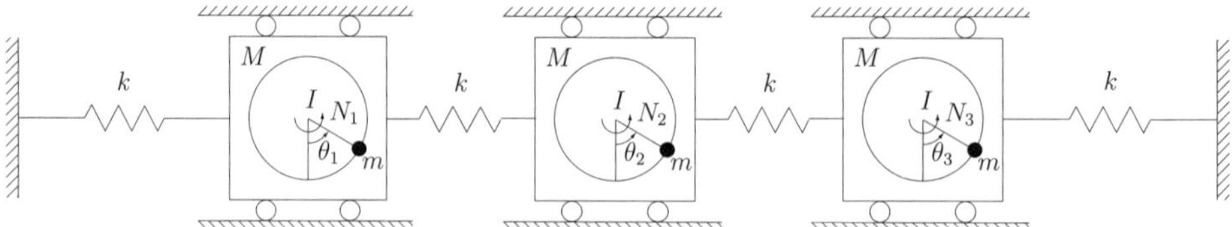

Figure 13.1 Multi-RTAC system.

tor (multi-RTAC) nonlinear system studied in [7,33]. This system is shown in Figure 13.1 and consists of three identical translational oscillating carts connected by linear springs along with three identical eccentric rotational inertias, which act as proof-mass actuators mounted on each cart. Rotational motion of each proof-mass is nonlinearly coupled with the translational motion of the corresponding cart that the proof-mass is mounted on, which provides the mechanism for control action. The oscillator carts, each with a mass M, are connected to each other as well as fixed supports via linear springs of stiffness k. The carts are constrained to one-dimensional motion. The rotational proof-mass actuators consist of a pendulum of mass m with mass moment of inertia I located at a distance e from the axis of rotation. In Figure 13.1, N_1, N_2, and N_3 denote the control torques applied to each proof-mass.

Letting q_i and $\dot{q}_i$, $i = 1, 2, 3$, denote the translational position and velocity, respectively, of each cart, letting θ_i and $\dot{\theta}_i$, $i = 1, 2, 3$, denote the angular position and angular velocity, respectively, of each rotational proof-mass, and using the total physical energy of the multi-RTAC system given by

$$
\begin{aligned}
V_\mathrm{s}(q_i, \dot{q}_i, \theta_i, \dot{\theta}_i) = {}& k(q_1^2 + q_2^2 + q_3^2 - q_1 q_2 - q_2 q_3) \\
& + \frac{1}{2}(M + m)(\dot{q}_1^2 + \dot{q}_2^2 + \dot{q}_3^2) + \frac{1}{2}(I + me^2)(\dot{\theta}_1^2 + \dot{\theta}_2^2 + \dot{\theta}_3^2) \\
& + me(\dot{q}_1\dot{\theta}_1 \cos\theta_1 + \dot{q}_2\dot{\theta}_2 \cos\theta_2 + \dot{q}_3\dot{\theta}_3 \cos\theta_3) \\
& + mge[(1 - \cos\theta_1) + (1 - \cos\theta_2) + (1 - \cos\theta_3)], \quad (13.69)
\end{aligned}
$$

the nonlinear dynamic equations of motion are given by

$$
\begin{aligned}
(M + m)\ddot{q}_1 &= -me(\ddot{\theta}_1 \cos\theta_1 - \dot{\theta}_1^2 \sin\theta_1) - 2kq_1 + kq_2, & (13.70) \\
(I + me^2)\ddot{\theta}_1 &= -me\ddot{q}_1 \cos\theta_1 - mge \sin\theta_1 + N_1, & (13.71) \\
(M + m)\ddot{q}_2 &= -me(\ddot{\theta}_2 \cos\theta_2 - \dot{\theta}_2^2 \sin\theta_2) + kq_1 - 2kq_2 + kq_3, & (13.72) \\
(I + me^2)\ddot{\theta}_2 &= -me\ddot{q}_2 \cos\theta_2 - mge \sin\theta_2 + N_2, & (13.73) \\
(M + m)\ddot{q}_3 &= -me(\ddot{\theta}_3 \cos\theta_3 - \dot{\theta}_3^2 \sin\theta_3) + kq_2 - 2kq_3, & (13.74) \\
(I + me^2)\ddot{\theta}_3 &= -me\ddot{q}_3 \cos\theta_3 - mge \sin\theta_3 + N_3, & (13.75)
\end{aligned}
$$

with the decentralized control inputs $u_i = N_i$, $i = 1, 2, 3$, and outputs $y_i = \theta_i$, $i = 1, 2, 3$.

Table 13.1 Problem data for the RTAC system.

Description	Parameter	Value	Units
Cart mass	M	1.7428	kg
Eccentric mass	m	0.2739	kg
Arm eccentricity	e	0.0537	m
Arm inertia	I	0.000884	kg·m^2
Spring stiffness	k	200	N/m
Controller parameter	m_{c}	0.0017	—
Controller parameter	k_{c}	0.166	—

To design a decentralized state-dependent hybrid controller for (13.70)–(13.75), let $n_{ci} = 1$, $V_{ci}(q_{ci}, \dot{q}_{ci}, \theta_i) = \frac{1}{2}m_{\mathrm{c}}\dot{q}_{ci}^2 + \frac{1}{2}k_{\mathrm{c}}(q_{ci} - \theta_i)^2$, $\mathcal{L}_{ci}(q_{ci}, \dot{q}_{ci}, \theta_i) = \frac{1}{2}m_{\mathrm{c}}\dot{q}_{ci}^2 - \frac{1}{2}k_{\mathrm{c}}(q_{ci} - \theta_i)^2$, $y_{qi} = \theta_i$, and $\eta_i(y_{qi}) = y_{qi}$, where $m_{\mathrm{c}} > 0$ and $k_{\mathrm{c}} > 0$, and $i = 1, 2, 3$. In this case, the decentralized state-dependent hybrid controller has the form

$$m_{\mathrm{c}}\ddot{q}_{ci} + k_{\mathrm{c}}(q_{ci} - \theta_i) = 0, \quad (q_{ci}, \dot{q}_{ci}, \theta_i, \dot{\theta}_i) \notin \mathcal{Z}_i, \quad i = 1, 2, 3, \quad (13.76)$$

$$\begin{bmatrix} \Delta q_{ci} \\ \Delta \dot{q}_{ci} \end{bmatrix} = \begin{bmatrix} \theta_i - q_{ci} \\ -\dot{q}_{ci} \end{bmatrix}, \quad (q_{ci}, \dot{q}_{ci}, \theta_i, \dot{\theta}_i) \in \mathcal{Z}_i, \quad (13.77)$$

$$u_i = k_{\mathrm{c}}(q_{ci} - \theta_i), \quad (13.78)$$

with the resetting set (13.67) taking the form

$$\mathcal{Z}_i = \left\{ (q_{ci}, \dot{q}_{ci}, \theta_i, \dot{\theta}_i) \in \mathbb{R}^4 : k_{\mathrm{c}}\dot{\theta}_i(q_{ci} - \theta_i) = 0 \text{ and } \begin{bmatrix} \theta_i - q_{ci} \\ -\dot{q}_{ci} \end{bmatrix} \neq 0 \right\}.$$

$$(13.79)$$

For the closed-loop system (13.70)–(13.75) and (13.76)–(13.79), the k_i-transversality condition given in Definition 13.1 is sufficiently complex that we have been unable to show it analytically. However, condition (13.19) was verified numerically. Hence, by Theorem 13.4, the closed-loop system (13.70)–(13.75) and (13.76)–(13.79) is globally asymptotically stable.

For the following simulation, we use system parameters shown in Table 13.1 with the system initial conditions $q_1(0) = -0.06$ m, $q_2(0) = 0.04$ m, $q_3(0) = 0.05$ m, $\dot{q}_i(0) = 0$, $\theta_i(0) = 0$, $\dot{\theta}_i(0) = 0$, $q_{ci}(0) = 0$, and $\dot{q}_{ci}(0) = 0$, $i = 1, 2, 3$. Figures 13.2 and 13.3 show, respectively, positions and velocities of the carts for the *uncontrolled* system with the proof-masses fixed. It can be seen that since the open-loop system is lossless, and hence, Lyapunov stable, it exhibits persistent oscillations. Alternatively, for the *controlled* system, Figure 13.4 shows positions of the carts versus time, and Figure 13.5 shows the cart velocities versus time. Figure 13.6 shows the angular positions of the pendulums versus time, and Figure 13.7 shows their angular velocities versus time. Figures 13.8 and 13.9 show the time history of each subcontroller state. Note that each subcontroller state is discontinuous. The

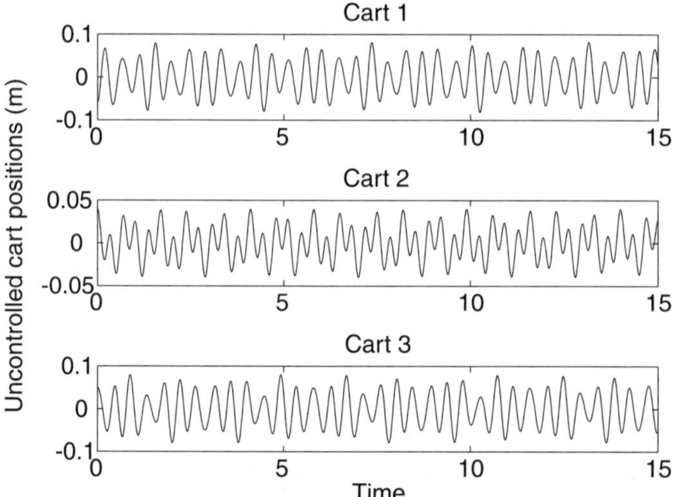

Figure 13.2 Positions of the carts for the uncontrolled system versus time.

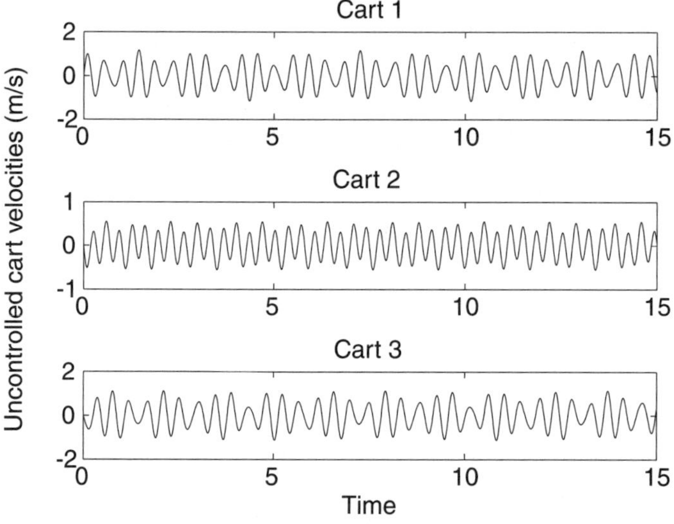

Figure 13.3 Velocities of the carts for the uncontrolled system versus time.

control torques versus time are shown in Figure 13.10 and are discontinuous at the resetting times as follows from (13.77) and (13.78). Figure 13.11 shows the physical energy of the plant, combined emulated energy of all subcontrollers, and the total energy of the multi-rotational/translational proof-mass actuator system. It can be seen that the total energy of the multi-RTAC system remains constant between subcontroller state resettings, which is in agreement with (13.64), whereas the physical energy of the plant is constantly decreasing.

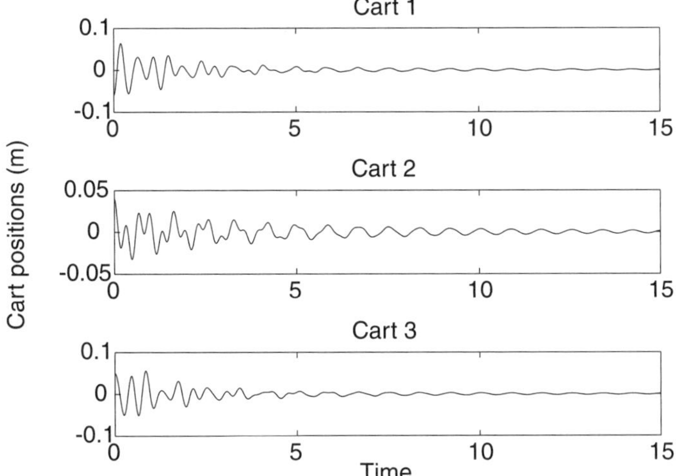

Figure 13.4 Positions of the carts versus time.

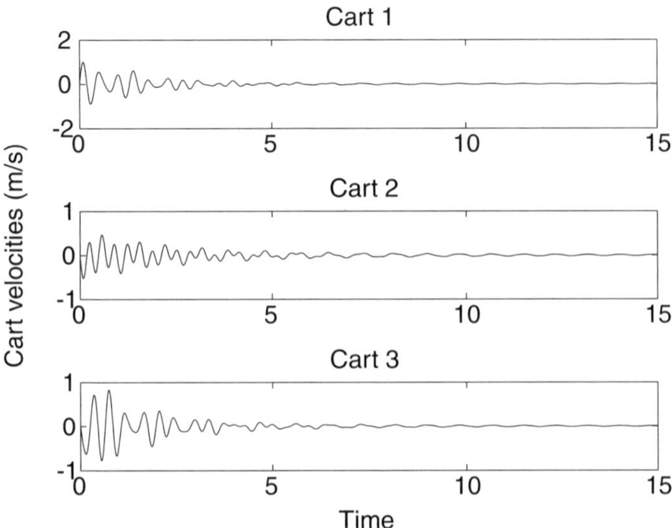

Figure 13.5 Velocities of the carts versus time.

13.6 Quasi-Thermodynamic Stabilization and Maximum Entropy Control

In this section, we use the recently developed notion of system thermodynamics [81] to develop thermodynamically consistent hybrid decentralized controllers for large-scale systems. Specifically, since our energy-based hybrid controller architecture involves the exchange of energy with conservation laws describing transfer, accumulation, and dissipation of energy between the subcontrollers and the plant subsystems, we construct a modified

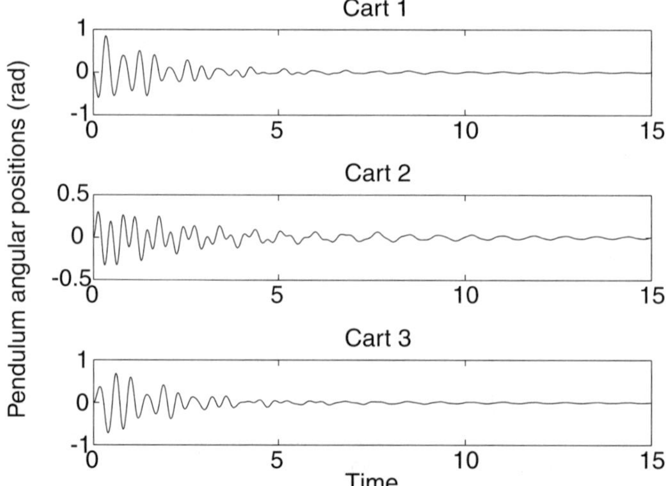

Figure 13.6 Angular positions of the pendulums versus time.

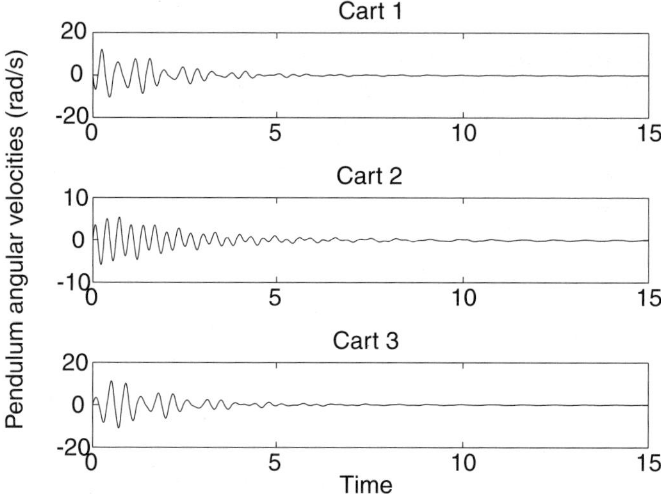

Figure 13.7 Angular velocities of the pendulums versus time.

hybrid controller that guarantees that each subsystem-subcontroller pair $(\mathcal{G}_i, \mathcal{G}_{ci})$ is consistent with basic thermodynamic principles after the first resetting event. To develop thermodynamically consistent hybrid decentralized controllers, consider the closed-loop subsystem-subcontroller pair $(\mathcal{G}_i, \mathcal{G}_{ci})$ given by (13.30) and (13.31) with $\tilde{\mathcal{Z}}_i$ given by

$$\tilde{\mathcal{Z}}_i \triangleq \left\{ \tilde{x}_i \in \tilde{\mathcal{D}}_i : \phi_i(\tilde{x}_i)(v_{\mathrm{p}i}(\tilde{x}_i) - v_{ci}(\tilde{x}_i)) = 0 \text{ and } v_{ci}(x_{ci}, h_i(x_i)) > 0 \right\},$$
$$i = 1, \ldots, q, \quad (13.80)$$

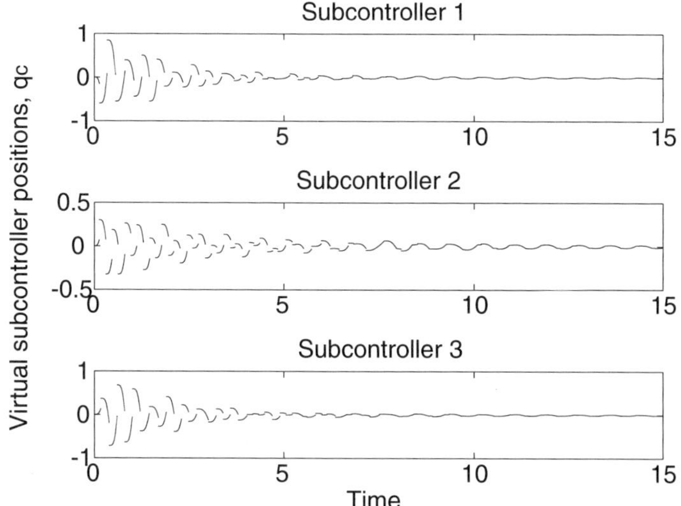

Figure 13.8 Subcontroller positions versus time.

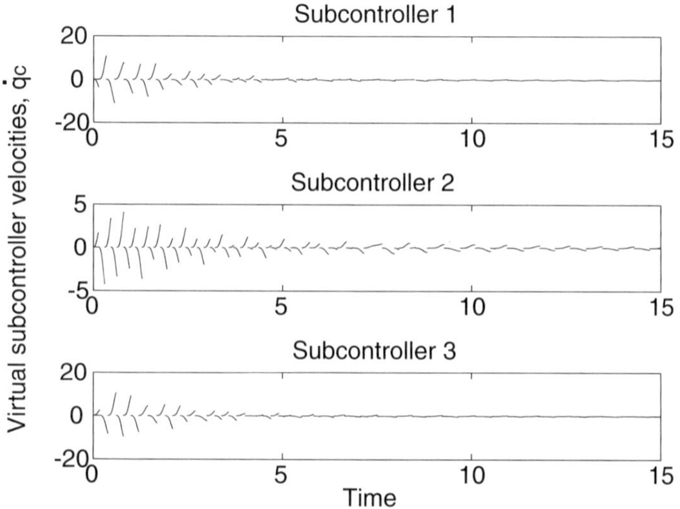

Figure 13.9 Subcontroller velocities versus time.

where $\phi_i(\tilde{x}_i) \triangleq s_{ci}(h_i(x_i), h_{ci}(x_{ci}, h_i(x_i)))$, $v_{pi}(\tilde{x}_i) \triangleq v_{si}(x_i)$, and $v_{ci}(\tilde{x}_i) \triangleq v_{ci}(x_{ci}, h_i(x_i))$. We refer to $\phi_i(\cdot)$ as the *net energy flow* function.

We assume that the energy flow function $\phi_i(\tilde{x}_i)$ is infinitely differentiable and the k_i-transversality condition (13.19) holds with

$$\mathcal{X}_i(\tilde{x}_i) = \phi_i(\tilde{x}_i)(v_{pi}(\tilde{x}_i) - v_{ci}(\tilde{x}_i)), \quad i = 1, \ldots, q. \tag{13.81}$$

To ensure a thermodynamically consistent energy flow between the subsys-

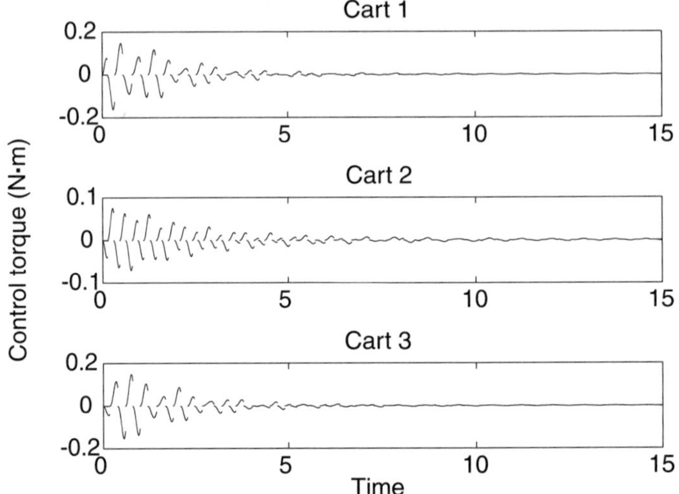

Figure 13.10 Control torques versus time.

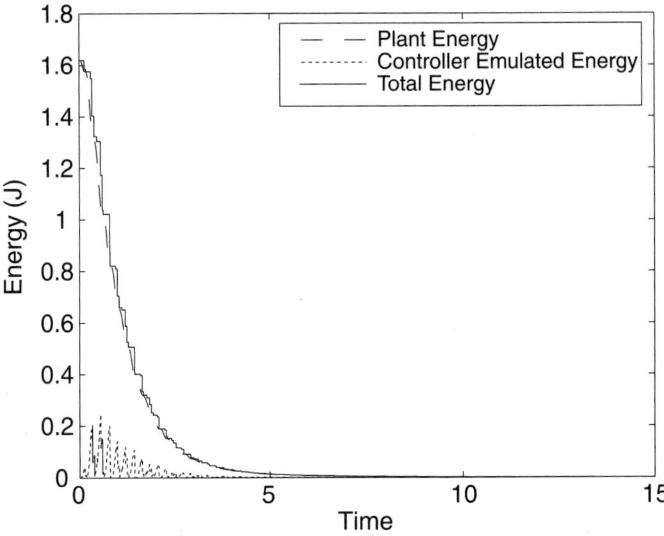

Figure 13.11 Plant, combined subcontroller, and total energies versus time.

tem $\mathcal{G}_i$ and subcontroller $\mathcal{G}_{ci}$ after the first resetting event, each subcontroller resetting logic must be designed in such a way as to satisfy three key thermodynamic axioms. Namely, between resettings the energy flow function $\phi_i(\cdot)$ must satisfy the following two assumptions [79, 81]:

Assumption 13.4. The connectivity matrix $\mathcal{C} \in \mathbb{R}^{2 \times 2}$ [81, p. 56] associated with the subsystem $\tilde{\mathcal{G}}_l$ is defined by

$$\mathcal{C}_{(i,j)} \triangleq \begin{cases} 0, & \text{if } \phi_l(\tilde{x}_l(t)) \equiv 0, \\ 1, & \text{otherwise,} \end{cases} \quad i \neq j, \quad i, j = 1, 2, \quad l = 1, \ldots, q, \quad t \geq t_1^+,$$

$$(13.82)$$

and

$$\mathcal{C}_{(i,i)} \triangleq -\mathcal{C}_{(k,i)}, \quad i \neq k, \quad i, k = 1, 2, \tag{13.83}$$

and satisfies rank $\mathcal{C} = 1$. Moreover, for every $i \neq j$ such that $\mathcal{C}_{(i,j)} = 1$, $\phi_l(\tilde{x}_l(t)) = 0$ if and only if $v_{\mathrm{p}l}(\tilde{x}_l) = v_{\mathrm{c}l}(\tilde{x}_l)$, $\tilde{x}_l(t) \notin \tilde{\mathcal{Z}}_l$, $l = 1, \ldots, q$, $t \geq t_1^+$.

Assumption 13.5. $\phi_i(\tilde{x}_i(t))(v_{\mathrm{p}i}(\tilde{x}_i) - v_{\mathrm{c}i}(\tilde{x}_i)) \leq 0$, $\tilde{x}_i(t) \notin \tilde{\mathcal{Z}}_i$, $i = 1, \ldots, q$, $t \geq t_1^+$.

Furthermore, across resettings the energy difference between the subsystem and the subcontroller must satisfy the following assumption [83, 84]:

Assumption 13.6. $[v_{\mathrm{p}i}(\tilde{x}_i + \tilde{f}_{\mathrm{d}i}(\tilde{x}_i)) - v_{\mathrm{c}i}(\tilde{x}_i + \tilde{f}_{\mathrm{d}i}(\tilde{x}_i))][v_{\mathrm{p}i}(\tilde{x}_i) - v_{\mathrm{c}i}(\tilde{x}_i)] \geq 0$, $i = 1, \ldots, q$, $\tilde{x}_i \in \tilde{\mathcal{Z}}_i$.

The fact that $\phi_i(\tilde{x}_i(t)) = 0$ if and only if $v_{\mathrm{p}i}(\tilde{x}_i(t)) = v_{\mathrm{c}i}(\tilde{x}_i(t))$, $\tilde{x}_i(t) \notin \tilde{\mathcal{Z}}_i$, $t \geq t_1^+$, implies that the ith subsystem and the ith subcontroller are *connected*; alternatively, $\phi_i(\tilde{x}_i(t)) \equiv 0$, $t \geq t_1^+$, implies that the ith subsystem and the ith subcontroller are *disconnected*. Assumption 13.4 implies that if the energies in the ith subsystem and the ith subcontroller are equal, then energy exchange between the ith subsystem $\mathcal{G}_i$ and the ith subcontroller $\mathcal{G}_{\mathrm{c}i}$ is not possible unless a resetting event occurs. This statement is consistent with the *zeroth law of thermodynamics*, which postulates that temperature equality is a necessary and sufficient condition for thermal equilibrium of an isolated system. Assumption 13.5 implies that energy flows from a more energetic subsystem to a less energetic subsystem and is consistent with the *second law of thermodynamics*, which states that heat (energy) must flow in the direction of lower temperatures. Finally, Assumption 13.6 implies that the energy difference between the ith subsystem $\mathcal{G}_i$ and the ith subcontroller $\mathcal{G}_{\mathrm{c}i}$ across resetting instants is monotonic, that is, $[v_{\mathrm{p}i}(\tilde{x}_i(t_k^+)) - v_{\mathrm{c}i}(\tilde{x}_i(t_k^+))][v_{\mathrm{p}i}(\tilde{x}_i(t_k)) - v_{\mathrm{c}i}(\tilde{x}_i(t_k))] \geq 0$ for all $v_{\mathrm{p}i}(\tilde{x}_i) \neq v_{\mathrm{c}i}(\tilde{x}_i)$, $\tilde{x}_i \in \tilde{\mathcal{Z}}_i$, $i = 1, \ldots, q$, $k \in \overline{\mathbb{Z}}_+$.

With the resetting law given by (13.80), it follows that each ith subsystem $\tilde{\mathcal{G}}_i$ of the closed-loop dynamical system $\tilde{\mathcal{G}}$ satisfies Assumptions 13.4–13.6 for all $t \geq t_1$. To see this, note that since $\phi_i(\tilde{x}_i) \neq 0$, the connectivity matrix $\mathcal{C}$ is given by

$$\mathcal{C} = \begin{bmatrix} -1 & 1 \\ 1 & -1 \end{bmatrix}, \tag{13.84}$$

and hence, rank $\mathcal{C} = 1$. The second condition in Assumption 13.4 need not be satisfied since the case where $\phi_i(\tilde{x}_i) = 0$ or $v_{\mathrm{p}i}(\tilde{x}_i) = v_{\mathrm{c}i}(\tilde{x}_i)$, corresponds to a resetting instant. Furthermore, it follows from the definition of the resetting set (13.80) that Assumption 13.5 is satisfied for each closed-loop subsystem pairs $(\mathcal{G}_i, \mathcal{G}_{\mathrm{c}i})$ for all $t \geq t_1^+$. Finally, since $v_{\mathrm{c}i}(\tilde{x}_i + \tilde{f}_{\mathrm{d}i}(\tilde{x}_i)) = 0$ and $v_{\mathrm{p}i}(\tilde{x}_i + \tilde{f}_{\mathrm{d}i}(\tilde{x}_i)) = v_{\mathrm{p}i}(\tilde{x}_i)$, $\tilde{x}_i \in \tilde{\mathcal{Z}}_i$, it follows from the definition of the resetting set that

$$[v_{\mathrm{p}i}(\tilde{x}_i + \tilde{f}_{\mathrm{d}i}(\tilde{x}_i)) - v_{\mathrm{c}i}(\tilde{x}_i + \tilde{f}_{\mathrm{d}i}(\tilde{x}_i))][v_{\mathrm{p}i}(\tilde{x}_i) - v_{\mathrm{c}i}(\tilde{x}_i)]$$
$$= v_{\mathrm{p}i}(\tilde{x}_i)[v_{\mathrm{p}i}(\tilde{x}_i) - v_{\mathrm{c}i}(\tilde{x}_i)]$$
$$\geq 0, \quad \tilde{x}_i \in \tilde{\mathcal{Z}}_i, \quad i = 1, \ldots, q, \tag{13.85}$$

and hence, Assumption 13.6 is satisfied across resettings. Hence, each ith closed-loop subsystem $\tilde{\mathcal{G}}_i$ of the closed-loop system $\tilde{\mathcal{G}}$ is thermodynamically consistent after the first resetting event in the sense of [79, 81, 83, 84]. Note that this statement is only true for each closed-loop subsystem $\tilde{\mathcal{G}}_i$. For the hybrid closed-loop system $\tilde{\mathcal{G}}$, Assumptions 13.4–13.6 may not hold since the interconnection function $\mathcal{I}(x)$ defining $\mathcal{G}$ may not necessarily correspond to a thermodynamically consistent model.

If $\tilde{\mathcal{D}}_{\mathrm{c}i} \subset \tilde{\mathcal{D}}$ is a compact positively invariant set with respect to the closed-loop dynamical system $\tilde{\mathcal{G}}$ given by (13.34) and (13.35) such that $0 \in \overset{\circ}{\tilde{\mathcal{D}}}_{\mathrm{c}i}$, and the k_i-transversality condition (13.19) holds with $\mathcal{X}_i(\tilde{x}_i) = \phi_i(\tilde{x}_i)(v_{\mathrm{p}i}(\tilde{x}_i) - v_{\mathrm{c}i}(\tilde{x}_i))$ for all $i = 1, \ldots, q$, then it follows from Theorem 13.3 that the zero solution $\tilde{x}(t) \equiv 0$ to the closed-loop system $\tilde{\mathcal{G}}$, with resetting set $\tilde{\mathcal{Z}}_i$ given by (13.80), is asymptotically stable. Furthermore, in this case, the hybrid decentralized controller (13.25) and (13.26), with resetting set (13.80), is a *quasi-thermodynamically stabilizing* compensator.

Finally, we show that the hybrid decentralized controllers developed in this section and Section 13.3 are maximum entropy controllers. To do this, the following hybrid definition of entropy is needed.

Definition 13.2. For each decentralized subcontroller $\mathcal{G}_{\mathrm{c}i}$ given by (13.25)–(13.27), a function $S_{\mathrm{c}i} : \overline{\mathbb{R}}_+ \to \mathbb{R}$, $i = 1, \ldots, q$, satisfying

$$S_{\mathrm{c}i}(v_{\mathrm{c}i}(\tilde{x}_i(T))) \geq S_{\mathrm{c}i}(v_{\mathrm{c}i}(\tilde{x}_i(t_1))) - \frac{1}{c_i} \sum_{k \in \mathbb{Z}_{[t_1,T)}} v_{\mathrm{c}i}(\tilde{x}_i(t_k)), \quad T \geq t_1,$$
$$i = 1, \ldots, q, \tag{13.86}$$

where $k \in \mathbb{Z}_{[t_1,T)} \triangleq \{k : t_1 \leq t_k < T\}$, $c_i > 0$, is called an *entropy* function of $\mathcal{G}_{\mathrm{c}i}$, $i = 1, \ldots, q$.

The next result gives necessary and sufficient conditions for establishing the existence of an entropy function of $\mathcal{G}_{\mathrm{c}i}$, $i = 1, \ldots, q$, over an interval $t \in (t_k, t_{k+1}]$ involving the consecutive resetting times t_k and t_{k+1}, $k \in \mathbb{Z}_+$.

Theorem 13.5. For each hybrid decentralized subcontroller $\mathcal{G}_{ci}$ given by (13.25)–(13.27), a function $S_{ci} : \overline{\mathbb{R}}_+ \to \mathbb{R}$, $i = 1, \ldots, q$, is an entropy function of $\mathcal{G}_{ci}$ if and only if

$$S_{ci}(v_{ci}(\tilde{x}_i(\hat{t}))) \geq S_{ci}(v_{ci}(\tilde{x}_i(t))), \quad t_k < t \leq \hat{t} \leq t_{k+1}, \quad i = 1, \ldots, q, \quad (13.87)$$

$$S_{ci}(v_{ci}(\tilde{x}_i(t_k) + \tilde{f}_{di}(\tilde{x}_i(t_k)))) \geq S_{ci}(v_{ci}(\tilde{x}_i(t_k))) - \frac{1}{c_i} v_{ci}(\tilde{x}_i(t_k)), \quad k \in \mathbb{Z}_+,$$
$$i = 1, \ldots, q. \quad (13.88)$$

Proof. Let $k \in \mathbb{Z}_+$ and suppose $S_{ci}(v_{ci})$ is an entropy function of $\mathcal{G}_{ci}$. Then, (13.86) holds. Now, since for $t_k < t \leq \hat{t} \leq t_{k+1}$, $\mathbb{Z}_{[t,\hat{t})} = \varnothing$, (13.87) is immediate. Next, note that

$$S_{ci}(v_{ci}(\tilde{x}_i(t_k^+))) \geq S_{ci}(v_{ci}(\tilde{x}_i(t_k))) - \frac{1}{c_i} v_{ci}(\tilde{x}_i(t_k)), \quad i = 1, \ldots, q, \quad (13.89)$$

which, since $\mathbb{Z}_{[t_k,t_k^+)} = k$, implies (13.88).

Conversely, suppose (13.87) and (13.88) hold, and let $\hat{t} \geq t \geq t_1$ and $\mathbb{Z}_{[t,\hat{t})} = \{i, i+1, \ldots, j\}$. (Note that if $\mathbb{Z}_{[t,\hat{t})} = \varnothing$, then the converse result is a direct consequence of (13.87).) If $\mathbb{Z}_{[t,\hat{t})} \neq \varnothing$, then it follows from (13.87) and (13.88) that

$$S_{cl}(v_{cl}(\tilde{x}_l(\hat{t}))) - S_{cl}(v_{cl}(\tilde{x}_l(t))) = S_{cl}(v_{cl}(\tilde{x}_l(\hat{t}))) - S_{cl}(v_{cl}(\tilde{x}_l(t_j^+)))$$
$$+ \sum_{m=0}^{j-i} S_{cl}(v_{cl}(\tilde{x}_l(t_{j-m})$$
$$+ \tilde{f}_{dl}(\tilde{x}_l(t_{j-m})))) - S_{cl}(v_{cl}(\tilde{x}_l(t_{j-m})))$$
$$+ \sum_{m=0}^{j-i-1} S_{cl}(v_{cl}(\tilde{x}_l(t_{j-m})))$$
$$- S_{cl}(v_{cl}(\tilde{x}_l(t_{j-m-1}^+)))$$
$$+ S_{cl}(v_{cl}(\tilde{x}_l(t_i))) - S_{cl}(v_{cl}(\tilde{x}_l(t)))$$
$$\geq -\frac{1}{c_l} \sum_{m=0}^{j-i} v_{cl}(\tilde{x}_l(t_{j-m}))$$
$$= -\frac{1}{c_l} \sum_{k \in \mathbb{Z}_{[t,\hat{t})}} v_{cl}(\tilde{x}_l(t_k)), \quad l = 1, \ldots, q,$$

$$(13.90)$$

which implies that $S_{ci}(v_{ci})$ is an entropy function of $\mathcal{G}_{ci}$, $i = 1, \ldots, q$. $\qquad \square$

The next theorem establishes the existence of an entropy function for $\mathcal{G}_{ci}$, $i = 1, \ldots, q$.

Theorem 13.6. Consider the hybrid decentralized subcontrollers $\mathcal{G}_{ci}$ given by (13.25)–(13.27), with $\tilde{\mathcal{Z}}_i$ given by (13.39) or (13.80). Then the function $S_{ci} : \overline{\mathbb{R}}_+ \to \mathbb{R}$, $i = 1, \dots, q$, given by

$$S_{ci}(v_{ci}) = \log_e(c_i + v_{ci}) - \log_e c_i, \quad v_{ci} \in \overline{\mathbb{R}}_+, \quad i = 1, \dots, q, \qquad (13.91)$$

where $c_i > 0$, is an entropy function of $\mathcal{G}_{ci}$, $i = 1, \dots, q$. In addition, for $i = 1, \dots, q$,

$$\dot{S}_{ci}(v_{ci}(\tilde{x}_i(t))) > 0, \quad \tilde{x}_i(t) \notin \tilde{\mathcal{Z}}_i, \quad t_k < t \le t_{k+1}, \qquad (13.92)$$

$$-\frac{1}{c_i} v_{ci}(\tilde{x}_i(t_k)) < \Delta S_{ci}(v_{ci}(\tilde{x}_i(t_k))) < -\frac{v_{ci}(\tilde{x}_i(t_k))}{c_i + v_{ci}(\tilde{x}_i(t_k))},$$

$$\tilde{x}_i(t_k) \in \tilde{\mathcal{Z}}_i, \quad k \in \mathbb{Z}_+. \qquad (13.93)$$

Proof. Since $\dot{v}_{ci}(\tilde{x}_i(t)) > 0$, $\tilde{x}_i(t) \notin \tilde{\mathcal{Z}}_i$, $i = 1, \dots, q$, $t \in (t_k, t_{k+1}]$, $k \in \mathbb{Z}_+$, it follows that

$$\dot{S}_{ci}(v_{ci}(\tilde{x}_i(t))) = \frac{\dot{v}_{ci}(\tilde{x}_i(t))}{c_i + v_{ci}(\tilde{x}_i(t))} > 0, \quad \tilde{x}_i(t) \notin \tilde{\mathcal{Z}}_i, \quad i = 1, \dots, q. \qquad (13.94)$$

Furthermore, since $v_{ci}(\tilde{x}_i(t_k) + \tilde{f}_{di}(\tilde{x}_i(t_k))) = 0$, $\tilde{x}_i(t_k) \in \tilde{\mathcal{Z}}_i$, $i = 1, \dots, q$, $k \in \mathbb{Z}_+$, it follows that, for $i = 1, \dots, q$,

$$\Delta S_{ci}(v_{ci}(\tilde{x}_i(t_k))) = \log_e\left[1 - \frac{v_{ci}(\tilde{x}_i(t_k))}{c_i + v_{ci}(\tilde{x}_i(t_k))}\right] > -\frac{1}{c_i} v_{ci}(\tilde{x}_i(t_k)),$$

$$\tilde{x}_i(t_k) \in \tilde{\mathcal{Z}}_i, \quad k \in \mathbb{Z}_+, \qquad (13.95)$$

and

$$\Delta S_{ci}(v_{ci}(\tilde{x}_i(t_k))) = \log_e\left[1 - \frac{v_{ci}(\tilde{x}_i(t_k))}{c_i + v_{ci}(\tilde{x}_i(t_k))}\right] < -\frac{v_{ci}(\tilde{x}_i(t_k))}{c_i + v_{ci}(\tilde{x}_i(t_k))},$$

$$\tilde{x}_i(t_k) \in \tilde{\mathcal{Z}}_i, \quad k \in \mathbb{Z}_+, \qquad (13.96)$$

where in (13.95) and (13.96) we used the fact that $\frac{x}{1+x} < \log_e(1+x) < x$, $x > -1$, $x \ne 0$. The result is now an immediate consequence of Theorem 13.5. $\square$

Using (13.94), the resetting set $\tilde{\mathcal{Z}}_i$, $i = 1, \dots, q$, given by (13.39) can be rewritten as

$$\tilde{\mathcal{Z}}_i \triangleq \left\{ \tilde{x}_i \in \tilde{\mathcal{D}}_i : \frac{\mathrm{d}}{\mathrm{d}t} S_{ci}(v_{ci}(\tilde{x}_i)) = 0 \text{ and } v_{ci}(\tilde{x}_i) > 0 \right\}, \quad i = 1, \dots, q, \qquad (13.97)$$

where $\frac{\mathrm{d}}{\mathrm{d}t} S_{ci}(v_{ci}(\tilde{x}_i(t))) \triangleq \lim_{\tau \to t^-} \frac{1}{t - \tau}[S_{ci}(v_{ci}(\tilde{x}_i(t))) - S_{ci}(v_{ci}(\tilde{x}_i(\tau)))]$ whenever limit on the right-hand side exists, and $S_{ci} = \log_e(c_i + v_{ci}) - \log_e c_i$ denotes the continuously differentiable ith subcontroller entropy. Hence, each decentralized controller $\mathcal{G}_{ci}$ corresponds to a maximum entropy controller. Alternatively, for $i = 1, \dots, q$, the resetting set $\tilde{\mathcal{Z}}_i$ given by (13.80) can be rewritten as $\{\tilde{x}_i(t_k) : k \in \mathbb{Z}_+\}$, where t_k is the maximum final time such that $S_{ci}(v_{ci}(\tilde{x}_i(t))) \leq S_{ci}(v_{ci}(\tilde{x}_i(t_1)))$ (or $S_{ci}(v_{ci}(\tilde{x}_i(t))) \geq S_{ci}(v_{ci}(\tilde{x}_i(t_1)))$) holds under the constraint $v_{pi}(\tilde{x}_i(t)) \geq v_{ci}(\tilde{x}_i(t))$ (or $v_{pi}(\tilde{x}_i(t)) \leq v_{ci}(\tilde{x}_i(t))$) for $0 \leq t < t_1$, and $S_{ci}(v_{ci}(\tilde{x}_i(t))) \leq S_{ci}(v_{ci}(\tilde{x}_i(t_{k+1})))$ holds under the constraint $v_{pi}(\tilde{x}_i(t)) \geq v_{ci}(\tilde{x}_i(t))$ for all $t_k \leq t < t_{k+1}$ and $k \in \mathbb{Z}_+$. Hence, each decentralized controller $\mathcal{G}_{ci}$ corresponds to a constrained maximum entropy controller.

13.7 Hybrid Decentralized Control for Combustion Systems

In this section, we apply the results developed in this chapter to the control of *thermoacoustic instabilities* in combustion processes. We stress that the combustion model we use can be stabilized by conventional nonlinear control methods. The aim here, however, is to show that hybrid decentralized control provides an extremely efficient mechanism for dissipating energy in the combustion process with far superior performance than any conventional control methodology. In particular, we show that the proposed hybrid decentralized controller provides finite-time stabilization.

To design a decentralized hybrid controller for combustion systems we concentrate on a two-mode, nonlinear time-averaged combustion model with nonlinearities present due to the second-order gas dynamics. This model is developed in Section 6.5 and is given by

$$\dot{x}_1(t) = \alpha_1 x_1(t) + \theta_1 x_2(t) - \beta(x_1(t)x_3(t) + x_2(t)x_4(t)) + u_1(t),$$
$$x_1(0) = x_{10}, \quad t \geq 0, \qquad (13.98)$$

$$\dot{x}_2(t) = -\theta_1 x_1(t) + \alpha_1 x_2(t) + \beta(x_2(t)x_3(t) - x_1(t)x_4(t)) + u_2(t),$$
$$x_2(0) = x_{20}, \qquad (13.99)$$

$$\dot{x}_3(t) = \alpha_2 x_3(t) + \theta_2 x_4(t) + \beta(x_1^2(t) - x_2^2(t)) + u_3(t), \quad x_3(0) = x_{30},$$
$$(13.100)$$

$$\dot{x}_4(t) = -\theta_2 x_3(t) + \alpha_2 x_4(t) + 2\beta x_1(t)x_2(t) + u_4(t), \quad x_4(0) = x_{40},$$
$$(13.101)$$

where $\alpha_1, \alpha_2 \in \mathbb{R}$ represent growth/decay constants, $\theta_1, \theta_2 \in \mathbb{R}$ represent frequency shift constants, $\beta = ((\gamma + 1)/8\gamma)\omega_1$, where γ denotes the ratio of specific heats, ω_1 is frequency of the fundamental mode, and u_i, $i = 1, \ldots, 4$, are control input signals. For the data parameters $\alpha_1 = 5$, $\alpha_2 = -55$, $\theta_1 = 4$, $\theta_2 = 32$, $\gamma = 1.4$, $\omega_1 = 1$, and $x(0) = [1, 1, 1, 1]^T$, the open-loop ($u_i(t) \equiv 0$, $i = 1, 2, 3, 4$) dynamics (13.98)–(13.101) result in a limit cycle instability.

Next, note that (13.98)–(13.101) can be rewritten in the form of (13.1) and (13.2) with $f_1(x_1, x_2) = [\alpha_1 x_1 + \theta_1 x_2, -\theta_1 x_1 + \alpha_1 x_2]^T$, $f_2(x_3, x_4) = [\alpha_2 x_3 + \theta_2 x_4, -\theta_2 x_3 + \alpha_2 x_4]^T$, $\mathcal{I}_1(x) = [-\beta(x_1 x_3 + x_2 x_4), \beta(x_2 x_3 - x_1 x_4)]^T$, $\mathcal{I}_2(x) = [\beta(x_1^2 - x_2^2), 2\beta x_1 x_2]^T$, $G_1(x_1, x_2) = I_2$, $G_2(x_3, x_4) = I_2$, $y_1 = h_1(x_1, x_2) = [x_1, x_2]^T$, and $y_2 = h_2(x_3, x_4) = [x_3, x_4]^T$. Here, we take $v_{s1}(x_1, x_2) = \frac{1}{2}(x_1^2 + x_2^2)$ and $v_{s2}(x_3, x_4) = \frac{1}{2}(x_3^2 + x_4^2)$ as our subsystem

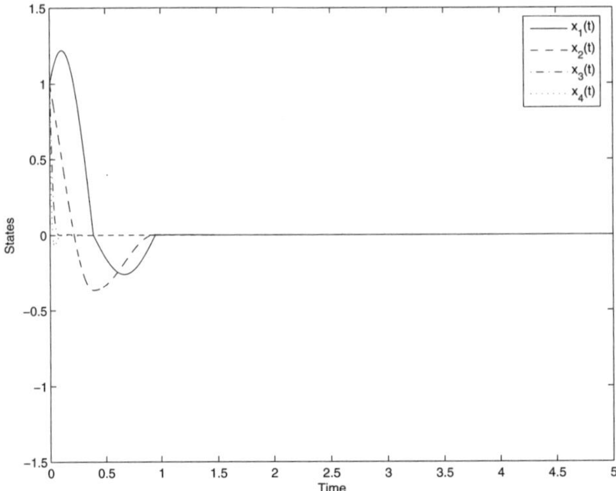

Figure 13.12 Plant state trajectories versus time.

energies. Now, it can be shown that the ith disconnected subsystem of (13.98)–(13.101) is lossless with respect to the supply rate $\hat{u}_i^{\mathrm{T}} y_i$, $i = 1, 2$, where $\hat{u}_1 = [u_1 + \alpha_1 x_1, u_2 + \alpha_1 x_2]^{\mathrm{T}}$ and $\hat{u}_2 = [u_3 + \alpha_2 x_3, u_4 + \alpha_2 x_4]^{\mathrm{T}}$. Furthermore, it can also be shown that (13.98)–(13.101) is lossless with respect to the supply rate $\hat{u}_1^{\mathrm{T}} y_1 + \hat{u}_2^{\mathrm{T}} y_2$.

Next, consider the decentralized dynamic compensator given by (13.25)–(13.27) with $f_{\mathrm{c}1}(x_{\mathrm{c}1}, y_1) = A_{\mathrm{c}1} x_{\mathrm{c}1} + B_{\mathrm{c}1} y_1$, $\eta_1(y_1) = 0$, $h_{\mathrm{c}1}(x_{\mathrm{c}1}, y_1) = B_{\mathrm{c}1}^{\mathrm{T}} x_{\mathrm{c}1}$, $f_{\mathrm{c}2}(x_{\mathrm{c}2}, y_2) \equiv 0$, $\eta_2(y_2) = 0$, and $h_{\mathrm{c}2}(x_{\mathrm{c}2}, y_2) \equiv 0$, where

$$A_{\mathrm{c}1} = \begin{bmatrix} 0 & 1 \\ -1 & 0 \end{bmatrix}, \quad B_{\mathrm{c}1} = \begin{bmatrix} 0 & 0 \\ 4 & 0 \end{bmatrix}, \tag{13.102}$$

and subcontroller energy is given by $v_{\mathrm{c}1}(x_{\mathrm{c}1}) = \frac{1}{2} x_{\mathrm{c}1}^{\mathrm{T}} x_{\mathrm{c}1}$. Furthermore, the resetting set (13.39) is given by

$$\mathcal{Z}_1 = \left\{ (x_1, x_2, x_{\mathrm{c}1}) : x_{\mathrm{c}1}^{\mathrm{T}} B_{\mathrm{c}1} \begin{bmatrix} x_1 \\ x_2 \end{bmatrix} = 0,\ x_{\mathrm{c}1} \neq 0 \right\}. \tag{13.103}$$

To illustrate the behavior of the closed-loop impulsive dynamical system, we choose the initial condition $x_{\mathrm{c}1}(0) = [0, 0]^{\mathrm{T}}$. For this system a straightforward, but lengthy, calculation shows that Assumptions 13.1 and 13.2 hold. However, the k_i-transversality condition is sufficiently complex that we have been unable to show it analytically. This condition was verified numerically, and hence, Assumption 13.3 appears to hold. Figure 13.12 shows the state trajectories of the plant versus time, and Figure 13.13 shows the state trajectories of the compensator versus time. Figure 13.14 shows the control inputs u_1 and u_2 versus time. Note that the compensator states are the only states that reset. Furthermore, the control force versus time is

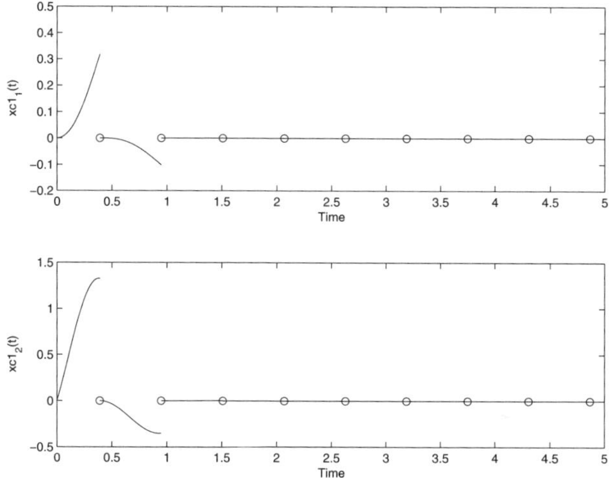

Figure 13.13 Compensator state trajectories versus time.

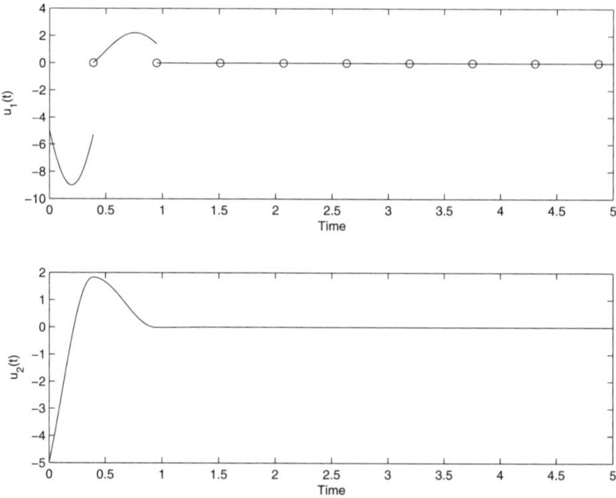

Figure 13.14 u_1 and u_2 versus time.

partially discontinuous at the resetting times. A comparison of $v_{s1}(x_1, x_2)$, $v_{s2}(x_3, x_4)$, $v_{c1}(x_{c1})$, and $v(x, x_{c1}) \triangleq v_{s1}(x_1, x_2) + v_{s2}(x_3, x_4) + v_{c1}(x_{c1})$ is shown in Figure 13.15. Note that the proposed hybrid decentralized controller achieves finite-time stabilization.

Next, we consider the case where $\alpha_1 = 0$ and $\alpha_2 = 0$. The other parameters remain as before. In this case, the decentralized dynamic compensators are given by (13.25)–(13.27) with $f_{ci}(x_{ci}, y_i) = A_{ci}x_{ci} + B_{ci}y_i$, $\eta_i(y_i) = 0$, $h_{ci}(x_{ci}, y_i) = B_{ci}^{\mathrm{T}}x_{ci}$, $i = 1, 2$, where A_{c1} and B_{c1} are given by

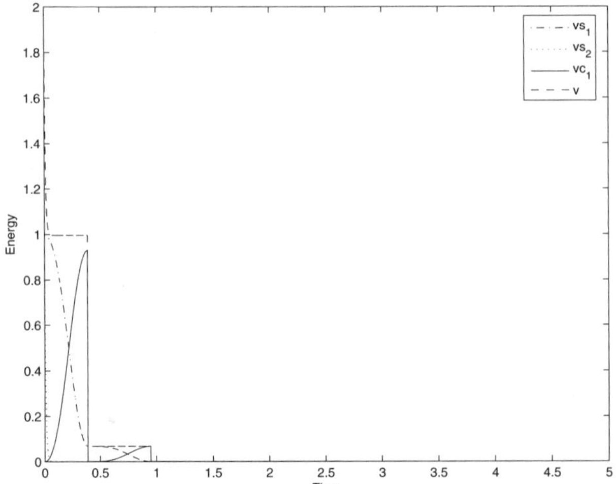

Figure 13.15 v_{s1}, v_{s2}, v_{c1}, and v versus time.

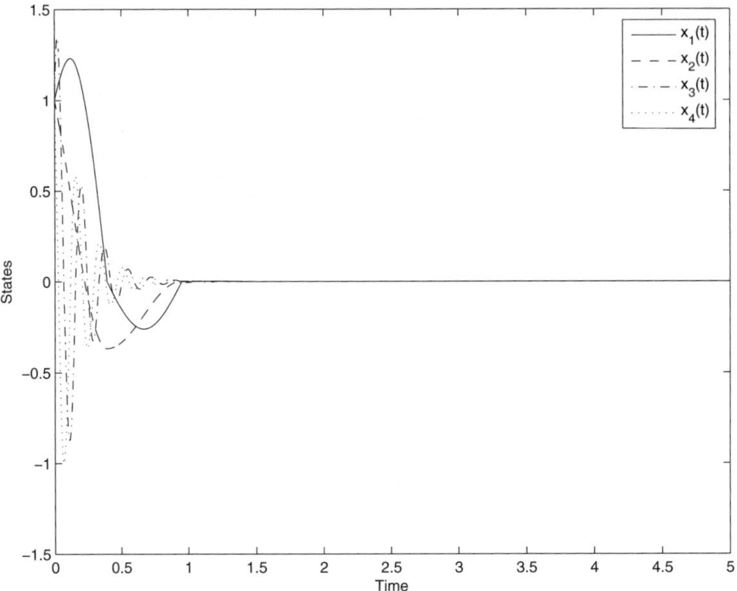

Figure 13.16 Plant state trajectories versus time.

(13.102),

$$A_{c2} = \begin{bmatrix} 0 & 1 \\ -1 & 0 \end{bmatrix}, \quad B_{c2} = \begin{bmatrix} 0 & 0 \\ 16 & 0 \end{bmatrix}, \tag{13.104}$$

and subcontroller energies are given by $v_{c1}(x_{c1}) = \frac{1}{2}x_{c1}^{\mathrm{T}}x_{c1}$ and $v_{c2}(x_{c2}) =$

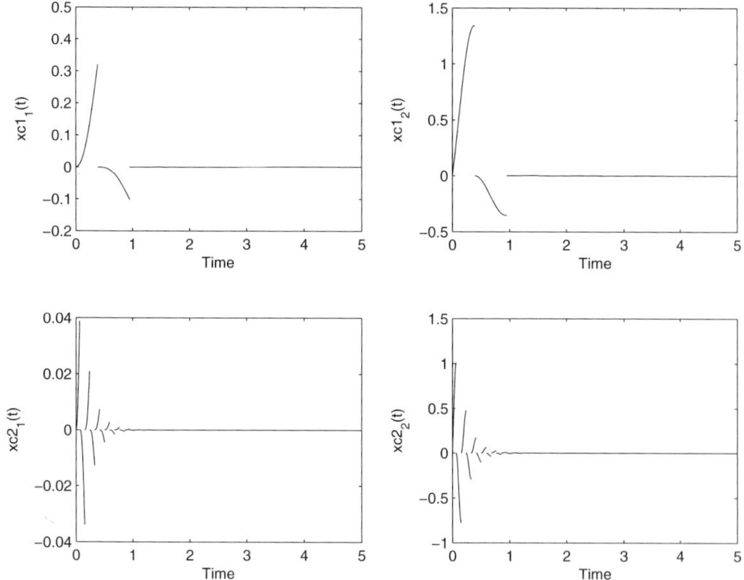

Figure 13.17 Compensator state trajectories versus time.

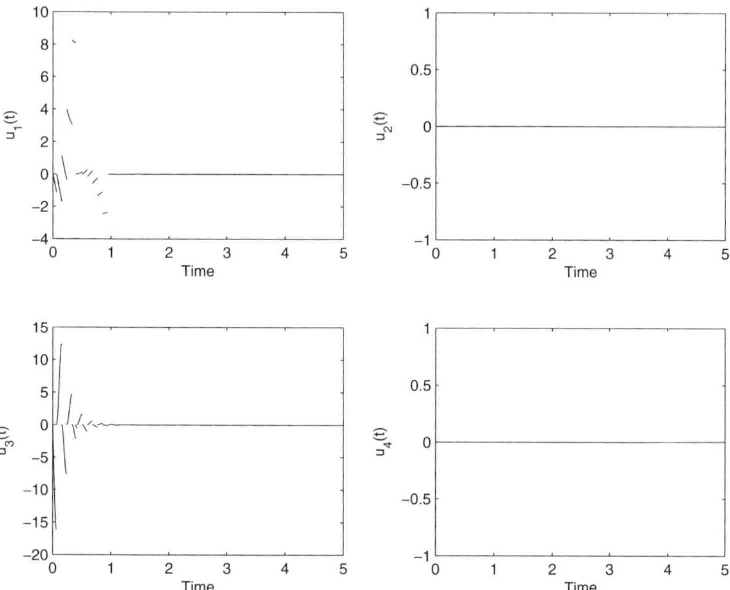

Figure 13.18 Control input versus time.

$\frac{1}{2}x_{\mathrm{c}2}^{\mathrm{T}}x_{\mathrm{c}2}$. Furthermore, the resetting set (13.39) is given by (13.103) and

$$\mathcal{Z}_2 = \left\{ (x_3, x_4, x_{\mathrm{c}2}) : x_{\mathrm{c}2}^{\mathrm{T}} B_{\mathrm{c}2} \begin{bmatrix} x_3 \\ x_4 \end{bmatrix} = 0, \ x_{\mathrm{c}2} \neq 0 \right\}. \qquad (13.105)$$

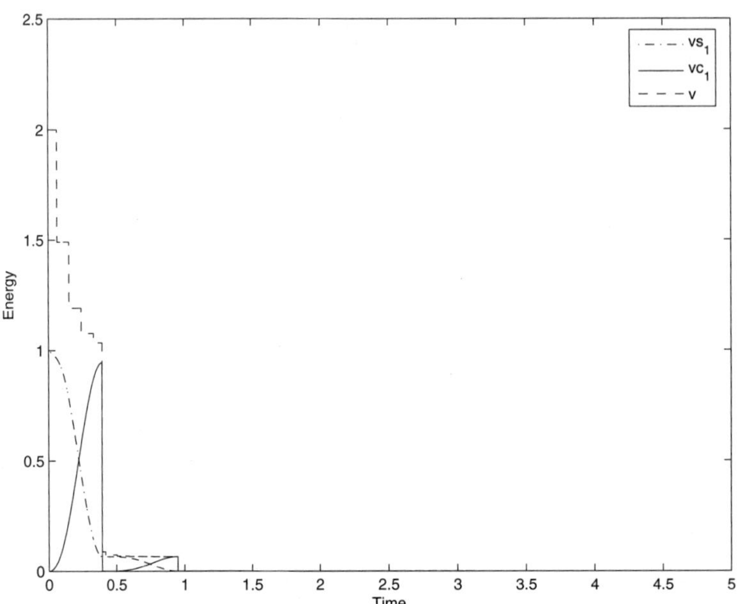

Figure 13.19 v_{s1}, v_{c1}, and v versus time.

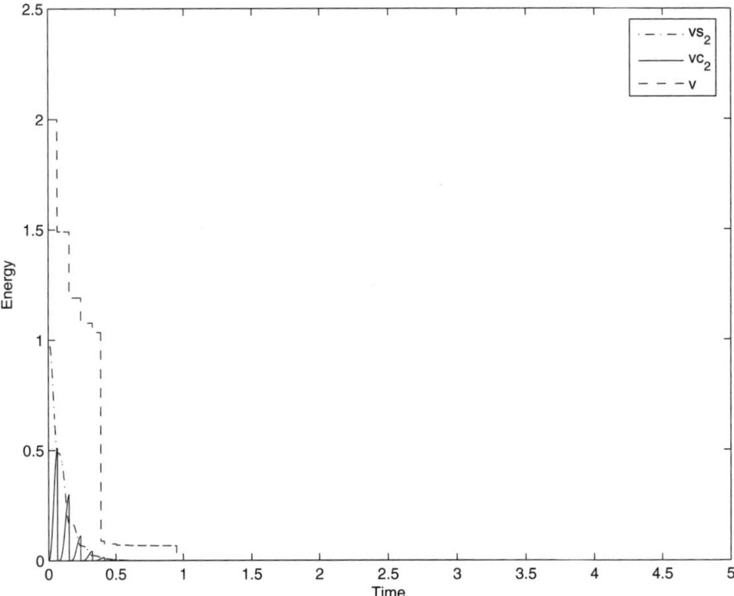

Figure 13.20 v_{s2}, v_{c2}, and v versus time.

Finally, the entropy functions $S_{c1}(v_{c1})$ and $S_{c2}(v_{c2})$ are given by $S_{ci}(v_{ci}) = \log_e[1 + v_{ci}(x_{ci})]$, $i = 1, 2$.

To illustrate the behavior of the closed-loop impulsive dynamical sys-

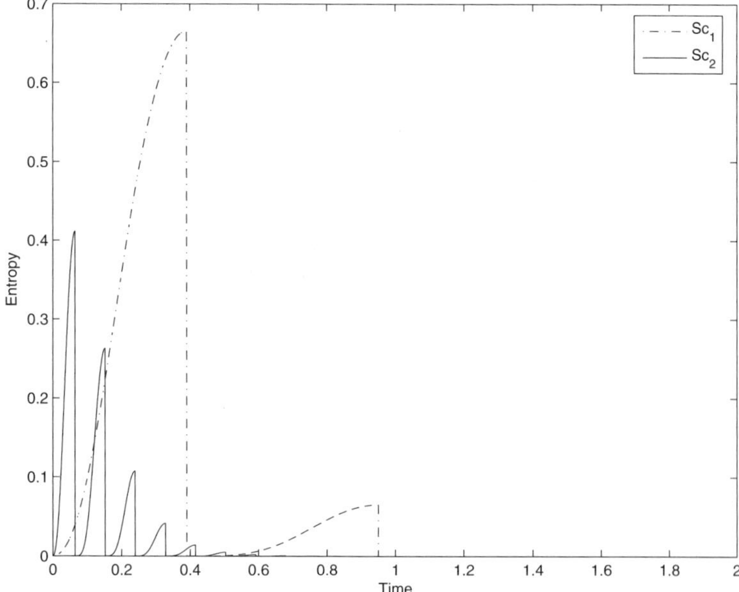

Figure 13.21 Controller entropy versus time.

tem, we choose initial conditions $x_{c1}(0) = [0,0]^{\mathrm{T}}$ and $x_{c2}(0) = [0,0]^{\mathrm{T}}$. For this system a straightforward, but lengthy, calculation shows that Assumptions 13.1 and 13.2 hold. However, the k_i-transversality condition is sufficiently complex that we have been unable to show it analytically. This condition was verified numerically, and hence, Assumption 13.3 appears to hold. Figure 13.16 shows the state trajectories of the plant versus time, and Figure 13.17 shows the state trajectories of the compensator versus time. Figure 13.18 shows the control input versus time. Note that the compensator states are the only states that reset. Once again, the proposed hybrid decentralized controller achieves finite-time stabilization. Furthermore, the control force versus time is partially discontinuous at the resetting times. A comparison of $v_{s1}(x_1, x_2)$, $v_{c1}(x_{c1})$, and $v(x, x_{c1}, x_{c2}) \triangleq v_{s1}(x_1, x_2) + v_{s2}(x_3, x_4) + v_{c1}(x_{c1}) + v_{c2}(x_{c2})$ is shown in Figure 13.19, and a comparison of $v_{s2}(x_3, x_4)$, $v_{c2}(x_{c2})$, and $v(x, x_{c1}, x_{c2})$ is shown in Figure 13.20. Finally, Figure 13.21 shows the controller entropy versus time. Note that the entropy of the controller strictly increases between resetting events.

13.8 Experimental Verification of Hybrid Decentralized Controller

In this section, we provide experimental results to show the efficacy of the hybrid decentralized controller architecture developed in this chapter. The experimental test bed constructed to implement the decentralized energy-based hybrid control technique is shown in Figure 13.22 and involves the

Figure 13.22 RTAC test bed.

Table 13.2 Model and manufacturer information of hardware used.

Description	Manufacturer	Model
Air Bushing	New Way Air Bearings	S301201
Laser sensor	Micro-Epsilon	ILD1300-200
DC motor	MicroMo	3863H012C
Shaft Encoder	MicroMo	HEDM5500J12
Motor Controller	Advanced Motion Controls	12A8
DAQ board	National Instruments	NI-6024E
Encoder/Timer	National Instruments	NI-6601

multi-RTAC system discussed in Section 13.5. The test bed consists of an aluminum base with air bushings floating on two rails providing translational motion for the carts with very low friction. Rotary actuators with eccentric arms and masses are fixed to the carts providing the control torques. The actuation is provided by DC motors driven by a set of linear motor controllers, and the measurements of the eccentric arm angles and cart positions are performed with a quadrature encoder on each motor and a laser displacement sensor for each cart, respectively. The controller is implemented with the MathWorks MATLAB$^{\textregistered}$, Simulink$^{\textregistered}$, and xPC Target^{TM} software using National Instruments PCI cards for input/output (I/O). The hardware used for the test bed is listed in Table 13.2.

Translational motion for each cart is provided by four air bushings mounted into aluminum blocks. These blocks are mounted to an aluminum plate to form the platforms of the carts. These platforms are also constructed to deliver air to the bushings through internal passageways to eliminate excessive air fittings. The air bushings float on two stainless steel precision shafts 0.5 inches in diameter that are affixed to the aluminum base. Negligible damping effects result from the motor and rail friction, resistance from hoses and wires, and air resistance. Supports are attached to the platforms to facilitate mounting of the rotational actuators and the proof-masses. The supports are designed in such a manner that they can be mounted either vertically or horizontally. This enables the experiment to be carried out with and without gravitational effect on the proof-masses.

Four pretensioned extension springs attach the carts to each other and to fixed supports mounted on the base. The springs are easy to remove so that springs with different stiffnesses can be used. The spring stiffness constant used for the testbed was measured to be 170 N/m and the springs are shown to be linear throughout the usable range. The control torque for each cart in the system is provided by means of a proof-mass attached to an actuator by an eccentric arm. The arms are constructed in such a way that various proof-masses may be used. The actuators are 12-volt DC motors that generate a continuous torque of 0.11 N·m each with a stall torque of 1.2 N·m and have a thermally limited continuous current of 7.6 A. Driving the motors are a set of pulse width modulated (PWM) servo amplifiers that can supply a peak current of 12 A and a continuous current of 6 A. The maximum continuous switching rate for each amplifier is 36 kHz. The units are operated in current mode, producing currents that are proportional to the input voltages. The motor controllers have built-in current limiters to protect the motors from high torque commands.

System state measurements are obtained with quadrature encoders and laser displacement sensors. The quadrature encoders measure the angular positions and velocities of the proof-masses. The encoders are attached to the back of the motors and have 1,024-line-per-revolution resolution. This gives an angular resolution of $0.09°$ when used in quadrature mode. The positions of the translational masses are measured with laser sensors that use optical triangulation to measure displacement while velocities are obtained by numerical differentiation of position data. The sensors measure position with a static resolution of 100 μm and dynamic resolution of 200 μm at a rate of 500 Hz and with a measurement range of 200 mm. Laser sensors were selected over other linear measurement sensors since they do not influence the motion of the carts. Note that the information of the position and velocity of the translational carts was only used to obtain a plot of the plant energy versus time and *not* to compute the energy-based hybrid control as follows from (13.78) and (13.79).

During experimentation slight noise in the cart position measurement was observed. This led to a substantial distortion of the cart velocity information obtained from numerical differentiation of the position data. To resolve this issue, an accelerometer was incorporated with the laser sensor. By numerically integrating the cart acceleration and using the laser sensor to eliminate the accumulated error resulting from numerical integration, a much smoother and more accurate cart velocity was determined. This problem does not appear to be as pronounced with the angular velocity determined from the pendulum's angular position due to a higher resolution of the quadrature encoder.

To implement the decentralized, energy-based hybrid control in real time we used the MathWorks MATLAB®, Simulink®, and xPC Target™

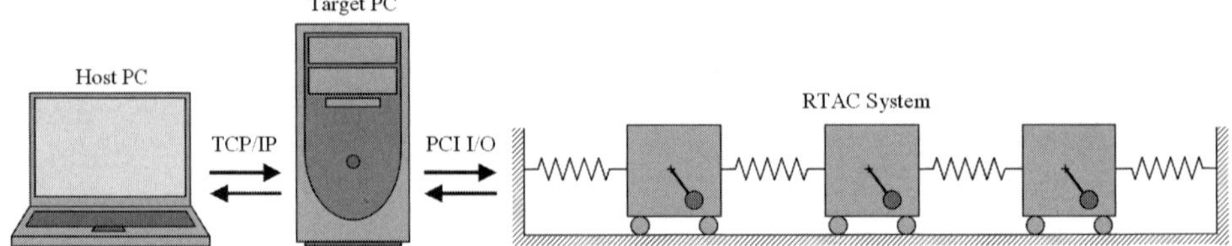

Figure 13.23 Diagram of real-time target implementation.

Table 13.3 Problem data for the multi-RTAC system.

Description	Parameter	Value	Units
Cart mass	M	1.7428	kg
Eccentric mass	m	0.2739	kg
Arm eccentricity	e	0.0537	m
Arm inertia	I	0.000884	kg·m^2
Spring stiffness	k	170	N/m
Controller parameter	m_c	0.0012	—
Controller parameter	k_c	0.0811	—

software. The diagram in Figure 13.23 illustrates the hardware layout. The control law is created in Simulink®, compiled into C code, and then downloaded onto the target PC. The target PC runs a real-time operating system that executes the Simulink® block diagram. The input/output for the target PC consists of National Instruments PCI-6024E and PCI-6601 cards. The PCI-6024E has a maximum input rate of 200 kS/s at 12-bit resolution. The minimum analog input is ±50 mV, and the maximum input is limited to ±10 V. The maximum analog output rate is 10 kS/s at 12-bit resolution with a maximum value of ± 10 V. The PCI-6024E cards are used to acquire the distances measured by the laser sensors and to send voltages to the motor controllers to generate the required control torques, whereas the PCI-6601 cards are used to read the encoders to obtain the angles and directions of rotation of the proof-masses.

Next, we show experimental results obtained by implementing the decentralized, energy-based hybrid control presented in Section 13.4 on the multi-RTAC test bed. The system parameters are shown in Table 13.3 with initial conditions $q_1(0) = -0.074$ m, $q_2(0) = 0.012$ m, $q_3(0) = 0.0055$ m, $\dot{q}_i(0) = 0$, $\theta_i(0) = 0$, $\dot{\theta}_i(0) = 0$, $q_{ci}(0) = 0$, and $\dot{q}_{ci}(0) = 0$, $i = 1, 2, 3$. Figure 13.24 shows positions of the carts versus time, and Figure 13.25 shows the cart velocities versus time. Figure 13.26 shows the angular positions of the pendulums versus time, and Figure 13.27 shows their angular velocities versus time. Figures 13.28 and 13.29 show the time history of each subcontroller state. Note that each subcontroller state is discontinuous. The control torques versus time are shown in Figure 13.30 and are discontinuous at the resetting times as follows from (13.77) and (13.78). These discontinu-

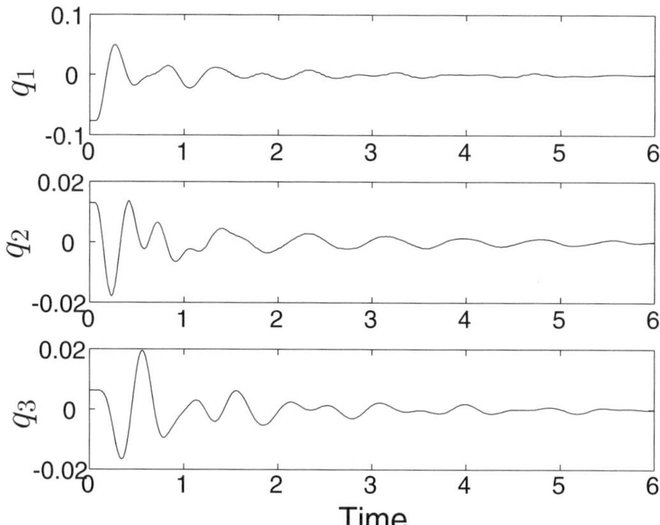

Figure 13.24 Positions of the carts in meters versus time.

ities indicate that the proof-masses suddenly change the direction of motion to prevent the flow of real physical energy back to the translational carts. Figure 13.31 shows the physical energy of the plant, combined emulated energy of all subcontrollers, and the total energy of the multi-RTAC system. Although the sum of the plant energy and controller emulated energy is supposed to remain constant between resettings as shown in (13.64), in the experimental setup the slight decreases in total energy are the result of damping effects that are always present in a physical system.

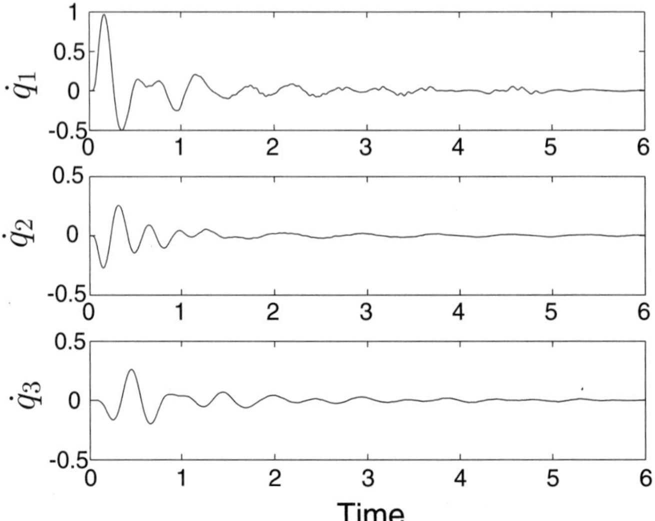

Figure 13.25 Velocities of the carts in m/s versus time.

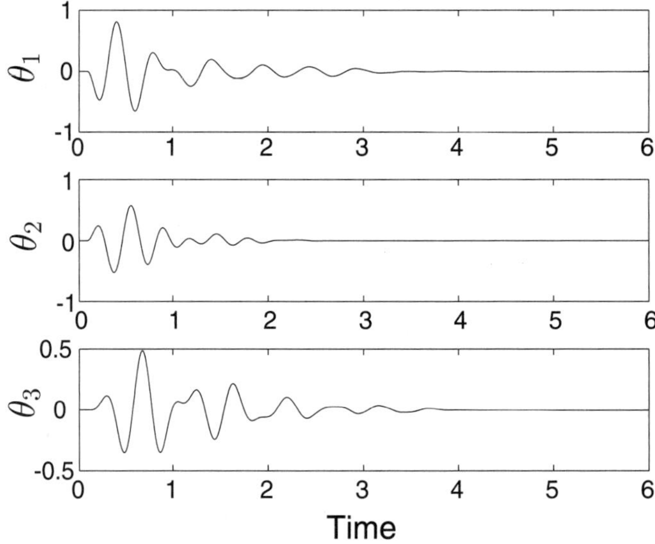

Figure 13.26 Angular positions of the pendulums in radians versus time.

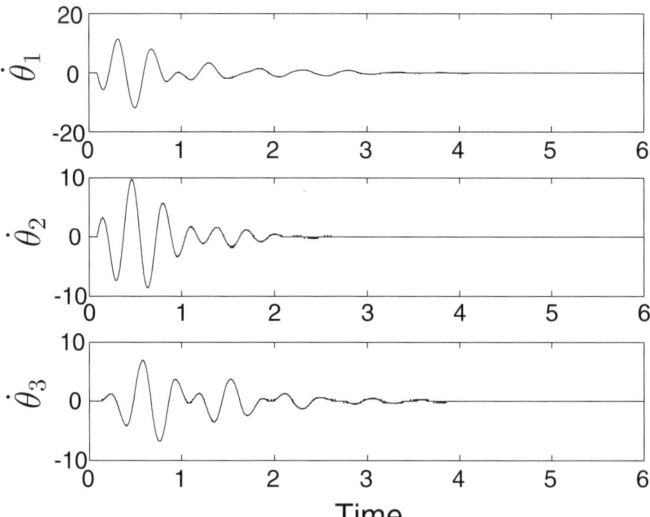

Figure 13.27 Angular velocities of the pendulums in rad/sec versus time.

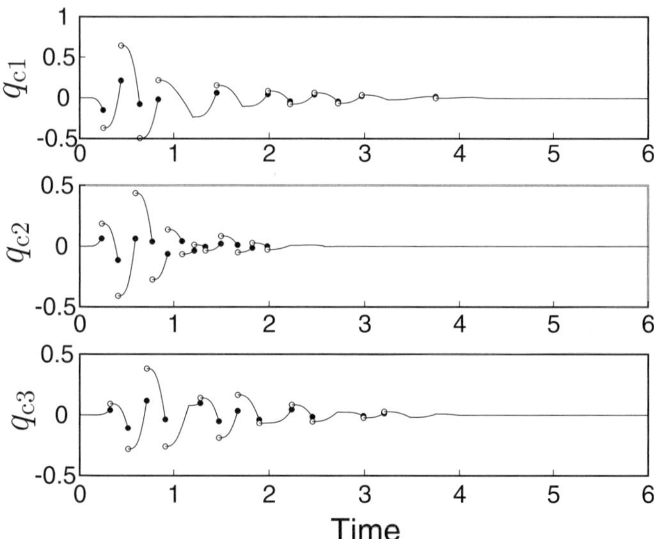

Figure 13.28 Subcontroller positions versus time.

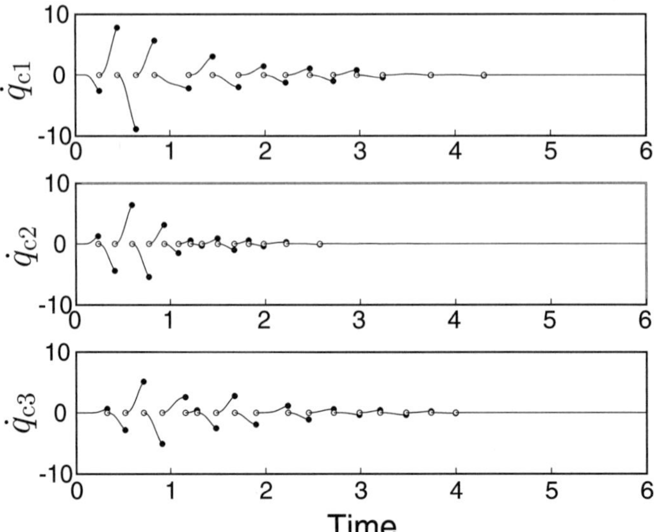

Figure 13.29 Subcontroller velocities versus time.

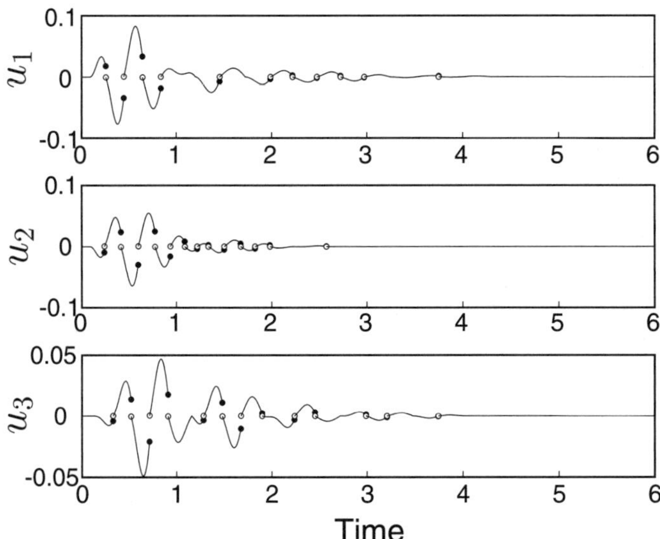

Figure 13.30 Control torques in N·m versus time.

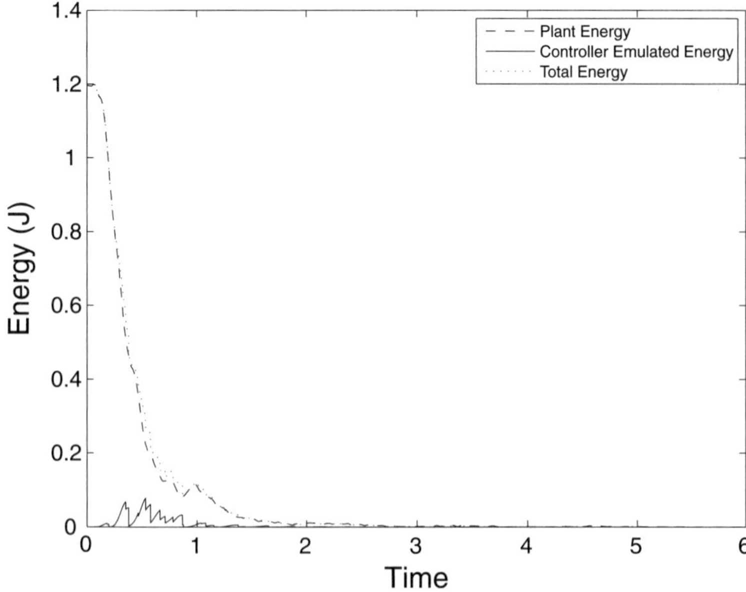

Figure 13.31 Plant, combined subcontroller, and total energies versus time.

Chapter Fourteen

Conclusion

In this monograph, we have developed a stability analysis and control design framework for large-scale complex dynamical systems. These systems are composed of interconnected subsystems and include air traffic control systems, power and energy grid systems, manufacturing and processing systems, aerospace and transportation systems, communication and information networks, integrative biological systems, biological neural networks, biomolecular and biochemical systems, nervous systems, immune systems, environmental and ecological systems, molecular, quantum, and nanoscale systems, particulate and chemical reaction systems, economic and financial systems, cellular systems, metabolic systems, planetary ecosystems (e.g., Gaia), and galaxies, to name but a few examples. The relationships between the subsystems is often circular—in an abstract and not necessarily spatial sense—giving rise to feedback interconnections. This leads to nonlinear models that can exhibit rich dynamical behavior, such as multiple equilibria, limit cycles, bifurcations, jump resonance phenomena, and chaos.

The complexity of large-scale dynamical systems is due to the natural scale of these systems and often necessitates a hierarchical decentralized architecture for analyzing and controlling these systems due to high system dimensionality and global communication connection constraints. The role of system uncertainty is also critical in the analysis and control design of large-scale, interconnected systems. Uncertainty and variations in large-scale systems may change constantly and unpredictably, resulting in rapid and often catastrophic transitions. To address the stability analysis and control design of large-scale systems in the presence of uncertainties, decentralized robust control architectures that combine logical operations with continuous dynamics are needed. These hybrid architectures provide hierarchical coordination and autonomy by utilizing higher-level reasoning and decision making.

The underlying intention of this monograph has been to present a general analysis and control design framework for large-scale dynamical systems, with an emphasis on vector Lyapunov function methods, vector dissipativity theory, and decentralized control architectures. It is hoped that this monograph will help stimulate increased interaction between engineers, economists, biologists, ecologists, physicists, computer scientists, and dynamical systems and control theorists. The potential for applying and extending this work across disciplines is enormous. For example, in economic systems the interaction of raw materials, finished goods, and financial resources can be modeled as large-scale systems with subsystem interconnections representing various interacting sectors in a dynamic economy. Similarly, network systems, computer networks, and telecommunication systems are amenable to large-scale modeling with interconnections governed by nodal dynamics and routing strategies that can be controlled to minimize waiting times and optimize system throughput.

Large-scale system models can also be used to model the interconnecting components of power grid systems with energy flow between regional distribution points subject to control and possible failure. Road, rail, air, and space transportation systems also give rise to large-scale systems with interconnections subject to failure and real-time modification. Modern heating, ventilating, and air conditioning (HVAC) systems in large commercial buildings are characterized by a large number of interconnected zones that require heating, ventilation, and cooling, and thus, also constitute large-scale systems. In particular, the automated operation of a network of HVAC systems for a regional collection of smart buildings involves a large number of interacting subsystems with multiple zones and components, nonlinear heat transfer models, multiple spatial and temporal timescales, and model uncertainties and disturbances involving changes in weather and solar radiation, varying heat loads, humidity, computers and lab equipment, and people and other latent heat sources. In all of the aforementioned applications, reliable system analysis and decentralized robust control system design, with integrated verification and validation, are essential for providing high system performance and reconfigurable system operation in the presence of system uncertainties and system component failures.

Bibliography

[1] R. P. Agarwal and V. Lakshmikantham, *Uniqueness and Nonuniqueness Criteria for Ordinary Differential Equations*. Singapore: World Scientific, 1993.

[2] N. Aggarwal and K. Fujimura, "Motion planning amidst planar moving obstacles," in *IEEE Inter. Conf. Robot. Autom.* (San Diego, CA), pp. 2153–2158, 1994.

[3] G. Amicucci, S. Monaco, and D. Normand-Cyrot, "Control Lyapunov stabilization of affine discrete-time systems," in *Proc. IEEE Conf. Dec. Contr.* (San Diego, CA), pp. 923–924, 1997.

[4] D. H. Anderson, *Compartmental Modeling and Tracer Kinetics*. New York: Springer-Verlag, 1983.

[5] M. Araki, "Input-output stability of composite feedback systems," *IEEE Trans. Autom. Contr.*, vol. AC-21, pp. 254–259, 1976.

[6] Z. Artstein, "Stabilization with relaxed controls," *Non. Anal. Theory, Meth. Appl.*, vol. 7, pp. 1163–1173, 1983.

[7] J. M. Avis, S. G. Nersesov, and R. Nathan, "Decentralized energy-based hybrid control for the multi-RTAC system," *Int. J. Contr.*, vol. 83, no. 8, pp. 1701–1709, 2010.

[8] E. Awad and F. E. C. Culick, "On the existence and stability of limit cycles for longitudinal acoustic modes in a combustion chamber," *Combust. Sci. Technol.*, vol. 9, pp. 195–222, 1986.

[9] E. Awad and F. E. C. Culick, "The two-mode approximation to nonlinear acoustics in combustion chambers I. Exact solution for second order acoustics," *Combust. Sci. Technol.*, vol. 65, pp. 39–65, 1989.

[10] A. Bacciotti and L. Rosier, *Liapunov Functions and Stability in Control Theory*. London: Springer-Verlag, 2001.

[11] D. D. Bainov and P. S. Simeonov, *Systems with Impulse Effect: Stability, Theory and Applications*. Chichester, UK: Ellis Horwood Limited, 1989.

[12] D. D. Bainov and P. S. Simeonov, *Impulsive Differential Equations: Periodic Solutions and Applications.* Chichester, UK: Longman Scientific & Technical, 1993.

[13] D. D. Bainov and P. S. Simeonov, *Impulsive Differential Equations: Asymptotic Properties of the Solutions.* Singapore: World Scientific, 1995.

[14] R. Bellman, "Vector Lyapunov functions," *SIAM Journal of Control,* vol. 1, pp. 32–34, 1962.

[15] A. Berman and R. J. Plemmons, *Nonnegative Matrices in the Mathematical Sciences.* New York: Academic Press, Inc., 1979.

[16] A. Berman, R. S. Varga, and R. C. Ward, "ALPS: Matrices with nonpositive off-diagonal entries," *Linear Algebra Appl.,* vol. 21, pp. 233–244, 1978.

[17] D. S. Bernstein, "Sequential design of decentralized dynamic compensation using the optimal projection equations," *Int. J. Contr.,* vol. 46, pp. 1569–1577, 1987.

[18] D. S. Bernstein, *Matrix Mathematics.* Princeton, NJ: Princeton University Press, 2005.

[19] D. S. Bernstein and S. P. Bhat, "Nonnegativity, reducibility, and semistability of mass action kinetics," in *Proc. IEEE Conf. Dec. Contr.* (Phoenix, AZ), pp. 2206–2211, 1999.

[20] D. S. Bernstein and S. P. Bhat, "Energy equipartition and the emergence of damping in lossless systems," in *Proc. IEEE Conf. Dec. Contr.* (Las Vegas, NV), pp. 2913–2918, 2002.

[21] D. S. Bernstein and D. C. Hyland, "Compartmental modeling and second-moment analysis of state space systems," *SIAM J. Matrix Anal. Appl.,* vol. 14, pp. 880–901, 1993.

[22] J. Bernussou and A. Titli, *Interconnected Dynamical Systems: Stability, Decomposition, and Decentralisation.* New York: North-Holland, 1982.

[23] S. P. Bhat and D. S. Bernstein, "Continuous finite-time stabilization of the translational and rotational double integrators," *IEEE Trans. Autom. Contr.,* vol. 43, pp. 678–682, 1998.

[24] S. P. Bhat and D. S. Bernstein, "Finite-time stability of continuous autonomous systems," *SIAM J. Control Optim.,* vol. 38, no. 3, pp. 751–766, 2000.

[25] S. P. Bhat and D. S. Bernstein, "Geometric homogeneity with applications to finite-time stability," *Math. Control Signals Systems*, vol. 17, pp. 101–127, 2005.

[26] S. Boyd, L. E. Ghaoui, E. Feron, and V. Balakrishnan, *Linear Matrix Inequalities in System and Control Theory*. Philadelphia, PA: In SIAM studies in applied mathematics, 1994.

[27] R. W. Brockett and J. C. Willems, "Stochastic control and the second law of thermodynamics," in *Proc. IEEE Conf. Dec. Contr.* (San Diego, CA), pp. 1007–1011, 1978.

[28] B. Brogliato, *Nonsmooth Mechanics*, 2nd ed. London: Springer-Verlag, 1999.

[29] R. F. Brown, "Compartmental system analysis: State of the art," *IEEE Trans. Biomed. Engineering*, vol. 27, pp. 1–11, 1980.

[30] J. Brunet, "Information theory and thermodynamics," *Cybernetica*, vol. 32, pp. 45–78, 1989.

[31] F. Bullo, J. Cortéz, and S. Martinéz, *Distributed Control of Robotic Networks*. Princeton, NJ: Princeton University Press, 2009.

[32] R. T. Bupp, D. S. Bernstein, V. Chellaboina, and W. M. Haddad, "Resetting virtual absorbers for vibration control," *J. Vibr. Contr.*, vol. 6, pp. 61–83, 2000.

[33] R. T. Bupp, D. S. Bernstein, and V. T. Coppola, "A benchmark problem for nonlinear control design," *Int. J. Robust and Nonlinear Control*, vol. 8, pp. 307–310, 1998.

[34] F. M. Callier, W. S. Chan, and C. A. Desoer, "Input-output stability of interconnected systems using decomposition: An improved formulation," in *Proc. IEEE Conf. Dec. Contr.* (New Orleans, LA), pp. 1249–1255, 1977.

[35] F. E. Cellier, *Continuous System Modeling*. New York: Springer-Verlag, 1991.

[36] P. R. Chandler, M. Pachter, and S. Rasmussen, "UAV cooperative control," in *Proc. Amer. Contr. Conf.* (Arlington, VA), pp. 50–55, 2001.

[37] V. Chellaboina, S. P. Bhat, and W. M. Haddad, "An invariance principle for nonlinear hybrid and impulsive dynamical systems," *Nonlinear Analysis: Theory, Methods, and Applications*, vol. 53, pp. 527–550, 2003.

[38] V. Chellaboina and W. M. Haddad, "Stability margins of discrete-time nonlinear-nonquadratic optimal regulators," in *Proc. IEEE Conf. Dec. Contr.* (Tampa, FL), pp. 1786–1791, 1998.

[39] V. Chellaboina and W. M. Haddad, "Stability margins of nonlinear optimal regulators with nonquadratic performance criteria involving cross-weighting terms," *Syst. Contr. Lett.*, vol. 39, pp. 71–78, 2000.

[40] V. Chellaboina and W. M. Haddad, "Stability margins of discrete-time nonlinear-nonquadratic optimal regulators," *Int. J. Syst. Sci.*, vol. 33, pp. 577–584, 2002.

[41] V. Chellaboina and W. M. Haddad, "A unification between partial stability and stability theory for time-varying systems," *Contr. Syst. Mag.*, vol. 22, no. 6, pp. 66–75, 2002, Erratum, vol. 23, p. 101, 2003.

[42] V. Chellaboina and W. M. Haddad, "Exponentially dissipative dynamical systems: A nonlinear extension of strict positive realness," *J. Math. Prob. Engin.*, vol. 2003, pp. 25–45, 2003.

[43] E. A. Coddington and N. Levinson, *Theory of Ordinary Differential Equations.* New York: McGraw-Hill, 1955.

[44] W. A. Coppel, *Stability and Asymptotic Behavior of Differential Equations.* Boston, MA: D. C. Heath and Co., 1965.

[45] J. P. Corfmat and A. S. Morse, "Decentralized control of linear multivariable systems," *Automatica*, vol. 12, pp. 479–495, 1976.

[46] F. A. Cosio and M. P. Castaneda, "Autonomous robot navigation using adaptive potential fields," *Mathematical and Computer Modelling*, vol. 49, no. 9, pp. 1141–1156, 2004.

[47] F. E. C. Culick, "Nonlinear behavior of acoustic waves in combustion chambers I," *Acta Astronautica*, vol. 3, pp. 715–734, 1976.

[48] E. J. Davison and W. Gesing, "Sequential stability and optimization of large scale decentralized systems," *Automatica*, vol. 15, pp. 307–324, 1979.

[49] J. P. Desai, J. P. Ostrowski, and V. Kumar, "Modeling and control of formations of nonholonomic mobile robots," *IEEE Trans. Robot. Automat.*, vol. 17, pp. 905–908, 2001.

[50] A. R. Dieguez, R. Sanz, and J. Lopez, "Deliberative on-line local path planning for autonomous mobile robots," *Journal of Intelligent and Robotic Systems: Theory and Applications*, vol. 37, no. 1, pp. 1–19, 2003.

[51] R. Diestel, *Graph Theory*. New York: Springer-Verlag, 1997.

[52] Z. Drici, "New directions in the method of vector Lyapunov functions," *J. Math. Anal. Appl.*, vol. 184, pp. 317–325, 1994.

[53] M. Egerstedt, X. Hu, and A. Stotsky, "Control of mobile platforms using a virtual vehicle approach," *IEEE Trans. Autom. Contr.*, vol. 46, pp. 1777–1782, 2001.

[54] L. P. Ellekilde and J. W. Perram, "Tool center trajectory planning for industrial robot manipulators using dynamical systems," *The International Journal of Robotics Research*, vol. 24, no. 5, pp. 385–396, 2005.

[55] F. Fahimi, C. Nataraj, and H. Ashrafiuon, "Real-time obstacle avoidance for multiple mobile robots," *Robotica*, vol. 27, no. 2, pp. 189–198, 2009.

[56] J. A. Fax and R. M. Murray, "Information flow and cooperative control of vehicle formations," *IEEE Trans. Autom. Contr.*, vol. 49, pp. 1465–1476, 2004.

[57] A. Ferrara and M. Rubagotti, "Second-order sliding-mode control of a mobile robot based on a harmonic potential field," *IET Control Theory and Applications*, vol. 2, no. 9, pp. 807–818, 2008.

[58] A. F. Filippov, *Differential Equations with Discontinuous Right-Hand Sides*. Mathematics and its Applications (Soviet Series), Dordrecht: Kluwer Academic Publishers, 1988.

[59] A. T. Fuller, "Optimization of some nonlinear control systems by means of Bellman's equation and dimensional analysis," *Int. J. Control*, vol. 3, pp. 359–394, 1966.

[60] S. S. Ge and Y. J. Sui, "Dynamic motion planning for mobile robots using potential field method," *Autonomous Robots*, vol. 1, pp. 207–222, 2002.

[61] J. C. Gentina, P. Borne, C. Burgat, J. Bernussou, and L. T. Grujic, "Sur la stabilité des systémes de grande dimension normes vectorielles," *R.A.I.R.O. Autom. Syst. Anal. Control*, vol. 13, pp. 57–75, 1979.

[62] K. Godfrey, *Compartmental Models and Their Applications*. New York: Academic Press, 1983.

[63] C. Godsil and G. Royle, *Algebraic Graph Theory*. New York: Springer-Verlag, 2001.

[64] J. W. Grizzle, G. Abba, and F. Plestan, "Asymptotically stable walking for biped robots: Analysis via systems with impulse effects," *IEEE Trans. Autom. Contr.*, vol. 46, pp. 51–64, 2001.

[65] L. T. Grujić, A. A. Martynyuk, and M. Ribbens-Pavella, *Large Scale Systems: Stability Under Structural and Singular Perturbations.* Berlin: Springer-Verlag, 1987.

[66] L. T. Gruyitch, J. P. Richard, P. Borne, and J. C. Gentina, *Stability Domains.* Boca Raton: Chapman & Hall/CRC, 2004.

[67] E. P. Gyftopoulos and E. Çubukçu, "Entropy: Thermodynamic definition and quantum expression," *Phys. Rev. E*, vol. 55, no. 4, pp. 3851–3858, 1997.

[68] W. M. Haddad and V. Chellaboina, "Dissipativity theory and stability of feedback interconnections for hybrid dynamical systems," *J. Math. Prob. Engin.*, vol. 7, pp. 299–355, 2001.

[69] W. M. Haddad and V. Chellaboina, "Stability and dissipativity theory for nonnegative dynamical systems: A unified analysis framework for biological and physiological systems," *Nonlinear Analysis: Real World Applications*, vol. 6, pp. 35–65, 2005.

[70] W. M. Haddad and V. Chellaboina, *Nonlinear Dynamical Systems and Control. A Lyapunov-Based Approach.* Princeton, NJ: Princeton University Press, 2008.

[71] W. M. Haddad, V. Chellaboina, and E. August, "Stability and dissipativity theory for discrete-time nonnegative and compartmental dynamical systems," *Int. J. Contr.*, vol. 76, pp. 1845–1861, 2003.

[72] W. M. Haddad, V. Chellaboina, and Q. Hui, *Nonnegative and Compartmental Dynamical Systems.* Princeton, NJ: Princeton University Press, 2010.

[73] W. M. Haddad, V. Chellaboina, Q. Hui, and S. G. Nersesov, "Energy- and entropy-based stabilization for lossless dynamical systems via hybrid controllers," *IEEE Trans. Autom. Contr.*, vol. 52, no. 9, pp. 1604–1614, 2007.

[74] W. M. Haddad, V. Chellaboina, and N. A. Kablar, "Nonlinear impulsive dynamical systems. Part I: Stability and dissipativity," *Int. J. Contr.*, vol. 74, pp. 1631–1658, 2001.

[75] W. M. Haddad, V. Chellaboina, and N. A. Kablar, "Nonlinear impulsive dynamical systems. Part II: Stability of feedback interconnections and optimality," *Int. J. Contr.*, vol. 74, pp. 1659–1677, 2001.

[76] W. M. Haddad, V. Chellaboina, and S. G. Nersesov, "On the equivalence between dissipativity and optimality of nonlinear hybrid controllers," *Int. J. Hybrid Syst.*, vol. 1, pp. 51–65, 2001.

[77] W. M. Haddad, V. Chellaboina, and S. G. Nersesov, "Hybrid nonnegative and compartmental dynamical systems," *J. Math. Prob. Engin.*, vol. 8, pp. 493–515, 2002.

[78] W. M. Haddad, V. Chellaboina, and S. G. Nersesov, "A unification between partial stability of state-dependent impulsive systems and stability theory for time-dependent impulsive systems," *Int. J. Hybrid Syst.*, vol. 2, no. 2, pp. 155–168, 2002.

[79] W. M. Haddad, V. Chellaboina, and S. G. Nersesov, "A system-theoretic foundation for thermodynamics: Energy flow, energy balance, energy equipartition, entropy, and ectropy," in *Proc. Amer. Contr. Conf.* (Boston, MA), pp. 396–417, 2004.

[80] W. M. Haddad, V. Chellaboina, and S. G. Nersesov, "Thermodynamics and large-scale nonlinear dynamical systems: A vector dissipative systems approach," *Dyn. Cont. Disc. Impl. Syst.*, vol. 11, pp. 609–649, 2004.

[81] W. M. Haddad, V. Chellaboina, and S. G. Nersesov, *Thermodynamics. A Dynamical Systems Approach.* Princeton, NJ: Princeton University Press, 2005.

[82] W. M. Haddad, V. Chellaboina, and S. G. Nersesov, *Impulsive and Hybrid Dynamical Systems. Stability, Dissipativity, and Control.* Princeton, NJ: Princeton University Press, 2006.

[83] W. M. Haddad, Q. Hui, S. G. Nersesov, and V. Chellaboina, "Thermodynamic modeling, energy equipartition, and nonconservation of entropy for discrete-time dynamical systems," in *Proc. Amer. Contr. Conf.* (Portland, OR), pp. 4832–4837, 2005.

[84] W. M. Haddad, Q. Hui, S. G. Nersesov, and V. Chellaboina, "Thermodynamic modeling, energy equipartition, and nonconservation of entropy for discrete-time dynamical systems," *Adv. Diff. Eqns.*, vol. 2005, no. 3, pp. 275–318, 2005.

[85] W. Hahn, *Stability of Motion.* Berlin: Springer Verlag, 1967.

[86] V. T. Haimo, "Finite-time controllers," *SIAM J. Control Optim.*, vol. 24, pp. 760–770, 1986.

[87] J. K. Hale, *Ordinary Differential Equations.* New York: Wiley, second ed., 1980. reprinted by Krieger, Malabar, 1991.

[88] S. R. Hall, D. G. MacMartin, and D. S. Bernstein, "Covariance averaging in the analysis of uncertain systems," in *Proc. IEEE Conf. Dec. Contr.*, (Tucson, AZ), pp. 1842–1859, 1992.

[89] D. J. Hill and P. J. Moylan, "Stability results of nonlinear feedback systems," *Automatica*, vol. 13, pp. 377–382, 1977.

[90] Y. Hong, "Finite-time stabilization and stabilizability of a class of controllable systems," *Syst. Contr. Lett.*, vol. 46, pp. 231–236, 2002.

[91] Y. Hong, J. Huang, and Y. Xu, "On an output feedback finite-time stabilization problem," *IEEE Trans. Autom. Control*, vol. 46, pp. 305–309, 2001.

[92] R. A. Horn and R. C. Johnson, *Matrix Analysis.* Cambridge, UK: Cambridge University Press, 1985.

[93] R. A. Horn and R. C. Johnson, *Topics in Matrix Analysis.* Cambridge, UK: Cambridge University Press, 1995.

[94] S. Hu, V. Lakshmikantham, and S. Leela, "Impulsive differential systems and the pulse phenomena," *J. Math. Anal. Appl.*, vol. 137, pp. 605–612, 1989.

[95] Q. Hui and W. M. Haddad, "Distributed nonlinear control algorithms for network consensus," *Automatica*, vol. 44, pp. 2375–2381, 2008.

[96] M. Ikeda and D. D. Šiljak, "Overlapping decompositions, expansions, and contractions of dynamic systems," *Large Scale Systems*, vol. 1, pp. 29–38, 1980.

[97] M. Ikeda and D. D. Šiljak, "Generalized decompositions of dynamic systems and vector Lyapunov functions," *IEEE Trans. Autom. Contr.*, vol. 26, pp. 1118–1125, 1981.

[98] M. Ikeda, D. D. Šiljak, and D. E. White, "Decentralized control with overlapping information sets," *J. Optim. Theory Appl.*, vol. 34, pp. 279–310, 1981.

[99] M. Ikeda, D. D. Šiljak, and D. E. White, "An inclusion principle for dynamic systems," *IEEE Trans. Autom. Contr.*, vol. 29, pp. 244–249, 1984.

[100] J. A. Jacquez, *Compartmental Analysis in Biology and Medicine,* 2nd ed. Ann Arbor, MI: University of Michigan Press, 1985.

[101] J. A. Jacquez and C. P. Simon, "Qualitative theory of compartmental systems," *SIAM Rev.*, vol. 35, pp. 43–79, 1993.

[102] A. Jadbabaie, J. Lin, and A. S. Morse, "Coordination of groups of mobile autonomous agents using nearest neighbor rules," *IEEE Trans. Autom. Contr.*, vol. 48, pp. 988–1001, 2003.

[103] C. C. Jahnke and F. E. C. Culick, "Application of dynamical systems theory to nonlinear combustion instabilities," *J. Propulsion Power*, vol. 10, pp. 508–517, 1994.

[104] M. Jamshidi, *Large-Scale Systems*. New York: North-Holland, 1983.

[105] V. Jurdjevic and J. P. Quinn, "Controllability and stability," *J. of Diff. Eqs.*, vol. 28, pp. 381–389, 1978.

[106] E. Kamke, "Zur Theorie der Systeme gewöhnlicher Differential - Gleichungen. II," *Acta Mathematica*, vol. 58, pp. 57–85, 1931.

[107] D. Karnopp, D. L. Margolis, and R. C. Rosenberg, *System Dynamics: A Unified Approach*. New York: Wiley-Interscience, 1990.

[108] M. Kawski, "Stabilization of nonlinear systems in the plane," *Syst. Contr. Lett.*, vol. 12, pp. 169–175, 1989.

[109] A. J. Keane and W. G. Price, "Statistical energy analysis of strongly coupled systems," *J. Sound Vibr.*, vol. 117, pp. 363–386, 1987.

[110] H. K. Khalil, *Nonlinear Systems*. Upper Saddle River, NJ: Prentice-Hall, 1996.

[111] D. H. Kim and P. Chongkug, "Limit cycle navigation method for mobile robot," in *27th Chinese Control Conference* (Kunming, Yunnan, China), pp. 320–324, 2008.

[112] D. H. Kim and J. H. Kim, "A real-time limit-cycle navigation method for fast mobile robots and its application to robot soccer," *Robotics and Autonomous Systems*, vol. 42, no. 1, pp. 17–30, 2003.

[113] Y. Kishimoto and D. S. Bernstein, "Thermodynamic modeling of interconnected systems I: Conservative coupling," *J. Sound Vibr.*, vol. 182, pp. 23–58, 1995.

[114] Y. Kishimoto and D. S. Bernstein, "Thermodynamic modeling of interconnected systems II: Dissipative coupling," *J. Sound Vibr.*, vol. 182, pp. 59–76, 1995.

[115] Y. Kishimoto, D. S. Bernstein, and S. R. Hall, "Energy flow modeling of interconnected structures: A deterministic foundation for statistical energy analysis," *J. Sound Vibr.*, vol. 186, pp. 407–445, 1995.

[116] N. N. Krasovskii, *Problems of the Theory of Stability of Motion*. Stanford, CA: Stanford University Press, 1959.

[117] V. Lakshmikantham, D. D. Bainov, and P. S. Simeonov, *Theory of Impulsive Differential Equations*. Singapore: World Scientific, 1989.

[118] V. Lakshmikantham, V. M. Matrosov, and S. Sivasundaram, *Vector Lyapunov Functions and Stability Analysis of Nonlinear Systems*. Dordrecht: Kluwer Academic Publishers, 1991.

[119] R. S. Langley, "A general derivation of the statistical energy analysis equations for coupled dynamic systems," *J. Sound Vibr.*, vol. 135, pp. 499–508, 1989.

[120] J. P. LaSalle, "Some extensions to Lyapunov's second method," *IRE Trans. Circuit Theory*, vol. 7, pp. 520–527, 1960.

[121] J. P. LaSalle, "An invariance principle in the theory of stability," in *Differential Equations and Dynamical Systems* (J. Hale and J. P. LaSalle, eds.), Mayaguez, PR: Proceedings of the International Symposium, 1965.

[122] E. L. Lasley and A. N. Michel, "Input-output stability of interconnected systems," *IEEE Trans. Autom. Contr.*, vol. AC-21, pp. 84–89, 1976.

[123] E. L. Lasley and A. N. Michel, "L_∞- and l_∞- stability of interconnected systems," *IEEE Trans. Circuits and Syst.*, vol. CAS-23, pp. 261–270, 1976.

[124] N. E. Leonard and E. Fiorelli, "Virtual leaders, artificial potentials, and coordinated control of groups," in *Proc. IEEE Conf. Dec. Contr.* (Orlando, FL), pp. 2968–2873, 2001.

[125] D. K. Lindner, "On the decentralized control of interconnected systems," *Syst. Contr. Lett.*, vol. 6, pp. 109–112, 1985.

[126] A. Linnemann, "Decentralized control of dynamically interconnected systems," *IEEE Trans. Autom. Contr.*, vol. 29, pp. 1052–1054, 1984.

[127] J. Lunze, "Stability analysis of large-scale systems composed of strongly coupled similar subsystems," *Automatica*, vol. 25, pp. 561–570, 1989.

[128] A. M. Lyapunov, *The General Problem of the Stability of Motion*. Kharkov, Russia: Kharkov Mathematical Society, 1892.

[129] R. H. Lyon, *Statistical Energy Analysis of Dynamical Systems: Theory and Applications*. Cambridge, MA: MIT Press, 1975.

[130] J. A. Marshall, M. E. Broucke, and B. A. Francis, "Formations of vehicles in cyclic pursuit," *IEEE Trans. Autom. Contr.*, vol. 49, pp. 1963–1974, 2004.

[131] A. A. Martynyuk, *Stability by Liapunov's Matrix Function Method with Applications*. New York: Marcel Dekker, Inc., 1998.

[132] A. A. Martynyuk, *Qualitative Methods in Nonlinear Dynamics. Novel Approaches to Liapunov's Matrix Functions*. New York: Marcel Dekker, Inc., 2002.

[133] V. M. Matrosov, "Method of vector Liapunov functions of interconnected systems with distributed parameters (Survey)," *Avtomatika i Telemekhanika*, vol. 33, pp. 63–75, 1972 (in Russian).

[134] M. Mesbahi, "On state-dependent dynamic graphs and their controllability properties," *IEEE Trans. Autom. Contr.*, vol. 50, pp. 387–392, 2005.

[135] M. Mesbahi and M. Egerstedt, *Graph Theoretic Methods in Multiagent Networks*. Princeton, NJ: Princeton University Press, 2010.

[136] A. N. Michel and R. K. Miller, *Qualitative Analysis of Large Scale Dynamical Systems*. New York: Academic Press, Inc., 1977.

[137] A. N. Michel, K. Wang, and B. Hu, *Qualitative Theory of Dynamical Systems*, 2nd ed. New York: Marcel Dekker, 2001.

[138] V. D. Mil'man and A. D. Myshkis, "On the stability of motion in the presence of impulses," *Sib. Math. J.*, vol. 1, pp. 233–237, 1960.

[139] V. D. Mil'man and A. D. Myshkis, "Approximate methods of solutions of differential equations," in *Random Impulses in Linear Dynamical Systems*, pp. 64–81, Kiev: Publ. House. Acad. Sci. Ukr. SSR, 1963.

[140] R. M. Murray, "Recent research in cooperative control of multi-vehicle systems," *J. Dyn. Syst. Meas. Contr.*, vol. 129, pp. 571–583, 2007.

[141] S. G. Nersesov and W. M. Haddad, "On the stability and control of nonlinear dynamical systems via vector Lyapunov functions," *IEEE Trans. Autom. Contr.*, vol. 51, no. 2, pp. 203–215, 2006.

[142] R. Olfati-Saber, "Flocking for multi-agent dynamic systems: Algorithms and theory," *IEEE Trans. Autom. Contr.*, vol. 51, pp. 401–420, 2006.

[143] R. Olfati-Saber and R. M. Murray, "Distributed cooperative control of multiple vehicle formations using structural potential functions," in *IFAC World Congr.* (Barcelona, Spain), 2002.

[144] P. Örgen, M. Egerstedt, and X. Hu, "A control Lyapunov function approach to multiagent coordination," *IEEE Trans. Robot. Automat.*, vol. 18, no. 5, pp. 847–851, 2002.

[145] U. Ozguner, "Near-optimal control of composite systems: The multi time-scale approach," *IEEE Trans. Autom. Contr.*, vol. 24, pp. 652–655, 1979.

[146] K. Pathak and S. K. Agrawal, "An integrated path-planning and control approach for nonholonomic unicycles using switched local potentials," *IEEE Transactions on Robotics*, vol. 21, no. 6, pp. 1201–1208, 2005.

[147] M. Pavon, "Stochastic control and nonequilibrium thermodynamical systems," *Appl. Math. Optim.*, vol. 19, pp. 187–202, 1989.

[148] R. K. Pearson and T. L. Johnson, "Energy equipartition and fluctuation-dissipation theorems for damped flexible structures," *Quart. Appl. Math.*, vol. 45, pp. 223–238, 1987.

[149] C. Qian and W. Lin, "A continuous feedback approach to global strong stabilization of nonlinear systems," *IEEE Trans. Autom. Control*, vol. 46, pp. 1061–1079, 2001.

[150] A. Ramakrishna and N. Viswanadham, "Decentralized control of interconnected dynamical systems," *IEEE Trans. Autom. Contr.*, vol. 27, pp. 159–164, 1982.

[151] W. Ren and R. W. Beard, *Distributed Consensus in Multi-Vehicle Cooperative Control.* London: Springer-Verlag, 2008.

[152] E. P. Ryan, "Singular optimal controls for second-order saturating systems," *Int. J. Contr.*, vol. 30, pp. 549–564, 1979.

[153] E. P. Ryan, "Finite-time stabilization of uncertain nonlinear planar systems," *Dynamics and Control*, vol. 1, pp. 83–94, 1991.

[154] R. Saeks, "On the decentralized control of interconnected dynamical systems," *IEEE Trans. Autom. Contr.*, vol. 24, pp. 269–271, 1979.

[155] A. M. Samoilenko and N. A. Perestyuk, *Impulsive Differential Equations.* Singapore: World Scientific, 1995.

[156] W. Sandberg, "On the mathematical foundations of compartmental analysis in biology, medicine and ecology," *IEEE Trans. Circuits and Systems*, vol. 25, pp. 273–279, 1978.

[157] R. Sepulchre, M. Janković, and P. Kokotović, *Constructive Nonlinear Control.* New York: Springer-Verlag, 1997.

[158] M. E. Sezer and O. Huseyin, "On decentralized stabilization of interconnected systems," *Automatica*, vol. 16, pp. 205–209, 1980.

[159] D. D. Šiljak, *Large-Scale Dynamic Systems: Stability and Structure*. New York: Elsevier North-Holland Inc., 1978.

[160] D. D. Šiljak, "Complex dynamical systems: Dimensionality, structure and uncertainty," *Large Scale Systems*, vol. 4, pp. 279–294, 1983.

[161] D. D. Šiljak, *Decentralized Control of Complex Systems*. San Diego, CA: Academic Press, 1991.

[162] M. G. Singh, *Decentralised Control*. New York: North-Holland, 1981.

[163] P. W. Smith, "Statistical models of coupled dynamical systems and the transition from weak to strong coupling," *Journal of the Acoustical Society of America*, vol. 65, pp. 695–698, 1979.

[164] T. R. Smith, H. Hanssmann, and N. E. Leonard, "Orientation control of multiple underwater vehicles with symmetry-breaking potentials," in *Proc. IEEE Conf. Dec. Contr.* (Orlando, FL), pp. 4598–4603, 2001.

[165] E. D. Sontag, "A universal construction of Artstein's theorem on nonlinear stabilization," *Syst. Contr. Lett.*, vol. 13, pp. 117–123, 1989.

[166] H. G. Tanner, A. Jadbabaie, and G. J. Pappas, "Flocking in fixed and switching networks," *IEEE Trans. Autom. Contr.*, vol. 52, pp. 863–868, 2007.

[167] J. Tsinias, "Existence of control Lyapunov functions and applications to state feedback stabilizability of nonlinear systems," *SIAM J. Control Optim.*, vol. 29, pp. 457–473, 1991.

[168] M. Vidyasagar, *Input-Output Analysis of Large-Scale Interconnected Systems*. Berlin: Springer-Verlag, 1981.

[169] T. Ważewski, "Systèmes des équations et des inégalités différentielles ordinaires aux deuxièmes membres monotones et leurs applications," *Annales de la Société Polonaise de Mathématique*, vol. 23, pp. 112–166, 1950.

[170] J. C. Willems, "Dissipative dynamical systems. Part I: General theory," *Arch. Rational Mech. Anal.*, vol. 45, pp. 321–351, 1972.

[171] J. C. Willems, "Dissipative dynamical systems. Part II: Linear systems with quadratic supply rates," *Arch. Rational Mech. Anal.*, vol. 45, pp. 352–393, 1972.

[172] J. C. Willems, "The behavioral approach to open and interconnected systems: Modeling by tearing, zooming, and linking," *Contr. Syst. Mag.*, vol. 27, no. 6, pp. 46–99, 2007.

[173] J. Woodhouse, "An approach to the theoretical background of statistical energy analysis applied to structural vibration," *Journal of the Acoustical Society of America*, vol. 69, pp. 1695–1709, 1981.

[174] W. Xiaohua, V. Yadav, and B. S. N., "Cooperative uav formation flying with obstacle/collision avoidance," *IEEE Transactions on Control Systems Technology*, vol. 15, no. 4, pp. 672–679, 2007.

[175] T. Yang, *Impulsive Control Theory*. Berlin: Springer-Verlag, 2001.

[176] T. Yoshizawa, *Stability Theory by Liapunov's Second Method*. Tokyo: Mathematical Society of Japan, 1966.

[177] P. G. Zavlangas and S. G. Tzafestas, "Motion control for mobile robot obstacle avoidance and navigation: A fuzzy logic-based approach," *Systems Analysis Modelling Simulation*, vol. 43, pp. 1625–1637, 2003.

[178] A. I. Zečević and D. D. Šiljak, *Control of Complex Systems: Structural Constraints and Uncertainty*. New York: Springer, 2010.

[179] J. M. Ziman, *Models of Disorder*. Cambridge, UK: Cambridge University Press, 1979.

Index

A

admissible controls, 94
anti-Clausius inequality, 82
  discrete-time system, 191
asymptotically stable, 13
  discrete-time system, 35
asymptotically stable matrix, 13, 228
  discrete-time system, 35
asymptotically stable set, 129
asymptotically stable with respect to z, 18
  discrete-time system, 39
  impulsive system, 216
asymptotically stable with respect to z uniformly in x_0, 18
  discrete-time system, 39
  impulsive system, 216
available storage, 50
  discrete-time system, 158
  impulsive system, 234

B

beating, 214
bond-graph modeling, 3

C

class $\mathcal{W}$ functions, 11
class $\mathcal{W}_\mathrm{d}$ functions, 14
Clausius' inequality, 82
  discrete-time system, 191
combustion control, 122
combustion processes, 122
combustion systems, 122
comparison system, 14
compartmental matrix, 11
  discrete-time system, 35
compartmental model, 2, 181
completely reachable, 48
  discrete-time system, 156
  impulsive system, 226
confluence, 214
connective stability, 2
connectivity matrix, 79
conservation of energy, 75
  discrete-time system, 182
consistency property, 11
control law, 94
control of networks, 127
control over networks, 127
control vector Lyapunov function, 6, 94, 95
  candidate, 95
  finite-time stabilization, 114
  impulsive system, 272, 273
controllable, 186
controller emulated energy, 315
cooperative control, 127

D

decentralized control, 4, 102
  hybrid systems, 284
decentralized energy-based controller, 320
dilation, 119
directed graph, 79
disconnected system, 48
  impulsive system, 229
dissipation inequality, 46
dissipative system, 46
  discrete-time system, 154
  impulsive system, 226
dissipativity theory, 3
distributed control, 4

E

ectropy, 6, 81
  discrete-time system, 190
emulated energy, 321
energy balance equation, 76
  discrete-time system, 182
energy equipartition, 86
  discrete-time system, 197
energy similar subsystems, 89
  discrete-time system, 199

energy storage function, 184
energy supply rate, 184
entropy, 6, 80
    discrete-time system, 188
equilibrium point
    continuous-time system, 11
    impulsive system, 213
essentially nonnegative function, 12
essentially nonnegative matrix, 11
Euler-Lagrange equation, 319
existential statement, 9
exponentially dissipative system, 46
    impulsive system, 226
exponentially stable with respect to z
    uniformly in x_0, 19
    impulsive system, 216
exponentially vector dissipative system, 48, 62
    impulsive system, 228, 250

F

feedback control law, 94
feedback interconnections, 71
    discrete-time system, 177
finite-time convergence, 109
    impulsive system, 291
finite-time stability, 107
finite-time stabilization
    impulsive system, 297
finite-time stable, 108
    impulsive system, 290
first law of thermodynamics, 77

discrete-time system, 183
flow, 11

G

gain margin, 102
    impulsive system, 283
generalized energy balance equation, 65
    discrete-time system, 170
    impulsive system, 253
generalized momenta, 320
generalized positions, 319
generalized velocities, 319
geometrically dissipative system, 154
geometrically stable with respect to z
    uniformly in x_0
    discrete-time system, 39
geometrically vector dissipative system, 156, 168
geometrically vector nonexpansive system, 171
geometrically vector passive system, 171
globally asymptotically stable, 13
    discrete-time system, 35
globally asymptotically stable set, 129
globally asymptotically stable with respect to z, 19
    discrete-time system, 39
    impulsive system, 216
globally asymptotically stable with respect to z uniformly in x_0, 19

discrete-time system, 39
impulsive system, 216
globally exponentially stable with respect to z uniformly in x_0, 19
    impulsive system, 217
globally finite-time stable, 109
    impulsive system, 291
globally geometrically stable with respect to z uniformly in x_0
    discrete-time system, 39
globally uniformly asymptotically stable set, 129
globally uniformly exponentially stable set, 130
graph, 79
graph theory, 3

H

homogeneous function, 119
hybrid decentralized control, 313
hybrid decentralized control design, 323
hybrid decentralized controller, 306
hybrid dissipation inequality, 226
hybrid entropy, 332
hybrid feedback control, 273
hybrid supply rate, 226
hybrid vector dissipation inequality, 212

I

impulsive differential equations, 7, 271
impulsive dynamical system, 271

input-dependent impulsive system, 226
input/state-dependent impulsive system, 224, 226
irreducible matrix, 79
irreversible thermodynamics, 82
isolated system, 82

K

Kalman-Yakubovich-Popov equations, 63
  discrete-time system, 169
  impulsive system, 251
Kamke condition, 12
Krasovskii-LaSalle theorem, 27

L

Lagrangian function, 320
Legendre transformation, 320
Lie derivative, 310
Lipschitz condition, 14
lossless system, 47
  discrete-time system, 154
  impulsive system, 226
Lyapunov stability, 109
  impulsive system, 291
Lyapunov stable, 13
  discrete-time system, 35
Lyapunov stable set, 129
Lyapunov stable with respect to z, 18
  discrete-time system, 38
  impulsive system, 216
Lyapunov stable with respect to z uniformly in x_0, 18

discrete-time system, 38
impulsive system, 216

M

M-matrix, 11
maximal interval of existence, 11
maximum entropy, 83
  discrete-time system, 191
maximum entropy controller, 335
maximum entropy-based control, 327
minimum ectropy, 83
  discrete-time system, 192
monotemperaturic system, 86
multi-rotational/translational proof-mass actuator (multi-RTAC), 323
multiagent network coordination, 127
multiagent systems, 127
multivehicle coordinated motion control, 135

N

net energy flow function, 329
nonconservation of entropy, 79
  discrete-time system, 187
nondecreasing function, 14
nonnegative function, 34
nonnegative matrix, 10, 11
nonnegative orthant, 10
nonnegative vector, 10
nonsingular M-matrix, 11
null space, 10

P

partial stability, 18
plant energy, 315, 321
positive matrix, 10, 11
positive orthant, 10
positive vector, 10
power balance equation, 77

Q

quasi-continuous dependence, 308
quasi-monotone increasing function, 11
  time-varying, 130
quasi-thermodynamically stabilizing compensator, 332

R

range space, 10
reachable, 186
required supply, 55
  discrete-time system, 163
  impulsive system, 240
resetting set, 213
resetting times, 214
reversible thermodynamics, 82

S

second law of thermodynamics, 80
  discrete-time system, 187
  hybrid control, 331
sector margin, 102
  impulsive system, 283
semigroup property, 11
semistable, 13
  discrete-time system, 35

semistable matrix, 13,
228
discrete-time system,
35
sequential continuity, 310
sequential optimization,
5
settling-time function,
109
solution
impulsive system, 213
solution curve, 108
solution of a differential
equation, 11
spectral abscissa, 10
spectral radius, 10
spectrum, 10
stability of feedback
large-scale systems,
264
stability of the feedback
interconnection, 72
discrete-time system,
178
impulsive system, 266
state-dependent
differential equations,
271
state-dependent
impulsive system, 213
statistical energy
analysis, 1
storage function, 46
impulsive system, 226
strong coupling, 92, 210
strongly connected
graph, 79
subcontroller emulated
energies, 315
subcontroller
Hamiltonian, 321
subcontroller kinetic
energy, 321
subcontroller Lagrangian,
321
subcontroller momentum,
321
subcontroller potential

energy, 321
subsystem
decomposition, 5
subsystem energies, 315
supply rate, 46
discrete-time system,
154
system kinetic energy,
319
system Lagrangian, 319
system potential energy,
319

T

thermoacoustic
instabilities, 122, 123
hybrid control, 335
thermodynamic
modeling, 75
third law of
thermodynamics, 82
discrete-time system,
191
time-dependent
differential equations,
271
time-dependent
impulsive system, 224
total energy, 315, 321
total kinetic energy, 320
total potential energy,
320
total subsystem energies,
315
trajectory curve, 108
transversal point, 311

U

uniformly asymptotically
stable set, 129
uniformly exponentially
stable set, 130
uniformly Lyapunov
stable set, 129
universal statement, 9

V

vector available storage,
48
discrete-time system,
156
impulsive system, 229
vector comparison
principle, 36
vector dissipation
inequality, 45, 48, 62
discrete-time system,
153, 156, 169
vector dissipative system,
48, 62
discrete-time system,
156, 168
impulsive system, 228,
250
vector dissipativity, 4
discrete-time system,
153
vector exponentially
nonexpansive system,
66
impulsive system, 254
vector exponentially
passive system, 66
impulsive system, 254
vector hybrid dissipation
inequality, 229
impulsive system, 250
vector hybrid
dissipativity, 212
vector hybrid supply
rate, 212, 228
vector invariant set
theorem, 25
vector lossless system,
48, 62
discrete-time system,
156, 169
impulsive system, 229,
251
vector Lyapunov
function, 2, 22
component decoupled,
23

discrete-time system,
42
impulsive system, 220
vector Lyapunov
theorem, 19
converse, 33
discrete-time system,
39
discrete-time,
time-varying system,
43
finite-time stability, 294
impulsive system, 217
set stability, 130
time-varying system, 30
vector nonexpansive
system, 66
discrete-time system,
171

impulsive system, 254
vector passive system, 66
discrete-time system,
171
impulsive system, 254
vector required supply,
53
discrete-time system,
161
impulsive system, 237
vector storage function,
4, 62
discrete-time system,
156, 169
impulsive system, 229,
250
vector supply rate, 4, 47
discrete-time system,
155

virtual subcontroller
position, 320
virtual subcontroller
velocity, 320

Z

Z-matrix, 11
Zeno solutions, 214
zero-state observable, 49
discrete-time, 156
impulsive system, 226
zeroth law of
thermodynamics, 80
discrete-time system,
187
hybrid control, 331

PRINCETON SERIES IN APPLIED MATHEMATICS

Chaotic Transitions in Deterministic and Stochastic Dynamical Systems: Applications of Melnikov Processes in Engineering, Physics, and Neuroscience, Emil Simiu

Selfsimilar Processes, Paul Embrechts and Makoto Maejima

Self-Regularity: A New Paradigm for Primal-Dual Interior-Point Algorithms, Jiming Peng, Cornelis Roos, and Tamás Terlaky

Analytic Theory of Global Bifurcation: An Introduction, Boris Buffoni and John Toland

Entropy, Andreas Greven, Gerhard Keller, and Gerald Warnecke, editors

Auxiliary Signal Design for Failure Detection, Stephen L. Campbell and Ramine Nikoukhah

Thermodynamics: A Dynamical Systems Approach, Wassim M. Haddad, VijaySekhar Chellaboina, and Sergey G. Nersesov

Optimization: Insights and Applications, Jan Brinkhuis and Vladimir Tikhomirov

Max Plus at Work, Modeling and Analysis of Synchronized Systems: A Course on Max-Plus Algebra and Its Applications, Bernd Heidergott, Geert Jan Olsder, and Jacob van der Woude

Impulsive and Hybrid Dynamical Systems: Stability, Dissipativity, and Control, Wassim M. Haddad, VijaySekhar Chellaboina, and Sergey G. Nersesov

The Traveling Salesman Problem: A Computational Study, David L. Applegate, Robert E. Bixby, Vasek Chvátal, and William J. Cook

Positive Definite Matrices, Rajendra Bhatia

Genomic Signal Processing, Ilya Shmulevich and Edward R. Dougherty

Wave Scattering by Time-Dependent Perturbations: An Introduction, G. F. Roach

Algebraic Curves over a Finite Field, J.W.P. Hirschfeld, G. Korchmáros, and F. Torres

Distributed Control of Robotic Networks: A Mathematical Approach to Motion Coordination Algorithms, Francesco Bullo, Jorge Cortés, and Sonia Martínez

Robust Optimization, Aharon Ben-Tal, Laurent El Ghaoui, and Arkadi Nemirovski

Control Theoretic Splines: Optimal Control, Statistics, and Path Planning, Magnus Egerstedt and Clyde Martin

Matrices, Moments, and Quadrature with Applications, Gene H. Golub and Gérard Meurant

Totally Nonnegative Matrices, Shaun M. Fallat and Charles R. Johnson

Matrix Completions, Moments, and Sums of Hermitian Squares, Mihály Bakonyi and Hugo J. Woerdeman

Modern Anti-windup Synthesis: Control Augmentation for Actuator Saturation, Luca Zaccarian and Andrew W. Teel

Graph Theoretic Methods in Multiagent Networks, Mehran Mesbahi and Magnus Egerstedt

Stability and Control of Large-Scale Dynamical Systems: A Vector Dissipative Systems Approach, Wassim M. Haddad and Sergey G. Nersesov